9th International Space Safety Conference 2017

Know Safety, No Pain

Toulouse, France
18 - 20 October 2017

ISBN: 978-1-5108-7531-9

Printed from e-media with permission by:

Curran Associates, Inc.
57 Morehouse Lane
Red Hook, NY 12571

Some format issues inherent in the e-media version may also appear in this print version.

Printed by Curran Associates, Inc. (2019)

For permission requests, please contact International Association for the Advancement of Space Safety (IAASS) at the address below.

International Association for the Advancement of Space Safety (IAASS)
Kapteynstraat 1
2201BB Noordwijk
The Netherlands

Phone: +31(0)712020023

spacesafety@iaass.org

Additional copies of this publication are available from:

Curran Associates, Inc.
57 Morehouse Lane
Red Hook, NY 12571 USA
Phone: 845-758-0400
Fax: 845-758-2633
Email: curran@proceedings.com
Web: www.proceedings.com

PROCEEDINGS OF THE 9TH INTERNATIONAL SPACE SAFETY CONFERENCE

KNOW SAFETY, NO PAIN

TOULOUSE (FRANCE) - 18-20 OCTOBER 2017

9TH INTERNATIONAL SPACE SAFETY CONFERENCE

KNOW SAFETY, NO PAIN

18-20 October 2017
Toulouse (France)

Meteo-France

42 Avenue Gaspard Coriolis

31100 Toulouse, France

Conference Organizing Committee

Chair/Co-Chairs

I. Rongier	K.-U. Schrogl
J. Nguyen	N. Takeuchi

Members

W. Ailor	M. Glissman	M. Nogami
N. Al Rashedi	D. Isakeit	P. Omaly
F. Alby	S. Kaul	J. Pelton
C. Arbinger	M. Kezirian	T. Pfitzer
T. Beard	P. Kirkpatrick	A. Quinn
F. Bittner	B. Lazare	L. Ren
G. Boy	T. Lips	N. Ridzuan
C. Cazaux	D. McKnight	T. Sgobba
A. Del Bianco	A. Menzel	P. Wilde
G. Gafka	M. Miki	T. Yoshihara

Publication: Proceedings of the 9th IAASS Conference “Know Safety, No Pain”
Toulouse (France) 18-20 October 2017
ISBN: 978-90-828378-0-3
Cover: BENNU'S JOURNEY - Europa - Credit: NASA Goddard Space Flight Center

Table of Contents

PROCEEDINGS OF THE 9TH INTERNATIONAL SPACE SAFETY CONFERENCE

PLENARY SESSION

KEYNOTE SPEECH
DI REIMOLD - DIRECTOR OF STRATEGIC OPERATIONS
OFFICE OF COMMERCIAL SPACE TRANSPORTATION

FAA-OCST, 800 Independence Avenue, SW Washington, DC 20591, USA, Email: Dorothy.Reimold@faa.gov

1. INTRODUCTION

On behalf of the FAA Administrator and my boss, Dr. George Nield, I'd like to thank the IAASS for the chance to be here in the beautiful city of Toulouse and to have the opportunity to talk about the commercial space transportation industry and the role that the FAA Office of Commercial Space Transportation plays therein.

To set the stage, this has been one of the busiest years ever for U.S. commercial space transportation. Last week, we licensed the seventeenth commercial launch of 2017. Fourteen of these launches have been done by SpaceX with their Falcon 9 and we've had one commercial launch each by Orbital ATK with the Minotaur IV, Rocket Lab USA with the Electron from New Zealand, and United Launch Alliance with the Atlas V. And there are still more commercial launches are planned for this calendar year.

One of our biggest challenges is keeping pace with industry. It's more than just launches—it's applications for new vehicles, reentries from orbit, reusable first stages that return to the launch site, new missions in low Earth orbit and beyond, new spaceports, and the fact that many of these new ideas are coming to us all at once.

Having said that, these are great challenges to have, based on an industry that continues to grow and innovate. And as we are being challenged to keep pace, its compelled us to think about doing things differently. So, I'd like to give you a short update on some of the current areas of interest that will further our safety pursuits and they include Standards related work, Research and Development, and Space Traffic Management.

2. SETTING STANDARDS

Earlier this morning we had a lively panel discussion on the role of standards in commercial human spaceflight safety governance. I think everyone agrees that standards are an important element of any safety framework that achieves our goal of continuous improvement in the safety performance of space operations. It is also clear that the U.S. Congress has long taken the view that we cannot burden the industry as they have clearly continued to hold the line that

"standards governing human space flight must evolve as the industry matures so that regulations neither stifle technology development nor expose occupants to avoidable risks as the public comes to expect greater safety from the industry."

Many of you may be familiar with a new industry-led commercial space standards development effort at ASTM called "Committee F47 on Commercial Spaceflight." The ASTM committee had its first organizational meeting in October 2016 and the first face-to-face meeting in Washington, DC in May 2017. A second face-to-face meeting is scheduled in Houston, Texas in December.

The Committee has already established five technical subcommittees to focus on:

1) Occupant Safety of Suborbital Launch Vehicles;
2) Occupant Safety of Orbital Launch Vehicles;
3) Unoccupied Launch and Reentry Vehicles;
4) Spaceports; and
5) Cross-Cutting (which covers items such as risk assessments, emergency plans and radiation protection.

In addition, there are subcommittees on terminology, standards road mapping, a regulatory liaison, and an Executive committee.

This ASTM effort builds on U.S. standards work of the Commercial Spaceflight Federation along with AIAA and SAE that dates back to 2012 and we are encouraged by the work being done to develop and maintain voluntary consensus standards and recommended practices for the commercial spaceflight industry.

You may also know that the U.S. Congress asked the Department of Transportation to report every 30 months (until December 31, 2021) on the progress of the commercial space transportation industry in developing voluntary industry consensus standards that promote best practices. They clearly understand the importance of this work and the commitment of industry to this approach.

3. RESEARCH AND DEVELOPMENT

There is a modest budget for space R&D within the FAA, at about $2 million a year, plus some additional funding related to air traffic control and a request of $4.5 million for facilities and equipment. Yet, we have a dynamic program for Research and Development which includes four key areas:

1) Space Traffic Management and Spaceport Operations
2) Space Transportation Vehicles
3) Human Spaceflight
4) Space Transportation Industry Viability

In order to ensure that we are targeting our limited resources correctly we have approached work under these four areas by asking key research questions:

1) What criteria, tools, and documents can optimize public safety and efficiency as Commercial Space Transportation is integrated into the National Airspace?
2) What advanced technologies and analysis techniques can optimize public safety and efficiency during on-orbit activity?
3) What criteria, tools, and documents can facilitate the continuous improvement of occupant safety for Commercial Space Transportation?
4) What criteria, tools and documents can improve the understanding of evolving Commercial Space Transportation market activities?

In answer to these questions we have prioritized some of our Research and Development activities to support standards development, an assessment of the current U.S. regulatory framework, determine the feasibility of using surveillance technologies designed for commercial aviation for commercial space, and use of empirical data as benchmarks for flight safety analysis tools. While this is by no means an exhaustive list of our ongoing R&D activities it does reflect our ongoing emphasis on safety. And if any of you would like to know more about the entirety of our R&D work please let me know.

4. COMMERCIAL SPACE TRAFFIC MANAGEMENT

It's a great time for investment in commercial space. An updated study called "Start-Up Space" by Bryce Space and Technology found that investment in commercial space activity in 2016 increased slightly compared to 2015 and that from 2000-2016, Start-up space ventures have attracted over $16.6 billion of investment in both satellite and launch companies. To give you an idea of recent growth, in the early 2000s, an average of three funded space companies were started per year… by comparison, in the last five years, the number of funded new space companies has averaged nearly 17 per year.

You are all familiar with the many new small satellites that have been launched in the past few years and new proposed "mega" constellations for satellite communications, remote sensing, and other applications. At the FAA we've see a large increase in the number of applicants for new launch vehicles, especially for small launch vehicles that are trying to meet the demand to launch all of those satellites.

All of this leads to a trend of space becoming more congested. This can also lead to more collisions and debris in space, something that is in everyone's interest to minimize and avoid.

With no widely accepted "Rules of the Road" for safety of space operations, the time is right for Space Traffic Management.

As discussed this morning in the panel session and as many of you will likely agree there is no common definition for what Space Traffic Management is… but we use the one from the 2006 COSMIC study used by the International Academy of Astronautics that said:

"STM is the set of technical and regulatory provisions for promoting safe access into outer space, operations in outer space, and return from outer space to Earth, free from physical or radio-frequency interference."

Although the U.S. Department of Defense has historically kept track of objects in space, they would like to be able to focus on national security issues rather than on the safety of civil and commercial space operations – or in other words-- be space warriors instead of space traffic cops.

As part of a 2015 law, the U.S. Congress asked DOT to deliver several reports. One of those, the Section 110 Report on Space Surveillance and Situational Awareness Data completed in September 2016, concluded that:

"It is feasible for a civil agency, specifically the Department of Transportation (DOT) acting through the Federal Aviation Administration (FAA) Office of Commercial Space Transportation, to release safety-related SSA data..."

The report also stated that:

- "Statutory authority would be required..."
- "A civil agency would require immunity from lawsuits similar to the DoD immunity provision..."
- "Additional resources would be necessary..."

We've given this a lot of thought for many years within our organization and with input from other key stakeholders. We believe that running STM from FAA has some distinct advantages:

- The FAA already has responsibility for overseeing commercial launches and reentries, and has a knowledgeable and experienced workforce
- Our dual mission-- to ensure public safety and to encourage, facilitate, and promote the industry-- provides an appropriate regulatory balance
- The FAA knows a bit about situational awareness and traffic management....

So what could a Civil Space Traffic Management System (CSTMS) look like? We think it would consist of the hardware, software, data and communications connectivity and it would have personnel to:

- Accept and process observations;
- Create and maintain a catalog of space objects;
- Perform safety-related analyses, and
- Develop and distribute safety-related products and services.

The FAA could manage and operate the system and carry out tasks for civil, commercial, and international space operators like:

- Conjunction Assessment (predict close approaches for orbital objects)
- Collision Avoidance (support maneuver planning)
- Launch Deconfliction (identify and resolve predicted conjunctions during launch windows, launches, and initial spacecraft orbits)
- Deorbit/Reentry Support (provide conjunction assessment support for maneuvers and reentries).

The overall reason for us to pursue this is to enhance the safety of space operations and to preserve the space environment. Specifically, the system will provide safety products and services as a public good.

While DOT/FAA continues to work closely with the U.S. Department of Defense and other stakeholders to determine how best to implement a Space Traffic Management system, the FAA has proposed conducting a Pilot Program to answer feasibility questions and lower programmatic risks.

This Pilot Program would be done jointly with the Department of Defense. It would demonstrate the ability of the FAA to provide orbital safety services. The objective is to stand up the capability to record observations, generate at least a partial catalog of space objects, and analyze potential conjunctions. Data produced by the program could be compared and validated with DoD data.

From a U.S. view, we would start with domestic approach and involve international partners as soon as practical. Clearly though, Space Traffic Management is an international issue, and one that is of interest to all space-faring nations. However, if we look a little bit deeper we can readily recognize that space based assets are already providing capabilities that we all cherish: for example, the ability to navigate, the ability to prepare for the weather that we will face in the near future, and the ability to instantly communicate between even the most remote places. And when we couple that with an understanding of how fragile the space environment really is, then we can see the universal value of efforts to preserve the space environment and defend it from anything that would prevent us from reaping the rewards space exploitation and space exploration.

The two major orbital collision events we have all witnessed has already made space debris impacts the prime threat to the people and spacecraft that travel through the space that surrounds our small planet. Let us not dwell on the mistakes of the past any longer than is necessary for us to learn the lessons we need to face the challenges posed by the future. Instead, out of the crucible of crisis may we form a common sense of purpose and accept the challenge to preserve the space environment and defend it from anything that would prevent us from reaping the rewards space exploitation and space exploration.

This challenge, like the one issued by President Kennedy more than 50 years ago, will not be easy to achieve. It is imperative that we cooperate and collaborate to the greatest extent possible, because only then will we truly succeed. Responsible elements in various governments have a role to play. For example, the US Air Force is now willing to pay more for the products they operate and procure to ensure the protection of the space environment. That is the kind of leadership by example that we all benefit from. But we need more than good governance to face our biggest challenges in the 21st century. We need to educate and inspire the public about how precious and fragile is the space that surrounds us. We need to challenge the best and brightest among us to find technical solutions. In the past, prizes have been effective in helping us develop and demonstrate solutions to major challenges, perhaps most famously when Charles Lindbergh won the Orteig by flying across the Atlantic, and more recently when Scale Composites won the Ansari X-Prize. Those two examples show that increasingly difficult challenges require larger teams.

We have done well when we set grand goals that "serve to organize and measure the best of our energies and skills." And to achieve those grand goals, we need to identify the critical incremental steps along the way. Before Neil Armstrong took that "last small step," teams of professionals working in academia, industry, and the US government had to identify each technical hurdle between us and the moon.

The FAA's Office of Commercial Space wants to be a part of a coalition of the willing to accept the challenge to preserve the space environment and defend it from anything that would prevent us from reaping the rewards space exploitation and space exploration. We know it won't be easy, but we also have reasons to believe that we can achieve great things when we work together as team that includes industry and academia. When the FAA was formed just over 50 years ago, commercial air-travel was perhaps the most risky mode of transportation and unaffordable for many among us. Clearly flying through the air presents significant technical challenges and has inherent risks, yet in less than one lifetime, by working together, by studying and sharing the safety lessons from each accident, commercial aviation has become the safest mode of transportation and the most practical way to travel between continents.

Moving to an operational international Space Traffic Management system will require some transition periods and not happen overnight. We call, this a "crawl, walk, run" approach with the idea that all of the key stakeholders gain comfort with progress and with the products and services that are provided. We think that such a move can be made in a relatively short period of time, and with a relatively modest impact in terms of the necessary resources.

Starting with a Pilot Program could yield an Initial Operating Capability in about two years with full operational capability a year later.

We are certainly open to feedback from the entire space community on how to make Space Traffic Management successful.

I think we'd all agree that perhaps the hardest part is to stop talking about Space Traffic Management and actually do something about it.

I'd be remiss if I didn't mention our growing international collaboration on all fronts: safety, operational norms, and infrastructure. We believe that the work the U.S. has done in terms of regulatory approach, airspace access, and now a renewed focus on spaceport implications needs to be shared with our global colleagues and a great deal of my time is spent in this regard. We are in the process of executing several bilateral agreements...and there's many more in the planning stage. We are also having an international roundtable later this month and planning for our first ever regional commercial space transportation workshop in which we will invite all countries with an interest to come in and listen and ask questions.

There's so much going on in the industry and that means we are moving at warp speed in AST. I'd like to thank you all for the opportunity to talk with you today and I look forward to being with you all again.

Thank you.

PROCEEDINGS OF THE 9TH INTERNATIONAL SPACE SAFETY CONFERENCE

SESSION 01
RE-ENTRY SAFETY – I

OBLATE-EARTH EFFECTS ON THE CALCULATION OF E_C DURING SPACECRAFT REENTRY

John B. Bacon[(1)], Mark J. Matney[(1)],

[(1)]NASA Johnson Space Center XI-4, 2101 NASA Pkwy. Houston TX 77058 USA john.bacon-1@nasa.gov
[(1)]NASA Johnson Space Center XI-4, 2101 NASA Pkwy. Houston TX 77058 USA mark.matney-1@nasa.gov

ABSTRACT

The bulge in the Earth at its equator has been shown [1] to lead to a clustering of natural decays biased to occur towards the equator and away from the orbit's extreme latitudes. Such clustering must be considered when predicting the Expectation of Casualty (E_c) during a natural decay because of the clustering of the human population in the same lower latitudes. This study expands upon prior work [1, 2], and formalizes the correction that must be made to the calculation of the average exposed population density as a result of this effect. Although a generic equation can be derived from this work to approximate the effects of gravitational and atmospheric perturbations on a final decay, such an equation averages certain important subtleties in achieving a best fit over all conditions. The authors recommend that direct simulation be used to calculate the true E_c for any specific entry as a more accurate method. A generic equation is provided, represented as a function of ballistic number and inclination of the entering spacecraft over the credible range of ballistic numbers.

1. BACKGROUND

In the final stages of orbital decay, the density scale height shrinks to only a few kilometers. At the same time, the rapidly-circularizing orbit encounters two effects of the oblate Earth, each of which (at moderate to high inclinations) permute the "effective altitude" to a degree comparable to or exceeding such typical scale heights. The primary influence is the elevation of the atmosphere relative to Earth's center in the equatorial regions, simply rising in lock step with the geoid. Because the Earth's radius at the equator is 21 kilometers (km) higher than at its poles, the effective density (and thus, the local deceleration) could hypothetically increase by a factor of nearly 50 at the equator relative to values at the poles under an idealized perfect circular Kepler orbit.

Simultaneously, the J2 gravitational orbit perturbation works in the opposite direction, raising the local radius of the orbit at extreme latitudes, and dropping it at the equator. There are additional J3 effects noted by Fremeaux, *et al.*, [3] which emerge as a small correction factor in the present study. Fig. 1 shows the progression of the local altitude and radius for a decaying polar orbit, and the progression in the Log_{10} of density during the decay. All values are plotted versus latitude, causing a rocking back and forth as time progresses. Clearly, the dominant nonlinear factor in the otherwise spiral decay is the physical uplift of the atmosphere into the orbit path.

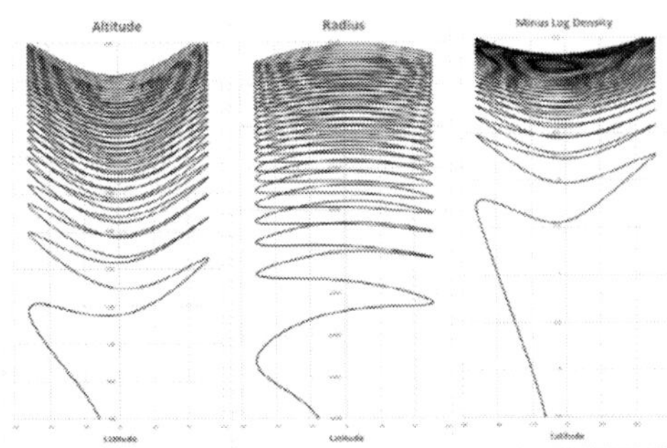

Figure 1. Altitude, radius, and Minus Log(Density) plotted vs. latitude for a decaying 50° orbit at 100kg/m2 ballistic number. While the J2 gravitational effect raises the local radius at the extreme latitudes, the elevation of the atmosphere above the physical bulge at the equator dominates the density variations encountered by the decaying spacecraft. Both the gravitational and atmospheric perturbations are evident in structure of the decay concentration curves.

With such strong cyclic variation in the drag profile, it should be expected that the distribution of natural decays around the orbit path (measured in Argument of Latitude [ArgLat with symbol Θ], which is the angle along an orbit in the orbit plane measured from the north-bound pass of the equator; generally computed as the true anomaly plus the argument of perigee) should instead show a statistical clustering that reflects this cyclic forcing function at one-half orbit period. Indeed, in earlier work, both simulation and analysis of well-documented real, natural polar orbit decays showed a cyclic clustering towards the equator. Preliminary assessments with the initial model revealed that the statistical biasing of natural decays to the more heavily populated lower latitudes could have up to a 13% effect on the calculated value of E_c. When the authors presented the preliminary model they acknowledged that further work would be necessary to completely define the statistical biases under different values of ballistic number (BN), inclination, and perhaps solar flux conditions. This paper reports the conclusion of that work, covering a full range of credible ballistic numbers and all orbit inclinations. Solar flux was shown in subsequent publication [2] to be inconsequential to the decay latitude bias problem. In the improved model the authors have reformulated the analysis in terms of ArgLat instead of latitude, which has allowed better comparison of the physics across cases.

2. APPROACH

Tens of thousands of natural decays were simulated in the NASA General Mission Analysis Tool (GMAT), in parametric studies every 5° of inclination for spacecraft with 50, 100, 150, and 200 kg/m^2 BN decaying for slightly more than 50 orbits each. The orbits were propagated to 90 km altitude. (Past research [2] has shown that at 90 km the residual time and distance to go to the surface is approximately independent of ArgLat for any given ballistic number. Therefore, to reduce computation time from weeks to days, the decays were truncated at 90 km). This truncation must be accounted for when considering actual expectation of casualty, because ground intercept will be several degrees downrange of the calculated entry point.

GMAT is a certified NASA mission planning tool with an adaptive step, ninth order Runge-Kutta integrator with eighth order error control that can numerically integrate the path of a decaying object through a selectable atmosphere. In this study, the MSIS-E 2000 atmosphere was modeled at F(10.7) identical daily and 90-day average values of 115 janskys, and the global geomagnetic index Kp constant was uniform at 3. Orbits were propagated with 4th-order spherical harmonics and full atmospheric responses to day-of-year and the orbit's Right Ascension of the Ascending Node (RAAN).

The mass of the modeled spacecraft was dithered in tiny steps to achieve a cumulative 2% spacecraft mass variation in 101 uniform steps per run. The mass dither is considered by the authors to be the subtlest and most linear way to induce a smooth variation in a very nonlinear process. This method allows the initial conditions to be as nearly identical as possible, while allowing the forces being studied to integrate in the nonlinear ways under study. The 2% total mass variation in a decay with more than 50 remaining orbits at inception is enough to observe one full cycle of the ArgLat of decay. (One hundred and one runs are needed to generate 100 gaps. It is the gaps that drive the compression data.).

Once a full cycle of ArgLat was observed, the relative compression (or "compression factor") C(Θ) of the decays was recorded at each value of Θ. C(Θ) is calculated by talking the ratio of the spacing between the ArgLat of any one iteration's decay and the next iteration's value, divided by the averaged step size over the full 360° cycle of ArgLat. (The non-integer number of steps is linearly interpolated between integer steps that bracket the 360° cycle.) A clustering of decays in one portion of the arc then yields a ratio less than one, and a rarefaction or sparseness of decays will lead to a number greater than one. Although this may seem counterintuitive, representing a variable <u>angle</u> per incremental decay scenario (vs. variable <u>decays</u> per unit angle) was the most easily immediately-normalized function when establishing common compression curves. When the function is inverted to a probability distribution function, each local point must be binomially expanded, as discussed later.

> *Example*: if 57.137 uniform increments of a satellite's mass caused the decay's ArgLat to vary by 360°, the average step size would be (360°/57.137) = 6.2751°. If the argument of decay latitude shifted only 5.72° from one iteration to the next at a particular region, the local compression rate would be (5.72°/6.2751) = 0.9115°

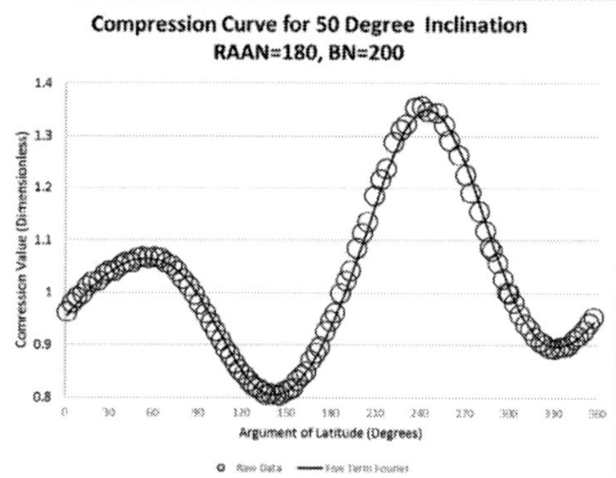

Figure 2. An example relative compression curve for BN=200 decays at 50° inclination and RAAN=180°. The curve illustrates the excellent fit of the five-term Fourier approximation. At the vernal equinox, this orbit has a beta angle of zero and passes directly under the subsolar point, creating a day-night asymmetry in the decay rate. The near mirror of this curve occurs at 90° RAAN (see Fig 6). The asymmetry is removed by plotting the rates at four orthogonal RAANs and averaging the curves. (This is best done <u>after</u> curve-fitting, because there is no way to align the individual statistical points to achieve the same values of ArgLat in all four plots.) Note in the curve how sparse the entry events are at the peaks relative to the valleys, in terms of spatial density of the points.

The resulting plot of relative compression vs. ArgLat shows a consistent cyclic behavior of concentrating entries as the decay location approaches the equator (ArgLat= 0 and 180), receding rapidly after passing it, and becoming most rarefied approaching the extreme latitudes (ArgLat = 90 and 270). An example plot is shown in Fig. 2. A 5-term Fourier analysis of the data is superimposed as the solid line, showing an excellent fit, with generally <0.7% average error of the fit from recorded values. The universal model of the compression curve over all conditions is generated by characterizing the trend of the first four best-fit Fourier coefficients under varying ballistic number, RAAN, and inclination. (The fifth term represents residual noise, but no implicit physics emerges from its trending.) Noise comparable to or greater than the fifth Fourier amplitude coefficient is introduced in the estimation of the curve in 256 discreet bins of ArgLat with nearest value in the true compression curve.

The entire series was repeated at four orthogonal orbit planes (RAAN=0, 90, 180, and 270) at the vernal equinox. The reason for RAAN variation was to average out the effects of the diurnal variation in the atmosphere, which presents variations in density of up to 25% of peak value, depending upon whether Θ lies near the terminator, subsolar point, or solar midnight.

In all then, (4 ballistic numbers) × (4 RAANs) × (101 runs/case) × (18 inclinations) led to 29,088 sample runs, leading to 288 curves of 100 data points each. An additional 30 special cases were run to explore conditions at the solstice, simplified gravitational models, and extreme ballistic numbers, for yet another 3030 runs.

Next, all 100 compression data pairs for each set of input conditions were spread into the closest of 256 uniform ArgLat bins. A Fourier analysis was performed on this 256-bin representation of the curve. The amplitude and phase of the first five oscillating terms were recorded, with the first term (the uniform average compression) defined as being one despite any curve-fitting residual.

The five Fourier terms were then explored for empirical trends, such that the amplitude and phase of each term can be derived for any input. The empirical formulation of the Fourier terms then, in theory, allows us to recreate the compression curve for any condition. Ultimately, chaotic trends in the phases of asymmetric Fourier terms, and low-inclination orbits in general were problematic to developing a fully-universal empirical prediction.

From the relative compression C(Θ) along such a reconstituted curve we can derive the relative likelihood of impacting at any particular latitude. For any given inclination (i) and ArgLat (Θ), we get that the latitude is ARCSIN(SIN(i)*SIN(Θ)), and the local statistical likelihood at that latitude relative to unit average will be a binomial expansion of 1/C. The coefficient 1/C must be binomially expanded to achieve the original true compression in relative decays per degree. This problem stems from the mathematical inequality that the most easily normalized initial compression curve is based upon N/Avg(X), which is not equal to N*Avg(1/X).

3. RESULTS

Each compression curve takes the general form of a sinusoidal curve in (two times the) ArgLat, (a 2Θ harmonic), perturbed by 1Θ, 3Θ, and 4Θ harmonic terms of generally lower amplitude, each term affected by a unique phase offset.

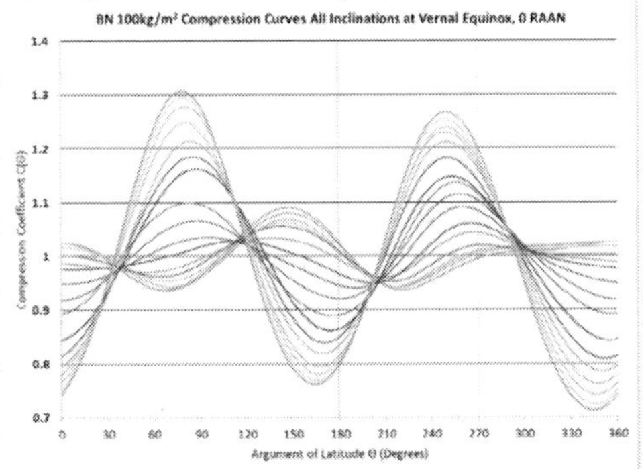

Figure 3. A family of compression curves for inclinations from 5° to 90° at RAAN=0° and BN=100 kg/m². Curve amplitude near 90° ArgLat increases with orbit inclination. The curves are more chaotic and invert the compression effect's locations at the lower inclinations, although at low amplitude and low inclination, the net effect on latitude clustering is small.

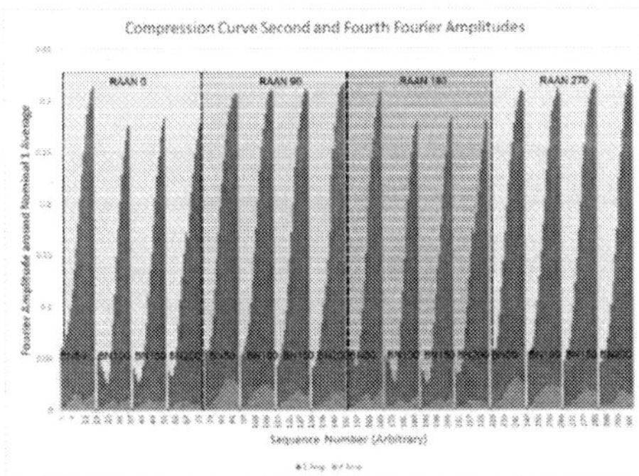

Figure 4a. The even-numbered (symmetric) Fourier term amplitudes of compression curves. Note that the 2Θ terms are typically a factor of 10 larger than the 4Θ terms and the patterns generally repeat with comparatively small variations for ballistic number and RAAN. In each of the 16 tooth-shaped curves, orbit inclination increases from 5 at the left to 90° at the right. Each of four major groupings shows the results for one RAAN value: 0°, 90°, 180°, and 270° from left to right. Within each group, each of four ballistic numbers are evaluated: 50, 100, 150, and 200 kg/m².

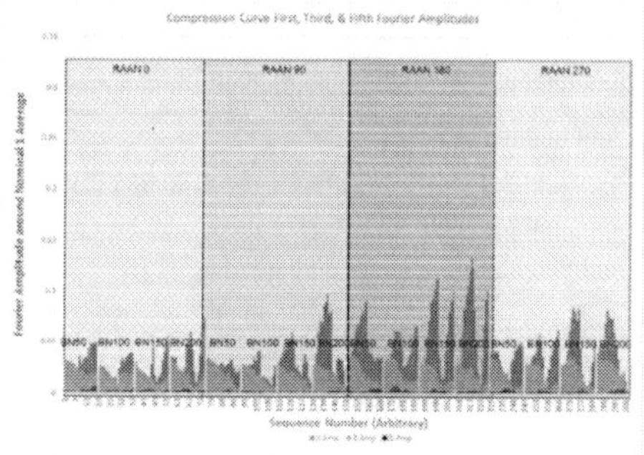

Figure 4b. The odd (asymmetric) Fourier term amplitudes of the compression curves. The groupings are as in Fig. 4a, and the vertical axis is to the same scale. The simple first harmonic wave in Θ is the largest but most irregular of all the secondary terms, and is believed to be heavily influenced by the exact path through or near the subsolar point, potentially containing some J3 gravitational information too. Note the interesting and as yet unexplained gap in most 1Θ amplitudes near 50°-55°. The 3Θ terms are very repeatable and are likely purely atmospheric in nature (see Fig 5c). The 5Θ terms are completely negligible, representing noise. (The gravitational model itself is only to four harmonic terms.)

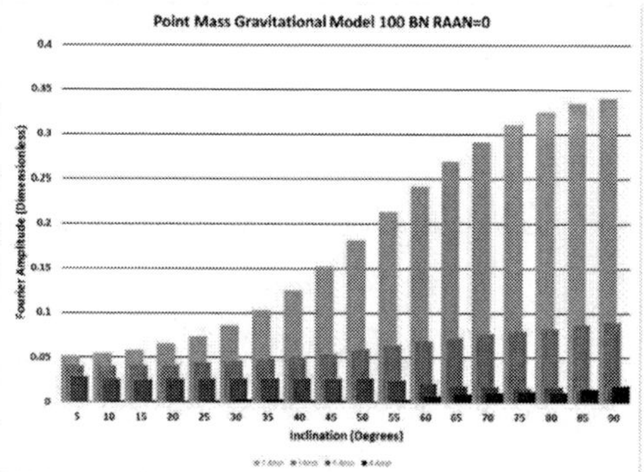

Figure 5a. The scale of atmospheric vs. gravitational effects is evident in the calculation of compression curves when Earth's simple point-mass gravitational model (0-order, 0-degree gravitational model) is used instead of the full 4-order, 4-degree gravitational model. Only atmospheric geometry—especially the solar heating effect—is driving the Θ, 3Θ, and 4Θ Fourier terms in this case, resulting in perturbing effects generally comparable in scale with the "wall of air" effect at low inclinations, dropping to about 25% influence at the highest inclinations for the 1Θ term, and <5% influence for the higher-order terms.

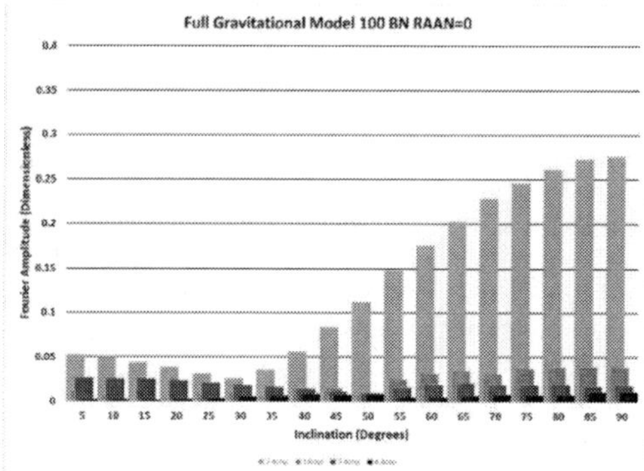

Figure 5b. The compression curve Fourier coefficients when the full gravitational model is used for the same conditions of 100 kg/m2 ballistic number and 0° RAAN. Note the reduction in nearly all of the coefficients, indicating that gravitational forces in many cases may mitigate the atmospheric effects.

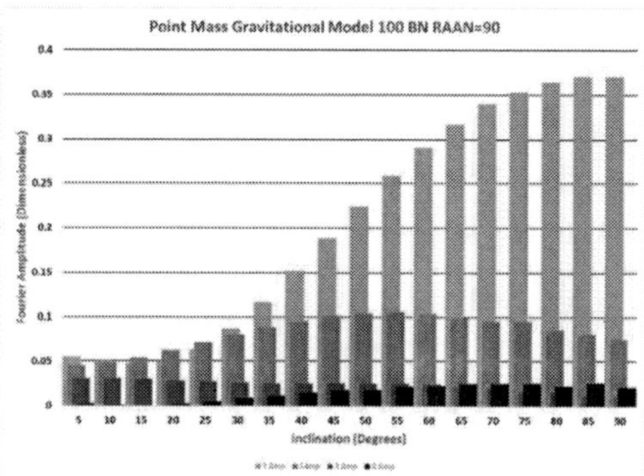

Figure 5c. The point mass gravitational model for a 100 kg/m² BN at 90° RAAN. Note the shift in height and shape of the 1Θ (blue) term, and the increase in the 2Θ (orange) term, all presumably due to atmospheric effects only. The path of the orbit relative to the subsolar point has a noticeable effect on the symmetry of the curve (Fig. 6). Note that the 3Θ (red) term is remarkably consistent across all the curves. This strongly implies that the J3 terms are atmospheric in nature, and real.

The basic 2Θ curve has an uprange offset of 15°-20° relative to nodal crossings at 0° and 180° of ArgLat at higher inclinations, with interesting trends in the opposite direction at the lower inclinations, where all the Fourier terms have comparable amplitudes. The shape of the curve is generally more disorganized at lower inclination orbits, but still exhibits this basic behavior (see Fig. 3).

The first five Fourier terms' amplitudes are shown in Figs. 4a and 4b grouped into terms that are respectively even and odd multiples of Θ, representing the effects that are symmetric north-to-south and those that are asymmetric.

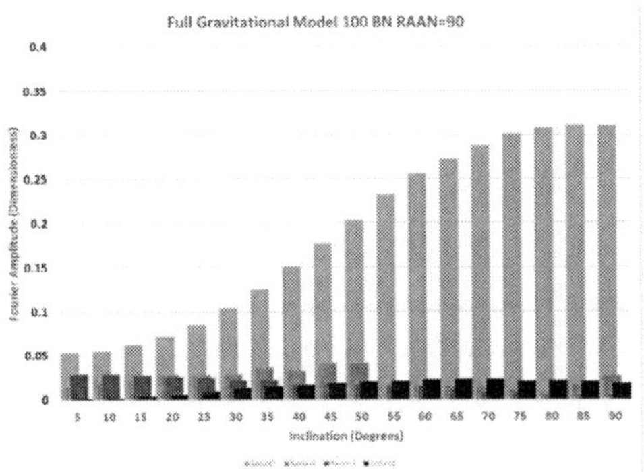

Figure 5d. The same conditions as in Fig. 5c above, with the full gravitational model employed. As with the RAAN=0 case pair, most Fourier terms have been reduced from the point-mass gravity model's case.

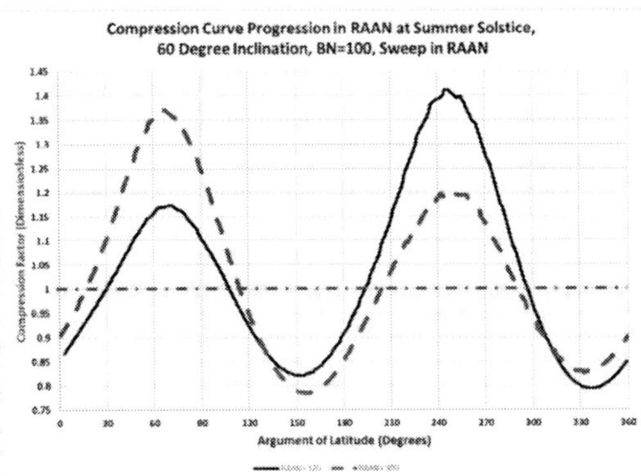

Figure 6a. The compression curve changes in character with passage of the orbit near the subsolar point at the summer solstice. Here a 60° inclination orbit of 100 kg/m2 ballistic number was propagated every 15° in RAAN, ultimately lining up nearly through maximum density a few degrees east of the subsolar point when the initial RAAN at the start of the propagation was 120°. At RAAN = 300° (+180° of RAAN from 120°), the curve family reached a nearly mirrored amplitude profile. It illustrates the effect on symmetry of major perturbing effects in the atmosphere other than the equatorial bulge that causes the basic SIN(2Θ) curve. When all 22 runs are sequenced, the curves exhibit a simple harmonic "sloshing" with cycling RAAN.

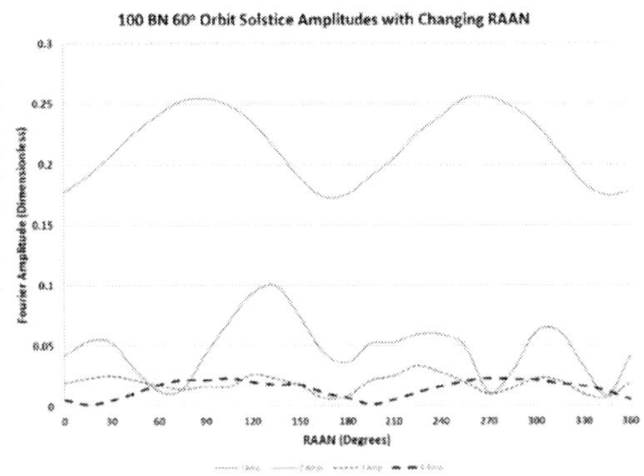

Figure 6b. The variation in Fourier term amplitudes with RAAN for the sweep of a 60° orbit. Note that the 2Θ amplitude term peaks well before the full compression curve does.

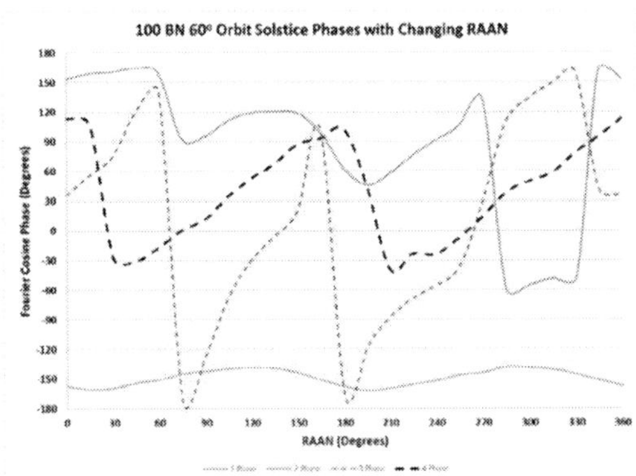

Figure 6c. The variation in Fourier term phases with RAAN for the sweep of a 60° orbit.

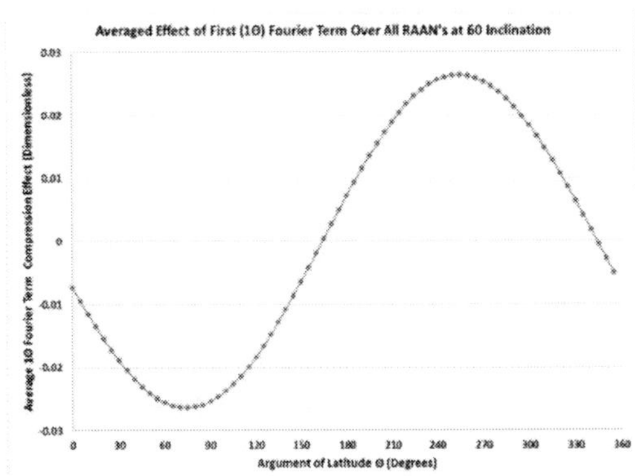

Figure 6d. When the first Fourier term in 1Θ is averaged at each value of Θ over all RAANs, a perfect sine wave emerges, representing the residue believed to be the J3 gravitational contribution. This slightly biases natural decays to occur in the northern (vastly more populated) hemisphere.

From the Fourier amplitudes in Fig. 1 (especially in the 90° and 270° RAAN cases) there is an apparent increasing trend in the peak 2Θ Fourier amplitude with increasing ballistic number. A special series of high ballistic number cases was run to explore this trend, with the peak amplitude at 90° inclination plotted in Fig. 7. From this we derive there is a simple quadratic trend in the symmetric compression effect as a function of ballistic number, while the asymmetric effects are increasingly less significant.

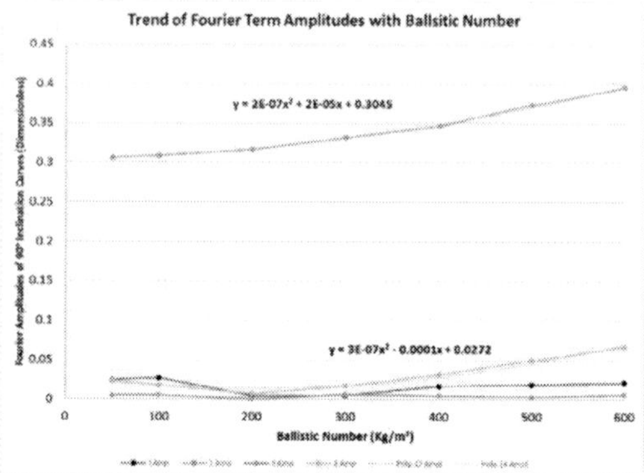

Figure 7. The trend of the first four Fourier term amplitudes for peak inclination orbits as ballistic number increases. The symmetric first and fourth terms show a simple quadratic growth, while the odd terms remain flat and comparatively small.

It should be noted that it is only the symmetric terms that progress monotonically in phase, while the asymmetric terms present a less ordered structure (see Figs. 8 and 9).

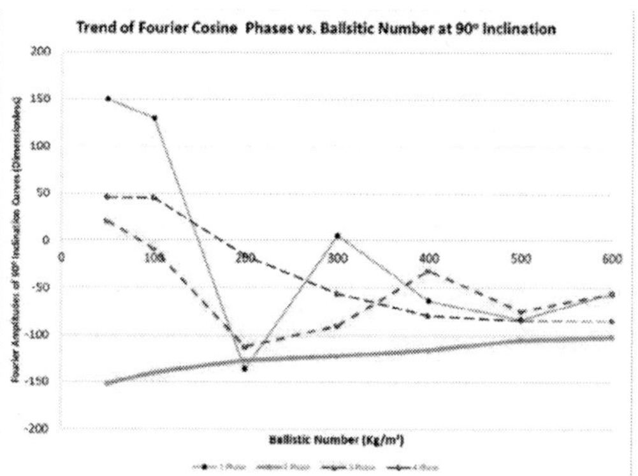

Figure 8. The phases of the 2Θ and 4Θ Fourier (cosine) terms are the only monotonically changing curves in the significant components of the compression curves.

The phases of the Fourier terms generally show a much noisier trend as a function of orbit inclination than do the amplitudes. A set of phases for one set of inclinations under common RAAN and BN conditions is shown in Fig. 9. As we have mentioned earlier, the compression curves do not exhibit particularly strong predictable structure in the lower inclinations. As in the high-ballistic-number study, only the even-numbered (symmetric) terms show a monotonic and convergent behavior with increasing inclination.

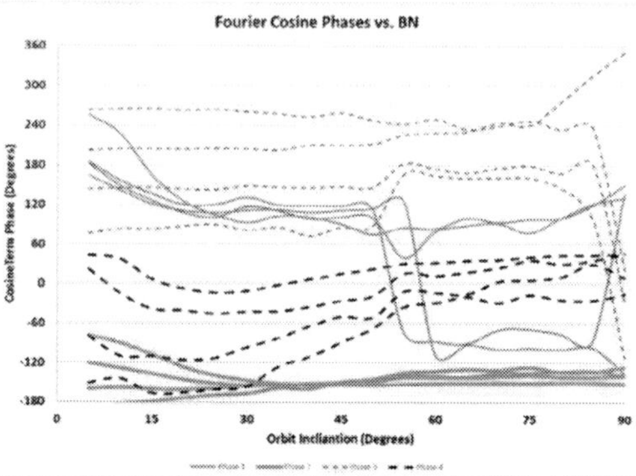

Figure 9. The phases of the first four cosine terms of the compression curves for all ballistic numbers at 90° RAAN. There are four curves for each Fourier phase term: one for each ballistic number (50, 100, 150, and 200 kg/m²). The curve for each successively higher ballistic number initiates farther lower on the graph than it did for the prior (smaller) ballistic number. Note the erratic behaviour of the asymmetric terms. (Phase 1 and Phase 3, implying the offsets from ArgLat = 0° of the 1Θ & 3Θ cosine terms.) The jump appears coincident in inclination with the gap in the amplitude trend, evident in Fig. 4b). Only the 2Θ family is very well behaved over all inclinations.

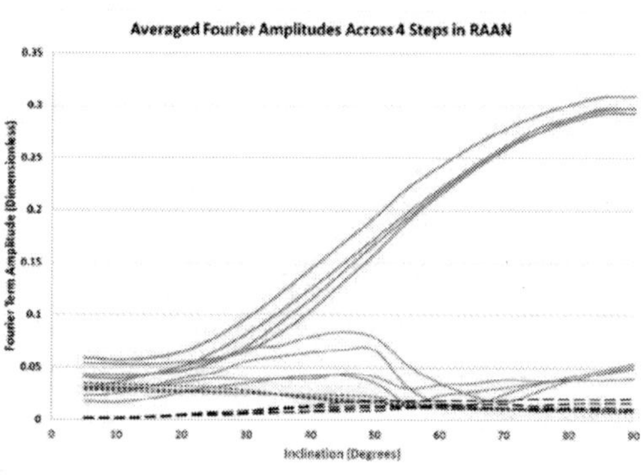

Figure 10. The amplitudes of the first four Fourier terms of the compression curves, each averaged across four orthogonal RAANs. As with Fig. 9, each of four parallel curves starts lower than the curve for the next lower ballistic number.

4. DISCUSSION

The physics of decay biasing are difficult to discern empirically in highly nonlinear and interacting gravitational and atmospheric density domains that are each modelled with many orders of comparable-scale and often competing perturbations. Even the 30,000+ simulations of the current study are insufficient to isolate all the pure physics, although some key effects are emerging.

The authors propose that the first four Fourier terms are sufficiently well-behaved in amplitude and phase that they can be used to generate a representative compression curve for any selected ballistic number and orbit inclination. The most problematic of these is the 1Θ term, which as seen in Fig. 6d is likely harboring real J3 gravitational effects observed by Fremeaux, *et al.* [3] along with the diurnal atmosphere effects. While the diurnal effects are large, these should be averaged out when estimating expectation of casualty for decays sufficiently far in the future that specific dates and RAANs are completely unknown.

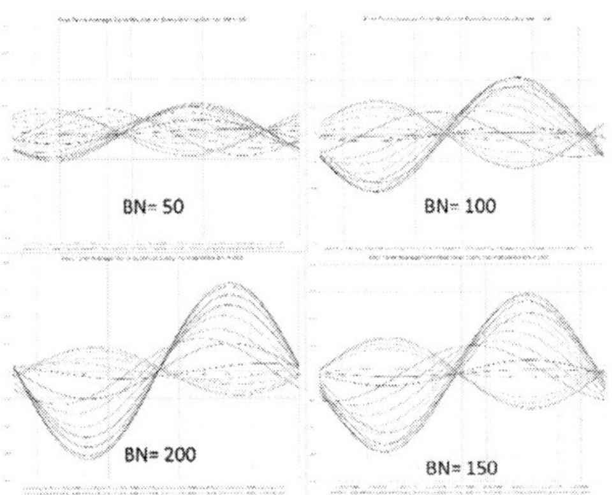

Figure 11. The average of the first Fourier term over four orthogonal RAANs at every value of ArgLat reveals a family of regular sine curves similar to that seen in Fig. 6d, whose amplitude and general regularity both grow with ballistic number. Each curve is a different inclination orbit, and all four plots are to the same scale. The curve amplitudes peak near 40° inclination and invert at the highest inclinations, but generally indicate a small preference (0.06 peak, vs 0.35 peak value in the two terms) to enter in the Northern Hemisphere for most orbits.

Although only four RAAN's were explored at the equinox (vs. the 22 at the solstice for one inclination shown in Figs. 6a-6d) the data can be similarly averaged to reveal (with some inherent noise) that the J3 effect seen in Fig. 6d is clearly at work. Fig. 11 is a compilation of the residue (average) curves of the 1Θ term as a function of ArgLat at the equinox for each of four ballistic numbers 50, 100, 150, and 200, clockwise from upper left.

The 3Θ phase term is sufficiently well behaved across all ballistic numbers and all inclinations below 85°, and its amplitude sufficiently small that it should be included, although its phase seems to be problematic. As evidenced by the point-mass gravitational study illustrated in Figs 5a-5d, the 3Θ term is expected to be atmospheric in nature. Fortunately, where the phase becomes noisy and suspect, the amplitude of the correction falls to near zero (perhaps an initial reason for the noise in estimating phase).

4.1 The Model

The authors continue to refine the model and the general understanding of the latitude biasing phenomenon in natural decays. The data presented has been used to build an *approximate* model of the compression curve for any set of conditions (see below). However, the cumulative errors in curve-fitting and then interpolating each Fourier term indicate that the most precise answer for any planned decay will be to perform a full simulation for the exact ballistic number and inclination of the vehicle in question, averaged over a continuum of RAANs (*i.e.*, every 15°) on a date near the equinox. This does not take long using the analytical tools developed for this study.

The next best model is to linearly interpolate the 8 Fourier coefficients in phase and amplitude for any given ballistic number and inclination between the available 288 data sets. This model is what will actually be rolled into screening software and quick-study tools, as it is a tabular lookup that can run well in a spreadsheet. This method takes many thousands of data elements that cannot be presented easily in a publication, and which will be under ongoing update and refinement.

The expedient, but least accurate approximate model presented here is an empirical derivation that—while far from perfect—is substantially more accurate than the prior assumption of a fully random decay along the ArgLat, and is useful for broad parametric studies to determine peak areas of concern in calculating E_c. This model uses a set of "eyeball fit" empirical curves to model the change in coefficients.

The model is built as follows:
Each Fourier term F_N of Nth order has an Amplitude A_N and a phase offset Φ_N to create a term of the form:

$$F_N = A_N * Cos\ (N*\Theta + \Phi_N)$$

In the equations that follow, ***BN*** is the ballistic number in kg/m^2, and ***INC*** is the inclination of the orbit in degrees.

a) $F_0=A_0=1$

b) The first Fourier term represents the averaged first-term contributions over all RAANs for any given ballistic number. The resulting average is expected to be the effect of J3 gravitational perturbation as the principal 1Θ driver into the nonlinear atmosphere. Until further simulations reveal more detail, the evident simplest trend from the preliminary 4-RAAN average is a family of curves with the same phase:
   a. The amplitude of the first Fourier Term is estimated as the empirical fit: $A_1=.0418+.000125*\boldsymbol{BN}*Sin((5.86*\boldsymbol{INC})+227°)$
      i. Note: the amplitude goes negative
   b. $\Phi_1 = 0°$
      i. *for low* ***BN***
      $\Phi_1 = 100°*sin(2*\boldsymbol{INC}+40°)$
      ii. *for BN >50*

c) The dominant second term F_2 follows a $\sin^3$ fit to inclination:
   a. $A_2= .0498 +0 .2576*sin^3(\boldsymbol{INC})$
   b. $\Phi_2 = 192°*\boldsymbol{BN} -170$
      i. *for INC >45°*
      $\Phi_2 = 70°*\boldsymbol{BN}$
      i. *for INC <45°*

d) The F_3 and F_4 terms should most correctly be estimated from the averaged curves in all RAANs. This work has not been completed, and will be better done once a broader set of RAANs have been studied. The two terms are estimated from trend data at all four RAANs as follows:
   F_3:
   a. $A_3= .0334 -.0003*\boldsymbol{INC}$
   b. $\Phi_3 =(\boldsymbol{BN})^{0.85}-60°$

e) The F_4 term follows
   a. $A_4=.0206-(8.5E-7)*(400-\boldsymbol{BN})*\boldsymbol{INC}$
      i. *for* ***INC*** *< 55°*
      $A_4=0.0206-(4.683E-5)*\boldsymbol{BN}$
      ii. *for* ***INC*** *>55°*
   b. $\Phi_4 =130 - 0.78*\boldsymbol{BN} + (\boldsymbol{BN}/50)*(\boldsymbol{INC}-50°)$
      i. *for* ***INC*** *< 55°*
      $\Phi_4 = (135-\boldsymbol{BN}/8)-\boldsymbol{INC}$
      i. *for* ***INC*** *> 55°*

The fits of these respective parametric representations to the averaged Fourier terms is illustrated in Fig. 12.

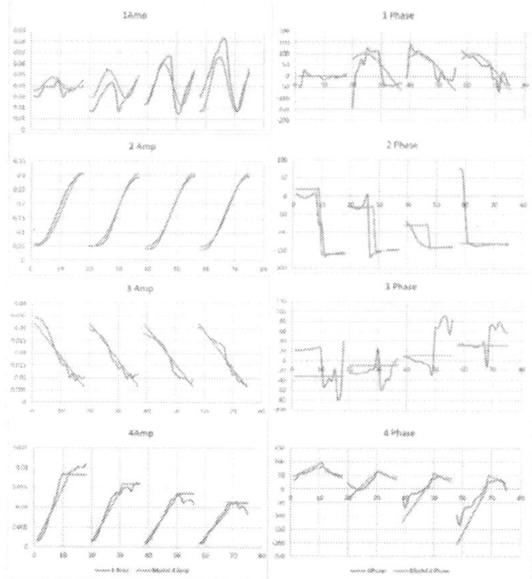

Figure 12. The empirical fits described in this section to the Fourier coefficients for compression curves at each inclination, averaged across four orthogonal RAANs at the Vernal Equinox. The four curves in each plot represent the coefficients for the four studied ballistic numbers in ascending order left to right. The jumps in the 2Θ and 3Θ phase curves do not lend themselves to simple models, and reconstructed curves in the regions of these breaks can show obvious qualitative differences.

Two representative fits of the empirical curve to the averaged scenario and to representative original simulation data are shown in Figs. 13a and 13b for orbits not situated within 5° of a break point in the empirical curve sets.

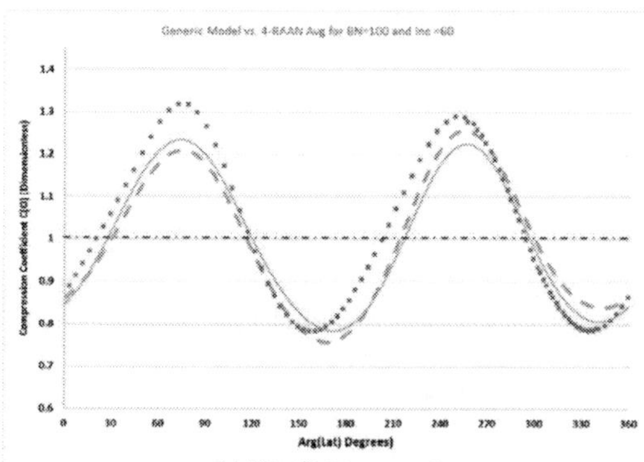

Figure 13a. Empirical (dashed) fit to the 4-RAAN average (solid) model for moderate BN and inclination. Data trace (points) from 270° RAAN is included to demonstrate how actual conditions will typically vary from the average on any given day.

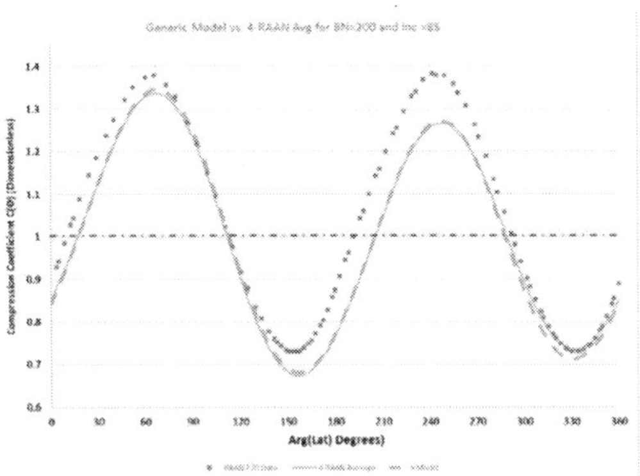

Figure 13b. The same comparison for high BN and inclination. The fit to the 4-RAAN average is excellent, but the statistical noise for any particular day of entry (in the form of the diurnal variation) is more evident.

4.2 Final E_c Calculation

With the averaged concentration factor available at each latitude, the E_c for a particular decay can finally be calculated with a slight modification to the traditional approach. The ArgLat at 90 km is incremented in small steps from 0° to 360°. The concentration factor is calculated for the ArgLat at 90 km for the specific orbit inclination and spacecraft ballistic number. A necessary pad of a few degrees is added to the ArgLat (variable with the ballistic number of the object in question) to propagate the debris farther downrange to the ground. (The calculation of this pad is beyond the scope of this paper.) The physical latitude on the planet associated with the adjusted impact ArgLat (Θ_{impact}) is calculated by the simple equation:

$$Latitude = ARCSIN(SIN(INC)*SIN(\Theta_{impact})) \quad (1)$$

The population in any given latitude band is extracted from the gridded population of the world, with a model of population growth applied. In the critical adjustment, this population is divided by the concentration factor associated with the 90-km ArgLat Θ to give an adjusted risk. The binomial correction that must be applied to correct for a normalized Probability Function is $P=(1/C(\Theta))-(1-C(\Theta))^2$. Note that the impact latitude will be encountered twice during this process (ascending and descending), with a different concentration factor applied. Each incremental risk is multiplied by the size of the incremental step in ArgLat divided by 360°, such that the sum of all steps is the total risk over the orbit.

The resulting mean population per square kilometer can be mapped as a function of inclination and ballistic number, as shown in Fig. 14, illustrating the difference in resulting risk under the new model, compared to the prior randomized-entry assumption. Expectation of casualty is a linear multiple of the mean exposed population per square kilometer, scaling as well with Debris Casualty Area. The ratio of the average populations under various inclination orbits is a weak function of ballistic number, as shown in Fig. 15.

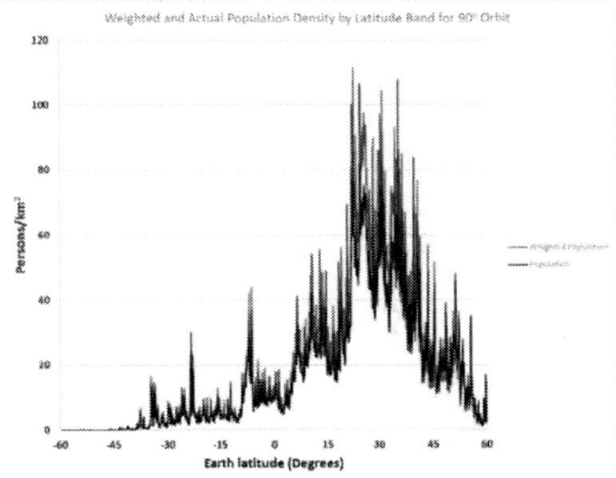

Figure 14. Weighted vs unweighted 2020 population under a 90° orbit

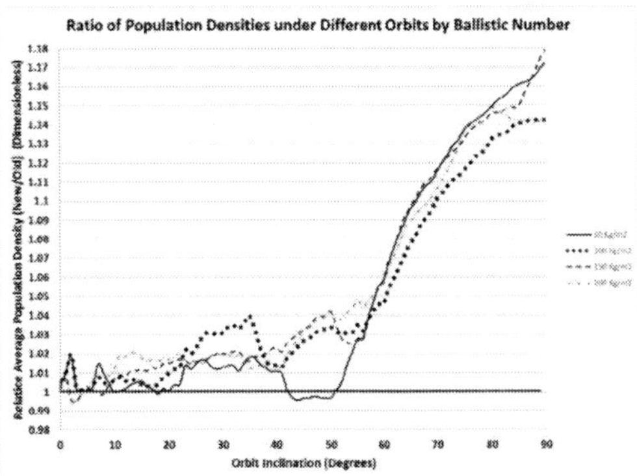

Figure 15. Ratio of mean population density under compression curve vs. the assumption of uniform decay around the orbit, as a function of orbit inclination for four ballistic numbers. A Θ_{impact} of 13.5 degrees downrange after the 90km altitude is included in this calculation.

4.3 Comparison with Prior Models

Prior work [1] identified a sinusoidal perturbation of the decay probability as a function of latitude, and implied a full symmetry of the effect. On an average case, this is approximately true, but for any particular orbit, the present work now shows the decay to be statistically clustered more strongly on the *approach* to the equator, as explained above. The physics of this has been discussed in [2], and there are significant ramifications for exceptionally low-dV intentional tactical de-orbit scenarios. In natural decays which usually occur over an indeterminate, broad timescale, many of these subtleties largely average out, but still have influence on the entry statistics.

In the current work, the representation has been shifted to a function of *Argument* of latitude, and not latitude, to achieve a more universal formulation, and to search for underlying physics. The current work also recognizes and accommodates remaining asymmetries in statistical decays, and refines the model for ballistic number differences.

The prior method of estimating E_c incorporated a rudimentary concentration factor based upon how long a circular orbit dwells over any given latitude. Typically, this correction is applied symmetrically around the equator in an algorithm to adjust the latitude-averaged population bands in strips only a few tens of kilometres wide to account for this circular "dwell time". In the method described within this paper, this correction is not applied independently, substituting instead a localized concentration factor in tiny numeric integration steps in ArgLat, applied and integrated band by band. Note that the prior method actually places stronger likelihood of decay at the latitude extremes, and minimizes the effect at the equator, in complete opposition to the effect explored here. The effective weighting of population is evident in Fig. 14.

To conclude the forward work described in the authors prior study [1], the influence of F(10.7) solar flux variations was shown in [2] to be negligible in the critical final eight orbits when the decay rate is geometrically increasing. This is because the expansion of the atmosphere results from radiant energy deposited above the altitudes where the lowest density scale heights (and thus, maximum effect of perturbations) occur.

Conversely, the global geomagnetic index (Kp) was shown to be the driving factor to disturb low altitude atmospheric density. Geomagnetic index (updated every 3 hours) determines the energy deposition of charged particles in the altitudes driving the final few orbits of decay. From a statistical standpoint, the Kp variation is rapid and random enough to only add noise to the more consistent geodetic effects of the wall of air and of J2 perturbations. That said, the effect of Kp is significant when attempting a drag-centric targeted decay on any particular day.

5. CONCLUSIONS

The statistical clustering of natural decays towards the equator has been reduced to an approximate parametric representation accounting for ballistic number, inclination, and gravitational perturbations, useful for parametric studies and potentially for insight into the physics. Some of the corrections to prior practice can lead to localized E_c changes of up to 36% over the random-decay assumption, and net integrated E_c's that are 18% or more different than prior practice would estimate. There is strong evidence that a slight bias of natural decays occurs in the northern hemisphere due to the J3 gravitational effect.

However, the chaotic nature of some parameters indicates that best precision will come from a dedicated simulation of each decay's ballistic number and orbit inclination (and especially, if known, the date and range of likely RAANs for the expected decay), rather than the inherently "fuzzy" approximation to the specific case in a set of global parameters that seek to optimally fit a broader range of cases.

The clustering is most conveniently rendered as a function of the ArgLat, which can then be transformed to latitude. With the local compression factor available, it is a minor adjustment to existing look-up tools to generate a statistically-weighted E_c that accounts not only for the inhomogeneity in the population density with latitude, but as well for the inhomogeneity in the distribution of decays around the orbit.

6. REFERENCES

1. Bacon, J.B. & Matney, M.J. (2016). *Statistical Issues for Calculating Reentry Hazards*, IAASS 8th Workshop on Space Safety, May 21 2016.

2. Bacon, J.B (2017). *Minimum dV for Targeted Spacecraft Disposal* Proc. 7th European Conference on Space Debris, Darmstadt, Germany, 18–21 April 2017 (http://spacedebris2017.sdo.esoc.esa.int)

3. Cid Borobia, E., Fremeaux, C., and Goester J.F.: *Fast Re-Entry Deorbitation with Acceptable Risk Level* Proc. 7th European Conference on Space Debris, Darmstadt, Germany, 18–21 April 2017 (http://spacedebris2017.sdo.esoc.esa.int)

IMPROVING ESTIMATION OF GROUND CASUALTY RISK FROM REENTERING SPACE OBJECTS

Chris L. Ostrom[(1)]

[(1)]*HX5 – Jacobs JETS Contract*
NASA Johnson Space Center, XI4-9E, 2101 NASA Pkwy., Houston, TX 77058, USA
christopher.l.ostrom@nasa.gov

ABSTRACT

A recent improvement to the long-term estimation of ground casualties from reentering space debris is the further refinement and update to the human population distribution. Previous human population distributions were based on global totals with simple scaling factors for future years, or a coarse grid of population counts in a subset of the world's countries, each cell having its own projected growth rate. The newest population model includes a 5-fold refinement in both latitude and longitude resolution. All areas along a single latitude are combined to form a global population distribution as a function of latitude, creating a more accurate population estimation based on non-uniform growth at the country and area levels.

Previous risk probability calculations used simplifying assumptions that did not account for the ellipsoidal nature of the Earth. The new method uses first, a simple analytical method to estimate the amount of time spent above each latitude band for a debris object with a given orbit inclination and second, a more complex numerical method that incorporates the effects of a non-spherical Earth. These new results are compared with the prior models to assess the magnitude of the effects on reentry casualty risk.

1. INTRODUCTION

The population database is used to estimate the casualty risk from surviving reentry debris [1]. The average population density under the satellite track is a critical factor in computing the probability of any reentering debris striking a person. Planning for uncontrolled reentries in a future year, the longitude dependency (both of the population distribution and of the reentry point) can be considered randomized, due to the precession of the orbit and the rotation of the Earth beneath the orbit. Thus, the population data presented here vary only as a function of latitude [2]. The orbital inclination of an orbiting object dictates the amount of time spent in each geographic latitude band. Each of these latitude bands has a different number of people living in it, so the density of people under a satellite's path is the average population density in each latitude band, weighted by the fraction of time the satellite travels over each band. The 'inclination-averaged' population density and casualty risk probabilities are computed outside the population database.

This population database required an update as the data currently used are considered obsolete, being from the Gridded Population of the World, version 2 (GPWv2), released in 2001, and the U.S. Census Bureau's International Programs Center (IPC) International Data Base (IDB) projections from May 2000. The new database is based on the Gridded Population of the World, version 4 (GPWv4), released in 2015, and the updated IDB from August 2016. As with the last update to the NASA reentry casualty risk model [2], there is a significant refinement in resolution of the global population (seen in Tab. 1).

Table 1. Comparison of Population Models

	Global Model	GPWv2	GPWv4
Year(s) of Estimation	1994	1990, 1995	2000, 2005, 2010, 2015, 2020
Grid Resolution	-	2.5 arc-minute (~5 km)	30 arc-second (~1 km)
Number of Input units (subnational geographic units)	1	127,000	~12,500,000

2. POPULATION DATA UPDATE

The updated global population database is a combination of two different data sources. The GPWv4 provides the total population and its distribution within each country's borders. The population growth projections from the IDB are then applied to each country's distribution and total population.

2.1. Gridded Population of the World, Version 4

The GPWv4 is produced by the Socioeconomic Data and Applications Center (SEDAC). SEDAC is a program at Columbia University's Center for International Earth Science Information Network and is part of NASA's Earth Observing System Data and Information System.

GPWv4 data are gridded in latitude and longitude with a resolution of 30 arc-seconds [4]. The publicly available data cover 243 countries and areas on six continents between the latitudes of 85° N and 60° S. The data are presented in a GeoTiff format, which allows georeferenced data to be included in an easy to read image file. Sample plots of the GPWv4 raw data can be seen in Figs. 1 and 2. For comparison, the same data from the previous model (GPWv2) are presented in Fig. 3 (reproduced from [5]) and Fig. 4. The population counts and densities are adjusted to the United Nations estimates for the years 2000, 2005, 2010, 2015, and 2020 [4]. All these data files, associated documents, and metadata are available for download on the internet from the SEDAC website: http://sedac.ciesin.columbia.edu/.

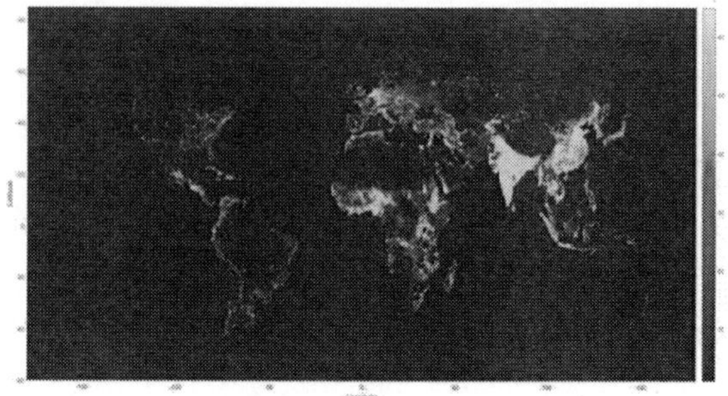

Figure 1. GPWv4 Data: U.N.-Corrected 2015 population grid of the world.

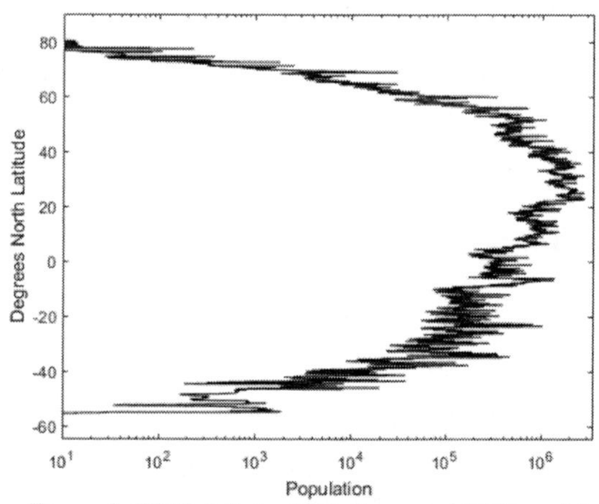

Figure 2. GPWv4 Data: U.N.-Corrected 2015 world population by latitude.

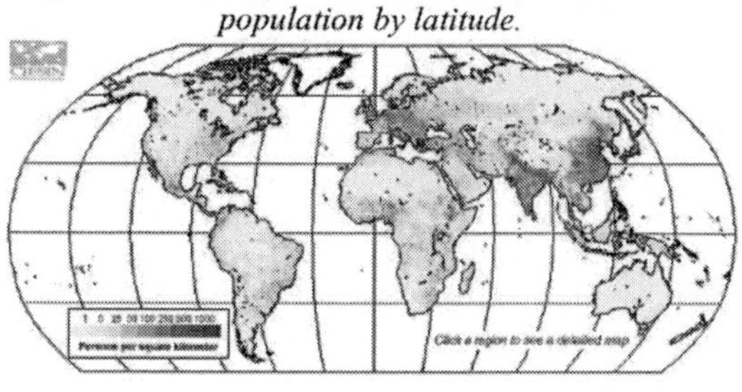

Figure 3. GPWv2 Data: U.N.-Corrected 1995 population grid of the world.

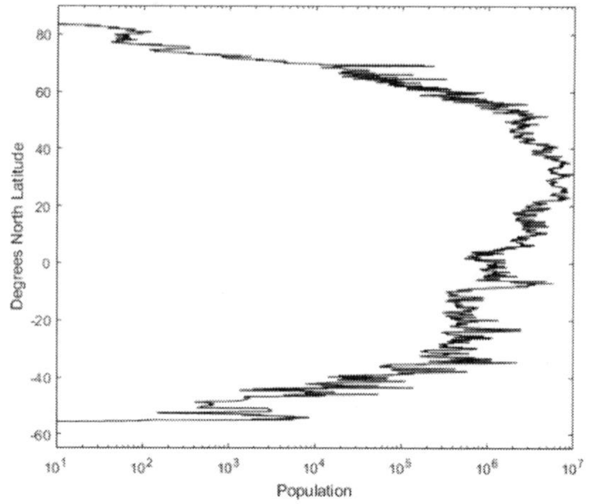

Figure 4. GPWv2 Data: U.N.-Corrected 1995 world population by latitude.

2.2. International Data Base

The IDB is produced by the IPC of the United States Bureau of the Census. The population projection data are based on projections from national statistics offices, the United Nations, and census estimates. The IDC data used here are from the latest update of August 2016 [6].

The IDB currently lists 228 areas with 2015 midyear populations of more than 5,000 [3]. The countries and areas are listed by name, and the database provides detailed annual information on all areas for the years 1950-2050. IDB data are available via Internet at www.census.gov/population/international/data/idb.

3. PREDICTING FUTURE POPULATION

The IDB and GPWv4 datasets do not use precisely the same countries and areas, so the populations had to be reconciled. This was done by examining the documentation for each source and isolating which areas were identified individually in one source but collected in the other. For example, the GPWv4 includes populations for French Guiana, Guadeloupe, and Martinique, but IDB combines them with metropolitan France [3,4]. There are a total of 227 countries and areas that the two databases have in common, plus the global totals. This process was done using a spreadsheet, comparing each dataset line by line. Once this was complete, the gridded population data were saved for later input into the data processing code.

The grid population predictions are generated by applying the annual projected population (given by IDB) to each country or area's fixed distribution (as given by GPWv4), cell by cell. This approximation does not allow for general urbanization trends worldwide, and forces the rural and urban populations to grow or shrink in unison within any country's borders. A future enhancement may be to apply a correction that accounts for the global urbanization trend. Such a correction is beyond the scope of the current work. Once this is complete, the cells are summed over longitude to re-create the latitude bands that are then used to compute the inclination-averaged population density. This process is summarized in Fig. 5.

This software creates tables of population and population density as a function of year and latitude, but the primary output is the annual global population distribution by 30-arcsecond latitude bin for the years 2000-2050 (with the years 1990-2000 available, but at a coarser resolution).

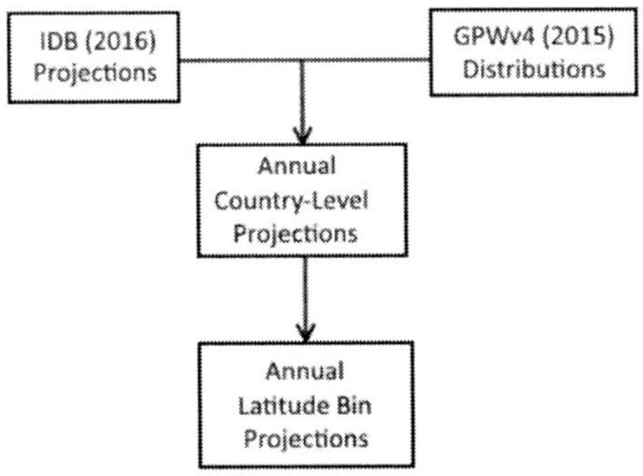

Figure 5. Data processing flow diagram.

There are two limitations to the output data produced by this scheme: the population prediction method, and the input data. The output data can only be as good as the sources of the input data (IDB and GPWv4). These are well-known and trusted data products, produced through research supported by the United Nations, the U.S. government (including NASA), and the individual countries and areas. GPW is the only source of global gridded population numbers (LandView is another gridded population dataset, produced by the U.S. Census Bureau, but does not cover non-U.S. populations), and since it does not provide a means of projecting population changes in the future, we rely on the IDB. There are many sources for population predictions; these sources (including IDB) typically rely on United Nations and individual countries and areas for data. The IDB was chosen because it uses these standard sources, and it is currently used by the previous version of this software.

The second limitation is the application of the population prediction method. The future population (by latitude bin) of each country is computed using the distribution found in GPWv4 in the year 2015. This method was previously used by Matney, *et al.*, in 2003 [2]. This method is useful, in that each country or area can have its population develop over time, but there is a potentially serious limitation, in that the relocation of population to areas not populated in the GPWv4 2015 baseline data set is not modeled. Fig. 6 shows the estimated change in world population density from 2000 to 2050, using the 2015 GPWv4 distribution. Fig. 7 shows the differences in the latitude-binned population densities predicted using the previous method (with 1995 baseline GPWv2 distribution and 2001 IDB population projections) and the baseline data sets provided in the GPWv4 product (the years 2005, 2010,and 2015).

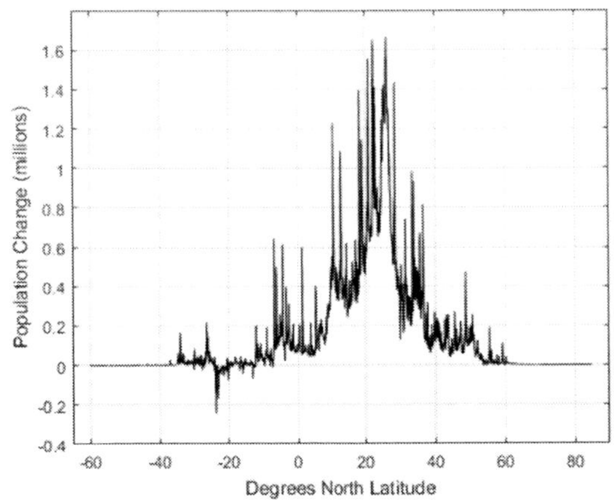

Figure 6. Predicted growth of global populations, 2000-2050 (2015 distribution)

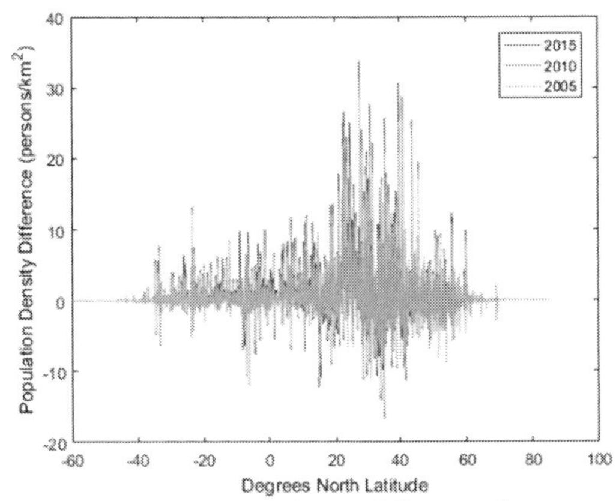

Figure 7. Population density differences between GPWv2 projections and GPWv4 baseline data.

4. INCLINATION-BASED, LATITUDE-AVERAGED POPULATION DENSITY

Objects that are decaying sufficiently far (months to years) in the future can only be predicted to reenter in a probabilistic manner. Because of the earth's rotation and the precession of the nodes, an object will not land preferentially on any longitude (which is why we have averaged the population density along each latitude band). However, the amount of time an object spends above each latitude is not equal, as is described by the probability distribution function (PDF) Eq. 1 [2], where $-\sin(i) \leq \sin(\lambda) \leq \sin(i)$ (λ is the geocentric latitude, and i the satellite orbit inclination). The satellite does not travel over latitudes where this relation does not hold, so the PDF will be zero outside of this range. The PDF is then multiplied by the latitude-binned population density and integrated to yield the inclination-based, latitude-average population density (see Eq. 2).

Current work by Bacon and Matney [7] has further refined the probability function to include a latitude bias towards the equator in final decay. Such bias derives from the atmosphere's approximately uniform profile above the geoid, creating an effective "wall of air" at equatorial crossings in the late decay stages. Because the correction for this phenomenon is a function of both the ballistic number and of the orbit's inclination, the correction is not modeled within the scope of this geographic study. Their correction algorithm substitutes for this more generic function when solving for a particular spacecraft and a particular orbit.

$$P(\lambda)d\lambda = \frac{\cos\lambda}{\pi\sqrt{\sin^2 i - \sin^2\lambda}}d\lambda \quad (1)$$

$$\overline{\rho_{pop}} = \int_{-i}^{i}\rho(\lambda)P(\lambda)d\lambda \quad (2)$$

Previous models assumed that the earth was spherical, and thus, that the geographic/geodetic latitude (as shown on maps, measured in the local horizontal on the surface of an elliptical Earth) of an object and its geocentric latitude (the angle measured at the Earth's center from the plane of the equator to the point of interest) are identical. There is however, a potentially significant difference (of up to 0.19° at 45° latitude) between these two latitudes, the geocentric (true) latitude being greater than the geographic latitude in the northern hemisphere, and the geographic latitude being greater than the geocentric latitude in the southern hemisphere. These two angles are connected through the relationship in Eq. 3, where λ_g is the geographic/geodetic latitude and λ is the geocentric latitude, and $e_\oplus = 0.08182$ is the eccentricity of the Earth's oblate ellipsoidal shape. Fig. 8 demonstrates the difference between these angles.

$$\tan\lambda = \left(1 - {e_\oplus}^2\right)\tan\lambda_g \quad (3)$$

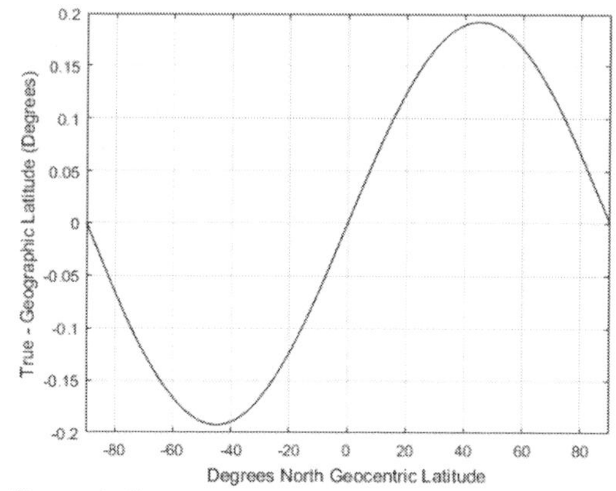

Figure 8. Geocentric (true) latitude minus geographic latitude as a function of geocentric (true) latitude.

The oblate nature of the earth not only causes the latitude angles to shift away from the equator, but also causes the area of each grid cell to change slightly. Eq. 4 gives the area of a grid cell on a spherical earth at a given geocentric latitude and longitude. Eq. 5 gives the grid cell area as a function of geocentric latitude and longitude on an ellipsoidal earth. The difference in areas was computed using these two methods as a function of geocentric latitude, and is seen in Fig. 9. This effect is much smaller than the change in latitude (this is, at worst, less than 0.2% error).

$$dA_{sph} = r^2 \sin\lambda \, d\lambda d\varphi \quad (4)$$

$$dA_{ell} = r(\lambda,\varphi)\cos\varphi\,[r(\lambda,\varphi) + r(\lambda,\varphi+\Delta\varphi)\cos(\varphi + \Delta\varphi)(1-\cos\Delta\lambda)]d\lambda d\varphi \quad (5)$$

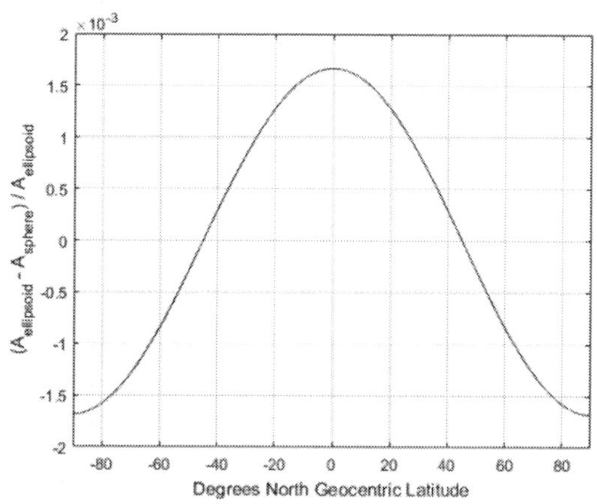

Figure 9. Relative difference in grid cell area in spherical and ellipsoidal Earth models.

Now that the population density has been corrected to geocentric latitude, and the areas have been adjusted (already completed and included in the GPWv4 data products), the inclination-based, latitude-averaged population density can be computed for each year (2015 and 2050 seen in Fig. 10).

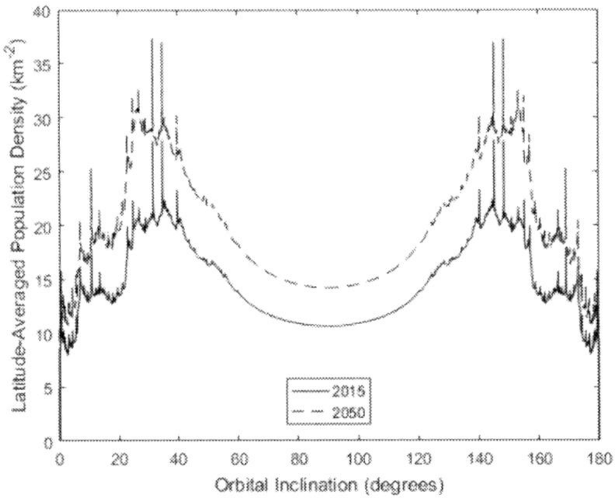

Figure 10. Latitude-averaged population density beneath a satellite, as a function of orbital inclination.

5. REENTRY CASUALTY RISK CALCULATION

The inclination-based population density can be used to compute the "acceptable" debris casualty area (DCA) for a reentry object, using Eq. 6. The factor of 10^{-4} in Eq. 6 is the level of risk specified in requirement 4.7-1 in NASA-STD 8719.14B [1]. Fig. 11 shows the difference in allowable DCA as calculated from Equation 6, applied to former population densities as calculated in Matney, *et al.* [2] and the derived densities under the new method, shown in Fig. 10.

$$DCA = \frac{0.0001}{\rho_{pop}} \quad (6)$$

6. CONCLUSIONS

There has been a significant increase in population density, globally, over the last 15 years, as shown in Fig. 7. This, along with finer resolution data, has contributed to a large reduction (as much as 67% in some inclination ranges) of "acceptable" DCA, seen in Fig. 11. Two effects of Earth asphericity (geocentric and geographic latitude difference, and grid cell size) have been examined, and have been incorporated into the new population-density averaging model and DCA calculation. The inclination based population density can differ by as much as 10% due to the difference in latitude, while the effect of unequal grid cells is much smaller, with a worst-case error of 0.2%.

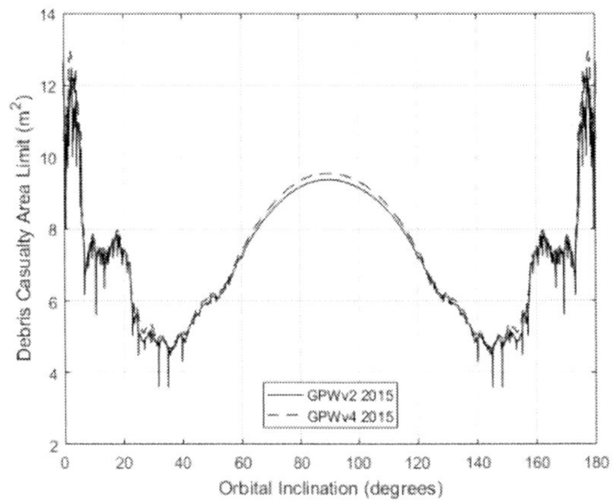

Figure 11. Maximum Debris Casualty Area to meet 1:10000 casualty risk, as a function of inclination. Comparison of GPWv2 densities on a spherical Earth, and GPWv4 densities on an ellipsoidal Earth.

7. FUTURE WORK

These estimates assume that the debris objects do not bounce off the ground, and the entire human population is assumed to be outdoors. Future augmentations to the reentry casualty risk model include accounting for global urbanization trends, the tendency for objects to reenter at latitudes nearer the equator than the poles, and any improvements to the calculation of DCA.

8. REFERENCES

1. NASA Technical Standard – Process for Limiting Orbital Debris, (May 2012). NASA-STD-8719.14A.

2. Matney, M.J., Opiela, J.N. (October 2003). "Improvements to NASA's Estimation of Ground Casualties from Reentering Space Debris," 54th International Astronautical Congress of the International Astronautical Federation, the International Academy of Astronautics, and the International Institute of Space Law, Bremen. Germany.

3. United States Bureau of the Census, "International Data Base." Available at: https://www.census.gov/population/international/data/idb/. Accessed 16 June 2017.

4. Center for International Earth Science Information Network - CIESIN - Columbia University. (2016). Gridded Population of the World, Version 4 (GPWv4): Population Count Adjusted to Match 2015 Revision of UN WPP Country Totals. Palisades, NY: NASA Socioeconomic Data and Applications Center

(SEDAC). Available at: http://dx.doi.org/10.7927/H4SF2T42. Accessed 16 June 2017.

5. CIESIN (Center for International Earth Science Information Network), Columbia University; International Food Policy Research Institute (IFPRI); and World Resources Institute (WRI). 2000. Gridded Population of the World (GPW), Version 2. Palisades, NY: CIESIN, Columbia University. Available at: http://sedac.cisin.columbia.edu/plue/gpw. Accessed 16 June 2017.

6. United States Bureau of the Census, "International Data Base: Population Estimates and Projections Methodology". Available at: http://www.census.gov/population/international/data/idb/estandproj.pdf. Accessed 16 June 2017.

7. Bacon, J.B., Matney, M.J.(October 2017). "Oblate-Earth Effects on the Calculation of Ec During Spacecraft Reentry", 9th International Association for the Advancement of Space Safety Conference, Toulouse, France.

OXIDATION LAWS AND EMISSIVITY DATA AT HIGH TEMPERATURE FOR IMPLEMENTATION IN DEBRISK CODE

L. Barka (1), M. Balat-Pichelin (1), E. Bêche (1), J-L. Sans (1), J. Annaloro (2), P. Omaly (2)

(1) *PROMES-CNRS laboratory, 7 rue du four solaire, 66120 Font-Romeu Odeillo, France*
marianne.balat@promes.cnrs.fr

(2) *CNES, 18 avenue Edouard Belin, 31401 Toulouse cedex 9, France*
julien.annaloro@cnes.fr

ABSTRACT

In order to mitigate debris in orbit and to avoid dramatic collisions on Earth after atmospheric re-entry at its end of-life, the spacecraft missions have to take into account the influence of re-entry on the spacecraft survivability. In this way, oxidation in air plasma conditions was studied on metallic alloys such as stainless steel 316L and Invar in order to obtain oxidation kinetic laws at high temperature and for short time duration (few minutes). Emissivity measurements at high temperature were also carried out on the pure alloys – in high vacuum to avoid oxidation – and on oxidized samples in air plasma and in standard air conditions. An important increase of the total emissivity for oxidized surfaces by a factor of nearly 3 or 4, depending on the material is obtained compared to the non-oxidized surfaces. Experimental results obtained for 316L and Invar with microstructural characterization using SEM, XRD and 3D roughness profilometry are presented.

1. INTRODUCTION

Materials used for future spacecraft missions must obey specific rules such as LOS (Loi relative aux Operations Spatiales, France), Clean Space Initiative of ESA… in order to be mainly or totally destroyed under the effect of severe aerothermodynamics and aero-thermochemical conditions encountered during the atmospheric re-entry. The non-equilibrium air plasma flow impinging the spacecraft materials enhances oxidation phenomena and also heat flux by species recombination and/or emissivity increase. Oxidation, ablation and melting phenomena can occur on the material leading to an important increase of the temperature and damage of the materials.
To study and to assess the debris survival risks during re-entry, CNES has developed its own engineering code, DEBRISK [1, 2]. It allows calculating the 3D trajectory of the satellite components, the surface heat flux and the debris disappearance altitudes during the re-entry. However, databases currently available in DEBRISK are limited for oxidation kinetics laws and for thermal radiative properties (emissivity) of materials at high temperature, parameters needed for the calculation of the final mass reaching the ground.
To increase the calculation accuracy of the DEBRISK code, CNES collaborates with PROMES-CNRS laboratory to perform oxidation in atmospheric re-entry conditions and to measure emissivity at high temperature. Oxidation kinetic laws and emissivity data obtained will be implemented in the DEBRISK code.
In the literature, there are few studies on oxidation of 316L and Invar at high temperature. Oxidation tests were mainly performed in standard air [3-6] and a little realized in air plasma or CO_2 conditions [7, 8] at lower temperature than the ones we are interested in. No emissivity data can be found in the literature for both the materials, hence the importance to perform oxidation in air plasma conditions and emissivity measurements at high temperature.

2. EXPERIMENTAL EMISSIVITY MEASUREMENT AND DEVICE

The method developed at PROMES-CNRS laboratory for emissivity measurements is a direct one. It consists of the measurement of two parameters, the true temperature and the total or spectral directional radiance of the sample in a given wavelength range [9]. The total or spectral directional emissivity (ε') is the ratio of the measured radiance (L) of the material to the part of the blackbody radiance (L_0) at the same temperature and wavelength range according to the following relation (1) expressed as:

$$\varepsilon' = L(T) / L_0(T) \qquad (1)$$

The hemispherical corresponding emissivity (ε) is calculated by integration of the directional values.
The experimental set-up MEDIASE (Moyen d'Essai et de DIagnostic en Ambiance Spatiale Extreme) shown in Fig. 1 is placed at the focus of the 1 MW Odeillo solar furnace in order to reach very high temperature in a few seconds. It allows studying the physico-chemical behavior of materials under extreme conditions at very high temperature (up to 2500 K). The set-up consists of a stainless steel vessel with a capacity of 6.10^{-2} m^3, equipped with a turbo-molecular pumping system allowing working in high vacuum conditions (up to 10^{-5}

Pa). A hemispherical silica-glass window (35 cm diameter) allows the heating of the sample (40 mm diameter, 2 mm thickness) with concentrated solar radiation by the gradual opening of the doors of the solar furnace, the high vacuum being preserved inside the vessel. The true temperature of the sample is measured on the back face using a two-color pyro-reflectometer developed in our laboratory [10] and calibrated on a blackbody before and after each measurements series up to 1900 K. The directional radiance measurements are performed using the radiometer equipped with a Cassegrain optics (KT4 Heimann, Germany). According to the needed wavelength range, the radiometer is used without filter (0.6-40 μm) or with narrow spectral filters (2.7; 5 or 5.5 μm) or spectral band filters (0.6-2.8; 3-5; 7-10; 8-14 μm). Directional radiance measurements are performed with an original motorized three-mirrors system goniometer, developed in our laboratory, in such a way the sample and the radiometer stay fixed. The directional data are acquired from 0 to 80° for a sample diameter of 40 mm. The window placed between the sample and the radiometer is adapted for the used wavelength range. It is composed of thallium iodobromide (KRS5) in order to perform measurements from 0.6 to 40 μm. The whole system radiometer-optical window-goniometer is calibrated before and after each series of measurements on a blackbody as for the optical pyro-reflectometer. The accuracy on emissivity is around 5% between 1500 and 2300 K for the spectral range of 0.6 to 40 μm.

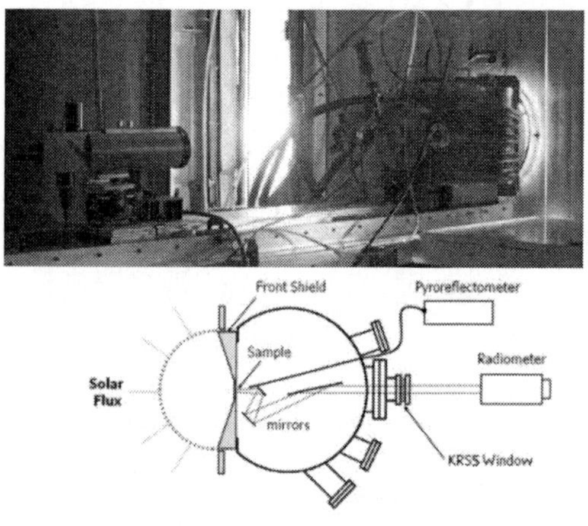

Figure 1. Picture and scheme of the MEDIASE device in the emissivity measurement configuration.

3. OXIDATION IN AIR PLASMA CONDITIONS AND SET-UP

The experimental set-up MESOX (Moyen d'Essai Solaire d'OXydation) shown in Fig. 2 is placed at the focus of the 6 kW solar furnace equipped with a variable opening shutter. It allows simulating on Earth the encountered conditions during the atmospheric re-entry (air plasma, high temperature and low pressure) of materials [11]. The set-up is equipped with a microwave generator (2450 MHz, 1200 W max) and a waveguide for creating non-equilibrium air plasma. A regulator, a gauge and a vacuum pump allow controlling precisely the total air pressure during the experiment. The incident concentrated solar flux can reach 5 $MW.m^{-2}$. The temperature measurement is performed on the front face of the sample by using a monochromatic (5 μm) optical pyrometer with a system of two mirrors. In this study, air plasma is created by a 300 W discharge surrounding the sample (15 x 15 x 2 mm^3 for oxidation studies and 25 mm diameter, 2 mm thickness for emissivity measurement) and the total air pressures are fixed at 300 and 2000 Pa. The air flow is set at 4 l/h to obtain a maximum oxygen dissociation rate of 80% [12]. The mass variation of the sample before and after test was measured to monitor the influence of the oxidation on the material. Microstructural characterizations were also carried out by different techniques such as SEM, 3D roughness profilometry and XRD before and after oxidation to know the morphology and the composition of the oxide layer formed on the sample surface.

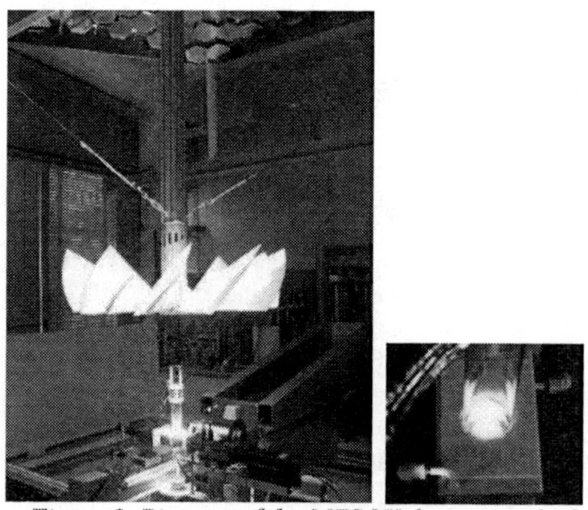

Figure 2. Pictures of the MESOX device (on the left) and air plasma (on the right).

4. EXPERIMENTAL RESULTS

In this section, the metallic alloys selected and the experimental results obtained for the study of the oxidation in air plasma conditions, and for the emissivity measurement are reported and discussed.

4.1. Materials

The materials used in this study are metallic alloys such as 316L and Invar provided by Goodfellow. The compositions of each material are the following for the main components:
- 316L: Fe 70%; Cr 17%; Ni 10 %; Mo 2%; Mn 1%
- Invar: Fe 64% and Ni 36%.

4.2. Oxidation in air plasma conditions

316L and Invar were oxidized in air plasma conditions at 300 and 2000 Pa up to their melting temperatures (Tm respectively 1670 and 1700 K) during short duration.
Oxidation tests were performed on parallelepipedic samples (15 x 15 x 2 mm^3) in order to obtain precise kinetic laws due to the homogeneity temperature on their surface. The Invar surfaces before and after oxidation at 300 Pa during 300 s for different temperatures are presented in Fig. 3.

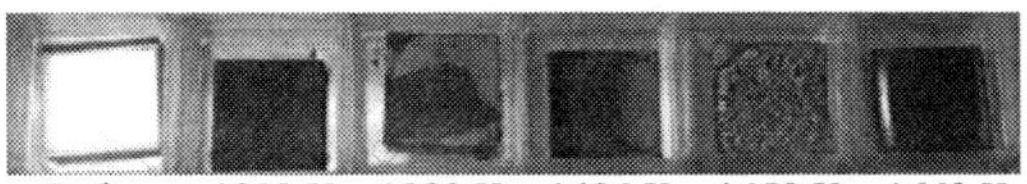

Ref. 1233 K 1380 K 1496 K 1652 K 1663 K

Figure 3. Pictures of Invar samples non-oxidized (ref.) and oxidized in air plasma at 300 Pa for different temperatures.

The influence of exposure to air plasma at high temperature is clearly visible on the Invar oxidized surfaces. It can be observed the color and the structure changes of the oxide layer, depending on the temperature. At low temperature, 1233 and 1380 K, the oxide layer is brown and non-adherent. With the temperature increasing at 1496 K, the oxide layer is adherent and changes its color, brown to gray, and becomes totally gray for higher temperatures. Moreover, holes and porosity are visible in several places on the surface at 1652 K that disappeared around 1663 K (close to Tm).
The mass variation (in $g/m^2.s$) for the materials after oxidation at 300 Pa versus temperature during 300 s at a temperature plateau is reported in Fig. 4. The mass gain for Invar is clearly huge compared to those of 316L from 1400 K. The significant difference of the mass gain for both the materials at around 1500 K and 300 Pa for duration 300 s is reported in Table 1.

Table 1. Mass gain (in g/m^2) for oxidized 316L and Invar samples around 1500 K at 300 Pa.

316L	15
Invar	134

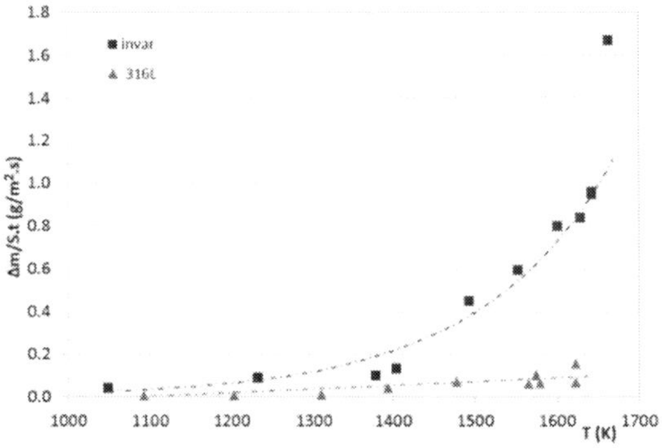

Figure 4. Mass gain versus temperature for Invar and 316L samples oxidized in air plasma at 300 Pa.

Oxidation in air plasma at 300 and 2000 Pa was also carried out for 316L around 1500 K for different duration times. The oxidized 316L surfaces in such conditions are presented in Fig. 5. It can be observed a slight peeling of the oxide layer regardless the duration time.

90 s 130 s 240 s 300 s 360 s

Figure 5. Pictures of 316 L samples oxidized in air plasma at 300 Pa for different duration times.

The mass variation laws versus time for 316L are reported in Fig. 6. Despite the duration time increase, the oxide layer keeps thin, this is why, linear oxidation kinetic laws are observed for the two total air pressures at around 1500 K. The mean law obtained is $\Delta m/S = 0.03\ t$, where $\Delta m/S$ is the mass variation per unit area in g/m^2, and t is the oxidation duration.

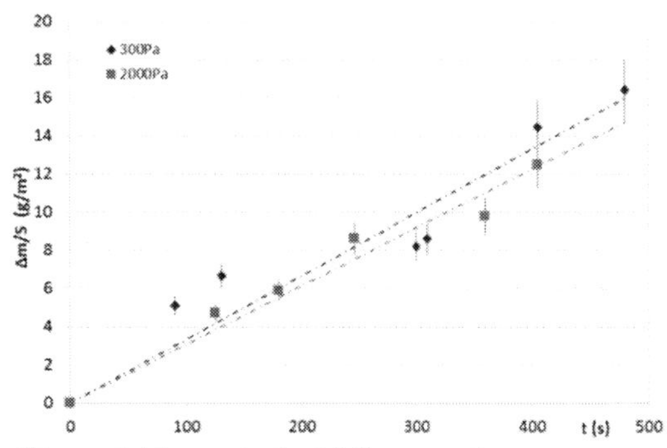

Figure 6. Mass gain for 316L versus duration time at 300 and 2000 Pa around 1500 K.

Scanning Electron Microscopy (SEM) was performed on oxidized 316L and Invar surfaces at different temperatures. SEM images are presented in Figs. 7 and 8. The growing of the crystalline grains is observed for both the materials with the increasing temperature. Moreover, for Invar, SEM images are corresponding to the oxidized surfaces presented in Fig. 3 and different morphologies have been observed when the temperature increases. The surface of the oxide layers is rough at low temperatures 1233 and 1380 K (brown color) and the crystal grain size increases with temperature (Figs. 8a and b). At 1496 K (brown-gray color), the oxide layer presents grains of larger size (Fig. 8c). Finally, around 1650 K (gray color), the oxide layer appears smooth with the presence of twins on the surface (Fig. 8d). Significant surface changes were encountered for Invar that could finally act on the emissivity.

Figure 7. SEM images of oxidized 316L samples at 300 Pa for different temperatures (scale bar 4 µm).

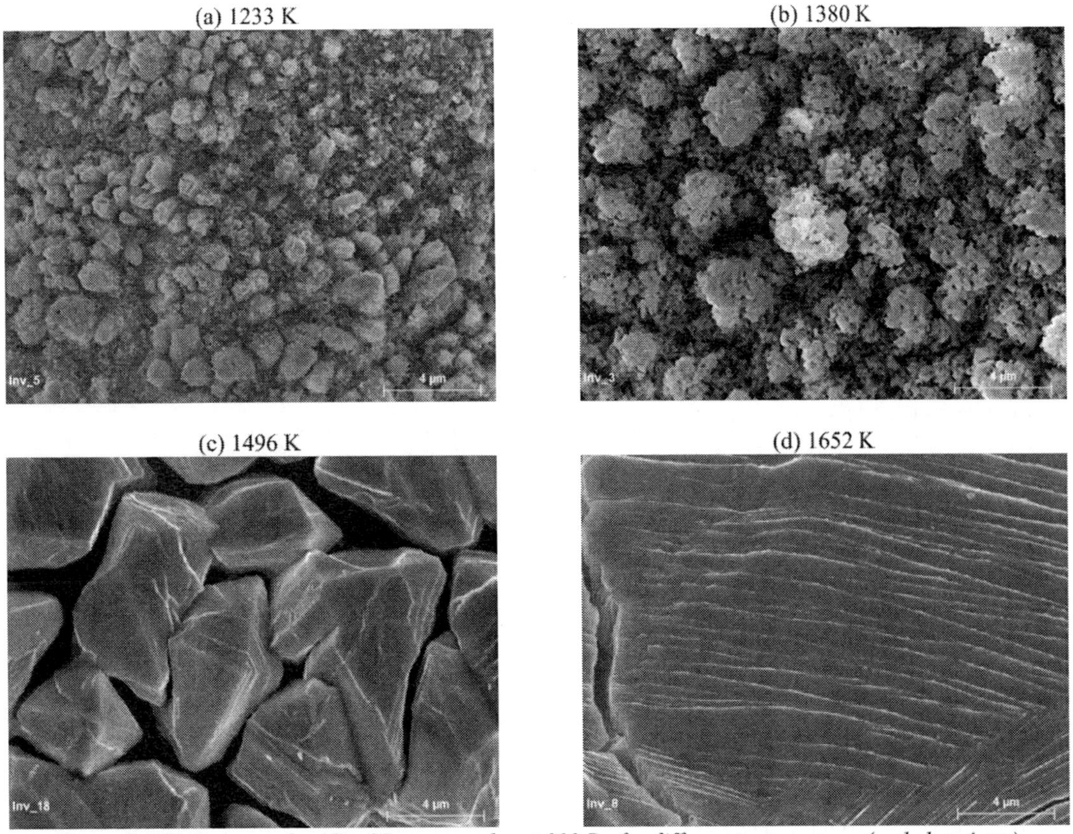

Figure 8. SEM images of oxidized Invar samples at 300 Pa for different temperatures (scale bar 4 µm).

X-Ray Diffraction (XRD) analyses were performed on the oxidized surfaces. Figure 9 and Table 2 respectively report the identified phases for oxidized 316L and Invar surfaces at different temperatures. For 316L, two iron oxides hematite Fe_2O_3 (ICDD 71-5088) and magnetite Fe_3O_4 (ICDD 75-0033) have been identified. The predominance of Fe_2O_3 (black round, Fig. 9) can be observed compared to Fe_3O_4 (black star) regardless the oxidation temperature. Moreover, some peaks of 316L substrate (black square) are also present except at 1586 K, probably due to the low thickness of the oxide layer formed at these temperatures.

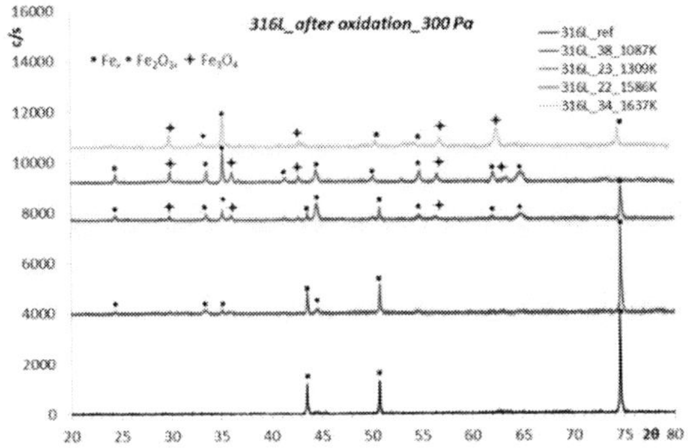

Figure 9. XRD patterns of oxidized 316L samples at 300Pa

For Invar, the oxides listed in Table 2 are following a decreasing order of the main oxide present in the layer. It can be noticed the predominance of Fe_2O_3 in the temperature range 1230-1650 K. At 1233 K, the additional contribution of $Ni_{0.4}Fe_{2.6}O_4$ (ICDD 87-2335) and Fe_3O_4 has been observed. The disappearance of Fe_3O_4 at 1380 K is noticed, due to the oxidation to Fe_2O_3. Moreover, at 1233 and 1380 K, the $FeNi_3$ (ICDD 71-8323) crystalline phase and the Invar substrate (ICDD 47-1405) were identified, probably due to the low thickness of the formed oxide layers. With the increasing temperature at 1496 K, the two iron oxides Fe_3O_4 and Fe_2O_3 are mainly present and $FeNi_3$, $Ni_{0.4}Fe_{2.6}O_4$ and Invar were not detected. Finally at 1652 K, Fe_2O_3 and Fe_3O_4 are also identified with $FeNi_3$, and no Invar was detected due to the important thickness of the oxide layers.

Table 2. Identified phases by XRD for oxidized Invar samples at 300 Pa for different temperatures.

Invar	
T (K)	Phases
1233	Fe_2O_3, $Ni_{0.4}Fe_{2.6}O_4$, Fe_3O_4, $FeNi_3$, Invar
1380	Fe_2O_3, $Ni_{0.4}Fe_{2.6}O_4$, $FeNi_3$, Invar
1496	Fe_2O_3, Fe_3O_4
1652	Fe_2O_3, Fe_3O_4, $FeNi_3$

4.3. Emissivity in high vacuum

In order to know the thermal radiative properties of 316L, total directional emissivity measurements and total hemispherical emissivity calculation - obtained by integration of the directional values - were performed in high vacuum (10^{-4} Pa). Emissivity of materials is dependent on the wavelength, the angle of emission, the temperature and also the surface morphology (rough or smooth) and composition. In this way, total directional emissivity measurements were performed on rough (sand-blasted) and smooth (as-received) 316L samples for several temperature levels in the total wavelength range from 0° to 80° incidence angle. Figure 10 reports the evolution of the total directional emissivity of a smooth 316L sample for several temperatures. It can be observed the decrease of the emissivity with the increasing temperature from 1100 to 1384 K and also, the increase of the emissivity for higher angles characteristics of metallic materials.

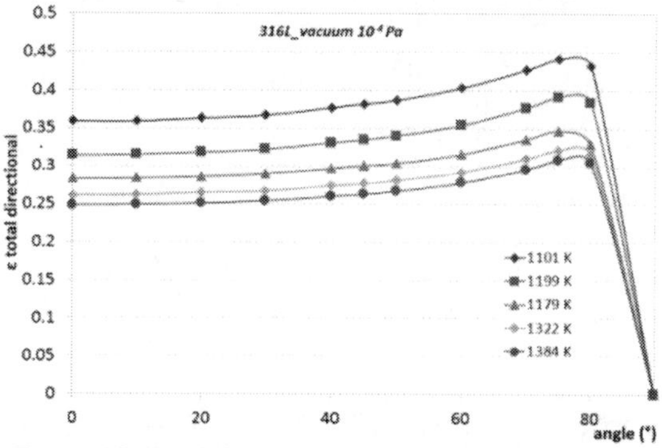

Figure 10. Total directional emissivity versus angle of incidence for a smooth 316L sample in high vacuum for several temperatures.

Scanning Electron Microscopy (SEM) was performed on the samples before and after emissivity measurement. SEM images for – smooth and rough – 316L samples before and after emissivity measurement in high vacuum and temperature conditions are presented in Figs. 11 and 12.

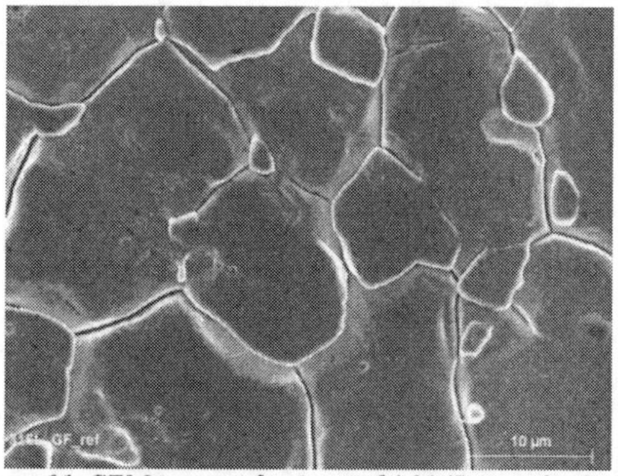

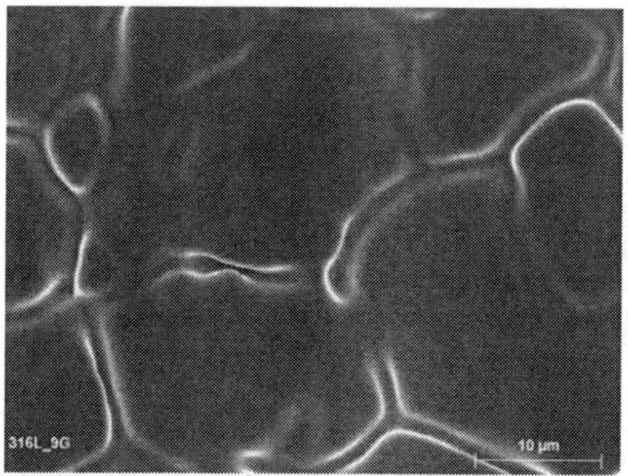

Figure 11. SEM images for 'smooth' 316L samples before (left) and after (right) emissivity evaluation in high vacuum at 1373 K (scale bar 10 µm).

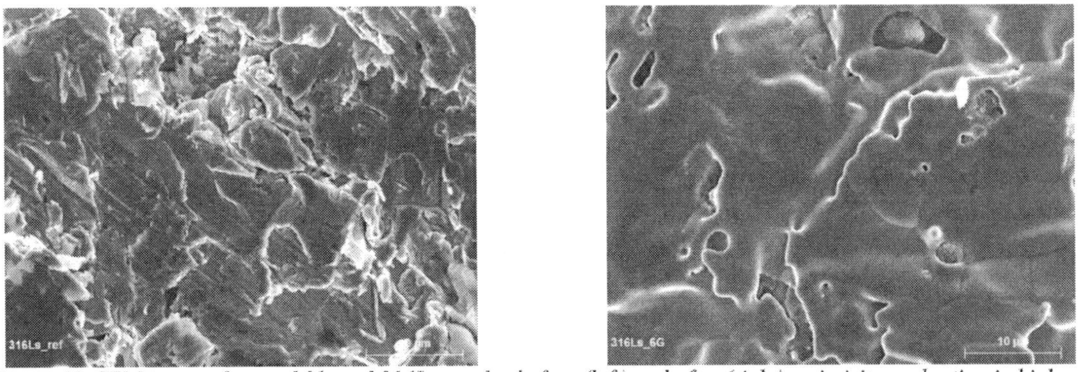

Figure 12. SEM images for sand-blasted 316L samples before (left) and after (right) emissivity evaluation in high vacuum at 1375 K (scale bar 10 µm).

For both the surfaces, it can be observed, after reaching the highest temperature level, a growing of the grains, leading to a global smoothing of the surface.
The evolution of the total hemispherical emissivity for 'smooth' (round dots) and sand-blasted (square open dots, 's' label) 316L samples in high vacuum versus temperature is reported in Fig. 13. Regardless the surface morphology, the total hemispherical emissivity of 316L samples decreases in the temperature range 1100-1200 K and becomes nearly constant – around 0.25-0.30 – from 1200 up to 1450 K. Nevertheless, the effect of the surface roughness is mainly visible for the lower temperature range.

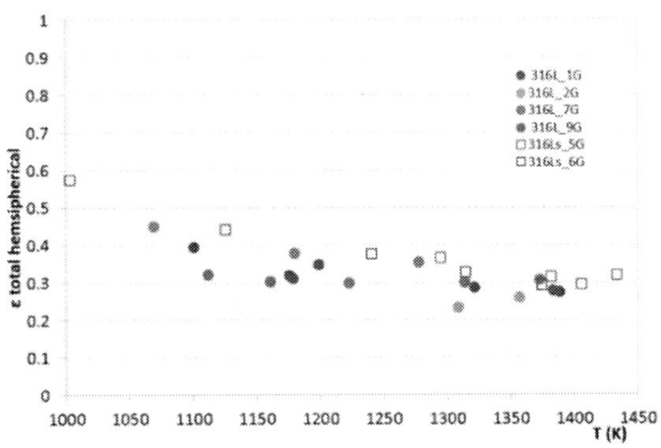

Figure 13. Evolution of the total hemispherical emissivity versus temperature for – 'smooth' and rough - 316L samples in high vacuum

Table 3 reports the surface roughness parameters for the samples before and after emissivity measurement: Sa is the arithmetic value and Sq the mean square surface roughness.

Table 3. Surface roughness parameters for 316L samples before and after emissivity measurement.

3D roughness parameters	'smooth' before ε	'smooth' after ε	sand-blasted before ε	sand-blasted after ε
Sa (µm)	0.11	0.01	0.40	0.36
Sq (µm)	0.24	0.17	0.57	0.48

The surface smoothing after the emissivity measurement at high temperature, due to the grain growth, can be noticed in Table 3 regardless the initial surface morphology by the decrease of the roughness parameters (Sq = 0.24 to 0.17 µm for 'smooth' sample and Sq = 0.57 to 0.48 µm for sand-blasted sample).

4.4. Emissivity for oxidized materials

During the atmospheric re-entry, materials from spacecraft are oxidized in air plasma, and the emissivity of oxides formed is very high compared to the non-oxidized materials. This is why, emissivity measurements on oxidized 316L and Invar samples were carried out.
In a first time, measurements of the total normal emissivity (as the samples were 25 mm in diameter) were performed on pre-oxidized samples in air plasma conditions at 300 or 2000 Pa up to their maximum temperature reached in the MESOX device: 316L_7R (1650 K, 300 Pa); 316L_3R (1720 K, 300 Pa); 316L_2R (1525 K, 2000 Pa); 316L_8R (1645 K, 2000 Pa); 316L_9R (1670 K, 2000 Pa); Inv_2R (1545 K, 300 Pa); Inv_3R (1635 K, 300 Pa); Inv_5R (1520 K, 2000 Pa) and Inv_8R (1649 K, 2000 Pa). Figures 14 and 15 respectively report the data of the total (0.6-40 µm) normal emissivity versus temperature for pre-oxidized 316L and Invar samples and also, those of 316L_9G and Inv_1gf samples treated in high vacuum in order to show the difference between oxidized and non-oxidized surface.
A global increase of the total normal emissivity is observed for 316L with temperature. At low

temperature, around 1100 K, the total normal emissivity for pre-oxidized samples is already high (≈ 0.80) and is still increasing for T ≥ 1400 K (≈ 0.90). Finally, it can be noticed a significant increase of the total emissivity for pre-oxidized surfaces by a factor of nearly 3 compared to the non-oxidized surface (316L_9G, pink dots).

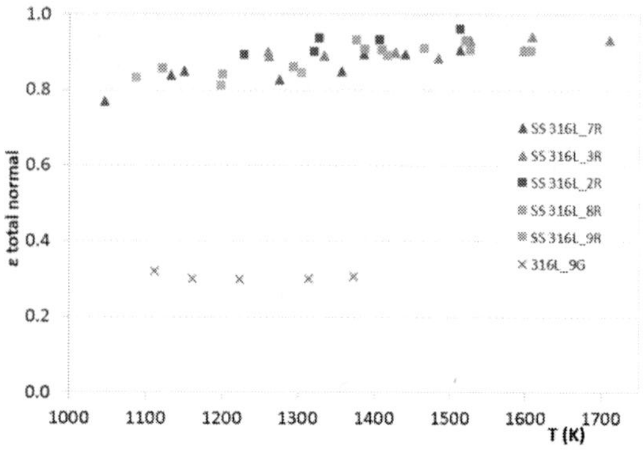

Figure 14. Evolution of the total normal emissivity versus temperature for some samples of 316L pre-oxidized in air plasma and in high vacuum (9G).

For Invar, two levels of emissivity can be noticed regardless the total air pressure, depending on the temperature, and more precisely below and above 1450 K. The total normal emissivity value is nearly constant around 0.80 up to 1400 K, then decreases around 0.75 at 1450 K and stabilizes around 0.70 up to 1600 K. The differences observed in the total normal emissivity values are due to the microstructural changes with temperature, observed in Table 2. Moreover, the total hemispherical emissivity value obtained for the non-oxidized Invar (Inv_1gf) is around 0.20, 4 times smaller than the emissivity of the oxide layers, hence the importance to perform emissivity measurement on pre-oxidized samples in air plasma conditions.

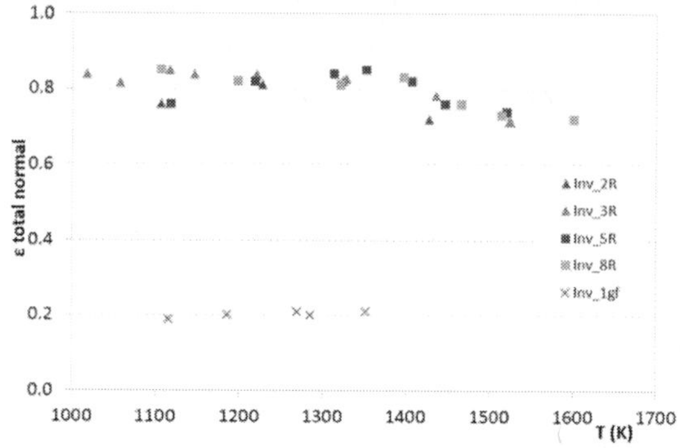

Figure 15. Evolution of the total normal emissivity versus temperature for some samples of Invar pre-oxidized in air plasma and in high vacuum (1gf).

SEM images of pre-oxidized samples, respectively 316L at 1642 K and Invar at 1649 K, at 2000 Pa before and after emissivity measurement are presented in Figs. 16 and 17. Before emissivity measurement, it can be observed the presence of an oxide layer on both the 316L (Fig. 16 left) and Invar surfaces (Fig. 17 left) with different morphologies. After emissivity measurement, the crystalline grain growth for 316L (Fig. 16 right) and for Invar (Fig. 17 right) can be noticed. For 316L, the oxide grains looking as 'flowers' are enlarged after emissivity measurement due to the continuous growing of the oxide grains. For Invar, intra-granular micro-cracks are more visible with an enlargement of the cracks and also the twins appear more spaced.

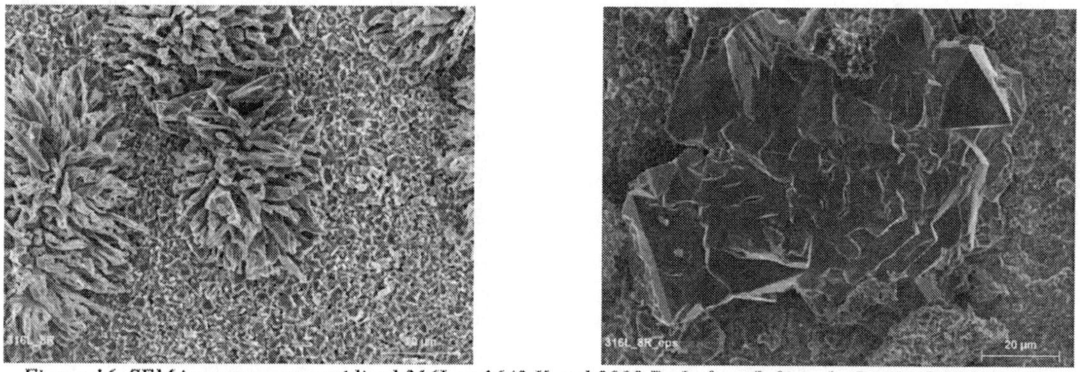

Figure 16. SEM images on pre-oxidized 316L at 1642 K and 2000 Pa before (left) and after (right) emissivity measurement (scale bar 20 µm).

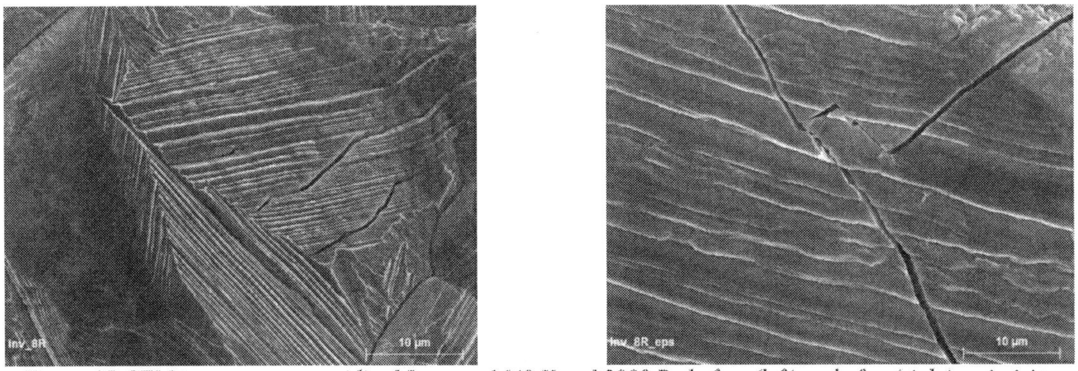

Figure 17. SEM images on pre-oxidized Invar at 1649 K and 2000 Pa before (left) and after (right) emissivity measurement (scale bar 10 µm).

In a second time, emissivity measurements were performed during the in situ oxidation in standard air at 300 or 2000 Pa. The total hemispherical emissivity (as the samples were 40 mm in diameter), obtained by integration of the directional values is respectively reported in Figs. 18 and 19 for 316L and Invar.

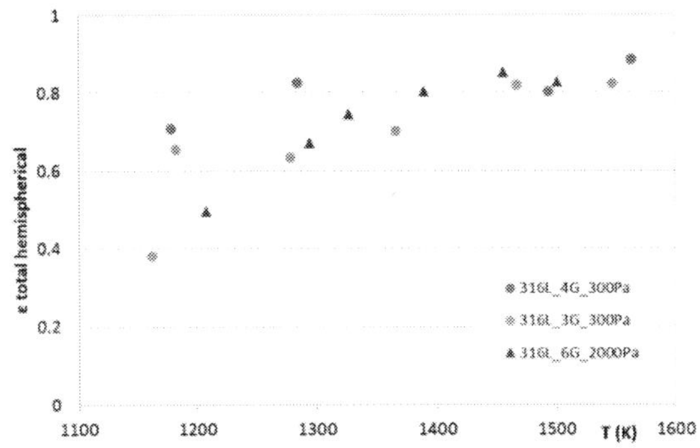

Figure 18. Evolution of the total hemispherical emissivity versus temperature for 316L samples in situ oxidized in standard air.

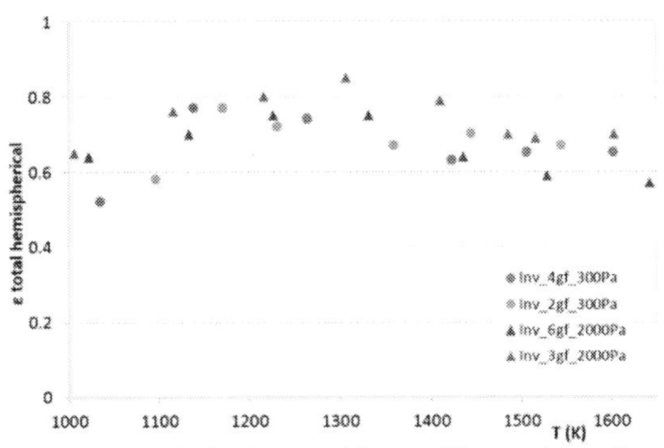

Figure 19. Evolution of the total hemispherical emissivity versus temperature for Invar samples in situ oxidized in standard air.

A global increase of the total hemispherical emissivity is observed for 316L with the increasing temperature at 300 and 2000 Pa. More precisely, the emissivity increases up to 1450 K due to the formation of oxides and becomes nearly constant around 0.85 at high temperature.

For Invar, as for the 316L oxidized in situ, the total hemispherical emissivity increases from 0.60 to 0.75 up to 1300 K, due to the formation of oxide layers, and then decreases to finally stabilize around 0.65 from 1450 K up to 1650 K due to the modification of the surface morphology.

Finally, it can be noticed for 316L and Invar, the higher total normal emissivity of pre-oxidized samples in air plasma compared to the emissivity of oxidized samples in standard air.

4.5. Emissivity implementation in the DEBRISK code

In order to calculate the surface heat flux of space debris and more precisely to better model their survival risks during the atmospheric re-entry, it is necessary to implement in the DEBRISK code the total hemispherical emissivity (ε^{Ω}) of pre-oxidized samples in air plasma conditions. However, the total normal emissivity (ε^{0}) was measured because the pre-oxidized samples were only 25 mm in diameter. In order to decide what value must be implemented, a comparison of the normal and hemispherical values obtained for in situ oxidized samples and normal values for pre-oxidized samples, for 316L and Invar around 1500 K, was done and reported in Table 4.

It is interesting to notice the similitude of the normal and hemispherical emissivity values for both the materials in situ oxidized taking into account the accuracy on the emissivity that is around 5%.

Compared to the normal values obtained on pre-oxidized samples, a different behavior is observed. Around 1500 K, for 316L, the emissivity value to take into account is equal to 0.94 ± 0.05 and it is very

different from the one obtained in standard air (0.83 ± 0.05). On the contrary, for Invar, all the values are very close and a mean emissivity of 0.72 ± 0.05 around 1500 K can be implemented.
Finally, for the implementation in the DEBRISK tool, we can consider that the total normal emissivity is equivalent to the total hemispherical emissivity for 316L and Invar samples oxidized in air plasma conditions.

Table 4. Total normal (ε^0) and hemispherical (ε^Ω) emissivity values for 316L and Invar samples obtained on in situ oxidized samples and on pre-oxidized ones.

	In situ	In situ	Pre-ox
	ε^0	ε^Ω	ε^0
316L-6G, 1500 K, 2000 Pa	0.86	0.83	-
316L-3G, 1548 K, 300 Pa	0.84	0.82	-
316L-9R, 1520 K, 2000 Pa	-	-	0.94
316L-7R, 1513 K, 300 Pa	-	-	0.90
Inv-3gf, 1516 K, 2000 Pa	0.70	0.69	-
Inv-2gf, 1545 K, 300 Pa	0.68	0.67	-
Inv-8R, 1514 K, 2000 Pa	-	-	0.73
Inv-3R, 1525 K, 300 Pa	-	-	0.71

5. CONCLUSION

Oxidation in air plasma conditions for 316L and Invar was studied and kinetic laws at high temperature and for short time durations were determined, features of the atmospheric re-entry. The most important mass gain was observed for Invar (134 g/m^2) compared to those of 316L (15 g/m^2) around 1500 K at 300 Pa.
Scanning Electron Microscopy has identified two iron oxides, Fe_2O_3 and Fe_3O_4 for both the materials with different crystalline orientations. Other oxides were detected for Invar, depending on the temperature, such as $Ni_{0.4}Fe_{2.6}O_4$ and $FeNi_3$.
Emissivity measurements were also performed on 316L and Invar in high vacuum (10^{-4} Pa) and on pre-oxidized samples in air plasma and in situ in standard air conditions. A significant increase of the total normal emissivity by a factor of nearly 3 or 4 was measured, depending on the material, for pre-oxidized samples compared to the non-oxidized ones. So, it clearly appears that the emissivity values obtained on pre-oxidized samples are important for the implementation in the DEBRISK code to better model the mass reaching the ground and the casualty area.

6. ACKNOWLEDGMENTS

This work was partly supported by the Programme "Investissements d'Avenir" (Investment for the Future) of the Agence Nationale de la Recherche (National Agency for Research) of the French State under awards numbers ANR-10-LABX-22-01-SOLSTICE through the co-funding of L. Barka PhD position and ANR-10-EQPX-49-SOCRATE.

7. REFERENCES

1. Omaly P., Spel M. (2012). DEBRISK, a tool for re-entry risk analysis, *Proc. 5th IAASS Conf. 'A Safer Space for a Safer World'*, Versailles (Fr), 17-19 oct. 2011, ESA SP-699.
2. Omaly P., Magnin Vella C., Galera S. (2013). DEBRISK, CNES tool for re-entry survivability prediction: validation and sensitivity analysis, *Proc. 6th IAASS Conf. 'Safety is not an option'*, Montréal (Ca).
3. Guo X., Kusabiraki K., Saji S. (2002). High-temperature scale formation of Fe-36%Ni bicrystals in Air, *Oxid. Met.* **58**, 589-605.
4. Wang X., Wang L., Zhu M., Zhang J., Lei M. (2006). Influence of high-intensity pulsed ion beams irradiation of oxidation behaviour of 316L stainless steel at 700°C, *Trans. Non Ferrous Met. Soc.* **16**, 676-680.
5. Buscail H., El Messki S., Riffard F., Perrier S. (2008). Role of molybdenum on the AISI 316L oxidation at 900°C, *J. Mater. Sci*, **43**, 6960-6966.
6. Labrador N., Lefort P. (1999). Oxydation de l'invar sous air, *J. Chem. Phys.* **96**, 810-829.
7. Vessel A., Mozetic M., Drenik A., Hauptman N., Balat-Pichelin M. (2008). High temperature oxidation of stainless steel AISI 316L in air plasma, *Appl. Surf. Sci*, **255**, 1759-1765.
8. Menecier S., Valette S., Denoirjean P. (2012). Invar oxidation in CO_2, kinetics and mechanism of formation of a wustite layer, *Therm. Anal. Calorim.* **107**, 607-616.
9. Balat-Pichelin M., Robert J.F., Sans J.L. (2006). Emissivity measurements on carbon-carbon composites at high temperature under high vacuum, *Appl. Surf. Sci.* **253**, 778-783.
10. Hernandez D., Sans J.L., Netchaieff A., Ridoux P., Le Sant V. (2009). Experimental validation of a pyroreflectometric method to determine the true temperature on opaque surface without hampering reflections, *Measurement.* **42**, 836-843.
11. Balat-Pichelin M., Charpentier L., Panerai F., Chazot O., Helber B., Nickel K. (2015). Passive/active oxidation transition for CMC structural materials designed for the IXV vehicle re-entry phase, *J. Eur. Ceram. Soc.* **35** (2), 487-502.
12. Balat-Pichelin M., Vesel A. (2006) Neutral oxygen atom density in the MESOX air plasma solar furnace facility, *Chem. Phys.* **327** (1), 112-118.

PROCEEDINGS OF THE 9TH INTERNATIONAL SPACE SAFETY CONFERENCE

SESSION 02 HUMAN PERFORMANCE FOR SAFETY & ORGANIZATIONAL CULTURE – I

STRATEGIC EMPLOYEE DEVELOPMENT IN THE GOVERNMENT SECTOR

Johnny Nguyen[1], Nathalie Guevara[2], Rebecca Barnett[3], and Barbara Thorpe[4]

[1] *National Aeronautics and Space Administration (NASA), Technical Assistant to Director of Safety and Mission Assurance, John F. Kennedy Space Center mail code SA, Kennedy Space Center, FL 32899, USA;* johnny.t.nguyen@nasa.gov

[2] *National Aeronautics and Space Administration (NASA), Human Resources Specialist, John F. Kennedy Space Center mail code BA-E, Kennedy Space Center, FL 32899, USA;* nathalie.guevara@nasa.gov

[3] *National Aeronautics and Space Administration (NASA)(formerly), Human Resources Specialist, John F. Kennedy Space Center, FL 32899, USA*

[4] *National Aeronautics and Space Administration (NASA), Management Analyst, John F. Kennedy Space Center mail code BA-E, Kennedy Space Center, FL 32899, USA;* barbara.p.thorpe@nasa.gov

ABSTRACT

As with many other U.S. agencies, succession planning is becoming a critical need for NASA. The primary drivers include (a) NASA's higher-than-average aged workforce with approximately 50% of employees eligible for retirement within 5 years; and (b) employees who need better developmental conversations to increase morale and retention. This problem is particularly concerning for Safety & Mission Assurance (S&MA) organizations since they traditionally rely on more experienced engineers and specialists to perform their organizations' functions.

In response to this challenge, the Kennedy Space Center (KSC) S&MA organization created the Strategic Employee Development (SED) program. The SED program's goal is to provide a proactive method to counter the primary drivers by creating a deeper "bench strength" and providing a more comprehensive developmental feedback experience for the employee. The SED is a new succession planning framework that enables customization to any organization, and in this case, specifically for an S&MA organization. This is accomplished via the identification of key positions, the corresponding critical competencies, and a process to help managers have relevant and meaningful development conversations with the workforce.

As a result of the SED, several tools and products were created that allows management to make better strategic workforce decisions. Although there are opportunities for improvement for the SED program, the most important impact has been on the quality of developmental discussions for employees.

1. ISSUE/OBJECTIVE

The issue of an aging workforce is felt throughout NASA and other Federal agencies, especially at the higher leadership positions [1]. The average age of Federal civilian non-postal employees is 47.4, with 14.8% eligible for retirement as of September 2016 [2]. At KSC, the average age is 47, with 15% eligible for retirement today and 34% eligible for retirement within 5 years. In comparison, the S&MA organization at KSC (i.e., a subset of KSC) has an average age of 49, with 25% eligible for retirement today and 43% eligible for retirement within 5 years (Figure 1).

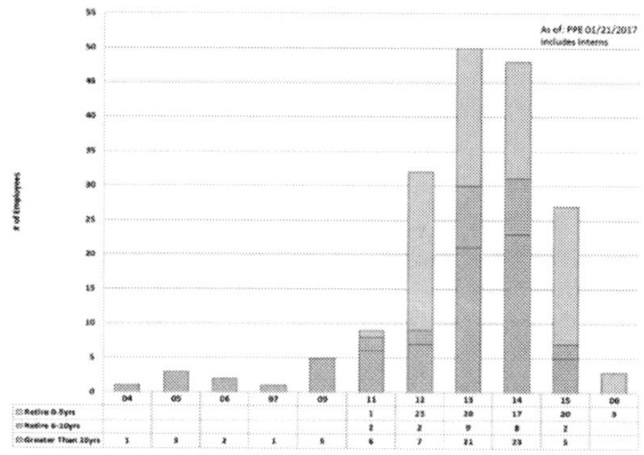

Figure 1: NASA KSC S&MA Workforce Retirement Eligibility Timeframe

A problem arises when so much of the workforce leaves in a short timeframe, taking with them highly-specialized skills and deep knowledge without allowing the organization enough time to train competent successors.

A separate but also relevant issue is the quality of conversation between the supervisor and employee on personal development. In the Federal government, the Office of Personnel Management administers a yearly Federal Employee Viewpoint Survey (FEVS) to assess employees' experiences in the workplace. Some of the questions included in the survey measure employees' perception of training and development (Table 1). Although the responses are quantitative, the interpretation can be highly subjective. The issue is not whether the results are low or high; the point is the evaluation scores for S&MA can be increased, and that is an opportunity for improvement. Given that NASA

S&MA has an aging workforce and many of them are leaving, it is more important than ever to retain the remaining workforce while preparing them for higher-level positions.

Table 1: Federal Employee Viewpoint Survey Items Related to Employee Development

Questions	SA 2016	KSC 2016
My training needs are assessed	74.0%	74.1%
My supervisor/ team leader provides me with constructive suggestions to improve my job performance	66.8%	74.9%
Supervisors/ team leaders in my work unit support employee development	67.2%	82.3%
How satisfied are you with the training you receive for your present job?	73.3%	68.7%

In response to these concerns, the KSC S&MA organization created the Strategic Employee Development (SED) program. SED program's goal is to provide a proactive method to counter the primary drivers of this issue by creating a deeper "bench strength" and providing a more comprehensive developmental feedback experience for the employee. The SED is a new succession planning framework that enables customization to any organization, and in this case, specifically for an S&MA organization. This is accomplished via the identification of key positions, the corresponding critical competencies, and a process to help managers have relevant and meaningful development conversations with the workforce.

The objectives of the Strategic Employee Development (SED) program are primarily to address and improve the above two primary issues. Additionally, there are numerous secondary objectives, including:

- Develop a succession plan (not a replacement plan that adheres to human resources policies and best practices) [3]
- Provide additional tools for supervisors in developmental discussion with their employees
- Increase awareness of employees' career goals across the entire management team

2. DESIGN/METHOD/APPROACH

2.1 Team Composition

The planning team that develops the overall succession planning strategy, products and materials needed, and change management requires certain skills and authority. Our team consisted of the following:

- Deputy Director, KSC S&MA: Authority and technical knowledge of the organization to lead major changes and indicates management buy-in
- Organizational Development Specialist: Expert advice on change management, succession planning theory, and facilitating conversations
- Human Resources Specialist: Expert on human resources policies and requirements
- Workforce Planner: Expert on providing detailed analysis of the organization's skills and demographics

2.2 Communication & Change Management Plan

The overarching communication and change management plan is a key tool developed at the very beginning of the SED process that should be constantly revisited and adjusted (Figure 2 and 3). This plan outlines the overall timeline to communicate with the various stakeholders (employees, supervisors, other interested organizations), purpose of communication (informational, solicitation for input, decisional), who should do the communication (senior manager, supervisor, human resources), and when to communicate.

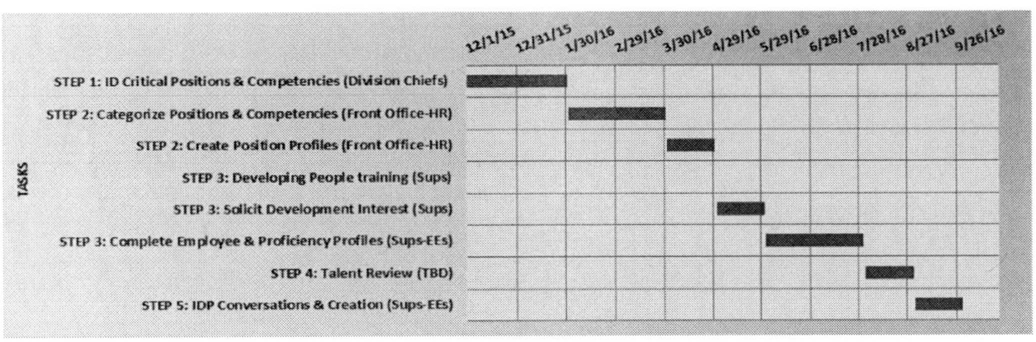

Figure 2: SED Program Schedule

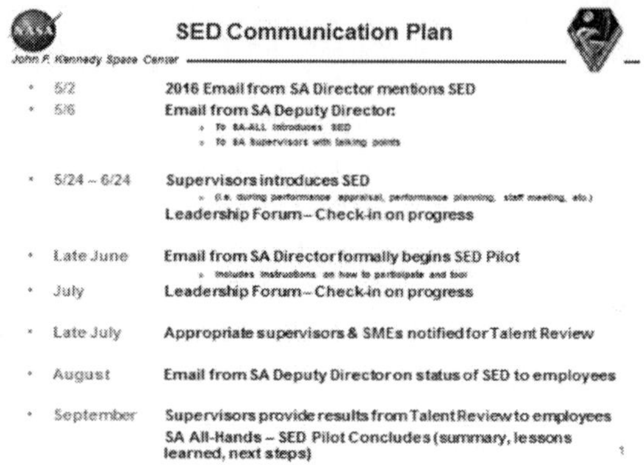

Figure 3: SED Communication Plan

2.3 Process Map

The team developed an overarching process map to guide the sequence of events (Figure 4). This outlines the major steps from start to finish. The process map provides a visual guide to the supervisors and participants regarding the overall process for the SED.

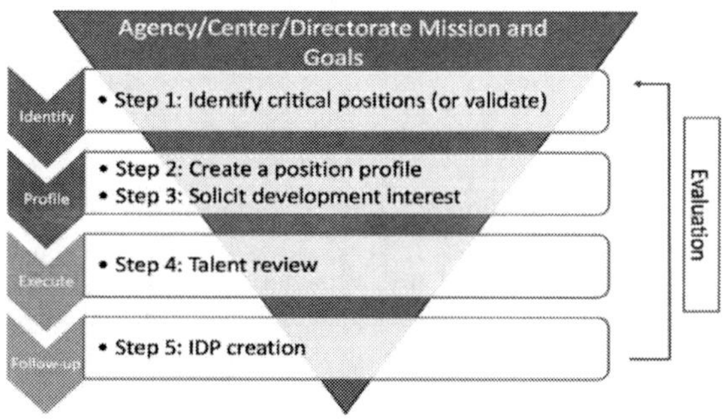

Figure 4: Strategic Employee Development Process Map

2.4 Identifying the Critical Positions

Depending on the size of the organization, perhaps all critical positions or only the most urgent positions should be considered as part of the SED. For our initial effort, we limited it to eight critical positions because it was our first run through of the process. We utilized a tool called Litmus Test for Identifying Key Positions which asks a series of questions to filter which positions should be deemed "critical."

2.5 Identifying the Key Competencies

For each of the critical positions, we engaged the current person in the position to help identify approximately ten key competencies in order to perform that position successfully and subdivided them into technical competencies and leadership competencies. We define key competency as a conceptual representation of a body of knowledge.

2.6 Employee Profiles

Once the management team identified and agreed upon critical positions and key competencies, we began the step to engage the entire workforce to voluntarily participate in the SED program through the completion of an employee profile. Profiles like these are typically used during succession management efforts to identify candidates' development needs [4]. It was important to communicate this was a voluntary program, and not participating would not adversely affect anyone's career opportunities. Conversely, by participating in this program, they would benefit from more detailed and specific personal development feedback, but the normal process for promotions and other opportunities would not be affected.

SED employee participants completed a profile identifying critical positions they were interested in and were asked to self-assess their key competencies. It is important to understand this is not a performance review that evaluates the employee to the skills they should have now in their current position. This is their own assessment of how ready they are to assume a different position in the organization. They are not expected to meet all the key competencies.

2.7 Brainstorming Developmental Ideas with Subject Matter Experts

Since the SED process will eventually involve discussing with each employee their career aspirations toward critical positions, we proactively brainstormed with the Subject Matter Experts (SME) currently in the critical positions to obtain their advice and suggestions on what skills and opportunities would be beneficial to prepare someone for their position. We compiled the information and turned it into a development reference tool for use with the supervisor and employee. Development reference tools are commonly used by organizations to include targeted tips and suggestions for developing specific competencies and skills [5]. See Figure 5 for a sample development reference tool.

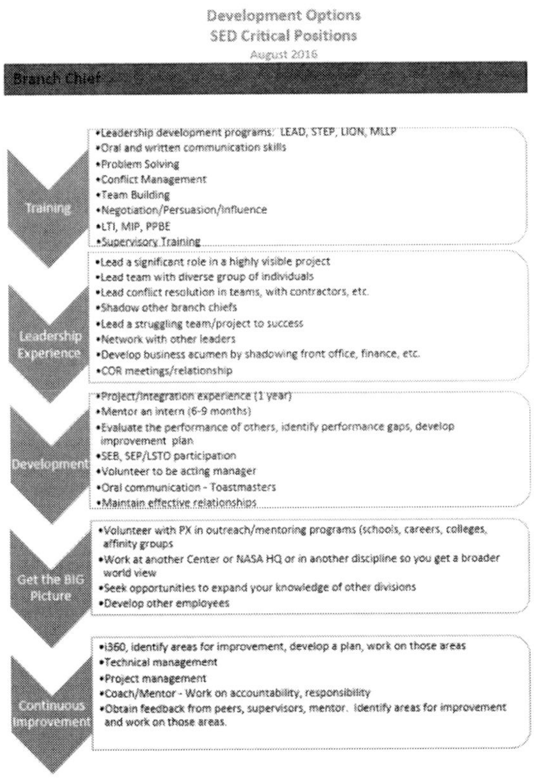

Figure 5: Sample Development Reference Tool for Branch Chief Position

2.8 Training for Supervisors on how to Deliver Feedback

Critical to the success of SED, and perhaps the most important step, is the successful delivery of personal development feedback to motivate the employee and to articulate clearly what the employee can do to prepare for advancement. Therefore, we conducted several workshops with our Organizational Development Specialist to discuss effective feedback delivery, and we procured a one-day training class called *Developing People* for all our supervisors to attend.

2.9 Review for Reconciliation and Readiness Level

Once the employees completed their employee profile, it was submitted to their supervisor. The supervisor then performed their own assessment of the employee per the same key competencies. The purpose of this step is to reveal any differences in assessments to discuss with the employee. For example, if an employee believes he/she is currently an expert in quality assurance engineering, but the supervisor believes the employee is at a journeyman level, there is a gap in perceived skill-level that will need to be discussed between the employee and supervisor at the end of the SED program. The supervisor will also make an initial recommendation on the employee's readiness level, which is defined as:

- Ready Now: Employee has the adequate proficiency level in all competencies to perform the position. Employee can hit the ground running now.
- Ready Later: Employee has some level of proficiency to perform the position but development is needed in some areas.

2.10 Talent Review

The Talent Review involves a senior management team meeting to discuss each of the key positions and the employees who self-identified an interest in them. Using the employee profiles, the supervisor's assessment, developmental ideas from SMEs, and an open conversation on how best to develop the employee, senior management is able to accomplish a few objectives:

- Common understanding among the management team on the readiness of the employee for certain critical positions
- An organizational-level perspective on the depth and readiness of the bench-strength
- A set of developmental feedback and suggestions specific to the individual participants that will aid them in developing for future opportunities

2.11 SED Employee Feedback Session

In this last step of the SED process, the employees who participated receive from their immediate supervisor the personalized feedback specific to the critical position they elected. Typically done one-on-one, this includes use of the products generated from the SED, such as the key competencies identified for the critical positions, the development brainstorming results with the SMEs, and the comprehensive feedback from the Talent Review.

3. EXECUTION

In executing the SED for KSC S&MA, there were 44 participants out of 170 possible employees, representing 26% participation. The official SED program took 5 months to execute. The planning phase took 5 months but should be significantly reduced for each successive year SED is implemented.

The SED resulted in the Bench Strength Table, a management tool that enables a holistic view of the organization for the critical positions identified in this SED (Figure 6). It captures the retirement eligibility of the person currently in the position, the number of employees who participated in the program that management deems ready now or ready later, and a brief description of the organization's strategy for the future. In other words, this tool captures management's strategy going forward regarding succession for these positions.

Positions	Retirement Eligibility	Ready Soon	Recruitment & Development Strategies	Overall Health	Progress
Agency Metrology/Calibration	5 or more	1	Look for external candidates Develop internal pool		
LSP CSO	Within 2 years	6	Continue to develop internal pool		
GSDO CSO	Within 2 years	4	Continue to develop internal pool		
Branch Chief	Within 2 years	9	Continue to develop internal pool (focus on leadership skills)		
Quality Engineer	Within 2 years	2	Re-direct/emphasize development of internal pool		
KSC Explosive Safety	2 to 5 years	1	Continue to develop and widen internal pool		
Lead Quality Assurance Specialist	5 or more	1	Continue to develop internal pool (focus on leadership skills)		
Administrative Officer	5 or more	1	Develop internal pool KSC pool available		

Overall Health Legend

Small bench strength relative to projected attrition and uniqueness of skill
Moderate bench strength relative to projected attrition and uniqueness of skill
Strong bench strength relative to projected attrition and uniqueness of skill

Figure 6: Bench Strength Table

For example, a supervisor position became vacant and management had a choice between directly reassigning someone to that vacancy or announcing a job vacancy. Management decided to choose the latter due to the Bench Strength Table indicating a deep talent pool for potential supervisors.

Training themes for our organization are another outcome of the SED. Based on information obtained from SED participants and the feedback coming out of the Talent Review, management was able to identify themes on the type of training from which the employees would benefit. This type of information allows organizations to focus training dollars toward on the most beneficial training. The training themes identified for NASA KSC S&MA are Assertiveness, Innovation, Business Acumen, Crucial Conversations, and Leadership.

4. COMPARISON TO TRADITIONAL SUCCESSION PLANNING METHODS

Compared to other traditional succession planning methods, the Strategic Employee Development initiative attempts to circumvent a major issue in government succession planning: pre-selection. In private companies, succession planning for critical positions involves selecting candidates with high potential and focusing company's resources and time to develop these candidates into even better performers and better candidates. In the government, very strict rules prohibit the employer from giving unequal opportunities for development or training to a group of employees. To avoid the issue of pre-selection, we put out a call to all employees to fill out an employee profile form that reflects their interest in critical positions. With this information, management is able to discuss development opportunities for those employees who are committed to growing and supporting the organization. Since the call went out to all employees, pre-selection was addressed.

Traditional succession planning methods are typically one-for-one. This means the organization is only planning on a per position level. The difference that SED brings is, although critical positions are identified, the discussion of the employee and their development feedback is not specific to that position. Rather, the feedback is to develop the skills necessary in order to perform in any like-position and is therefore not as limiting. Traditional succession planning is also usually done secretly, where management discusses potential successors without their knowledge, and sometimes this feedback gets back to the person. The SED process brings transparency to succession planning where participants opt-in to the program. The biggest drawback for a voluntary program like SED is that high potential employees may not participate giving management a less than comprehensive understanding of their organization's bench strength.

5. FINDINGS

5.1 Post Focus Groups Results and Opportunities for Improvement

- Directorate culture
  - Having full support from the top is pivotal
  - Ensure buy-in from supervisors from the beginning
  - Involve all supervisors throughout initiative

- Execution
  - Consider having broader areas of work (e.g., technical vs management track) rather than specific critical positions
  - Emphasize a similar approach across supervisors so that the quality of the employee experience is consistent
  - Ensure there is clear development feedback and options for each employee at the talent review, and get agreement on expected communication to the employee after talent review
- Change management
  - Identify supervisors who are implementing best practices and leverage their experience
  - Over communicate at every level. Intent can be lost in translation
  - Have a forum to share ideas on how to have critical conversations

5.2 Research Limitations/Implications

- Dependent on number of participants in the program. With more participants, the higher the fidelity of organizational data and more impact on the organization. With fewer participants, the organization will get skewed results.
- Duration of program requires significant preparation and may take several months to complete
- Requires some manual upkeep after program to continue reaping benefits of the SED

5.3 Practical Implications (i.e., Benefits of SED)

- SED can assist organizations with aging workforces
- Scalable to size of organization
- Framework applicable to almost any industry
- Customized for S&MA, with opportunity to add more S&MA specific data for increased impact
- Increases supervisor competency
- Increases employee morale
- Alignment of management team on workforce strategy
- Strategic insight into organization's training needs

5.4 Originality/Value

- Bench strength concept versus traditional one-for-one backfill
- Linking the "identify key competencies" to "brainstorming for developmental ideas" to "personalized individual feedback"
- Complies to HR policies (i.e., opt-in) while enabling management to have honest developmental conversations regarding their employees
- Leverages the SME that currently occupies the critical positions for developmental ideas on how people can develop the skills required and to the appropriate level

6. Conclusion and Recommendations for Further Research

Management always views succession planning as important, but it is usually only performed when the need for it becomes critical. Even then, it is usually conducted privately without the knowledge of the employees being considered for one-to-one backfills.

The SED process is holistic, where employees opt-in, and it gives management a continual method to reap the benefits of succession planning. The SED addresses the aging workforce issues that many organizations face, as well as improving morale by demonstrating management is actively involved in the employee's long-term career goals and in providing more substantial developmental feedback. The SED process can be customized for any organization just as it was customized and performed for the NASA Kennedy Space Center Safety & Mission Assurance organization. Although participation was at 26%, the process still yielded management tools and strategy, substantial employee development feedback, and an overall improvement on survey feedback regarding employee development and morale.

As a final note, the following are recommendations for further enhancements or research:

- Align the SED with the organization's natural performance feedback timeline
- Over-communicate with management and employees on the steps and benefits of the SED to ensure alignment of expectations
- Research industry trends to identify any "new skills" that do not currently exist in the organization but may be needed to stay competitive

7. REFERENCES

1. *Preparing for an Aging Workforce.* Retrieved from the Society for Human Resource Management (SHRM) website. https://www.shrm.org/hr-today/trends-and-forecasting/research-and-surveys/Documents/Preparing_for_an_Aging_Workforce-Government_Industry_Report.pdf
2. U.S. Office of Personnel Management (OPM). (September 30, 2016). *Data, Analysis & Documentation, Federal Employee Reports, Profile of Federal Civilian Non-Postal Employees. Retrieved September 20, 2017, from* https://www.opm.gov/policy-data-oversight/data-analysis-documentation/federal-employment-reports/reports-publications/profile-of-federal-civilian-non-postal-employees/

3. Ellehuus, C., Rowen, T., Connolly, N., Martin-Weinstein, J., Schmidt, C., Freire, P. (2007). *High Impact Succession Management: From Succession Planning to Strategic Executive Talent Management*. Washington D.C: U.S. Corporate Leadership Council.
4. Corportate Leadership Council (2009). *Identify and Prepare Your Next Generation of Leaders*. Washington D.C: U.S. Corporate Leadership Council.
5. GoEbase. (2005). *Building a Development Resource Guide*. www.goebase.com.

HUMAN ERROR ASSESSMENT AND REDUCTION TECHNIQUE (HEART) and HUMAN FACTORS ANALYSIS AND CLASSIFICATION SYSTEM (HFACS)

Tiffaney Miller Alexander, PhD

NASA/Kennedy Space Center, John F. Kennedy Space Center, FL 32899, USA, Tiffaney.M.Alexander@nasa.gov

ABSTRACT

Research results have shown that more than half of aviation, aerospace and aeronautics mishaps/incidents are attributed to human error. As a part of Safety within space exploration ground processing operations, the identification and/or classification of underlying contributors and causes of human error must be identified, in order to manage human error.

This research provides a framework and methodology using the Human Error Assessment and Reduction Technique (HEART) and Human Factor Analysis and Classification System (HFACS), as an analysis tool to identify contributing factors, their impact on human error events, and predict the Human Error probabilities (HEPs) of future occurrences. This research methodology was applied (retrospectively) to six (6) NASA ground processing operations scenarios and thirty (30) years of Launch Vehicle related mishap data. This modifiable framework can be used and followed by other space and similar complex operations.

1. INTRODUCTION

NASA Kennedy Space Center (KSC) Ground Processing Operations (GPO) deals with processing launch vehicles and space-craft used in the earth's atmosphere and beyond [1].

In 1993, NASA found that 78% of the incidents related to Space Shuttle Ground Processing Operations, since April 1991, was a result of human error [4].

In this research, human error contributing factors are identified and a comparative analysis of two existing tools are conducted to determine the human factors responsible for the failures, mishaps, etc. [10].

The research hypotheses are: 1) Contributing factors: unsafe acts of operators, preconditions for unsafe acts, unsafe supervision, and/or organizational influences (multiple causes) do not have an impact on human error events (i.e. mishaps, close calls, incident or accidents) in NASA ground processing operations. 2) The HFACS framework conceptual model can be proven to be a viable analysis and classification system to help classify both latent and active underlying contributors and causes of human error in NASA ground processing operations. 3) The development of a model using the HEART assessment can be used as a tool to help determine the probability of human error occurrence in NASA ground processing operations. This research will focus on the methodology and framework developed for analyzing the quantitative and qualitative aspects of human error contributing factors [1].

In order to display relationships between variables, a conceptual model, such as a theoretical framework was used. The Generic Error Modeling System (GEMS) was used to relate the basics (generic) basic error types presented by James Reason to the three performance levels presented by Jens Rasmussen [5].

Below is a Table outlining this relationship.

Table 1: Reason's three Basic Error Types in relation to Rasmussen's three Performance Levels [5]

Performance Level	Error Type	Description
Skill-based level	Slips and lapses	Automated non-cognizant errors of automatic processing (attention/memory) during regular routine actions that are identified quickly [5].
Rule-based level	Rule-based mistakes	Errors of rule-based behavior. For example: applying the wrong rule for a give situation (often with a tendency to keep repeating the same wrong actions "strong but wrong").
Knowledge-based level	Knowledge-based mistakes	Errors of cognitive (knowledge-based) processing whereby a problem is not analyzed correctly (or not at all) and this results in an error (e.g. wrong response to a multitude of alarms based on an incomplete understanding of the actual problem).

This research objective provides a framework and methodology using the Human Error Assessment and Reduction Technique (HEART) and Human Factor Analysis and Classification System (HFACS), as an analysis tool to identify contributing factors, their impact on human error events, and predict the Human Error probabilities (HEPs) of future occurrences.

Due to HEART's quantitative ergonomic approach through analysis of ergonomic factors that may have substantial and negative effects on human performance, it was selected for this research. The technique is capable of providing human factor specialists with quantitative supported data for design and other recommendations for overall improvement [2].

HFACS is considered a comprehensive analysis of human error that takes into account multiple causes of human failure [10] and for this reason, it was selected for this research. The HFACS method has generic terms and descriptors that allow its use for an array of industries and activities, thus providing an advantage [6].

2. PRELIMINARY ANALYSIS

Prior to the development of a Kennedy Space Center (KSC) Ground Processing Operations (GPO) Human Error framework, a preliminary analysis was performed. This was performed to help identify and categorize examples of historical Launch Vehicle specific ground processing operations tasks generally performed during Ground Processing Operations. The specific tasks in this research were matched with the associated HEART nominal human "unreliability probabilities," in order to calculate the Human Error Probability (HEP). The HEART Error Producing Conditions (EPC) were matched with the HFACS conditions (human error effects) to determine what category each fell under: unsafe acts, preconditions for unsafe acts, unsafe supervision and organization influences.

For this effort, informal discussions were conducted with three (3) Subject Matter Experts (SME) (3 SMEs with 34, 31 and 30 years KSC GPO experience). For the HEART Generic Tasks, the SMEs identified and categorized examples of historical shuttle specific ground processing operations. For the HEART Surveys used in this research, the SMEs identified general tasks and error producing conditions associated with three (3) NASA KSC Ground Processing Area Scenarios. The SMEs were also asked to modify the HFACS' four (4) Levels of Human Error examples with the specific KSC examples for categorization.

The framework below encapsulates the three (3) scenarios in this research, relating to Ground Processing Operations, Human error, the GPO tasks performed at KSC, HFACS category levels and sublevels with specific Error Producing Conditions, as it leads to Human error related NASA mishaps.

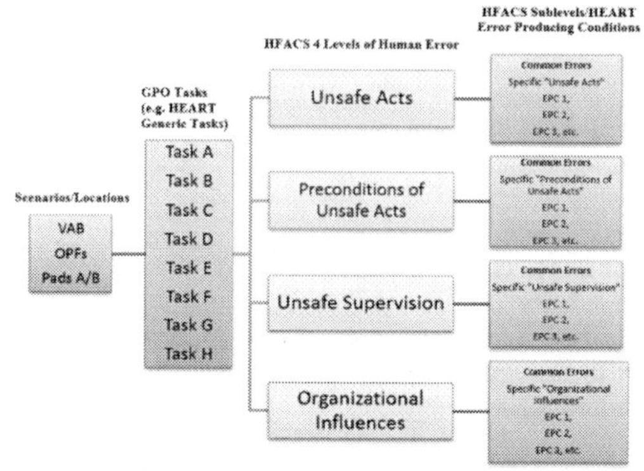

Figure 1. KSC Ground Processing Operations Human Error Framework [1]

3. METHODOLOGY

The Methodology used for this research consists of eight (8) stages: Literature Review, Existing Gap, Experiment Overview (Assessment approach), Data Collection (Mishap Data and Survey Participant data), Qualitative Study (HEART Method, SME Preliminary analysis and HFACS Method), Surveys, Quantitative Study (HEART Data analysis, HEP Calculation, and Binary Logistics Regression), and Data Analysis – Validation (Triangulation).

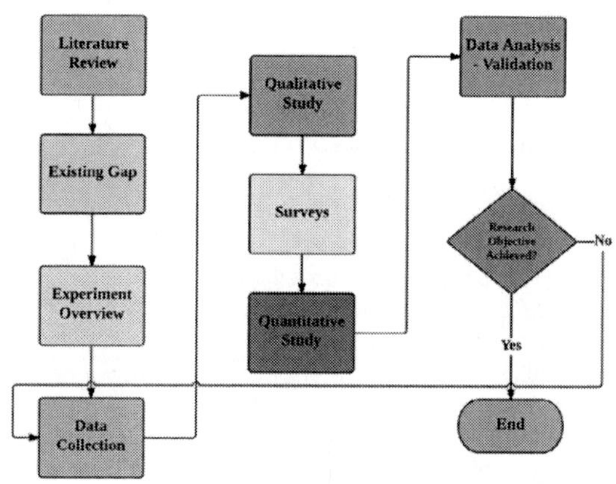

Figure 2. Research Methodology [1]

4. RESULTS AND DISCUSSION

For the Human Error Assessment and Reduction Technique (HEART), survey subjects were asked to answer a total of twenty one (21) questions. The questions consisted of three (3) job related questions for statistical purposes and eighteen (18) evaluation questions involving three (3) NASA KSC GPO Scenarios (VAB, OPFs and Pads A/B).

The assessment is based on the level of affect an Error Producing Condition (EPC) may have on a specific GPO task (very low, low, moderate, high, and very high). For this research, fortyone (41) KSC team members with Ground Processing Operations experience participated in the research survey. Participants' average years working at KSC is 23.9875 years (ranging 10 to 37 years). Participants' average years working at KSC supporting Ground Processing Operations is 19.8125 years (ranging 3 to 33 years). Using the HFACS tool to categorize the Ground Processing Operations (GPO) human error related mishaps from October 1984 – May 2014, two binary logistics regression models were generated. The HEART Method was used to calculate predicted Human Error Probabilities (HEP). NASA Shuttle Mishap data from October 1984 – May 2014 was categorized and sorted into the HFACS Four Levels of Human Error categories. The HFACS four levels are shown in Fig. 3.

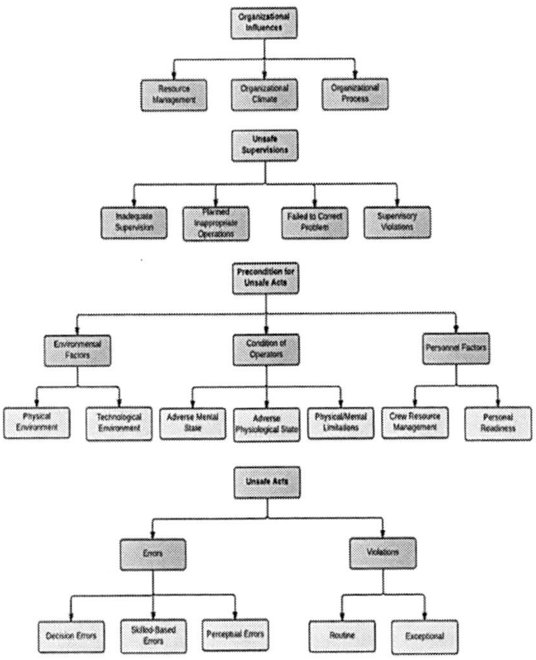

Figure 3. Human Factors Analysis and Classification System (HFACS) Overview [7]

The mishap data was further categorized into the eight (8) HFACS Sublevels: Skilled Based, Routine violations, Crew Resource Management, Perceptual Errors, Exceptional Violations, Decision Based, Physical Environment and Supervisory Violation.

For this research the HFACS human error probability results are used for a Binary Logistics Regression Analysis. The results from the analysis are represented in Tab. 2 below.

The occurrence of a Mishap was identified as a "response/event" and represented by a "0" or "1". The Regression Analysis was used to generate a regression model in which the predicted probability of each occurrence was calculated. The Binary Logistic Regression Expression for this Model is below:

$$P(1) = e^{(Y')} / (1 + e^{(Y')}) \quad (1)$$

Table 2. Identified HFACS Regression Model Values

HFACS Human Error Factor	Fitted Probability	P value	Odds Ratio
Skilled Based	27%	0.070	10.15
Decision Based	41%	0.037	18.72
Perceptual Errors	47%	0.009	24.25
Routine Violation	9%	0.444	2.67
Exceptional Violation	36%	0.052	15.67
Crew Resource Management	14%	0.283	4.38
Physical Environment	33%	0.138	13.72
Supervisory Violation	24%	0.226	8.61

In order to simplify the model, a Binary Logistics Regression Analysis with Stepwise Backward Elimination was performed. The analysis results are below.

The Binary Logistic Regression Expression for this Model is:

$$P(1) = e^{(Y')} / (1 + e^{(Y')}) \quad (2)$$

Table 3. Identified HFACS Regression Model with Backward Elimination P Values

HFACS Human Error Factor	Fitted Probability	P value	Odds Ratio
Skilled Based	27%	0.000	3.54
Decision Based	41%	0.005	6.60
Perceptual Error	52%	0.000	10.33
Exceptional Violation	36%	0.013	5.38

Of the calculated HEART Human Error Probability (HEP) values, three (3) values had the highest probability (22%, 22% and 26%). These three HEP values are from three (3) survey questions, which all had an Error Producing Condition (EPC) commonality. These commonalties are:

- Accessibility Limitations, Physical Stress and Tiredness.
- Poor lighting dealing with the Physical environment.

4.1 Binary Logistic Model Validation

For the Binary Logistic Regression model with Stepwise Backward Elimination, all of the p value factors are less than 0.05, thus indicating that they are statistically significant.

Three Goodness-to-Fit tests are performed for this research and all three tests' p values are greater than 0.05, indicating that the null hypothesis for the first hypothesis is rejected (Tab. 4). Therefore, HFACS contribution factors do have an impact on human error events in NASA GPO.

Table 4. HFACS Goodness-of-Fit Tests with Backward Elimination P Values

Goodness-of-Fit tests	P value
Deviance	0.725
Pearson	0.458
Hosmer-Lemeshow	0.795

5. MODEL VALIDATION

Model validation of the HFACS model was confirmed by consistency and comparison to other research conducted with the HFACS and data within the aeronautics field.

Human Factors research conducted by Remotely Piloted Aircraft (RPA) Operations [9], found that in prior HFACS RPA mishap studies, 3 of the 5 studies' largest percentage of mishaps fell into the "Unsafe Acts" category [10].

Another, Remotely Piloted Aircraft Mishap HFACS analysis on recurrent error pathways on 95 Remotely Piloted Aircraft Mishaps [8], identified that "Perceptional" and "Skilled" Based Error pathways (under HFACS "Unsafe Acts") had common associated latent failures. Together these are accountable for the majority of crewmember related mishaps.

Lastly, a study of the Republic of China (ROC) Air Force study [3] revealed that the ten (10) highest ranking frequencies of occurrence fell within the "Unsafe Acts (Level 1)" and "Preconditions for unsafe acts (Level 2)" HFACS Categories. When comparing the ROC study and KSC GPO, the highest frequencies of occurrence are consistent with the Level 1 and Level 2 majority Human error occurrence.

5.1 HEART Human Error Probability (HEP) Validation

As stated in the results and discussion (section 4.0), the three highest HEART HEP values from the survey data had Error Producing Condition commonalities between them ("Accessibility Limitations," "Physical Stress" and "Tiredness").

By comparison, these HEART EPCs to the NASA KSC Modified Levels of the HFACS, are best matched with HFACS Preconditions of Unsafe Acts (Level 2) sublevels "Physical Environment Confined Space" and "Adverse Physiological States - Physical Fatigue."

Through evaluation, there is some correlation between the three (3) highest HEART HEP values (22%, 22%, and 26%) from the survey data to the HFACS Physical Environment Fitted Probability of 33% from the binary logistics regression (before Backward Elimination) to draw statistically valid conclusions.

It is noted that all three (3) SMEs contributions to the developed Scenarios, tasks, subtasks and identified EPCs all fell under the HFACS (Level 2) Preconditions of Unsafe Acts Levels (Physical Environment Factor and Adverse Physiological States) category. Also, in the Preliminary Analysis of this study, many of the survey participants stated that they believed one of the biggest potential influences to human error mishaps in Ground Processing Operations (GPO) is confined spaces, which is a physical environment limitation.

Therefore the HEP probability values are consistent with previous studies in this research that identify the highest frequency and majority of human error occurrences falling under the HFACS Level 1 and Level 2 Category.

6. CONCLUSION

The HFACS and HEART results and validation, confirm that the KSC Ground Processing Operations Framework is a valid approach for human error analysis.

The KSC GPO framework is flexible in that it allows modification, for other operations. This capability was demonstrated in this research, by modifying the HEART's "Generic Tasks" to specific tasks performed in the Operation.

The research can be modified and used for complex operations, such as other Space Operations and Space Programs on an International Level.

Stages within the research methodology can be modified for use in recorded or documented Safety surveillance data over a period of time, as it relates to Operations and failure occurrences. The final stage of the KSC GPO framework encapsulates all of the error producing conditions and covers contributing factors for both models.

The central contribution of this research is a unique complex operations framework that incorporates a Human Reliability Analysis, Human Error Taxonomy and Human Error Framework, as it related to Space Operations.

Fig. 4 displays a set of steps that provides a research Methodology approach that Space Operations and other complex organizations may use to modify and apply to their unique processes.

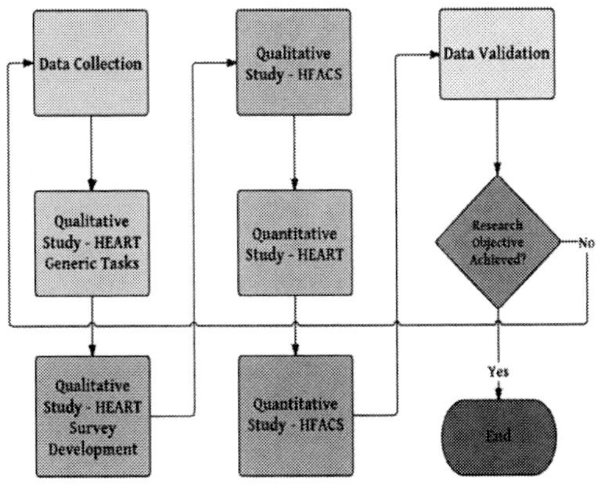

Figure 4. Research Methodology Contribution [1]

7. ACKNOWLEDGMENTS

I would like to thank the NASA/Kennedy Space Center for funding my research. I value and thank my UCF PhD Advisor, Dr. Pamela McCauley and committee, Dr. Luis Rabelo, Dr. Jose Nunez, Dr. Robert Hoekstra, Dr. Waldemar Karwowski and statistician, Dr. Liqiang Ni for the expertise and insight they provided me during my research. I appreciate my NASA KSC family, Safety and Mission Assurance family, mentors and leaders who have encouraged me throughout my entire research process. To all of the remarkable Shuttle Quality Engineers and Quality Assurance Technicians who participated in my surveys and my Subject Matter Experts; I truly appreciated all of your valuable knowledge, expertise, perspectives and input during my research. Finally, I would like to thank my husband, Ervin D. Alexander and PhD colleague, Dr. Joylene Ware for proofreading my manuscript.

8. REFERENCES

1. Alexander, T. (2016). A Systems Approach to Assessing, Interpreting and Applying Human Error Mishap Data to Mitigate Risk of Future Incidents in a Space Exploration Ground Processing Operations Environment. (Doctoral dissertation, University of Central Florida Orlando, Florida)
2. Eastman Kodak Company Staff. (2004). *Kodak's ergonomic design for people at work.* John Wiley & Sons.
3. Li, W., 2006 Li, W., & Harris, D. (2006). Pilot error and its relationship with higher organizational levels: HFACS analysis of 523 accidents. *Aviation, Space, and Environmental Medicine, 77*(10), 1056-1061.
4. Perry, R.U. and Committee (1993). Report of the Shuttle Processing Review Team. NASA. Kennedy Space Center
5. Reason, J. (1990). *Human error.* Cambridge university press.
6. Reinach, S., & Viale, A. (2006). Application of a human error framework to conduct train accident/incident investigations. *Accident Analysis & Prevention, 38*(2), 396-406.
7. Shappell, S. A., & Wiegmann, D. A. (2012). *A human error approach to aviation accident analysis: The human factors analysis and classification system.* Ashgate Publishing, Ltd.
8. Tvaryanas, A. P., & Thompson, W. T. (2008). Recurrent error pathways in HFACS data: analysis of 95 mishaps with remotely piloted aircraft. *Aviation, space, and environmental medicine, 79*(5), 525-532.
9. Tvaryanas, A. P., Thompson, W. T., & Constable, S. H. (2006). Human factors in remotely piloted aircraft operations: HFACS analysis of 221 mishaps over 10 years. *Aviation, space, and environmental medicine, 77*(7), 724-732.
10. Wiegmann, D. A., & Shappell, S. A. (2001). Human error analysis of commercial aviation accidents: Application of the human factors analysis and classification system (HFACS). *Aviation, Space, and Environmental Medicine, 72*(11), 1006-1016.

PROCEEDINGS OF THE 9TH INTERNATIONAL SPACE SAFETY CONFERENCE

SESSION 03
COMMERCIAL SPACEFLIGHT – I

EVALUATING AND COMMUNICATING ACCEPTABLE LEVELS OF RISK IN SPACE TOURISM

Chris Beauregard

Space Policy Institute | 1957 E St. NW Washington, DC 20008 | beauregard@gwu.edu

ABSTRACT

In the advent of commercial human spaceflight, there exist a number of ventures that aim to serve the public sector through government contracts, as well as some that seek to provide human spaceflight capabilities to paying private customers. A human presence in space is inherently complicated, dangerous and relatively novel compared to the human existence on Earth and, accordingly, risks must be properly calculated for such missions. Such risks, including those of the technical, safety and business sense, must be properly communicated between organizations in the spaceflight sector and regulators, policymakers, customers and general public. However, the risk tolerance acceptable to different parties, such as government entities and the private sector, can vary quite widely. As new technologies are developed and demonstrated, an increasing amount of excitement surrounds the fledgling industry, which has already experienced a series of accidents. Future mission failures in commercial human spaceflight are a near certainty, but through observation of precautions, communication of risk and creation of effective policies, such incidents can be mitigated, as will potential stagnation of an industry where acceptance of risk is a requirement to participate.

Keywords: Risk Management, Human Spaceflight, Space Tourism

1. INDUSTRY OVERVIEW

The remarkable era of human spaceflight is wrought with the excitement of exploration, scientific discovery, and challenges to the limits of the human body and engineering prowess. In terms of national pride and global impact, there is something about human exploration that robotics and imagery of deep space cannot quite match. Humans were first launched into orbit within sixty years of first achievement of manned flight and to the moon and back within another decade. In the present day, a near-permanent human presence in low Earth orbit has been established and it is important to recognize just how much has been accomplished. Such feats, however, do not come easily or inexpensively, in terms of financial capital and human life. In the pursuit of extending the human presence, over five hundred individuals have risked their lives by venturing to space, including nearly two dozen who made the ultimate sacrifice, not to mention scores of non-participant losses, due to accidents, such as engineers and members of the public. The lessons learned thanks to their expeditions, much like those from the deep sea, high altitude, and uncharted wilderness, will not soon be forgotten and have been used to improve the odds of success in future missions. The risks for these explorers, however, were understood and accepted, as a space program's ability to communicate risk and mission expectations is no accident. It is through decades of careful research, design, and precaution that organizations are able to limit and express risk within complex mission profiles to those who willingly accept them. In this advent of commercial human spaceflight, which includes orbital and suborbital private tourism, there are a new set of policy and safety concerns that must be addressed.

Historically, human access to space has been reserved for national space programs and a select few high-net-worth individuals that have paid sums over $20 million for orbital access aboard the International Space Station. Before the age of paying tourists, several other private individuals had flown for business purposes and as payload specialists in the United States Shuttle Program, including scientists, politicians, journalists, teachers, and civilians. For those with the financial means and willingness to accept the risk, private spaceflight is perhaps the most exciting venture accessible. Space tourism is a market often compared to climbing Mount Everest or trekking to the South Pole, as it is generally reserved for deep-pocketed risk-takers who will likely experience it once in a lifetime to challenge themselves, check it off of a bucket list or impress peers. With the added novelty of the adventure, more comfortable accommodations and lower activity requirement than that of a long-distance or high-altitude excursion, commercial spaceflight is poised to become a viable market thanks to successful technology demonstrations and a strong indication by willing participants. There currently exist several private ventures that aim to create the market for widely accessible space tourism for suborbital flights, orbital adventures and beyond. In fact, it is estimated that the space tourism may swell to a billion-dollar industry by 2022 [1].

Suborbital trips are expected to cost between $15,000 and $300,000 and would bring tourists to the edge of space by means of several different launch vehicles, most of which would include a period of weightlessness and a view of the Earth's curvature. At the lowest levels of complexity, cost and risk, Worldview and Zero2infinity, among others, aim to provide balloon-based experiences for passengers, which would provide experiences at the edge of Earth's atmosphere. At the suborbital level, companies are seeking to provide short-duration trips via spacecraft such as Blue Origin's New Shepard rocket and Virgin Galactic's SpaceShipTwo spaceplane. Orbital flights, due to higher velocity and altitude, will continue to be vastly more expensive and carry a greater number of qualifications for passengers. There are, however, several companies pursuing orbital excursions for private passengers, including Boeing's CST-100 Starliner and Bigelow Aerospace through their inflatable habitats. With their vision set beyond earth orbit, SpaceX, Space Adventures and several others seek to provide voyages to the Moon, with price tags in the tens of millions of dollars.

"Space transportation is inherently risky and the future of the commercial human spaceflight industry will depend on its ability to continually improve its safety performance.

- Commercial Space Launch Amendments Act

---

2. RISK INVOLVED

Earth is a risk-filled environment, though most risks that comprise daily life are accounted for and effectively mitigated. Travel via motor vehicles and aircraft, consumption of substances and medications, failure to observe warnings and lack of exercise all contribute to an individual's everyday health risk profile, but a number of other activities fall outside of such risk accounting. For travel, entertainment, and sporting, the following activities give a perspective of the approximate mortality risk undertaken by willing participants:

Activity	Fatality rate
Commercial air travel	1 in 7,000,000 [2]
Skydiving	1 in 101,083 [3]
Travel by motor vehicle	1 in 48,000 [4]
Scuba diving	1 in 34,000 [5]
Base jumping	1 in 2,317 [6]
Hang gliding	1 in 560 [7]
Human spaceflight	1 in 91 [8]
Climbing above 6000m in the Himalayas	1 in 10 [9]

Given the hype surrounding the next-generation thrill ride industry, such excitement should be somewhat tempered with the realization that the ventures still have some obstacles to overcome in their varied stages of development. Not the least of these concerns is the risk profiles that will be undertaken by the commercial space tourists, the companies providing flights and the regulatory entities certifying their missions. Human spaceflight subjects the body to a number of forces, and a lack thereof, that would otherwise not be experienced on earth. The launch of objects into space is an extremely accelerative process that imparts significant G-forces on the human body. A typical launch will subject upon a passenger three times the force of Earth's gravity, which is typically compensated for through reorientation of the human body to a reclined position to best tolerate the force, though the effects are only partially mitigated. The noise and vibration can also affect passengers and create a stress-inducing experience that may impair communication and awareness. For missions that include the experience of microgravity, passengers should anticipate some spatial disorientation with a possible onset of space-induced motion sickness [10]. For flights that allow passengers to leave their seats, the environment may also be conducive to personal injury, especially if the passenger is unable to re-fasten themselves in their seat before reentry. In orbital flights and high-altitude suborbital flights, the human body will also experience an increased exposure to radiation which, while not an immediate concern, is also a factor worth noting. Finally, during the reentry period, passengers will once again be exposed to high G-forces and a significant amount of noise and vibration. In addition to relatively minor health complications, participants in human spaceflight expose themselves to more catastrophic risks that could occur at any time, including decompression of a capsule, resulting in deprivation of breathable oxygen and exposure of the cabin to environmental conditions intolerable to life. During launch and reentry especially, atmospheric pressure and extreme heat on the craft can also create a risk of catastrophic loss of vehicle and crew should the structural integrity of the craft be compromised.

3. REGULATORY LANDSCAPE

With these risks in mind, mitigation of them can be partially achieved through observation of precautions such as the establishment of flight requirements for hopeful tourists. While US government entities such as the Federal Aviation Administration (FAA) have been developing their own guidelines for space tourism, health requirements have not yet been established, leaving many to believe the companies will self-regulate astronaut health guidelines [11]. Though FAA's Office of Commercial Space Transportation (FAA AST) currently oversees the licensing process for all US-based launches, including tourism, their focus is generally limited to public safety, property damage and supervision of the training processes for crew members [12]. FAA's ability to regulate human spaceflight was prohibited in the Commercial Space Launch Amendments Act of 2004 in order to encourage the growth of the industry. Thus, FAA AST is only able to regulate minor facets of commercial human flights, such as informing passengers of risks and training of participants "to respond to emergency situations, including smoke, fire, loss of cabin pressure, and emergency exit." In fact, the Federal Aviation Administration Office of Commercial Space Transportation *Recommended Practices for Human Space Flight Occupant Safety* lists "Medical Limits for Space Flight Participants" as a "Notable Omission." [13] In the case of non-US passengers, the case becomes slightly complicated, as foreign participants must be informed in the same way as us citizens, but discrepancies exist between US export control regulations and FAA informed consent regulations [14].

The criteria for astronauts as public servants have historically been very scrutinizing, often requiring high aptitude and fitness levels, advanced degrees, countless flight hours, extensive personal and family health history and strict personality requirements, but future commercial passengers will not be expected to meet many of those requirements. With regard to medical certification, Space Shuttle pilots were required to meet FAA Class I guidelines, while mission specialists adhered to slightly less restrictive Class II guidelines and payload specialists, Class III. From a risk perspective, the government tends to adhere to more restrictive standards of assurance, as organizations such as NASA under federal oversight cannot, in good faith, certify public servants to fly a mission with a high likelihood of failure. The initial risk profile target for the Shuttle program, for example, was a 1 in 1,000, while engineers estimated 1 in 100 and some management estimates cited a 1 in 100,000 chance of catastrophic loss of crew. Realistically, however, the chances of mission failure were far greater, as some experts cited retrospective statistics as grave as 1 in 9 [15].

Similarly, the current regulations state that commercial pilots are required to obtain an FAA pilot certificate with instrument rating, possess the necessary knowledge, experience and skills to control the vehicle and demonstrate an ability to withstand the stresses of spaceflight. Additionally, any safety-critical members of the flight crew must obtain an FAA Class II medical certificate. Finally, participants are required to agree to accept the risks involved by providing informed consent in writing. The FAA also encourages operators to comply with the "Recommended Practices for Human Space Flight Occupant Safety," though adherence is not required [16]. The FAA is currently most interested in maintaining public safety and prevention of third party harm and are relying on commercial spaceflight operators to create safe operating environment for its participants. However, as the industry matures, the scope of FAA AST may include occupant safety and mission assurance.

Some of the appropriate risks for spaceflight tourism are outlined in "Acceptable reusable launch vehicle mission risk," which specifies that the maximum risk level, which is measured in expected average number of casualties, to the general public shall be no more than 1x10-4 and those to any individual shall be no more than a 1x10-6 probability of casualty per mission [17]. However, these requirements only provide guidance for the general public. Spaceflight participants must also be properly informed, in writing, of the safety record of the individual vehicle, of the historical record of human spaceflight, that the vehicle has not been certified by the government, and of all hazards, with the understanding that some are unknown. Consent must then be issued in writing in addition to a waiver of claims against the US government [18]. To facilitate industry growth, a moratorium on additional space tourism-related regulations was established and has since been extended to 2023, providing a "learning period for operators." [19]

4. COMMUNICATION OF RISK

The risks associated with human spaceflight, as with other high-stakes, complex systems, are a complicated concept to effectively communicate with the general population and cannot always be accurately simplified to numbers. It is simple enough to emphasize that missions are risky, and reminders of what can happen when things go awry are ever present. Normalization of risk, though inherent in human nature, can lead the population to believe that there are never close calls and can create a perception of over-simplification. However, NASA, as an organization, has always appeared to operate as risk-averse and overly cautious, especially with the lives of its personnel. The risk profiles with crew versus cargo payloads differ quite a bit, as separate rating criteria are used to evaluate the two mission profiles. Similarly, the risk that the government is willing to take with taxpayer dollars and lives of public servants differs from the risk that commercial entities and private individuals are willing to take. Some risk is understood as a prerequisite for success, especially in business ventures, but with highly-publicized proceedings and human lives in the balance, how much risk are private ventures willing to take and how much *should* they be allowed to take? Communicating such a concept to the public is no easy task. According to NASA Associate Administrator for Human Exploration and Operations, Bill Gerstenmaier, the topic of risk "is one that we spend a lot of time and energy on within NASA, but we probably do not spend as much time with the public." [20] With so many emerging spaceflight systems born from companies with varying degrees of successful track records, it is important to recognize that the public must understand and accept the risk of failure-organizational, operational, mechanical and catastrophic

While risk can never be completely removed from a system, it can be mitigated. However, even through careful reduction and use of statistical modeling, a simple figure, though easy for a consumer to understand, will never tell the full story. Soft data is a universal truth, especially with a limited data set from which to draw conclusions. Unfortunately, it may require some missteps to uncover insights and concrete statistics, though the hope is for miraculous near misses that allow for adjustments to systems and processes to ensure safer flights in the future. With the added complication of communication such concepts to policy makers holding pens by which entire industries live and die, such risks need to be accurately, simply, and carefully communicated. The intricate balance in this principle is one that must be struck correctly, and the outcome will determine jobs, bottom lines, human lives, and chances of reelection.

5. POLICY RECOMMENDATIONS

As there currently exist few industry standards for paying customers, there is a case to be made for additional regulations, as long as they do not excessively prohibit further growth of the industry. As with any effective legislation, it should be created in anticipation of industry developments and occurrences to foster growth without inhibiting progress in the best interest of the public. Preemptive legislation, if correctly formulated and instituted, can prove vital to an industry's survival after mishaps, which are a near certainty in spaceflight. For example, in a 2011 hearing before the US House Subcommittee on Space and Aeronautics, Bill Gerstenmaier reminded policymakers that failure was a fact of life in spaceflight and it would be a major step backward to significantly pause the Commercial Orbital Transportation Services missions in light of accidents, or to over-regulate in response to them [21] In a similar fashion, it would be shortsighted to place unnecessary restraints on a budding commercial tourism industry due to accidents, unless of course, they are frequent, catastrophic and immutable. Industry survivability could ultimately rest in the hands of consumers and their willingness to follow through with their ticketed flights after such incidents. It could also be argued that, even after an accident, the industry would recover, based on ongoing non-tourism-based commercial missions that would reinforce the reliability of launch vehicles.

With that said, it is worth examining prudent policies that could be instituted to better ensure the safety and success of commercial space tourism. From a passenger and crew health perspective, it would be unwise to allow companies to institute subjective health requirements for their passengers without standards being set by an advisory body and the playing field should be relatively even for companies performing similar services. An example of such regulation should include medical certification of passengers via screened medical history for presence of disqualifying conditions, such as pulmonary, cardiovascular, vestibular or central nervous system risks, certain medications, mental illness, previous surgeries, and injuries. Prospective passengers should also be subjected to basic flight training and acclimated to the experience through means of simulations, centrifuge testing and other flight testing, specifically for orbital trajectories and beyond. Additionally, the certifying flight surgeon should be an employee of an entity other than the one with which the passenger will be flying to ensure an accurate evaluation with independent recommendations. A potential solution would be use of FAA designates, as currently employed by the aviation industry, to certify flightworthiness of participants and personnel. Such qualifications could ensure fewer adverse responses in flight, unfortunately, at additional cost and risk of self-deselection from their ticketed flight. Crew certification should be much more substantial and equivalent to an FAA Class I medical certificate, as well as advanced training in emergency situations, anti-g suits, medical response and other personal safety measures for themselves and passengers. An additional measure that has been advocated by the Office of Commercial Space Transportation to ensure participant readiness is the use of military support vehicles by commercial companies. The use of such test vehicles are not currently permitted, but could potentially be useful for training and familiarization of personnel and participants alike.

Such programs could potentially be overseen by FAA AST, NASA's Office of the Chief Health and Medical Officer, or a similar entity. To support this endeavor, as well as compensate for other shortcomings, funding must be increased for FAA AST, as the office will require additional resources to perform its workload and is not sufficiently equipped to expand its role [22]. In the same breath, it is also worth considering whether or not it is appropriate to have a single entity concurrently promoting and scrutinizing companies for commercial spaceflight undertakings. The Federal Aviation Administration's Office of Commercial Space Transportation's mission is "to ensure protection of the public, property, and the national security and foreign policy interests of the United States during commercial launch or reentry activities, and to encourage, facilitate, and promote U.S. commercial space transportation." Though not inherently counterproductive or a conflict of interest, the two missions may, at times, contradict one another.

Rather than creating new jurisdictional authority and in addition to an increased budget, the Office of Commercial Space Transportation could also be promoted to a higher level as an office under the Secretary of Transportation, perhaps in parallel with the Federal Aviation Administration or as a standalone entity. The present hierarchy does not delegate the appropriate authority or funding to AST and entangles space transportation within the same realm as traditional aviation. This restructuring would be beneficial for industry and the government, as it would assist in enabling transparent certification processes, consolidation of authority, better planning within the industry and government and simplification of development efforts [23]. In the interest of addressing several of these concerns, a review of domain definition should be conducted by the International Civil Aviation Organization with the intention of redefining space as a domain separate from aviation. This is a logical step for many reasons, especially considering the difference in environment, velocity and vehicle design. Additionally, space tourism should not fall under the profile of a common carrier as the present intention of space tourism is not to transport participants between locations. The outcome of this determination may also impact the entity tasked with investigation of any accidents that occur.

6. FUTURE CONSIDERATIONS

While it may be impossible to anticipate the nuances of an industry that hasn't quite taken shape, there are a number of considerations that might need to be addressed in the near future. Acknowledging that the opportunity of spaceflight might be used for purposes other than personal enjoyment, it should be asked what participant activities should be permitted and which should be banned. Could potential participants seek other tangible benefits during their flight, such as promotion of a brand or themselves? Will smartphones, cameras or other handheld technology be allowed? With the anticipation that participants could potentially have their vitals monitored, it may also be worth exploring what can be done with the data, especially considering the relatively small number of human spaceflight datasets. How would the sharing of that data take place, how would it benefit future studies of human health in space and how would participants be protected? When considering the context in which space tourism is marketed to consumers, is spaceflight likened more to a thrill ride like a roller coaster, an extreme sport like skydiving, an extraordinary tourist destination like summiting Mount Everest? The image portrayed may determine how the industry will be regulated and operators must carefully consider the perception of regulators and the public.

The industry is still in its infancy, but, as it develops, how should regulators stratify the different industry segments? Identifying orbital versus suborbital segments and piloted versus unpiloted craft is a logical starting point, but other unique identifiers will become apparent as the market matures and each will deserve specific attention and regulation, or lack thereof. Would a passenger flight half-way across the planet in an orbital-class vehicle be considered a tourism venture or a common carrier? When considering the differences between providers and lack of industry-wide standards, it should also be explored how these gaps in standards should be addressed. For example, because Blue Origin does not have a pilot to alter the trajectory of the vehicle, they do not need to abide by the same FAA regulations as a craft that is piloted.

The establishment of industry standards can be expected to occur as the market further develops, but there are a multitude of important considerations that should be posed in the process and examined in advance. In environments where actors are rushing to bring a product to market, there may be left with gaps in standards, which are established, in part, to create duties for providers to their customers. These standards could potentially be set up through working groups within a standards body and in conjunction with the FAA's Commercial Space Transportation Advisory Committee. It's far more advantageous to have a coalition of operators determine standards than imposition of rules from a federal level.

7. CONCLUSION

Within the next decade, a new era of tourism in space is set to begin and a delicate regulatory balance must be struck to allow the market to flourish. Organizations such as FAA AST have been instrumental in encouraging innovation and championing an initial hands-off approach to many facets of the industry that could have otherwise be strictly regulated, though they continue to operate in the best interest of public safety and have established a realistic and effective informed consent regime and licensing process.

As the industry develops, coordination among actors, regulators and standards bodies is necessary to ensure cooperation in safety issues and sharing of information regarding mishaps. These mishaps, while they may affect public perception of the industry, should be anticipated and abated to ensure increases in flight frequency. Regulations that are instituted should be done so with the understanding of their temporality to compensate for changes in the industry, technology maturation and evolution of other interrelated markets and business practices.

8. ACKNOWLEDGEMENTS

This paper is published with a sincere appreciation for the support of the Space Policy Institute in Washington, DC and the European Space Policy Institute in Vienna, Austria.

Special thanks is due for the insight of Scott Pace (Space Policy Institute), Jean-Jacques Tortora (European Space Policy Institute), Kris Lehnhardt (George Washington University Hospital), George Nield (FAA Office of Commercial Space Transportation), Frank Slazer (Aerospace Industries Association), and Chris Hearsey (Bigelow Aerospace).

9. REFERENCES

1. "Statement of Dr. George C. Nield, Associate Administrator for Commerclal Space Transportation" 20 Mar. 2012, https://science.house.gov/sites/republicans.science.house.gov/files/documents/hearings/HHRG-112-SY16-WState-GNield-20120320.pdf.
2. Waycaster, G. C., Matsumura, T., Bilotkach, V., Haftka, R. T. and Kim, N. H. (2017), Review of Regulatory Emphasis on Transportation Safety in the United States, 2002–2009: Public versus Private Modes. Risk Analysis. doi:10.1111/risa.12693
3. Richard L. Celsi, Randall L. Rose, Thomas W. Leigh; An Exploration of High-Risk Leisure Consumption through Skydiving. J Consum Res 1993; 20 (1): 1-23. doi: 10.1086/209330
4. Leonard Evans. Traffic Fatality Reductions: United States Compared With 25 Other Countries. American Journal of Public Health: August 2014, Vol. 104, No. 8, pp. 1501-1507.
5. Vann, RD; Lang, MA, eds. (2011). "Recreational Diving Fatalities." (PDF). Proceedings of the Divers Alert Network 2010 April 8–10 Workshop. Durham, N.C.: Divers Alert Network. ISBN 978-0-615-54812-8. Retrieved 24 May 2016.
6. Kjetil Soreide, Christian Lycke Ellingsen, Vibeke Knutson. How dangerous is BASE j umping? An analysis of adverse events in 20,850 jumps from the Kjerag Massif, Norway. J Trauma. 2007 May; 62(5): 1113–1117.
7. Feletti, F., & Goin, J. (2014). Accidents and injuries related to powered paragliding: a cross-sectional study. BMJ Open, 4(8), e005508. http://doi.org/10.1136/bmjopen-2014-005508
8. "A look at people killed during space missions - Phys.org." 1 Nov. 2014, https://phys.org/news/2014-11-people-space-missions.html
9. Windsor JS, Firth PG, Grocott MP, et al Mountain mortality: a review of deaths that occur during recreational activities in the mountains Postgraduate Medical Journal 2009;85:316-321.
10. Abadie, L., Lloyd, C., & Shelhamer, M. (2015, November 18). Gravity, Who Needs It? NASA Studies Your Body in Space. Retrieved April 29, 2017, from

https://www.nasa.gov/sites/default/files/atoms/files/your_body_six_month_in_space_11_18_15_0.pdf

11. Space.com Staff (2012, March 21). Space Tourism and Private Space Travel Must Be Safe. Retrieved April 29, 2017, from http://www.space.com/14982-private-space-travelrisks-faa.html
12. U.S. Code of Federal Regulations. (n.d.). Retrieved May 01, 2017, from https://www.govregs.com/regulations/title14_chapterIII
13. FAA AST. (2014, August 27). Recommended Practices for Human Space Flight Occupant Safety. Retrieved May 1, 2017, from https://www.faa.gov/about/office_org/headquarters_offices/ast/media/Recommended_Practices_for_HSF_Occupant_Safety-Version_1-TC14-0037.pdf
14. "The Foreign Space Flight Participant Problem: Can a ... - SSRN papers." 2 Jun. 2012, https://papers.ssrn.com/sol3/papers.cfm?abstract_id=2072479.
15. "CxP Presentation Template [SBU] - NASA Technical Reports Server ...." 26 Oct. 2010, https://ntrs.nasa.gov/archive/nasa/casi.ntrs.nasa.gov/20100036684.pdf.
16. Nield, George C. 26 September, 2017. "50 Ways To Leave Your Earth". Panel, International Astronautical Congress, Adelaide, Australia.
17. "14 CFR 431.35 - Acceptable reusable launch vehicle mission risk ...." 3 May. 2017, https://www.law.cornell.edu/cfr/text/14/431.35.
18. "14 CFR 460.45 - Operator informing space flight participant of risk. ..." 3 May. 2017, https://www.law.cornell.edu/cfr/text/14/460.45.
19. "House aviation subcommittee shows renewed interest in commercial ...." 24 Jun. 2016, http://spacenews.com/house-aviation-subcommittee-shows-renewed-interest-in-commercial-spaceflight.
20. Foust, J. (2017, February 13). Recalculating Risk. Retrieved May 01, 2017 from http://www.thespacereview.com/article/3171/1
21. NASA'S commercial cargo providers: Are they ready to supply the space station in the post-shuttle era?, 112th Cong. (2011) (testimony of William Gerstenmaier).
22. Foust, J. (2017, February 09). FAA commercial space office faces budget squeeze. Retrieved May 01, 2017, from http://spacenews.com/faa-commercial-space-office-faces-budget-squeeze/
23. *Restructuring Commercial Space Regulation*. 2017. Aerospace Industries Association. aia-aerospace.org

ADDITIONAL RSOURCES

24. Chapter III, Parts 400 to 460, of Title 14 Code of Federal Regulations. https://www.faa.gov/about/office_org/headquarters offices/ast/licenses_permits/media/Part_400_compilation.pdf
25. Ethics of Human Spaceflight: What is an Acceptable Level of Risk? (2016, November 8). Retrieved April 29, 2017, from https://deathbycosmos.com/blog/iss-ethics-of-spaceflight
26. Bensoussan, Denis (06/2010). Space tourism risks: A space insurance perspective. Acta astronautica. , 66 (11-12), p. 1633 - 1638. (ISSN: 0094-5765)
27. Kahn J, Liverman CT, McCoy MA, editors. Health Standards for Long Duration and Exploration Spaceflight: Ethics Principles, Responsibilities, and Decision Framework. Washington (DC): National Academies Press (US); 2014 Jun 23. 4, Risk Acceptance and Responsibilities in Human Spaceflight and Terrestrial Activities. https://www.ncbi.nlm.nih.gov/books/NBK222145/
28. Commercial Space Industry Developments and FAA Challenges, 16 (2016) (testimony of Gerald L. Dillingham, Ph.D). http://www.gao.gov/assets/680/677943.pdf
29. NASA Office of Safety and Mission Assurance (2003, June 19). Human-Rating Requirements and Guidelines for Space Flight Systems. http://www.dept.aoe.vt.edu/~cdhall/courses/aoe4065/NASADesignSPs/N_PG_8705_0002_.pdf
30. "Human-Rating Requirements for Space Systems" NASA Procedures and Guidelines NPG: 8705.2. http://www.dept.aoe.vt.edu/~cdhall/courses/aoe4065/NASADesignSPs/N_PG_8705_0002_.pdf
31. FAA "Recommended Practices for Human Space Flight Occupant Safety." https://www.faa.gov/about/office_org/headquarters offices/ast/news_announcements/media/Recommended_Practices_for_HSF_Occupant_Safety-Version1.pdf.

SOLAR ENERGY AND ELECTRIC PROPULSION FOR BETTER SAFETY DESIGN AND OPERATION OF NEAR SPACE AND SUBORBITAL VEHICLES

Norul Ridzuan Zakaria[1], Muhammad Amin Zakaria[2], Md Sayuti Ishak[3], Anass Hanafi[4], Ivan Cuzzi[4], Azahar Mat Hasan[5]

[1]*SOLVES, Corso Regio Parco, 24, Torino,Italy, ikam290200@hotmail.com*
[2]*Perak State.Government, Ipoh, Malaysia, m.aminzakaria@gmail.com*
[3]*University Science Malaysia, Trans-Krian, Malaysia,cesayuti@@usm.my*
[4]*Al-Biruni, University of Torino, Torino, Italy, anasshanafi.mail@gmail.com, ivancuzzi@gmail.com*
[5]*Radar Aero Marine Sdn Bhd, Kuala Lumpur, Malaysia, radarsdnbhd@gmail.com*

ABSTRACT

There is an obvious trend in considering the usage of solar energy and electric propulsion in designing aircrafts. This trend is expanding into the design of near space and suborbital vehicles, as the same principle is applicable in the design of both aircrafts and vehicles for near space and suborbital operations.

Solar energy and electric propulsion not only may increase the economics for the production and operation of the new space and suborbital vehicles, but also its safety, by designing safety into the vehicles when employing the technology of solar energy and electric propulsion in both the design and operation of the vehicles. Considering solar energy and electric propulsion in designing near space and suborbital vehicles may lead to the usage of lighter, lesser, smaller and less expensive materials and engineering systems for propulsion and control surface and less complex configuration, which may lead to better aerodynamics, ergonomics and safety. The innovative configuration of electric propulsion and exclusion of conventional engineering system essential to the conventional turboprop and turbofan propulsion in near space and suborbital vehicles may also lead to better operational safety of the vehicles. Such is possible because unlike conventional aircraft propulsion, electric propulsion may be installed in nonconventional configuration and operated with lesser components without handicap in basic significant aircraft engineering principles such as minimum frontal cross section, minimum weight and clean design for better aerodynamics.

The application of solar energy and electric propulsion, which is so much influenced by new material and energy transfer technology in the design and operation of near space and suborbital vehicles may also lead to application of quantum physics and engineering to maximize the efficiency of the design and engineering of the systems involved particularly to ensure very high conversion of light into energy and transfer of the energy to the subsystems. Another characteristic that most probably will result due to the application of solar energy and electric propulsion in the design and operation of such vehicles will be autonomy, where in this case the autonomous piloting will not eliminate the human piloting but will increase their efficiency and therefore enhance the safety of the operation of such vehicles. Both quantum engineering and autonomy is actually the common exploratory pathway that will emerge from engineering operation that try to maximize the efficiency of solar energy and electric propulsion, and this may lead to new, innovative and interesting design and operation of new space and suborbital vehicles.

The positive trend of considering solar energy and electric propulsion as the critical factor in designing and operating near space and suborbital vehicles most probably will lead not only to more interesting and more economic, but also better safety in the design and operation of such vehicles. The authors have involved in few study exercises in Italy and Malaysia in trying to design systems and subsystems of near space and suborbital vehicles employing solar energy and electric propulsion, and this paper will discuss few of the experiences and explain how such will lead to better safety design and operation of such vehicles.

1. INTRODUCTION

There is a collective effort in realizing a spaceport for suborbital flight in Malaysia for the past decades, and recently with the emergence of near space flight, now there is also an effort to realize a nearspaceport in the country as a mean to establish a facility that is less complex and less expensive than the spaceport, yet create an opportunity for flight "to the edge of space", as the region of near space flight is below the altitude of space flight, but above the altitude of conventional flight, commercially known as "the edge of space".

The objective of near space flight is to provide services similar to that of services provided by orbital satellites. To do this effectively, the aircraft is not only required to fly at very high altitudes, but also to remain airborne at such altitudes for very long time. For this, the aircraft requires practical propulsion that is light and uses fuel that is easily renewable at those altitudes, and it is concluded that solar rechargeable electric propulsion is probably the best.

Solar energy is a "commodity" for the new space industry in tropical countries at and near the equator that receive high intensity of solar energy all year, yet it need to be smartly managed and exploited (1). This commodity is very useful for near space flight and justifies the plan to create a nearspaceport in Malaysia, which is a tropical country, in the state of Perak on the west coast of Peninsular Malaysia (2).

Figure 1. A layout of the proposed nearspaceport in Perak, Malaysia.

Solar rechargeable near space aircrafts do not have to land on Earth or descend to lower altitudes to be refueled. It can remain airborne at these high altitudes for months The high altitudes is also above the unstable atmospheric regions, therefore there is no risk of encountering bad weather as in the lower altitudes, and the aircraft can ascend or descend through the lower altitudes when those altitudes are free of atmospheric disturbances.

It is true that there are significant atmospheric disturbances in the tropical countries like Malaysia, but as explained in the above paragraph, near space aircraft can easily be programmed to avoid such disturbances and remain in operation even during the storms. This is another justification for the planned nearspaceport in Malaysia.

Being in operation at near space altitudes, which is generally concluded to be above 20km, but below 100km from sea level, enables the aircraft to perform not only commercial services in communication and observation commonly provided by satellites, but also scientific missions which can effectively be done from a stable platform at those altitudes.

Among the scientific missions will be solar observation including for space weather monitoring. Near space aircraft can be used effectively for such solar observation because it will be airborne for a significant period of time above the layers of atmosphere that have been filtering the "solar signatures", and to operate it from a tropical nearspaceport with maximum the solar intensity, will only enhance its effectiveness. Space weather monitoring using a near space platform in Malaysia will be a major scientific activity since a small disturbance in the tropical South East Asia weather can create a "domino effect" in the global weather that will result in hurricanes storming north America for example, and monitoring the solar activities can provide significant understanding of such.

An economic-scientific field that will be an interest for the operation of near space platform in Malaysia will be forestry management, as the stability of the tropical forest of South East Asia is a significant global environmental determinant. Near space aircrafts can easily perform deforestation and forest fire monitoring, for example.

As for the commercial observation and communication, near space aircrafts will be able to provide a simpler and more affordable operation than orbiting satellites particularly for tropical countries. These have further justified the nearspaceport project in Perak, Malaysia. The nearspaceport, will also serve as the MCC (mission control center) for those near space missions.

As for suborbital flight, the subject has been around within the Malaysian space community for years since there is a spaceport project in the country. Therefore it is natural that near space flight is being looked upon as a catalyst for the emergence and development of suborbital flight in the country. It is even a viable concept to look upon near space flight as part of the process in the maturity of the suborbital flight initiative in Malaysia. There were studies completed on suborbital vehicles for operation in the country, and on-going studies to equip those vehicles with solar rechargeable electric propulsion.

2. INNOVATIVE DESIGN AND ENGINEERING

Economic factors may limit innovative design and engineering as it is always economic to use existing major components in designing and constructing aircraft, particularly the propulsion system, as there are large selections of turboprop and turbofan propulsions available in the market to power variety of aircrafts. This is demonstrated by many occurrences where a model of turboprop or turbofan is produced in many series to power many models of aircraft of similar class. Therefore for economic reason, aircraft design and engineering are limited by the availability of the turboprop or turbofan propulsions in the market.

However, for electric propulsion, the situation is different because there is not yet a common electric propulsion in the market that is available to power few aircrafts of a category and probably it is even economic if a new electric propulsion is designed and developed together with a new aircraft, albeit using common electrical components and technology. Therefore, electric propulsion aircrafts can be more innovative in its design and engineering.

The conventional configuration of electric propulsion is that the propulsion is mounted in the YZ-plane, or "vertically" facing the front as in the mounting of conventional turboprop and turbofan propulsions.

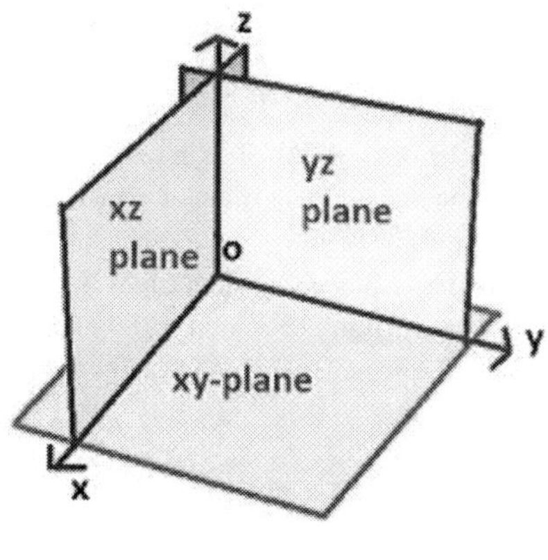

Figure 2. The XY, XZ and YZ-planes.

When such configuration involve only the mounting of an engine at the centerline, it does not significantly affect the frontal cross section area and wing span of the aircraft, but when the mounting involve 2 or more engines and off the center, as in the wing mounted twin engine aircrafts, it significantly increase the frontal cross section area and wing span of the aircraft and decrease the aerodynamics of the aircraft.

The configuration where the propulsion is installed within the fuselage of an aircraft is called centerline thrust (CLT), and one or more units of propulsion can be mounted in centerline thrust configuration. A twin engine CLT aircraft is more aerodynamic than a twin engine wing mounted aircraft of the same class because the earlier can achieve up to 30% reduction in frontal cross section area and wing span than the later (3).

The author has designed a CLT aeroplane with zero torque due to having an electric propulsion with contra rotating propellers called, ANTS (Aeroplane of Neutral Thrust & Synchronization), where "Neutral" and "Synchronization" refers to "centerline" and "torque cancellation" respectively.

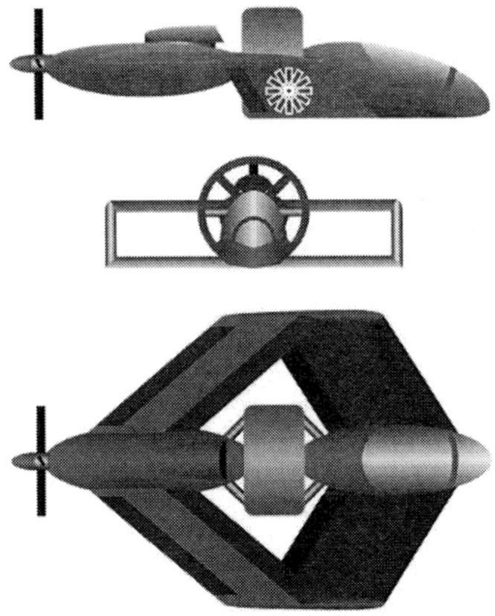

Figure 3. Illustrations of ANTS.

There is an ongoing exercise by SOLVES where an aircraft with electric propulsion mounted unconventionally in the XY-plane or "horizontally" facing the bottom is being simulated. This aircraft which is designed to look like a fighter jet is meant to be a high-speed electric light passenger aircraft and is called, SELF (Solar Electric Light & Fast).

SELF is designed to carry 2 passengers, but with its design and capability, it can be converted into a fighter-bomber with the advantages of solar-electric propulsion and large passenger cockpit for military payloads. Therefore SELF will be able to serve both the civilian and military markets.

The propulsion of SELF consists of 3 ducted fans installed inline at the XY-plane within the fuselage. If such propulsion is to be mounted conventionally in the YZ-plane, it might have ended up in the design of wing or off-center mounted propulsion which is less aerodynamic.

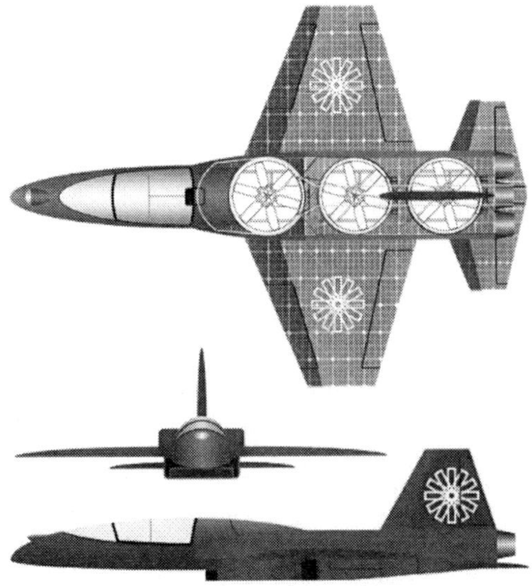

Figure 4. Illustrations of SELF. The top view shows its XY-plane mounted propulsion of 3 ducted fans (illustrated as exposed for visibility).

The XY-plane mounted electric propulsion of SELF resulted in a clean wing design and improved aerodynamics. The propulsion system is layered in 3 levels where the lower level is meant for air inlets, the middle level is occupied by the ducted fans and the upper level is dedicated for air outlets.

Figure 5. An illustration of SELF showing the airflow through its 3-level propulsion system.

The electric propulsion of SELF is solar rechargeable, hence the solar cells on top of its fuselage and wings. As there is a plan on developing near space flight capability in Malaysia, besides planning for a nearspaceport, there is also an exploration to design a near space aircrafts based on SELF. The result is another version of SELF with different wing design for near space flight, demonstrating further flexibility of SELF into another market segment; near space.

This aircraft which has longer wing to fly at higher altitudes and accommodate more solar cells is called, INSECT (Integrated Near Space Electric Carrier Transporter), where "Carrier" refers to "carrying scientific payloads", and "Transporter" refers to "transporting passengers". It may lead to a strategic plan of developing a near space aircraft for near space tourism, since it is able to carry 2 passengers, besides providing all the common commercial sciencific services able to be provided by near space aircraft.

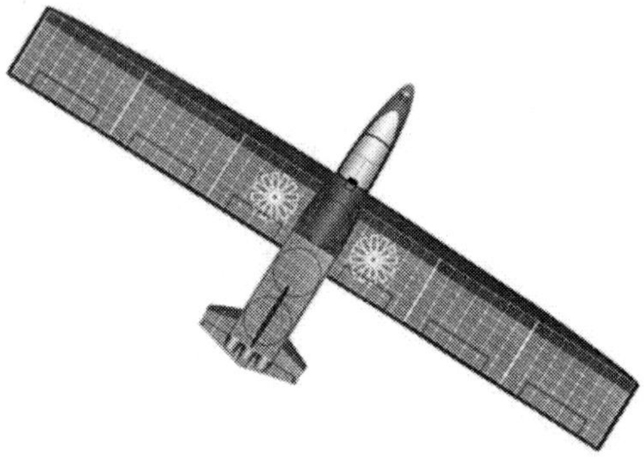

Figure 6. An illustration of INSECT.

There is also an effort to tactically operate an existing near space aircraft in Malaysia by Radar Aero Marine. The aircraft, which is Solar Stratos, has successfully been flown in Switzerland albeit not at a near space altitude and with a conventional configuration in propulsion mounting. The aircraft can also carry a passenger beside scientific payloads.

Figure 7. An illustration of Solar Stratos.

As for suborbital vehicles, the author has presented a paper in the 6th IAASS Conference which described a design of VTOL (vertical take off & landing) suborbital vehicle called, SOLVES (Sub-Orbital Low cost & low risk Vehicle to the Edge of Space), capable of safe vertical descend using an electric propulsion system equipped with vectoring nozzle. The electric propulsion is not to provide lift, but only to provide orientation good enough to land the vehicle at a predetermined location (4).

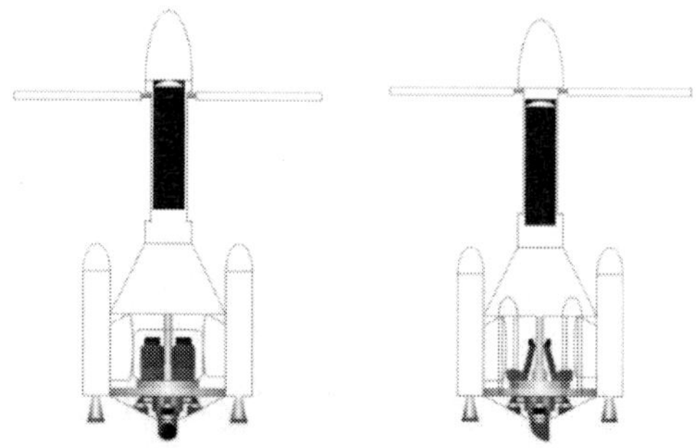

Figure 8.Illustrations of SOLVES showing its electric propulsion (in green) with vectoring nozzle.

The SOLVES paper also proposed HANZAR (Horizontal Aero Neutral Z-Axes Rotating) wings, where "Neutral" refers to "rotate freely naturally or passively due to air pressure". The HANZAR wings are to provide the lift to land SOLVES safely.

During descends, the HANZAR wings will be automatically extended and rotate naturally due to air pressure. This is possible due to the symmetrical aerofoil design of the wings with its axes of rotation at their center.

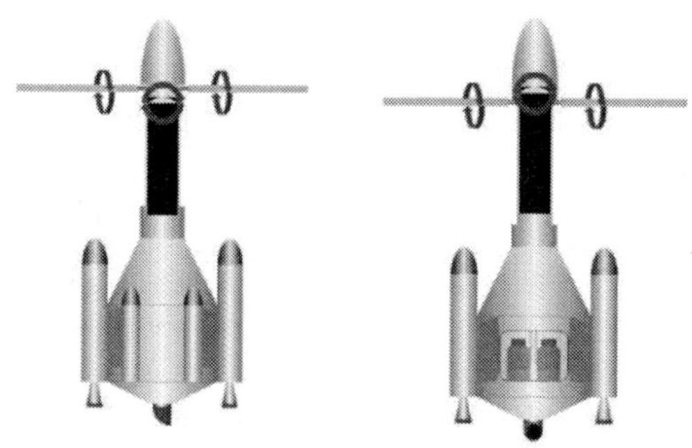

Figure 9. Illustration of SOLVES with HANZAR wings in operation.

If HANZAR wings are to equip an aircraft and activated when the aircraft lose its thrust, the aircraft can be landed safely because the HANZAR wings will compensate the lost lift. However, the aircraft need to have a perfect tube fuselage since HANZAR wings need to be symmetrical to effectively functional. For such operation, HANZAR wings will occupy large spaces on the sides of the fuselage and may not be economic.

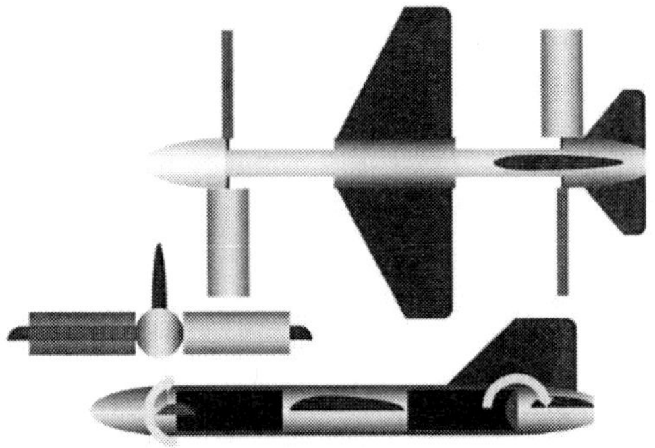

Figure 10. Illustrations of an aircraft with perfect tube fuselage and 4 HANZAR wings in operation.

If a wing with effective variable cross section can be realized, and the wing can is convertible into HANZAR wing, such wing can function both for conventional flight and HANZAR wing operation. In this case, the aircraft does not need to have a perfect tube fuselage to carry HANZAR wings. Such wings may create "crashless" aircrafts and will be safe for emergency landing during total propulsion failure.

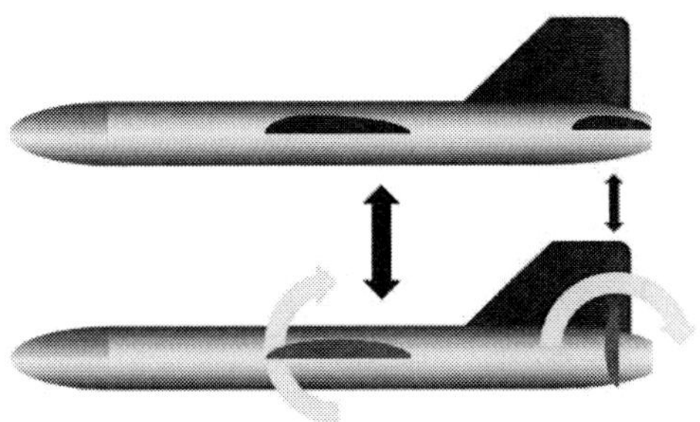

Figure 11. An illustration of a "crashless" aircraft with its wings convertible to HANZAR wing.

In October 2015, ALTEC SpA, Italy completed a study of a VTOL suborbital aircraft called, "Aerocosmos Spacejet" for a private company in Malaysia. When the suborbital tourism industry is matured, the capability and flexibility of the suborbital vehicle will be significant, and VTOL suborbital aircraft will have a definite advantage (5).

The study however did not describe in detail the VTOL propulsion system, except it did say that the VTOL is to be provided by air-breathing engines, but focused more on its rocket propulsion that is to be dedicated to ascend the suborbital aircraft to an altitude of 100km from sea level. Actually the VTOL propulsion system was yet to be determined.

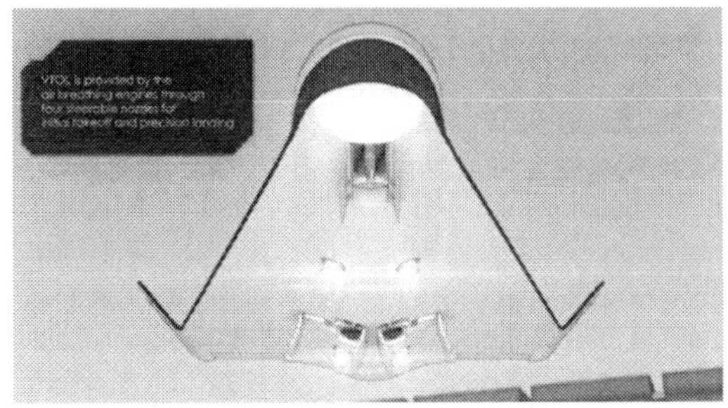

Figure 12. An Illustration of Aerocosmos Spacejet indicating its VTOL capability.

Now, SOLVES is proposing the Malaysian company to consider installing an electric propulsion to provide the VTOL capability of Aerocomos Spacejet. Most probably the electric propulsion is to be as simple and as light as possible only for limited vertical motion, while the rest of the flight including earlier non-vertical ascend and will be performed using rocket engine. Therefore a further study of the electric propulsion Aerocosmos Spacejet can be expected.

This is a unique example where turbofan propulsion is being replaced by electric propulsion for a suborbital vehicle design. Obviously this will affect the safety of the propulsion system, where gone are all the safety risks related to the turbofan. The electric propulsion too has its safety risks, but being a new and improving engineering system, there is much more room for the improvement of its safety elements. The electric propulsion is meant to increase the safety of Aerocomos Spacejet besides making it less complex, lighter and also less expensive.

This trend on solar electric propulsion for near space and suborbital vehicles will lead to more study on the quantum efficiency of the propulsion system, as quantum efficiency is defined as, the ratio of the number of carriers collected by the solar cells to the number of photons of a given energy incident on the solar cells. There are 2 types of quantum efficiency; the external quantum efficiency and internal quantum efficiency. The external quantum efficiency equals to the magnitude of photocurrent electrons divided by number of incident photons, while the internal quantum efficiency is the magnitude of photocurrent electrons divided by number of photons absorbed. Quantum efficiency can be expressed as a function of wavelength or energy.

A study has proven that the external quantum efficiency can exceed 100% with the usage of quantum dot solar cells. The quantum dots were of 5 nanometer diameter dots made of lead selenide (PbSe) tightly packed into a film of solar cells (6). Such is the beauty of Quantum Physics, where efficiency can exceed 100%, which is not possible under Classical Physics.

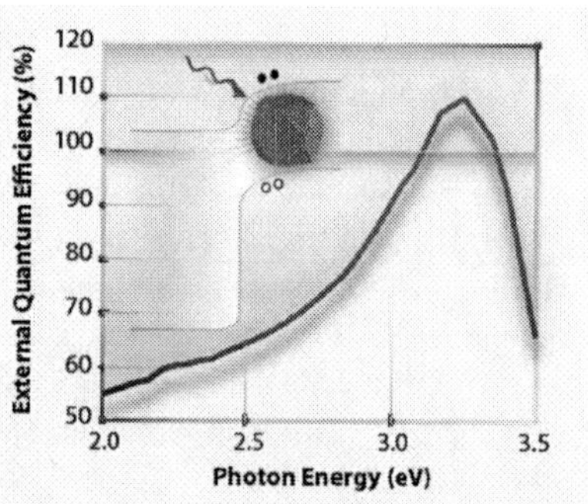

Figure 13. A simulation of external quantum efficiency as the function of energy, where the efficiency exceeds 100%.

Scientific study of such subjects is very proper to be done in Malaysia that owns a tremendous amount of this solar energy "new space industry commodity", and has plans for solar electric propulsion near space and suborbital vehicles.

Autonomy is an engineering that will affect the design, operation and safety of near space and suborbital vehicles because it is the next logical step in the development of solar electric propulsion aircraft. This is due to the simplification and potentials of employing autonomy in electric-powered aircraft.

Recharging of electric-powered near space and suborbital vehicles using high power microwaves beaming is another technology that may become the next logical step in the development of those vehicles. Efforts and development of such technology have been recorded, particularly in military application.

3. BETTER SAFETY FROM SOLAR ELECTRIC PROPULSION

Basically the usage of electric propulsion on near space and suborbital vehicles has eliminated turboprop and turbofan propulsion system including its fuel storage and feeding system in the design and engineering of such vehicles. This eliminates all the safety risks related to turboprop and turbofan propulsion system, as far as high flammable fuel is concerned.

As electric propulsion is still at a discovery and rapid improvement stage very unlike the already matured turboprop and turbofan, there is so much opportunity to include safety features at the pre-design stage of electric propulsion. There is even an opportunity to make compulsory of procedures and certification at the product definition level in the process of developing electric propulsion, which will be a huge step forward in safety design and engineering.

There should be an inclination toward the study of safer materials, methodology and design in the early stages of realizing solar electric propulsion. For example, in selecting the materials, methodology and design to increase the quantum efficiency, there should be emphasize on using the safest materials, methodology and design.

Some conventional components will not be safe for usage in near space flight, particularly moving parts because they wear more rapidly in the near-vacuum environment at near space altitudes (7). Therefore the design should emphasize on the usage of simple components with minimum moving parts.

Being aerodynamic is significant in reducing the power requirement for flight as the ratio of lift to drag depends on aerodynamics. The reduction in power requirement will enable the application of simpler and safer propulsion system. Right materials and design can be employed for better aerodynamics as demonstrated by the XY-plane ducted fans of SELF.

A near space aircraft accident in 2003 which was carefully investigated by NASA, has provided some insights into the vulnerability of such aircraft. The aircraft was Helios HP03-2, a HALE-UAV (High Altitude Long Endurance Unmanned Aerial Vehicle), a solar powered aircraft of flying wing design. Photos of the last moment before it crashed into the sea were recorded, showing that its wing has been shredded into pieces (8).

Figure 14. A photo of Helios HP03-2 in flight.

Figure 15. A photo of Helios HP03-2 showing its wing has been shredded into pieces before it crashed into the sea.

Among others, the investigation has revealed that the accident was caused by the inability of the design and engineering of its wing structure to endure atmospheric turbulence at low level and was not caused by any failure of its propulsion system. However since its propulsion consists of wing mounted propellers, the propulsion does affect its wing design.

Such design and engineering can be improved if its propulsion is in the CLT (centerline thrust) configuration. Not only CLT design will be more aerodynamic, but if any of its propulsion unit fails, it will not suffer any loss of symmetry, albeit loss of power.

Loss of symmetry due to engine fail for near space aircraft which has relatively very long wing will cause extra wing load and stress to the side that is suffering the engine fail. This absolutely will increase safety risk as reported by the Helios accident investigation.

Taking the Helios accident investigation as a lesson, the author has designed a HALE-UAV similar to Helios, but with CLT configuration called, WAU (Wing dominated Aircraft design for Unmanned flight), and "wau" is also the traditional Malay big kite This CLT HALE-UAV has 4 electric engines with individual, but synchronized propellers installed in-line at the fuselage. The 4 propellers of contra-rotating pairs cancel any possibility of torque that may result from the propulsion. This will produce a very stable well-balanced static and dynamic design.

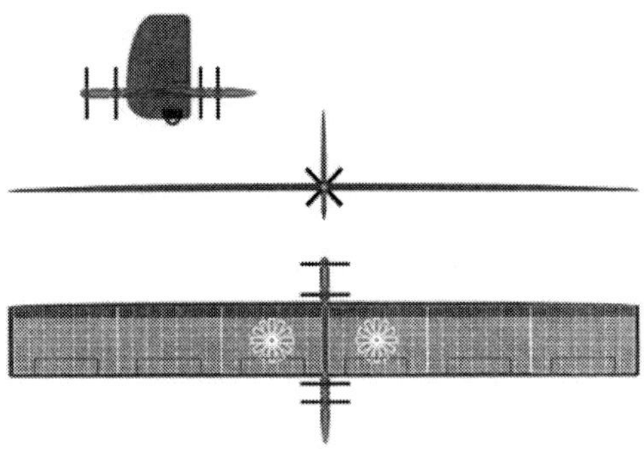

Figure 16. Illustration of WAU CLT HALE-UAV.

The clean wing design resulted from the CLT configuration will significantly reduce the stress in the wing particularly during high wing dihedral flight. Since high wing dihedral is common for HALE-UAV, this capability is very significant. This should be of concern because the Helios accident report did mention about the stress suffered by the wing of HP03-2 during high wing dihedral which led to the accident.

Diagram 17. A photo Helio HP03-2 in flight with high wing dihedral.

The quality and quantity of solar cells affect the safety of near space and suborbital vehicles. The efficiency of solar cells affects the design, size and mass of the aircrafts, as the surface area where the solar cells are exposed to sunlight is very significant.

High quantum efficiency such as that achieved by the employment of PbSe quantum dots will enable simplification of propulsion system and reduction in size and weight of the wing and other surfaces meant to collect the sunlight. This will positively affect the safety of the aircraft.

Autonomy is an advanced concept that is easily developed with the application of electric propulsion. Due to the standard protocols and availability of electrical subsystems for autonomy, the "language" for being autonomous is "electric", as almost all signals and commands for autonomous system are of electrical.

In this case, autonomy is not to replace human pilots, but to further enhance the capability of human pilots acting to remind, correct and justify any actions by them. Autonomy also is to take over the critical tasks of a human pilot during emergency when he or she is not able to function as required. Therefore, autonomy that can be expected to become a common practice in the operation of electric powered near space and suborbital vehicles will only increase the safety of those vehicles.

There were studies to establish the capability to recharge electric-powered UCAV (unmanned combat aerial vehicle) using high power microwave. This UCAV will use battery and solar cell to power its propulsion in hostile area to avoid detection during microwave recharging, but it will receive high power microwave energy beamed up from microwave transmitter when at safer airspace. A mobile transmitter using 4m diameter antenna beaming microwave of 75kw at 94GHz was modeled in this study (9).

Recharging of electric propulsion using high power microwave will further enhance the safety of near space aircraft which employ such recharging technique, because mobile transmitter aboard ground, sea and air vehicles can be employed for such emergency. The aircraft however need to be equipped with rectenna or receiving antenna.

4. CONCLUSIONS

1. The trend toward the application of electric propulsion and solar energy in the design of near space and sub orbital vehicle is also the elimination of turboprop and turbofan propulsion system in such design, and all the safety risks related to the turboprop and turbofan system.
2. Electric propulsion is not yet as matured as turboprop or turbofan. Therefore there are many rooms for safety improvement even at the product definition and pre-design stage, a perfect case for "prevention is better than cure".
3. The employment of electric propulsion and solar energy in the design and operation of near space and suborbital operation encourages innovative design and engineering. This in turn may produce more effective and safer design for near space and suborbital vehicles.
4. Near-vacuum environment at near space altitudes causes rapid wear and tear of moving parts. Therefore simple components with no or less moving parts should be used in the design of near space aircraft.
5. Near space aircraft commonly have long wing span to fly effectively at high altitude and to accommodate maximum number of solar cells, meaning there is an emphasize in the design and engineering of the wing. Therefore the design and engineering of the wing of near space aircraft is a major safety issue.

6. CLT configuration for electric propulsion on near space aircraft will produce clean wing design significant to cause less stress to the wing particularly during high wing dihedral, which is common for near space aircraft. This will increase the safety of the aircraft as far as the wing span is concern.
7. CLT configuration also reduce the frontal cross section and wing span of near space aircraft, improving aerodynamic and decreasing the power requirement from the electric propulsion. This will increase the safety of the propulsion as far as the power production is concerned.
8. HANZAR wing is worth to be considered particularly for near space flight safety, where long wing is a great concern.
9. Near space flight employing electric propulsion and solar energy maybe an area where Quantum Physics can contribute significantly to its safety, as the quantum efficiency of solar cells that determine the magnitude of the power production can be effectively increased by the science of quantum physics. The high quantum efficiency achievable by using new materials, processes and quantum techniques will enable the design of more efficient and safer near space aircraft.
10. Autonomy is not to replace human pilot, but is to enhance the effectiveness of human pilot and to compensate his or her tasks which are not able to be performed as required. Autonomy can easily be developed with the application of electric propulsion due to standardization and simplicity of processes. It will only enhance the safety of near space and suborbital flight.
11. An electric propulsion recharging method of high power microwave beaming-up can further enhance the safety of near space flight because such recharging will be very useful for emergency. As it has been studied, most probably it will be employed initially to recharge UCAV equipped with rectenna.
12. Near space flight may be a step closer to suborbital flight and space tourism as near space flight is "the flight to the edge of space", and its experience and learning curve particularly in the employment of electric propulsion and solar energy will be very useful for similar application in suborbital flight.

5. REFERENCES

1. Zakaria, M.A. (2017). The keynote speech by the Chief Minister of Perak, Malaysia for the 9th IAASS Conference.
2. Radar Aero Marine & Urban Scale. (2017). Pangkor Nearspace Ecoresort Proposal, Malaysia.
3. Zakaria, N.R., Zakaria, N.Z. (2006). Advantages of Dornier Do-335 Centre Line Thrust (CLT) Aeroplane Compared to Wing Mounted Twin Engines Aeroplanes of the Same Classification and a Proposed Design for CLT Aeroplane. Seminar on Engineering and Technology, Putrajaya, Malaysia.
4. Zakaria, N.R., Nasrun, N., Zulkifli, M.R.,Yamin, M.I., Othman, J., Zakaria, N.R. (2013). The Design and Operation of Suborbital Low Cost and Low Risk Vehicle to the Edge of Space (SOLVES), 6th IAASS Conference.
5. Zakaria, N.R., Nasrun, N., Abu, J., Jusoh, A., Azim, L., Said, M.A., Ishak, M.S., Zakaria, N.R. (2011). The Advantages, Potentials and Safety of VTOL Suborbital Space Tourism Operations. 5th IAASS Conference.
6. Semonin, O., Luther, J., Choi, S., Chen, H.Y., Gao, J., Nozik, A., Beard, (2001). Peak External Photocurrent Quantum Efficiency Exceeding 100% via MEG in a Quantum Dot Solar Cell. Science.
7. Noll, T.E., Brown, J.M., Perez-Davis, M.E., Ishmael, S.E., Tiffany, G.E., Gaier, M. (2004). Investigation of the Helios Prototype Aircraft Mishap. NASA.
8. Pathfinder and the Development of Solar Rechargeable Aircraft. (1994). E&TR July 1994.
9. Board of Army Science and Technology Commission on Engineering and Technical Systems, National Research Council. (1992). Strategic Technologies for the Army of the 21st Century. National Academy Press.

ARGUING THE (SAFETY) CASE(S) FOR SPACE

Andy Quinn MSc PhD MRAeS CEng

Saturn Safety Management Systems Ltd, Bath, BA2 9BR, UK, Email:

ABSTRACT

Space safety demands that appropriate levels of rigour are applied to license applications in order to demonstrate that the 'system' (spaceport, operator, vehicle etc.) is compliant to the stated requirements. As there are different and novel space vehicles, it is clear that prescriptive requirements are not appropriate and that performance (and risk) based requirements are more suitable. In the US, the FAA-AST regulations are not prescriptive and require an applicant to submit a Safety Review Document to cover the key regulatory requirements. The document's main headings relate to relevant Code of Federal Regulations Part 400 requirements and therefore the applicant's descriptions and 'solutions' to each requirement are essentially the compliance statements (or evidence). The document could therefore be described as the 'safety case'; or can it?

A safety case can mean different things to different people, in different countries. The ICAO definition is '*A document which provides substantial evidence that the system to which it pertains meets its safety objectives'*. In the UK, the term 'Safety Case' is used in respect of a set of one or more documents that include claims, arguments and evidence that a system is safe. Another UK definition is that a safety case is defined as '*a structured argument supported by a body of evidence that provides a compelling, comprehensible and valid case that a system is safe for a given application in a given operating environment.*

The supporting arguments are the spine of a safety case as these explain the rationale of how lower-level goals (or claims) can help substantiate the overall goal (of safety/compliance) and as well as explaining how the evidence can be interpreted to meet the goals.

This paper presents different safety case formats and the rationale why, if properly argued (constructed), safety cases can assist, rather than hinder, regulators and operators alike. The author of this paper contends that regulations for space activities should be performance based (and not overly prescriptive) and that a formal and *structured argument* can assist regulators in assessing the compliance to appropriately rationalised requirements and a structured argument, with justifications and assumptions can assist the operator's case for award of a launch license.

1. INTRODUCTION

Primarily in the UK, but adopted (and adapted) within other countries, a safety case, and associated report, is a means to convey the attributes of a system such that a regulator can approve its use. A safety case can be defined as [1] '*a structured argument supported by a body of evidence that provides a compelling, comprehensible and valid case that a system is safe for a given application in a given operating environment'.*

Although originating in the UK in the early 1970's, the safety case concept was elevated due to disastrous events and the subsequent inquiry recommendations that prompted a dramatic change in regulation and approaches to safety management.

The Deepwater Horizon accident occurred in 2010 [2] and lessons identified included taking a more performance and risk-based approach and by articulating this within a safety case: *Recommendation A2: The Department of the Interior should develop a proactive, risk-based performance approach specific to individual facilities, operations and environments, similar to the "safety case" approach in* ***the North Sea****.*

The Deepwater disaster occurred after the UK's North Sea disaster in **1988** (Piper Alpha) whereby Lord Cullen [3] made 106 specific recommendations to initiate a new and improved safety regime and in particular, the safety case regime.

As well as recommending less prescriptive approaches, another important recommendation transferred oversight of the offshore oil & gas industry from the Department of Energy to the Health & Safety Executive (HSE). This then brought the risks under the ALARP (As Low As Reasonably Practicable) principal whereby residual risks of a system are to be 'As Low As Reasonably Practicable' (within the HSE's Tolerability of Risk framework).

So when should we use a safety case regime? Is it just for complex systems or safety critical systems or should everything have a safety case? What about cost and what is proportionate?

Figure 1 attempts to illustrate the 'new product, new market = high complexity/high risk' approach, per the Ansoff Matrix.

		Solution	
		Familiar	**Unfamiliar**
Problem	**Familiar**	Minimal argument and standard evidence from the domain i.e. certificates of design	Focused argument on reasons for novel solution, plus the appropriate evidence
Problem	**Unfamiliar**	Minimal argument and standard evidence from another domain i.e. use of recognised standards	Extensive argument and evidence, with substantial independent scrutiny and application of novel standards and technology i.e. Space Launch Operations

Figure 1: Safety Case Complexity – based Ansoff Matrix

During the lifecycle of a system (from concept, development, manufacturing, testing, in-service operation, disposal) the point of the safety case is to challenge whether the system is 'acceptably safe' for instance. During the SC development (and by using rigor and challenge) this would undoubtedly provide some aspects which are not fully substantiated; hence this would mean a recommendation can be made to either improve the design or processes, or having identified a residual risk, then the point is to highlight this (not to hide this). The design team would then have to provide evidence that this has been analyzed (or state that this will be analyzed during the next phase) and hence may require approval of a concession (also called waiver). This may be (or may not) acceptable for this stage of the safety case because of the risk classification i.e. if a risk was identified as intolerable or a design requirement had not been substantiated and further re-design is not practicable, then the point of the safety case would be to say 'stop' An example could be the original X37 project – whereby the design seemed to be fine, but then the programme was cancelled at NASA after the technical assurance report provided evidence that the design was not robust, mainly due to poor reliability (of the design and of the reliability processes). In this example, the safety engineer constructing the safety case would have identified this during the concept or development stages, using a properly constructed argument to challenge the design, and processes and hence would have saved a lot of time and money being invested in a novel but non-starting project (the project was stopped in 2001, but could have been stopped 2 years previous using the Safety Case methodology).

2. TYPES OF SAFETY CASE ARGUMENT

Qualitative Argument; this will be used throughout the SC as the author details compliance to standards and requirements; some of which may have an indirect link to the system's desired attributes (competency of personnel's qualifications and experience, certain Quality Assurance aspects, etc.).

Quantitative Argument; here the author will set a probabilistic target (as a goal) and then argue by quantitative statistical reasoning, to establish whether the analysis meets the objective (through reliability, availability, maintainability predictions and performance analysis and modelling etc.).

Deterministic Argument; this relies on axioms, logic and proof.

	Attribute	
	Fault-Tolerance	**Reliability/ Availability**
Goal/Claim	Safety is maintained under stated failure conditions	System reliability meets the stated objective
Requirement	Subsystem 'X' shall be 2 Fault Tolerant	Subsystem 'X' Loss of Function shall be $1x10^{-4}$ per mission
Design Features	Redundant channel Zonal separation	Hardware component reliability
Evidence	FMECA, FTA Design Review	FTA CMF

Figure 2: Example Claims, Arguments, Evidence Attributes

3. SAFETY CASE CONSTRUCT

The SC is best constructed as a team effort starting with the Top Goal/Claim i.e. System 'X' is acceptably safe to operate in context 'Y'. What we are trying to communicate is a clear, comprehensive and *defensible* argument (case) with supporting evidence. So this sounds like legal jargon – exactly the point! The SC is attempting to justify a system is safe to operate (for complex systems with severe risks) and that the system has met the legislative requirements. Why? – such that, in the event of a catastrophic/critical outcome, the operator can provide a defensible argument (in a court of law if appropriate) that the legislation was complied with and that all that was reasonably practicable to prevent the accident was done.

After the Top Goal/Claim is established then the context in which the system is operating in needs to be defined, along with assumptions.
The strategy (or argument) then evolves by *'arguing over effective design in accordance with 'X' standards and 'Y' means of compliance and that the system is operated and maintained effectively and that all hazards have been identified and managed within an effective Safety Management System'* for instance.

Then the next level down provides the goals/claims to support this strategy, and so on, until the lowest goals/claims are as far as you can derive (with the arguments), whereby the analyst/team provide the evidence. Here we are looking for a true/false solution (evidence) but of course there will be conclusions whereby partial substantiation has been achieved – thus prompting re-design or operating limits/procedures etc.

There is **Direct Evidence**: this is evidence which is clearly linked directly to the goal/requirement itself (test evidence, analysis, historical data, compliance matrices – to standards, Verification Cross-Reference Matrix, etc.) i.e. evidence could be a test result showing that a safety critical parameter is within tolerance.

Then there is **Indirect (Backing) Evidence** that supports the direct evidence, giving it credibility i.e. a qualified and experienced engineer using certified test equipment.

The evidence is presented at the lowest level in the safety case as depicted in Figure 3:

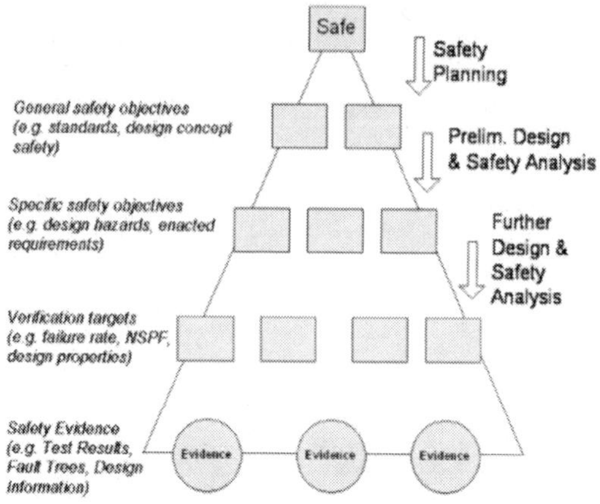

Figure 3: Safety Case Construct

An aircraft, for example, has many systems and subsystems that combine to present the overall Safety Case (SC). So for instance, propulsion system requirements are flowed down to the engine manufacturer to produce the SC of the engine. The engine system will then have a specific Top Goal such that the engine operation is acceptably safe on the aircraft 'X' and meets its safety objective of $<1x10^{-4}$ per flying hour (per engine) resulting in a Hazardous Event; note in aviation an engine cannot result in a catastrophic event by itself (per Certification Specifications). Figure 4 below depicts a Total Safety case hierarchy diagram for an air-launched suborbital vehicle with similar systems/ subsystems to that of an aircraft (with the obvious differences of the rocket engine etc.).

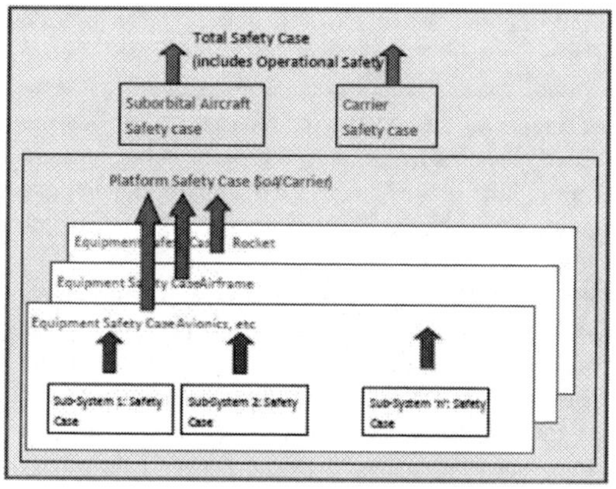

Figure 4: Safety Case Hierarchy

3.1 Safety Case Notations

There are different ways to present a safety case and this depends on the complexity of the system under consideration. A simple system can be described in a document report and have a Tabular safety argument. For more complex systems, in the UK, we recognise the following two notations:

Goal Structuring Notation (GSN). The GSN notation was developed at York University and further developed there by Professor Tim Kelly. A generic GSN example is in Figure 5:

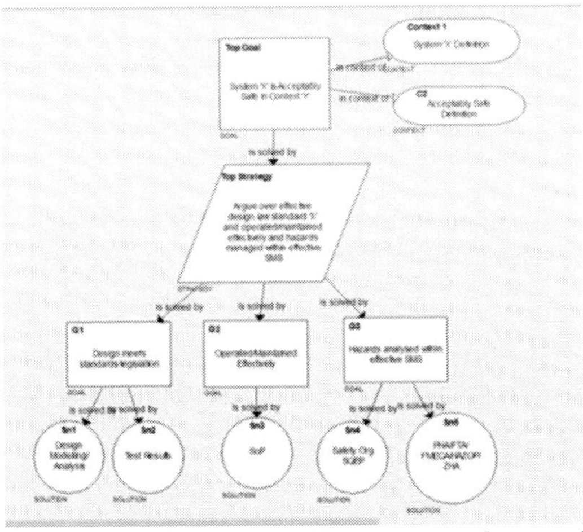

Figure 5: Goal Structuring Notation – simple example using the Adelard Assurance and Safety Case Environment Tool [5]

Claims, Argument, Evidence (CAE). The CAE notation is in Figure 6.

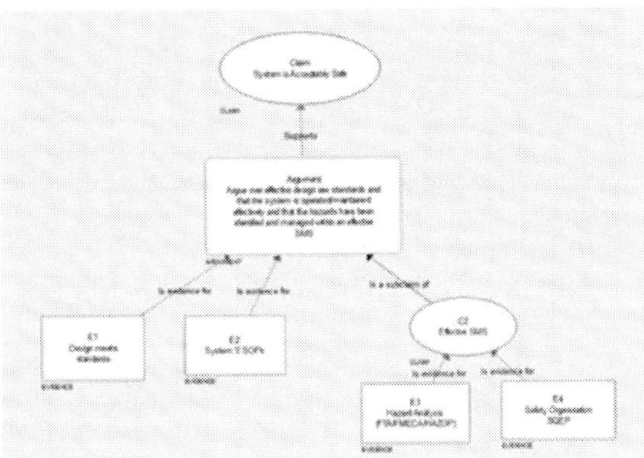

Figure 6: Claim Argument Evidence – simple example using the Adelard Assurance and Safety Case Environment Tool [5]

The CAE methodology was developed in the UK by Adelard in the Assurance and Safety Case Environment (ASCE) tool. This is an interactive tool whereby the author can 'open' the nodes and write the claims, arguments and evidence. Diagrams and tables can be imported and the nodes can be linked to each other and references can link to reports (analysis/test results etc.) directly when viewing in the ASCE 'Browser'. This is considered a very useful tool for a regulator for instance when interrogating the evidence and arguments.

3.2. Spine of The Safety Case

The supporting arguments are the spine of a safety case as these explain the rationale of how lower-level goals (or claims) can help substantiate the overall goal (of safety/ compliance) and as well as explaining how the evidence can be interpreted to meet the goals.

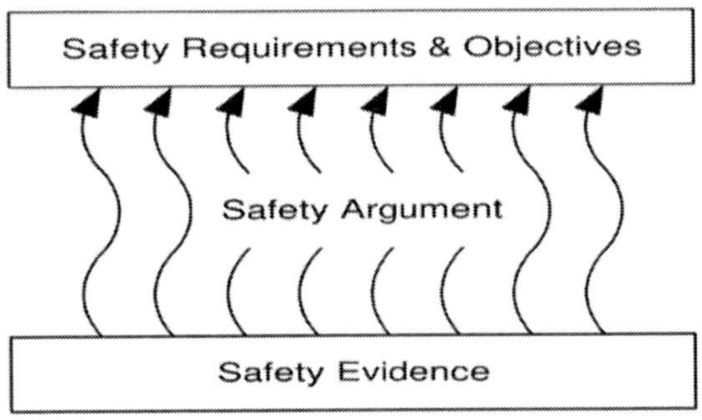

Figure 7: Argument – Spine of the SC

3.3 The Life of The Safety Case

The safety case remains live throughout the lifecycle of the system and should start as early as possible i.e. at the concept phase:

Preliminary/Initial (Concept Phase) Safety Case. This captures:

- The scope of the SC and definition of the system (and boundaries/interfaces to other systems or dependent SCs i.e. space vehicle to spaceport to range safety command and control etc.).
- Defines the SC approach; arguing functional attributes, product attributes, process attributes.
- Identifies the key safety requirements and applicable standards/means of compliance (and references the Safety Management Plan and Safety Program. Plan).
- Identifies the key safety issues (from the associated hazard analysis) with high-level derived safety requirements for the subsystems and system level. So here we follow the Design

For Minimum Risk (DMR) approach with fail-safe, damage tolerance requirements etc.

- The Preliminary Report (SCR) then sets out the argument (story) and provides the evidence to substantiate the Top Goal; hence it summarises the SC and should provide recommendations to continue work to gather the evidence detailed OR provides an early opportunity to find out if there are fundamental flaws in the design or intended operation. Hence the SC Statement should clearly detail whether the project is viable to continue to the next phase.

Interim (Detailed Design) Safety Case. This is an expansion of the Preliminary Safety Case and should increase confidence that the initial SC can be supported i.e. based on the detailed design and further hazard analysis using appropriate safety tools & techniques. The lower-level goals and supporting evidence 'nodes' can be populated with the design architecture, some modelling to provide validation of requirements, the Failure Mode Effect & Criticality Analysis and Fault Tree Analysis will have derived additional requirements for the design (interlocks, protection measures), including apportionment of risk (reliability) budgets etc. The Interim Safety Case Report (SCR) will state that the design is acceptable to now progress to Manufacture and then Test (or will detail flaws in the design which should either be re-designed or be accepted with operating limitations/procedures [if acceptable to the Authorities]).

Operational Safety Case. This is the SC that demonstrates to the Regulator that the agreed requirements have been met, in accordance with the appropriate standards, and that where necessary the alternate acceptable means of compliance demonstrate the systems is safe for its intended use. This can therefore be seen as the Initial Operating Capability SC. Going forward throughout the operating life of the system, the Operational SC can be updated (where required due to modifications or changes in operation) or a more bespoke Operating SC could be developed focusing on operations, maintenance and modifications etc. Back to the Operational SC, this covers:

- Sufficient assurance that the system as constructed is acceptably safe to be allowed to operate in its intended operational context for the first time.
- Completed arguments (story of the design, analysis, build, test) with associated assumptions and justifications where needed.
- Completed evidence (solutions nodes) that substantiate the associated goals and hence the Top Goal. This includes compliance matrices for design, safety, testing etc. and hence verification of all requirements per the Preliminary SC (and additional requirements added throughout the project).
- Descriptions of any applied for concessions or 'work arounds' on specific issues. This will provide analysis of the problem and associated operating limitations or procedures.
- Conclusion that the system is acceptably safe with the list of limitations and operating procedures for any issues.

3.4. Maintenance of The Safety Case

Throughout the operating life of the system, the safety case should remain 'live' as an important part of the Safety Management System (SMS). Part of an effective SMS is Change Management and changes to the design either due to mission requirements or safety issues need to be managed effectively i.e. considering the effects to existing hazards (and does the change introduce new hazards) and impact to other subsystems etc.

4. SC PITFALLS (& TIPS TO AVOID THEM)

4.1. Proving the system is 'safe'

It is easy/straight-forward to develop a safety case to provide a positive 'story' (argument) with all evidence being present and correct. This is especially true when the author/reviewer has a mindset (or cognitive bias) to favour information that confirms their preconceptions and hence are not looking to challenge the information. The key (and added value) is to develop a structured argument that challenges the Top Goal and provides arguments and justification where required i.e. demonstrating an equivalent level of safety or design means of compliance. At the start of the safety case construction, it is important to establish the requirements and standards as part of a team effort. The safety specialist should lead the workshop with the aim of developing the safety case argument (through the lower level goals and justification notes/assumptions).

4.2. Retrospective Safety Case

Starting a safety case after the design is completed is problematic. The safety rationale and argument will be less robust, trying to link the 'as designed' evidence (including probabilistic results) to meet some safety objective, as well as design standards. This could lead to a re-design of parts of the system with associated overrun in costs and schedule. The Safety Case should begin at the concept phase, along with preliminary hazard analysis, such that appropriate safety requirements and safety objectives can be derived i.e. that are realistic and achievable. The Safety Case construct can be a useful tool in developing a strong argument and then detail the sort of evidence that should be sought i.e. the initial safety case 'solution's (evidence) will actually be requirements; for instance detailing appropriate standards or regulatory requirements) that should be satisfied in order to substantiate the safety case goals. In relation to safety and reliability, the solutions can detail the activities that will need to be completed as the project develops.

4.3. Lack of SQEP and Independent Assurance

The Safety Case author should be a safety specialist with knowledge of the system under scrutiny i.e. Suitably Qualified and Experienced Person (SQEP). Where this is not the case, then, as detailed above, confirmation bias will creep in as well as lack of understanding of appropriate requirements and standards. The UK Ministry of Defence Nimrod aircraft SC was 'a lamentable job from start to finish ..... riddled with errors....tick box exercise ..... open hazards.....' etc. [6]. With complex and safety critical systems that have severe consequences, it is advisable to employ an Independent Safety Advisor to provide assurance that the SC author (and those conducting the hazard analysis) are conforming to best practice and that the system has appropriate requirements and standards to follow.

4.4 Interfaces (Inter-system or between systems)

Interfaces present weak areas i.e. boundaries between subsystems, systems and external systems. Check the assumptions and dependency statements. Ensure the interface failure modes have been appropriately considered and derived safety requirements recorded (and properly transferred to suppliers or external stakeholders).

4.5. Arguments

Arguments without evidence are unfounded. Construct arguments correctly – get independent checks.

4.6. Evidence

Evidence without arguments are unsubstantiated (unexplained).

4.7. Shelf-bound Safety Case

Once constructed and a Safety Case Report written (and a launch license granted for instance) then 'job done'. The SCR and diagram (GSN/CAE) then sits on the shelf gathering dust. Wrong!

5. SAFETY CASE FOR SPACE

The NASA Safety Center [7] provides further rationale in a system safety failure case on the UK's Piper Alpha disaster, as mentioned in the Introduction. The NASA 'Case for Safety' states: *Fortunately, the UK's movement toward Safety Cases after the Piper Alpha disaster finds a parallel in NASA system safety engineering and methodology as Risk Informed Safety Cases (RISC). The RISC is developed beginning early in the systems development lifecycle and reviewed at each major milestone. Then it plays a key role in NASA's acceptance and possibly certification (if applicable) of the newly developed system.*

There are different and novel space vehicles and hence it is clear that prescriptive requirements are not appropriate and that performance (and risk) based requirements are more suitable. This non-prescriptive approach is how the UK HSE manages safety and this is how the prospective UK Space Agency (and UK Commercial Spaceflight CAA team) intend to handle space launch risks. In the US, the FAA-AST regulations are also not too prescriptive and require an applicant to submit a Safety Review Document to cover the key regulatory requirements. The document's main headings relate to relevant Code of Federal Regulations Part 400 requirements and therefore the applicant's descriptions and 'solutions' to each requirement are essentially the compliance statements (or evidence). The document could therefore be described as the 'safety case'; or can it? On the one hand, requirements have been identified, but on the other hand, to what standards and to which alternate acceptable means of compliance?

5.1. Benefits of Safety Case for Space

Space systems are complex and novel; hence how does an operator substantiate their launch license application without justifying and arguing their evidence (and

where required, arguing an equivalence or alternate acceptable means of compliance – to the regulator's requirements)?
A well-constructed safety case, based on expert knowledge of the system provides the operator with an opportunity to tell the regulator the story of the design, including what requirements were identified and how these have been met – and where issues were identified, how these were mitigated to an equivalent or alternate means of compliance i.e. by further analysis and testing and demonstrating the system functions safely. Rockets are risky, and consequences of failures are generally severe resulting in Loss of Vehicle on the launch pad or during the ascent. The Safety Case provides an opportunity to challenge the design and operation in order to influence the design.

5.2. Operators Safety Case

Whether a spaceport operator or space launch vehicle designer/operator, a safety case will assist in providing the 'glue' between your system, subsystems and externally to interfacing systems. It will help drive the design by compiling a structured argument based on the agreed requirements and complying to the agreed standards and means of compliance.

The operator's SC safety engineer has a chance to challenge the design and operations. An example would be to challenge why prelaunch chill-down times are reduced (procedural change); here the SC author could quote best practice and then raise a recommendation that the procedure is changed to reflect best practice or that additional mitigation is introduced to monitor the component temperatures to ensure the shorter chill-down times actually provide the desired component operating temperatures.

5.3. Regulators 'Assurance' Case

One could argue that, for an operator to submit a safety case as part of their launch license application, the Regulators should have their own assurance case – which would essentially be a structured framework (the argument) based on the legislation (requirements) that, if followed, would provide confidence that the applicant's 'case' would meet the regulations for launch.

In essence, this is where the regulators can define the credible bounds of an acceptable safety argument and provide guidance on the level of evidence required to meet the individual goals and hence overall safety case (top goal).

Taking the FAA-AST CFR Part 415 Appendix B, the list of required Safety Review Documents in theory could be the evidence (that the FAA-AST are looking for i.e. they have defined the boundary for the applicant with these headings). This SRD outline should include:

1.0. Lunch Description (§415.109)
2.0. Launch Operator Organization (§415.111)
3.0. Launch Personnel Certification Program (§415.113 and §417.105 of this chapter)
4.0. Flight Safety (§415.115 & §417 Subpart C)
5.0. Ground Safety (§415.117)
6.0. Launch Plans (§415.119 and §417.111 of this chapter)
7.0. Launch Schedule (§415.121)
8.0. Computing Systems and Software (§415.123)
9.0. Unique Safety Policies, Requirements and Practices (§415.125)
10.0. Flight Safety System Design and Operation Data (§415.127)
11.0. Flight Safety System Test Data (§415.129)
12.0. Flight Safety System Crew Data (§415.131)
13.0 Safety at End of Launch (§415.133)

Hence the operator should provide descriptions and justifications and supporting analysis for each of the Chapters above.

Perhaps a future exercise, for the Regulators, would be to compile their own assurance case to evaluate whether the above list of 'evidence' is sufficient/complete. Additionally, this approach may assist them (the US FAA-AST) in expanding their regulations to include safety of those on board and then the airworthiness/spaceworthiness of the launch vehicles.

4. CONCLUSIONS

Safety Cases should add value to your business whether a space launch vehicle designer/operator or spaceport operator. If constructed properly, by safety SQEP (and checked with Independence) the SC will provide the story (the argument) that the 'system' is acceptably safe to operate in the defined environment. This argument, based on deterministic, qualitative and quantitative claims/goals will be supported by the evidence – these should be true/false statements that summarise the documented evidence (design reports, safety analysis, test results etc.) to provide a level of confidence in the associated goals i.e. the evidence should be substantiated. Where this is not the case, then this will be partially substantiated (or at worst unsubstantiated)

and a level of confidence assigned; in this instance additional measures should be provided to allow the system to operate i.e. operating limitations, operating procedures etc.
The SC will then be presented as a SC Report as part of a launch license application with the aim of assisting the Regulators understand the 'story' of your system's conception, design, build, testing and how it will operate. The next step will be for Regulators to be aware of Safety Cases and to know how to 'read' and 'interrogate' the robustness of the SC to enable them to approve the system. Regulators should use recognised tools to aid in the SC review and provide guidance to operators in compiling safety cases (defined boundaries and guidance on depth of evidence required).

5. REFERENCES

[1] United Kingdom Ministry of Defence, Defence Standard 00-56, Issue 4

[2] National Commission on the BP Deepwater Horizon Oil Spill and Offshore Drilling, Deepwater: The Gulf Oil Disaster and the Future of Offshore Drilling, p252, January 2011

[3] The Hon. Lord Cullen, The Public Inquiry into the Piper Alpha Disaster, Vols 1 and 2 (Report to Parliament by Secretary of State for Energy by Command of Her Majesty, 1990)

[4] National Aeronautics and Space Administration, Independent Assessment of X37 Safety & Mission Assurance Processes and Design Feature, Headquarters Office of Safety & Mission Assurance June 18, 2001

[5] Adelard, Assurance and Safety Case Environemnt (ASCE) tool, Available at: Https://www.Adelard.com

[6] Charles Haddon-Cave, The Nimrod Review, HC 1025, London: The Stationery Office Limited, 2009

[7] National Aeronautics and Space Administration, NASA Safety Centre, The Case for Safety, May 2013 Volume 7 Issue 4

PROCEEDINGS OF THE 9TH INTERNATIONAL SPACE SAFETY CONFERENCE

SESSION 05
SPACE DEBRIS – I

FURTHER STUDY OF SPACE DEBRIS COLLISION WARNING TECHNIQUES

WANG Rong-lan, LIU Wei, Yan Rui-dong, Liu Si-qing

National Space Science Center ,Chinese Academy of Sciences,Beijing,100190,China

ABSTRACT

Collision warning and avoidance is the main method of the satellite risk mitigation to the catalog debris. The precision improvement and collision warning confidence analysis are the main problems of the current collision warning work. This paper first introduced the collision warning work flow and development in NSSC (National Space Science Center, Chinese Academy of Sciences) . Then some recent work were introduced on the TLE precision improvement, drag coefficient analysis, the covariance analysis and the Space Environment Effect to the prediction accuracy, which improved the warning confidence to some extent.

1. INTRODUCTION

The number of space debris is increasing with more and more aerospace activities. According to SSN (Space Surveillance Network) , the number of the catalog objects has reached to over 17000.The collision between the catalog object and the spacecraft is catastrophic for the spacecraft. The collision warning is the main method of predicting the dangerous intersection, the avoidance is the main means of mitigating the risk.

Many studies have been made because of the importance of the collision avoidance. Many units have founded the collision warning and analysis system which has been applied to their satellites based on the collision warning flow and specification. However the warning techniques and application have some problems. For example, the 2009 collision between Iridium 33 and Cosmos 2251 satellite, after the event, many research departments studied the events and made the collision warning analysis for the event. But the results were different on the intersection epoch, the intersection distance and the collision probability. The differences were mainly due to the three factors: the data source, the orbit determination and prediction model and the error analysis method. So the warning precision and confidence improvement are the main problem in the current warning work[1].

This paper first introduced the collision warning work flow and the critical techniques, then some main work were given on TLE precision improvement, the drag parameter analysis, the covariance analysis, which improved the warning confidence to some extent.

2. COLLISION WARNING FLOW AND CRITICAL TECHNIQUES

Collision warning is to analyze the collision risk between the satellite and space debris. The collision warning flow see Fig.1, it is as following:

(1) Data collection is the first thing, including the orbital data, the environment parameter, the characteristic data, the error data and so on. Then these data would be imported to the database and the calculation would begin.

(2) The warning criterion should be determined based on the warning task and the negotiation with the engineering department, then the warning task would be triggered, and the parallel computing would begin.

(3) After the warning calculation finished, the dangerous intersection results would be filtered and the intersection results would be analyzed in detail.

(4) Precise warning analysis would be made after determining the risk with update data and improved method.

(5) After the danger confirmation, the information need report the corresponding departments.

(6) After the predicted intersection epoch, the warning evaluation need be made to improve the collision warning work.

Now the collision warning work has developed for many years, some critical techniques, such as error analysis, probability calculation, warning threshold have been grasped. Many warning practices were made. Warning products can be provided for the different warning units. However the warning accuracy and confidence are the critical factors to limit the development of the warning project application. Their improvements need consider two main factors: one is the data accuracy, the other is the atmospheric model. The data improvement is to improve the ground-based and space-based surveillance ability. The improvement of the atmospheric model need more data. However the space environment changes are the main factor of affecting the atmospheric drag.

Besides the surveillance ability improvement and the support of the high-precision model, how to improve the warning ability by improving the current method with the current data is the main problem in the current warning work.

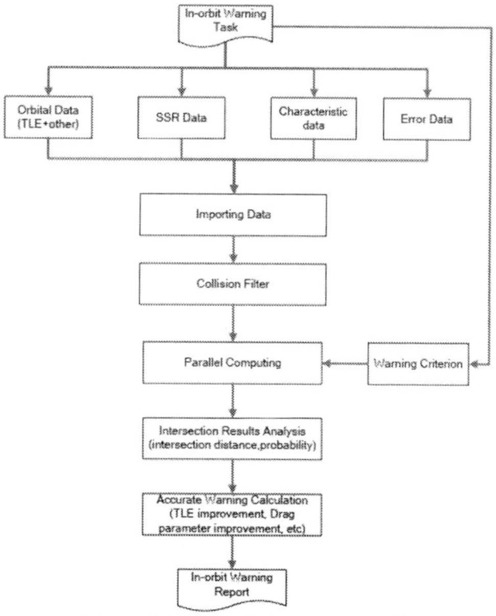

Figure 1. The collision warning flow

3. WARNING TECHNIQUE DEEP ANALYSIS

3.1. TLE precision improvement analysis[2]

TLE is the only public and useful data in the collision warning work, it is the main data source in the warning filter. But it has low precision. Without the support of the high-precision data, how to make the collision warning analysis with TLEs is the main problem for many warning departments. TLE improvement is a valid method.

TLE data is matched with SGP4/SDP4 model. The simplified SGP4/SDP4 model ignored the random errors produced by the high-order periodic terms perturbation, which can be removed by the statistical method. Prediction errors would decrease with the improved orbit and the high-precision prediction model.

So the process of TLE precision improvement is as following: TLEs are regarded as the "true" data, the pseudo observation data with public TLE were produced. The improved orbit data were gotten with the high-precision numerical model by the orbit determination. The orbit predictions were made with the improved data and high-precision model to improve the prediction accuracy and warning confidence.

The pseudo observation data were produced by refitting the orbit with the historical data. The fitting period determination method was given for the objects at the different orbit altitude. The prediction errors were given for the new orbit elements. The results demonstrated that the prediction accuracy had the large improvement, especially for > 1d prediction period. Fig.2 gave the Tiangong-1(TLE no 37820) statistical position errors with TLE+SGP and the modified TLE+ the numerical method. From the position errors, the modified TLE+ the numerical method was better. The position errors with TLE+SGP were 65km for 6d prediction period. The position errors with the modified TLE+ the numerical method were only 13km for 6d prediction period.

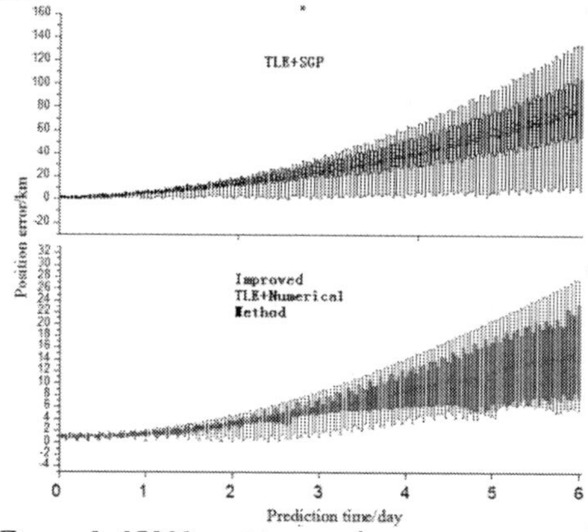

Figure 2. 37820 position prediction error statistic

3.2. Satellite drag coefficient analysis[3]

Atmospheric drag is the main error source of orbit determination and prediction for the LEO objects. The errors of the atmospheric model usually are 15%~30%, the errors would be larger in the solar maximum or magnetic storm which can be modified by resolving the satellite orbit data. The atmospheric model errors are correlated with the space environment parameter. The drag coefficient can compensated for it. The drag coefficient is the important parameter affecting the orbit prediction of LEO satellites which is gotten by the orbit determination. The drag coefficient was usually thought as invariable in the process of the orbit determination which would resulte in the errors of the prediction results.

The relationship between the drag coefficient and space environment parameter were analyzed with the wavelet analysis. Then the correlation coefficients were gotten by the sliding analysis and the prediction period was determined. The relationship between the drag coefficient and environment parameter was gotten by the regression analysis method. The time period for the chosen data was determined in the regression model. The results represented that the gotten equation of the drag coefficient could improve the orbit prediction accuracy.

Cd series were gotten with 55 day GPS orbit data of TG-1 at the different arc length. The environment parameter such as F107 and ap,Ap at the same period were made the wavelet decomposition at the high and low frequency different levels. The sliding correlation of the different level signal for the space environment parameter and Cd series was analyzed, the results represented that when the sliding value was 1d,2d, the drag coefficient of the wavelet decomposition and the corresponding F10.7 and Ap had the good lag correlation. So the Cd prediction by the F10.7 and Ap was possible.

The analysis results represented that the geomagnetic index Ap and the drag coefficient series Cd was the negative correlation which meaned that the NRLMSIS-00 model at the zone of the low,medium latitude and 350km altitude, the ascent stage of the geomagnetic activity had the excessive response, the descent stage had the lack of the response. See Fig.3.

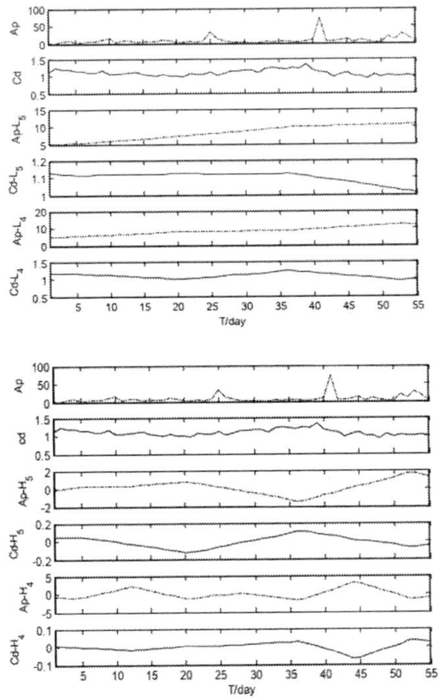

Figure 3. The boxplot of $A_p, \overline{C_d}$ *sliding correlation*

The solar radiation flux F10.7 and the drag coefficient Cd had the strong positive correlation. At the low frequency and F107 ascent stage Cd series is stable, at the descent stage the Cd series decreased which showed that the NRLMSIS-00model responded appropriately at the rise stage and responded excessively at the descent stage. At the high-frequency part the strong positive correlation was given which showed the NRLMSIS-00 model responded insufficiently at the rise stage and stage the excessive response was showed at the decrease. See Fig.4.

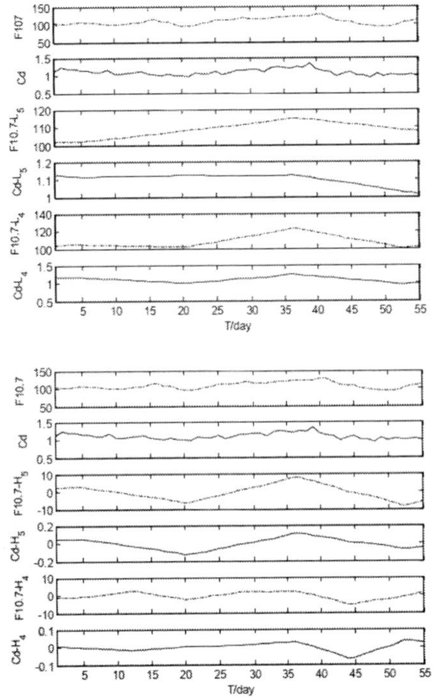

Figure 4. The boxplot of $f_{10.7}, \overline{C_d}$ *sliding correlation*

Sliding analysis found that the drag coefficient Cd had the lag correlation with the geomagnetic index Ap and the solar radiation flux. Comparing with Ap, the correlation of F107 was stronger. The sliding value was 2d,1d for the drag coefficient Cd to the geomagnetic index Ap and the solar radiation flux F107. So in the actual project application, Cd could be predicted for 1-2d with the observed space environment index.

At last the multielement nolinear regression analysis was made for the different layer signals. The model was:

$$C_d = \sum_{i=1}^{5} \alpha_{i0} + \alpha_{i1}x_{i1} + \alpha_{i2}x_{i1}^2 + \alpha_{i3}x_{i2} + \alpha_{i4}x_{i2}^2$$

The regression analysis were made with the quadratic polynomial, in which, x_{i1}, x_{i2} was the layers of the signal; $\alpha_{i0}, \alpha_{i1}, \alpha_{i2}, \alpha_{i3}, \alpha_{i4}$ were the regression coefficient. Cd was gotten by adding the high-frequency and low frequency signal.The sample series length n was analyzed for the different regression model, Cd prediction residuals showed that Cd was the optimal when n=20d for the F107 single-element regression model. see Fig.5 and Tab.1.

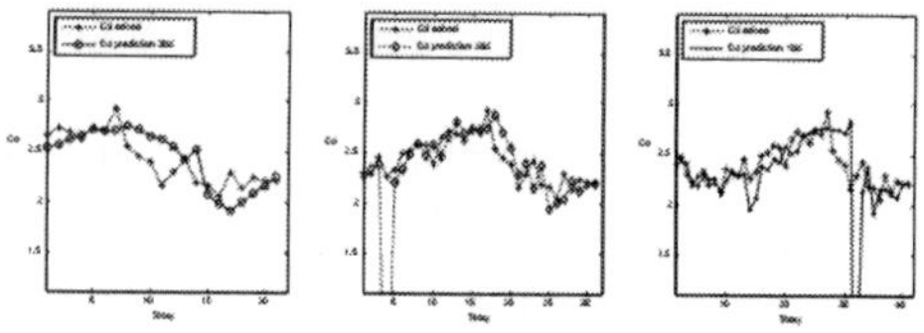

Figure 5. The statistic of regression analyses and verification

Table1. The prediction errors of position with different C_d mode for 1 day prediction

Prediction method	Cd	X /km	Y /km	Z /km	Position Errors /km
Fixed value	2.2	5941.8	3075.3	-790.6	2.7
Solved by OD	2.9	5939.7	3078.5	-794.1	2.5
Solved by regression analysis	2.6	5940.8	3076.9	-792.3	0.1

The relationship between satellite drag coefficient and space environment parameter was analyzed based on the wavelet analysis. The method was given for the Cd prediction which solved the problem in the project application that the drag coefficient gotten by the orbit determination with the forepart data was used to the subsequent orbit prediction. That would induce the large errors in space environment volatile period.

3.3. Ballistic Coefficient Improvement Analysis [4]

The drag model is the largest error in the dynamics model of the LEO satellite. The errors of the ballistic coefficient and atmospheric density dominate the drag errors. The ballistic coefficient is very critical for the space object orbit determination and prediction which is especially prominent for the non-cooperative space object. The ballistic coefficient mean was calculated with TLE and the atmospheric density which solved the large error problem. The calculation equation was given with orbit perturbation acceleration equation and semi-major axis change equation, combined with the numerical integral, the derived equation was:

$$BC = -\frac{\mu \Delta a_{t_1\ \mathrm{drag}}^{t_2}}{\sum_{t_1}^{t_2} \rho a_t^2 v_t v_r^2 \Delta t}$$

Of which , $BC = C_d A/M$,Cd is the drag coefficient, A/M is the area-mass ratio, v_r is the velocity vector of satellite relative to the atmosphere. $\Delta a_{t_1\ \mathrm{drag}}^{t_2}$ is the semi-major axis change between t1 and t2, ρ is the atmospheric density at the position of some epoch, v_t is the satellite velocity at some epoch.

The density in the above equation did not use the density at the perigee, but use the model calculated density discretized with some step, which reduced the errors of the model and made the ballistic coefficient more accurate.

To verify the analysis, the two methods were used to calculate the ballistic coefficient BC in the process of TLE improvement.(1) the estimated BC which used the TLE simulated observation data, the ballistic coefficient was estimated in the process of the TLE improvement.(2) the BC mean, which meaned the ballistic coefficient of the 10d TLE series was calculated with the above method, the mean was gotten after deleting the abnormal values. The improved results would be compared to evaluate the two kinds calculation methods of the ballistic coefficient. The published TLEs was regarded as the reference orbit, E_{T-N} was defined as the prediction position errors with TLE+SGP4 minus to the position errors with the improved elements+ the numerical method. $E_{T-N} > 0$ represented the prediction accuracy improved, the values were larger, the improved results were better. Fig.6 gave the prediction error E_{T-N} changes with the two methods. From Fig.6,the prediction accuracy decreased with the ballistic coefficient gotten by the first method, at the same situation, the prediction accuracy increased with the ballistic coefficient gotten by the second method compared with SGP4.

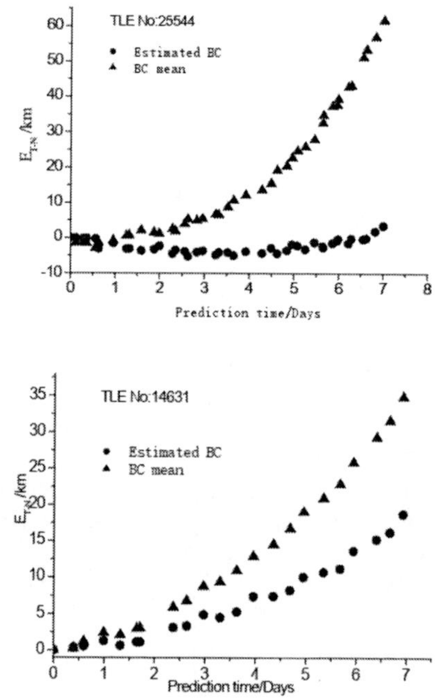

Figure 6. Errors comparison of two ballistic coefficient calculation methods

3.4. Covariance Calculation and Effect Analysis[5]

In the current collision warning work, the collision probability is often the important criterion of space object dangerous intersection. The prediction covariance accuracy affected the collision probability directly. So the reasonable covariance calculation method is the very meaningfully for improving the collision warning accuracy.

Two kinds of prediction covariance were analyzed. (1) The orbit was predicted with the high-precision orbit data and numerical model, then the predicted elements and the high-precision elements were compared. (2). The orbit was predicted with TLEs and SGP4/SDP4 prediction model. The prediction position and velocity was compared with TLEs. The prediction covariance was gotten statistically with 1-7 day prediction period. Combined with the TG-1 dangerous intersection in 2012, the effect of the covariance accuracy to the warning was analyzed with the different covariance. See Tab.2 and Tab.3.

The position covariance comparison found with the two kinds of methods that the precision by the high-precision data combined with the numerical methods is highest. When the prediction period is 1 day, the U sigma is about 2 km.

The two kinds of position covariance were used to the collision warning calculation. From the collision probability when the prediction model is fixed, the covariance is the main factor which would lead to the different collision probability. Different orbit data and prediction model can calculate the different covariance. So it is meaningful to improve the accuracy of space debris collision warning based on the covariance with the different prediction model and orbit data.

Table 2. Forecast standard deviation of different forecast period

Errors /km		1d	2d	3d	4d	5d	6d	7d
TLE error	U	2.0	7.0	12.7	21.9	33.5	48.1	68.7
	N	0.1	0.2	0.3	0.4	0.6	0.8	1.0
	W	0.1	0.3	0.4	0.6	0.7	0.9	1.0
High precision data error	U	1.2	3.6	7.8	13.6	20.3	28.1	37.6
	N	2.0	7.0	12.7	21.9	33.5	48.1	68.7
	W	0.1	0.2	0.3	0.4	0.6	0.8	1.0

Table 3. Collision probability of different covariance

Data	Pc of satellite dangerous intersection	
	The calculated covariance matrix	The correlation coefficient of the covariance matrix
High precision data	1.387398×10^{-5}	1.442304×10^{-5}
TLE data	1.438180×10^{-6}	1.390196×10^{-6}

3.5. Space Environment Effect to the prediction accuracy

Solar activity burst induced the geomagnetic disturbance which made the atomaspheric density change and further affect the orbit prediction accuracy. The complex space environment was classified as four types according to the F10.7 and Ap index. See Tab.4

Table 4. Space environment types

Type	f10.7>140	Ap>40
00	N/0	N/0
01	N/0	Y/1
10	Y/1	N/0
11	Y/1	Y/1

Tab.5 gave the biggest prediction errors of different prediction periods under the four typical space environment. For the 00 and 10 type, the prediction errors were small and usually less than 2km for 1day prediction within 3 days of events. But for type 01 and 11 of the geomagnetic disturbance, the prediction errors were large whose effect could last 3 days.

Table 5. Predcition Errors(km) of different space environment types

Type	0~1d				1~2d				2~3d			
	0	1	2	3	0	1	2	3	0	1	2	3
00	0.5	0.8	1.4	0.7	1.8	3.5	1.8	3.1	2.0	7.7	1.1	3.3
01	10.0	4.9	4.5	--	38.7	22.1	--	--	77.4	--	99.2	--
10	2.0	1.7	1.6	0.6	6.2	7.8	4.3	2.0	9.7	18.2	7.8	3.0
11	6.0	4.1	5.2	1.0	29.3	4.4	18.1	2.2	61.2	6.9	37.1	4.0

4. CONCLUSION

This paper gave the work flow and studied development for the current collision warning work. The main problems were put forward. Then some work were introduced in the recent years including the TLE precision improvement, the drag parameter improvement analysis, the covariance computation analysis and space Environment Effect to the prediction accuracy, which improved the warning accuracy and confidence to some extent.

5. REFERENCES

1. Wang RL,Liu W, et al.,Thinking problems of the present collision warning work by analyzing the intersection between Cosmos 2251 and Iridium 33.Space debris research and application, 2012(3),11-17

2. Liu W,Wang RL,Liu SQ,et al. TLE Prediction Accuracy Improvement and its Application in Collision Warning, Chinese Journal of Space Science,34(4), 449-459.

3. Liu W, Wang RL, Liu SQ,et al. Analysis of satellite drag coefficient based on wavelet transformation. Journal of Astronautics, 2015,36(2),142-150

4. Liu W, Wang R L,Liu SQ,et al.A valid calculation method of ballistic coefficient, Journal of Atrodynamics,2014,4(3),7-9

5 Yan R D,Wang R L, Liu SQ,et al., Study of Covariance Calculation in Space Objects Collision Warning, Chinese Journal of Space Science,2014,34(4): 441-448.

ON THE END-OF-LIFE DISPOSAL OF SPACECRAFT AND ORBITAL STAGES OPERATING IN INCLINED GEOSYNCHRONOUS ORBITS

Luciano Anselmo, Carmen Pardini

Space Flight Dynamics Laboratory, Institute of Information Science and Technologies (ISTI)
National Research Council (CNR), Via G. Moruzzi 1, 56124 Pisa, Italy
Email: Luciano.Anselmo@isti.cnr.it; Carmen.Pardini@isti.cnr.it

ABSTRACT

The current internationally accepted guideline for the appropriate end-of-life disposal of spacecraft and orbital stages, in order to guarantee no further long-term interference with the geosynchronous protected region, was developed when the nearly sole function of the geosynchronous region was to host geostationary satellites, maintained close to the Earth equator as much as possible. Significantly inclined geosynchronous orbits are nevertheless becoming increasingly attractive for various applications, so the original re-orbiting guideline was re-assessed to evaluate its application limits. It was found that an extension of the original formula would be possible to cover disposal inclinations up to ≈ 30°, with an additional velocity variation penalty ≤ 11.5 m/s. However, this approach will not be generally applicable for higher initial inclinations, were the disposal strategy should be addressed on a case by case basis, being strongly dependent on the initial orbital and celestial conditions.

1. INTRODUCTION

Having in mind the goal of preserving the utilization of the geostationary orbit, the Space Debris Mitigation Guidelines, compiled and issued by the Inter-Agency Space Debris Coordination Committee (IADC), defined a toroidal geosynchronous protected region, extending 200 km below and above the geosynchronous altitude of 35 786 km, and spanning in latitude ± 15° with respect to the Earth equatorial plane, and introduced a simple formula for the appropriate end-of-life disposal of geostationary spacecraft, in order to guarantee no further interference with the protected region over the long-term [1]. The requirement of remaining outside the protected region after mission completion was applicable not only to spacecraft, but also to any separable propulsion system (if any) or orbital stage used for reaching the geosynchronous region.

The "IADC formula" requests the fulfillment of the following two conditions for a successful disposal orbit above the geosynchronous protected region [1]:

1. A minimum increase of the perigee altitude h_p of:

$$\Delta h_p[\text{km}] \geq h_0 + (1000 \cdot C_R \cdot A/M) \qquad (1)$$

where C_R is the solar radiation pressure coefficient, A is the average aspect area (in m^2) of the re-orbited object, and M its mass (in kg). h_0 is set equal to 235 km, representing the sum of the upper limit of the protected region (+ 200 km) and of the maximum descent of the re-orbited object due to luni-solar & geopotential perturbations (35 km).

2. An initial eccentricity:

$$e_0 \leq 0.003 \qquad (2)$$

Having the possibility of implementing an appropriate choice of the initial conditions of the disposal orbit, the IADC guideline, i.e. avoiding the interference with the protected region over the long-term, might be fulfilled even with different and slightly more convenient, from the propellant consumption point of view, re-orbiting strategies [2], but the purpose of the IADC formula was just to offer a simple and straightforward approach irrespective of initial conditions, aside from perigee altitude and eccentricity.

The geosynchronous protected region definition and the end-of-life disposal formula were however elaborated when the nearly exclusive utilization of the geosynchronous region consisted of geostationary satellites placed and maintained, during their operational lifetime, close to the Earth equatorial plane. When finally abandoned, due to the concurring action of geopotential and luni-solar perturbations, such objects display a characteristic periodic orbit plane evolution, with a period of about 53 years and a maximum inclination of about 15°.

More specifically, the angular momentum of the orbit, as seen from the North Pole, is characterized by a counterclockwise precession around an axis lying between the rotational axis of the Earth and the pole of the ecliptic, approximately 7.4° away from the polar axis of the Earth. Being one half of the aperture of the precession cone nearly 7.4° as well, the orbital inclination with respect to the equatorial plane of the Earth goes from zero to about 15° and then again to zero in approximately 53 years, completing a full precessional cycle. It is the superimposition of three precessions: around the rotational axis of the Earth, mainly due to the planet oblateness; around the ecliptic pole, due to the third body attraction of the Sun; and around the angular momentum of the Moon orbit, due to the third body attraction of the natural satellite [3]. The plane normal to the pole of the total precession is the Generalized Laplace Plane (GLP) of the system made of Sun, Moon and oblate Earth.

In 2017, the situation in and around the geosynchronous region, with semi-major axis 37 948 km $\leq a \leq$ 46 380 km and eccentricity $e \leq 0.25$, was still largely dominated by objects with orbital inclination with respect to the Earth equatorial plane $i \leq 18°$. On 1 January 2017, among a total tally of 1533 objects, only 19, i.e. 1.2%, had $i > 18°$, i.e. a Highly Inclined Geosynchronous Orbit (HIGSO) [4]. As of 31 July 2017, their number had grown by just one more, to 20, 19 spacecraft and 1 mission related object, i.e. the telescope dust cover of the International Ultraviolet Explorer (IUE).

However, significantly inclined geosynchronous orbits are now becoming increasingly popular for various applications, like satellite navigation systems, as the IGSO component of the Chinese Beidou ($i = 53°$-$58°$), the Indian IRNSS ($i = 28°$-$30°$), and the Japanese QZSS ($i = 40°$-$45°$), science (e.g. the NASA's Solar Dynamics Observatory, with $i = 29°$), telecommunications and intelligence, and this trend may rapidly increase in the future. In the light of these developments, the aim of this paper is to review the current definition of the IADC geosynchronous mitigation guideline, assessing if it would need an extension. Special attention is paid to the end-of-life disposal, in order to check the potential weaknesses of the current IADC formula and re-orbiting recommendations, focusing on the consequences of having operational orbits characterized by medium or high inclinations.

2. STABILITY VS. INSTABILITY OF NEARLY CIRCULAR GEOSYNCHRONOUS ORBITS

The orbital plane and eccentricity evolution of nearly circular geosynchronous orbits have been deeply investigated over many years, and a good recent review of some of the relevant literature can be found in [5] and [6]. The possibility of eccentricity instability for some inclinations was already pointed out back in the 1960's [7][8].

A mechanism, discovered for artificial satellites by Lidov in 1961 [9], and for asteroids by Kozai in 1962 [10], can strongly affect the eccentricity of objects with a high inclination with respect to the generalized Laplace plane [11]. Driven by the conservation of the orbit angular momentum component normal to the generalized Laplace plane, it consists in an exchange of eccentricity with inclination, leading to large-amplitude inclination and eccentricity oscillations accompanied by librations of the perigee around either 90° or 270°. Therefore, due to the Kozai-Lidov effect, the eccentricity of high inclination near geosynchronous orbits might grow significantly, irrespective of its initial value [12].

However, the picture is actually much more complex than that derived from the classical Kozai-Lidov mechanism, due to fact that luni-solar perturbations also give rise to resonances dependent only on inclination, and others dependent on semi-major axis, eccentricity and inclination [13][14][15]. The resonances depending only on inclination occur when there is commensurability between the secular rates of change of the right ascension of the ascending node $\dot{\Omega}$ and of the argument of perigee $\dot{\omega}$ [13][16][17]. The main inclination resonance conditions for prograde orbits (i.e. with $i \leq 90°$) are summarized in Tab. 1.

Table 1
Inclination resonance conditions for Earth's oblateness (J_2) and luni-solar perturbations ($\dot{\Omega}/\dot{\omega} \approx k$)

Value of k	Resonant term	Inclination (°) prograde solution
$-\infty$	$\dot{\omega} \approx 0$	63.44
-2	$2\dot{\omega}+\dot{\Omega} \approx 0$	56.06
-1	$\dot{\omega}+\dot{\Omega} \approx 0$	46.38
0	$\dot{\Omega} \approx 0$	90
1	$\dot{\omega}-\dot{\Omega} \approx 0$	73.15
2	$2\dot{\omega}-\dot{\Omega} \approx 0$	69.01
∞	$\dot{\omega} \approx 0$	63.44

If the initial inclination i_0 of nearly circular and geosynchronous orbits is around 30° or less, these resonances are not met during the long-term orbit plane evolution and the Kozai-Lidov mechanism does not affect the orbit behavior. Therefore, the eccentricity displays periodic variations with amplitude depending on the initial conditions, but remains in any case bounded for a very long time, with no significant systematic growth.

For inclinations $i_0 > 30°$, certain combinations of initial conditions lead, instead, to a significant eccentricity growth, often with a complex interplay between resonances depending only on inclination, resonances depending on semi-major axis, eccentricity and inclination [18], and, when applicable, the Kozai-Lidov mechanism. However, for the purposes of the present work, what must be remarked is that for nearly circular and geosynchronous orbits with $i_0 \leq 30°$ it is always possible to avoid a significant eccentricity growth over 200 years, irrespective of the other initial conditions (i.e. epoch, right ascension of the ascending node Ω and argument of perigee ω), while with $i_0 > 30°$ only specific initial conditions would be able to guarantee a relative stability of the eccentricity.

3. A POSSIBLE EXTENSION OF THE IADC FORMULA

In addition to the initial inclination, the eccentricity evolution of nearly geosynchronous orbits also depends on the relative positions of the Sun and the Moon. The Moon motion, in particular, is quite complex. With respect to the Earth equator and equinox reference coordinate system, the Moon argument of periapsis has a precession period of 8.85 years, while the right ascension of the ascending node oscillates about 0°, from approximately – 13° to + 13°, with a period of 18.6 years. With the same period, the Moon inclination with respect to the Earth equator varies from 18° to 29° [19].

In order to investigate the eccentricity stability of nearly circular and geosynchronous orbits, the full range of initial conditions concerning the Sun, the Moon, the right ascension of the ascending node and the argument of perigee was explored using the STELA (version 3.1.1) and SATRAP (version 2.1) orbit propagators. The Semi-analytic Tool for End of Life Analysis (STELA) was designed by the Centre National d'Études Spatiales (CNES) to support the French Space Act and allows, among other things, efficient and accurate long-term propagations of nearly geosynchronous orbits [20]. The SATellite Reentry Analysis Program (SATRAP) is instead a numerical code for the trajectory propagation around the Earth, including all the major perturbations [21].

In addition to the full range of orbital and celestial initial conditions, the IADC formula (Eq. 1, with $h_0 = 235$ km) and any possible extension were tested over a time interval of 200 years assuming $e_0 = 0.003$, i.e. the maximum initial eccentricity recommended by the IADC, and $C_R A/M = 0.1$ m^2/kg, i.e. a quite conservative value, compared with those found for current spacecraft and upper stages, where $C_R A/M < 0.05$ m^2/kg. With such assumptions, it was found that the IADC guideline, i.e. Eq. 1 with $h_0 = 235$ km, would be able to guarantee no further interference with the geosynchronous protected region over 200 years, irrespective of initial Sun and Moon position, ascending node, and argument of perigee, only for initial disposal inclinations $i_0 \leq 2°$. In other words, with $e_0 = 0.003$, $C_R A/M = 0.1$ m^2/kg and $i_0 \leq 2°$, the IADC formula would be able to guarantee a perigee altitude ≥ 200 km above the geostationary height of the disposal orbit, over 200 years, for any initial right ascension of the ascending node, argument of perigee, day of the year and phase of the Moon orbit plane cycle.

With higher initial disposal inclinations, this would be no longer guaranteed, even though the applicability of the IADC formula might be extended up to $i_0 \leq 10°$, for any initial condition, if sporadic crossings of the geosynchronous protected region by ~ 10 km were deemed acceptable. Concerning still higher initial inclinations, up to $i_0 = 30°$, where the amplitude of the eccentricity oscillations progressively increases, but remains anyway upper bounded, a relationship with the form of Eq. 1 might be maintained by just appropriately increasing the value of h_0, as detailed in Tab. 2. With $i_0 > 30°$, for the reasons discussed in the previous section, the eccentricity may grow to quite large values, for certain initial conditions, due to resonances and/or the Kozai-Lidov mechanism. Consequently, it is not possible to define end-of-life disposal strategies being simple, universal and inexpensive at the same time, at least with current technologies.

Table 2

Modification of the parameter h_0 in the IADC formula, as a function of the initial disposal inclination i_0, to guarantee no further interference with the geo-synchronous protected region over 200 years, irrespective of initial Sun and Moon position, ascending node, and argument of perigee

h_0 (km)	i_0
235	≤ 2°
285	≤ 18°
300	≤ 20°
350	≤ 26°
400	≤ 28°
500	≤ 29°
550	≤ 30°

It should be remarked that the results obtained, summarized in Tab. 2, do not mean that much less expensive solutions, in terms of velocity variation (ΔV) or propellant consumption, do not exist. On the contrary, for many sets of initial conditions, quite smaller values of h_0 would be able to avoid the long-term crossing of the geosynchronous protected region for a given value of i_0. However, this should be checked with a case by case analysis, being not a general result valid for any possible combination of initial conditions. Moreover, when a geosynchronous mission ends, some initial disposal conditions may be adjusted and optimized, as the eccentricity and the argument of perigee, but changing the inclination and the right ascension of the ascending node would be too expensive and, of course, nothing could be made to change the Sun and Moon position.

In general, the worst initial conditions, i.e. those needing values of h_0 close to those listed in Tab. 2, occur around the spring equinox, with the perigee of the disposal orbit lying on the equatorial plane opposite to the Sun, i.e. with $\Omega_0 \approx 180°$ and $\omega_0 \approx 0°$. The additional Moon contribution depends on the initial inclination i_0. If i_0 is small, the worst case is found when the right ascension of the Moon ascending node is maximum (i.e. $\approx 13°$). When i_0 assumes intermediate values (10°-20°), the worst initial conditions correspond to a right ascension of the Moon ascending node around 0° and to the maximum inclination with respect to the Earth equator (i.e. $\approx 29°$). Finally, for $i_0 \approx 26°$-30°, the worst initial conditions correspond again to a right ascension of the Moon ascending node around 0°, but with the minimum inclination with respect to the Earth equator (i.e. $\approx 18°$).

The extreme values of Tab. 2, i.e. the first three and the last two ones, can be accurately fitted (Fig. 1) with the following equation, where h_0 is expressed in km and i_0 in degrees:

$$h_0 = 5.11 \times 10^{-5} (i_0 - 2°)^5 - 1.71 \times 10^{-3} (i_0 - 2°)^4 + 1.65 \times 10^{-2} (i_0 - 2°)^3 + 0.16 (i_0 - 2°)^2 \quad (3)$$

Eq. 3 might be therefore used to generalize Eq. 1 up to $i_0 = 30°$, even though for $i_0 \approx 26°$-28° would lead to h_0 values significantly higher than those found and listed in Tab. 2.

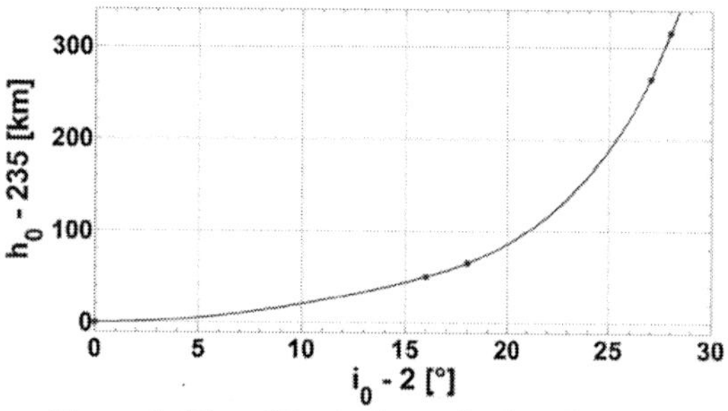

Figure 1. Plot of Eq. 3, fitting the first three and the last two values of Tab. 2

Regarding the additional ΔV needed to implement the extended IADC formula, it can be evaluated as follows. The end-of-life re-orbiting according to Eqs. 1 and 2, from a near circular geosynchronous orbit, leads to a semi-major axis increase Δa given by:

$$\Delta a = (1 + e_0)[h_0 + (1000 \cdot C_R \cdot A/M)] + e_0 a_{GEO} \quad (4)$$

where a_{GEO} is the geostationary semi-major axis (i.e. 42 164.14 km). Compared with a re-orbiting carried out using the IADC formula, where $h_0 = 235$ km, the additional semi-major axis increase Δa_+ entailed by the application of the extended formula is given by:

$$\Delta a_+[\text{km}] = (1 + e_0)(h_0 - 235) \approx h_0 - 235 \quad (5)$$

Being applicable, for $\Delta a \ll a$ and nearly circular orbits, the following expression, where n represents the mean motion:

$$\Delta a = \frac{2}{n} \Delta V \quad (6)$$

the additional velocity variation ΔV_+ to be applied for implementing the extended IADC formula, compared with the original one, can be expressed as (Fig. 2):

$$\Delta V_+[\text{km/s}] \approx 3.646 \times 10^{-5} (h_0[\text{km}] - 235) \quad (7)$$

For $h_0 = 550$ km, the maximum value given in Tab. 2 and applicable for $i_0 \leq 30°$, this would imply $\Delta V_+ = 11.5$ m/s, i.e. an additional velocity variation equivalent to that needed to re-orbit, with $e_0 = 0$, a spacecraft with $C_R A/M = 0.08$ m^2/kg, using the original IADC formula with $h_0 = 235$ km.

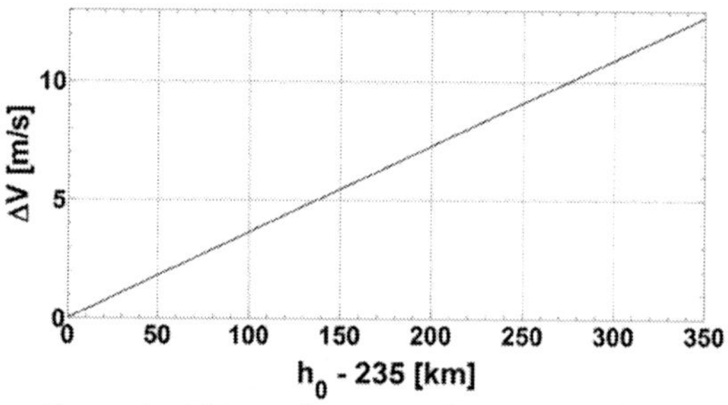

Figure 2. Additional ΔV, as a function of h_0, to be delivered compared with the original IADC formula

Therefore, even assuming the most unfavorable initial conditions, an extension of the IADC formula following either Tab. 2 or Eq. 3 might be practicable up to $i_0 = 30°$, with a maximum ΔV penalty of 11.5 m/s. This, of course, would have a not negligible cost, but could be economically affordable and technically feasible. Moreover, as already pointed out, much less expensive solutions, in terms of additional ΔV, would be often available for appropriate combinations of orbital and celestial initial conditions.

4. VERY HIGH INITIAL INCLINATIONS

As explained in Section 2, when $i_0 > 30°$ a significant growth of eccentricity cannot be avoided without a careful selection of the initial conditions. A simple and universal end-of-life re-orbit strategy, like that represented by Eqs. 1 and 2 with a varying value of h_0, is therefore no longer applicable.

Considering, for instance, the IGSO component of the Chinese Beidou navigation system ($i = 53°$-$58°$), the long-term eccentricity behavior is extremely sensitive to the initial conditions, in particular the right ascension of the ascending node Ω_0, for an assigned epoch [5]. Taking, as an example, the Beidou IGSO-1 spacecraft, assuming no further orbital control since 7 June 2013, with $\Omega_0 = 0°$ would have been possible to maintain a small eccentricity for nearly three centuries, for just one century with $\Omega_0 = 30°$, but for less than 25 years with $\Omega_0 = 120°$ or $\Omega_0 = 225°$ [5]. On the other hand, selecting the initial ascending node in the intervals $90° \leq \Omega_0 \leq 180°$ or $210° \leq \Omega_0 \leq 250°$ would have led to a growth of the eccentricity above 0.83, causing the reentry of the satellite in the atmosphere after intervals of time from several decades to a few centuries [5].

These examples clearly show that a case by case analysis would be needed to determine the outcome of the orbit long-term evolution when the initial inclinations are so large. Ignoring the possibility to reduce the initial disposal inclination below 30°, too expensive and unfeasible in practice, certain initial conditions might be compatible with the re-orbiting in a super-synchronous nearly circular orbit maintaining a small eccentricity over 200 years. In other cases, the initial ascending nodes might lead, after several decades or a few centuries, to large eccentricities and the consequent spacecraft decay. However, in general, the change of the right ascension of the ascending node would be too much expensive and unfeasible as an inclination reduction, and having a satellite reentering on the Earth after crossing the geosynchronous protected region for several decades or a few centuries would not be much different, in practice, from a non reentering object crossing the protected region as well. Even choosing an appropriate launch time, as proposed in [5] to obtain the desired disposal conditions, might conflict with mission constraints and operations, in exchange of uncertain potential benefits.

In principle, at least for very high inclination orbits, the Kozai-Lidov effect might offer a very elegant and effective disposal mechanism. In fact, as recalled in Section 2, the progressive amplitude growth of the eccentricity oscillations is accompanied by librations of the perigee around either 90° or 270°. This means that the crossings of the geosynchronous altitude would occur far from the nodes, i.e. considerably above or below the equatorial plane and the geostationary ring. However, this beautiful "protection mechanism" derives from the hierarchical and restricted three-body Kozai-Lidov formalism. In the real world, adding to the Sun the Earth's oblateness, the Moon, other relevant perturbations as direct solar radiation pressure, and the intricate web of luni-solar resonances, the simple pattern of the classical Kozai-Lidov mechanism is strongly affected and, during certain phases of the orbit evolution, the geosynchronous altitude can be crossed also near the equatorial plane, depending on the initial conditions.

Examples of this behavior are provided by a couple of long-term propagations, carried out with SATRAP, taking into account all relevant perturbations, including atmospheric drag for perigee heights below 1000 km. A fictitious passive satellite with $C_R A/M = 0.05$ m^2/kg, $a_0 = a_{GEO}$ and $i_0 = 75°$ was propagated for more than 200 years, starting on 20 March 2019, with two sets of initial conditions:

1. Assuming $e_0 = 0.0001$, $\Omega_0 = 180°$, $\omega_0 = 0°$;
2. Assuming $e_0 = 0.001$, $\Omega_0 = 0°$, $\omega_0 = 0°$.

The results obtained in the first case are summarized in Figs. 3 and 4, those obtained in the second one in Figs. 5 and 6. In the first case, the inclination increases up to 84° and the eccentricity exceeds 0.7 (Fig. 3), but the satellite is still in orbit after 220 years. As in the classical Kozai-Lidov effect, the eccentricity growth and its successive large amplitude oscillation is accompanied by a libration of the perigee around a "stable" value (Fig. 4). However, such a value is neither 90° nor 270°, but ~0°. This means that when the eccentricity is large, the perigee is close to the node, i.e. to the equatorial plane, and the Kozai-Lidov mechanism is therefore not able to keep safe the geosynchronous protected region.

Looking at the second case, the eccentricity growth is so large to induce satellite decay after 128 years (Fig. 5). Concerning the coupled evolution of eccentricity and argument of perigee (Fig. 6), again the simple pattern of the classical Kozai-Lidov effect is disrupted, with no libration around a specific value of the argument of perigee. Nevertheless, during the first 100 years, when the perigee is close to the node, and to the equatorial plane, the eccentricity is small, while when the eccentricity is maximum (~0.8), the perigee is above the equator at the greatest distance as well, so a dynamical protection mechanism still plays a role (Fig. 6). Such protection mechanism is lost in the following orbit evolution, when the eccentricity goes down to 0.35

and then rebounds to 0.84 and the consequent decay, but this flight phase is too short (< 30 years) to involve significant interactions with the geosynchronous protected region.

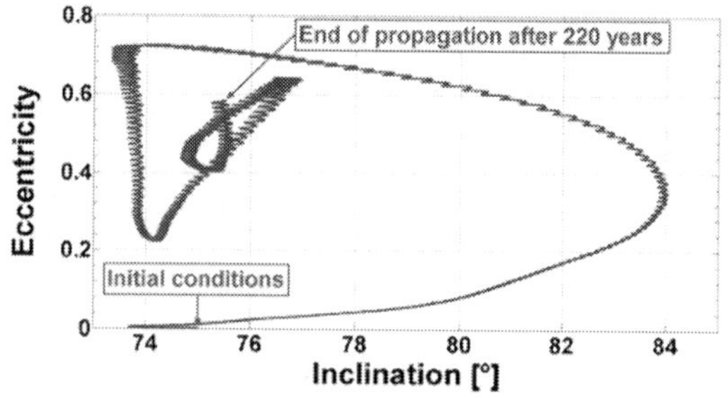

Figure 3. Eccentricity vs. inclination for the first case with $i_0 = 75°$

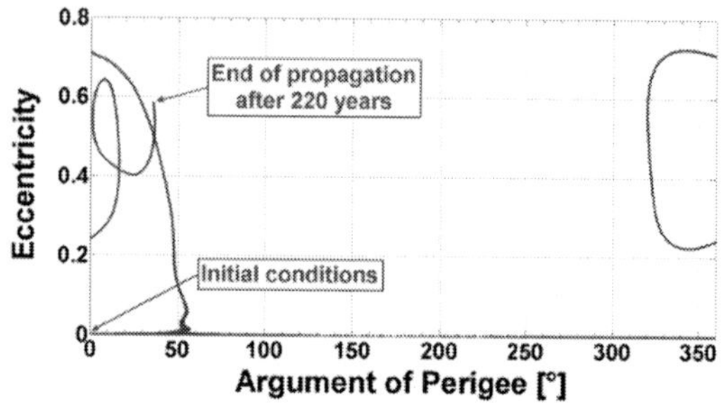

Figure 4. Eccentricity as a function of the argument of perigee for the first case with $i_0 = 75°$

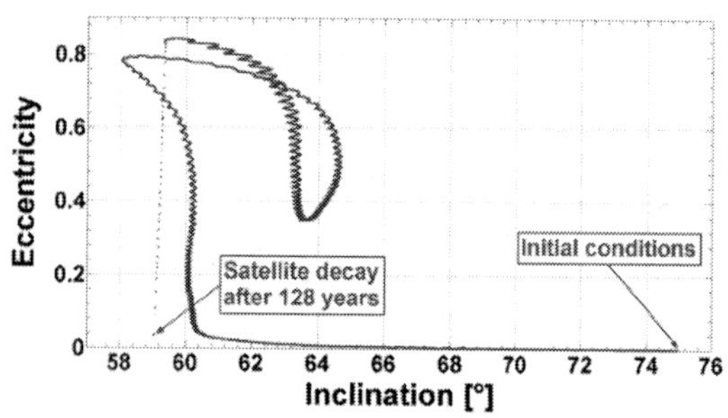

Figure 5. Eccentricity vs. inclination for the second case with $i_0 = 75°$

It is therefore quite evident that no simple, universal and effective end-of-life disposal guideline can be devised for the re-orbiting of near circular geosynchronous objects with $i_0 > 30°$. On the other hand, if the total number of these objects will remain constrained to a few thousand during the next century, even leaving them totally uncontrolled at the end-of-life would not increase significantly the collision probability in the geostationary ring and in the geosynchronous protected region, because they would be spread in a huge volume of space by the perturbations, spending only occasional and relatively short periods of time in the most populated geosynchronous belt. However, an end-of-life re-orbiting above the geosynchronous altitude of a few hundred km might be in any case beneficial, even with an unbounded eccentricity, for further reducing the average long-term collision probability with objects residing in the geosynchronous protected region.

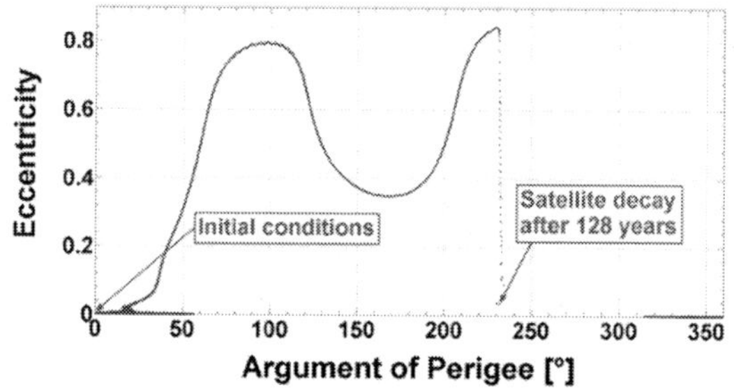

Figure 6. Eccentricity as a function of the argument of perigee for the second case with $i_0 = 75°$

5. CONCLUSIONS

The original IADC guideline for the end-of-life re-orbiting of spacecraft and orbital stages above the geosynchronous altitude, summarized by Eqs. 1 and 2, with $h_0 = 235$ km, was devised for objects close to the geostationary ring and resulted strictly valid, irrespective of the orbital and celestial initial conditions, for $i_0 \leq 2°$, and applicable in most cases for $i_0 \leq 10°$.

The potential growing use of geosynchronous orbits with higher inclinations has raised the problem of what strategy to adopt at the end-of-life for these new classes of objects. For $i_0 \leq 30°$, it was shown that an effective extension of the IADC formula would be possible by just increasing the value of h_0 as a function of i_0, with a maximum of 550 km for $i_0 = 30°$. This possibility arises from the fact that, despite the complexity of the perturbations acting on such orbits, the eccentricity would remain bounded to sufficiently low values for at least 200 years, irrespective of the orbital and celestial initial conditions. With $i_0 = 30°$ and $h_0 = 550$ km, no crossing of the geosynchronous protected region would occur, over 200 years, even in the less favorable combinations of orbital (Ω_0 and ω_0) and celestial (Sun and Moon position) initial conditions. It should be however pointed out that very often it would be possible to avoid any further long-term interference with the protected region adopting significantly lower values of h_0, being those listed in Tab. 2, or computed with Eq. 3, just the values able to guarantee the fulfillment of the guideline goal even with the worst initial conditions.

The cost of implementing the extended IADC formula, in terms of additional ΔV to be delivered by the propulsion system and mission impact, would not be negligible, but affordable. With $i_0 = 30°$, the maximum additional ΔV penalty would be 11.5 m/s, approximately doubling the cost currently incurred to apply the original IADC formula to nearly geostationary satellites.

Regarding the present geosynchronous satellite systems operational at high inclination, the Indian IRNSS ($i = 28°$-$30°$) and the NASA's Solar Dynamics Observatory ($i = 29°$) are the obvious candidates for the implementation of the extended IADC formula at the end-of-life. In other words, depending on the initial conditions, for each satellite to be disposed above the geosynchronous altitude, an appropriate value of $h_0 \leq 550$ km might be found to guarantee no further interference with the geosynchronous protected region for more than 200 years.

This approach will not be instead generally applicable for $i_0 > 30°$, i.e. for the Chinese Beidou ($i = 53°$-$58°$) and the Japanese QZSS ($i = 40°$-$45°$), just to mention the couple of systems currently operational. Reducing the initial inclination and/or changing the initial right ascension of the ascending node of the disposal orbit would be of course too much expensive and unfeasible, not to mention the Moon and Sun positions. Even choosing appropriate launch times, provided it were feasible from a mission point of view, might not be so beneficial, being quite tricky the accurate prediction of end-of-life epochs several years in advance.

The problem should be therefore addressed on a case by case basis, strongly dependent as it is from the initial orbital and celestial conditions. In certain cases, it should be possible to constrain the eccentricity growth from several decades to a few centuries, in others the eccentricity would grow to values so large to cause orbital decay in several decades. For very high inclinations and appropriate initial conditions, the interaction between the Kozai-Lidov effect and the other perturbations might lead to the emergence of a dynamical protection mechanism of the geosynchronous protected region, effective at least for several decades.

As a consequence of this situation, no easy to apply, general and cost-effective end-of-life disposal solution can be recommended for the re-orbiting of near circular geosynchronous objects with $i_0 > 30°$. Nevertheless, if the total number of these objects will not grow by more than two orders of magnitude in the next century, even simply abandoning them at the end of their missions would not lead to a significant increase of the collision probability in the geostationary ring and in the geosynchronous protected region, because they would be spread in a huge volume of space by the perturbations, spending only occasional and relatively short periods of time in the most populated geosynchronous belt. However, in order to further reduce the average long-term collision probability with objects residing in the geosynchronous protected region, an end-of-life re-orbiting above the geosynchronous altitude of a few hundred km might be in any case beneficial, even with an unbounded eccentricity.

6. ACKNOWLEDGMENTS

The work described in this paper was carried out in the framework of the Agreement No. 2015-028-R.0 between the Italian Space Agency (ASI) and the National Institute of Astrophysics (INAF) on "Space Debris: Support to IADC Activities and Pre-Operational Validation for SST".

7. REFERENCES

1. Steering Group & Working Group 4 (2004; 2007). Space Debris Mitigation Guidelines, Document IADC-02-01 & Revision 1, Inter-Agency Space Debris Coordination Committee (IADC).
2. Anselmo, L. & Pardini, C. (2008). Space Debris Mitigation in Geosynchronous Orbit. Advances in Space Research 41(7), 1091–1099.
3. Beutler, G. (2005). Methods of Celestial Mechanics, Volume II: Application to Planetary System, Geodynamics and Satellite Geodesy, Springer, Berlin, Germany, pp153–155.
4. Lemmens, S. (Ed.) (2017). Classification of Geosynchronous Objects, Issue 19, Revision 1, ESA's Space Debris Office, European Space Agency, Darmstadt, Germany.
5. Zhao, C.-Y., Zhang, M.-J., Wang, H.-B., Xiong, J.-N., Zhu, T.-L. & Zhang, W. (2015). Analysis on the Long-term Dynamical Evolution of the Inclined Geosynchronous Orbits in the Chinese BeiDou Navigation System. Advances in Space Research 56(3), 377–387.
6. Zhu, T.-L., Zhao, C.-Y. & Zhang, M.-J. (2017). Analytic Model for the Long-term Evolution of Circular Earth Satellite Orbits Including Lunar Node Regression. arXiv:1703.00655v1.
7. Allan, R.R. & Cook, G.E. (1964). The Long-period Motion of the Plane of a Distant Circular Orbit. Proceedings of the Royal Society A: Mathematical, Physical and Engineering Sciences 280(1380), 97–109.
8. Breiter, S. (2001). Lunisolar Resonances Revisited. Celestial Mechanics and Dynamical Astronomy 81(1-2), 81–91.
9. Lidov, M.L. (1961) Evolution of Artificial Planetary Satellites under the Action of Gravitational Perturbations Due to External Bodies. Iskusstviennye Sputniki Zemli (in Russian) 8, 5–45.
10. Kozai, Y. (1962) Secular Perturbations of Asteroids with High Inclination and Eccentricity.

The Astronomical Journal 67(9), 591–598.

11. Tamayo, D., Burns, J.A., Hamilton, D.P. & Nicholson, P.D. (2013). Dynamical Instabilities in High-obliquity Systems. The Astronomical Journal 145(54), 12pp.
12. Wytrzyszczak, I., Breiter, S. & Borczyk, W. (2007). Regular and Chaotic Motion of High Altitude Satellites. Advances in Space Research 40(1), 134–142.
13. Cook, G.E. (1962). Luni-solar Perturbations of the Orbit of an Earth Satellite. Geophys. J.R. Astron. Soc. 6, 271–291.
14. Hughes, S. (1980). Earth Satellite Orbits with Resonant Lunisolar Perturbations. I. Resonances Dependent Only on Inclination. Proc. R. Soc. Lond. A 372, 243–264.
15. Hughes, S. (1981). Earth Satellite Orbits with Resonant Lunisolar Perturbations. II. Resonances Dependent on the Semi-major Axis, Eccentricity and Inclination. Proc. R. Soc. Lond. A 375, 379–396.
16. Yokoyama, T. (1999). Dynamics of Some Fictitious Satellites of Venus and Mars. Planetary and Space Science 47, 619–627.
17. Anselmo, L. & Pardini, C. (2011). Orbital Evolution of the First Upper Stages Used for the New European and Chinese Navigation Satellite Systems. Acta Astron. 68, 2066–2079.
18. Bordovitsyna, T.V., Tomilova, I.V. & Chuvashov, I.N. (2014). Secular Resonances as a Source of Dynamic Chaoticity in the Long-term Orbital Evolution of Uncontrolled Satellites. Solar Syst. Res. 48, 259–268.
19. Roncoli, R.R. (2005). Lunar Constants and Models Document. Report JPL D-32296, Jet Propulsion Laboratory, California Institute of Technology, Pasadena, California, USA.
20. Anonymous (2016). STELA User's Guide, Version 3.1.1. Centre National d'Études Spatiales (CNES), Toulouse, France.
21. Pardini, C. & Anselmo, L. (1994). SATRAP: Satellite Reentry Analysis Program. Internal Report C94-17, CNUCE Institute, Consiglio Nazionale delle Ricerche (CNR), Pisa, Italy.

MITIGATION MEASURES FOR ORBITAL DEBRIS: NO MORE RELEASE OF ARIANE SYLDA

S. Heinrich[1], K.Mathis [2]
(1) ALTRAN France, 4 Allée des Cormorans, F-06150 Cannes,
Email: stephane.heinrich@altran.com
(2) CNES, 52 rue Jacques Hillairet – F-75612 Paris cedex,
Email:kevin.mathis@cnes.fr

ABSTRACT

One of the historical assets of the ARIANE launcher is to be able to provide a dual payload launch in a same mission. This is due to high performance structural frame SYLDA (SYstème de Lancement Double Ariane) carrying the upper payload and enveloping the lower payload. Up to now, this double launch structure was released during flight operations and deorbited (re-entry) naturally.

The Space Debris Mitigation policy of IADC (Inter-Agency Space Debris Committee) promote all spacecraft operations to be terminated via a controlled re-entry in the earth atmosphere above a non-populated area such as SPOUA (South Pacific Ocean Unpopulated Area) or via a non-controlled re-entry within 25 years.

All evolutions of Ariane launcher (formerly A5ME and now A6) have been investigated to provide this capability with the implementation of a re-ignitable upper stage with the cryogenic VINCI engine.
Although Ariane launcher complies wih the regulation, to go further, the case of the dual launch was studied in order to no release SYLDA structure during operations.

In this perspective, CNES initiated an internal brainstorming on innovative SYLDA concepts able to remain attached to the launcher upper stage and able to release the two payloads. In this process, CNES worked with ALTRAN in order to collect the innovative concepts investigated by industry and initiate a preliminary study of baselined concepts.

The intention of this paper is to describe the outcomes of the ALTRAN / CNES study during brainstorming phase on innovative concepts and preliminary study phase on baselined concepts.

1. INTRODUCTION

The European launcher industry introduced since decades the use of dual launch of payloads to give the capability to handle two payloads for the same orbit in the same launch.
The up-to-date status for Ariane 6 is based on a reuse of Ariane 5ME studies leading to same upper stage technologies, a longer SYLDA and fairing, and the re-ignitable VINCI engine able to perform controlled re-entries.

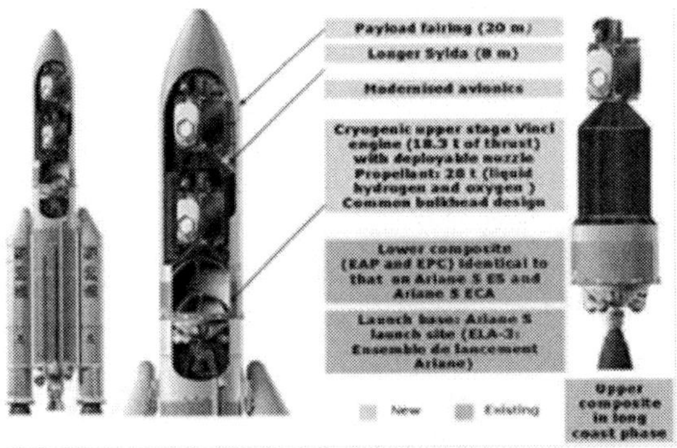

Figure 1: Upper Stage evolution for ARIANE 5ME

The design of SYLDA is based on a middle cylinder section and two end cones. Each element is based on 4 carbon sandwich panels with aluminium honeycomb glued-assembled together. The mechanical interfaces are provided via metallic rings glued-assembled with panels. A pyrotechnical cut joint at lower section is assembled with interface rings via rivets. The SYLDA provides venting holes in the upper cone and the capability of human intervention holes in cylindrical part.

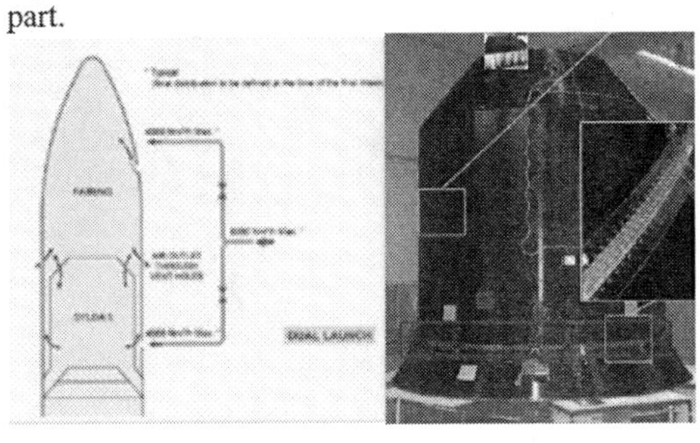

Figure 2: SYLDA5 design details

2. STATE OF THE ART

This section will introduce a dedicated focus on the state of the art concerning the dual launch systems.

Historically, the dual launch system on Ariane 5 was achieved via the SPELTRA concept (external structural frame), cancelled later in favour of SYLDA5 internal frame under fairing.

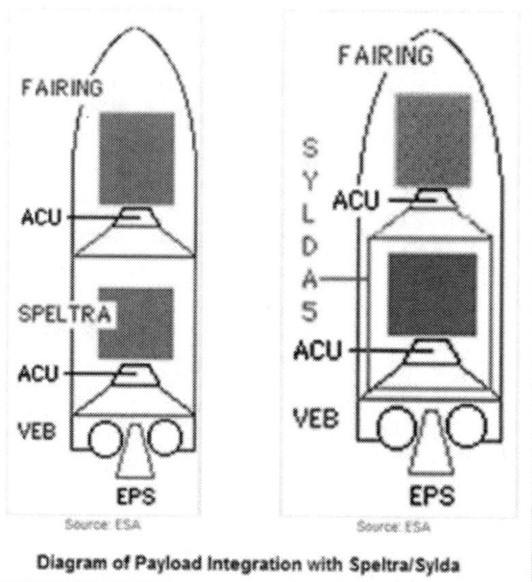

Figure 3: SPELTRA / SYLDA comparative concepts

An equivalent concept has been developed in the USA industry and implemented at ULA (United Launch Alliance) with the Dual Spacecraft System (DSS) [8].

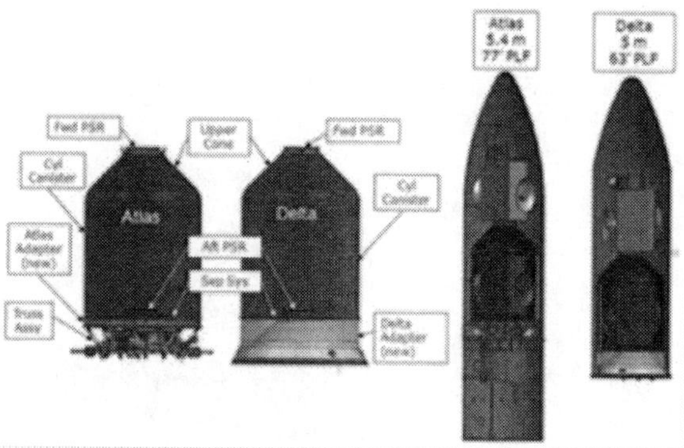

Figure 4: DSS-5 concept for DELTA / ATLAS [9]

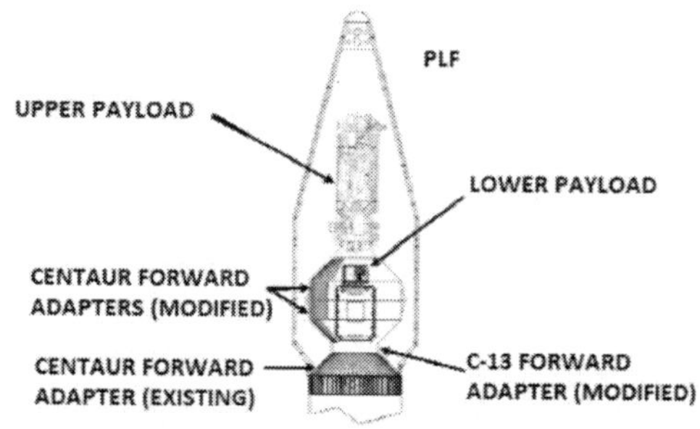

Figure 5: DSS-4 concept for ATLAS 4m fairing [8]

Other designs are available on smaller launchers such as:

- DSS-4 on 4m ATLAS fairing
- VEGA VESPA
- PSLV DLA (Dual Launch Adapter) [12]

Figure 6: VEGA (left) & PSLV (right) DLS concepts

Following variations can be found between concepts:

- Materials: metallic / carbon,
- Processes: sandwich panel / partial overwrap,
- Pyro joint location: lower, middle or upper section.

Historically, an attractive concept has already been implemented as Dual Launch System in the lunar APOLLO missions between Command Module (CM) and Lunar Module (LM).

The upper stage was offering a 4 panels pyrotechnical cut and a 4 flower petals aperture and expansion in order to allow the release of LM from the upper stage.

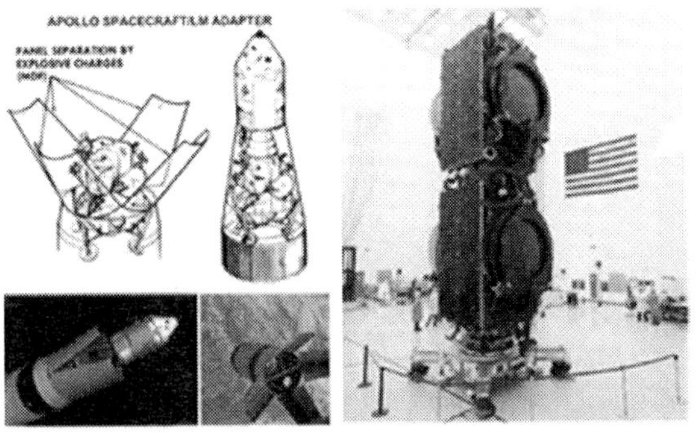

Figure 7: (Left) Dual launch concept for Apollo LM & CM; (Right) BOEING satellites 1st stack in 2014 [10]

More recently, the BOEING Satellite Company developed their new generation of full electric platform BSS-702SP bus including a special cylinder adapter able to stack another BOEING satellite. Initial ABS-3A and Eutelsat 115 West B satellites were designed to be launched in pairs to bring down launch costs on the first dual launch never operated on a Falcon 9 (1st of March 2015) [10] [11].

This attractive solution is in fact an initiative from satellite manufacturer/operator offering a dual launch of a composite independently of launch provider.

3. CONSTRAINTS

The study had the constraints to investigate concepts of structural frames able to carry the upper payload and encapsulate the lower payload. This concept had the same requirements that were used on current SYLDA. Nevertheless, the brainstorming phase needed to avoid moving towards the current SYLDA concept in order to re-examine all potential concepts for a global trade-off.

Requirements were defined according to following data:

- **Geometry**: interfaces and allowable volume
    - Allowable volume under fairing for external volume;
    - Allowable volume inside for internal payload;
    - NOTA: Those volumes must be respected even under mechanical constraints (including mechanical displacements).
    - Allowable volume for payload release;
    - Lower interface with upper stage;
    - Upper Interface with upper adapter cone;
    - NOTA: Alternative interfaces must not be discarded and can be proposed.
    - Height between lower & upper interfaces;
    - NOTA: Capability of the concept to be easily adapted in height is an asset;
- **Load Cases**: input data & success criteria
    - Upper payload maximum mass;
    - Upper payload maximum CoG;
    - Minimum lateral eigen mode;
    - Interface flux;
    - Maximum acceptable overflux to upper stage;
    - Maximum acceptable overflux to payloads;
- **Environmental Aspects**: input data
    - Service temperatures;
    - Upper stage mass;
    - Upper stage maximum thrust and accelerations;
    - Maximum acceptable impact on U/S CoG offset (balancing);

Other requirements were added to define operations constraints (including controlled re-entry phase).

4. BRAINSTORMING PHASE

This section will present the results of brainstorming phases performed separately at CNES & ALTRAN levels.

4.1 At CNES Level

The solutions investigated at the French Space Agency (CNES) before the beginning of work with ALTRAN were mostly based on:

- Usual SYLDA "tethered" with the upper stage;
- SYLDA opening the upper section as a "hat";
- "Lift" concept to slide the P/L in upper position;
- SYLDA opening as "flower petals";
- SYLDA with "retractable" concentric sections;
- Standard I/F frame able to "stack" 2 P/Ls;
- "Origami" retracting frame;
- Usual SYLDA translated along P/L via a "telescopic arm" and whisked away.

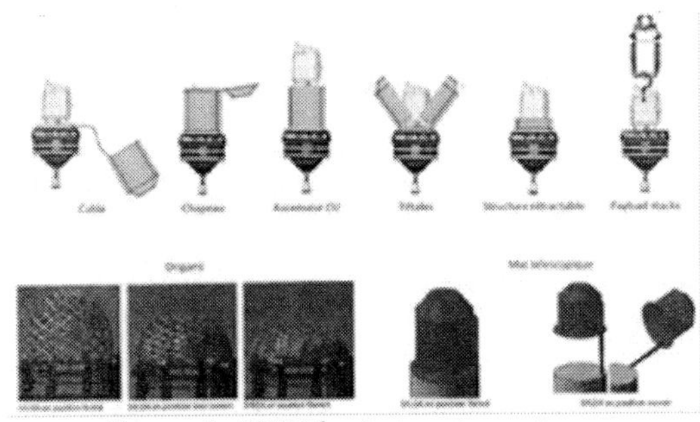

Figure 8: CNES brainstorming outcomes

4.2 AT ALTRAN Level

The solutions investigated at ALTRAN Cannes Space Design Office during a preliminary phase without knowing CNES solutions were mostly based on 2 major structural frame options:

- Usual SYLDA shell (based on sandwich frame);
- Innovative gantry (based on carbon bars).

Several variants adapted to those structural concepts lead to the following SYLDA technical solutions, illustrated from figures 9 to 14:

- Opening as "petals' of flower "2 or 4,
- Opening its upper cone as a "hat",
- Opening as "hinged door" panels,
- Retracting via "sliding" sections,
- Widening to a conical shape.

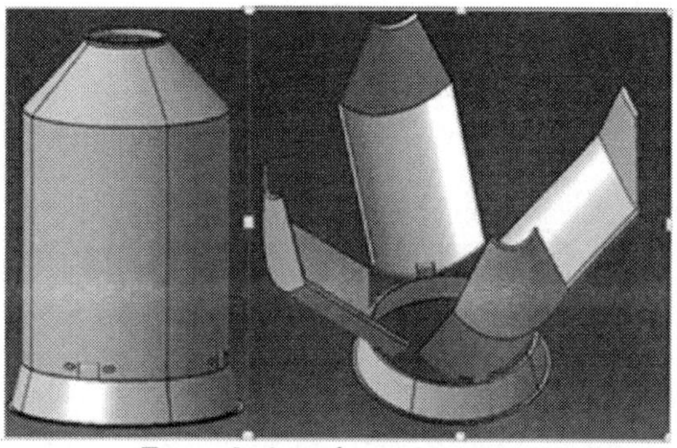

Figure 9: 4 petals cut & aperture

At this step, the consideration of additive pyrotechnical cuts raised a concern due to increase of induced pyro shocks and added mass. An alternative concept was taken in to consideration versus usual SYLDA shell concept.

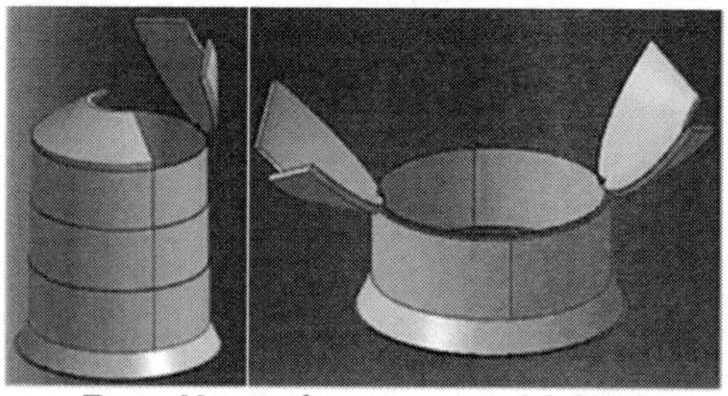

Figure 10 conical sections cut & slided back

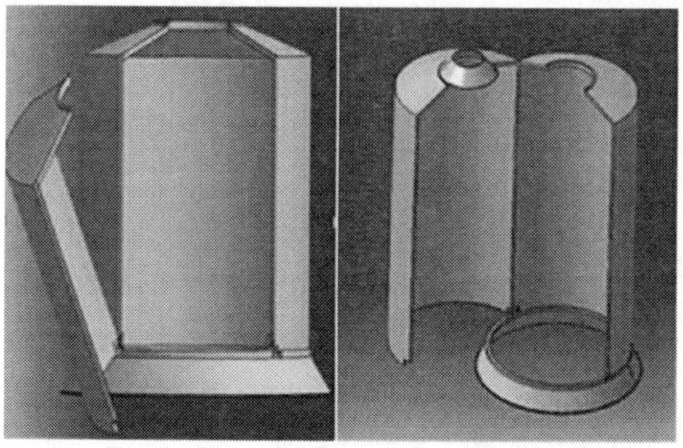

Figure 11: 4 & 2 hinged doors aperture

This concept was based on an innovative gantry made of tubular carbon bars and allows considering less dismantlement points and potentially less added mass and alternative dismantlement technologies. ALTRAN proposed a reuse of dismantlement joints investigated in parallel on a ESA CLEANSAT study [6] based on SMA (Shape Memory Alloys). This study concerned satellite dismantlement mechanisms activated by temperature in atmospheric re-entry to improve their demise.

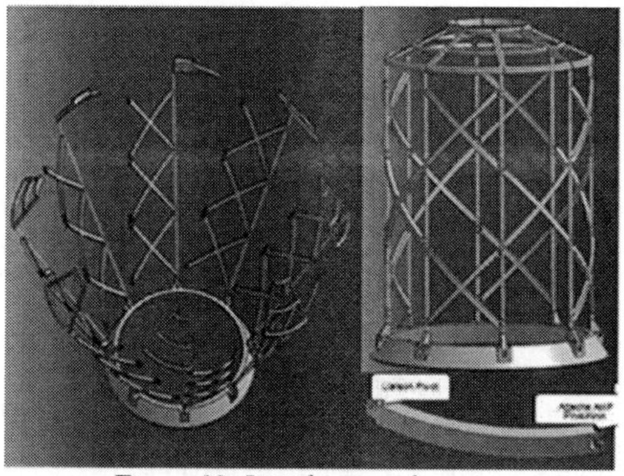

Figure 12: Initial gantry design

The preliminary mechanical analysis of the gantry concept versus shell demonstrated a real performance of the solution. The cylindrical part of the SYLDA is in fact not the more operating part in terms of stresses and stiffness and could be clearly attractive. At the opposite, all conical sections are loaded and clearly inappropriate for the gantry concept. Thus, it was decided to further investigate the gantry concept as offering a lot of flexibility such as being able to slide and articulate bars to achieve an expandable capability to a conical shape.

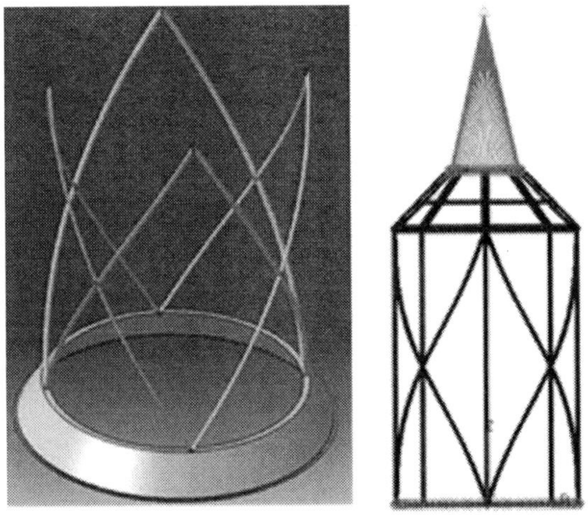

Figure 13: Simplified gantry design & FEM

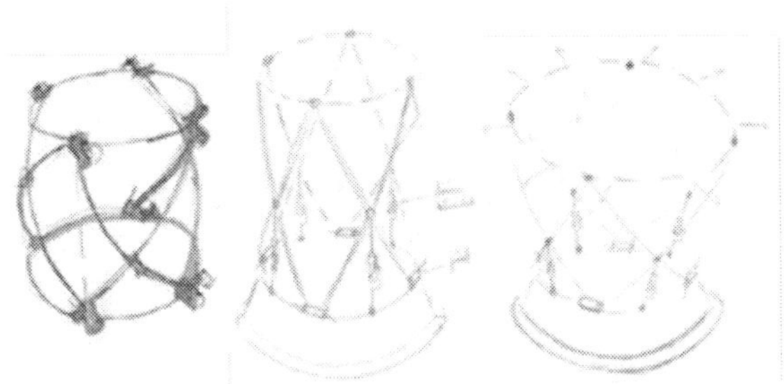

Figure 14: Potential cinematic behaviour of the gantry

5. INNOVATIVE PHASE

The second phase of this activity was the occasion to merge the two brainstorming phases and initiate an innovative phase in order to enlarge and consolidate the range of solutions. The objectives of this phase were to trade-off each concept following several parameters in order to prepare a ranking to select the more attractive concepts.

The first step of this phase was to initiate a trade-off for the best way to handle payload adapter cones (standard parts to remain unchanged in the study, "out of DLS").

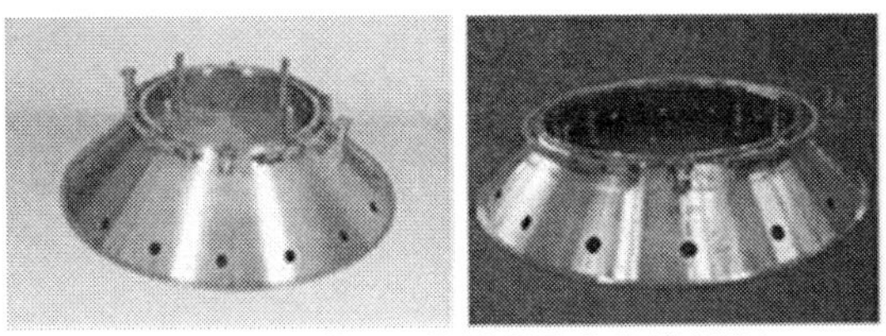

Figure 15: Cones PAS 937 & PAS 1194

In most scenarios, the case of the upper SYLDA cone has to be managed in the same trade-off which led to the following outcomes:

- DLS cone + PAS deployed on the same side.
  - ⇨ It is non-compliant in term of balancing.
- DLS cone on one side and PAS on other side.
  - ⇨ Marginal but acceptable in balancing.
- 75% DLS vs 25% DLS + PAS.
  - ⇨ Best option in term of balancing.
  - ⇨ Ideal option for 4 vertical cuts.
- 50% DLS vs 50% DLS + PAS (see Figure 16).
  - ⇨ Acceptable in balancing with less pyro cuts.
  - ⇨ Best option for 2 vertical cuts.

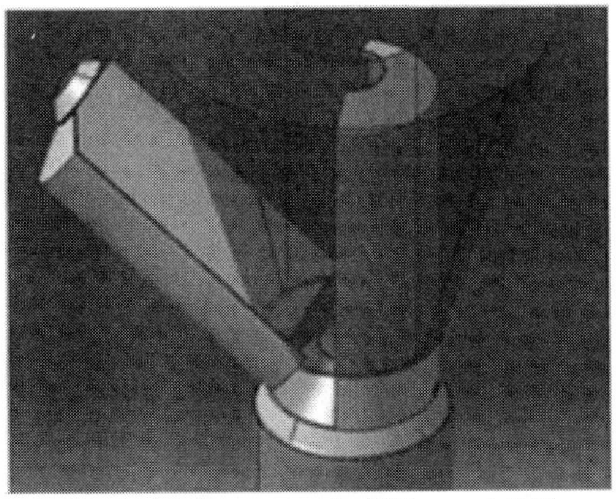

Figure 16: Best option for DLS cone + PAS

Main scenarios addressed by CNES were mostly based on a reuse of the entire SYLDA so as to limit industrial impact. Those concepts were further investigated in order to comply with the balancing requirement needed during rotation rates induced during payload release.
In those cases, after upper payload release, the SYLDA was supposed to be stacked on the opposite side of the upper stage and the system shall be reversible to initial position to allow and resist to final engine boosts:

- Tethered version (see Figure 17: *Upgraded tethered concept*) was modified with 2 links based on "carpenter strips" enrolled in a winch (allowing control in traction / compression). The tilt of the SYLDA being achieved by upper stage tilting via attitude control system.
- SYLDA translation via a boom or a robotic arm (see Figure 18) was reinvestigated in a reversible mode. This led to consider a very long telescopic arm.
- An upgraded version was then investigated based on semi-circular robotic arms with standard 2- axis rotators (see Figure 19). The use of semi-circular rigid arms (without telescopic device) were attractive to fit shape and volume of the upper stage

around the payload and needed length to realign SYLDA at the opposite side of the upper stage.

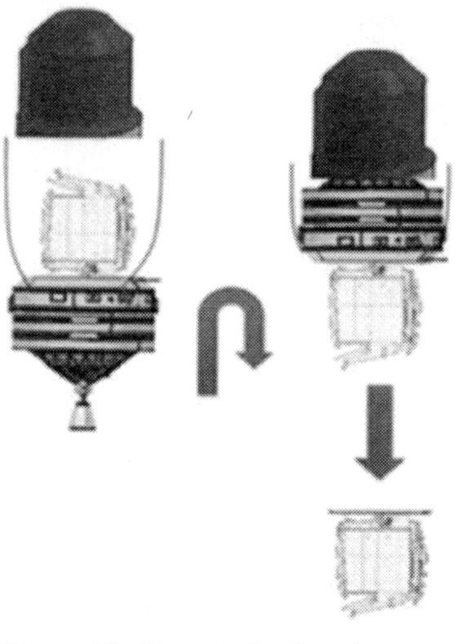

Figure 17: Upgraded tethered concept

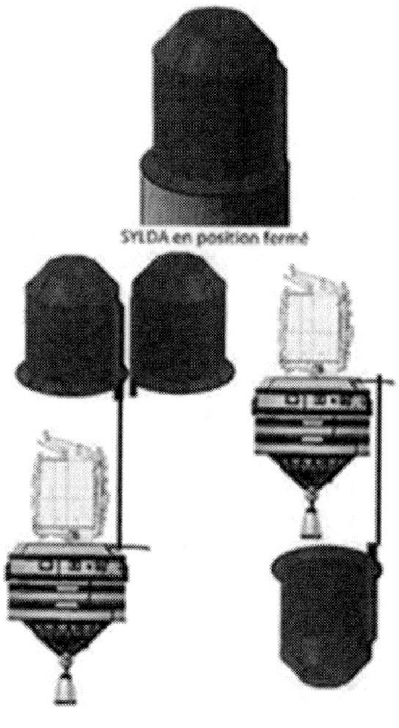

Figure 18: Telescopic boom or robotic arm concept

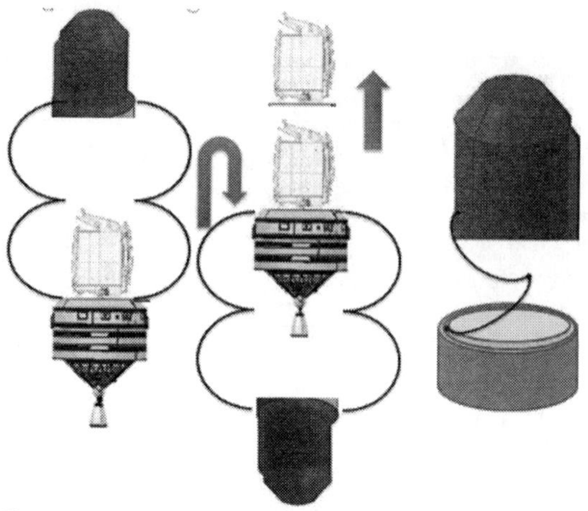

Figure 19: Circular robotic arm (2 or 1 arm version)

All those concepts were assessed pretty risky with regard to upper stage environment and operations. Moreover, the safety of those operations would require costly robotic arms and mechanisms for an upper stage which has to be considered as expendable.

Alternative ALTRAN concepts proposed were as well further investigated:

- A "flanked" version was proposed aiming to dock SYLDA panels along the upper stage body. It offers the benefit of a only 2 vertical cuts, a simple deployment arm and mechanism.

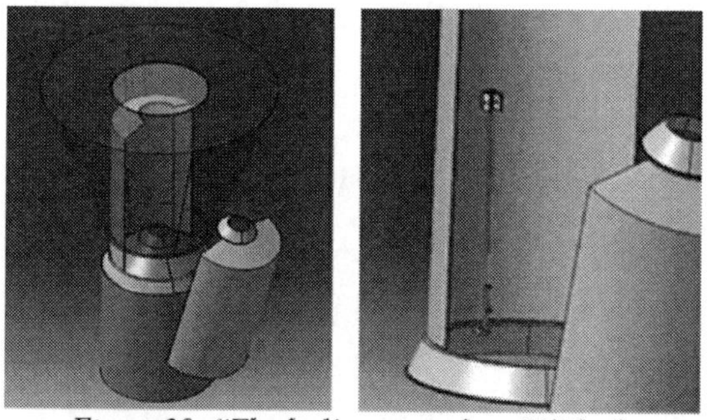

Figure 20: "Flanked' version along U/S body

- The "slided" version was further investigated in order to meet all requirements. The forbidden clearance cone and the intention to minimize the number of horizontal section cuts (costlier in pyro length and mass than vertical cuts) were hard to be achieved. The best compromise found on the concept was 4 sections cut with the upper conical section added to a fifth cylindrical section cut in vertical and fold away as petal flower.

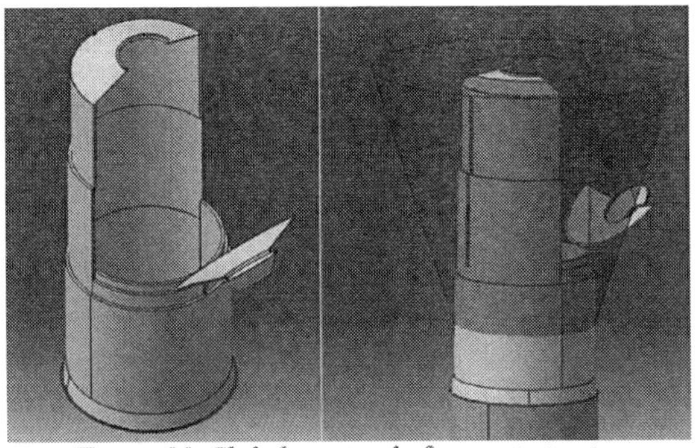

Figure 21: Slided version before optimization

- The "lift" concept was further investigated and raised the following preliminary outcomes. It forces to reconsider the usual design of mechanical attachment of the lower payload with regard to the upper stage and the opportunity of a "SPELTRA" or an external frame coming from the upper stage. In addition, the integration of the lower payload on the ground was consequently fully re-examined. It forces as well to consider sliding, and lifting technological device (not frequently used in space in such dimensions). On technological concepts raised by CNES was to even consider a floating internal device with direct or helicoidal translation induced by upper stage attitude control system to avoid motorization. This concept was clearly found attractive but requiring deeper investigation for a proper assessment with regard to alternative concepts.

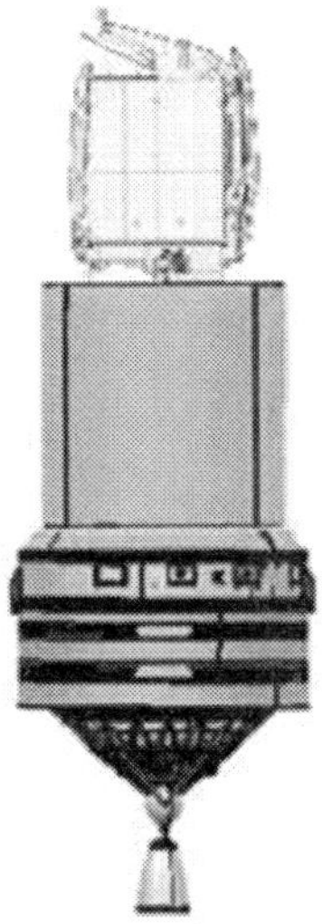

Figure 22: "Lift" concept

Another innovative concept proposed by ALTRAN was shorty addressed and discarded as out the objectives of the study restricted to the SYLDA concept. The intention was to re-use of Fokker Space concept named "ConeXpress" – ESA CxOLEV (an autonomous space tug aimed for maintenance and deorbit purpose using electrical propulsion offering huge delta V capabilities).

The idea came from the observation during study that:

- Conical sections are usually empty / inert mass,
- Conical section PAS + DLS are inconvenient frames in most scenarios leading to complexification of the dismantlement and fold away mechanisms.

The idea still remains at ALTRAN that this concept and volume could be declined in several versions of "Space Tugs" based on a generic avionic and propulsion family:

- Short PAS cone,
- Large DLS upper cone + PAS,
- Entire SYLDA frame.

In parallel of ALTRAN research studies based on sustainable used of space "GREENSPACE", this concept is further investigated to address all potential missions:

- Capability to deliver 2 payloads in different orbits in the same launch,
- Capability to deliver cargo to ISS,
- Capability to be reused for another space infrastructure (ISS, propulsion module for a shuttle, a space exploration mission).

Figure 23: PAS/DLS cone / SYLDA used as Space Tug

6. CONCEPTS SELECTION

Preliminary evaluation was performed by using the following criteria:

- Launcher aspects (L/V or launch operator):
    - Launcher performance: mass/resistance (transverse stiffness) ratio;
    - Modularity (availability in several sizes);
    - Flight operations constraints (balancing, inertia, needed energy);

- Spacecraft aspects (S/C or P/L operator):
  - Volume (compliance with regard to current SYLDA environment);
  - Shocks level (capability to include low-/no-shocks devices);
- SYLDA aspects (attractivity of DLS concept):
  - Recurrent cost (simple / complex, assembly technologies);
  - Ground operation constraints (simple / complex operation during launcher integration phases);
  - Reliability and safety;
- Technological aspects (maturity, DLS & U/S development)
  - Heritage, technological maturity;
  - TRL (Technology Readiness Level);
  - Non-recurring costs, R&D time;

And the ranking was done with the following scoring:

- Initial 2 criteria are for ranking at "System" level.
  - DLS concepts are assessed with regard to historical current SYLDA with mark "7".
- Final 2 criteria are for ranking at "Technological" level.
  - DLS concepts are assessed with regard to historical current SYLDA with mark "9".

Criteria were ranked with grades from 1 to 9. Consequently, products were ranked from 1 to 81. Global ranking was assessed by the product of sub-rankings 150 % "Systems" and 100% "Technological".

As a conclusion of the innovative phase, the evaluation process raised the following outcomes:

- Current SYLDA is the best ranked by far.
  - No innovative concept provided more benefits to compensate induced drawbacks.
- Gantry concept still appears as the most promising in term of mass saving, but further analyses were necessary to verify some requirements like created overfluxes.
- Space Tug concept is the best "System" ranked (in terms of benefits for operators), the worst ranked of technological ranking (very high lead time and non-recurring cost).

With the ranking assumptions used, the evaluation process identified the most attractive concepts as:

- SYLDA shell concept opened as flower petals.
  - Baselined as the best compromise.
- Lift concept (additional innovative concept).
- SYLDA gantry (additional attractive concept but need to be confirmed with regard to mass and overfluxes requirements).

In the next section the first concept.

7. BASELINED CONCEPT 1

Based on assessment performed in previous phase, the concept baselined is as usual SYLDA cut in vertical and fold away like "flower petals".

At this step of the study, the following questions have still to be investigated in parallel:

- The number of panels (2, 3, 4) remains to be confirmed for selection after dedicated trade-off.
- The best process used for the structural frame of the cylindrical section remains to be confirmed:
  - Usual SYLDA made of carbon skins / aluminium honeycomb;
  - SYLDA shell made of Fibres Metal Laminate (FML);
  - Innovative Gantry made of carbon bars.
- The strength of deployed structural panels and locking devices with regard to the acceleration generated during re-entry boost or the need of a reversible deployment mechanism back to lock in initial position.

The trade-off concerning the number of cut petals highlighted that 2 petals configurations always led to an interference with the forbidden clearance volume for the payload release (only removed at high angle of aperture).

Moreover, the distance between panels at this position became very problematic to achieve an efficient support device to withstand the boost acceleration in open-deployed configuration.

At the opposite, 4 petals configuration has very short angle of aperture and distance between panels in open configuration, but the saving of length of pyro-cut in a 3 panels configuration was considered as the best compromise.

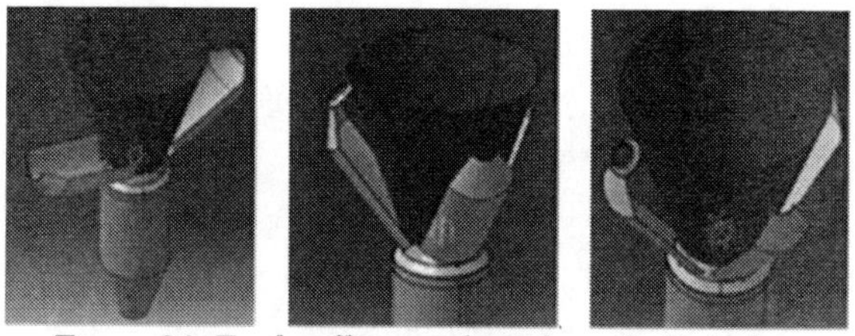

Figure 24: Trade-off on number of panels (2, 3, 4)

FML (Fibre Metal Laminate) is an alternative process for a shell concept. It is based on sandwich layers of following structural materials:

- GLARE: Glass/Aluminium
- ARALL: Kevlar (Aramid)/Aluminium
- CARAL: Carbone/Aluminium
- CentrAl: GLARE / HC Aluminium
- HTCL: Carbone / Titane

Based on previous ALTRAN study on those materials with potential application on space pressurized habitat, it was proposed to CNES to address the stiffness versus mass efficiency of those materials in the study case of a SYLDA shell.

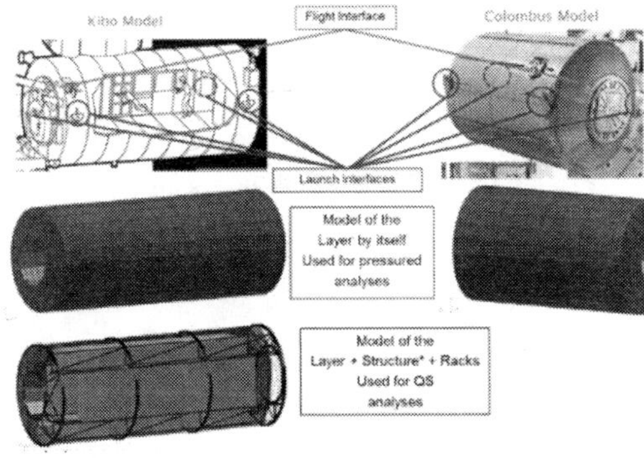

Configuration			Layer Al	Thickness Al (mm)	Layer Fibers/Epoxy	Thickness Fibers/Epoxy (mm)	Total Thickness	Mass (Kg)	[illegible] (%)
Metal Alloy	Al 2024 T3						1.40	[illegible]	
FMLs	Glare 2	3/2-0.3	3	0.3	2	0.25	1.40	493.2	9%
	Arall 2	3/2-0.3	3	0.3	2	0.25	1.40	458.7	-15%
	Carall 2	3/2-0.3	3	0.3	2	0.25	1.40	447.8	-17%

Figure 25: Outcomes of FML study on ISS habitat

After iteration, the optimisation of the different processes with objective to cope with the transverse rigidity eigen mode, the best FML identified is ARALL in a double disposition of 3/2 – 0.3 (i.e. 6/4 – 0.3mm). This preliminary study has been compared with carbon skins 2mm thick / aluminium honeycomb with initial panel thickness of 40mm. The optimisation in that study came with optimisation of honeycomb surface via rough perforations. Considering the potential of honeycomb with regard to FML with equivalent mass/stiffness, it was decided to remain in the historical process (sandwich honeycomb panel). Then in the study, SYLDA shell panels were set at usual historical thickness and local stiffening adaptations to local stresses were set in carbon skin thickness.The first FEM simulations with this last definition were performed in deployed configuration (engine boost re-entry environment). Thus, these first detailed FEM simulations positively confirmed the assumption that deployed panels could withstand engine boost acceleration (and therefore comfort the baselined concept). Nevertheless, it was already identified that 3 support devices would be need for design-optimisation as supporting high concentration loads.

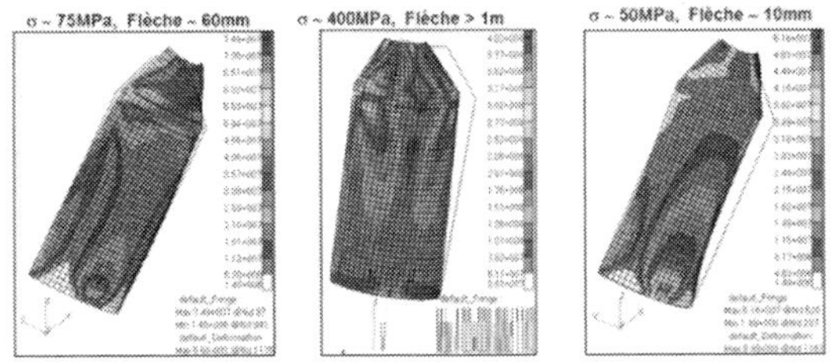

Figure 26: FEM analysis at maximum re-entry acceleration (according to X, Y or Z direction) on worst case panel

The design of the shell was then optimised in mass to cope with the main requirement (transverse eigen frequency) with all usual elements (human and ventilation holes) complying with standard space design rules.

Figure 27: DLS design optimised with skin evolutive thickness (each colour is associated to a specific thickness)

The following zones required intensive care:

- Lower conical section,
- Upper conical section,
- I/Fs with pyrotechnical joints,
- Panel hinges,
- Panel supports (in deployed configuration).

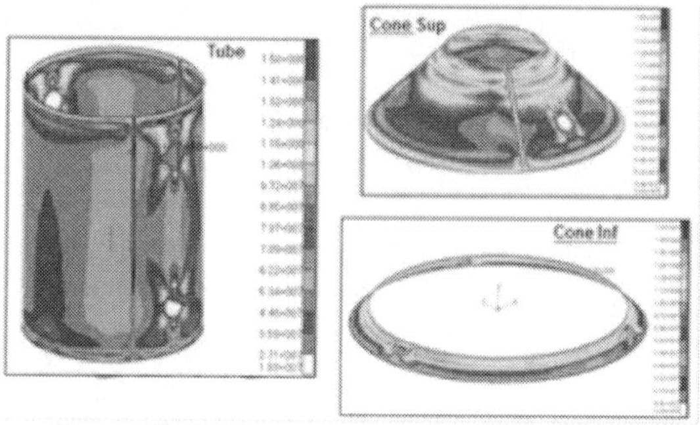

Figure 28: DLS stress in launch configuration at maximum flux

The design for the deployment, locking and supporting device was achieved via:

- A single hinge on panel;
- A simple spiral spring (almost constant torque);
- End pins at panel edges interfacing with supporting device;
- A honeycomb sandwich supporting panel with grooves to guide, brake and lock the panel pins;
- A braking device with a rotating came around the groove catching panel pins with resistive torque made with linear springs;
- A locking device made with the same came device achieving blocking / locking after came rotation under tension of the springs.

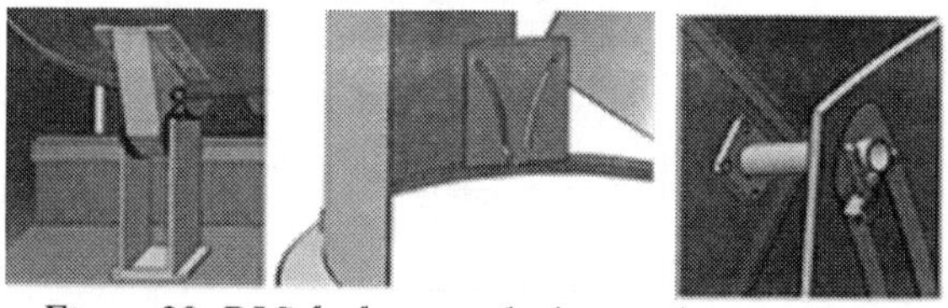

Figure 29: DLS deployment, locking and supporting design

The optimisation of the panels gave the inertia data allowing investigating the kinematic aspects.
For a common opening time of the panels selected around 20 seconds, the difference of inertia of 2 panels (with / without PAS cone) gave 2 values of motor torque compatible of the selected spring. The breaking and locking system was designed with a resistive torque not higher than ten times the motor torque.

The present study leads to overall mass of the complete system 13% in excess of the initial mass target. The supporting system and mechanisms were considered heavy and marginal in design, thus putting in question the concept selected.

An alternative idea was found, to get rid of the supporting panels system with circular panel end pins guided and locked at lower conical section. This last evolution allowed to achieve an excess mass of 7% with respect to the target mass.

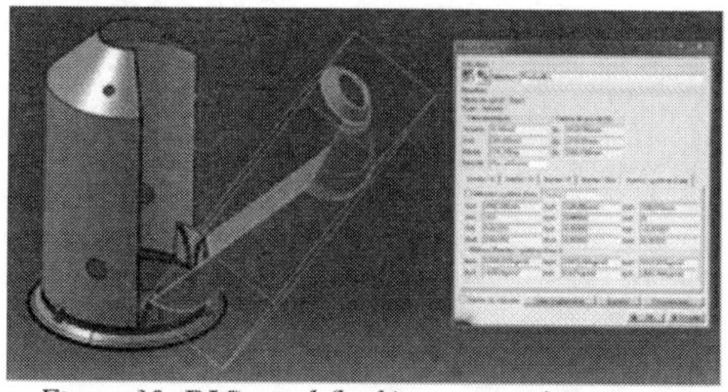

Figure 30: DLS panel (high) inertia and cinematics

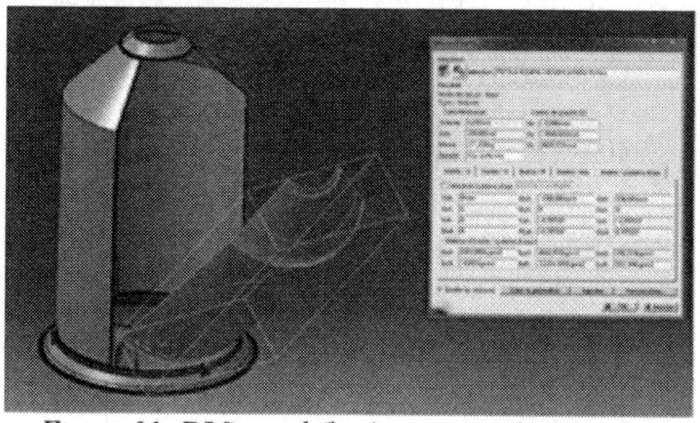

Figure 31: DLS panel (low) inertia and cinematics

The reasons can be explained as follow:

- The load case of re-entry boost implies local stiffeners close to interfaces;
- Petal separation implies as well local stiffeners to transmit those loads all along pyrotechnic joints and close to upper conical section (for the panel carrying the PAS cone).

8. CONCLUSION

Despite a very detailed design optimisation performed for an intended preliminary phase, some uncertainties remain at this step on mass budget due to assumptions made. Nevertheless, the envelope of mass impact of the concept is quantified, and it can be concluded that a no release SYLDA has a mass around 30% higher than its equivalent release SYLDA.

Concerning costs, all elements selected in the concept have put priority in terms of simplicity and maturity to maximise cost savings. A rough estimation of cost impacts with regard estimation of usual SYLDA cost leads to increase the cost by a factor of 1.5 – 2.

9. REFERENCES

[1] B. O'Connor, Process for Limiting Orbital Debris, NASA-STD-8719-14ACHG-1, NASA 2012

[2] EU, European Code of Conduct for Space Debris Mitigation, 28.6.2004

[3] ESA, ESA/ADMIN/IPOL(2014)2 – Space Debris Mitigation Policy for Projects, ESA DG, 28/03/2014

[4] CNES, "French Space Operations Act – Technical regulations", 31.03.2011

[5] ESA Space Debris Office, ESA and Space Debris, BR-309, March 2013

[6] ESOC 2017, ESA CLEANSAT SMA dismantlement mechanisms, April 2017

[7] Ariane 5 Users Manual (Issue5, July 2011)

[8] Catherine Andrulis and Benjamin Colvin, Dual Spacecraft System, AIAA 2008-7658

[9] Roy Miller, Dual Spacecraft System, AIAA 2011 http://www.ulalaunch.com/uploads/docs/Published_Papers/Commercial_Space/AIAA_DSS_4_DSS_5_2011-09-12_Final.pdf

[10] http://www.americaspace.com/2015/02/27/spacex-ready-for-first-dual-satellite-mission-to-geostationary-orbit-on-net-sunday/

[11] https://www.nasaspaceflight.com/2016/06/spacex-falcon-9-dual-satellite-launch/

[12] http://spaceflight101.com/pslv-c35/indias-pslv-rocket-counts-down-to-first-multi-orbit-mission-with-eight-satellites/

OPTIMIZATION TECHNIQUES FOR FEATURE DETECTION OF ORBITAL DEBRIS

Helia Sharif[(1)], **Christian Pfaab**[(2)], **Matthew Hoelzel**[(2)]

[(1)]*German Aerospace Center (DLR), Robert-Hooke-Straße 7,28359 Bremen, Germany, Email: Helia.Sharif@dlr.de*
[(2)]*Universitat Bremen, Bibliothekstraße 5,28359 Bremen, Germany*

ABSTRACT

Space debris is exponentially increasing and becoming a pressing concern, as demonstrated by the recent malfunctioning of the European Space Agency Envisat [1] and the Japanese Space Agency Hitomi [2]. In this study, we propose a novel vision-based feature detection technique for the next generation of satellites to perform autonomous obstacle avoidance.

The study evaluates the strength and shortcomings of thermal and visible imaging to detect an approaching subject. For this purpose, the subject was imaged in various lighting conditions and contrasts of temperature relative to the ambient temperature.

Preliminary results suggested that thermal imaging at distances greater than 2m offers a higher detection accuracy in contrast to using visible imaging. However, outdoors test proved that both images offered similar poor detection readings. Fluctuations of wind and turbulence may have contributed to the low performance of the thermal images, as well as extreme illuminations respectively impacted the detection results of the visible images which were observed.

1. INTRODUCTION

Due to the proliferation of orbital debris, it is now more crucial for spacecraft to be equipped with autonomous obstacle avoidance capabilities. Vision-aided techniques have been utilized for obstacle navigation onboard spacecraft, unmanned aerial vehicles (UAV)s, and even self-driving cars [3–7]. However, due to the harsh lighting conditions in space, visible imaging is often faced with the challenge of false detection. The feature detector's performance via thermal and visible images were analyzed.

Due to lack of access to both thermal and visible datasets of fractionated systems in orbit, self-assembled quadcopters were used for the simulated tests.

2. DATA ACQUISITION

The simulation experiments were carried out by flying two DJI Flame F450 quadcopters in parallel Fig. 1 and using an Oriented Fast and Rotated Brief (ORB) feature detector to identify an approaching target at various distances and speed.

Onboard one quadcopter, the feature detection algorithm was tested utilizing a CAT S60 Android device equipped with integrated thermal and visible cameras of 80x60 pixel resolution each, and a Qualcomm MSM8952-3 (4 x 1.5GHz + 4 x 1.2GHz) processor.

The second quadcopter carried a thermal target equipped with a custom electric heater, which maintained the target at a constant temperature, ensuring thus consistent temperature readings at various distances from the cameras throughout the experiment. This is because the quadcopters' propellers were not producing consistent heat for a fair thermal analysis. The electric heater, a 4x7cm^2 aluminium enclosure as shown in Fig. 2, maintained a consistent 50°C temperature and was imaged against a clear sky with average temperature of 12°C and when the background was trees, large field of grass, buildings, and cars with average temperatures of about 48°C which allowed the quadcopter to be camouflaged.

A temperature sensor which was installed on the electric heater verified the consistent temperature measurements throughout the test, while another one measured the acquired the ambient temperature.

The imaging quadcopter recorded thermal and visible images at 4Hz along with the acceleration, GPS, and pressure readings for each image to later correlate the altitude distance of the camera from the target per image. The Arduino sensors also recorded the gyroscopic, humidity, as well as the relative and ambient temperature which was paired with each image. Specifically, one temperature sensor was placed directly next to the electric heater and the other was placed on the other quadcopter in order to measure the temperature of the surrounding environment.

In order to evaluate the robustness of the feature detector, the quadcopters collected data in 8 consecutive days at various times during the day to ensure a wide range of lighting condition as well as ambient temperature was present during the test. The target was imaged across a range of 1m to 10m distance from the imaging cameras.

A total of 502 thermal images with 60x80 pixel resolution, and 502 visible images with 640x480 pixels were collected in parallel at 4Hz rate during the flight. In order to simulate a spacecraft's limited memory and processing power, the feature detection algorithm was applied via Sony Xperia Z1 Compact Android device for the analysis of the data.

Figure 1: Illustration of target and imaging quadcopter during outdoors field test.

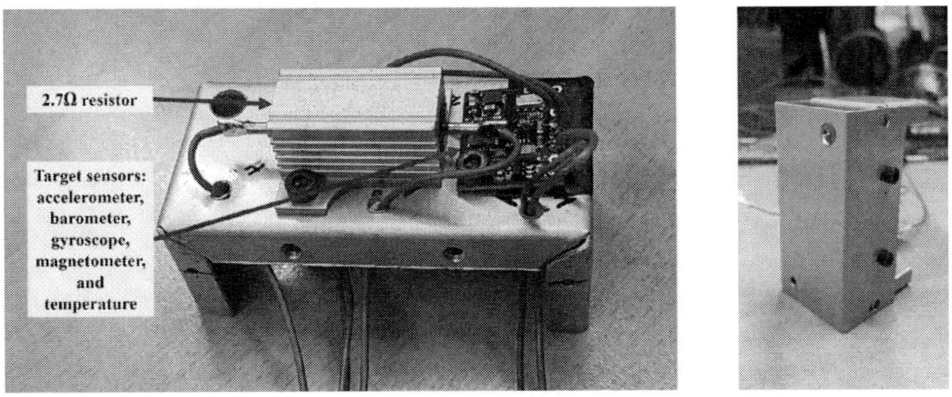

Figure 2: Custom thermal heater

3. ORB FEATURE DETECTION

The Oriented Fast and Rotated Brief (ORB) feature detection was utilized for extracting features of the target from the test image for both thermal and visible images. This is because ORB is a robust feature detector that's rotation and orientation invariant, which allows us to track the target at various distances and rotation angles [8]. ORB applies Feature from Accelerated Segment Test (FAST) [9] to extract the features from the image Fig. 3. In order to optimize the number of features, 10,000 features and a FAST threshold of 5 were utilized instead of the 500 features and 20 FAST threshold. Harris corner detection was further applied to filter out features of low corner ranking [10]. The remaining features were then converted to 256-bits binary descriptors.

In order to establish whether the target was present in the test image, the extracted features from a reference image were compared with the features extracted from the positive and negative test images. Brute-Force matching was then applied to compare the most similar features in the two images. The similarity of the features was established by the minimum Hamming distance which is the sum of the differing bits of the binary descriptors. For instance, a 3bit reference descriptor of "011" and the test descriptor of "000", yields a Hamming distance of 1. The smallest Hamming distance was then marked as the minimum Hamming distance between the two images. In other words, the more similar the descriptors of the two images, the smaller their Hamming distances.

4. HOMOGRAPHY ESTIMATION

In this study, we utilize 2D images which limits our ability to visually estimate the depth of the target in realtime [12]. Homography estimation calculations were utilized for evaluating the rotation angle and translation vector to estimate the trajectory of the target relative to the cameras. This is of interest to analyze the performance of vision-based

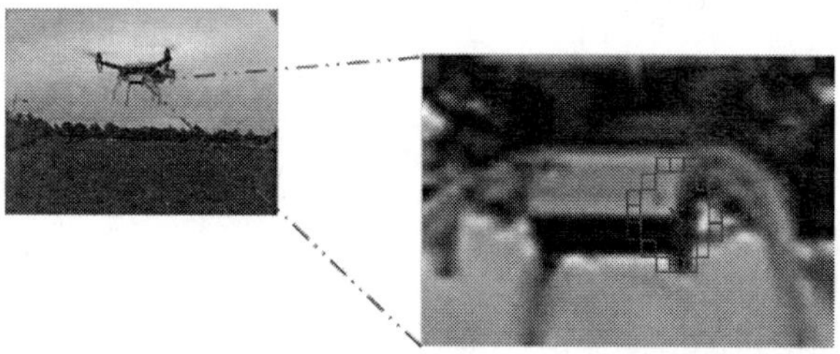

Figure 3: An illustration of an ORB feature along with its neighbors used to create the binary descriptor. Although the illustration depicts a Bresenham circle of radius 3, we applied the default radius of 9 for our analysis, as recommended by Rublee et al. [11].

orientation and positioning estimator, when an oblong camouflaged tumbling debris approaching the spacecraft Fig. 4.

Rotation and translation vectors were extracted from the homography matrix to help further estimate the pose of the target from the test image relative to the reference [13].

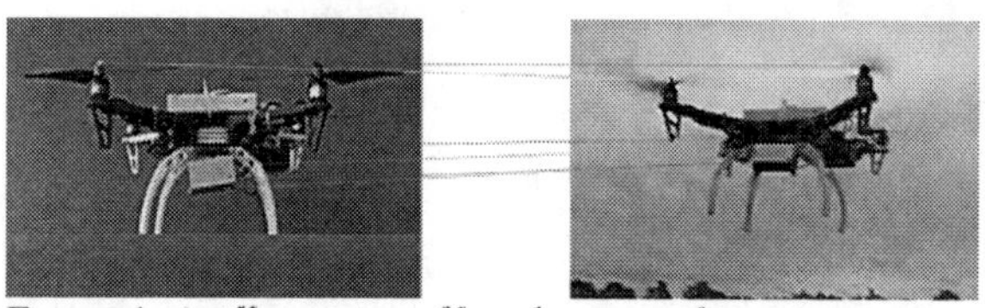

Figure 4: An illustration of how homography estimation was performed to evaluate the positioning of the target via select key features.

5. ANALYSIS OF RESULTS

The images were processed into two main categories to assess the robustness of the technique when the target is camouflaged and when it was not, for both thermal and visible images Fig. 5a and 5b.

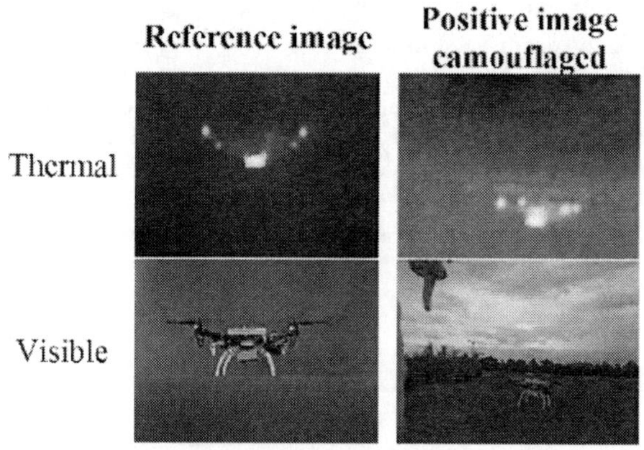

Figure 5a: Demonstration of thermal and visible images camouflaged

Figure 5b: Demonstration of thermal and visible images uncamouflaged within the background

Based on the preliminary results obtained indoors at the laboratory, we were confident that the thermal imaging device would outperform the visible imaging target. However, we observed that the feature detector was unable to differentiate between a positive and negative image for both thermal and visible images. As it can be seen in the results, there is a significant overlap between negative images when the target was absent in the scene and its respective positive images.

Motion blur and poor illumination conditions are likely the contributing factors to the poor results, and further tests will be carried out to enhance the feature detector under such extreme imaging conditions.

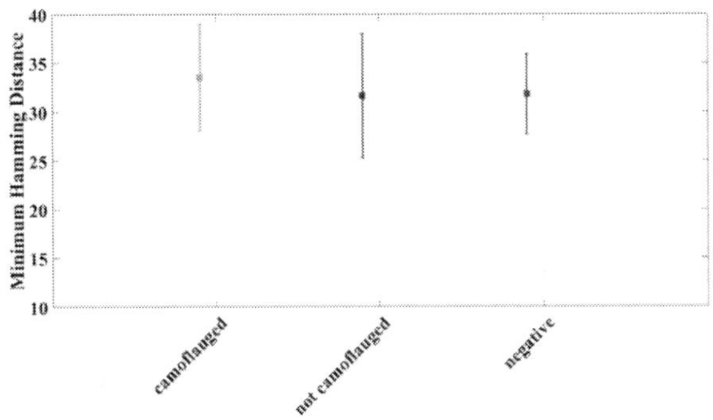

Figure 6: Minimum Hamming distances for visible test images.

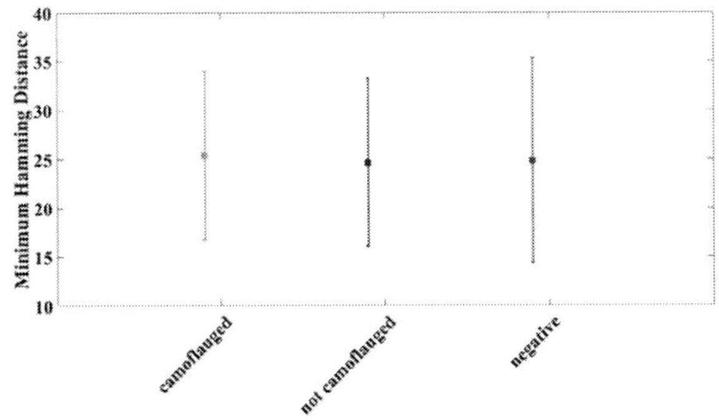

Figure 7: Minimum Hamming distances for thermal test images.

Moreover, to further assess the thermal and visible images' accuracy in estimating the pose of the target were carried out. Images with least blurring and ideal lighting conditions were hand selected from the dataset and processed into three categories for both thermal and visible images: the electric heater was facing the camera at -45°, 0°, and +45° rotation angle along the x-axis (yaw) as shown in Fig. 8.

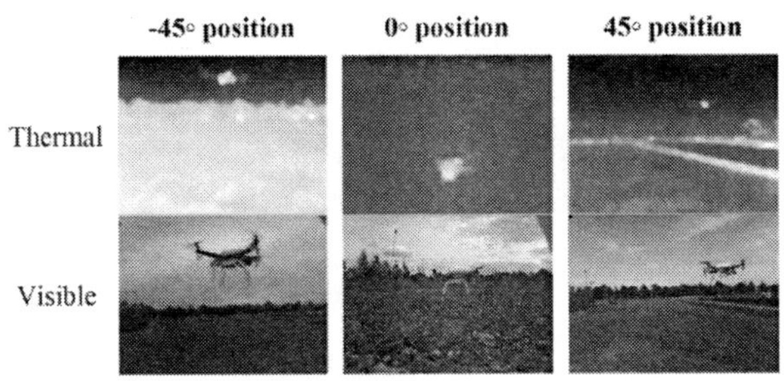

Figure 8: A sample images of the quadcopter at about -45°, 0°, and 45° rotation along the x-axis (yaw).

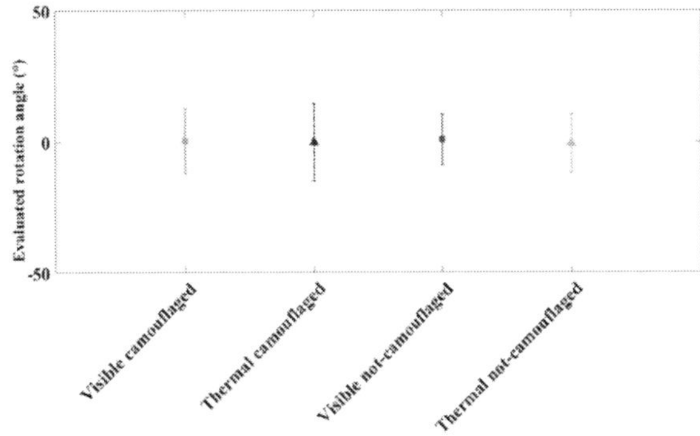

Figure 9: Estimation of rotation angles for 0° rotation along yaw.

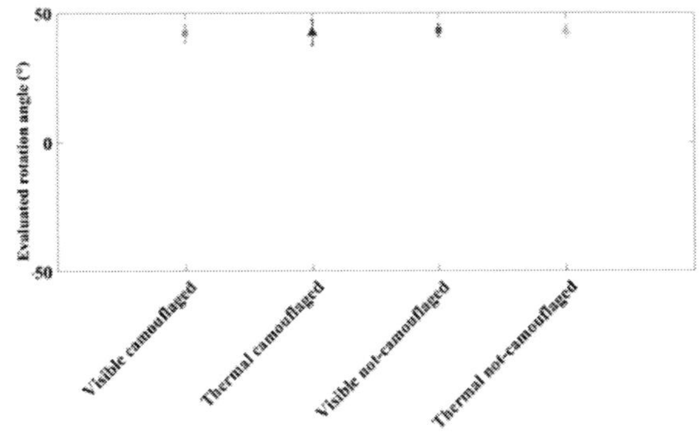

Figure 10: Estimation of rotation angles for 45° rotation along yaw.

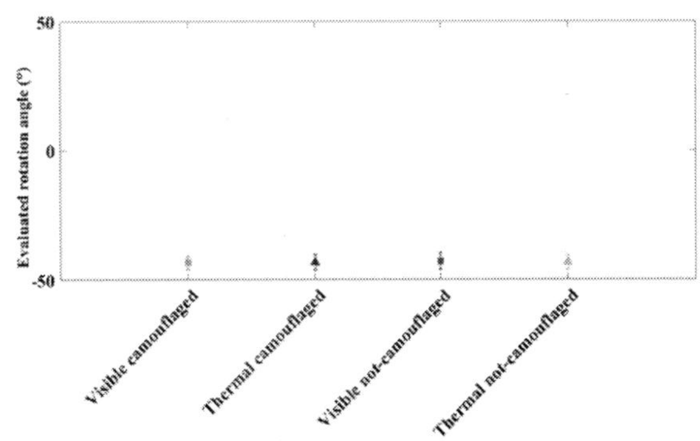

Figure 11: Estimation of rotation angles for -45° rotation along yaw.

Extensive tests suggested that by randomly selecting five features with the smallest minimum Hamming distance per test, an optimized pose estimation of the target could be achieved. Although for the 0° rotation results of camouflaged images, the error bar is larger, overall, the mean of all results illustrated that the pose estimation was effectively evaluating the rotation angles.

The preliminary results confirmed that not only orientation was able to be obtained, but illumination and motion blur significantly reduced the performance of the feature detector.

6. CONCLUSION

The results illustrated that both thermal and visible images behaved similarly and that the feature detector was unable to differentiate between when the target was present and absent in the test images. Further tests will be carried out to confirm which external factors impacted the performance to further enhance the performance.

Furthermore, the pose estimation results illustrated that both thermal and visible images were less accurate in camouflaged environments. We intend to carry out further tests in more controlled environments such as in a thermal vacuum chamber to better image the thermal signatures of the target and evaluate its appearance in lower atmospheric pressures like in orbit. In this way, the feature detector can be better tuned for the ultimate application of detecting and tracking fractionated objects in orbit.

7. ACKNOWLEDGEMENTS

Special thanks to Borja Martinez Calvo at OHB, Michael Lund at Universitat Bremen's Pfaab Lab, Joseph Howse at Nummist Media Corporation Ltd., and Santosh Raju at Universitat Bremen for providing support and guidance in carrying out the experiment. This research was made possible with financial support of the German Aerospace Center (DLR) and the Excellence Initiative of German Research Foundation (DFG).

8. REFERENCES

[1] P. Dubock, F.Spoto, J.Simpson,D.Spencer, E. Schutte and H. Sontag, "The Envisat satellite and its integration," *ESA bulletin*, vol. 106, pp. 26–45, 2001.

[2] T. Takahashi, M. Kokubun, K. Mitsuda, R. Kelly, T. Ohashi, F. Aharonian, H. Akamatsu, F. Akimoto, S. Allen, N. Anabuki *et al.*, "The Astro-H (Hitomi) xray astronomy satellite," pp. 99050U–99050U, 2016.

[3] Y. Cheng, A. E. Johnson, L. H. Mattheis, and A. A. Wolf, "Passive imaging based hazard avoidance for spacecraft safe landing," 2001.

[4] A. Huertas, Y. Cheng, and R. Madison, "Passive imaging based multi-cue hazard detection for spacecraft safe landing," in *Aerospace Conference, 2006 IEEE*. IEEE, 2006, pp. 14–pp.

[5] S. Zhao, F. Lin, K. Peng, X. Dong, B. M. Chen, and T.H. Lee"Vision-aided estimation of attitude, velocity,and inertial measurement bias for uav stabilization," Journal of Intelligent & Robotic Systems, vol. 81, no. 3-4, p. 531, 2016

[6] U. Lee, S. Yoon, H. Shim, P. Vasseur, and C. Demonceaux, "Local path planning in a complex environment for self-driving car," in *Cyber Technology in Automation, Control, and Intelligent Systems (CYBER), 2014 IEEE 4th Annual International Conference on*. IEEE, 2014, pp. 445–450.

[7] C. Hane, T. Sattler, and M. Pollefeys, "Obstacle detection for self-driving cars using only monocular cameras and wheel odometry," in *Intelligent Robots and Systems (IROS), 2015 IEEE/RSJ International Conference on*. IEEE, 2015, pp. 5101–5108.

[8] H. Sharif and M. Holzel, "A comparison of prefilters in orb-based object detection," *Pattern Recognition Letters*, vol. 93, pp. 154–161, 2017.

[9] N. Cornelis and L. Van Gool, "Fast scale invariant feature detection and matching on programmable graphics hardware," pp. 1–8, 2008.

[10] C. Harris and M. Stephens, "A combined corner and edge detector." in *Alvey vision conference*, vol. 15. Citeseer, 1988, p. 50.

[11] E. Rublee, V. Rabaud, K. Konolige, and G. Bradski, "Orb: an efficient alternative to sift or surf," in *Computer Vision (ICCV), 2011 IEEE International Conference on*. IEEE, 2011, pp. 2564–2571.

[12] H. Wang and C. Schmid, "Action recognition with improved trajectories," pp. 3551–3558, 2013.

[13] F. Lu and E. Milios, "Robot pose estimation in unknown environments by matching 2d range scans," *Journal of Intelligent and Robotic systems*, vol. 18, no. 3, pp. 249–275, 1997.

PROCEEDINGS OF THE 9TH INTERNATIONAL SPACE SAFETY CONFERENCE

SESSION 06 REGULATIONS & STANDARDS – I

CREATION OF COMPREHENSIVE GLOBAL SPACE RISK SCALE (SRS)

Joseph N. Pelton

International Association for the Advancement of Space Safety (IAASS), Kapteynstraat , 2201BB Noordwjk, The Netherlands, E-mail: joepelton@verizon.net

ABSTRACT

There would be value in developing, on a global and systematic basis, what might be called a Space Risk Scale (SRS). Such a scale broken down into the six major categories of: (i) Public Launch and Re-entry Safety; (ii) Crew on Board Safety; (iii) Environmental Protection; (iv) Ground Personnel Safety; (v) Space Traffic Management (including the Protozone); and (vi) Cosmic Threats and Planetary Defense. Through liaison and risk updating this could be internationally agreed and applied on an on-going basis. Such as global Space Risk Scale (SRS) could aid scientists, engineers and policy makers in understanding and calibrating the many safety risks that are associated with space-based activities that could also include quite diverse public hazards such as cosmic threats, orbital debris, environmental pollution, or in-orbit or sub-orbital collisions. Various space safety engineers, including Darren McKnight, in particular, have begun efforts to characterize a risk scale that might allow comparison of a wide range of potential safety risks. This is a difficult task for many reasons. Economic, engineering, policy, and especially life-threatening risks associated with various space-related activities are hard to compare.

Safety, risk minimization, and program assurance processes and related activities can vary widely in terms of cost, staff time, and program impact and their relative payoff can vary greatly. Some efforts to minimize risk can involve a minimum of time and effort and provide significant safety margins. Alternatively, other efforts can require major investments of time and money, yet only add minor safety margins. This suggests that others might undertake efforts to develop an algorithm to assess relative "cost-effectiveness" of alternative safety-enhancing strategies. Clearly this is also an important concern and interest, but it should be clearly understood this is quite different issue, that is beyond the scope of this paper. Here the prime focus is only on the feasibility of developing a space risk scale (SRS) and it potential applications.

There are literally thousands of risks that involve space applications, space science, and space exploration, including cosmic hazards. Not only national space agencies, but a growing number of private space concerns are investing billions of dollars to minimize risk and enhance safety for space-related systems. This paper thus seeks to explore a new methodology that would allow a universal process for seeking quantitative measure of space risk on a scale that can translate across a wide spectrum of space activities. The space insurance industry might find such a methodology of particular interest.

1 INTRODUCTION

There is not currently an effective methodology for measuring the degree of risk associated with particular space systems involving human flight or particular cosmic space hazards. There is neither a quantitative nor even qualitative scaling process that might allow comparative indexing of such risks. Likewise, there is no way of comparing their relative importance and urgency. Hazard scales have been developed for asteroid strikes known as the Palermo Scale and the Torino Impact Hazard Scale, but there is no comparable system related to such aspects as orbital space debris, coronal mass ejections, solar radiation flares, or even changes to the Earth's magnetosphere. It would seem apparent that astrophysicists and those that model space debris behavior would provide a useful service if they were to collaborate to develop probability predictions or computer simulation programs that would help to predict the likelihood of these various events and the relative size and relative probability of these occurrences. This effort might be carried out in collaboration with the scientists that developed both the Palermo and the Torino Impact Hazard Scale.

A different undertaking would be required with regard to space programs involving human space flight, but this could also be useful. This effort to create a space risk scale (SRS) would presumably seek to cover manned space flight to Earth orbit and beyond, so-called space tourism flights, hypersonic transportation systems, and public and/or private space station systems. This effort should also seek to cover unmanned launch vehicles, and unmanned spacecraft. This would presumably be developed by those involved with the design and deployment and operation of unmanned spacecraft and associated launch systems.

At this time there is clearly no "risk equations" that provide answers as to how safe these space systems are or tradeoff formulas that allows an assessment as what types of safety measures or risk reduction processes might be prudently applied.

In short, there is not now any systematic way for policy makers to sort out levels of risk, any significant shifts in the level of risk that might be anticipated, nor a way to assess the relative level of risk with regard to a wide spectrum of various types of potential hazards or accidents that might occur. The simplified system that Darren McKnight has suggested, see Figure 1 is a worthwhile start.

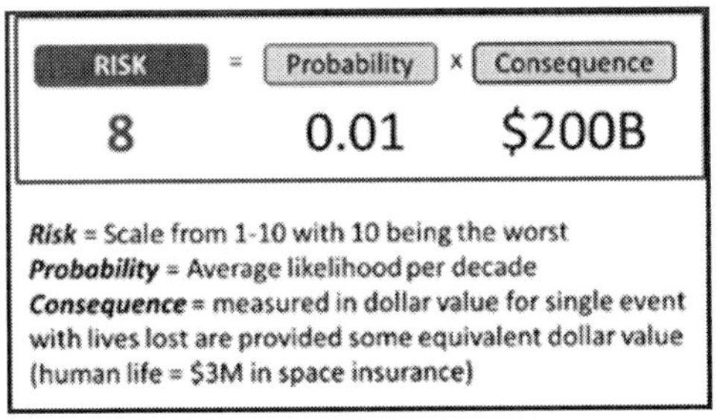

Figure 1: A Risk Assessment Process

This effort that is related to space projects involving crew and passengers is in need of refinement as well as international coordination. This scale, nevertheless, has the merit of simplicity. It considers only three factors. It also allows the possibility of a comparison of risks that span space missions involving human space (or suborbital) flight, various cosmic space hazards and orbital space debris, and even the reliability of various components in spacecraft or launcher systems that carry humans or even launchers carrying unmanned spacecraft of known value and well-established reliability of performance statistics.

These three factors are: (i) "Risk" (on a scale of 1 to 10); (ii) Probability of occurrence per year or decade; and (iii) Consequence (expressed in dollar terms) The largest increment in determining consequence would include valuing human life on the basis of a $3 million payout for each personal fatality in terms of life insurance. (This is an arbitrary figure which would have to be agreed within a panel from the insurance industry and it would also have to be indexed to inflation over time. There are obvious concerns about the qualitative nature of who determines the "risk factor" and establishes the "probability factor". A reasonably large panel with expertise might be able to provide useful assessment of these factors. This process might be developed on as control process that is commonly applied to conducting a Delphi Survey. These controls would include respondents providing these assessments independently without discussion and with a three-round process to produce a precise number with high level of agreement.

A refinement of this system would be to create different "classes of events" with different levels of probability of occurrence. There could be a Class 1 event involving a major accident with a large loss of life but with a minor probability of occurrence (e.g. an off-course launcher that hits a large airliner). There could be a progression down through Class 2, to Class 3 to a Class 4 event where the probability of occurrence is much higher, but the loss is not as extreme (e.g. a mission abort where the downed crew of a manned launch lands in the ocean and one of the crew is drowned).

There could also be other types of refinements that would be to divide the methodology based on the type of hazard that is being analyzed where the "probabilities" and "classes of events" could be attuned to the type of hazard and the probability of occurrence based on statistical data that could be collected and refined over time. The following structural breakdown for the Space Risk Scale as to types of hazards that might be categorized in progressively more detail over time is provided below in Appendix. The probabilities indicated in this chart are not based on statistical evidence, but rather are intended to be illustrative in nature. Actual development of probabilities and clear-cut definition of "Class 1 events" (i.e. from most severe down to "Class 4 events" or least severe) would need be perfected and agreed over time as statistical evidence or panels of experts consult to agree on actual numbers.

In short it would be useful for space scientists and engineers as well as national and international policy makers and safety and security officials to develop a systematic system for understanding of nature of risks for all types of space hazards and on a scale that allowed at least so form of comparative analysis. This could be followed by a means for assessment of their probability of occurrence, potential consequences (measured in a quantitative way), key contributory factors, and factors that might lead to significant increases or decreases in the level of risk and a minimization of "risk consequences".

A ten-level risk scale, supplemented by a "Type 1" down to a "Type 4" event significance could be useful to policy makers understanding the significance of true "black swan" events such as a major asteroid strike or catastrophic solar coronal mass event (CME) hit on Earth at one extreme, down to something like a reparable leak in an astronaut suit might have utility.

This paper argues for the logic of developing a process to place space-related "risks" into a scaled system

This would allow for various scales to be developed to analyze a wide variety of space-related risk concerns. These could be combined into a computer-based global space risk evaluation system. Such a system would need to be capable of constant evolution to add new concerns, technologies and capabilities as they emerge.

Clearly there is a problem of so-called "apples" to "oranges" comparison in such a risk analysis assessment. Equating the basic risk factor to the loss of human life that is quantified at $3 million per life as a metric is problematic, but some quantitative measures for qualitative threats is ultimately necessary to creating a scale that can be used by practitioners of risk assessment.

The value—and novelty--of such a "global space risk scale" would be some form of systematic measure that would allow some form of crude comparison and risk analyses. This paper addresses this issue and seeks to define a systematic process way forward to develop an internationally agreed process for systematic space hazard and risk indexing.

The Chart provided in Appendix was only intended to demonstrate that different types or classes of risk could be assessed based on their probability of occurrence. To be of real value this type of chart would have to include probability of occurrence or probability of failure supported by detailed measurements or data points noting many different types of risk concerns. This supporting data would include all of the various factors that are associated with true safety engineering. These aspects of a risk analysis and space risk scale would of course include such factors as: (i) key component safety margins; (ii) back up escape system or fault tolerance; (iii) safety of fuel, propulsion and fuel storage and delivery systems; (iv) survivability and performance of life support systems; (v) rigorous testing of system and component performance and resilience; (vi) and many other factors.

For launch systems with crew and passengers aboard the following subsystems as provided in Table 1 can in each instance be critical to life and death and thus ultimately safety standards, performance and risk assessment should be included for all these key component aspects of a fully integrated launch system.

Table 1: Key subsystems associated with human spaceflight

Key Subsystems for Testing, of Failure Analysis, and Backup Availability
○ Environmental control and life support system
○ Main propulsion system and fuels
○ Guidance
○ Navigation and control system
○ Avionics and software
○ Main structural system
○ Thermal protection system
○ Thermal control system
○ Health monitoring system
○ Electrical power system
○ Mechanical system
○ Flight safety system, and crew systems

In short it would be useful for space scientists and engineers as well as national and international policy makers and safety and security officials to develop a systematic system for understanding of nature of risks for all types of space hazards and on a scale that allowed at least so form of comparative analysis. This could be followed by a means for assessment of their probability of occurrence, potential consequences (measured in a quantitative way), key contributory factors, and factors that might lead to significant increases or decreases in the level of risk and a minimization of "risk consequences".

A ten-level risk scale, supplemented by a "Type 1" down to a "Type 4" event significance could be useful to policy makers understanding the significance of true "black swan" events such as a major asteroid strike or catastrophic solar coronal mass event (CME) hit on Earth at one extreme, down to something like a reparable leak in an astronaut suit might have utility.

This paper argues for the logic of developing a process to place space-related "risks" into a scaled system

The value, and novelty, of such a "global space risk scale" would be some form of systematic measure that would allow some form of crude comparison and risk analyses. This paper addresses this issue and seeks to define a systematic process way forward to develop an internationally agreed process for systematic space hazard and risk indexing.

2. SPACE RISK SCALE (SRS) ADDED VALUE

The "why do it?" question is of course key. What value would it bring to the table? Would space safety be improved and space hazards reduced? This is difficult to answer in the abstract and many of the benefits would only come to light once this new tool was developed and refined. One can, however, speculate on a number of possible useful outcomes, particularly if the SRS were color coded and some "bright red" indicators that signal major risk elements with reasonably high levels of occurrence being clearly signaled. Here are just a few possible examples of improved steps that might be taken to reduce clearly indicated space-related risks if such a Space Risk Scale (SRS) were formulated within the international space safety community.

- Deficiencies of relative risk information pertaining to asteroid survey capabilities. Currently the U.S. Congress has instructed NASA to survey the skies for asteroids that are 140 meters or larger in size. We now know from research on the 1908 Tunguska event in Siberia that a 40 meter (or so) asteroid impacted and destroyed an area equivalent to the size of the San Francisco and Silicon Valley area. This suggests an inadequate guidance to NASA by the Congress because a lack of understanding of the risks and probability of occurrence. This is because 35 meter asteroids although perhaps 50 times less massive than 140 meter asteroids, they are probably 50 to 100 times more predominant in the number of potentially hazardous near earth objects. This would suggest that an Infrared telescope such as the Courier as conceived by the B612 Foundation, capable of finding asteroids down to 35 meters in size, would be perhaps 50 to 100 times more likely to locate an asteroid city killer than the NASA NEOCAM that is being designed to conform to current guidance 140 meter guidelines provided by the US Congress.
- Inadequate understanding of the changing risk related to solar storms. Currently there is not a clear picture of how changes to the Earth's magnetosphere and its protective shielding of our planet from massive coronal mass ejections (CMEs) plus urbanization, population growth, and increased reliance on vital infrastructure has shifted the probability of a major solar storm strike that could create trillions of dollars (U.S.) of harm to electrical grid networks, pipeline systems, telecommunications networks, and satellites. CMEs are much more common than asteroid strikes by orders of magnitude. Again the "bright red" indicators of a major CME strike that would show up on a global space risk scale could help alert the world to this potential hazard and help define an improved strategy to cope with such hazards.
- Improved safety designed by hypersonic transportation systems and space planes. Currently there is intensive design work being carried out to design new space planes and even hypersonic transportation that would perhaps fly at velocities in the Mach 2 to even Mach 7 range. Since such systems might eventually carry very large number of passengers there are obvious concerns to develop systems that are as safe as possible. The difficulty of designing craft that are 80% safe, or 95% safe, or 99% safe, or 99.9% safe on a scale of difficulty and expense to achieve these levels of safety probably involve spending ten times more money to reach each threshold. In some ways it is like the cost to purify water to higher and higher degrees of purity. At some point the analysis associated with a space risk scale might suggest that the key would not be to engineer a system that was ten times more safe. Rather the key would be to find an emergency safety system that would allow the crew and passengers to escape with safety from a mission critical failed system. Concepts such as parachutes for aircraft or other rescue systems might thus provide the solution to higher levels of passenger safety. Again the SRS could aid in analyzing the best and most efficient pathway to safety.

3. CONCLUSIONS

These are but three examples, but clearly there are many more. Seeing risks, quantifying risks, and looking at relative costs for alternative pathways forward to improve safety and hazard minimization, could produce many benefits over time to governmental regulators, designers of new craft, business investors, insurance interests, and even consumers and the general public. Clearly there are two steps in this process. One step is to create a "Space Risk Scale" that identifies the probability and consequence of a space hazard occurring. The second step that is perhaps much more difficult is to identify strategies, and their anticipated cost and time to implement, that can be employed to reduce the probability of occurrence or lower the degree of the expected consequence. The first step is proposed to be taken first within the international space safety community. Once there are at least rudimentary forms of the SRS in place, then work on the second step that seeks to identify strategies that improve reliability, resilience, and safety and to highlight those processes that might be the most cost effective and/or quickest and easiest to achieve.

The real challenge is whether there is an institution that is willing to take the first step to develop the first iteration of a comprehensive space risk scale (SRS)?

4. REFERENCES

1. Darren McKnight, "Orbital Debris Remediation: A Risk Management Problem" Space News, Jan. 28, 2015, http://spacenews.com/op-ed-orbital-debris-remediation-a-risk-management-problem/

2. "Hanspeter, Schaub, Jasper, Lee E. Z., Anderson, Paul V., McKnight,, Darren S. "Cost and risk assessment for spacecraft operation decisions caused by the space debris environment" Acta Astronautica, Volume 113, p. 66-79. http://adsabs.harvard.edu/abs/2015AcAau.113...66S

3. Privat correspondence with Darren McKnight in January 2017.

APPENDIX

Straw Man Representation of Probabilities for Different Types of Space-Related Risks

The table here below is only intended to demonstrate that different types or classes of risk could be assessed based on their probability of occurrence. To be of real value this type of table would have these probability of occurrence or probability of failure supported by detailed measurements or data points noting many different types of risk concerns.

Key Structural Divisions by Type of Hazard to be Included in the Space Risk Scale				
Type of Hazard	Class 1 (most severe)	Class 2 (severe)	Class 3 (with key fatalities)	Class 4 (minor fatalities)
Variety of Cosmic or Space-Based Hazards				
Asteroid or Comet Impacts (from 35 meters to 1 km or larger in size) (including air burst, water impact and land impact)	$P = 10^{-10}$ 1km or larger	$P=10^{-8}$ Over 200m	$P= 10^{-6}$ 35-200m	$P=10^{-4}$ Under 35m
Coronal Mass Ejection hitting Earth's magnetic field (The probability of major fatality calculations will need to be altered based on changes in urbanization, infrastructure vulnerability, and changes to the geomagnetosphere that protects earth from CME destructive potential to populated regions of the world.	$P=10^{-6}$ Relative Impact velocity of over 2.5x106 Km/hr	$P=10^{-5}$ Relative Impact velocity of 1 to 2.5x106 Km/hr	$P=10^{-3}$ Relative Impact velocity of 0.5 to 1x106 Km/hr	$P=10^{-2}$ Relative Impact velocity below 0.5 x106 Km/hr
Solar radiation flare (X-class and higher)	$P=10^{-6}$	$P=10^{-5}$	$P=10^{-3}$	$P=10^{-2}$
Changes to the Earth's Magnetosphere	Does not apply	Does not apply	Does not apply	Does not apply
Orbital space debris collision	$P=10^{-7}$ High velocity impact with spacecraft over 1 metric ton in size	$P=10^{-5}$ Moderately high velocity impact with spaceraft over 500 kg in size	$P=10^{-3}$ Moderate velocity impact with spaceraft 50-500 kg in size	$P=10^{-1}$ Low velocity collision involving small units
Human Space Flight Launch Vehicles				
Liquid fuel launch vehicles with instant escape evacuation systems	TBD	TBD	TBD	TBD
Air-launched vehicles with carrier craft to allow launch from high altitude (e.g. Stratolaunch).	TBD	TBD	TBD	TBD
Solid fuel launch vehicles without instant escape evacuation	TBD	TBD	TBD	TBD
Nuclear propulsion craft	TBD	TBD	TBD	TBD
Spaceplanes for sub-orbital flights				
Vehicles with hybrid fuel systems capable of being throttled off and carrier vehicle launch (i.e. SpaceshipTwo)	TBD	TBD	TBD	TBD
Vehicles with vertical takeoff and vertical landing (i.e. New Shepard)	TBD	TBD	TBD	TBD
Hypersonic Transport				
Craft with liquid Hydrogen Peroxide (H_2O_2) fueled motors	TBD	TBD	TBD	TBD
Space stations				
International Space Station	TBD	TBD	TBD	TBD
Cislunar space station	TBD	TBD	TBD	TBD

Note: TBD = To Be Defined

THIRD-PARTY CERTIFICATION FOR SPACE SYSTEMS SAFETY AND ENVIRONMENTAL SUSTAINABILITY

Tommaso Sgobba, Isabelle Rongier

International Association for the Advancement of Space Safety, Kapteynstraat , 2201BB Noordwjk, The Netherlands, E-mail:iaass.president@gmail.com

ABSTRACT

For the past half century humans have flown to space on systems developed by national space (research) agencies, while unmanned launch and satellite operations have been carried out either by government organizations or by few specialized commercial operators. In the near future, privately owned and operated vehicles will begin suborbital and orbital operations with paying passengers on board, while a variety of newcomers are swelling the group of satellite and launch operators on a scale never seen before. The long and sometimes painful process through which safety rules and processes matured in space agencies lead to the separate organizations and required access to independent highly specialized technical teams. The same happened for space operations namely in the fields of space debris mitigation/remediation, and collision avoidance procedures.

Traditionally, when licensing commercial risky activities there is a government regulatory body that has the responsibility for levying safety requirements, perform surveillance activities, and certify compliance. However, considering the ever-widening technical competence gap between high-tech industry and traditional government regulatory organizations, such way of operations is becoming impractical in many technological fields. A different set up is advisable in which space industry collectively takes the lead in systems and operations certification, while the government keeps responsibility for overall policies, direction, and international coordination. This paper will discuss how such set-up can be implemented for commercial space developments and operations, based on examples from other fields, like classification societies in ship and offshore structures, IAQG in aerospace third-party quality certification.

1. INTRODUCTION

Traditionally, when licensing commercial risky activities there is a government regulatory body that has the responsibility for levying safety requirements, performing surveillance activities, and certifying compliance.

However, considering the ever-widening technical competence gap between high-tech industry and traditional government regulatory organizations, such way of operations is becoming impractical in many technological fields. Space is one of those fields, which has some further peculiarities. Namely, an additional gap between countries with long tradition of government programs, and countries which are newcomers in the space arena. In few cases, specialized knowledge may reside in a different government body (e.g. national space agency), in many cases it may be partially or totally missing. We will discuss in this paper some possible scheme of third-party certification which may be an effective way of addressing the issue. We will discuss possible third-party certification schemes (as an alternative to regulatory bodies or national space agencies) in the field of commercial human spaceflight, and in the field of space operations and space debris mitigation.

2. COMMERCIAL SPACE SAFETY AND TECHNICAL STANDARDS

The starting point for any safety certification process is the availability of standards.

2.1. Experience in United States

The U.S. commercial spaceflight industry is developing under specific constraints on the establishment of government regulations and international activities: namely legal moratorium until 2024 on safety regulations for humans on board, and export control rules. The moratorium was spearheaded by proponents of suborbital spaceflight fearing imposition of extensive (and expensive) aviation-like certification test programs. The export control rules limit instead the dissemination of technical information outside the country because of space technologies dual-use.

Current U.S. commercial space transportation legislation limits FAA licensing to the protection of public safety during launch and re-entry operations. However, there is one remarkable exception that applies to operators providing commercial transportation services to the International Space Station under the terms of NASA's Commercial Crew Program (CCP).

They are required to obtain a NASA safety certificate for the safety of humans on board, as foreseen by the original agreements signed by governments participating to the ISS program.

Differently from all other NASA programs, companies involved in the CCP program own and operate their spacecraft and infrastructure and are free to design the system they think is best, and to use the manufacturing and business operating techniques they choose. However, the NASA program has to implement the safety policy outlined in a dedicated document (ESMDCCTSCR-12.10), and companies must meet or exceed a pre-determined set of NASA technical and safety requirements. In the initial phase of the CCP program, NASA made an inventory of such standards and recommended them either as reference baseline (meet or exceed) or as good practices.

The safety policy document ESMD-CCTSCR-12.10 "Commercial Crew Transportation System Certification Requirements for NASA Low Earth Orbit Missions" includes four parts:

a) Certification: outlining scope and elements of the certification process: validation of the technical and performance requirements/standards; verification of compliance with requirements/standards; consideration of operational experience; and acceptance of residual technical risk due to hazards, waivers, non-compliances, etc.
b) Documentation: compilation of plans and documents required for submittal at project milestones to collectively prove that the system meets technical requirements and is safe.
c) Safety Requirements: system capabilities in three primary categories of system safety, crew/human control of the system, and crew survival/aborts.
d) Standards: in the fields of engineering, safety, and medical/health, subdivided in those that must either be met as written, or equivalent alternate proposed to NASA Technical Authority for approval, or representing recommended best practices.

Safety policy and technical standards used by NASA Crew Commercial Program represent an excellent reference from which U.S. commercial human spaceflight industry could start developing policies and standards to be used on non-NASA suborbital and orbital commercial spaceflight programs. Furthermore, industry can use the NASA CCP certification program as model for developing their own independent third-party certification process.

2.2 Experience in Europe

Commercial space safety and technical standards have been issued in Europe since the early 1990's as part of an initiative called ECSS (European Cooperation for Space Standards).

In 1987, Europe started an ambitious human spaceflight program including Hermes spaceplane, Ariane 5 human-rated launcher, and Columbus Man-Tended Free-Flyer. At that time the European Space Agency (ESA) had a safety standard, PSS-01-40, based on Shuttle/Spacelab experience. Because of the close involvement in those new programs of other European space agencies, in particular the French Space Agency CNES, and because of industry interest to establish common standards for their commercial satellite business, a joint standardization initiative was launched to create a single system of European space standards. The aim was to improve industrial efficiency and competitiveness, by making available a single set of technical, safety/QA, and management standards to satisfy government and commercial contractual needs without differentiation. In the autumn of 1993, European agencies and industry partners signed the ECSS terms of reference which defined the framework and the basic rules of the system, and concurrently the space agencies committed to gradually discard and replace their standards with ECSS standards in all future contracts. In accordance to such terms, the European space industry assumed from the outset an equal rank in the direction and development of ECSS standards, and held an equal share of voting rights.

Several standards in the ECSS system were an adaptation of previous agencies standards, but many new standards were developed in the course of the years. The ECSS technical standards were extensively used for designing European elements, facilities and payloads of the International Space Station, including in particular Columbus module and ATV cargo transportation vehicle. The previous ESA PSS-01-40 safety standard was the basis for the ECSS-ST-Q-40 "Safety". This standard is essentially a statement of safety policy. It includes four main parts:

a) Safety Program: planning, organization, roles and responsibilities, safety review process, approval authority, training/awareness, documentation;
b) Safety Engineering: policy and principles, risk reduction and control, failure tolerance, design for minimum risk, safety-critical functions, operations safety;
c) Safety Analysis Requirements and Techniques: hazard analysis, safety risk assessment, supporting analyses

d) Safety Verification: hazard reporting, verifications planning and methods, hazard close-out

The ECSS-ST-Q-40 standard identifies three roles: supplier, customer, and safety authority. The supplier is the system developer. The safety authority is the relevant government organization having ultimate safety responsibility. The customer could be an agency, an industrial customer, a space operator, or a recognized third party.

In addition to ECSS, CNES, on behalf of the French government who is the launching state of European launchers, has developed a specific flight and reentry safety regulation (French Space Operation Act), based on the same principles. This regulation was transformed in a French law in 2008, and applied since December 2010 to protect people not only in orbit (ISS, Zhengzhou) but also on Earth.

3. THIRD-PARTY CERTIFICATION FOR COMMERCIAL HUMAN SPACEFLIGHT

Because it is simply impossible to prescribe detailed safety design solutions for something that never existed before, space agencies use to issue safety and technical requirements that are goal/performance oriented. Because of their genericity, verification of compliance with those requirements is a crucial task demanding highly-skilled interdisciplinary competence and independent authority. During safety reviews of a CCP spacecraft, NASA will identify issues and make recommendations, but cannot impose design or operational solutions. However, NASA Technical Authority can refuse to accept a risk.

The competence of NASA's multi-disciplinary safety review panels, and of the specialist teams and labs that support them is well known, but those are rather unique circumstances that have no match in traditional regulatory organizations. There is currently no obvious substitute for NASA's technical capabilities for certifying (non-NASA) commercial human orbital and sub-orbital space systems. If and when FAA would be required to regulate crew and passengers safety of commercial spaceflight, the problem would become apparent. However, as long as the prerequisites of competence, independence and authority are fulfilled, safety certification can be performed also by a non-governmental organization.
This is also fully applicable to safety reviews performed by CNES and ESA at European level for launches and spacecraft operations, even without human on board, where technical expertise is mandatory to support Safety Office decisions.

Safety is a strategic business development driver in many industries. An unsafe design may kill (sooner or later) a unique design and operational concept. Sometimes a single major disaster can jeopardize the future of an entire industrial sector. Cooperating on safety to develop and enforce best-practices can be a matter of business survival, but often awareness is gained only after catastrophic accidents. That's why our association IAASS was created and was very soon fruitful for all the safety experts (from US, Canada, Europe, Japan, …) who could therefore share their skills and experiences on space safety.

We will review now examples of non-government third-party certification in some industries.

3.1 Classification Societies

In the second half of the 18th century, marine insurers based at Lloyd's coffee house in London, developed a system for the independent inspection of ships presented to them for insurance coverage. In 1760, a committee was formed for with the express purpose of classifying the condition of each ship on an annual basis. In 1834, the Lloyd's Register of British and Foreign Shipping was reconstituted as a self-standing 'classification society'. Following the example, a number of Classification Societies were established worldwide.

In 1948, the United Nations International Maritime Organization (IMO) was established to deal with maritime safety, traffic and environmental issues within an international framework. In 1968, the major classification societies established the International Association of Classification Societies (IACS) to provide coordinated technical support and guidance to the International Maritime Organization (IMO) and to national maritime organizations.

Legally, Classification Societies act as a "Recognized Organizations" carrying out statutory surveys and certification as delegated by national maritime administrations (flag administrations). In particular:

a) Developing technical standards for design and construction of ships
b) Approving designs against those standards
c) Conducting technical surveillance during construction
d) Performing in-service inspection and periodic survey during operation
e) Reviewing and approving in-service modifications
f) Performing safety research and development programs

The standards published by Classification Societies, together with the requirements set down in the various International Conventions of the International Maritime Organization (IMO) and the marine legislation of the flag states, form a comprehensive and coherent set of standards for design, construction and maintenance in operation of ships. Classification Societies maintain a leading role in all matters related to technical standards and certification activities, while national and international governmental organizations concentrate primarily on operational and environmental matters.

3.2 Center for Offshore Safety

On April 20, 2010, the Deepwater Horizon oil rig located in the Gulf of Mexico exploded and subsequently sank killing 11 people, injuring 17, and causing the largest marine spill and environmental catastrophe in history. The report of the U.S. Presidential Commission that investigated the disaster made, among others, the recommendation to establish a safety institute. In early 2011, following the recommendation, industry created the Center for Offshore Safety (COS), with the stated mission to "promote the highest level of safety" for offshore operations through "leadership and effective management systems". The Center for Offshore Safety helps operators to develop safety programs based on best-practices that are rigorously assessed by well-trained third-party auditors. The COS is a branch of a trade organization, thus attracting some criticism about its independence.

3.3 FIA & Global Institute for Motor Sport Safety

In the first three decades of the Formula 1 World Championship, inaugurated in 1950, a racing driver's life expectancy could often be measured in fewer than two seasons. It was accepted that total risk was something that went with the job (Tremayne, 2000). The turning point was the Imola Grand Prix of 1994 with live coverage of Roland Ratzenberger and Ayrton Senna deaths that forced the car racing industry to look seriously at safety or risk the loss of television rights. In the days after the Imola crashes the FIA (Fédèration Internationale de l'Automobile) established the Safety Advisory Expert Group, later renamed FIA Safety Institute, to identify innovative technologies to improve car and racetrack safety. Currently replaced by the Global Institute for Motor Sport Safety, the nstitute works in close collaboration with the FIA, in particular on matters associated with the development of safety regulations, however the regulatory, homologation and enforcement processes are the exclusive responsibility of the FIA, its World Council and specialized Commissions.

3.4 Institute of Nuclear Power Operations

In March 1979, a series of mechanical and human errors at the Three Mile Island nuclear generating power plant in Pennsylvania, caused an accident that profoundly affected industry. A combination of stuck valves, misread instruments, lack of information and poor decisions led to a partial meltdown of the reactor core and the release of radioactive gases into the atmosphere. The Kemeny Commission, which president Jimmy Carter formed to investigate the accident recommended that "The (nuclear power) industry should establish a program that specifies appropriate safety standards including those for management, quality assurance, and operating procedures and practices, and that conducts independent evaluations" and that "There must be a systematic gathering, review and analysis of operating experience at all nuclear power plants, coupled with an industrywide international communications network to facilitate the speedy flow of this information to affected parties" (Kemeny, 1979). In response to the recommendations, industry established in December 1979, the Institute of Nuclear Power Operations (INPO).

3.5 Building a Commercial Space Safety Institute

No matter how much the cost of spaceflight will be cut, without the advancement of space safety there will be no growth, no expansion and perhaps no future at all for the commercial human spaceflight industry. Advancing space safety engineering applications, research, and education should be a primary strategic business goal of industry as a whole. Industry should build on the experience accumulated in more than 50 years of government space programs. The space industry should develop a notion of space safety and sustainability as their own collective strategic goal. Industry should support the establishment and operations of a commercial space safety institute, as an independent entity aimed at developing and enforcing, in cooperation with all stakeholders, standards of excellence to ensure continuous safe access and operational integrity of space operations.

Table 1 Commercial Space Safety Institute (CSSI) set-up

	COMPANY	SAFETY INSTITUTE	REGULATORY BODY	INT. ORG
POLICIES	-	*advise*	*develop*	*coordinate*
STANDARDS	*implement*	*develop*	*validate*	-
CERTIFICATION	*data*	*perform*	*oversight*	-
PROCESSES	*establish/execute*	*establish/execute*	*establish/execute*	-
AUDITS	-	*Company*	*Safety Institute*	-
	COMPETENCE	*COMPETENCE*		
		INDEPENDENCE	*INDEPENDENCE*	
			AUTHORITY	*AUTHORITY*

The CSSI would provide standardization and safety certification services commercial human spaceflight industry. The SSI should be a "recognized organization" approved by and operating under oversight of a national regulatory entity, with further international coordination. See Figure 1. The commercial space safety institute would also coordinate, support and promote research in the field of space safety engineering, support educational programs, and provide professional training opportunities to members.

4. ISO SPACE DEBRIS STANDARDS

In the early times of automobile, speeding, reckless driving, crashes, and fatalities were new issues. Serious debates were held over whether the automobile was inherently evil. In US, the Georgia's Court of Appeals wrote: *"Automobiles are to be classed with ferocious animals and ... the law relating to the duty of owners of such animals is to be applied ..."*.

The first remedies comprised collection of statistics, and a social response focused on controlling and improving driver behavior. Conferences, and safety campaigns were held hoping that increased awareness would promote careful driving. Manufacturers took initiatives to improve the vehicle response (e.g. 4 wheels braking), but they contended that improving roads, licensing drivers, and regulating traffic was the key to prevent accidents. In England road rules were initially left to municipality to define. Some required driving on the right side, others on the left side, and others left the choice to the driver. Clearly the rules of the road were needed, and they had to be nationally enforced. Nowadays in space operations we are still in the early steps of a similar path of promoting awareness, collecting statistics, and issuing guidelines. In 1999, a working group of IADC (Interagency Debris Coordination Committee) developed the first set of consensus international space debris mitigation guidelines, which were adopted in October 2002, and presented to the Scientific and Technical Committee of the United Nations' Committee on the Peaceful Uses of Outer Space (COPUOS) the following year. In 2004, the European Code of Conduct for Space Debris Mitigation was issued. In 2007, the United Nations General Assembly endorsed the COPUOS Space Debris Mitigation Guidelines.

In 2011, ISO standard 24113 was issued which was still very close to the original IADC guidelines and truly a policy guidelines that says 'what' needs to be done and not 'how' to do it. Although called standard, until enforced by some instrument ISO 24113 is still a set of guidelines. While some countries and organizations have already taken the step of mandating its implementation in their space programs (as the French Space Operation Act), worldwide implementation is still pending. As a minimum, we can say that the progress made until now does not match the seriousness of the problem.

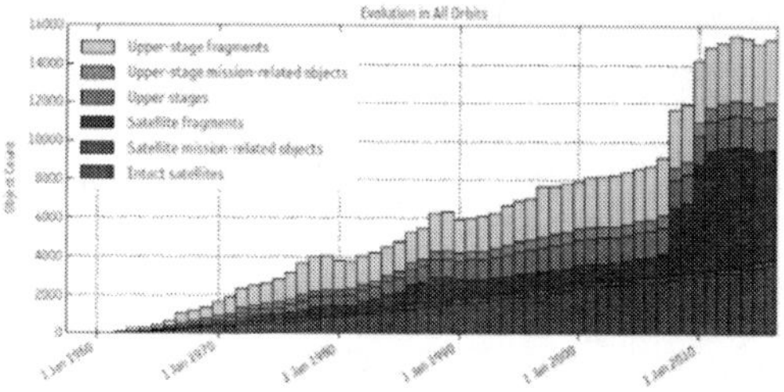

Figure 1 Space Debris population as of January 2017 (credit: ESA)

A growing problem that the arrival of newcomers, launches of CubeSats, and plans for mega constellation will further exacerbate.

5. VERIFICATION OF COMPLIANCE WITH SPACE DEBRIS MITIGATION STANDARDS

The establishment of an international body similar to ICAO (International Civil Aviation Organisation) for regulating civil and commercial space activities is a political matter not yet on the table. The alternative is to look for an industry initiative, and IAQG explained here below is an interesting model.

5.1 International Aerospace Quality Group (IAQG) as model

One of the key rules of quality assurance is to have 'trustable' suppliers, which would deliver goods compliant with specification and keep performing/improving. The suppliers must have a 'quality system' in place which is initially and then periodically audited by the customer. Most suppliers have multiple customers and specialize in one or more fields like aerospace, automotive, etc. In the past, suppliers would receive multiple audits per year which represented a cost and a distraction for them as well as for their customers. When in 1987, ISO launched a 'universal' quality system standard later known as ISO 9001, almost immediately specialized companies appeared which were offering third-party certification services. Overall it was an almost ideal solution for everybody. But there were weaknesses, the following two in particular. The first was that the ISO requirements were too generic to satisfy the needs of the aerospace industry, the second was that the competence and rigor of the auditors where often not up to the task. ISO 9001 auditing became mainly a commercial market more than a testimony of effective quality system.

At that point in time, major aerospace primes recognizing that they were often sharing the same supply chain, decided in 1998 to set up a non-profit organization, the International Aerospace Quality Group (IAQG), with the responsibility to adapt the ISO standard to the need of the aerospace industry and to establish a unique system of shared auditing and measuring processes. Nowadays AS/EN 9100, the IAQG aerospace quality standard, is used worldwide, recognized by major government regulatory body, and implemented by suppliers and primes as well.

5.2 Institute for Space Debris Prevention and Control (ISDPC)

The issue of space debris mitigation is not only a matter of political will but also of access to skilled human resources which are rare outside major industries and agencies. In 2013, the IAASS presented at the 50th COPUOS STSC session the concept of setting up a non-profit organization (on the model of IAQG), called Institute for Space Debris Prevention and Control (ISDPC). It would be a collaborative consortium of industry, open to the participation of experts from government organizations, implementing processes for verifying compliance with voluntary space debris standards directly or through recognized certification services providers.

The Institute ultimate goal would be to build through experience and lessons learned collaborative mechanisms for more safe and productive space activity.

In particular, the Institute for Space Debris Prevention and Control would:

a) Make available to all stakeholders in space activity, whether large or small, expert assessments for safe space operations.
b) Perform third party certification of space systems and operations procedures against ISO Space Debris Requirements (ISO 24113, etc.)
c) Produce additional technical and management implementation guidelines
d) Verifies that best practices are known and systematically applied
e) Assess risks of debris events. (Risks to all rather than risks to the operator's own satellites only)
f) Examine planned repurposing or ad hoc major maneuvers and repositioning
g) Assess uncertainties in maneuvers and plans near stated end of life
h) Enable proactive accommodation of circumstances that would eventually become serious issues requiring ad hoc responses.

6. CONCLUSIONS

Future space governance should move away from classical models of separation of roles between government regulatory bodies and industry, and move towards a closer collaboration of all stakeholders. The space industry should move towards developing a notion of space safety and sustainability as their own collective national and international strategic goal. Under the oversight of relevant regulatory bodies and with the benefit of international cooperation, stakeholders should establish specialized institutes, as independent entities aimed at support the development and enforcing implementation of standards of excellence to ensure continuous safe access and operational integrity of space operations.

7. REFERENCES

1. NASAFacts "Crew Commercial Program", www.nasa.gov/pdf/608888main_CCP.pdf

2. IACS International Association of Classification Societies, "Classification Societies-What, Why and How". www.iacs.org.uk/media/3785/iacs-class-what-why-how.pdf

3. ECSS-ST-Q-40 "Safety".

4. A/AC.105/C.1/2011/CRP.14, R. Jakhu, "Towards Long-term Sustainability of Space Activities: Overcoming the Challenges of Space Debris", UN Committee on the Peaceful Uses of Outer Space Scientific and Technical Subcommittee Forty-eighth session Vienna, 7-18 February 2011.

5. 9104/1 AEROSPACE STANDARD (R) Requirements for Aviation, Space, and Defence Quality Management System Certification Programs, 4 December 2011. www.sjac.or.jp/jaqg/info/2012/doc/PB_9104-1_EN.pdf

COMMONALITY BETWEEN CHICAGO CONVENTION AND OUTER SPACE TREATY: MERGING AIR AND SPACE TRAFFIC MANAGEMENT FOR AVIATION AND NEAR-SPACE

Dr Sanat Kaul

International Foundation for Aviation and Development (IFFAAD), 602, Ashadeep Building, 9-Hailey Road, New Delhi, Delhi 110001, India, Email: iffaadindia@hotmail.com

1. INTRODUCTION

While there is no legal separation between where Airspace ends and Outer Space starts, there is a general acceptance that 100 km vertical could be the limit of Airspace (based on von Karman delimitation proposal). The major difference between the two is that while Airspace above a land territory (including territorial waters) of a country is the Sovereign Airspace of that country as per Article 1 of Chicago Convention of 1944, the Outer Space Treaty of 1967 defines Outer Space (beyond that undefined limit of Territorial Air Space) as the 'common heritage of mankind'. Similarly, under Art 12 of the Chicago Convention the Air Space over the High Seas (except for territorial waters) will be as established by the Convention. Accordingly, International Civil Aviation Organisation (ICAO), which was created by the Chicago Convention, also treats the Airspace over High Seas like a common heritage of mankind as it is open for all to use. Therefore, both Air Space over High Seas (70% of airspace) and Outer Space can be considered as "Global Commons". There is, thus, a great commonality between the Chicago Convention and Outer Space Treaty.

Article 28 of the Chicago Convention provides that each contracting State shall provide in its territory navigational aids to facilitate international travel by air. However, since Airspace over High Seas is with ICAO, it has to provide the navigational facility. To overcome this responsibility, ICAO has divided the entire Airspace over High Seas into many Flight Information Regions (FIR). After seeking the assent of the adjoining country it has allotted FIRs to them. However, no sovereignty is passed on to the state.

Space Traffic Management in the near space will become a challenge as more and more small satellites enter near space and sub-orbital flights become a reality. United States, the leading Air and Space power in the world today, has already designated its Federal Aviation Authority with the responsibility of laying down rules for commercial space flights. Similarly, twinning the role of aviation safety management and Outer Space safety management in a single international authority would bring the desired impact of managing air and space traffic in a harmonious manner without a legal hitch through up to say about 100 km vertical and beyond. With the ADS-B surveillance technology soon becoming mandatory, management of Air and Space traffic over land up to about 100 km vertical (and beyond) will become realizable within the ICAO requirement of *Communications*, Navigation and Surveillance/Air Traffic Management (CNS/ATM).
Merging the responsibility of Air and near Space traffic management in atmosphere and near space can, therefore, be best left to ICAO with minor tweaking of the Annexes to Chicago Convention as neither the upper limit of Airspace nor the beginning of Outer Space has been defined in any of the Treaties mentioned above.

This paper will examine and show why it is now necessary for ICAO to take over Air and near Space Traffic Management as technology is similar for both and atmosphere extends up to 100 km for aircraft to fly. Further it becomes obligatory for ICAO to do this and Outer Space Treaty of 1957 is silent on the issue of space traffic management.

2. NEW SPACE

Travel into upper atmosphere and Outer Space will no longer remain an occasional launch when the Air Traffic Controller informs all aviation traffic to stay clear by issue of a Temporary Flight Restriction (TFR). With the coming of Space Tourism, Sub-orbital Flights, hypersonic aircraft, high altitude balloons well as UAVs in near future there will be much greater requirement of better management of Upper Air and Near Space. Currently, countries are obliged to manage the air traffic over their sovereign airspace as well as over the Flights Information Region (FIR) over high seas allotted to them by ICAO. The existing technology based on radars is not able to provide good surveillance over High Seas since radars are land based and are unable to provide surveillance over High Seas. Radars find it difficult to cover terrain like Polar Regions or even hill features. Since 1960s with the introduction of Global Positioning System (GPS) by US Air Force the management of Air Traffic is undergoing a change. Global Navigation Satellite System (GNSS), a generic name for constellations like GPS of US, GLONASS of Russian Federation, BeiDou of China and local constellation like IRNSS or NAVIC of India have become popular method of navigation not only by aircraft but in general. ICAO intends to shift from terrestrial navigation to satellite based navigation for aviation. However, there are many issues with satellite based navigation. To begin with GPS signals are weak and subject to interference both natural and motivated. In addition GPS and GLONASS are military satellite constellations and are deliberately made inaccurate for civil purpose. For Aviation navigation purposes Augmentation satellites are needed like WAAS in the US, EGNOS in Europe and GAGAN in Asia. These satellites are supposed to correct the poor signal and take care of ionosphere disturbances which distort GPS signals

3. AIRSPACE MANAGEMENT ISSUES

Presently 70% of earth's Airspace does not have a continuous air traffic surveillance as radars cannot reach it. In this non-radar Airspaces (NRA), flight surveillance is managed procedurally. To ensure flight safety, large separation distances are mandatory inflight paths passing through NRA. But due to the rise in transoceanic air traffic, a smaller than normal separation distance is already being permitted in some cases. It may also be added that the present surveillance by radars is over only 30% of the earth's atmosphere that too up to a height of about 50,000 ft which is just over 15 km (vertical). Even military aircraft do not go beyond 100,000 ft or about 30 km vertical above the sea level.

In order to overcome the gaps in continuous surveillance over airspace under use the ICAO Council approved the Global Air Navigation Plan (GANP) with the ultimate goal of a fully harmonized global Air Navigation system with global interoperability as a minimum objective. In order to achieve this it was felt that there is a need to introduce new technology. Use of GNSS as a technology was the obvious answer

3.1. A new Technology is introduced: ADS-B

Automatic Dependent Surveillance Broadcast (ADS-B) is a new GNSS based technology that enables aircraft broadcast signals including flight related information i.e., identification, position, altitude, velocity and other relevant information i.e. surveillance on a regular basis to ATCs as well as to other aircraft. This technology ensures that a moving object in air regularly transmits its position and other details to all other aircraft and ATCs via the aircraft's Mode S transponder. However, for accuracy ADS-B also requires ground stations. US has already constructed over 640 ground stations as FAA has announced use of ADS-B (IN) as a compulsory equipment on all aircraft by 2020. Australia is also doing the same. Terrestrial-based ADS-B infrastructure, therefore, becomes an alternative (or add-on) to radars but not a complete solution to complete Air Traffic Management.

GNSS technology is available free to all aircraft and with the help of Augmentation satellites aircraft are able to navigate all parts of the earth. However, aircraft over high seas and polar regions are unable to keep the Air Traffic Control in the loop with this technology.

3.2. Shortcoming of ADS-B

While ADS-B technology is GNSS based and is able to broadcast the position of an object/ aircrafte to the ATC as well as to all other aircrfts in the region giving its fligt direction, speed etc

While ADS-B technology is considered better than land based Radar technology as it provides a better ground coverage, it also suffers from the same major deficiency as Radars. Both require ground station. As a result both technologies are unable to cover 70% of the earth's surface ie the High Seas and Polar regions and difficult terrains.

3.3. Solution to the problem: ADS-B over satellite

In order to overcome the issues of coverage mentioned above DLR, the German aerospace research agency, in 2008 investigated the possibility of monitoring global air traffic from space without ground stations as it could resolve the issues of surveillance in non-radar area (NRA) i.e. the High Seas and Polar regions outlined above. This, they felt, could be achieved by satellites intercepting the information broadcast by aircraft, in this case data from the Automatic Dependent Surveillance - Broadcast (ADS-B), which can then be transmitted to existing ground infrastructure with the help of Mode S transponders on commercial aircraft. This will ensure global surveillance of all air traffic. By 2010 DLR conducted successfully a series of high altitude test flights.

Under the aegis of the European Space Agency (ESA) Thales Alenia Space Deutschland GmbH led the "Space Based ADS-B In-Orbit Demonstration (IOD) Payload Development for Air Traffic Surveillance Project". This project was mainly aimed at demonstrating the functionality of this technology under representative space environment conditions, and to verifying the link budget of this payload for the various ADS-B compatible equipment onboard aircraft on different geo-locations.

This was followed on in May 2013 by European Space Agency's (ESA) Proba V experiment (Project for OnBoard Autonomy - Vegetation) on board the newest satellite of ESAs. Its main task was to provide actual vegetation data to the community and serve as gap filler between the SPOT-VGT and Sentinel-3 satellites.

This experiment of the ADS-B receiver on board the satellite was the first experiment of its kind, receiving 1090ES (Extended Squitter) ADS-B squitter signals transmitted from aircraft.

The primary goal of the DLR project ADS-B over Satellite was to demonstrate the feasibility of an orbital ADS-B system by means of an In-Orbit Demonstration (IOD) and to evaluate the characteristics and performances which may be important for future space based air traffic control systems. As preparatory missions DLR conducted in 2010 the first of a series of high altitude test flights. During a test flight in northern Sweden a terrestrial ADS-B receiver has been used and basic assumptions regarding the maximum reception distance in NRA could be verified. Based on these trials and pre-development studies a system concept for a space based ADS-B reception has been developed, mostly based on commercial of-the-shelf hardware.
The achieved results of the IOD should take into account the constraints under which this experiment was performed, as there were limitations in cost and time and in particular available resources on the satellite in terms of available power and geometry. Further, the reception of 1090 Extended Squitter ADSB messages on board of the PROBA-V satellite is mainly affected by the fact that the signal path between a LEO satellite and an aircraft is much longer as the LEO satellite flies at about 820 km of altitude while an aircraft flies at 15 km altitude, which may lead to a loss of signal quality.

In this connection the quality of antenna on board a satellite to receive the ADS-B signals is very important. The same goes with ADS-B receivers.

ADS-B over satellite is still in early stages of commercialization and will overcome the deficiencies to provide a full coverage of surveillance over High Seas and Polar regions of all moving objects in Airspace provided they carry an appropriate transponder.

3.4. Airspace No More Only Aircraft Domain

Chicago Convention has used the word “Aircraft” without defining it and assumed it as the sole user of airspace, there are now many other users of the airspace. However, these new users are not subject to the requirement imposed upon aircraft. With new users of airspace similar conditions need to be imposed since surveillance of other than aircraft is a must for all airspace users and therefore, similar requirements must be imposed on all. The commonality of the instrument falls on ADS-B as it can be put on all objects including balloons and platforms.

Simply put, ADS-B works wherever it is placed, be that on an aircraft, or on a space vehicle, or on a stratospheric balloon, or on a remotely piloted system. The technology follows the transponder. Space borne, the future hope.

Only a satellite constellation has the capability to provide a global coverage at any possible flight level, avoiding limitations imposed by terrestrial ADS-B. This could be implemented by receiving ADS-B signals, which are broadcasted regularly by each equipped aircraft, spacecraft or even a balloon giving information on its position, speed, direction etc. by LEO satellites. This data can then be made available to already existing ATC ground infrastructures.

Therefore, a satellite-based surveillance network will provide enhanced Air Traffic Services in areas where the traffic density, the location, or the cost of "conventional" ATC equipment would not justify any installation of radar and/or terrestrial ADS-B. It can also include VHF coverage in fringe areas and areas where existing radar is to be decommissioned, and where the replacement costs are not justified.

Results: “The Proof of Concept” experiment for ADSB has proved successful. It has successfully validated the principle of detecting even the weak Mode-S transponder transmission from a Low Earth Orbit satellite. With improvement in receivers and antenna the quality could improve.

Signal loss is due to the low signal level resulting from the distance between the receiving satellite at an altitude of approximately 820 km and the transmitting aircraft at an altitude of 12-15 km. Further, RF signal loss is also due to the shapes of the satellite antenna vertical radiation pattern and the aircraft antenna vertical radiation pattern. Corruption of messages by garbling, when several messages arriving at the ADS-B antenna onboard of the satellite at the same time overlap and thus cannot be decoded by the ADS-B receiver. Speed of the satellite of about 27000 km/h, leads to a limit the time of observation for each detected aircraft to about 3 minutes maximum.

Receiver: The ADS-B receiver developed for the PROBA-V in-orbit demonstration provides in its output data no direct signal level measurement values, which mainly would represent the noise level. Instead the receiver provides a correlation gain value for each successfully decoded message, which denotes the performance of the correlation process. It may be added that on June 2013 an Airbus aircraft flying over Scotland was first noticed from space by a new receiver, the ADS-B device of DLR on Proba. In another experiment by DLR when ADS-b over satellite was switched on over 12,000 ADS-b messages were received in two hours with the satellite at an altitude of 820 km.

Other advantages of ADS-B over satellite is that as the long-distance Air Traffic becomes global, it can bring considerable fuel savings, a reduction in greenhouse gases emissions and greater safety of aircraft and passengers.

3.5. Commercialization of ADS-B over Satellite for Aviation

Aireon is the first commercial attempt for ADS-B over Satellite for aviation sector was already started in 2014 with the announcement of "Aireon" a joint venture between Iridium Communications, Nav Canada, the Irish Aviation Authority, Italian air traffic services provider ENAV, and Danish ATS provider Naviair. Nav Canada also has signed on as Aireon's first customer, and plans to implement space-based ADS-B beginning with its busy North Atlantic airspace. NAV Canada has backed Space-based ADS-B as its major technology for ATM especially over high seas. By March 2014 Aireon had successfully completed qualification testing for the harsh environment of Space. The company produced 81 ADS-B 1090 Extended Squitters receiver payloads. It intends to launch 66 satellites into LEO along with six in orbit as spares

With the starting of Aireon as commercial venture for ADS-B-over-satellite, the beginnings of a new era surveillance of common Air and near space integration has, in fact, started. As more commercial companies are taking interest in use of small and nano satellites, only ADS-B transponder may be able to provide the task of integration of Air and Space Traffic Management.

4. THE INTEGRATION OF AIR AND SPACE TRAFFIC MANAGEMENT

While we describe the conventional limit of Airspace as 100 km in our introduction above, the current commercial air traffic is only up to about 40,000 feet or say 12 km.

Airspace is currently used by commercial traffic up to about 40,000 feet or say 12 km above the sea level. Further, from 60,000 (18.3 km) to 100,000 (30.5 km) feet it was traditionally used for military operations. Now, however, new commercial operations have started to take place in this airspace which include unmanned free balloons, high endurance UAS and commercial space operations. Air Traffic Services will need to be extended, out of necessity, enhance its role to include the new air based objects. This will lead to start of integration of aviation and space activities in the upper airspace. Even space tourism activities propose to go up to 100 km and beyond into a bit of outer Space. The current set of Radars will be inadequate and new radars for higher altitudes are very expensive. Therefore, new solutions have to be found. Here the new ADS-B over satellite will be a game changer for aerospace management.

The Space-Based ADS-B Antenna: The antenna used for ADS-B over Satellite is an antenna array of two elements. Each element is a capacitive fed, shorted patch antenna. There is no direct mechanical bonding between the feeding structure and the patch. This makes the patch assembly very easy. Except for the feeder no dielectric material is used. It has been found that this gives an extra 12% gain increase. The patch antenna is shorted in the center of the patch to the ground plane of the spacecraft structure. This avoids potential charging. The technology for space based antennas is still young and hopefully there will be more improvements

4.1. Temporary Flight Restrictions (TFR)

As more and more space launches take place from existing and new launch locations leading to greater demand for declaration of temporary flight restrictions (TFR) the use of TFR will need to be made more sophisticated. Other commercial Expendable Launch Vehicles include Sea Launch Vehicles which operates currently from a floating platform in a remote region of the Pacific Ocean and high-altitude Balloons being launched by amateurs and Universities. As more and more countries and private companies start their own launches from ground or high seas the issue of TFR will become more problematic as the conflict between a growing aviation industry with horizontal requirement of airspace and growing space launch industry with vertical requirement takes place. With Nano and other small satellites getting launch in the thousands per year in future this issue will become very important. For example, Planet Lab, a seven-year-old company, has already sent 233 satellites into space on a mix of American, Indian, Japanese and Russian rockets. India recently launched 104 satellites in one single launch. Another company One Web in Florida, USA wants to surround the planet with hundreds or even thousands of satellites in the low orbit and form a network accessible to 3 billion people on Earth who lack high-speed internet services. So far very little attention was paid to the airspace above 60,000 feet as this airspace was not particularly important to commercial space or other non-government users. But not anymore. More sea launches will a require more TFR over high seas. It is here that ICAO has to take a proactive stand as it is their mandate and obligation to maintain and encourage growth of safe and orderly traffic in airspace including over high seas.

4.2. Balloons

We are also witnessing frequent launches of unmanned free stratospheric balloons by amateurs and universities that reach the upper airspace for short periods and return. Some larger commercial unmanned free stratospheric balloons that can float above 60,000 feet for several days. These can also cause conflict with other modes of aerospace travel

In near future we expect supersonic and then hypersonic commercial aircraft will operate at or above 60,000 feet. As a matter of fact, Concorde, the supersonic aircraft flew for over 27 years until 2003. Once occupied by military and space users, this airspace from 60,000 to 100,000 feet is becoming increasingly attractive to commercial users. Further, ELV are giving way to reusable Launch Vehicles (RLV) which have their own requirements. Can the present state of Air Traffic Management based on terrestrial ground equipment like Radars manage it?

4.3. Facilitate Integration of Space Operations

Technically integration of air and space traffic management cannot take place unless a common technology is available for surveillance and guidance. Timely and accurate position information is needed to be made available to air and space traffic control systems from all airspace users. There is a particular need for surveillance information on airspace users in the commercial space industry. Further, the integration, even if technically possible, will need acceptance at political level as global standards will need to be made. ICAO is the obvious institution for it.

As we move towards commercial use of space from Expendable Launch Vehicles to a variety of diverse operational types sharing this upper airspace is a challenge. This is also a tremendous opportunity for ICAO to demonstrate leadership, as is mandated in Chicago Convention of 1944, in airspace integration of both horizontal and vertical kind and support innovation in these developing industries if they are equipped to operate in shared airspace. There is a need to *"the development of standards and recommended practices between aircraft and space vehicles in sharing of the airspace...This research needs to concentrate at developing more advanced concepts that separate launch and reentry vehicles from other aircraft based on separation standards as opposed to airspace restrictions."* Another aspect is that the presently followed practice of horizontal heights dividing the Airspace will need to adjust to near vertical requirements launch vehicles as well as of space tourism vehicles. In order to achieve this goal, space vehicles and other high-altitude operations will require onboard equipment that provides adequate communication, navigation and surveillance data to allow for the development of separation standards. ADS-B over satellite comes into play at this juncture as its transponder can be put on all including balloons and UAVs. More so as the efficacy of radars are getting reduced at higher altitudes.

4.4. Airspace Class and Requirements: both horizontal and vertical

The need to accommodate a variety of operational types aircraft, spacecraft, rockets, balloons, platforms etc. with different requirements in shared airspace will disrupt the existing system of Air Traffic Control. This requires accurate and reliable detection and tracking information. The reliability of this information is a critical factor in developing separation standards. However, the operational characteristics, including speed, maneuverability, trajectory and mission requirements must also be considered in determining separation requirements. Here the role of ADS-B over satellite becomes very important as this technology will provide full coverage of all objects including aircraft in Airspace provided they are equipped with a transponder. These will include free balloons, rocket powered vehicles, ELV, vertically launched reusable vehicles, controlled reentering vehicles, controlled reusable first stage, reentering vehicle in free fall, high endurance UAVs and others.
ADS-B equipment transmits both GPS and barometric altitude information. This allows for validation of the altimetry error that is acceptable in applying existing separation standards. It is important to note that separation standards are based on the position of aircraft and other objects relative to one another, not absolute position. Therefore it is important to ensure that the quality of data received from each airspace user is comparable. It has been confirmed that ADS-B altitude information is superior to barometric and is preferred over barometric even at a height of International Space Station

4.5. ADS-B in Outer Space up to ISS

Extending GNSS PNT right up to International Space Station(ISS) at about 350 km in the Space is actually being attempted. Once space grade receivers are placed on a satellite constellation, both CNS/ATM for both upper airspace and near outer space say up to ISS at a height of 350 KM from Earth, at least, seems feasible.
GNSS PTN is used by ISS to meet altitude determination and enables more precision for space flights operations like rendezvous and docking, station keeping, formation flying and GEO satellite servicing. Using GPS/GNSS on ISS and crew rescue vehicles is a challenge being looked into by scientists. The problems of blackouts and multipaths are yet another issue to be tackled. Once these issues are sorted out, it is hoped that ADS-B over satellite will be able to send signals to Earth of all moving objects up to that height, be they satellite or platforms, as long as they have an ADS-B compliant transponder on board.

When this is confirmed by trials then there will be full integration of Air Traffic Management between Airspace and near Space.

The major challenges for a crewed vehicle as at re-entry into the atmosphere remain. The GPS receiver has to go through the region referred to as the blackout region (between 82 to 67 km) where RF signals encounter a lot of disturbance due to the Earth's ionosphere. The RF signals received are much lower signal strength, and typically GPS tracking is reduced to less than 4 satellites for several minutes. The GPS receiver needs to be able to recover the GPS signal following blackout.

If we mandate all satellites and other space objects to equip with ADS-B, and all vehicles and platforms operations in and through upper airspace with ADS-B over satellite, it will enable all varieties of vehicles to perform more efficiently in upper airspace. ADS-B provides the ability to display these vehicles on controller workstations and provide surveillance data to spaceports and vehicle operators. Simply put, ADS-B works wherever it is placed, be that on an aircraft, or on a space vehicle, or on a stratospheric balloon, or on a remotely piloted system at any height. The technology follows the transponder. The transponder is the key, not the aircraft or the vehicle

4.6. International Space Station (ISS)

It may also be stated that GNSS PNT is already being used by International Space Station at a height of about 350 km to meet altitude determination and enables more precision for spaceflight operations like rendezvous and even docking, station keeping, formation flying and GEO satellite servicing. Further, by placing space grade receivers on a satellite constellation like Aireon, it will provide not only 100% global surveillance of all aircraft but also all other space vehicles (satellites, balloons, platforms etc) and space objects right up to 100 km and beyond. As receivers and antennas improve with quality the transmission of data will also improve.

The most important aspect for the reception of ADS-B messages in space are the receiving conditions for the 1090 MHz Extended Squitter signal on board the satellite. In comparison to ground based ADS-B surveillance with a range of up to 300 km maximum, the signal path between a LEO satellite orbiting at 820 km altitude and an aircraft is much longer, which results in a low signal level at the ADS-B receiver. Thus the Mode-S signals have to be detected nearly at noise level by a correlation process. The quality of these space based receivers the becomes very important

5. ROLE OF UNOOSA and ICAO

In 2007 the United Nations Office of Outer Space Affairs (UNOOSA) setup the International Committee on Global Navigation Satellite Systems (ICG) for compatibility/interoperability of GNSS systems and protection of GNSS spectrum, orbital debris mitigation, and orbit de-confliction. The Interagency Operations Advisory Group addresses strategic issues related to inter-agency interoperability, space communications, and navigation matters.

ICAO is another source of international governance of GNSS. Its Standards and Recommended Practices (SARPs) for GNSS are found in Amendment 76 to Annex 10 of the Chicago Convention and provide extensive guidance on the technical aspects of GNSS and its application in furtherance of international aviation safety.

Integration of commercial space users with aviation users is becoming highly desirable for the present but perhaps, necessary in future. Both ICAO and ONOOSA are seized of the need and have started meeting. Already three joint Conferences have taken place the last one in August of 2017.

Re-entering stages of a spacecraft (about 75 km), specially the new reusable type poses new challenges. For reusable launch vehicles, under operator control, there is a need to develop tactical separation standards between aircraft and spacecraft.

The objective of full integration of the aviation and rocket based system to maximize airspace efficiency right up to higher levels of airspace (100 km) and beyond into LEO is very desirable. In such a scenario, a space vehicle operation is a fully integrated with airspace user providing the space operator with assured access to the Airspace and does not disrupt the aviation operators.

5. CONCLUSIONS

The accommodation of diverse modes using shared airspace is now growing and integration of ATM and STM is desirable. The technical solution to this problem is becoming more and more obvious. Mandate the use of ADS-B-over satellite for all users of airspace and near-space.

Finally, integration of upper airspace and near-space will depend upon the coordination between ICAO and UNOOSA. Since ICAO and UNOOSA have already started a dialogue on the subject, it is hoped that the results will be positive.

6. REFERENCES

1. Convention on International Civil Aviation, signed at Chicago on 7th December 1944, known as Chicago Convention.

2. Treaty on Principles Governing the Activities of States in the exploration and use of Outer Space, including the Moon and Other Celestial Bodies, opened for signatures at Moscow, London and

Washington, on 27th January 1967 known as Outer Space Treaty.

3. Stilwell, R., N. Demidovich, The Bridge to Space: CNS Technology for High Altitude operations, Integrated Communications, Navigation and Surveillance Conference (ICNS), 2017. Herndon, VA, USA, 18-20 April 2017.

4. GNSS and Sustainable Access to Space, Diane Howard, J.D., LL.M., DCL Embry-Riddle Aeronautical University.

6. Collins, M., First Space-Based ADS-B Satellites in Orbit: Launch Customer Nav Canada has not set Mandate. www.aopa.org/news-andmedia/all-news/2017/january/18/first-space-based-adsb-satllite-in-orbit

7. K. Werner, K., J. Bredemeyer, T. Delovski ADS-B over Satellite, Global Air Traffic Surveillance from Space. http://elib.dlr.de/91303/1/Final-Paper_ESAV2014-AoS.pdf

8. Spacecraft Reentry Basics: The Aerospace Corporation: www.aerospace.org/cords/all-aboutdebris-and-rentry/spacecraft-reentry/

9. ADS-B over Satellite: eoPortal; https://directoru.eoportal.org/web/eoportal/satellite-missions/ a/ads-b

10. Vigier, C., Automatic Dependent Surveillance Broadcast (ADS-B) Surveillance development for Air Traffic Management, Airbus FAST Magazine http://www.jmargolin.com/sense3/ref14_FAST47_5-adsb.pdf

SAFETY REGULATION FOR UK LAUNCH

Andrew Kuh[(1)], Pete Lindsay[(1)], Amy Taylor Saunders[(1)]

[(1)]*UK Space Agency, Polaris House, Swindon, United Kingdom. Email: andrew.kuh@ukspaceagency.bis.gsi.gov.uk*

ABSTRACT

The UK is aiming to increase its share of the commercial spaceflight market by enabling competitive, commercial and safe spaceflight for UK-based satellite launch services and sub-orbital flights.

The UK Space Agency is working with the Department for Transport on the Space Industry Bill. Introduced to Parliament in June 2017, this Bill is intended to provide for regulations covering launch and associated activities from the UK. Safety is at the forefront; securing safety will be the primary duty of the regulator.

The Bill is subject to ongoing Parliamentary scrutiny and debate, but we expect that once enacted, the Bill will create a sustainable legal base for regulation of spaceports, ranges, suborbital spaceflight and launch to orbit – as well as in orbit operations, which are already regulated under the Outer Space Act (1986).

Designing a new regulatory model which protects safety and enables industrial growth represents both a significant challenge and a significant opportunity.

1. THE SPACE INDUSTRY BILL

The UK has been licensing satellite operations and overseas launch since 1986, but regulating domestic spaceports, ranges and launch operations is breaking new ground for us. As a newcomer in this arena, the UK has a significant challenge in designing proportionate regulation from scratch. Critically, in order to keep pace with the commercial pace of today's operators, regulators must develop new, responsive regulatory models. This creates a significant opportunity to create a proportionate and streamlined regulation and licensing framework with safety at its core.

Introduction of the Space Industry Bill is an important first step to enabling spaceflight activities in the UK, and a concrete indication to investors that the UK is serious about promoting growth in the space sector. The Bill itself aims to provide certainty and assurance to Parliament, regulators, industry and the general public on the Government's approach to regulating spaceflight, whilst having the flexibility to allow industry to grow. This reflects best practice for creating regulatory frameworks, whatever the industry: supporting the industry whilst protecting public safety and managing government risk. We aim to contribute to new global standards in space regulation, but cannot stand alone in this ambition and we are keen to work with likeminded partners.

1.1 General Approach and Principles

By working collaboratively and transparently with industry through a process of continuing engagement, we will demonstrate Government's commitment that the regulatory framework will be designed to take account of commercial practice, available technologies and future developments in the industry, whilst providing the necessary operational safety through existing best practice in comparable industries and the international space sector. This is, however, a novel and complex challenge, requiring detailed policy development and building internal expertise to then deliver the regulatory regime effectively.

Its novelty is twofold: firstly, for the UK, this activity has never been regulated before. The UK's directly comparable activities are mainly limited to the military sphere, which has separate mechanisms for assuring safety to that proposed here for private, commercial entities.

Secondly, our particular approach to allowing licensed entities to undertake critical safety roles is largely unprecedented in other countries. It will require a sophisticated approach to deliver sufficient regulatory oversight and assurance whilst allowing licensees the flexibility to decide the operational model which satisfies safety requirements and meets their commercial needs.

The complexity of this challenge is manifold, from technical, operational, organisational, regulatory and policy perspectives – all of which must be reflected in the legislation.

1.2 The Regulator

The Bill places duties and powers on the regulator for spaceflight and related activities. The regulator may be either the Secretary of State – which in practice, means a part of central government itself, or, in respect of some functions, persons appointed by the government – which in practice typically refers to arms-length public bodies.

The Government's intention is that the UK Space Agency (UKSA) and the Civil Aviation Authority (CAA) will carry out the functions of the regulator. Our intention is to appoint the CAA to license sub-orbital spaceflight activities, such as spaceplanes, and horizontal launch spaceports. The UKSA will license space activities and vertical launch spaceports – which will include launch of satellites into orbit.

The Health and Safety Executive will also play a regulatory role. The Health and Safety at Work Act (1974) will apply to the space industry in respect to safety of employees and the general public. The provisions in the Space Industry Bill relating to risk to health and safety only address issues particular to the risks involved in spaceflight; and not covered by the 1974 Act.

This ensures that the bodies with existing expertise lead on the relevant aspects of regulation. The licensing process will be developed jointly by the relevant regulators with clear lines of responsibility being drawn and agreed in advance.

The Bill also establishes the main regulatory functions, through which we aim to enable safe and sustainable commercial spaceflight.

2. REGULATION AND LICENSING

The regulations provided for in this legislation are also being developed. The Government intends to lay before Parliament a set of Statutory Instruments, and accompanying guidance material for industry, in the coming years (pending successful passage of the Space Industry Bill.)

2.1 Duties of the Regulator

The principal duties and powers of the regulator are set out in clause 2 of the Space Industry Bill. The primary duty of the regulator is to secure public safety. There are however other factors to which the regulator must also have regard:

(a) the interests of persons carried by spacecraft or carrier aircraft;

(b) the requirements of persons carrying out spaceflight activities;

(c) the interests of any other persons in relation to the use of land, sea and airspace;

(d) the requirements of persons with interests in property carried by spacecraft;

(e) any environmental objectives set by the Secretary of State;

(f) the interests of national security;

(g) any international obligations of the United Kingdom.

In other words: we wish to ensure that the regulator takes a rounded, holistic view – while nonetheless ensuring that safety remains the *primary* duty, which may not be 'traded off' against these other considerations.

2.2 Regulatory Functions

The main functions of the regulator are:

(a) licensing of spaceflight operations, spaceports, range control service providers;

(b) monitoring and enforcement;

(c) making orders relating to land;

(d) issuing directions;

(e) issuing guidance.

2.2 Licensing

Licensing will be critical in an environment where we expect all elements of a launch to be provided on a commercial basis. It is through licensing that the regulator will gain assurance that the correct protocols and capabilities are in place to secure safety.

There will be five types of licence: for a launch vehicle operator; for a spaceport; for a range provider (termed *range control service provider* in the Bill), for sub-orbital operations and for in-orbit operations.

The licensing process will be streamlined to offer operators as simple and navigable system as possible. To mitigate the potential challenge for applicants of having several regulators involved in any operation, we are developing a system which will offer a 'single point of entry' for licensing – we believe this will be welcome in the short term, and essential in the medium-to-long term as the frequency of launches increases.

2.3 Launch Vehicle Operator Licence

The launch license will consider risk at each stage of the launch process: pre-launch (suitability and worthiness of the launch vehicle, sub-assembly and assembly operations); launch (preparation, planning and procedure); and post-launch (retrieval, comparison of actual activity to that intended). Assessment of each stage will be conducted to the key principle of ensuring the risks to third party safety are as low as reasonably practicable (ALARP) – a conventional method of managing risk used across industrial sectors in the UK.

The key documentation to be considered here will be the Safety Management Case which will set out:

- agreed operator readiness and mission-rules criteria are acceptable to the regulator;
- safety limitations and safeguards, weather and climactic conditions suitable for flight and landing, adherence to planned trajectories and/or orbital parameters, serviceability of the spacecraft, trackability and mitigation of on orbit collision risk);
- Duties, responsibilities and authority of the launch vehicle and satellite operator;
- Duties, responsibilities and authority of the person responsible for range control service provision;
- Mandatory accident and safety occurrence reporting; emergency response plan;
- Minimum capability requirements for traffic surveillance, spacecraft tracking and communication, segregation from other traffic, orbital parameters, collision avoidance, flightpath clear of populated areas, etc.

A launch license will only be granted where all areas have been satisfied. There is a means by which to allow exemption or exceptions in the license (e.g. to take account of international agreements). Each element of the process will still be followed in such circumstances, but may be more simply assured.

2.4 Spaceport Licence

The intent is to be able to issue a spaceport licence in advance of a launch vehicle operator being identified. This may be commercially advantageous to a licence holder to help attract potential launch vehicle operators to that location. The issue of the licence does not mean that *any* launch vehicle would be suitable to operate from that location. It does mean that an applicant has demonstrated, *in principle*, a type of launch vehicle may be suitable to operate from that location. The licence would be issued limiting the use of the spaceport to a specific launch vehicle type as appropriate. If the spaceport operator wishes to operate another type of vehicle, an application will be needed to add that to their licence.

Awarding of a licence will be based around factors such as: individual risk analysis (including location, proposed operating vehicles, proximity to population centres etc.); assessment of physical infrastructure; ground safety analysis; suitability of the operating company; employment of suitably qualified and experienced personnel and any other requirements that may be set out in the licence conditions.

Specific licence conditions with regards to safety will include factors such as: management of the risks posed by spaceflight activity through a Safety Management System (SMS); implementation of an emergency response plan for space operations; provision of rescue and fire-fighting commensurate with the risk to people and property; implementation of an explosive site plan, and configuration of the site in accordance with the plan.

A licence when awarded is likely to be granted in perpetuity, but subject to regular review and assessment that the operator continues to meet conditions therein.

2.5 Range Control Service Provider Licence

We intend to license range control as a *provision of services* – emphasising a) that the entity (or entities) operating a range is ultimately providing a service (or services) to the launch vehicle operator; b) that range control consists of many aspects, each of which is considered a *range control service*, and these services may be provided by one or several entities in respect of any one launch activity.

A range control service provider will be licensed to monitor and track restricted areas to allow spaceflight activities to be conducted safely. The licensing will allow for the range control service provider to be a separate entity than the launch vehicle operator, or for the two entities to be the same legal person – though in the latter case, conditions to secure independence of operations would be required to manage any conflicts of interest (for example, through providing for separate chains of command, and identification of certain roles which may not be performed by the same individuals). The licence conditions will require that the operator can safely and securely undertake:

- Identification of the range, including defining exclusion, restriction and warning zones for specific missions;
- Co-ordinating arrangements for activation and operation of the range;
- Ensuring appropriate notifications are put in place;
- Surveillance of the range area;
- Tracking of the launch vehicle during flight;

- Communication of any failure to comply with range safety procedures.

The primary focus for the UK government is how to regulate the provision of ranges in such a way that safety is assured. We will remain open to different organisational models and technical solutions for how range services may be provided, as long as safety critical requirements are adhered to.

3. THIRD PARTY RISK ASSESSMENT METHODOLOGY

3.1. General Approach

Critical to assessing safety of third parties is implementing a coherent and robust risk assessment methodology. We will build on the approach of other countries whilst reflecting the unique operating environment of the UK and making the most of the UK's strong heritage in aviation regulation, and health and safety in other sectors. We intend to align to existing practices from other countries where possible, since operators considering launching from the UK may already comply with these.

In particular, we have used the US Federal Aviation Authority Office of Commercial Space Transportation (FAA-AST) approach indicated in the *Flight Safety Analysis Handbook* as a starting point to develop a UK-specific third party risk assessment methodology – recognising that the geographical and legal context is somewhat different in the US to the UK.

Our intended methodology will include five fundamental steps:

- Identification of hazards associated with launching rockets or spaceplanes;
- Calculation of failure frequencies;
- Identification of areas which could be affected by failure at any point in the launch and flight;
- Estimation of the consequences of a failure;
- The risk of fatalities in these areas.

This outline methodology is not intended to be legally prescriptive: it will be recommended in guidance to operators, rather than mandated as a precondition for licensing. Following the method, however, should allow the individual risk to be calculated with an understanding of the approach applied. An assessment could be justified in using alternative methods, but these methods would need to be fully explained to provide the rationale behind the approach.

3.2 Individual Risk

There are some significant differences in our proposed methodology to that of the FAA-AST. These are not fundamental, but will require those assessing risk to follow a slightly different approach. The most significant of these is that the level of risk to the uninvolved general public (third parties) will be expressed in terms of *individual risk* (IR) as opposed to a collective risk expressed as 'expected casualties' (Ec) – IR meaning the individual deaths per year in zones around the spaceport sites and launch trajectories. IR is the conventional way of assessing risk in other industries in the UK, and so gives a useful point of comparison and consistency with other regulations.

Individual risk calculations will require the use of failure *frequencies* as opposed to failure *probabilities*, as used for the Ec calculation – which naturally introduces a temporal dimension to the calculation.

Further, we propose the calculation of probabilities of individuals being affected by the incident when assessing the IR, as opposed to population numbers affected by the incident for the Ec calculation.

Finally, the Ec methodology considers casualties rather than more specifically fatalities, which is the level of harm reflected in UK risk criteria. However, given the uncertainty over many of the calculations in this process, the use of US methodology derived casualty areas might add a 'safety margin' to the UK calculations.

One of the key points emerging from our studies in this area is that without reliable test data on which to base calculations, there will be a large degree of uncertainty in any calculations. The risk can nonetheless be managed by taking the most conservative estimates and ensuring that appropriate measures are in place to reduce the risk to as low as reasonably practicable (the most obvious measure being that one conducts launches in a remote area.)

4. CONCLUSION

Through a combination of best practice (both from countries with more mature launch capabilities and from other comparable industries), and an innovative, proportionate approach emphasising safety, the UK government will introduce flexible, proportionate regulations to secure the safety of spaceflight activities.

PROCEEDINGS OF THE 9TH INTERNATIONAL SPACE SAFETY CONFERENCE

SESSION 08 DESIGNING SAFETY – I

SAFETY VERIFICATION OF SOLAR ARRAY DRIVE ASSEMBLY STRENGTH BASED ON THE MISSION PROFILE

ZHU Xinggao, REN Liming, CHEN Fengxi

China Astronautics Standards Institute, Beijing 100071, China

ABSTRACT

It proposed a methodology of strength safety design and verification that can reveal structure response in the mission profile. Solar array drive assembly for example, A finite element model of solar array drive assembly is established using modern digital simulation software. The model is enhanced effect by vibration swept test. Random vibration and thermal analysis are implemented. The results make a reasonable judgment. It shows that the maximum of the response of the structure are all in the safety margin. Strength safety design of solar array drive assembly is reasonable.

1. INTRODUCTION

During the whole satellite life, the spacecraft has experienced severe vibration environment during the transportation, launching and flight of the ground. The random vibration excitation consists of the thrust pulsation caused by the unstable combustion of the launch vehicle engine, the unbalanced rotation of relevant equipment and the resonant frequency coupling between the structure and the system. The thermal environment consists of the external heat flow and the internal heat source from motor, disk ring body and other components.

The random vibration environment is an important consideration in the process of product development. The mission in spacecraft structure design is very important that includes determining random load, calculating structure response and optimizing structural safety design. In vacuum, the structure become worse and worse because of the poor heat dissipation conditions. High temperature make the non-metallic materials drop more excellent performance. The extreme high and low temperature lead to a lot of thermal stress of metal parts and damage the structure In addition, there is a large temperature gradient between the rotating part and the fixed part. The mechanism may cause the air gap or even to be stuck.

The finite element analysis is an effective method to predict the dynamic characteristics of the structure. The vibration form of the structure can be judge in the product design stage, which is beneficial to the improvement and optimization of the structure. The strength design verification is great significance that can reduce the product development cycle and cost of production. It can ensure the reliability of the spacecraft and the selection of the testing scheme of reliability qualification

2. PROPOSED METHOD

2.1 Methodology

Solar Array Driving Assembly (SADA)is a typical space mechanical and electrical product. There are many structures in the bolt connection and special parts which is the aerospace materials. The structural characteristics lead to structural analysis of the model damping be no determinacy. It is difficult to provide accurate modal damping parameters in the design stage. Therefore, a methodology of strength safety design and verification is proposed as Figure 1.

The modal damping parameters of the finite element

model are modified by the vibration swept test. The random vibration analysis and the thermal analysis of the modified model are carried out. The simulation results can verify the structural design safety, at the same time it can guide the implementation of experiment test.

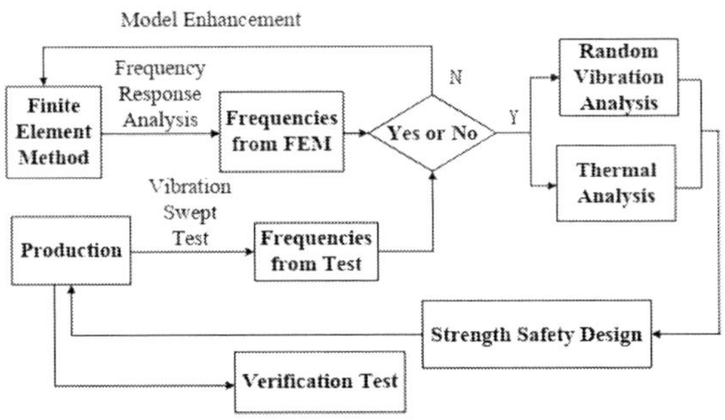

Figure 1 Methodology of strength design safety verification

2.2 Establishment of Finite Element Model

2.2.1 Establishment of three-dimensional model

SADA is a typical space mechanical and electrical product, it consists of drive shaft, a front bearing assembly (bearing and front bearing), a rear bearing assembly (bearing and rear bearing), a motor assembly (motor stator and motor rotor), a resolver assembly (rotary stator and rotary rotor), a slip ring assembly (disk ring body and brush tray), shell. (main shell, rear shell, brush tray and back cover) and other components.

Figure 2 Solar Array Drive Assembly

Figure 3 Three-dimensional model

The motor receiving the signal from drive controller can output the rotating speed and torque. SADA drives solar wing and plays a function role in capturing, tracking, stopping and keeping orientation. SADA can transmit the signal and power between solar wing and satellite.

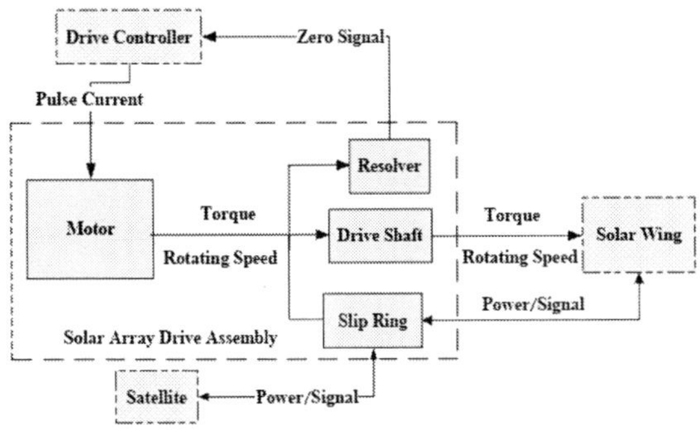

Figure 5 The working principle of SADA

2.2.2 Characterization of the model

According to SADA structural characteristics, the hexahedral element is be selected. The structural of SADA is discredited into a series of units which is jounced at the node adjacent units through the tie connection. It can transmit force and torque by the bolts which are connected. There are two connecting flanges in SADA, one is on the shell of the product connecting on the star, the other is on the end of the SADM output shaft connecting the solar wing. when the SADA turns, the solar wing is driven, so the SADA main housing 12 bolt holes are fully constrained, SADM finite element discrete model is shown in Figure 6.

Figure 6 Finite element model

In the parts of SADA, the material of drive shaft is TC4, the material of motor assembly and resolver assembly are 9Cr18, the material of shell (main shell, rear shell, brush tray and back cover) are 2A12. the material of disk ring body is Polyimide, be shown in Table 1 and Table 2.

Table 1 Material Properties(Mechanics)

Material	Elastic Modulus (MPa)	Density (kg/m^3)	Poisson's ratio	Allowable Stress (MPa)
9Cr18	2.06E5	7750	0.30	980
TC4	1.09E5	4400	0.33	895
2A12	7.1E4	2800	0.33	420
Polyimide	5.03E3	1380	0.37	400

Table 2 Material Properties(Thermal)

Material	Thermal conductivity W/(m K))	Thermal expansion coefficient (α(/K))	Specific heat capacity J/(Kg K)	Allowable Stress (MPa)
9Cr18	15.2	1.66E-5	460	980
TC4	117	2.36E-5	920	895
2A12	7.2	8.8E-6	565	420
Polyimide	0.3	5E-5	1100	400

2.2.3 Model Enhancement

There is a vibration swept test, the acceleration frequency is 0.5g (g is the acceleration of gravity, the value is 9.8m/s^2), the frequency range is 10~2000Hz.The base frequency of the SADM structure is 318.6Hz and the acceleration is 14.962g, therefore, the amplification factor is:

$$Q=14.926g/0.5g=29.924$$

Because the modal damping ratio and the dynamic amplification factor are reciprocal to each other, a modal damping ratio is 0.033. The finite element model needs to be modified according to the experimental data. In the simulation results of SADA, frequencies from FEM is 310Hz and acceleration is 14.592g. The results of test and FEM is shown in table 3. It is verified that the modified model is an effective finite element model, which can be followed by random vibration analysis and thermal analysis.

Table 3 The Results of Test and FEM

	Test	FEM	Error (%)	Model Verification Result
Frequency(Hz)	318.6	310	2.7	√
Response(m/s^2)	14.962	14.592	2.47	√

3. NUMERICAL SIMULATION

3.1 Random Vibration Analysis

The random vibration excitation consists of the thrust pulsation caused by the unstable combustion of the launch vehicle engine, unbalanced rotation of relevant equipment and the resonant frequency coupling between the structure and the system. Random vibration analysis confirms to the nation standard, reveal structure response and supply a reference to design group. The random vibration excitation of SADA is determined by the mechanical environmental test. It is shown in Figure 7.

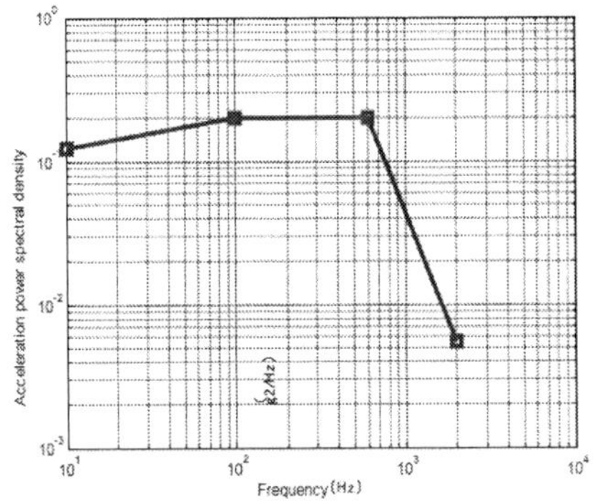

Figure 7 Random vibration condition of SADA

The results from random vibration analysis of SADA is shown in table 4. The stress cloud and acceleration response curves are shown in figure 8 and figure 9.

Table 4 The results from random vibration analysis of SADA

	Frequency (Hz)	Stress (MPa)	Acceleration (m/s^2)	Magnification Factor
X	760	59.43	742.86	5.92
Y	560	10.45	783.67	6.24
Z	760	58.76	796.94	6.35

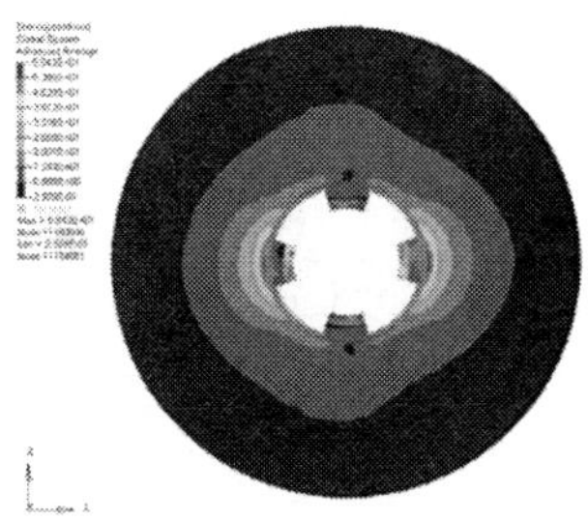

Direction X

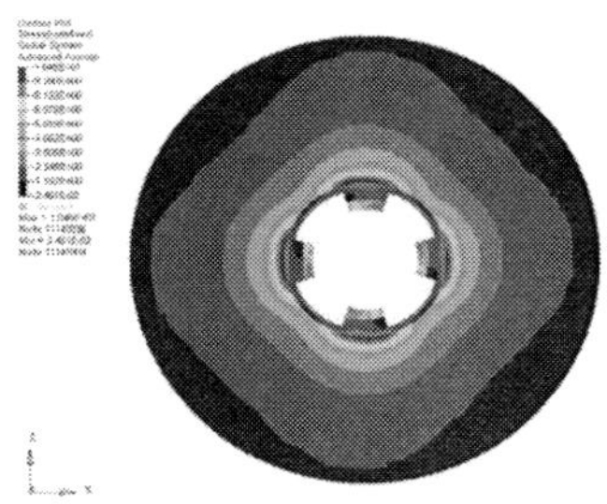

Direction Y

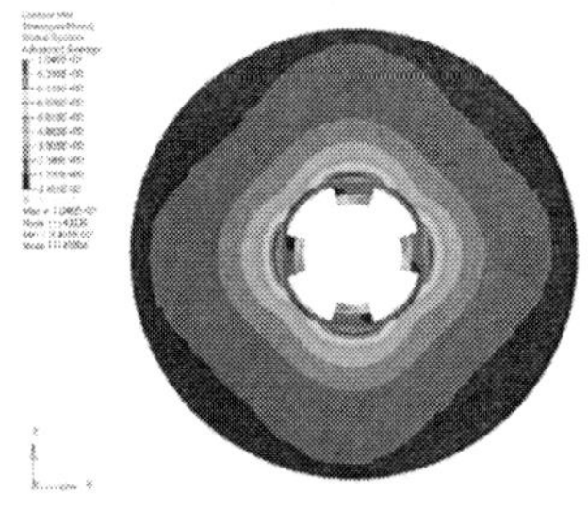

Direction Z

Figure 8 The stress cloud of SADA

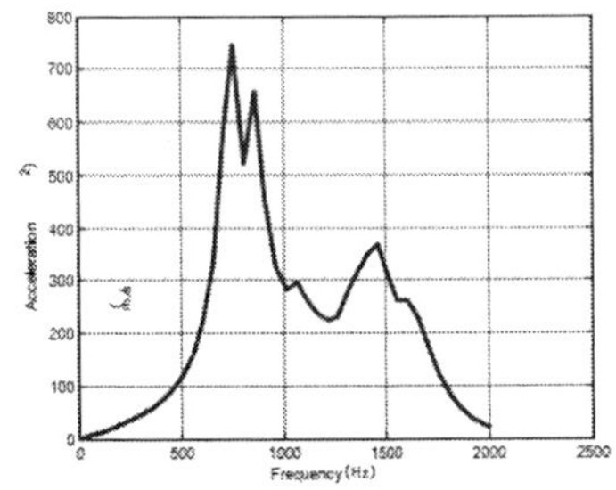

Direction X

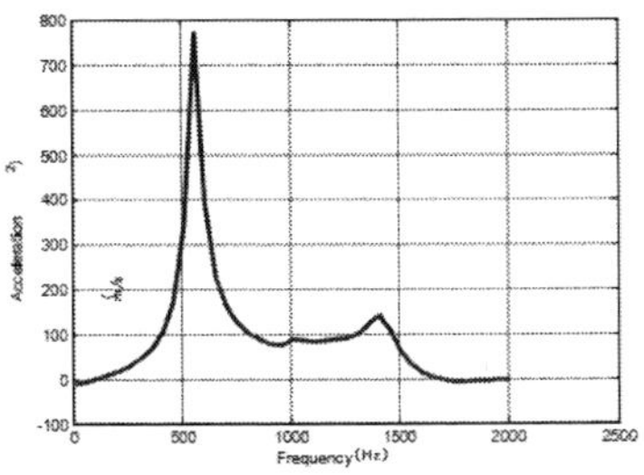

Direction Y

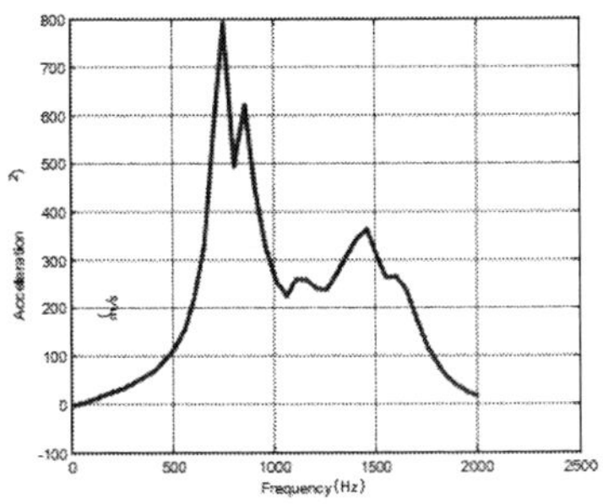

Direction Z

Figure 9 The acceleration response curves of SADA

The maximum stress and acceleration response occur at the disk ring body. X is the direction of the maximum stress. The maximum stress value is 59.43MPa which is less than the allowable strength of the material. Z is the direction of the maximum acceleration. The RMS of the acceleration is 796.94m/s^2. Magnification factor is 6.35 which is less than 10. The structural strength is reasonable.

3.2 Thermal Analysis

The thermal excitation consists of the external heat flow and the internal heat source from motor, disk ring body and other components. The thermal analysis confirms to the nation standard including thermal cycling analysis and thermal vacuum analysis. The is a high and low temperature cycle. The thermal vacuum is a steady state condition at high and low temperature. The high temperature is 75°C and low temperature is -35°C. They are shown in Figure 10

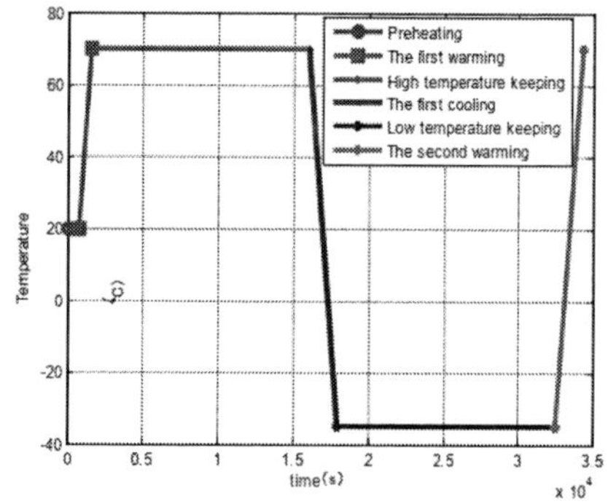

Thermal cycling

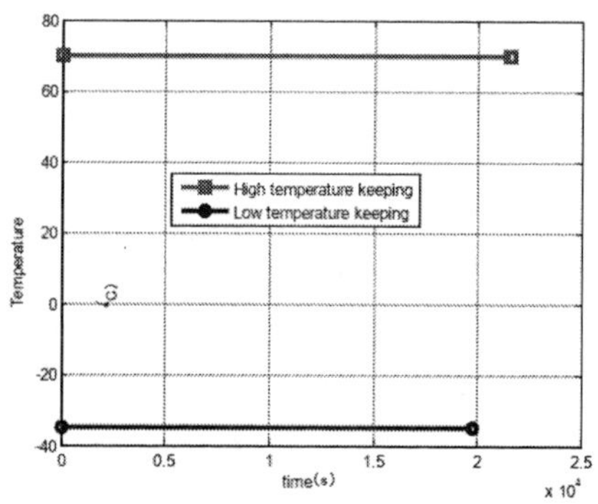

Thermal vacuum

Figure 10 The thermal condition of SADA

The thermal distribution, thermal stress and thermal deformation of the thermal cycling analysis are shown in figure 11.

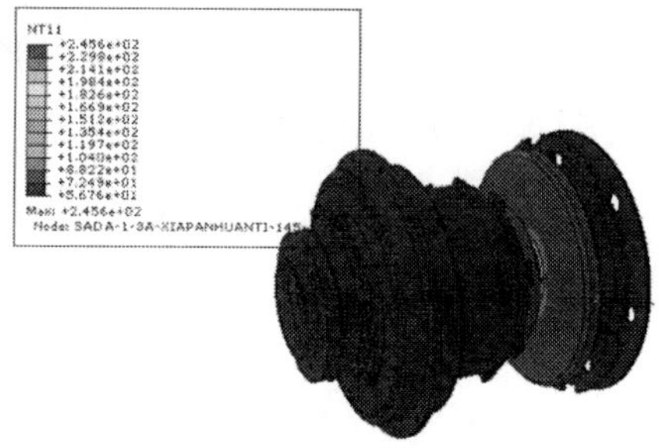

(a) Thermal distribution

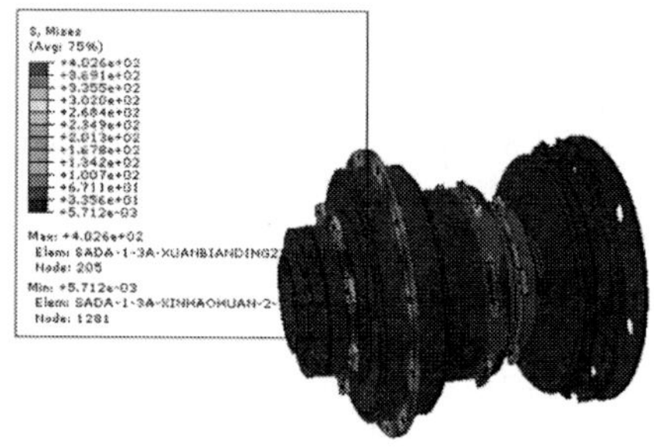

(b) Thermal stress

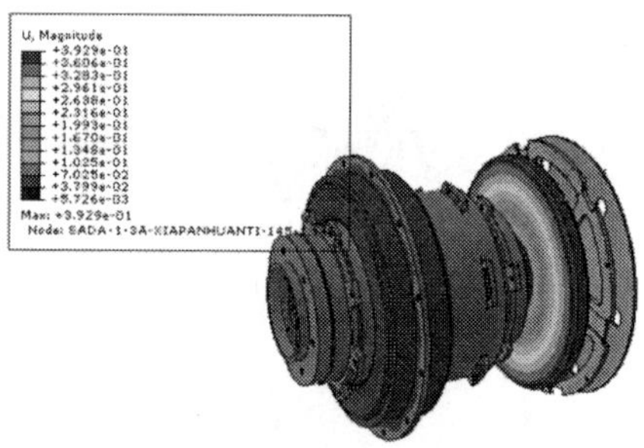

(c) Thermal deformation

Figure11 The results of thermal cycling analysis

During the thermal cycling, the highest temperature occurs at the disk ring body. The high temperature is 245.6°C, and the low temperature is 148.5°C.

The resolver of maximum thermal stress is 402.6MPa, which occurs at high temperature. The bearing of maximum thermal stress is 372.5MPa, which occurs at low temperature. Both are less than the allowable stress of the material.

The maximum thermal deformation occurs at the disk ring. The maximum thermal deformation of the disk ring body is 0.3929 mm.

The thermal distribution, thermal stress and thermal deformation of the thermal vacuum analysis are shown in figure 12.

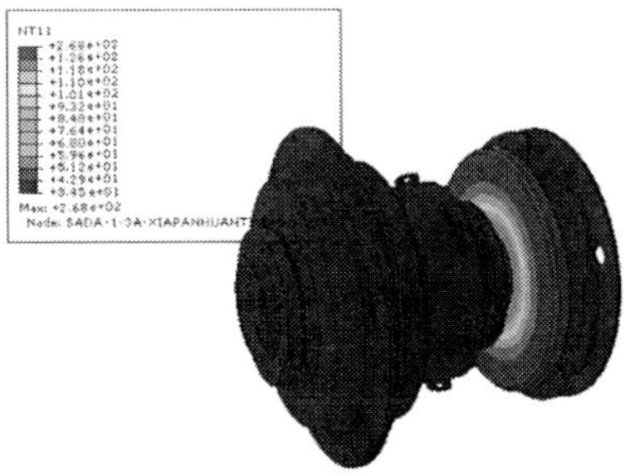

(a) Thermal distribution

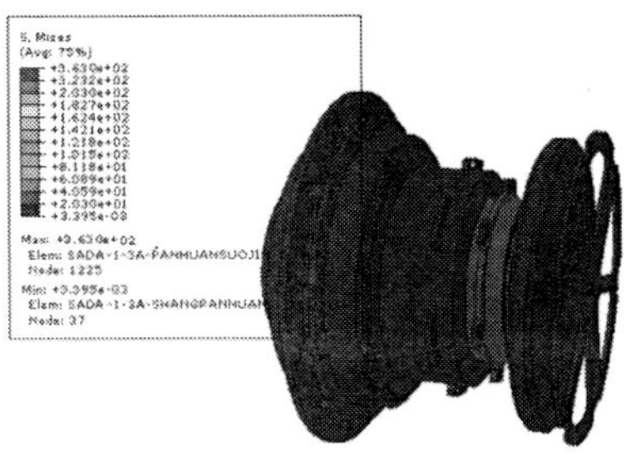

(b) Thermal stress

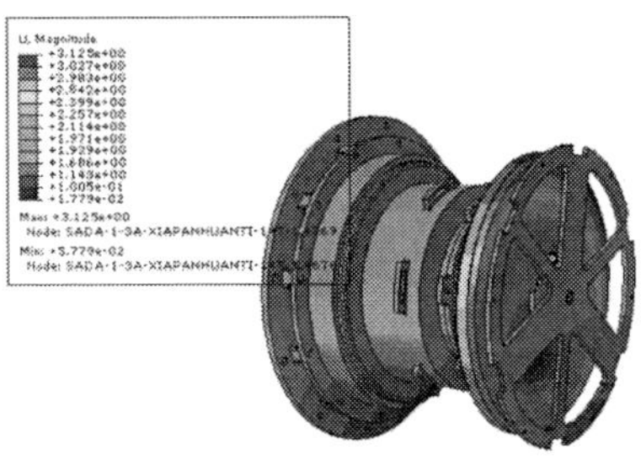

(c) Thermal deformation

Figure12 The results of thermal vacuum analysis

The highest temperature occurs at the disk ring body. The temperature of the disk ring body is 268 °C at 70 °C. The temperature of the disk ring body is 183 °C at -35 °C.

The maximum thermal stress occurs at the disk ring body. The thermal stress of the disk ring body is 363MPa at 70°C. The thermal stress of the disk ring body is 262.1MPa at -35°C. Both are less than the allowable stress of the material.

The maximum thermal deformation occurs at the disk ring body. The thermal deformation of the disk ring body is 3.125 mm at 70°C. The thermal deformation of the disk ring body is 2.444 mm at -35°C.

4. CONCLUSIONS

In this paper, a methodology of strength safety design and verification is proposed. The finite element model of SADA is established. The model is enhanced by vibration swept test. The structure response simulation of Solar Array Drive Assembly is calculated by using finite element analysis software. It show that the risk ring body is the weak part of SADA, which is found by random vibration analysis and thermal analysis. Strength safety design methodology is verified and proved to be reasonable. The methodology of strength safety design supplies a reference to design group that can reduce the design cycle and cost. It can also provide some reference for the development of similar products. The modeling precision depends heavily on the computer performance and the expertise of the analyst. It is about that finite element model need to be established more exactly, and more and more experiment data need to be conducted to make the model more correct

5. REFERENCES

[1] MA Xing rui. The Researching Evolvement of Spacecraft Deployment and Driving Mechanism Journal of Astronautics,2006,27(6):1123-1126

[2] YU Dengyun Mechanism Technology of Spacecraft [M]. Science and Technology of China Press,2011

[3]. HOU Zengqi Thermal Control Technology of Spacecraft [M]. Science and Technology of China Press,2008

[4] YANG Baoning. Engineering Methods for Determining Quasi-static Limit Load for Structure under Random Vibration [J]. Spacecraft Engineering, 2006,15(3):33-37

[5] DONG Deyi. Random Vibration Analysis and its Verification for Large Aperture Mirror Component Journal of Vibration and Shock, 2011,30 (11): 74-78

LESSONS LEARNED FROM NASA SPACE LAUNCH SYSTEM (SLS) EXPLORATION MISSION 1 (EM-1) PAYLOAD SAFETY REVIEW PANEL (PSRP) FOR SECONDARY PAYLOADS

Takashi Goto[(1)], Masami Miki[(1)], Masako Kikuchi[(1)], Koji Oga[(1)], Manami Nogami[(2)], Toshinori Ikenaga[(2)], Ryu Funase[(3)], and Tatsuaki Hashimoto[(4)]

[(1)]Japan Manned Space Systems Corporation (JAMSS), Tsukuba Center Inc, D, 2-1-6, Sengen, Tsukuba, Ibaraki, 305-0047, Japan, Email:goto.takashi@jamss.co.jp
[(2)]Japan Aerospace Exploration Agency (JAXA),Tsukuba Space Center 2-1-1, Sengen, Tsukuba, Ibaraki, 305-8505 Japan
[(3)]University of Tokyo, Tokyo, Japan
[(4)]Institute of Space and Astronautical Science, JAXA, Sagamihara, Japan

ABSTRACT

NASA has been developing a new heavy expendable launch vehicle called the Space Launch System (SLS). For the first SLS flight, JAXA and the University of Tokyo have been cooperatively assembling two CubeSats as secondary payloads: OMOTENASHI (Outstanding MOon exploration TEchnologies demonstrated by NAno Semi-Hard Impactor) and EQUULEUS (EQUilibriUm Lunar-Earth point 6U Spacecraft). On SLS program, unique requirements for safety review process and safety technic for secondary payloads, with which OMOTENASHI and EQUULEUS must comply has been provided. As one of the preparation for NASA PSRP, we, JAXA Safety and Mission Assurance (S&MA), compared SLS safety requirements with JAXA's those. Furthermore, we have been continuously discussing safety requirements with NASA SLS Payload Safety Review Panel (PSRP). This paper will explain the summary of safety requirements for secondary payload of SLS and show our safety activities.

1. INTRODUCTION

NASA has been developing a new heavy expendable launch vehicle called the Space Launch System (SLS) (Figure 1). It will launch astronauts to an asteroid and eventually to Mars by the agency's Orion spacecraft in missions. The first SLS launch, Exploration Mission-1 (EM-1), has been planned in 2018 with an uncrewed Orion spacecraft to a stable orbit beyond the moon and bring it back to Earth to demonstrate the integrated system performance of the SLS rocket and Orion spacecraft and ground support teams prior to a crewed flight. Furthermore, for EM-1, SLS will also launch 13 secondary payloads to utilize the excessive capability.

Figure 2 illustrates the elements of the SLS first flight configuration. For EM-1, secondary payloads are mounted inside the Orion Stage Adaptor located between the Interim Cryogenic Propulsion System (ICPS), and the Orion Multi-Purpose Crew Vehicle (MPCV). Secondary payloads are stowed in deployers called dispenser, and dispensers are mounted on the brackets located around the inner surface of the Orion Stage Adaptor (figure 3). It is possible that at least some of these bracket can support a larger 12U payload, but a 12U payload will not be flown on EM-1.

Figure 1. Conceptual drawing of the SLS first flight configuration launch. [1]

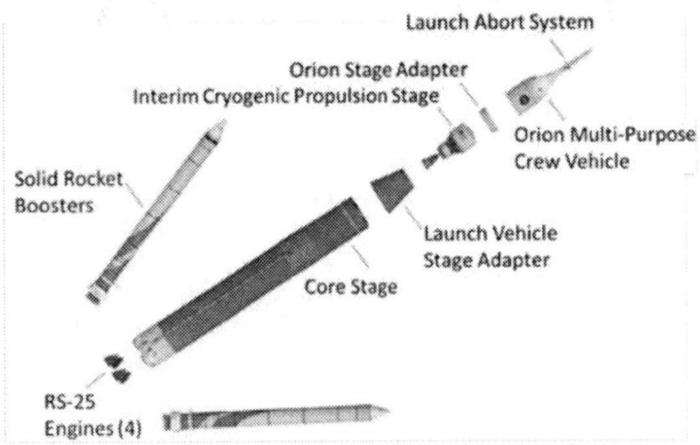

Figure 2. SLS EM-1 configuration. [1]

2. SECONDARY PAYLOADS

For the first SLS flight, JAXA and the University of Tokyo have been cooperatively assembling two 6U CubeSats as secondary payloads: OMOTENASHI

(Outstanding MOon exploration TEchnologies demonstrated by NAno Semi-Hard Impactor) (Figure 4) and EQUULEUS (EQUilibriUm Lunar-Earth point 6U Spacecraft) (Figure 5).

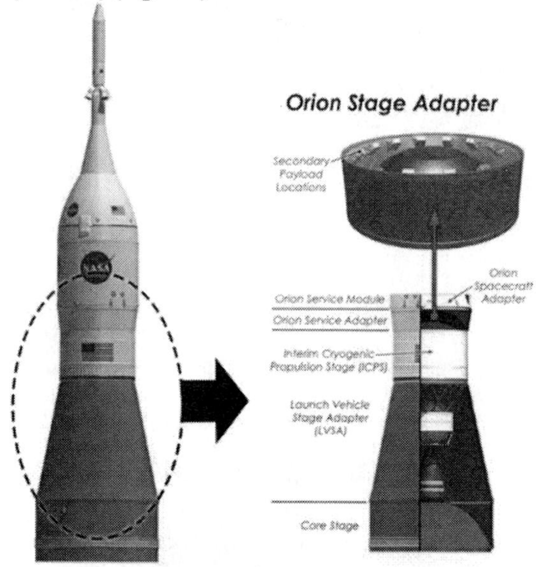

Figure 3. Location of secondary payloads for SLS EM-1. [1]

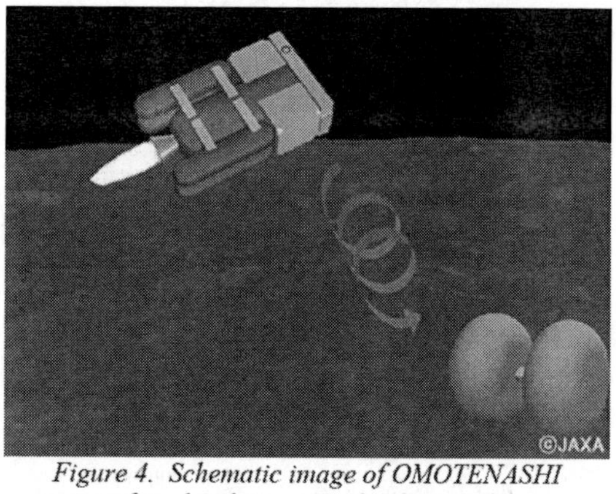

Figure 4. Schematic image of OMOTENASHI spacecraft and airbag system landing on the moon.

Figure 5. Schematic image of EQUULEUS.

After deployment from dispenser, OMOTENASHI will control its orbit to lunar impact orbit by gas-jet thrusters. Several minutes before the lunar impact, the spacecraft decelerates by solid rocket motor. After the solid motor is burn out, Surface Probe will be deployed by the deployment system using Non-Explosive Actuator (NEA) and its semi-hard impact to the moon with shock absorber using airbag system will be performed.

EQUULEUS has two wings of solar array panels that have a one-axis gimbal. For the launch configration, the solar array panels are folded to satisfy the envelope of the dispenser. After the deployment of the EQUUELUS, the panels will be automatically deployed and EQUULEUS fly to the Earth-Moon L2 libration orbit by Lunar flyby and water resist jet propulsion system consisted of pressurizing tank, water bladder, Vaporization chamber and so on.

3. COMPARISON OF SAFETY REQUIREMENT

In SLS program, unique requirements for safety review process (SLS-PLAN-217 "Space Launch Systems Plan Exploration Mission -1 Secondary Payload Safety Review Process")[2] and safety technics (SLS-RQMT-216A "Space Launch System Program Exploration Mission 1 Safety Requirements for Secondary Payload Hardware")[3] for secondary payloads, with which OMOTENASHI and EQUULEUS must comply has been provided. Furthermore, their safety assessments have to be reviewed by NASA Payload Safety Review Panel (PSRP) for SLS secondary payloads, and those will be conducted based on each design phase.

Meanwhile, JAXA System Safety Review Panel (SSRP) usually reviews small secondary payload of HII-A based on JMR-002B[4]. JMR-002B is the JAXA SSRP requirement including technics and process for rocket payload. Therefore, this is the first time for JAXA Safety and Mission Assurance (S&MA) to review the hazard analysis based on SLS safety requirements. For further understanding of SLS requirements, we compared SLS safety requirements to JMR-002B. Comparing and clarifying the differences between both requirements make it possible to understand the big picture of SLS safety requirements and the intent of SLS requirements easier. The comparison results of NASA PSRP requirements and JAXA SSRP those are shown below.

3.1. Safety Process Requirement

The summary of scope comparison of the safety review process requirements of NASA PSRP and JAXA SSRP is shown in Table 1. There is not so much difference in the scope between both requirements. As for its coverage, launch site operation is in the scope in JMR-002B, however, is out of the scope in SLS-PLAN-217A, while Kennedy Space Center (KSC) requirement is applied to launch site operation for secondary payload and reviewed by KSC Ground Safety Review Panel (GSRP).

Then, as for the review method, there is not either much difference between SLS-PLAN-217A and JMR-002B as shown in the followings. The safety assessments of both panels are accomplished by using:

- A phased safety process (phases 0, I, II, III) that corresponds to the payload conceptual, preliminary, critical design, and final acceptance review phases.
- Hazard Reports (HR) (and safety waivers if applicable)

Category	SLS-PLAN-217A	JMR-002B
Apply to	Secondary Payload, dispenser	Payload, GSE
Hazard to	SLS EM-1 (SLS, Orion)	Launch Vehicle, Person, Public, Third Parties Private Property, Launch Site Facilities, environment
coverage	T-0 until Deploy +15 sec (Launch site operation is applied other requirement document)	Launch Site Operation, T-0 until Separation

Table 1. Summary of comparison of scope

3.2. Safety Technical Requirement

The definition of hazard severity of SLS-RQMT-216B is shown in Table 2. Comparing it with JMR-002B, it is found that there is not much difference in hazard severity between both requirements.

Furthermore, there is not much difference in basic policy of system safety including priority of safety design and Failure Tolerance/ Design for Minimum Risk (DFMR) and so on between SLS-RQMT-216B and JMR-002B.

While, it is found that some SLS requirements are more stringent than JAXA's in regard to safety technics. In particular, SLS requirements require items which rarely exist in unmanned rocket payload requirement such as fracture control, battery and mechanical system. It seems that this is because SLS requirements could be established with assuming the manned mission including next flight (EM-2) and after. For example, as for structural design requirement, secondary payloads should be implemented fracture control because those are applied NASA Technical Standard requirements which are similar to International Space Station (ISS) those[10], although it is not required in JMR-002B. Additionally, as for the battery requirement, the ISS requirement referred in SLS-RQMT-216B should be applied to second payloads for all batteries usage.

As described the above, it is found that SLS requirements are more similar to ISS requirements rather than JMR-002B. Therefore, we found that we should make the most use of the experience regarding ISS safety review rather than expendable launch vehicle payload safety review because we have experiences obtained through safety review of not only Japanese launch vehicle (HII-A) secondary payloads but also ISS payloads.

Category		SLS-RQMT-216B	JMR-002B
Catastrophic hazard	Personnel	Loss of life or permanently disabling injury.	Death or severe injury of third-party (including permanent damage), death or severe permanent damage of personnel
	Facilities, Equipment, Assets	Loss of vehicle prior to completing its mission, or loss of essential flight/ground assets	Loss of or severe damage to public or private property, loss of system (launch vehicle, payload/GSE) or launch site facilities, and/or serious environmental damage.
Critical hazard	Personnel	Injury or occupational illness requiring definitive/specialty hospital/medical treatment resulting in loss of mission.	Minor injury of third-party, severe personnel injury
	Facilities, Equipment, Assets	Loss of ESD mission, condition that requires safe-haven, or major damage to essential flight/ground assets	Minor damage to public or private property, severe damage to system (launch vehicle, payload/GSE) or launch site facilities, and/or reversible severe environmental damage.

Table 2. Definition of hazard severity of SLS-RQMT-216B and JMR-002B

4. INDIVIDUAL DISCUSSION FOR SAFETY REQUIREMENT

While it is found that SLS requirements are similar to ISS requirements, we found that there are some issues in the requirements of SLS-RQMT-216 based on our experience, also, that we could perform safety assessment more effective by implementing safety analysis with integrated configuration having dispenser thanks to its flight configuration.

Therefore, we have discussed the issue with NASA to implement the appropriate safety assessment to small secondary payloads which are composed of a lot of COTS due to limited costs and schedules. This chapter shows some of the results of the discussions.

4.1. Fracture Control

Use the decimal system in Arabic figures for the numbering of headings and subheadings. Major (or section) headings are to be in capitals and bold.

4.1.1. Classification

As stated above, because NASA-STD-5019A referred to SLS-RQMT-216B requires fracture control of secondary payloads, secondary payloads have to be performed fracture control classification for each component based on the requirement. If general approaches such as fail-safe or low-risk are selected as fracture control, a lot of verification corresponding to each approach are required and the volume of workloads may increase. However, since secondary payload is basically stowed in dispenser during launch phase, we considered that secondary payloads could use the dispenser as one of the control. Therefore we discussed with NASA that secondary payloads could employ the approach in integrated configuration with dispenser. As a result, some components could classify "Non Fracture Critical (NFC) contained" approach that volume of verification and workload are relatively less for secondary payload.

4.1.2. Leak Before Burst (LBB) Assessment

Pressure components should be basically classified into either NFC sealed containers or Non Hazardous Leak Before Burst (NHLBB) according to NASA-STD-5019A. EQUULEUS has some pressure components. And the project expected that these have been classified into NFC sealed containers because they were not high pressure systems. Also, the EQUULEUS project didn't expected to perform LBB because its assessment needs a lot of works to perform the analysis considering fracture toughness and plastic collapse pressure or testing.
Meanwhile, NASA-STD-5019A requires LBB assessment of NFC sealed containers for verification. However we consider that LBB assessment is not necessary for low pressure components because LBB assessment is not required for NFC sealed containers in ISS requirements. Therefore we proposed that LBB should not be required for the pressure component of EQUULEUS because they have low stored energy and enough strength margins for MDP. Also, we proposed that EQUULEUS implement proof test for NFC sealed containers instead of LBB assessment. As a result, we obtained the agreement that LBB assessment is not necessary for EQUULEUS.

4.2. Bonding Class

We also discussed about bonding requirement with NASA. SLS-RQMT-216 revision A requires class R (2.5 milliohms or less) bond of conductive surfaces including Multilayer Insulation (MLI) for a hazard control of explosion in flammable atmospheres. However, considering the intent of the requirement, class R bonding is to excess for MLI because it is for protecting electronics which requires a low noise from RF mission. Because conductive surface has no power, we believed that secondary payloads only had to remove ignition source such as electrostatic discharge of MLI in flammable atmospheres. Therefore we considered that class S (1.0 ohm or less) bonding for protecting against electrostatic discharge is adequate requirement for MLI. As a result of discussion with NASA, bonding requirements have been moderated to class H (0.1 ohm or less) for protecting against fire or shock to personnel for hazard control of explosion in flammable atmospheres in SLS-RQMT-216 revision B.
We consider that class S is adequate for flammable atmosphere hazard, and that we should have a continuously discussion with NASA.

4.3. For Phase II PSRP

OMOTENASHI and EQUULEUS have finished phase 0/I NASA SLS EM-1 PSRP. For phase II they will have to be determined those hazard controls and proceeded the verification regarding to design. The followings show the items which are particularly to be issues for phase II.

- According to the comparison result of present requirements and discussion with NASA, it is found that NASA SLS EM-1 PSRP has no special requirement for solid rocket motor which OMOTENASHI has employed. However, there is a possibility that new special requirement may be imposed on solid rocket motor. We should keep eyes on establishing requirement of solid recket motor when design of it regarding hazard control and verification are determined.

5. CONCLUSION

OMOTENASHI and EQUULEUS, 6U CubeSats as secondary payloads for the first SLS flight, which JAXA and the University of Tokyo have been cooperatively assemblingmust comply with unique requirement for NASA SLS EM-1. As one of the preparation for NASA PSRP, JAXA S&MA compared SLS safety requirements to JAXA's, and obtained a better understanding. Furthermore, As for the requirements in which there are some issues, we have proposed hazard controls and verifications equivalent to the safety requirements for small secondary payloads to NASA with consideration of the intent of the requirement. We have been building consensus about the proposals with NASA and we will have continuously discussed safety requirements with NASA SLS PSRP in order to get an appropriate safety design for next phase SLS PSRP.

6. REFERENCES

1. Space Launch System, https://www.nasa.gov/exploration/systems/sls/index.html (accessed March 1 2017).

2. Space Launch System Program (SLSP) Exploration Mission -1 Secondary Payload Safety Review Process, SLS-PLAN-217A, NASA, 2016.

3. Space Launch System Program (SLSP) Exploration Mission 1 (EM-1) Safety Requirements for Secondary Payload Hardware, SLS-RQMT-216B, NASA, 2017.

4. Launch Vehicle Payload Safety Standard, JMR-002B, JAXA, 2008.

5. Secondary Payload Interface Definition and Requirements Document (IDRD), SLS-SPIE-RQMT-018, NASA, 2016.

6. Fracture Control Requirements for Spaceflight Hardware, NASA-STD-5019A, NASA, 2016.

7. Electrical Bonding for NASA Launch Vehicles, Spacecraft, Payloads, and Flight Equipment, NASA-STD-4003A, NASA, 2016

8. Flammability, Offgassing, and Compatibility Requirement and Test Procedures, NASA-STD-6001B, NASA, 2011.

9. Standard Materials and Processes Requirements for Spacecraft, NASA-STD-6016, NASA, 2008.

10. Safety Requirements Document, SSP 50021, NASA, 1995

APQP+ METHODOLOGY FOR DEPENDABILITY & SAFETY ACTIVITIES IN DEVELOPMENT AND PRODUCTION OF THE NEW EUROPEAN LAUNCHER ARIANE 6

Isabelle Guerinel

ArianeGroup, 66 Route de Verneuil, 78130 Les Mureaux, France
Email:isabelle.guerinel@airbusafran-launchers.com

ABSTRACT

The ArianeGroup way, especially for the new Ariane 6 Program, is "Right-First-Time develop" and "Right-First-Time produce".

This means to master technical risks early in Development Phase, which covers Design and Industrialization (Production Engineering) activities.

This ArianeGroup approach to Program technical risk management relies on APQP (Advanced Product Quality Planning) methodology. By addressing relevant questions, it aims at properly identifying/assessing risks for Launcher System functions, products and processes, especially for Safety concerns. This risks assessment approach allows prioritizing and planning properly Dependability & Safety and Engineering activities in Design & Industrialization processes.

Several complementary sub-processes embedded in the APQP methodology are used in the Ariane 6 Program. They are deployed to support both Project Management, like Critical Items, and Development activities, like Critical Points and Key Characteristics.

This APQP+ methodology contributes to the Project Risk Management.

1. INTRODUCTION

This paper presents the implementation of APQP+ in Ariane 6 Program aimed to master technical Risks including those related to Safety. It identifies the different Risk Management processes, their characteristics and the links between all of them.

This paper explains how the Safety is managed from the beginning of the development up to the exploitation/operational phase of the Program through these different processes.

2. APQP+ TO SUPPORT PROJECT MANAGEMENT

Program Logic for risks management activities are driven by Program Risks and Opportunities Management process and by APQP+ implementation.

2.1 APQP+ Process

APQP+ means Advanced Product Quality Planning adapted to Ariane 6 (+). The APQP+ standard is a **high-level** Program management procedure whose main goal is to anticipate activities to reduce risks:

- Ensuring the execution of the necessary steps at the appropriate time throughout the product lifecycle based on cross functional approach.
- Providing effective early warning signals to drive on-time and on-quality delivery of products and processes by monitoring key deliverables.
- Allowing Proactive and Preventive Actions.

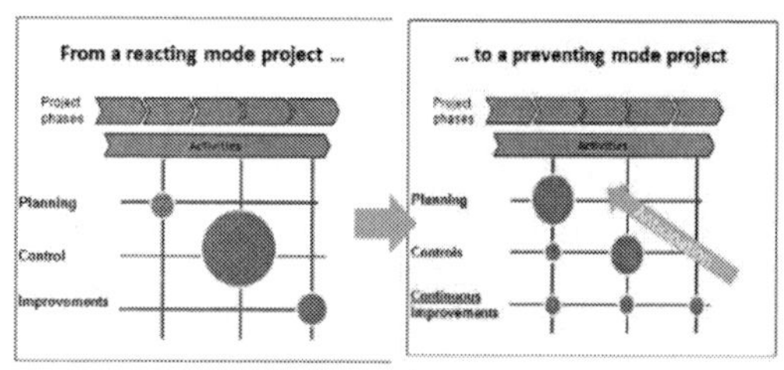

Fig. 1- APQP+Mindset

For each phase of the Program and for each element of it (Product/Process/Function), key deliverables are identified and kept under control.

In parallel, each Product, Process or Function of the Ariane 6 Program is evaluated using APQP+ Risk Analysis tool to determine its Criticality Level. The Safety is one of the evaluation criteria. Most critical items constitute the Ariane 6 Critical Item List.

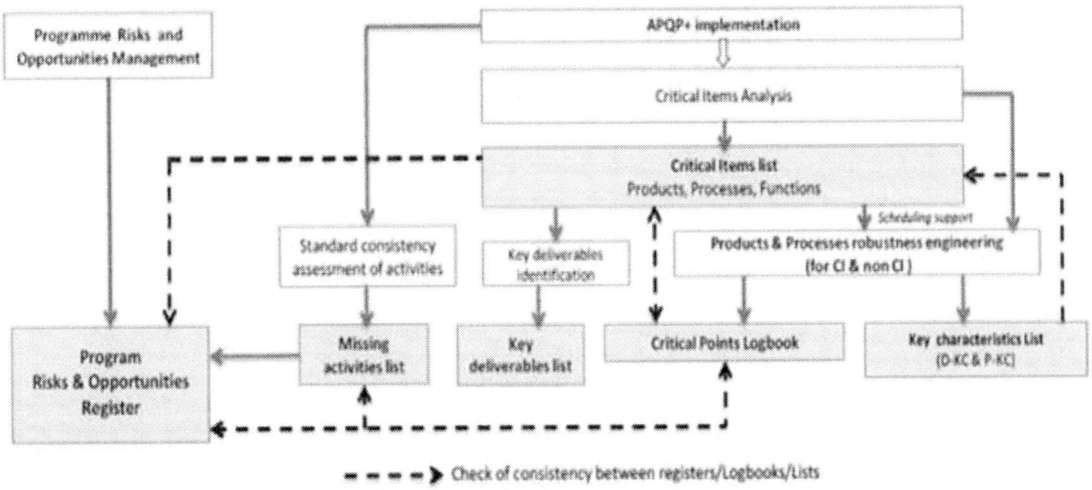

Fig.2 – Integrated Risk Management Process

2.2 Critical Items

Items are identified as critical after APQP+ Risk Analysis which is based on the assessment of a set of project and technical criteria built on ArianeGroup Lessons Learned. The aim is to anticipate risks of Launcher System development and exploitation phases. Depending on the kind of items (Product, Process or Function), the applicable criteria are slightly different.

For Ariane 6 functions:

Risks analysis criteria				
Operational use (Flight...) (security &performances)		In Development: minimize schedule & NRC (Risks in development on products)		
Safety (F1)	Expected performances (F2)	Lessons Learnt (F3)	Complex study/ Margin (F4)	Programmatic criticality (Over Cost/ Planning)

For Ariane 6 products:

Risks analysis criteria								
	Operational use (Flight...) (security & performances)		In Development: minimize schedule & NRC (Risks in development on products)				Manufacturing & Operation (Minimize RC, maximize availability, maximize OTD)	
Re-use Or New Development	Safety (D1)	Performances (D2)	Complex Study (D3)	Technology maturity (D4)	Programmatic criticality (Over Cost/ Planning) (D5)	Maturity of stakeholders (Confidence level to do; internal or external partner) (D6)	MAIT Maturity product/ process (D7)	Lessons Learned in operational life (D8)

For Ariane 6 processes:

Risks analysis criteria					
Operational use (Flight...) (security & performances)	In Development: minimize schedule & NRC (Risks in development on products)			Manufacturing & Operation (Minimize RC, maximize availability, maximize OTD)	
Special process (C1)	New process (C2)	Current process with NC during other programs developments (C3)	Programmatic criticality (Over Cost/ Planning) (C4)	Current Process with significant NC in operational life (C5)	Single unit of MAIT or sharing unit at its limit of production capacity (C6)

For products and processes, in industrialization phase criteria are analyzed in addition to operational performances and risks in development.

Risk identified that relate to Safety criteria, for product or process, lead directly to include the item in Critical Item List. Risk on Safety criteria, for function, has to be worsen by an additional risk in development to raise it to the Critical Item List.

On Ariane 6 Program, there are today 161 Critical Items, 42 CIs for functions, 56 CIs for products, and 63 Cis for processes. Among them, 56 Critical Items (11 functions, 26 products and 19 processes) have been highlighted for special Program management focus. These items have therefore potentially significant effect on product realization, performance and use, including Safety. The objective is then to strongly manage these items by identifying and implementing one or several mitigation plans.

For performance (Operational, Flight/Safety) risks, the Critical Points process (cf. §3.1) is used. For production, operation, maintenance risks (with potential Safety impact), the Key Characteristic selection process (cf. §3.2) is implemented. Finally, for programmatic risks (schedule, cost, development with potential Safety impact), the Project Risk process (cf. §4) is followed.

In addition, Key Deliverables associated to Critical Items are identified, and each one is monitored with a dedicated "APQP+ check-list". Such monitoring is started at the beginning of the activity in order to anticipate any potential deviation with reference to the standard process.

3. APQP+ RELIES ON DEVELOPMENT PROCESSES

After securing the Project by anticipation with APQP+ process, the "Right-First-Time for development" and "Right-First-Time for production" approaches on Ariane 6 are reinforced by specific development processes.

They are based on Lessons learned, Dependability & Safety analysis, Industrialization risks analysis and engineering design activities such as tests, numerical analysis, flight exploitations etc.

3.1 Critical Points Process

This process is a mean to help demonstrating the conformity between the Design and the Specification (theoretical reliability/safety). It can also be extended to help the demonstration of conformity between the manufactured products and its design (practical reliability/safety).

Critical Points are linked to Critical Items List in the sense that they contribute to explain the criticality of the products, processes and functions retained.

Critical Points are those elements bearing unacceptable risks regarding manufacturability or performances (operational performance, reliability, safety, etc.) Critical Points are fundamental for the management of the reliability and safety risks which are identified throughout the life cycle of the program. It gives the traceability of the risks and of the associated mitigation plans that are identified and managed by the project team. They are identified during development phase and are controlled over the whole life cycle of the product/system at top management level.

This approach mainly relies on dependability & safety analysis, on lessons learned from past programs ned) and on the analysis of technical risks by multi-disciplinary teams.

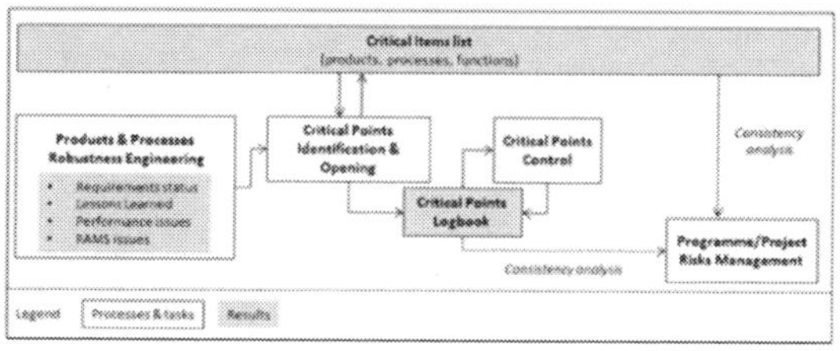

Fig. 3 Critical Points management

Meant for mastering program technical risks through product and processes robustness engineering, a Critical Point will be opened in the following cases:

- Major difficulty occurring during the development or during operational phase, which could have an impact on operational performance, reliability and safety requirements of the product.
- Violation of a design failure criterion or process which has consequences on the safety of persons or of key assets, on the environment or on the reliability of products.
- Risk situated in the unacceptability zone of the criticality matrix of the concerned product, because either the severity of the consequences and/or the probability of occurrence do not fulfil the applicable reliability and safety requirements. This method is applicable only if an FMECA is performed. Otherwise, criteria are based on the non-fulfillment of either quantitative or qualitative requirement separately.

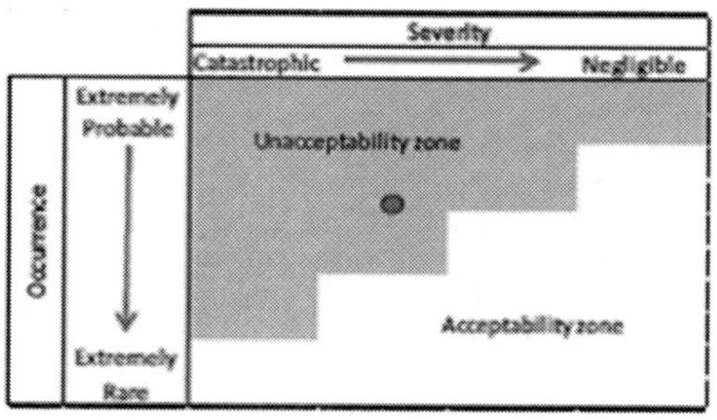

Fig.4 Risk situated in the unacceptability zone

The next step in the process is to control the Critical Points. The goal of this step is to define and manage mitigation actions/plans for each Critical Point to control its criticality (severity x probability) and to [try to] close it. For qualitative requirements (e.g. Fail-Safe, Fail-Operational), there are no notions of "severity x probability". Associated Critical Points are closed when the requirement is fulfilled.

According to mitigations actions implementation, Critical Points status can evolve from "*Open*" to:

- "*Accepted as it is*": mitigation actions are not efficient enough to reach the acceptable area in the criticality matrix. Those critical points must be agreed by upper-level management/ customer.
- "*Attenuated*": critical point has been brought in the acceptable area of the criticality matrix, with specific recurring action in serial production (detectability, imposed means, processes),
- "*Closed*": critical point has been brought in the acceptable area of the criticality matrix, without dedicated recurring action in serial production.

At the end of the development phase, the target is to have no Critical Point still *open*, no Critical Point *accepted as is*, and as few as possible Critical Points needing recurring actions ("*Attenuated*").

This Critical Point approach is applied to each equipment/product belonging to the system under development/production. Therefore, Critical Points must be considered for each item in the Product Breakdown Structure (PBS). As mentioned in the following figure, information on Critical Points on each level must be shared with upper level management.

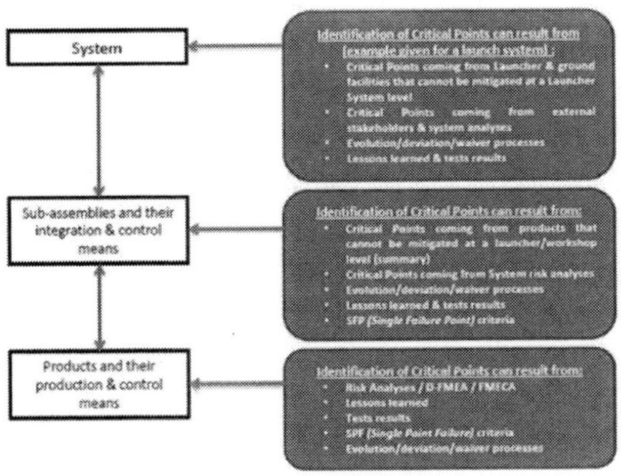

Fig.5 Critical Points in Product Breakdown

During the operational phase, after completion of development, Critical Points Management must also be ensured with a constantly re-evaluation of the criticality status of each Critical Point. Any modification affecting a Critical Point during the development phase or during the production phase must be managed on top level of the technical organization of the program (Chief Engineer). The need for a modification can triggered by:

- an anomaly or change disposition affecting the product/system with consequences on the probability of occurrence of the risk relevant to the Critical Point;
- an external event in the environment of the product that put into question the severity or the probability of the Critical Point.

Critical Points that may remain "*open*" and/or "*accepted as it is*" must be carefully monitored. In this case, they are equivalent to a request for deviation at the development phase and request for waiver for noncompliance to design.

3.2 Key Characteristics for Design and for Process (D-KC & P-KC)

This Key Characteristics process is a means to help assuring the conformity between the manufactured product and its design (practical reliability/safety) during the production phase.

The process moves from the identification of Design Key Characteristics (D-KC) and Process Key Characteristics (P-KC) in order to provide the necessary inputs to the elaboration of an efficient control policy and to meet reliability and safety requirements.

Figure 6 illustrates the link between the main documents/activities concerned by the process:

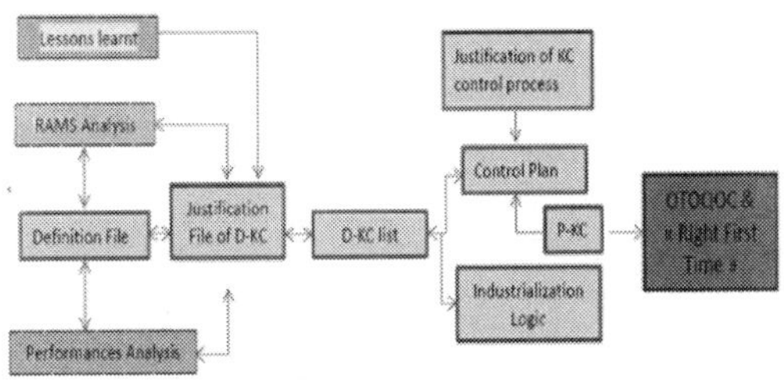

Fig.6 General KC Process

A Design Key Characteristic / Process Key Characteristics (D-KC/P-KC) is an attribute or feature of the product/realization process whose variation has a significant effect on product safety, fit, form, function, performance, or service life. In particular, D-KC/P-KC will be identified for Critical Items.

Key characteristics identification and ranking is a major co-engineering activity carried on with dependability & safety engineers that will guarantee that:

- Design characteristics having high sensitivity to requirements/performances are identified as KC,
- Design characteristics having high sensitivity to requirements/performances are identified as KC,

- These Design KC are driving Industrialization activities, including Control Plan definition,
- The full value chain (supplier, procurement, incoming inspection, design office, production, quality…) is aware and focused on these KC.

Regarding Critical Items analysis:

- For elements identified as Critical Items, identification and ranking of D-KC&P-KC is mandatory and must rely on a structured approach such as D-FMEA, Zonal Analysis, Performances analysis, etc.
- For elements which are not identified as Critical Items, identification and ranking of D-KC&P-KC could rely on simplified approaches (lessons learnt, engineering judgement, etc.)

Co-engineering approaches are necessary in order to:

- Allow the Design Office to satisfy technical requirements while taking into account production constraints,
- Allow MAIT to define an adapted process to the design tolerances taking into account programmatic criteria (cost, rate and schedule).

Design Characteristics:

D-KC identification is based on lessons learnt, systems/products/equipment performances analysis, and dependability & safety analysis such as D-FMEA which helps identifying and/or consolidating the list of D-KC, and associated ranking through the identification of causes leading to failure modes and their associated severity.

D-KC ranking aims at defining the criticality of design characteristics with reference to required performances. This ranking is a necessary input to determine the required level of the control plan and associated conformity demonstrations. A characteristic which is "*Critical*" requires a high level of conformity demonstration, a characteristic which is "*Major*" requires a reinforced level of conformity demonstration, and a characteristic which is "*Minor*" follows the conformity demonstration covered by standards procedures. D-KC are only those product characteristics which are classified as "*Critical*" and "*Major*".

To initiate the ranking, D-KC "*Critical*" are linked to Safety, D-KC "*Major*" are linked to mandatory performances mandatory (reliability, availability, maintainability, payload capacity, range, etc.)

The D-KC list will evolve during the product life cycle in accordance with the results of activities, tests, analysis, etc. At the end of development phase, the target should be to have as few D-KC as possible.

Actions shall be settled in order to optimize this list, for example studies to quantify margins and associated scattering, or definition evolutions.

Process Characteristics:

In the case of a D-KC which is not directly controllable, the process characteristics (P-KC) must be verified in order to guarantee the conformity of the D-KC.

The P-KC determination aims at proving product conformity and ensuring the stability of manufacturing process through statistical approaches. For example, in order to detect drifts and/or out-population points. Those P-KC are not necessarily linked to D-KC but could be identified for cost/planning aspects.

P-KC identification must be principally made through P-FMEA, which consists in identifying the risks of failure of the MAIT process of a part/product/system, with potential impact on safety, quality, D-KC, cost and planning. In case of a P-KC directly linked to a D-KC, P-KC and D-KC ranking shall be similar.

4. PROJECT RISK MANAGEMENT

The Ariane 6 Governance and the consequent high level of responsibility carried by the Launcher System contractor require to perform Risk Management at Program Level. This process aims to identify events with potential significant impact on quality, schedule, and cost, whatever the origin, and to take all necessary actions for risk mitigation such to secure the achievement of program objectives. One of the source of program risk can be a safety concern.

The Project Risk Management is the basic Program mastering which is reinforced on Ariane 6 Program by APQP+ methodology for technical matters with the dedicated processes presented above. This basic Program mastering is based on different steps:

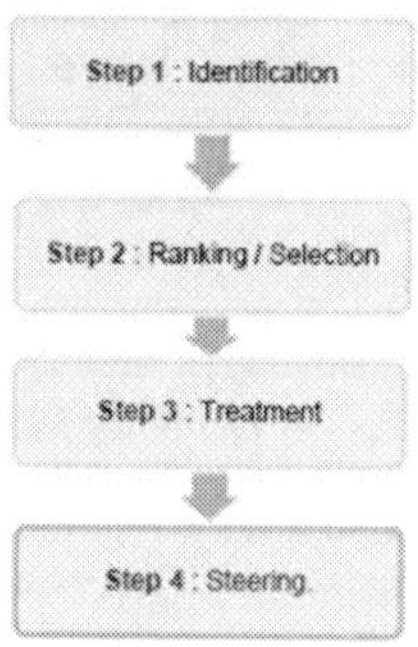

Fig.7 Major steps of the Risk Management Process

The different status of Project Risks are:

- "*Draft*": new risk proposal to be validated
- "*Open*": risk validated and treatment ongoing
- "*Traded/Accepted*": action plan performed, risk at targeted criticality
- "*Closed*": risk can no longer appear
- "*Occurred*": risk became an event
- "*Rejected*": new risk proposal not validated
- "*Transferred*": risk to be treated by another entity

The Identification step is based on lessons learned, on analysis of a group of risky situations, and on compliance analysis to contractual and technical requirements, including safety. It uses a standard checklist which addresses functional areas like political, financial, industrial, management, and technical. The identification of new risks is a continuous process during the program life cycle.

The ranking level of a risk depends on the combination of occurrence probability and consequence severity, which leads to an overall criticality as shown in the matrix in figure 8.

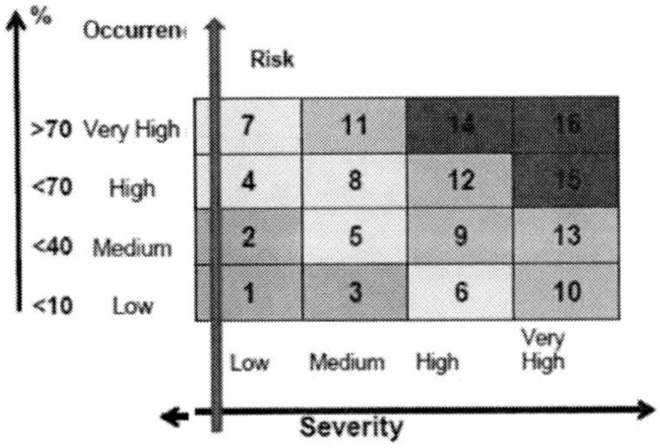

Fig.8 Project risk criticality matrix

Depending on criticality, the management of Project Risk with its associated action plans is handled either at Program top level or at Work Package level. The "*Treatment*" step consists in calculating the initial weighted financial impact, defining detailed action plans, calculating the target weighted financial impact, deciding action plan implementation, and build the Target Criticality matrix.

The "*Steering*" step consists in updating the Project risks. As the program is carried on, the risk register must be updated depending upon the occurrence of events and new inputs. Each risk owner has to update continuously his Risk portfolio. In order to consolidate the program objectives achievement, the aim of the steering step is to control the action plans implementation and to assess their efficiency, to update the list of action plans, to re-assess the corresponding risk criticality and to monitor risks evolution. At the end, the consistency between Project Risks registers, Critical Item list and Critical Points logbook is a task of the Project Risks Management process.

5. CONCLUSION

New or updated APQP+ Directives and Procedures of ArianeGroup concerning technical Risks Management, applied to the Ariane 6 Program, allow to integrate safety constrains from the beginning of the development up to the exploitation/operational phase of the system. Such APQP+ process contributes and reinforces Project Risk Management. It is particularly necessary with the new Ariane 6 Governance and the consequent high level of responsibility carried by the Launcher System contractor which makes mandatory Risks Management at Program level.

At the beginning of the Program, as part of the first APQP+ step, the safety risk is one of the criteria used to identify Critical Items of the Project (Products, Processes and Functions). During the development phase, safety risks are introduced into one of the following risk management processes:

- Critical Points if some major development or exploitation difficulties could impact safety performances, or if a qualitative or quantitative safety requirement is not fulfilled
- Key Characteristics for design or for process (D-KC/P-KC) if the safety risk is linked to a specific characteristic of a function or to a specific physical (hardware) characteristic of a product or of a production process,
- Project Risk Management if the safety risk impacts cost, schedule or performance of the project.

One safety risk can appear in several Risk Management processes depending on the origin and its impacts.

The consistency between Project Risks registers, Critical Item list and Critical Points logbooks is ensured by the Project Risks Management process.

6. REFERENCES

1. ArianeGroup Directive D-0016-E - Integration of Risks Management Processes within Programs

2. ArianeGroup Procedure P-0078-E - Advanced Product Quality Planning (APQP+)

3. A6-LI-1000000-X-0008-ASL – Ariane 6 Critical Item List

4. ArianeGroup Procedure P-0106 - Critical Points Management

5. ArianeGroup Procedure P-0100 - Ranking of Definition and Processes characteristics

6. ArianeGroup Procedure P-0003-E - Program Risks & Opportunities Management

7. A6-MG-1000000-X-0001-CORE-ASL – Ariane 6 Launcher System Program Management Plan

PROCEEDINGS OF THE 9TH INTERNATIONAL SPACE SAFETY CONFERENCE

SESSION 09 PROBABILISTIC RISK ASSESSMENT

UNCERTAINTY ESTIMATION CHEAT SHEET FOR PROBABILISTIC RISK ASSESSMENT

Paul Britton [1], Mohammad Al Hassan [2], Robert Ring [3]

(1) National Aeronautics and Space Administration (NASA), Marshall Space Flight Center, Huntsville, AL 35812, USA, Email: paul.t.britton@nasa.gov
(2) National Aeronautics and Space Administration (NASA), Marshall Space Flight Center, Huntsville, AL 35812, USA, Email: mohammad.i.alhassan@nasa.gov
(3) Bastion Technologies Incorporated, Marshall Space Flight Center, Huntsville, AL 35812, USA, Email:robert.w.ring@nasa.gov

ABSTRACT

"Uncertainty analysis itself is uncertain, therefore, you cannot evaluate it exactly," Source Uncertain.

Quantitative results for aerospace engineering problems are influenced by many sources of uncertainty. Uncertainty analysis aims to make a technical contribution to decision-making through the quantification of uncertainties in the relevant variables as well as through the propagation of these uncertainties up to the result. Uncertainty can be thought of as a measure of the 'goodness' of a result and is typically represented as statistical dispersion.

This paper will explain common measures of centrality and dispersion; and—with examples—will provide guidelines for how they may be estimated to ensure effective technical contributions to decision-making.

1. INTRODUCTION

When uncertainty estimates are expected to inform decision-makers, it is especially important to carefully consider, understand, and communicate the significance of the statistical parameters used in the characterization of failure probability distributions. This paper will focus on a few fundamental examples and the principles paramount in achieving successful uncertainty estimation. After establishing definitions and providing background material, we will illustrate key principles as we step through the quantification of uncertainty. Finally, the risk implications of uncertainty estimation are summarized in a convenient reference card: Uncertainty Estimation Cheat Sheet.

2. DEFINITIONS AND BACKGROUND MATERIAL

In an attempt to make later concepts accessible to a broader audience, this section provides a synopsis of some of the relevant concepts of probability theory. Here and in what follows, **boldface** indicates a word or phrase that is being defined or explained.

2.1. Probability Distributions

Informally, a **probability distribution** is a mathematical function that assigns probabilities to each element of the **sample space** (the set of all possible outcomes in an experiment). Probability distributions are either discrete, or continuous, or a mixture of both types. However, the topics, herein, require only some basic knowledge of continuous distributions.

A **random variable** is a function that maps outcomes of an experiment to numerical quantities. For a continuous distribution, the **probability density function (pdf)** is the function that is used to generate the probability that a random variable *X* lies within an interval [*a, b*]:

$$\Pr[a \leq X \leq b] = \int_a^b f(x)dx \tag{1}$$

In this paper we will discuss the following continuous distributions: The exponential, normal, and lognormal.

The probability density of the **exponential distribution** is:

$$f(t) = \begin{cases} \lambda e^{-\lambda t}, & t \geq 0 \\ 0, & t < 0 \end{cases} \tag{2}$$

Assuming an exponential failure model for a component's mission exposure time, the probability of failure of the component before time T is given by the cumulative distribution function calculated as:

$$F(T,\lambda) = \int_0^T \lambda e^{-\lambda t} dt = 1 - e^{-\lambda T} \tag{3}$$

The exponential is a simple model with one parameter, and its properties are widely understood. It is commonly used to model components operating within their service life because of the following characteristics:

1) The hazard function (instantaneous failure rate) is constant with respect to time and is equal to λ, which implies:

2) The memoryless property; that is, components do not wear out: they function as good as new regardless of how long they have been in service.

The pdf of the **normal (or Gaussian) distribution** is:

$$f(x) = \frac{1}{\sqrt{2\pi\sigma^2}} e^{-\frac{(x-\mu)^2}{2\sigma^2}} \tag{4}$$

Where the parameters μ and σ are the mean and the standard deviation, respectively. The normal pdf is symmetric about x = μ. An example of data that follows a normal pdf are repeated measurements of a physical characteristic of a part (such as, weight, length, thickness).

The pdf of the **lognormal distribution** parameterized with the mean (μ) and standard deviation (σ) of the underlying normal distribution is given as:

$$f(\lambda) = \frac{1}{\lambda\sigma\sqrt{2\pi}} \exp\left(-\frac{[\ln(\lambda)-\mu]^2}{2\sigma^2}\right), (0 < \lambda < \infty) \tag{5}$$

An important property of the lognormal is its relationship to the normal distribution. If Y is lognormal, then $X = \ln(Y)$ is normal. The lognormal is a good choice for representing failure rate uncertainty because it is strictly defined on the positive x axis, and the distribution has a heavy (stretched out) right tail (i.e., right skewed).

2.2. Central Tendency

A measure of **central tendency** (or **centrality**) is a single value that describes a central or typical value for a probability distribution or set of data. It may refer to the center of probability (**median**) or center of probability density (**mean**) or a most probable value (**mode**). The mean, median and mode are the most common examples and are defined below. In subsequent sections, we will look at the mean, mode and median, and explain some of the conditions for their appropriate usage.

In the case of a probability distribution $p(x)$ defined on a finite set $\{x_1, x_2, \ldots, x_n\}$, the **arithmetic mean** or expected value of x is a weighted sum:

$$E[X] = \sum xp(x) \tag{6}$$

For a continuous distributions, the arithmetic mean is:

$$E[X] = \int xf(x)\,dx, \tag{7}$$

where the weighting function $f(x)$ is the pdf of X. Noteworthy is the fact that the mean is susceptible to the influence of outliers. These are values that lie far from the central body of the distribution. The effect of large outliers is to pull the mean away from the median towards the outlier. This can be understood by considering that the mean is analogous to the center of mass.

The **median** or 50th percentile is the midpoint where half of the probability (area under the pdf) lies to either side.

$$\int_{-\infty}^{median} f(x)\,dx = \int_{median}^{\infty} f(x)\,dx = \frac{1}{2} \tag{8}$$

The **mode** is a local maximum or peak of the pdf.

2.3. Dispersion

Dispersion refers to the spread of the distribution. A measure of **dispersion** is a non-negative real number that quantifies the deviation from the central tendency. A large deviation (relative to the magnitude of the central tendency) is indicative of a distribution that is spread out or dispersed. Examples to be discussed are the variance, the standard deviation, and the error factor.

The **variance** is the expected value of the squared deviations about the mean:

$$Var[X] = E[(X - E[X])^2] \tag{9}$$

The square root of the variance is the **standard deviation**:

$$\sigma = \sqrt{variance} \tag{10}$$

A nice property of the standard deviation is that it has the same units as the quantity being measured.

A frequently used measure of dispersion for the lognormal is the **error factor (EF)**. The EF defines dispersion about the median. The EF is defined as the square root of the 95th percentile divided by the 5th percentile. Equivalently, the EF is equal to the 50th divided by the 5th and the 95th divided by the 50th as summarized in the Eq. 11:

$$EF = \sqrt{\frac{95^{th}\,percentile}{5^{th}\,percentile}} = \frac{95^{th}\,Percentile}{50^{th}\,Percentile} = \frac{50^{th}\,Percentile}{5^{th}\,Percentile} \tag{11}$$

The EF has a minimum value of one, which represents certainty.

2.4. Failure Rate Uncertainty

Component failure rates (λ) are not physical quantities; that is, they cannot be measured directly but must be inferred. Operational or test reliability data expressed as total exposure time and number of failures from which to infer the failure rate for highly reliable components, are scarce. Failure rates have to be inferred by similarity to generic sources, such as those published in component reliability databases. In some cases, estimates are developed using engineering judgment or by eliciting the estimates from subject matter experts.
In the Bayesian interpretation, the parameter value is random (synonymous with uncertain) and is represented by a probability distribution. Previous research evaluated different distributions to represent the uncertainty of the parameter λ [1]. They found the lognormal distribution was appropriate for simple components with a single failure mode. Basic Event quantification for PRA then becomes one of estimating the central tendency and dispersion of the lognormal distribution, and then using Monte Carlo methods to simulate the probability distribution for failure of the component. The lognormal failure rate uncertainty pdf is illustrated in Figure 1 below.

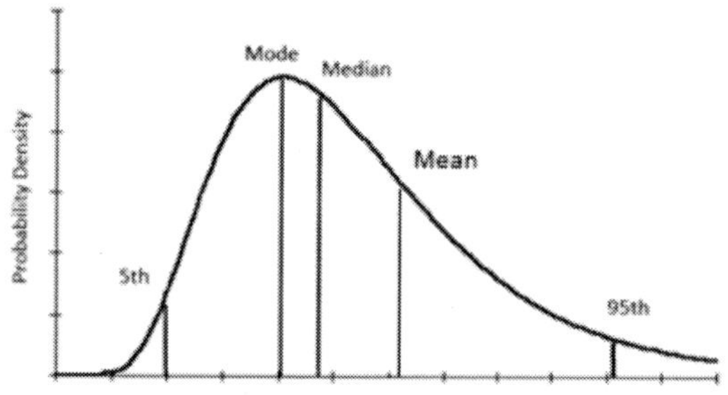

Figure 1.

Uncertainty has many sources in addition to variation among individuals within a population and lack of knowledge due to sparse data. However, this paper examines the implications of applying uncertainty around central tendency estimates in order to quantify degree of belief – in particular when expressing degree of belief via the shape of the lognormal pdf.

3. THE BAYESIAN APPROACH

Application of classical life data analysis requires component data in the form of failures and exposure time or number of demands. The data is fit to a distributional model of times to failure, goodness of fit tests are performed, and measures of centrality and dispersion are estimated. The model's parameters are assumed to be fixed (but unknown) quantities. The neighborhood in which the parameter value lies is estimated from the sampling distribution and is expressed as a confidence interval. Classical prediction methods rely solely on the data and do not permit prior knowledge to influence the estimate.

Highly reliable components produced in small quantities, such as in space applications, do not have enough operating time and failure history to yield useful confidence bounds using classical statistical data analysis methods.

Bayesian approach is able to address the challenges described above because it admits prior experience into the estimation procedure in the form of a prior degree of belief about the likely values of the parameter in the form of a prior distribution. Specific data in the form of a likelihood function is then applied through Bayes Theorem to update the prior belief to yield the posterior uncertainty pdf.

Bayesian updating produces normative results in that if one believes the prior distribution, then one ought to believe the posterior distribution. It is important to note the implication of the previous statement. If after Bayesian updating, one does not believe the posterior distribution, then the prior is likely wrong. Hence it is important for the prior distribution to be developed and reviewed in a deliberative process with the help of subject matter experts to assure credibility.

In our experience, engineers with specific discipline expertise are familiar with the shape and properties of the normal probability distribution, but have little direct experience with skewed distributions, such as the lognormal. Recall, the normal probability density function is symmetric about its global maximum (mode). The median (50th percentile), and mean coincide with the mode. Perhaps due to the coincidence of measures of central tendency of the normal distribution, subject matter experts do not understand the relationship of these measures when using the lognormal pdf. Subject matter experts who often assist Probabilistic Risk Assessment (PRA) analysts in the quantification of the prior failure rate distribution must be educated to develop an intuitive understanding of how the lognormal distribution morphs as its central tendency and dispersion measures are varied. One of the main purposes of this paper is to illustrate with specific examples the effects of varying one of the parameters, such as the dispersion while holding another fixed to show the effect on the remaining parameters.

Specifying any two parameter values completely specifies the lognormal distribution. Thus we can solve for μ and σ and then fill in the remaining parameter values using the formulas in Figure 2.

Parameter	As a function of μ and σ
Mean	$\exp\left(\mu + \frac{\sigma^2}{2}\right)$
Median	$\exp(\mu)$
Mode	$\exp(\mu - \sigma^2)$
Standard Deviation	$\sqrt{[\exp(\sigma^2) - 1]\exp(2\mu + \sigma^2)}$
Error Factor	$\exp(1.64485\sigma)$

Figure 2.

4. FUNDAMENTAL EXAMPLES OF UNCERTAINTY ESTIMATION

There is strong evidence that we perceive things logarithmically [2]. In other words, when we think something is twice as big, it might be more like eight times as big.

PRA analysts are often in the position of asking experts to estimate the percentiles of the uncertainty pdf. Experts are prone to overconfidence bias. For example, if we elicit the 5th and 95th percentiles, experts are likely to give answers that are closer to the 25th and 75th percentiles. This is especially true for rare events.

Component reliability data developed to support reliability allocation goals are an important source of data to help develop prior uncertainty estimates for use in the PRA. Reliability predictions are reported as point estimates. The PRA has to estimate uncertainty to create the probability distribution that represent degree of belief. Heuristic approaches have been used. These approaches consider the data source applicability with respect to similarity. Multipliers can be applied to convert the data from the reported operating environment to a more applicable one. The heuristic method proscribes using the provided point estimate, which is typically assumed to be an estimate of the mean failure rate and then, depending on data source applicability, apply an assumed error factor based on heuristic guidelines. The resultant prior distribution is then assumed to be a lognormal distribution with the provided mean and the heuristic error factor.

Clearly, any recipe for developing a prior needs to be augmented by visualization to confirm that the resulting distribution is credible. The discussion that follows illustrates a concern that when the point estimate is assumed to be the mean failure rate, the resultant pdf is counterintuitive – and is in fact non-conservative. PRA methodology tries to achieve the best estimate of the risk. But, in questionable cases should err on the side of conservatism.

The discussion that follows compares several modifications to the aforementioned heuristic approach. The comparison cases assume the point estimate represents one of the measures of central location (i.e. the mean, median or the mode) and is fixed, while the error factor is applied or varied. We then illustrate with a hypothetical trade study in which the recommended alternative, from a risk standpoint, depends on the measure of central location that is held fixed.

Figures 3-6 illustrate the effect on the resultant prior pdf for the three variations of the heuristic method using assumed measures of dispersion error factors of 3 and 9, respectively. What is given in the contractor's reliability analysis report is the point estimate for failure rate (λ) of 50 failures per million hours (FPMH) of exposure. The solid green curve is the base case assuming the point estimate is the mean; the dashed orange curve assumes the point estimate is the median; and the dotted purple curve assumes the point estimate is the mode. The square point on each of the resultant lognormal pdf curves is the median (50th percentiles) and the triangle is the mean.

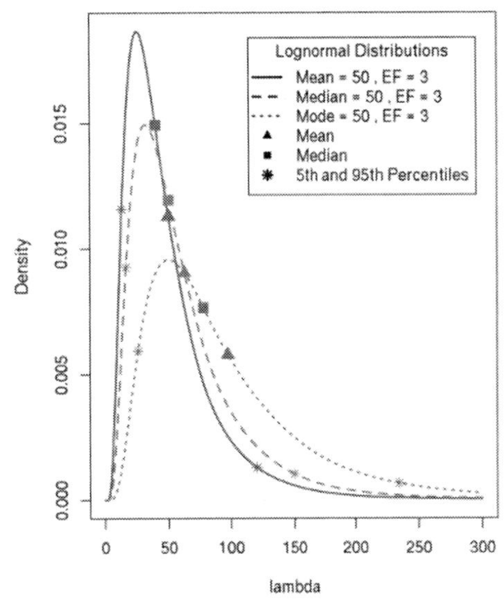

Figure 3

Central Value = 50, Error Factor = 3			
Fixed →	Mean	Median	Mode
Mean	50.00000	62.49422	97.62917
Median	40.00370	50.00000	78.11056
Mode	25.60710	32.00592	50.00000
5th	13.33457	16.66667	26.03685
95th	120.0111	150.0000	234.3317

Figure 4

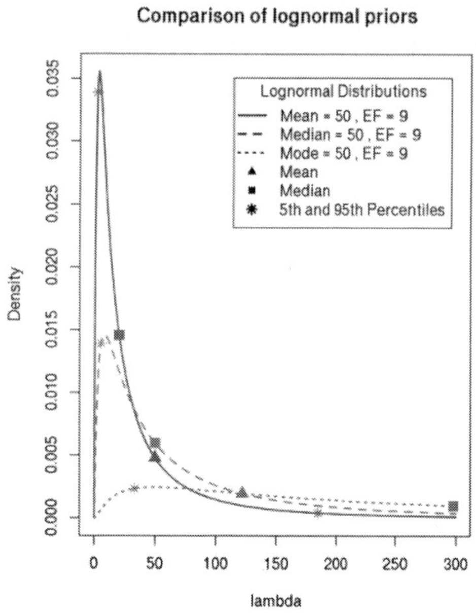

Figure 5

Central Value = 50, Error Factor = 9			
Fixed →	Mean	Median	Mode
Mean	50.00000	122.0252	726.7891
Median	20.48757	50.00000	297.80292
Mode	3.439787	8.394814	50.00000
5th	2.276397	5.555556	33.089213
95th	184.3882	450.0000	2680.2263

Figure 6

Notice that when the mean is assumed fixed, the resultant lognormal pdf is forced to the left. This may not be the desired result.

The next set of results (Figures 7-9) are similar comparisons, but allows us to view what is happening from a different perspective. The first case begins with holding the mean fixed to a value of 50 while varying the error factor between 3 and 9. In the other cases we hold the median and mode fixed while varying the error factor.

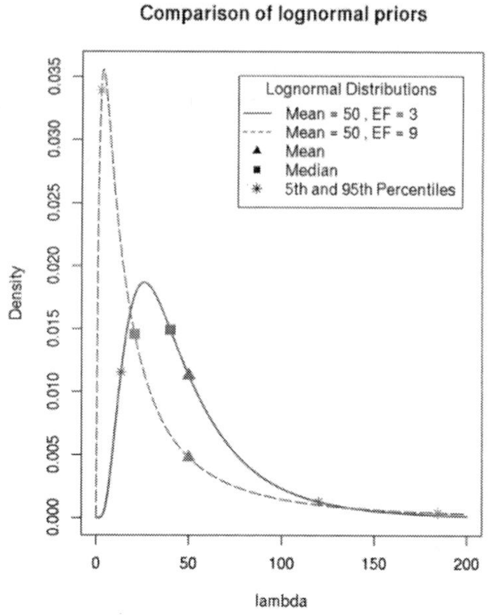

Figure 7

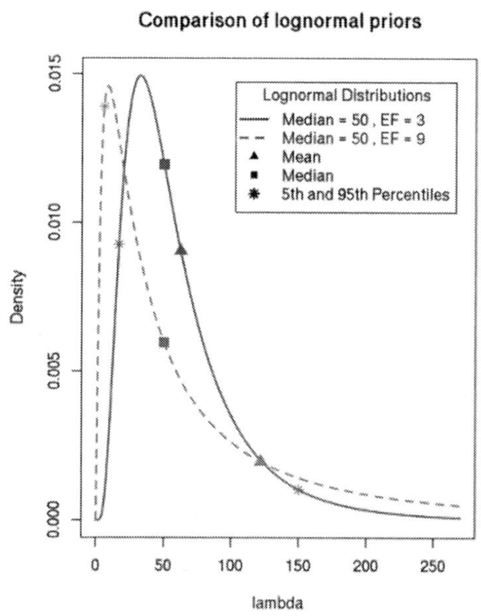

Figure 8

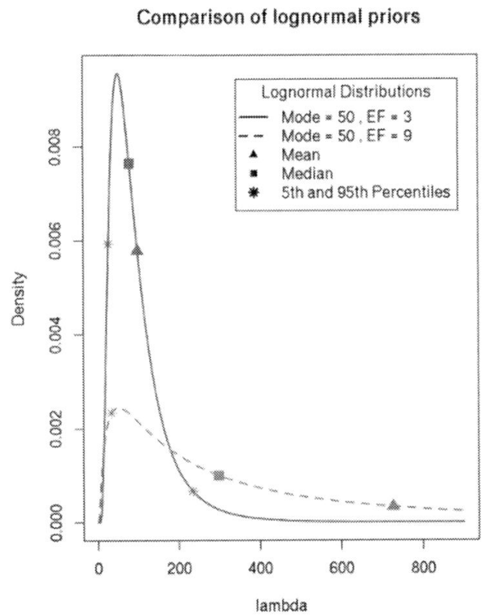

Figure 9

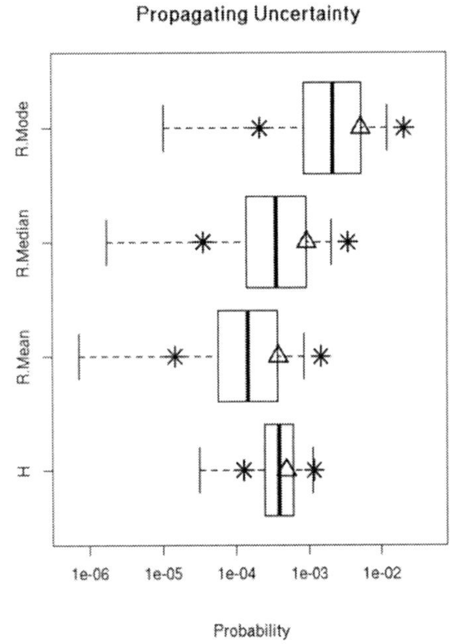

Figure 10

Option -->	Redundant			Heritage
Fixed -->	Mean	Median	Mode	NA
Mean	3.77E-04	9.18E-04	5.34E-03	4.80E-04
Median	1.46E-04	3.55E-04	2.14E-03	3.83E-04
5th	1.47E-05	3.59E-05	2.16E-04	1.28E-04
95th	1.45E-03	3.53E-03	2.07E-02	1.16E-03

Figure 11

We observe that the general shape of the density curves are quite sensitive to holding the mean and mode fixed. Therefore, we strongly discourage heuristics that do not involve steps that require analysists and subject matter experts to assess visualizations of the resulting density curves.

This final example illustrates a hypothetical trade study. It compares a highly reliable, heritage, zero failure tolerant design with a retrofitted redundant option that is not only susceptible to common cause failure modes but is such that each leg of redundancy is considered less reliable than the heritage design.

The heritage design has a well-established lognormal failure rate distribution with a mean failure rate of 50 FPMH and an error factor of 3. The legs of the redundant design are estimated (through engineering judgement) to have failure rates of 150 and 200; and associated error factors of 6 and 9, respectively. A common cause failure basic event is also assumed to follow a beta distribution with a 5th percentile of 0.1 and a 95th percentile of 0.4 [3]. The time both options are exposed to failure is 8 hours.

Figures 10-11 compare the risk of the heritage option with three alternative methods of quantifying the uncertainty of the risk of the redundant design option: by assuming the provided failure rates are fixed and represent either the mean, median or mode.

These results demonstrate that the effects on risk-informed decisions by the mere choice of the central parameter about which uncertainty is estimated can, in fact, be pivotal!

5. UNCERTAINTY ESTIMATION CHEAT SHEET (FOR LOGNORMAL UNCERTAINTY)

The purpose of the cheat sheet (Figure 12) is to reinforce an understanding of the cause and effect relationships between the adjusting of parameters (that measure central tendency and dispersion) and their risk implications. The cheat sheet is qualitative in nature and must be taken with a grain of salt. Ultimately, the choice of which measure of central tendency to hold fixed is subjective. However, it is important to understand and consider the risk implications of these choices within the context of the assumptions and beliefs of those involved in the estimation process.

Lognormal Uncertainty Estimation Cheat Sheet

Uncertainty Increased	
Fixed	**Risk**
Mean	Lower
Median	Neutral
Mode	Higher

Uncertainty Decreased	
Fixed	**Risk**
Mean	Higher
Median	Neutral
Mode	Lower

Figure 12

6. CONCLUSION

Although many cases are presented, the typical case for aerospace PRA is to assume the measure of central tendency is the mean and keeps it fixed while increasing the uncertainty (error factor). Unfortunately, this is often the case without strong rationale. Our recommended default for heuristic estimation of lognormal uncertainty is to quantify the median from the given data and then adjust the error factor accordingly. The median automatically remains fixed since it is independent of the error factor. Our recommendation holds even in the case where data uncertainty is absent and only a central value is given. In this case, the data tells us the mean, median and mode coincide. Therefore, we proceed by fixing the median to the given central value while estimating uncertainty about the median to obtain results.

Theoretical distributions do not always behave intuitively. Care must be taken when adjusting the parameters of a distribution as part of a heuristic or other method. One ought to understand the relationships and effects of all relevant parameters as well as the risk implications. It is our hope that the Uncertainty Estimation Cheat Sheet (for Lognormal Uncertainty) will help those involved in the PRA process (such as managers, subject matter experts and PRA analysts) make effective technical contributions to decision-making.

7. REFERENCES

1. Parameter Uncertainty for ASP Models, Knudsen, James K. and Smith Curtis L., Idaho National Engineering Laboratory, October 1995

2. Why do we perceive logarithmically? Authors Lav R Varshney, John Z Sun Publication date 2013/2/1 Journal Significance Volume 10 Issue 1 Pages 28-31

3. Common Cause Failure Modeling. 8th IAASS Conference. Authors: Frank Hark, Steven Novack, Rob Ring, Paul Britton

PROBABILISTIC RISK ASSESSMENT MODEL DEVELOPMENT & APPLICATIONS TO OPERATIONAL DECISION MAKING IN HTV

Hiraku Kudo[(1)], Toru Yoshihara[(1)], Tatsuya Shirai[(1)], Masami Miki[(2)], Satomi Takada [(2)], Takashi Goto[(2)], Koji Oga[(2)]

[(1)]*Japan Aerospace Exploration Agency (JAXA), Tsukuba Space Center 2-1-1, Sengen, Tsukuba-shi, Ibaraki, 305-8505 Japan*

[(2)]*Japan Manned Space Systems Corporation (JAMSS), Tsukuba Center Inc, D, 2-1-6, Sengen, Tsukuba, Ibaraki, 305-0047, Japan, Email:miki.masami@jamss.co.jp*

ABSTRACT

Japan participates in the International Space Station (ISS) program, and Japan Aerospace Exploration Agency (JAXA) provides an unmanned orbital carrier, H-II Transfer Vehicle (HTV). The initial HTV flight was launched in 2009. So far, six HTVs have been flown to ISS successfully.

JAXA has conducted Probabilistic Risk Assessment (PRA) for the HTV, while NASA also separately developed an HTV PRA model. In order to have common models for HTV, NASA and JAXA started their joint efforts to update their models and the common models were completed in 2016.

PRA is a powerful tool to make decision for the system at development phase and operation phase, however, JAXA currently cannot perform quantitative risk assessment of their space systems effectively; For almost all of space systems, quantitative assessment are performed using by conventional method (reliability assessment). While, for manned space systems like JEM and HTV, PRA was conducted. However, there was few feedback from PRA to system design or operation because PRA was performed only to show the quantification of likelihood in Hazard Reports so far.

To solve this issue, for the first step, JAXA focuses on effectiveness of trade study of PRA in order to support decision making. Now that NASA and JAXA completed to build their common models of HTV PRA, trade study by using this model can be performed.

As the first attempt, JAXA performed PRA trade study for HTV6 re-entry phase Safety Review Panel. Also JAXA performed PRA trade studies for the anomalies occurred on the previous HTV6 flights.

This paper will explain the result of PRA trade studies applied to the HTV6 flights.

1. INTRODUCTION

1.1. Introduction of the HTV

Japan participates in the ISS program providing the Japanese Experimental Module (JEM) and the HTV.

The HTV is an unmanned orbital carrier that can carry both pressurized and unpressurized cargo to the ISS. The HTV is composed of five sections; the Pressurized Carrier, Unpressurized Carrier, Avionics Module, Propulsion Module, and Exposed Pallet (Fig.1).

The HTV is launched by the H-IIB transfer vehicle, aboard with up to 6,000kg of system and experiment racks, and supplies. When the HTV approaches close to the ISS, the Space Station Remote Manipulator System (SSRMS) grapples the HTV and berths it to the ISS. While the HTV is berthed, the ISS crew can enter in it to unload cargo such as food, clothes, and experiment equipment. Then, the HTV is loaded with the wasted materials. After the transfer, the HTV is deberthed from the ISS and reenters the atmosphere (Fig.2).

The HTV initial flight vehicle was launched in 2009. So far, six HTVs have been flown to ISS successfully.

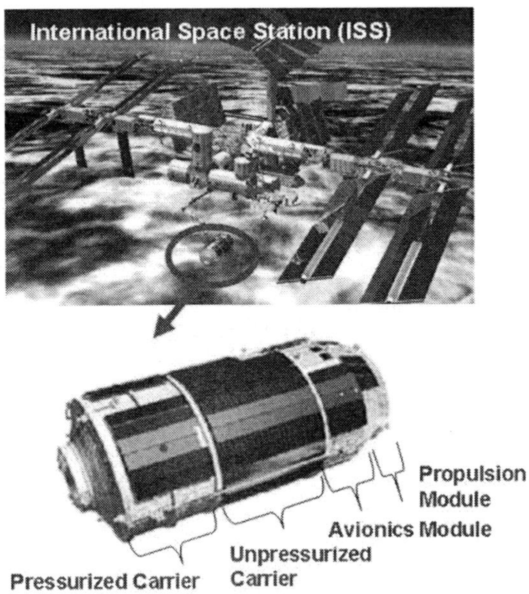

Figure 1. H-II Transfer Vehicle (HTV)

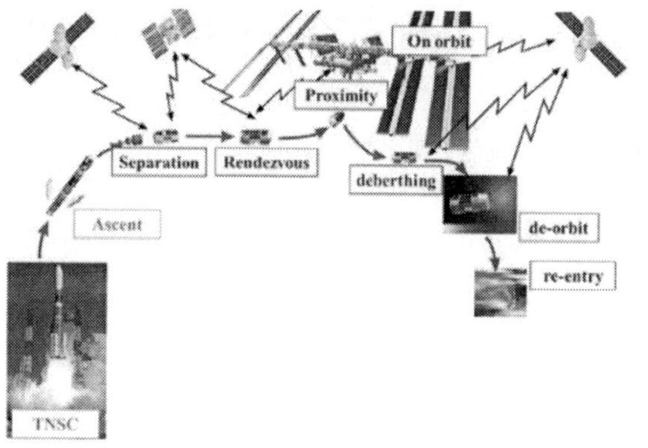

Figure 2. Overview of HTV operation

1.2. HTV PRA common model building

In JAXA, hazard analysis has been conducted for all HTV flights according to the ISS safety requirements. There has been no major incident so far and safety of the ISS crew and vehicle is assured by the adequate hazard analysis, controls and verification. In addition to this qualitative analysis, JAXA has conducted PRA for the HTV to get better safety assurance in the design/manufacturing phase.

While JAXA performed PRA for the HTV individually, NASA also separately developed the HTV PRA model mainly for calculating ISS generic risks, and for trade studies involving the HTV. However, NASA has several uncertainties in developing the HTV PRA model because NASA did not have enough data regarding the HTV. For this reason, among visiting vehicle PRA analysis in NASA, the analysis of the HTV has greater uncertainty than other visiting vehicles.

Therefore, NASA and JAXA started their joint efforts to update their models driven by the request from NASA. Through this activities, NASA and JAXA updated their model and finally the common models have been completed in 2016.

1.3. Motivation for trial of decision making using HTV PRA common model

PRA is a comprehensive, structured, and logical analysis method aimed at identifying and assessing risks in complex technological systems for the purpose of cost-effectively improving their safety and performance. Especially trade study of PRA is a powerful tool to make decision for the system at development phase and operation phase.

JAXA currently cannot perform quantitative risk assessment of the space systems effectively;

- For almost all of space systems, quantitative assessment are performed using by conventional method (reliability assessment), therefore a comprehensive, structured, and logical analysis like PRA has not been adopted.
- For manned space systems like JEM and HTV, PRA was conducted. However, there was few feedback from PRA to system design or operation because PRA was performed only to show the quantification of likelihood in Hazard Reports as mentioned above.

To solve this issue, for the first step, JAXA focuses on effectiveness of trade study of PRA in order to support decision making at development and operation phase. Now that NASA and JAXA completed to build their common models of HTV PRA, trade study by using this model can be performed.

As the first attempt, JAXA performed PRA trade study for HTV6 re-entry phase Safety Review Panel. Also JAXA performed PRA trade studies for the anomalies occurred on the previous HTV6 flights.

2. PRA TRADE STUDY FOR HTV6 RE-ENTRY SAFETY

2.1. Background

JAXA conduct the HTV re-entry safety review panel to assure public safety. JAXA has its own safety requirement for re-entry: JERG-0-047 Safety Standard for Controlled Reentry, and the HTV has to be complied with this standard.

JERG-0-047 requires that Expected number of casualty (Ec) shall satisfy the following criteria:

(1) Successful re-entry with all fragments within expected re-entry area. [Casualty of the people inside Footprint, if any.]

(2) Natural re-entry resulted from the loss of re-entry function. [System reliability for re-entry related functions are taken into account for calculation.]

(3) Unsuccessful (failed) re-entry that leads to Footprint slide out of the planned location. [Probability of failure during re-entry maneuver is taken into account for calculation.]

On the previous HTV flight, there was no deployable payloads like KITE (Kounotori Integrated Tether Experiment) (See section 2.2 overview of KITE system), therefore, Ec of natural re-entry could be calculated considering just component failures of the HTV system, and compliance with the above criteria could be shown with relative ease.

However, on HTV6, there were several discussion that the risk of casualty seemed to have more uncertainty because failure mode of KITE had some uncertainty. Especially, it was concerned that end-mass of KITE would collide HTV6 structure and might cause damage to controlled re-entry system of HTV6 if KITE could not cut off the tether. (Loss of controlled re-entry system results in natural re-entry.)

Therefore, PRA trade study was conducted to establish how much risk HTV6 with KITE had during re-entry phase.

2.2. Overview of KITE system

Major changes of HTV6 from the previous HTV series included build-in payloads to demonstrate new technologies, one of which was KITE (Kounotori Integrated Tether Experiment).
KITE was an experimental electrodynamic tether (EDT). The tether was equipped with a 20 kg end-mass, and would have been 700 m long when deployed (Fig.3).

Figure 3. KITE operation configuration

The EDT experiment was scheduled following Kounotori 6's departure from the ISS, with a planned duration of one week. After the experiment, the tether would have been separated before the spacecraft proceeds with the de-orbit maneuvers.

2.3. PRA Model updating

For the trade study of HTV6 re-entry, HTV PRA COMMON MODEL could not be used as it was, therefore the model was modified.
The end state of the original models was "collision to the ISS" as shown in the Event Sequence Diagram in Fig.4.

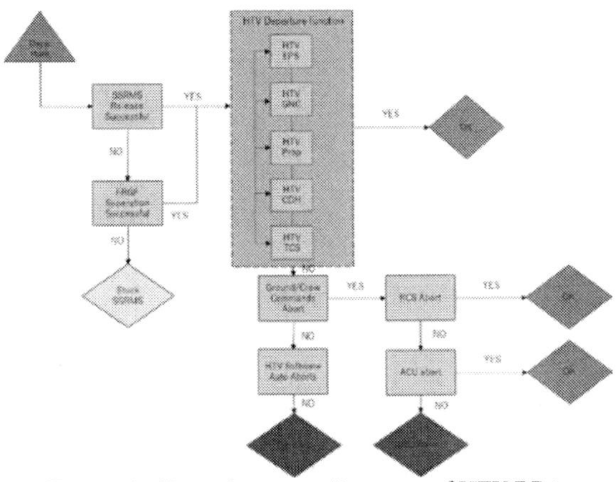

Figure 4. Event Sequence Diagram of HTV PRA COMMON MODEL

From re-entry safety perspective, it is important to assess whether HTV can perform controlled re-entry or not, therefore, the following modification was applied to the original model:

- The end state was changed to "loss of re-entry system".
- The pivotal events regarding "abort" were removed from the original model because the ability of abort was nothing to do with re-entry.
- The pivotal events regarding "KITE failure" was added. Major mode of loss of controlled re-entry system caused by KITE failure was assumed as "failure of cutting off the tether" (Even if KITE could not cut off the tether, loss of re-entry system of HTV6 might not occurred. However, in the PRA mode, it was assumed that failure of cutting-off the tether resulted in loss of re-entry system as worst case scenario.)

The modified ESD was shown in Fig.5.

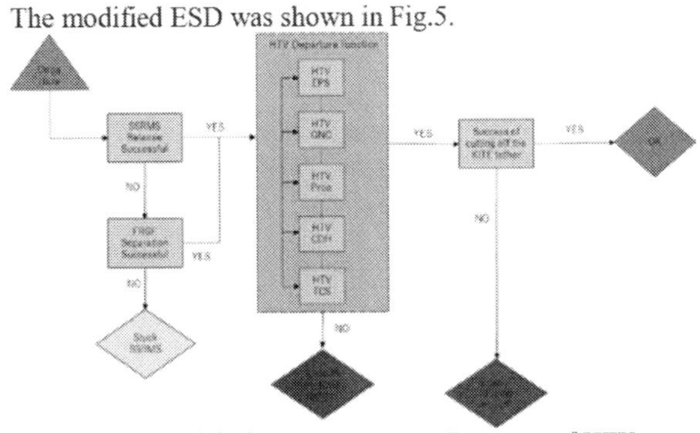

Figure 5. Modified Event Sequence Diagram of HTV PRA COMMON MODEL

2.4. Assumption

To conduct trade study for HTV6 re-entry by using the modified PRA COMMON MODEL, following assumption was established:

- Probability of loss of controlled re-entry for the HTV with KITE was compared with that for the HTV without KITE
  (This comparison means how much KITE contributes to loss of re-entry system)
- The pivotal event of "cutting off the KITE tether" sets to "always success" in order to calculate the probability of controlled re-entry for the HTV without KITE

2.5. Result of trade study

The result of this trade study is shown in Fig.6.
The upper probability shows loss of controlled re-entry system of the HTV with KITE (This probability means loss of re-entry system of HTV6 configuration), while the lower probability shows loss of controlled re-entry system of the HTV without KITE. The result shows that there is not so much differences between the two

probabilities and the probability of loss of controlled re-entry does not depend on the failure of KITE so much.

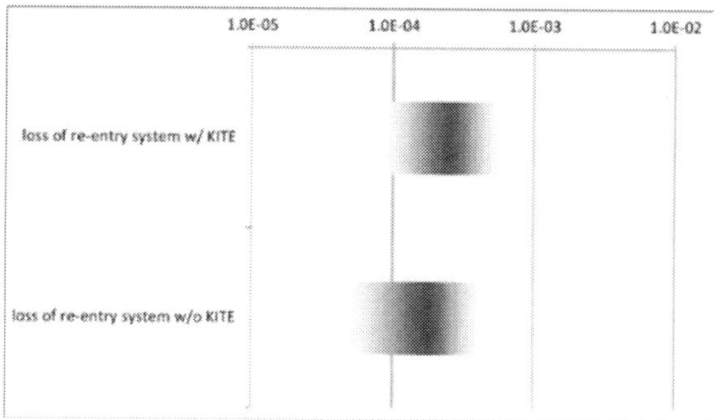

Figure 6. Comparison of loss of re-entry system

This result was explained at the HTV re-entry safety review panel. JAXA concluded that the risk of re-entry of HTV6 with KITE was acceptable based on the qualitative assessment for risk assessment of HTV6 controlled re-entry as well as the result of PRA trade study.

Thus, JAXA S&MA found that PRA trade study could support the rationale for acceptance of the risk at safety review panel.

3. PRA TRADE STUDY FOR ANOMALIES ON HTV6 ON-ORBIT OPERATION

3.1. Background

During HTV6 on-orbit operation, several anomalies were occurred, and HTV Flight Control Team (FCT) have performed real-time decision making (disposition) for each anomaly.

In order to study whether PRA can be used as support tool of decision-making by FCT in JAXA or not, JAXA conducted trade study using anomalies on HTV6 on-orbit operation of the following two cases:

3.2. <CASE1> Anomaly of ESA-A

3.2.1. Contents of anomaly

The HTV has several sensors to measure its position, and attitude as GNC (Guidance Navigation and Control). One of the sensors is Earth Sensor Assembly (ESA). The HTV has two strings for ESA; ESA-A and ESA-B. On Dec 10 2016, during on-orbit operation before proximity phase, ESA-A was failed and main sensor system of ESA was switched to ESA-B.

3.2.2. Overview of HTV GNC

Guidance, Navigation and Control (GN&C) subsystem is designed to control the trajectory and attitude through all flight phase. It consists of Guidance and Control Computer (GCC), Abort Control Unit (ACU), Earth Sensor Assemblies (ESA), Space Integrated GPS/INS (SIGI), Rendezvous Sensor (RVS), Valve Drive Electronics-1 (VDE-1) and Valve Drive Electronics-2 (VDE-2). Fig.7 shows the block diagram of HTV GN&C subsystem.

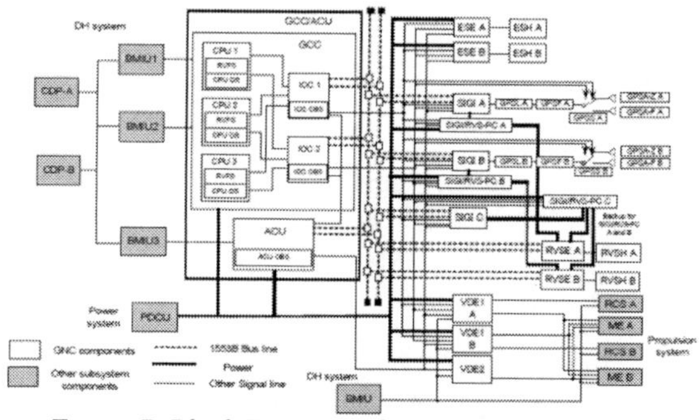

Figure 7. Block Diagram of GNC Subsystem

The component list of GNC subsystem is shown in Tab.1, which shows the major functions of GNC components.

Table 1. Major function of GNC Components

Component		Major Functions
GCC (Guidance and Control Computer)		Sensor Interface, actuator interface, GNC component power ON/OFF, guidance, navigation, control, CMD/TLM processing, FDIR functions
ACU (Abort Control Unit)		Sensor interface, actuator interface, VDE power ON/OFF, control
ESA (Earth Sensor Assembly)		Roll/Pitch attitude detection
SIGI (Space Integrated GPS / INS)	Gyro	Roll/Pitch/Yaw angular rate detection
	Accelerometer	Three axes acceleration detection
	GPSR (GPS Receiver)	GPS raw data receive, absolute navigation
RVS (Rendezvous Sensor)		Relative attitude and position detection, relative attitude rate and velocity detection
VDE (Valve Drive Electronics)		Valve drive to operate propulsion subsystem

GNC subsystem has some Fault Detection, Isolation and Recovery (FDIR) function. GNC subsystem detects failures of GNC sensors, actuators, Communication and DH subsystem, propulsion subsystem and Carrier. If Rendezvous Flight Software (RVFS) detects the failure of the primary sensor, RVFS masks it and use the output data of the secondary sensor. To detect the failure of the secondary sensor, RVFS also checks sensor status data, data content and compares the output data of the secondary sensor with the reference data. If RVFS detects the failure of the secondary sensor, RVFS executes abort. From the safety perspective, loss of GNC sensor will result in collision to the ISS.

3.2.3. Disposition of HTV FCT

HTV FCT decided to continue flight using the data from ESA-B (no use of ESA-A) and HTV 6 was successful to be docked to ISS.

3.2.4. PRA Model updating

For the trade study, HTV PRA COMMON MODEL was used as it was, because the end state of the original models was "collision to the ISS" and failure of sensors were already included in the model. The ESD is shown in Fig.8.

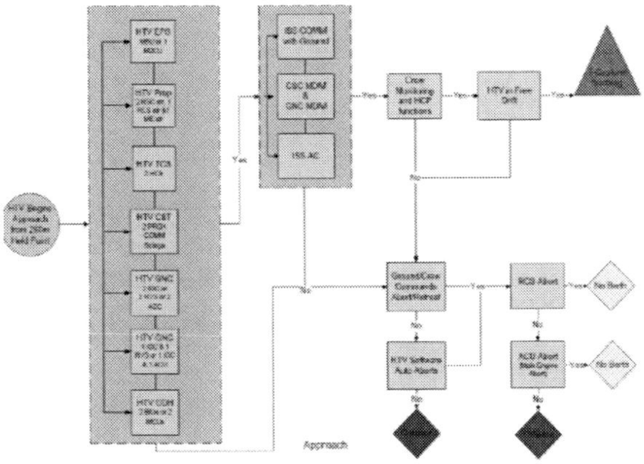

Figure 8. Event Sequence Diagram of HTV PRA COMMON MODEL

3.2.5. Assumption of HTV PRA trade study

To conduct trade study, following assumption was established:

- Probability of HTV collision to ISS without ESA-A (ESA-A is failed) was compared with that with normal condition (both ESA-A and ESA-B are normal).
- The basic event of "ESA-A" sets to "always failure" in order to calculate the probability of HTV collision to ISS without ESA-A

3.2.6. Result of trade study

The result of this trade study is shown in Fig.9.
The upper probability shows HTV6 collision with normal condition, while the lower probability shows HTV6 collision without ESA-A.

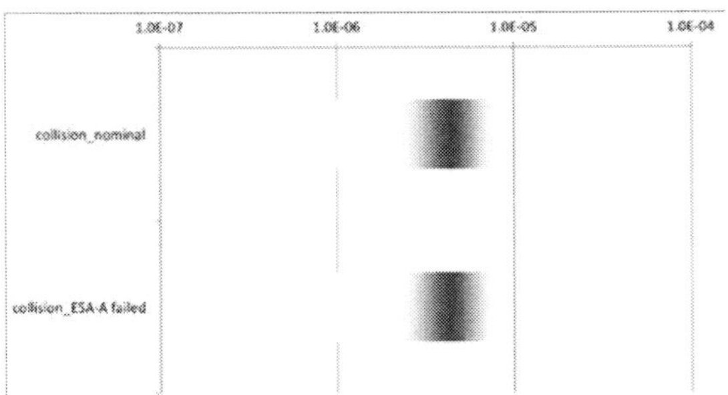

Figure 9. Comparison of normal v.s. ESA-A failure

The result shows that there is not so much differences between the two probabilities, therefore, collision risk did not become higher with no use of ESA-A.
Thus, JAXA S&MA found that PRA trade study could support the decision by HTV FCT regarding ESA anomaly.

3.3. <CASE2> Anomaly of status of heaters

3.3.1. Contents of anomaly

On Dec 14 2016, during on-orbit operation after docking to the ISS, several heater status of Avionic Module and Propulsion Module turned to ERROR. Also, switches of HCE-1A (Heater Control Electronics), which were upper side of a part of the heaters, became OFF.

3.3.2. Overview of HTV TCS

The HTV has many heaters to prevent components from freezing up. Especially, from safety perspective, freezing of propellant system such as plumbing will result in loss of attitude control of The HTV, therefore, redundant heater systems are established in The HTV. Fig.10 shows the block diagram of HTV TCS subsystem.

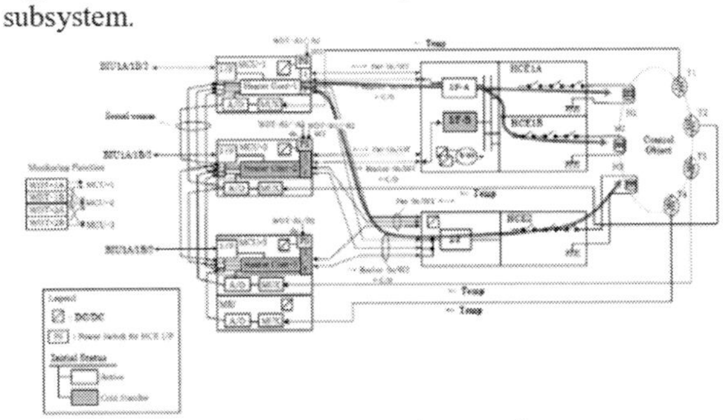

Figure 10. Block Diagram of TCS Subsystem

3.3.3. Disposition of HTV FCT

HTV FCT decided to continue flight as it is with real time monitoring of these heaters status.

3.3.4. PRA Model updating

For the trade study, HTV PRA COMMON MODEL was used as it was, because the end state of the original models was "collision to the ISS" and failure of HCE were already included in the model. (Same as shown in section 3.2.4)
However, The model of HTV PRA does not have the failure mode of a part of component such as "status error" or "switch failure", therefore, in this trade study, component failure of HCE-1A was used instead of the above failure mode. (This was very conservative assumption.)

3.3.5. Assumption of HTV PRA trade study

JAXA S&MA perform PRA trade study whether PRA can support HTV FCT decision regarding heater anomaly during operation.

To conduct trade study, following assumption was established:

- Trade study was performed by using HTV PRA COMMON MODEL
- Probability of HTV collision to ISS without HCE-1A (HCE-1A is failed) was compared with that with normal condition (both HCE-1A are normal).
- The basic event of "HCE-1A" sets to "always failure" in order to calculate the probability of HTV collision to ISS without HCE-1A. (This was very conservative assumption.)

3.3.6. Result of trade study

The result of this trade study is shown in Figure 11.
The upper probability shows HTV6 collision with normal condition, while the lower probability shows HTV6 collision without HCE-1A.

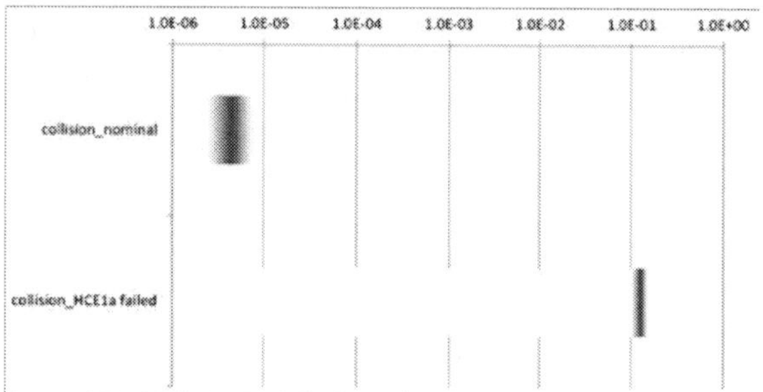

Figure 11. Comparison of normal v.s. HCE-1A failure

The result shows that there is much differences between the two probabilities. However, as mentioned the above, this result was very conservative because of the conservative assumption. Therefore, on one hand, PRA trade study cannot support this kind of anomaly because the HTV PRA models do not have such failure mode like status failure or switch failure. However, on the other hand, this result shows that there is much risk if failure of HCE-1A should occur during operation phase, therefore, it is shown that PRA have the potential to provide the information what kinds of next failure mode should be aware during operation.
Thus, JAXA S&MA could find the advantages and disadvantages of PRA trade study.

4. CONCLUSION AND FUTURE WORK

JAXA started studies to use PRA as applications to the decision making in JAXA internally. As the first attempt, JAXA performed PRA trade study for HTV6 re-entry phase Safety Review Panel. Also JAXA performed PRA trade studies for the anomalies occurred on the previous HTV6 flights.
As mentioned in section 3, JAXA S&MA found the advantages and disadvantages of PRA trade study. Based on these findings, JAXA S&MA will continue to try to use PRA as applications to the decision making in JAXA internally.
Also, JAXA is developing new transfer vehicle: the HTV-X. JAXA will study whether they can apply PRA to the HTV-X or not. If PRA is applied to the HTV-X, JAXA will start international collaboration to build a PRA model for the HTV-X with NASA.

5. REFERENCES

1. M.Stamatelatos, Ph.D., H.Dezfuli, Ph.D., NASA/SP-2011-3421 Probabilistic Risk Assessment Procedures Guide for NASA Managers and Practitioners.
2. T. Fukatsu, N. Takeuchi, T. Seike, K. Oga and Y. Sato, JEM Probabilistic Safety Assessment, PSAM 6, Vol.1, Elsevier Science Ltd., pp.949-955 (2002).
3. N. Takeuchi, T. Sato, T. Seike, K. Oga, S. Takada and Y. Sato, PSA for Depressurization Hazard in ISS module, PSAM7, Vol. X. (2004)
4. S. Takada, T. Fukatsu, T. Sato, K. Oga, S. Yoshikawa, M. Miki and Y. Sato, Probabilistic Safety Assessment directed to Manned Space System Operation, 1st IAASS(2005)
5. S. Takada, T. Fukatsu, T. Sato , M. Miki, K. Oga, S. Yoshikawa and Y. Sato, Improvement of JEM Operation by PSA, PSAM8 (2006)
6. S. Takada, N. Takeuchi, M. Miki, Y. Sato, Effective Probabilistic Risk Assessment Application for Japanese Manned Space Systems, PSAM9 (2008)
7. S. Takada, M. Miki, Y. Sato, Probabilistic Risk Assessment for H-II Transfer Vehicle, 3rd IAASS (2008)
8. MAPI contract, H-II Transfer Vehicle Analysis Notebook (2016)
9. M. Miki, S. Takada, T.Goto, K.Oga, T.Yoshihara, H.Kudo, N.Kamimori, J.Ni, J.Yasensky, P.Mortillaro, INTERNATIONAL COLLABORATION FOR HTV PRA ANALYSIS, 8th IAASS (2016)

A REAL-TIME LAUNCHING CALIBRATION SYSTEM HARDWARE DESIGN, AND FAILURE ANALYSIS APPROACH FOR THE REAL-TIME MEXICAN SATELLITE SPACE LAUNCH CENTER USING FTA AND MARKOV CHAINS

Omar Ariosto Niño Prieto(1), Francisco Ruiz Ciriaco(1), Vicente Guevara Ayala(1) Cuahutémoc Covarrubias Carranza(1), José Luis Sampayo García(1)

(1) OneSide Tech, Guadalajara Design and Invention Center, Research & Development; Circuito Lomas Altas Coto 1 MZ E-11. Serrada Bosques de STA Anita. C.P. 45645 Tlajomulco de Zuñiga, JAL.
Email:omar.ariosto@gmail.com

ABSTRACT

In this paper, the Real-Time Calibration System Hardware Design for the Mexican Real-Time Satellite Launch Platform is presented. This work describes the Space System Engineering Design, using Stuctured Analysis for Real-Time (SA-RT) methodology, the Hardware design and Probabilistic Risk Assessment using Fault Three Analysis (FTA) and Markov Chains. The software design is described in detail using LACATRE modeling methodology. This work was created at OneSide Tech Start-Up Headquarters, at Guadalajara, Mexico.

Keywords – Calibration System, Real-Time Sensors, Hardware Design, Software Design, Markov Chains, Fault Tree Analysis (FTA), Space Rockets, External factor, Quantitative data, SA-RT, Kalman filters, trajectories.

1. INTRODUCTION

The use and implementation of Kalman filters [34-38] with the purpose to align and calibrate complex inertial guidance systems based on altitude error equation adding model parameters from related systems like accelerometers and gyroscope included inside the satellites. This approach, allows the interaction between the Hardware installed in the Launch Pad, the one installed inside the Rocket, the satellites and the Control Center. Mexico holds the fourth best place in the world for building a Space Port to launch Space Satellites, since its geographic location is optimal for its construction (near the Equator)[23, 30]. It is essential to have the Probabilistic Risk Assessment and Failure Analysis Simulation using FTA (Widhchill Quality Software)[18] and Markov Chains in Space Systems Engineering from its design, in order to minimize risks and avoid any possible catastrophe or position rocket failure throughout any wrong orbit by means of a quantitative probabilistic approach and Simulation.

Mexico is planning to construct a Space Port [23] and a Rocket Space Launch Center in the near future in order to be part of the nations with Space Programs [19-22]. Some methodologies are used for the System Design, Modelling and Engineering (Hardware and Software), such as Structured Analysis for Real Time (SA-RT)[3,12], Detailed Block Diagrams for Hardware Design and LACATRE Real-Time Software Design [4 and 5]. This is the first step to propose and build the first Satellite Launch Platform in Mexico, including the Real-Time Calibration System Hardware Design. Related Researching Work had been presented at 6th IAASS Conference at Montreal Canada in May 2013 [26]. At 7th IAASS Conference at Friedrichshafen Germany in October 2014 [27]; researching work had been published at IJERA NASA Indexed magazine (Vol 5. Issue 11 2015) [29], and presented during the 67th International Astronautical Congress IAC 2016 at Guadalajara Mexico in September 2016 [31].

2. THE REAL-TIME LAUNCHING CALIBRATION SYSTEM

Since the beginning of space technology, human kind developed ad continue improved Telemetry, some examples are the use of radiosonde, developed early on 1930, that device recorded pressure and temperature which were used after the experiment. Today it communicates to a computer in Real-Time using radio [31]. In space launch systems, telemetry is used to monitoring extreme environment (temperature, acceleration, vibration,etc.), the energy supply, antenna alignment and (at long distances, e.g., in spaceflight) signal travel time. This data is used for decision support to determine the position and health of a launch vehicle and determine safety flight termination [31].

Problems may arise during launch, the system accuracy and reliability must be constantly controlled in Real-Time by use of multiple sensors and communications subsystems redundancy. Decreasing reliability of aging systems leads to higher operations and maintenance costs and increased risk of launch delay [31].

Noise is an unwanted effect and in the Space launch there are multiple sources that could lead to catastrophic

consequences [19, 20 and 21] . Kalman Filters are a powerful tool to deal with noise and they can act in Real-Time [34-36].
The Real-Time Launching Calibration System will monitor the launching orbital position in order to avoid launching errors that will eventually fail to position the satellites in the correct orbit after launching. Some launching errors are currently observed in some rockets, for example the Galileo Satellite launch, located in the wrong orbit on 22 August 2014 [9 and 10]. This Hardware design also allows preventing and minimizing the impact of a major catastrophic event, when activating a Rocket Leave module if an imminent failure is detected.

3. KALMAN FILTERS

RADAR (See Figure 1)[17 and 37] measurements with all kind of noises are get helped with Kalman filters, following basic sequence of tasks. First, the system object where Kalman filter is working will be initialized; later, the aircraft RADAR [17 and 37] system will read all data provided by the rocket. All kind of noise generated from the rocket will be passed to the Kalman filter system object [34-38].

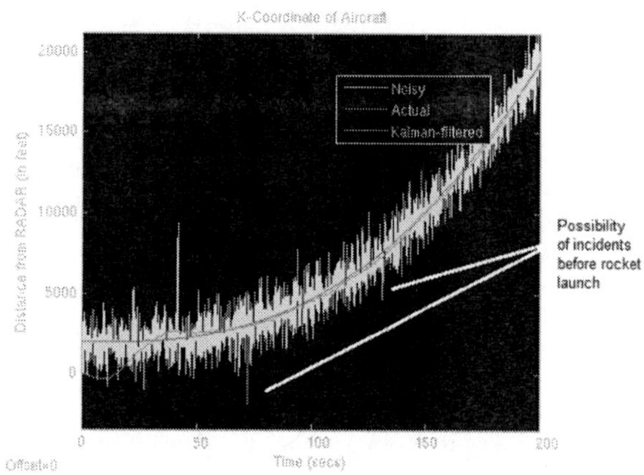

Figure 1. Probability of incidents before Rocket launch

At the launching moment of the rocket it is possible to identify some noises and we can compare against previous launching moments detecting in which case, failures had occurred. The Kalman filters need to be executed using MPU6050 3 Axis model (See Figure 2)[38].

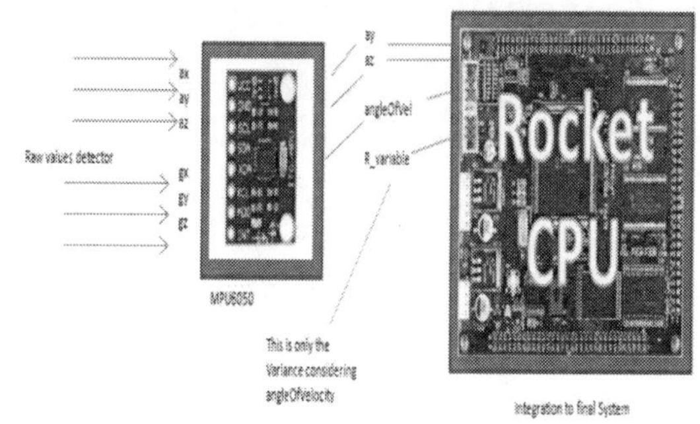

Figure 2. CPU Rocket

Inside of the Rocket CPU, the arcTan calculus implementing R variable will be deployed to get radian and rad/s values to generate Kalman filters [34-38]. In base of that value we can support the decision if the rocket is presenting failures during the Space Launch.

4. GENERAL SYSTEM

The proposed Real-Time Launching Calibration System Hardware design is located inside the general Rocket and Platform system design, and is the detailed design for the Real-Time System described in Figure 3. The complete system will send data to the Space Launch Center Monitoring System. Through the communication module, the Physical Variable Sensors will send data to the Space Launch Center Monitoring System, and then this information will go through a Network that will allow interfacing with the Real-Time System, which monitor the incoming events.
The wireless information from the Rocket and the Space Launch Center is bidirectional, as the Real Time Space Monitoring System receives information from the Sensors located on targeted components and systems, for example, those from the Rocket and Launch Pad in order to monitor the Physical Variables to avoid and minimize catastrophes while continous monitoring; the Real-Time Launching Calibration Hardware design continues the previous resarch and System creation for related Real-Time Systems, such as the Real-Time Volcanic Monitoring System, the Mexican Space Launch Center design and the Satellites failure Modeling [1, 2, 6, 8, 11, 26, 27, 29 and 31]. The complement and specialization that is covered on this paper includes a detailed Hardware and Software design described in the following sections.

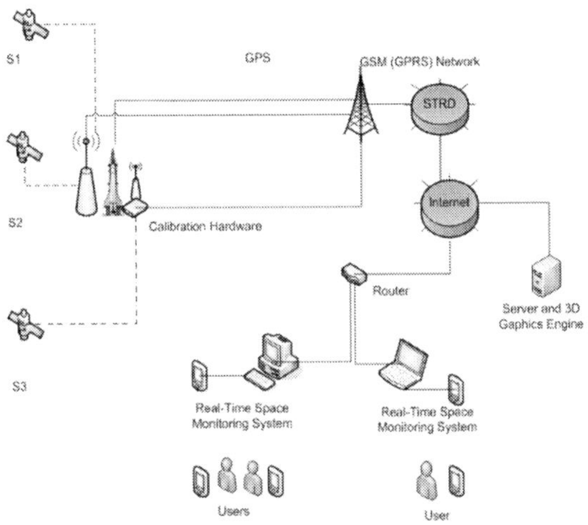

Figure 3. General System

5. GENERAL SYSTEM DESIGN

The previous design of the Real-Time Launching Calibration System [31] included a general validation state in the General State Transition Diagram [31], but with this new approach, the Validation Design evolved to become the Detailed HW/SW Validation and After Launch HW/SW Validation States (See Figure 4). They work together obtaining the validation results of the Rocket, during and after the Launching process. Those System States have a new design that allows describing the Calibration Process in a better way.

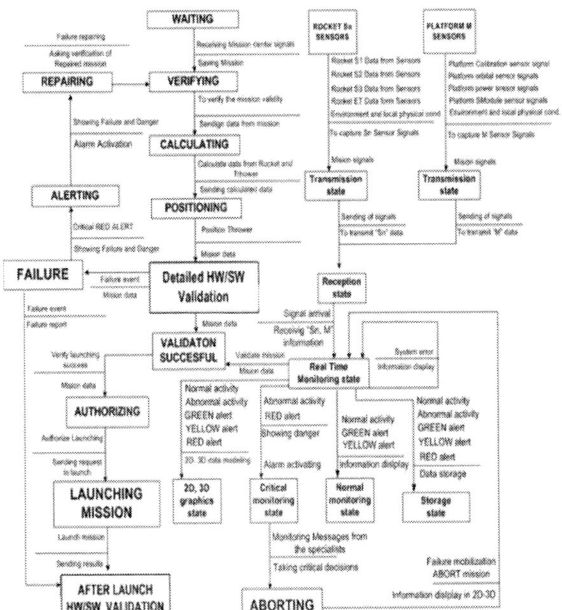

Figure 4. General State Transition Diagram

6. DATA FLOW DIAGRAM (DFD)

The validation that is done previous launching and after launching is described with the new DFD (See figure 5). All the information enter to the Control Bar, for example, the following events:

- The position Sensor after Launch (DATA).
- Mission Data NOISE.
- Real Mission Data (for all the components, for example, the Rocket Data, the Launch Pad Data and the External Factor Data).
- The Rocket Noise.
- The Platform Noise.
- The Environnemental Noise.
- Control Data.

All these data interact with all systems, so it needs validation in order to obtain the correct and succesful results. The Data that is out of the Control Bar, are the following events:

- Sending Data position for sensors after Launch.
- Sending all types of Noise.
- Sending all types of data from all the components.
- Sending the control data.

Then, there are several actions as follows:

1. To Receive sensor Information After Launch.
2. To Receive Noise.
3. To Receive Real Data.
4. To Receive Control Data.

All the data is transmited to the Hardware Validation (Kalman Filter) Action within the electronic components, so after this process, depending on the result from all the components will perform an action according to the information and data obtained; for example, the action failure, when the wrong orbit is detected after launching the rocket, will trigger an alert to activate the action of *Activating AUX propulsors*, integrated with the Rocket design in order to correct the wrong orbit of the Rocket, and the action *Correct Trajectory*, will eventually correct the wrong orbit. The same for the *Data success* and *Data failure* depending on the results after HW validation. All the information returns to the *Real-Time Data Monitoring Action* within all the system integration inside the Platform, the Rocket, and the External Factor. When the Rocket had launched there still exists a probability of failure or catastrophe, if happened, another action will be trigered with confirmation from the Sensor Data, for example, abnormal heat and turbulences, that will lead to the *Rocket Leave Activation* action; this means that the cargo module, where the most important systmes such as satellites or human are located, will leave the failing Rocket. With this action, lifes will be saved and enconomical damage will be reduced.

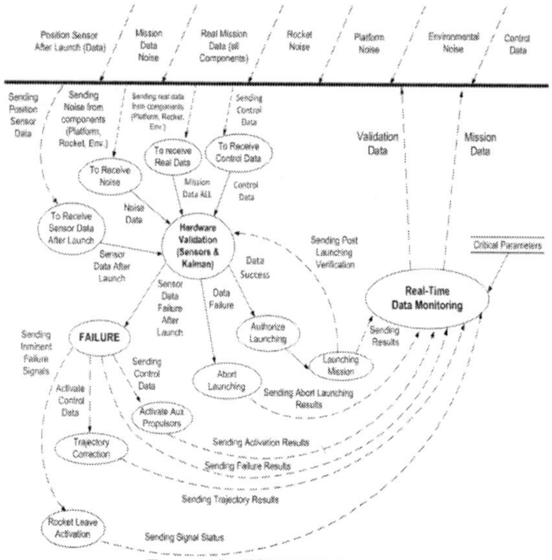

Figure 5. HW DFD Design

7. STATE TRANSITION DIAGRAM

The State Transition Diagram (See Figure 6) is logically obtained from the previous DFD (See Figure 5), and shows each state within the system while functioning. The different initial Sensors States, such as *Launching Platform and Mission Data*, will send data from the components that measure the trajectory of the Rocket and the launching parameters; at the same time, the *Generating Noise* state uses the Kalman filters to determine the noise, and the *Control State* to handle the controls of all the systems in Real-Time. The actions and System information from each state will determine the change of the next state. When all the sensor data, platform noise and control data is ready, then the system change to the *Transmission State* where the same information lead the change to the Reception State. All these states will transmit information such as the following:

- The position Sensor after launch (DATA)
- Mission Data NOISE.
- Real Mission Data (for all the components).
- The Rocket Noise.
- The Platform Noise.
- The Environnemental Noise.
- Control Data.
- Data signals.
- Failure Data.

All of this information after the *Reception State*, change to the *Real-Time Monitoring State* where the Hardware and Sofware is monitored in Real-Time. The Validation States such as *Validating Noise, Validating Launching Platform Data* and *Validating Control Data* states classify the incomming signals from the Launch Pad and Rocket Systems and sensors, so after receiving them, the state will change depending on the Sensors values. All the signals change in several sub-states inside validation state, as the Hardware validates all the signal information in the following states:

- Validating Sensor Information After Launch.
- Failure after launch
- Activate propulsors
- Imminent failure state
- Rocket leave
- Validating Noise.
- Validating Real Data from components.
- Validating Control Data.
- Success State.
- Failure State.

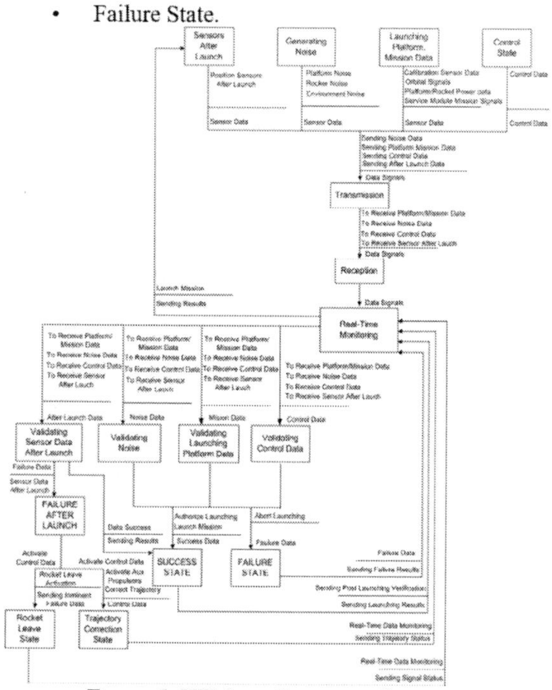

Figure 6. HW State Transition Diagram

After passing all the sub-states inside validation as mentioned, depending on the data parameters, will change as Success or Failure state. All the information returns to the Real-Time Monitoring State to continue with the mission monitoring. The data sensor resluts will describe if the mission sould be aborted or should proceed increasing the probability of success. In case of failure detection the mission will be aborted minimizing the impact of a catastrophe and if the rocket had launched, will be monitored. Some sub-states are found after the validation state, however if a dysfunctional behavior suggest a major catastrophic failure, then will have the following states: *Failure after launch, Activate auxiliar propulsors, Imminent failure, Rocket leave.*

If there is an imminent failure, the rocket will be lost, but the cargo or humans will be saved minimizing the catastrophe.

8. CALIBRATION SYSTEM SOFTWARE DESIGN (LACATRE)

The new Safety Software design, based on the previous Real Time Volcanic Monitoring System [1, 2, 6, 8 and 11], is represented by the LACATRE [4, and 5] Real-Time systems methodology.

The *main ( )* thread P1 will work with the following threads with descending priority (See Figure 7 and 8)[31]:

- P2 Rocket Sn Sensors.
- P3 Platform Sensors.
- P4 Environment and P. Variable Sensors.
- P5 Monitoring.
- P6 Alarm.
- P7 3D.
- P8 Position Thrower.
- P9 Validating.
- P10 Verify Data.
- P11 Execute Throwing.
- P12 Execute Mission.

The threads are synchronized by semaphores to ensure that all the processes have the correct data. The signals from the components of the Mexican Satellite Launch Platform, including those within the rocket and all the physical components that interact with the system. They are obtained with a communication data protocol and transmitted by the Data Resource (R.S.) to the threads P2, P3, P4 and P7 that collect data from the Satellite Launch Platform monitoring activity and the Rocket; this should include numerical data captured by the sensors from the components of the Platform. All this information will be received by the MsgA FIFO structure that will access the monitoring system. Furthermore, all the procedures will show numerical data on the screen, for example: Physical variables, heat information, electrical information, etc[31].

All the data are saved in a database, and the messages from each platform component and Rocket are sent independently to one FIFO data structure (MsgA), and the P5 thread receives them to process the information and data received by the functions included in the code of the P2, P3, P4 threads, (P5..N in case of more than 3 threads because only 3 are represented in the LACATRE Diagram), this is useful because if everything is all right the mission will proceed to the next state; whenever an abnormal activity exists, the function code sends a message to another FIFO structure (MsgB) to register the abnormality, and those messages are transmitted to the P6 thread, which writes down the abnormal data into the database, and activates an alarm and the mission will be aborted. Meanwhile, the function shows all the abnormal activity (or normal activity) on the screen of the Monitoring System and at the same time, it send another message with the status of the monitoring result to the MsgC, another FIFO for the validating Thread P9. At this moment, the Position Thrower P8, is generating also information with sensors, and it read data from the Calculation Resource (C. S). If any error found, it change the state, until it is corrected. Then the information of the Thread P8 is sent to the Launch Resource (L.S.)[31].

At this moment we have both Sensor and System information received on MsgC FIFO structure, (w data) and read from the Launch Resource (L.S.) (x data) to be validated in P9. This step is critical, because x and w data are validated, depending on the results: (x AND w) = True, (x OR w) = False to proceed to the next state. If any value is False, the mission will be aborted. If true, then the verification state with the P10 Verify Data Thread on Figure 8 will receive the information on MsgD FIFO structure [31].

The Second LACATRE diagram in Figure 8, will have the final mission states if x and w variables are TRUE. The Verify Data P10 and Execute Throwing P11 Threads has the control of the mission that could be aborted at any time if any unexpected behaviour or failure is still found in process. This is to be able to abort the mission when any failure is still detected that will produce an imminent catastrophe, however, the damages will be minimized and life will be saved. If everything still right, then the Mission will be launched, and the final P12 Execute Mission thread will be executed, showing the mission successful and the mission data will be saved[31].

The Kalman Filters measurements will generate the *Noise ( )* function and the *Noise Data Resource N.S.* pointed to the *Verify Data P10, Execute Throwing P11* and *Execute Mission P12* threads that will include the noise factor in the system as an external factor to be considered for the mission success (See Figure 8).

The complementary design has 3 new threads that will be working with the Kalman Filters results, and with the general system results working together with the HW and SW. The Noise (NS) Resource Data will be coming to the validating threads P9, P10, P11, P12, P13 and P14 with all the system and Kalman information in order to decide the action that will be done in case of catastrophe. *The Critical Noise Resource Data C.N.S* as well as the *Critical Noise ( )* Function is very critical whilst their data will be the parameter to activate the Emergency *Rocket Leave ( )* function if the sensors detect an explosion or an imminent failure. The new threads are (See Figure 9):

- P13: Emergency Propulsors.
- P14: Emergency trayectory correction.
- P15 Emergency Rocket Leave Activation.

They will receive information from the data structure FIFO named MsgG with XX data, taken from Kalman Filters results, afther this will be a process to decide if the data is between the safe parameters and if not, will trigger an alarm: AUX(Activate Auxiliar Propulsors), TC(Trajectory Correction), ELA (Emergency Leave Activation), if any of them happens, then the following actions and procecess will be executed (See Figure 9):

- AUX () -> Activate aux propulsors ().
- TC () -> Trajectory Correction ().
- ELA () -> Emergency Leave Activation ().

If there is no emergency action, the rocket will continue as lauched, and the mission will be successful as expected.

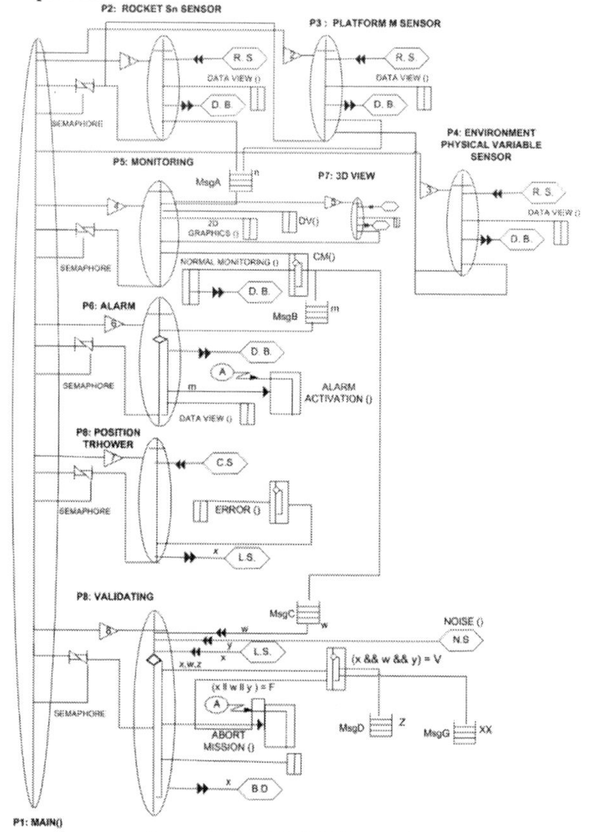

Figure 7. LACATRE Design 1

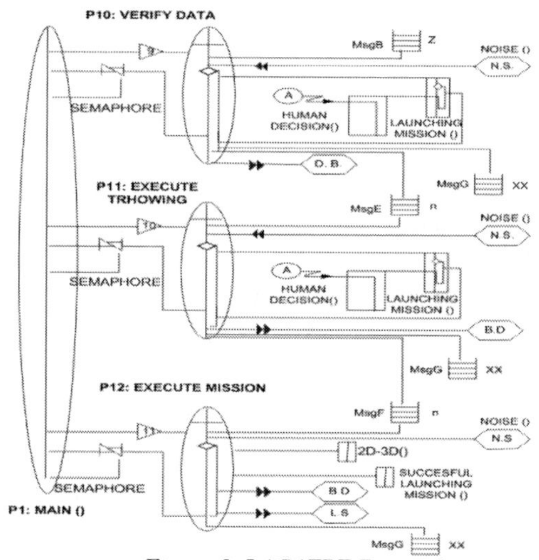

Figure 8. LACATRE Design 2

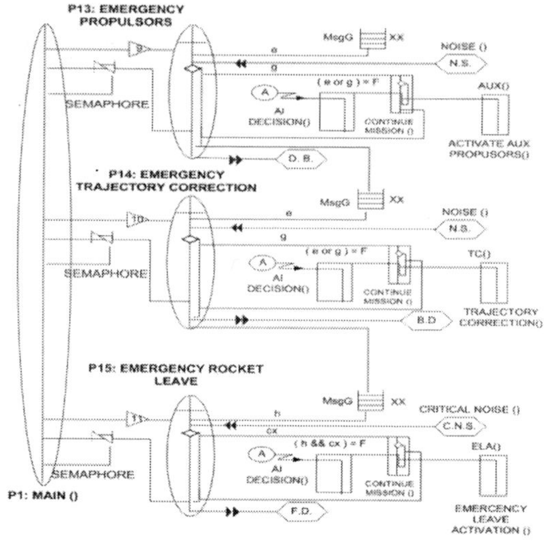

Figure 9. LACATRE 3

9. CALIBRATION SYSTEM HARDWARE DESIGN

The Real-Time Launching Calibration System has a Hardware (HW) design described below (see Fig. 10)[31]. The Space Physical Systems include the Launching Platform and the Rocket [31]. The Real-Time Calibration System has a group of sensors distributed in both, the rocket and the Launching Pad. The sensors will be able to receive information of the Rocket health, orbit status, and HW failures; they also will be sending wireless signals to the Real Time Hardware OS System in the Launching Pad in Real-Time. This electronic device have the HW system requirements such as CPU, Memory, optical and electronic components [31].

The type of sensor will be like the following, distributed to monitor all the Hardware systems[31]:

- Accelerometer sensors.
- Temperature sensors.
- Visual (camera) sensors.
- Infrared sensors.
- Inclinometer sensors.
- Electrostatic change sensor.
- Laser sensors.

The laser sensors will be peer to peer for remote monitoring the trajectory within the correct orbit, before and after the takeoff of the rocket with one or more satellites inside [31].

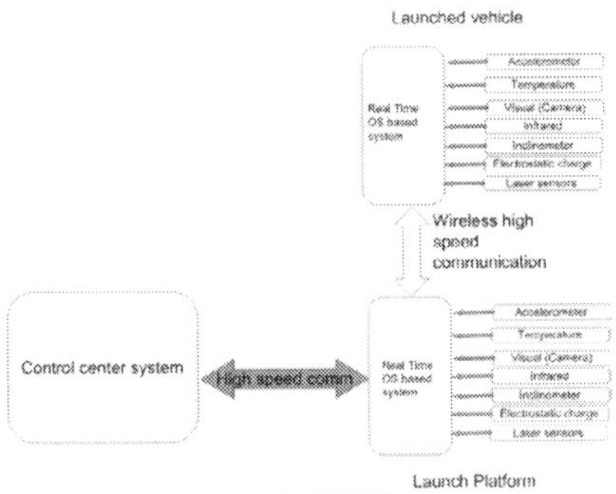

Figure 10. HW Design

The initiative describe the new detailed hardware model that detect failures in the Rocket (see Fig. 11).

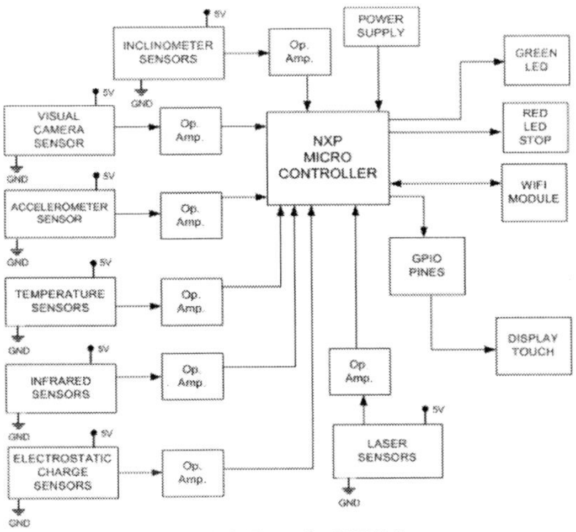

Figure11. Detailed HW Design

The proposal describes the data flow diagram for the new hardware model. The principal element is the microcontroller connected to 13 modules: *visual sensor, accelerometer sensor, temperature sensor, infrared sensor, laser sensor, inclinometer sensor, electrostatic charge sensors, Power supply, green and red led, the wifi module and display touch.*

The modules connected to microcontroller will send the data or information about the possible failures in the Rocket, the microcontroller has a routine that manages the failures, it will send the data using the Wifi module to the launching platform.

10. RISK ASSESSMENT: FTA & MARKOV CHAINS

The Risk Assessment and security model of the Real-Time Launhing Calibration System Hardware Design, is represented with Failure Tree Analysis (FTA) and Markov Chain modeling as a whole. The Fault tree analysis translates the Rocket Launcher System into a logical diagram to detect hazards and risks involved at the moment when the Rocket is launched. Each logical gate and small events are represented creating a visual model of how human factors, internal and external factors may intensify, to become a technical failure. AND gate [14] means both events are needed to become a failure while OR[14] gates means only one of the initial events are needed to create an eventual failure, the Fault Tree was constructed with Widhchill Quality Software [18]:

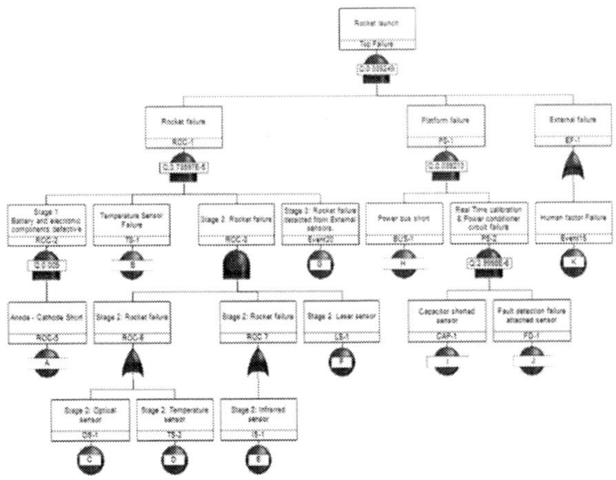

Figure 9. Simulation FTA Results

The Boolean algebra reduction is described as follows as Eqs. 1-4, and the components described below:

$$RocketLaunch = (RocketFailure \times PlatformFailure + ExternalFailure) \quad (1)$$

Where:

$$RocketFailure = (ROC5 \times TS1 \times (OS1 + TS2) \times IS1 \times LS1 \times Event20) \quad (2)$$

Then:

$$PlatformFailure = (BUS1 \times (CAP1 \times FD1)) \quad (3)$$

And:

$$ExternalFailure = EF1 \quad (4)$$

So the TOP Event equation results with the corresponding initial events description as Eq. 5:

$$TOP = \{ [(A * B * ((C + D) * E * F) * G)] * [(H * (I * J))] + [K] \} \quad (5)$$

Node A: Anode-Cathode common failures detected including delayed processes like primary issue in failures stats.

Node B: Temperature sensor detecting certain levels of temperature considered between ranges of safer levels.

Node C: Optical sensor to assimilate all kind of variations of color textures in many parts of the launch platform.

Node D: Temperature sensor to determine eligible levels to continue with the rocket launch.

Node F: Laser sensor to detect three indispensable targets from certain kind of distance.

Node G: External laser sensors using infra-red remote devices integrating IR sensitive elements.

Node H: Electrical portion from the system checking tightness.

Node I: Detection of passive electronic component where energy is stored in the form of electrostatic field.

Node J: Sensors are not working properly.

Node K: Including maintenance errors that lead to immediate or latent unsafe conditions.

The Fault tree analysis identify dependent failure potentials examining all possible events in the table showed below where it is possible to detect into the system which is the most potential hazard listing all the dependencies based on probabilistic index considering the most risk possibility until the latest.

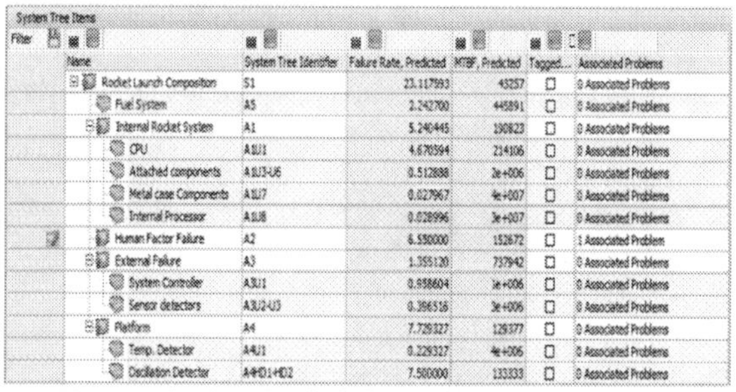

System Tree Items

Name	System Tree Identifier	Failure Rate, Predicted	MTBF, Predicted	Tagged...	Associated Problems
Rocket Launch Compositon	S1	23.117593	43257	☐	0 Associated Problems
Fuel System	A5	2.242700	445891	☐	0 Associated Problems
Internal Rocket System	A1	5.240445	190823	☐	0 Associated Problems
CPU	A1U1	4.670594	214106	☐	0 Associated Problems
Attached components	A1U3-U6	0.512888	2e+006	☐	0 Associated Problems
Metal case Components	A1U7	0.027967	4e+007	☐	0 Associated Problems
Internal Processor	A1U8	0.028996	3e+007	☐	0 Associated Problems
Human Factor Failure	A2	6.550000	152672	☐	1 Associated Problem
External Failure	A3	1.355120	737942	☐	0 Associated Problems
System Controler	A3U1	0.958604	1e+006	☐	0 Associated Problems
Sensor detectors	A3U2-U3	0.396516	3e+006	☐	0 Associated Problems
Platform	A4	7.729327	129377	☐	0 Associated Problems
Temp. Detector	A4U1	0.229327	4e+006	☐	0 Associated Problems
Oscilation Detecter	A4HD1+HD2	7.500000	133333	☐	0 Associated Problems

Figure 10. FTA Probabilistic Simulation Results

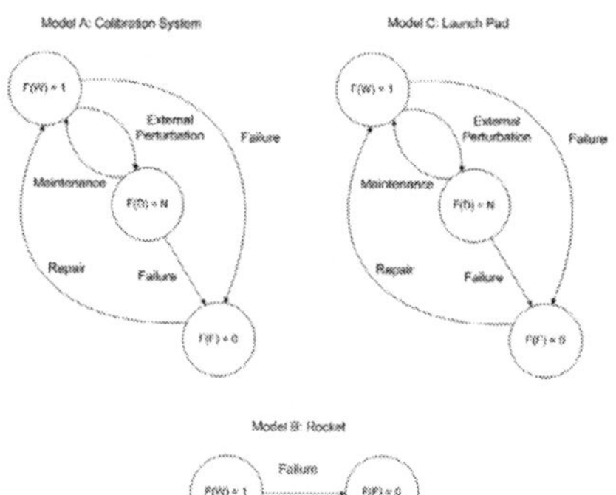

Figure 11. Initial MARKOV models

The Real-Time Launhing Calibration System Hardware Design, has the probabilistic Risk Assessment using Markov Chains. The Calibration System has the behavior of degradable Systems affected by Rocket lauching, environment and time (Figure 11). The failure model has tree faisable states: F(W)=1 when everything is working well, F(f)=0 when any system is in a failure state and F(D)=N when the components are in a degraded state. The Launch Pad model has also a degradable behavior and the Rocket model has a Non-Repairable system behavior. This mathematical approach is another way to see the dysfunctional states of the general system working as a whole. The Markov modeling representation is described with Alta Rica 3.0 failure modeling oriented programming language for critical systems aplying the composition with an Object/Model Oriented methodology [32 and 33]. The Calibration System, the Launch Pad and Rocket models are composed obtaining a general Markov system fault model (Figure 12) which has several states, but the semantics of the The Real-Time Launhing Calibration System Hardware Design is respected whilst there are some states which make no sense on persisting, for instance, the failure of all the systems at once, is not allowed because the complete system must remain working, otherwise, it will fail. However in case of emergency, for example if the Rocket fails, the following emergency states will be trigered: F(W,F,D) = X, F(W,F,F) = X, F(D,F,D) = X, F(D,F,W) = X, F(F,F,W) = X, F(W,F,W) = X, where X is the emergency action depending on the current failure state in the general Markov model (Figure 12).

Figure 12. The MARKOV general system model

The final model (Figure 12) describes the behavior of the system with the sensors placed on the launched vehicle, the Real-Time based OS System and the Control Center System; the degraded states within the system is proposed from 0.1 –the minimum degraded state- which can be completed from 0.9 to 0, when the addition must be: 0.1 + W = 1, where W = 0.9. The state changes depends on the transitions, for example: failure, External Perturbation, Repair, Maintenance, and their probability to occur; each state could have the value of F(A,B,C)= {a, b, c, d, e, 1} depending upon the state on the model.

11. CONCLUSION AND ONGOING RESEARCH

The Real-Time Launhing Calibration System Hardware Design are in their designing and conceptualization phase; this design will allow to minimize the consequences of catastrophes if occured, as history tell us that many hazards and failures had occured. The rocket Failure has several causes, for example the rocket engines use the pinciple of reaction. They are propelled/propulsed by igniting fuel which creates high pressure under control. If control of this process is lost, it will reach critical pressure, temperature or both. This usually results in an explosion. Another cause such as described for uper-cold oxygen reacting with the carbon fiber composites within the fuel tank; the proposal is to put several monitoring sensors that will inform the Rocket and the whole System state within the Launch Pad. The Hardware and Software within the Calibration System will monitor the components, and if a catastrophe is imminent, the system will reduce the consequences such as lives or economical loss. The construction of the Space Port in Mexico will have several technological benefits such as launching Satellites in a lower cost but with high reliability in the Space industry. If a Satellite could be saved, if a catastrophe occurs, for example when it leaves the rocket if an imminent failure is detected, this will have benefits to the clients and the cargo can be totally or partially saved.

12. REFERENCES

1. Niño O., Colmenares E, Martin M,. "Sistema de Monitoreo Volcánico en Tiempo Real", Congreso de IEEE, ANDESCON2008 (Ciudad del Cusco, Peru) Octubre de 2008. ISBN: 978-603-45345-0-6.
2. Niño O., Colmenares E, Martin M., "Propuesta para predecir eventos en el Sistema de Monitoreo Volcánico en Tiempo Real basado en la informática bio-inspirada y algoritmos de Redes Neuronales". Congreso SENIE Aguascalientes 2008 Universidad Autónoma de México. (UAM) Oct. de 2008. ISBN : 978-970-31-0944-9
3. E. Yourdon (1989). Modern Structured Analysis. Englewood Cliffs, New Yersey; Yourdon Press.
4. Ponsignon, J.-F. Maranzana, M.; Aubry, R (1997). Reuse and reengineering of multitasking Real-time applications in LACATRE. Real-Time Systems, 1997. Proceedings, Ninth Euromicro Workshop on, pp. 233 – 240, ISBN 0-8186-8034-2.
5. J. J. SCHWARZ, J. J. SKUBICH, Graphical programming for Real-Time Systems, Control Engineering. Practice, Vol. 1, No. 1, pp. 43–49, 1993.
6. Niño O., Colmenares E., Analysis and Risk Assessment of the Real Time Volcanic Monitoring System. IPN ISSN 1870-4069: Advances in Computer Science and Engineering, Volume 45, Antonio Alarcon EDS. Mexico DF, 2010.
7. Cantu-Paz, E. and Kamath C., An empirical comparison of combinations of evolutionary algorithms and neural networks for classification problems. IEEE Transactions on Systems, Man, and Cybernetics-Part B: Cybernetics, pages 915-927, 2005. UCRL-JC-151936.
8. Niño O., Colmenares E., Real Time Fire Reconnaissance Satellite Monitoring System, (2012), 2do Congreso Nacional y 1er Latinoamericano de Ciencia y Tecnología Aeroespacial; Sociedad Mexicana de Ciencia y Tecnología Aeroespacial (SOMECyTA), Septiembre 2012, San Luis Potosí, pp192.
9. Galileo: European Commission requests full details of launch problems from Arianespace and ESA http://europa.eu/rapid/press-release_IP-14-945_en.htm Recovered on September 2017.
10. Galileo Launch, Initially Hailed as Success, Is a Failure. http://spacenews.com/41650galileo-launch-initially-hailed-as-success-is-a-failure/ recovered on September 2017.
11. Niño O., Colmenares E., Huerta E., Aguila A. (2013) An Approach of Failure Analysis of the Real-Time Fire Reconnaissance Satellite-Monitoring System. Computer and Information Science Vol 6 no 4. Canadian Center of Science and Education. 139-153.
12. Roman G., Domer V., Matjaz C. Babau J.P. Skubich J.J. (2000) An Experiment in Design and Analysis of Real-Time Applications; Journal of Computing and Information Technology, CIT 8, 2000, 3. 181-195.
13. Rausand M., (1994) System Reliability Theory, Models and Statistical Methods, John Wiley and Sons Inc.
14. Gondran M. Pagès A. (1980). Fiabilité des Systèmes ; Collection de la Direction des Etudes et Recherches d'Electricité de France; 39, Eyrolles Editeur/Paris ; 1980.

15. Andrews J.D and Moss T.R. (1993); Reliability and Risk Assessment ; Longman Scientific & Technical; 1993.
16. J.A. McCall., (1977), Factors in Software Quality, General Electric no. 77C1502, jun 1977.
17. Stürmer, I., Travkin, D., (2007): Automated Transformation of MATLAB Simulink and Stateflow Models. Proc. of 4th Workshop on Object-oriented Modeling of Embedded Real-time Systems, pp. 57-62, 2007
18. Kultyshev, A. Y., Blagodarev, A. A., Gladkii, A. V., & Shanturov, D. N. (2013). The experience of implementing and using the Windchill product lifecycle management system at the energy machine building enterprise. Thermal Engineering, 60 (8), 559 - 566. http://dx.doi.org/10.1134/S0040601513080065
19. Programa del transbordador espacial. http://enciclopedia.us.es/index.php/Programa_del_transbordador_espacial Recovered on September 2017.
20. http://www.nasa.gov/mission_pages/shuttle/shuttlemissions/archives/sts-51L.html recovered on September 2017.
21. http://www.wired.com/2009/02/launchfailures/ Recovered on September 2017.
22. Lorenzi L., (1997) European Space Policies, Italy, Liberal, Democratic and Reformers Group;http://www.assembly.coe.int/nw/xml/XRef/X2H-Xref- ViewHTML.asp?FileID=7807&lang=EN Recovered on September 2017.
23. http://www.dgcs.unam.mx/boletin/bdboletin/2011_088.html Recovered on September 2017.
24. Peter Pin-Shan Chen: The Entity-Relationship Model--Toward a Unified View of Data. In: ACM Transactions on Database Systems, Vol 1, No 1, März 1976.
25. http://www.russianspaceweb.com/soyuz2_3_lv.html Recovered on September 2014.
26. Niño O., Colmenares E. (2013), Real-Time Fire Reconnaissance Satellite Monitoring System Failure Model. 6th IAASS Conference 21-23 May 2013, Montreal, Canada. "Safety is not an option", Proceedings European Space Agency (ESA SP-715, September 2013).
27. Niño O., Colmenares E., Carrera J. A., (2014), REAL-TIME MEXICAN SATELLITE SPACE LAUNCH CENTER FAILURE MODEL, SYSTEM DESIGN AND FAILURE ANALYSIS, 7th IAASS Conference 20-22 October 2014, Friedrichshafen, Germany "Space safety is no accident", Proceedings European Space Agency (ISBN 978-3-319-15981-2, Springer International Publishing, October 2014).
28. NOAA National Weather Service. A Brief History of Upper-air Observations 09 Aug. 16. http://www.ua.nws.noaa.gov/reqdahdr.htm
29. Niño O., Colmenares E., Carrera J. A., (2015) "A Mathematical Modeling Approach of the Failure Analysis for the Real-Time Mexican Satellite Space Launch Center". Journal of Engineering Research and Applications IJERA; ISSN: 2248-9622, Vol. 5, Issue 11, (Part - 1) November 2015, pp.38-45.
30. Gomez G. et al. Centro de Lanzamiento Espacial, Quintana Roo, México; Agencia Espacial Mexicana (AEM), Resúmenes de Ponencias; Foro y mesa de trabajo con el tema Investigación Científica y Tecnológica; Ensenada, Baja California 13-15 Octubre (2010).
31. Niño O. et al. (2016) A REAL-TIME LAUNCHING CALIBRATION SYSTEM AND FAILURE ANALYSIS APPROACH FOR THE REAL-TIME MEXICAN SATELLITE SPACE LAUNCH CENTER 67th International Astronautical Congress (IAC), Guadalajara, Mexico, 26-30 September 2016 IAC-16-34091.
32. Cherfi A. et al. (2014) AltaRica 3 Based Models for ISO 26262 Automotive Safety Mechanisms. F. Ortmeier and A. Rauzy (Eds.): IMBSA 2014, LNCS 8822, pp. 123–136, 2014. © Springer International Publishing Switzerland 2014.
33. FARAGO F., DECHEV A., (2014) DYSFUNCTIONAL MODELING OF SPACE LAUNCHERS 19e Congrès de Maîtrise des Risques et Sûreté de Fonctionnement – Dijon 21-23 octobre 2014.
34. Wu Y. et al (2017) Dynamic Magnetometer Calibration and Alignment to Inertial Sensors by Kalman Filtering. IEEE.
35. Cantinelli L. et al. (2016) Auto-Calibration Methods of Kinematic Parameters and Magnetometer Offset for the Localization of a Tracked Mobile Robot. Dip. di Ing. Elett. e Inf. (DIEEI), Unviversità degli Studi di Catania, Catania, Italy.
36. Beravs, T. et al. (2013) Three-Axial Accelerometer Calibration Using Kalman Filter Covariance Matrix for Online Estimation of Optimal Sensor Orientation. IEEE.
37. https://www.mathworks.com/help/dsp/examples/estimating-position-of-an-aircraft-using-kalman-filter.html?s_tid=gn_loc_drop. Recovered on September 2017.
38. http://blog.tkjelectronics.dk/2012/09/a-practical-approach-to-kalman-filter-and-how-to-implement-it/ Recov. September 2017.

PROCEEDINGS OF THE 9TH INTERNATIONAL SPACE SAFETY CONFERENCE

SESSION 10
LAUNCH SAFETY – I

NEW CONSENSUS STANDARDS FOR SHIP SAFETY DESIGN DURING LAUNCH AND REENTRY

Paul D. Wilde (1) , **Erik F. Larson** (2)

(1) *Federal Aviation Administration, 800 Independence Ave., Washington, DC 20591 USA, Email: paul.wilde@faa.gov*

(2) *ACTA Inc., 2790, Skypark Dr. Ste 310, Torrance, CA 90505, USA, E-mail: larson@actainc.com*

ABSTRACT

This paper describes recently revised safety standards for the protection of ships during launch and re-entry based on the consensus positions held by a broad cross section of the US range safety community, including the DoD, FAA, and NASA. Specifically, this paper presents innovative improvements published in the Range Commanders Council (RCC) 321-16 Standard, the foundational document that defines consensus standards for the range risk management process and risk criteria in the US. The new ship protection standards address all hazards from planned and unplanned events by specifying risk management criteria to define regions where Notices to Mariners shall be issued and where surveillance should be performed. The rationale for the RCC ship protection standards is also discussed, including recent updates.

This paper also describes an application of the recently revised safety standards for the protection of ships to commercial space transportation conducted under the authority of the FAA/AST. Specifically, the FAA granted SpaceX a waiver to regulatory requirements related to ship protection that enabled the CRS-8 mission: the flight of a Falcon 9 launch vehicle to deliver the cargo inside the Dragon capsule to the ISS. This paper discusses how the CRS-8 launch, which complied with the new RCC 321-16 Standard, satisfied the FAA's waiver criteria related to public health and safety, the safety of property, and the public interest.

1. INTRODUCTION TO RCC ORGANIZATION

The mission of the Range Commanders Council (RCC) is to serve the technical and operational needs of U.S. test, training, and operational ranges. RCC member organizations are U.S. government test, training, or operational ranges with a mission in the Research Development Test and Evaluation (RDT&E), Operational Test and Evaluation (OT&E), or training arenas. "Ranges" in the U.S. are where hazardous operations and tests are conducted such as missile flight, space launch and reentry, unmanned vehicle operation, explosive use, hazardous emissions, directed energy and laser operations. The term range is often shorthand for Department of Defense (DoD) national ranges, such as the Eastern and Western Ranges at Cape Canaveral, Florida and Vandenberg Air Force Base, California where most U.S. orbital launches originate, and the other Major Range and Test Facility Bases (MRTFB), such as those operated by the US Air Force, Army, and Navy. Non-military RCC members include the National Aeronautics and Space Administration (NASA) and the Department of Energy's National Nuclear Security Administration, Nevada Site Office. The Federal Aviation Administration (FAA) Office of Commercial Space Transportation (AST) is an associate member of the RCC. Associate Members are U.S. government agencies (other than member ranges) or agencies that are currently dedicated contractors working for the government, such as Sandia National Laboratory or Lawrence Livermore Laboratory, who have interests in common with the RCC and who have applied and been accepted for participation in RCC activities. Each representative of an associate member must possess technical knowledge of value and be willing to help solve problems within RCC areas of endeavor.

From its inception to the present, *the RCC has remained outside formally established DoD structures. This has been one of the organization's strengths as it has allowed frank and open dialogue* among the Range Commanders and other participants with the freedom to address a broad range of issues. The advantages of informal assistance and guidance on common problems have been salient factors in the RCC's development of widely-accepted technical standards for the ranges. The RCC also remains an effective consulting body to more formally-established military and civilian organizations in US.

The RCC provides a framework where:

1. Common needs are identified, and common solutions are sought,
2. Technical standards are established and disseminated,
3. Joint procurement opportunities are explored,
4. Technical and equipment exchanges are facilitated,

5. Advanced concepts and technical innovations are assessed, and promising applications are identified.

The RCC consists of groups that address various issues such as airspace operations, environmental stewardship, meteorology, telecommunications, timing, telemetry, and frequency management. The RCC's Range Safety Group (RSG) supports, through standardization, development, and continuous improvement, the safe conduct of hazardous operations on the test, training, and operational ranges and related facilities. This paper describes recently revised standards developed by the RSG, which includes three standing committees: Risk Committee, Flight Termination Systems Committee, and Unmanned Aerial Systems Committee.

2. INTRODUCTION TO RCC 321 STANDARD

Consensus based technical standards are a primary product of the RCC. All RCC Standards are advisory documents. However, all the major changes to the official public risk tolerability criteria applied to civil, commercial, and military space launch and reentry operations (i.e. adopted by the FAA, NASA, and USAF) in the last decade have followed changes to the RCC 321 Standard. Specifically, the USAF[1] and the NASA[2] officially adopted a collective risk criterion of 100E-6 Expected Casualties (EC) in 2010, following the precedent set in the RCC 321-07 (published in 2007)[3]. More recently, the FAA promulgated a similar change to the collective risk criteria in the Code of Federal Regulations (CFR) that governs US commercial launch and re-entry operations.[4] The RCC 321 safety criteria have also served as a benchmark, and have sometimes been adopted outright by international authorities that govern the safety of launch[5] and reentry[6] operations. Thus, the usefulness and credibility of RCC products are recognized throughout the U.S. Government (USG), industry, academia, and internationally.

RCC 321-16 Standard is the foundational document that defines consensus standards for the range risk management process and risk criteria. The RCC 321 Standard and Supplement are specifically intended to provide a common set of range safety policies, risk criteria, and guidelines for managing risk to people and assets during manned and unmanned flight operations. These safety criteria in the RCC 321 Standard apply to launch and reentry hazards generated by endo-atmospheric and exo-atmospheric range activities including both guided and unguided missiles and missile intercepts, space launches, and reentry vehicles. (RCC 321 does not address aviation operations or UAV operations; RCC 323 provides safety criteria for unmanned air vehicles.) Specifically, the RCC 321 Standard and Supplement establishes consensus standards and guidelines on:

a. Tolerable risk criteria for both the general public (involuntary acceptance) and mission essential personnel (voluntary acceptance), excluding people in the launch or reentry vehicle.
b. Debris injury thresholds for unprotected people.
c. Debris hazard thresholds for aircraft and ships.
d. Vulnerability models for commercial transport aircraft and business class jets, and an
e. Evaluation of flight hazards to critical assets.

With the publication of RCC 321-16 Standard (in 2016) comes the following:

f. Updates to the ship protection criteria, ship surveillance and vulnerability guidelines,
g. Updates to COLA standards and guidelines, and
h. Extensions to property protection standards.

RCC 321 content is based on the consensus positions held by the US range safety community. Consensus standards with the highest levels of priorities are generally provided in the main body of the document, while the supplement contains lower levels of priority requirements, guidelines, and example methods. The Supplement provides additional detailed information to assist in implementation of the standards, including the rationale for the standards. Updates to the RCC 321-10 Supplement are in progress, which will include updates to the ship vulnerability guidelines, models, rationale, etc. The standards in RCC 321 are not considered absolute; rather, this document and its supplement are intended to provide a baseline for defining acceptable risks for hazardous range operations and to assist in the development and implementation of robust risk assessments.

3. SHIP PROTECTION STANDARDS

The ship protection standards in RCC 321-16 address all launch and re-entry hazards from planned and unplanned events by specifying risk management criteria to define regions where Notices to Mariners shall be issued and where surveillance should be performed. The term "ship" includes boats and watercraft of all sizes, and separate standards apply to "non-mission" versus "mission-essential" ships.

3.1. Standards for Warning Areas

To protect non-mission ships, RCC 321-16 states that *"Notices to Mariners (NtMs) shall be issued to warn non-mission ships of regions defined by one of the following approaches:*

(1) where the probability of debris capable of causing a casualty impacting on or near a vessel exceeds 10E−6 (1E−5), accounting for all relevant hazards; or

(2) the union of the areas where the individual probability of casualty for any person onboard exceeds the criteria in a of Subsection 3.2.1 [i.e. individuals must not be exposed to a probability of casualty greater than 1E−6 for any single mission], the collective casualty expectation for an individual ship would exceed the criterion in b of Subsection 3.2.1 [i.e. collective risk for the general public must not exceed a casualty expectation of 100E−6 per mission], and the catastrophic risk for an individual ship would exceed the provisional criteria outlined in Section 3.6 [summarized below]." The words in square brackets are not in the Standard, but are added here to promote clarity. In addition, the remainder of this section explains the new standards.

The Standard recognizes that, in some cases, a range may choose to expand the warning area to provide additional mitigation so that the individual and collective risk criteria are met. On the other hand, extremely large ship warning areas may be ineffective because some ship captains might ignore warnings that would cause extreme departures from their planned route. Therefore, warnings may be optional when expected ship traffic in the affected area is low and adequate observation will be performed.

The first option, which is the more traditional approach employed by most US range safety organizations, requires formal warnings to waterborne vessels based on the probability of impact (PI) on a vessel that is "capable of causing a casualty." Specifically, people in ships must be warned of launch or reentry hazards, from planned events or malfunctions, anywhere the probability of impact on a vessel exceeds 1E-5 (one in one-hundred thousand). The previous version of the Standard included a footnote to state explicitly that "this includes any debris capable of producing a casualty to an unsheltered person." Thus, *debris "capable of causing a casualty" translates most simply into inert debris expected to impact with a kinetic energy greater than 15J (11 ft-lb).*

Determining the explosive debris impacts that are "capable of causing a casualty" is more complicated; for example depending on the explosive yield and the type of vessel exposed.[7] For a relatively small yield, the prime threat to ships in shallow water is often loss of hull integrity. However, for a large yield, the largest damage radius is often due to capsize from a surface wave for small boats in deep water. For explosions in the vicinity of small ships, the loss of ship hazard radius is larger than any hazard radius which directly leads to a casualty. The most recent ship vulnerability analysis indicates that *a 9 kPa (1.3 psi) incident overpressure threshold is a reasonably conservative and simple metric for "capable of causing a casualty" that applies to all yields in all water depths for smaller boats (15m, or 50 ft. long or less).*

Since (a) even a small boat is more than ten times larger in horizontal extent than an individual human being, and (b) the debris "capable of causing a casualty" are equated with thresholds for unsheltered people, the 1E-5 PI standard ensures than NtMs are always issued for regions where an exposed individual would exceed the 1E-6 probability of casualty criterion. Additional rationale is summarized below.

The second option in the RCC 321-16 Standard explicitly ensures that warnings are issued for areas where individual, collective, and catastrophic risks would exceed standard safety limits. Specifically, people in ships must be warned of hazards, from planned events or malfunctions, in the union (superset) of the areas where the individual probability of casualty for any person onboard exceeds the 1E-6 probability of casualty, the collective casualty expectation *for an individual ship* would exceed a 100E−6 EC, and the catastrophic risk for an individual ship would exceed the provisional criteria explained below. The probability of an adverse outcome (e.g. casualty) is the aggregated probability due to all types of hazards to people on ships, such as the loss of the vessel due to a sufficiently large inert fragment impact that penetrates the hull, an impact that ignites onboard fuel, air blast, or an underwater pressure wave.

The intent of the collective and catastrophic risk components of this standard is to ensure that warnings are always issued for areas where the presence of any given single ship would exceed the prescribed limits. For example, *the EC computed for an individual ship to comply with this standard assumes the presence of an individual ship for each type of waterborne vessel foreseen as potentially present.* Thus, this approach avoids the potential complications associated with estimates of the ship traffic and the effectiveness NtMs for the region where debris might fall.

RCC 321-16 includes a policy objective that "the risk of a catastrophic mishap should be mitigated," and provides catastrophic[1] risk criteria designed to protect against scenarios involving numerous casualties. RCC 321-16 includes provisional catastrophic risk criteria to supplement the collective and individual risk criteria. Specifically, catastrophe aversion limits are defined by the general formula,

$$P \times N^{1.5} \leq \text{criterion} \qquad (1)$$

where P is the cumulative probability of all events capable of causing N or more casualties, N is the number of casualties associated with a scenario, and

[1] The term catastrophic refers to multiple casualties with a minimum ranging from five to ten depending upon the particular regulation. The criterion presented here is "risk averse," a term that is used in academia and applies to all values of *N* above one.

the criterion is the maximum allowable collective risk for the event with various scenarios as feasible outcomes. The Standard recommends criterion of 1E-4 for the general public, and 3E-4 for mission essential and critical operations personnel. This formula is used to define the recommended catastrophe aversion criteria, or acceptable catastrophic risk profile, not to indicate how to compute the potential for catastrophic outcomes. The RCC 321-07 Supplement provides guidelines designed to facilitate evaluation of catastrophe potential for launch and reentry, and a forthcoming update will contain additional details needed to facilitate higher fidelity analyses of the probability of a catastrophic ship impact.

To fully implement the catastrophe aversion criteria put forward by the RCC requires computation of a risk profile, examples of which are publicly available [6,8,9], followed by comparison with the maximum acceptable risk profile shown in Fig. 1 However, any single scenario that corresponds to a point above the red line shown in Fig. 1 conclusively demonstrates that the hazardous operation exceeds the recommended catastrophe aversion criteria. Thus, the catastrophe aversion criteria may be used to identify individual scenarios or failure modes that present elevated catastrophe potential and practical mitigations

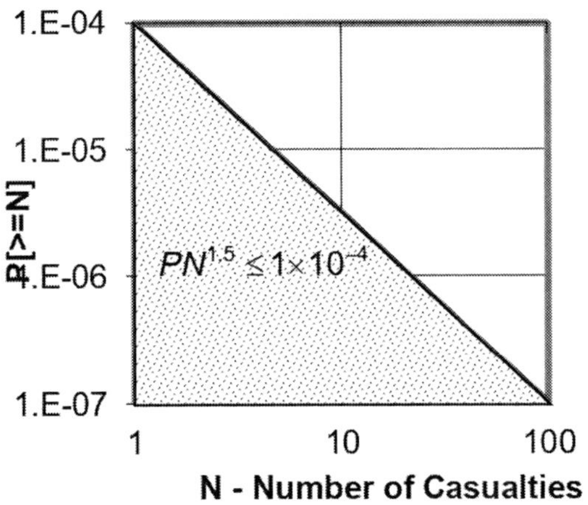

Figure 1: Provisional Catastrophic Public Risk Criteria in RCC 321-16

For example, an analysis could show that the presence of a high concentration of people (e.g. on a cruise ship) in a location subject to impact from large debris jettisoned during a launch corresponds to a point above the red line shown in Fig. 1. Thus, in that case surveillance would be required to ensure no cruise ship in that area during the spaceflight operation to demonstrate compliance with the 321 Standard as discussed in the next section. Examples of risk profile computations were previously published to demonstrate the application of the RCC 321 catastrophe criteria to the ATV-1 reentry hazard area for ships and aircraft [6], as well as a more generic explosion event[10]. More discussion appears below.

3.2. Standards for Surveillance Areas

The word "surveillance" did not appear in previous versions of RCC 321 Standard, and the 321-10 Supplement simply stated that "surveillance regions are typically functions of need (traffic density in a hazard region) and technical feasibility of monitoring." Indeed, US agency level requirements have typically delegated decisions regarding specific requirements for ship surveillance to more local authorities; for example, AFI 91-217 does not mention ship surveillance and NASA's NPR 8715.5 simply states that "launch/flight commit criteria shall provide for...surveillance of any established hazard areas." However, the latest version of the RCC 321 Standard addresses ship surveillance requirements explicitly and unambiguously, and thus fulfills a previous void in US consensus standards for launch and reentry safety.

The RCC 321-16 Standard requires the use of suitable observation techniques (i.e. surveillance) to locate non-mission ships during a launch countdown in those region(s) where the: (1) individual probability of casualty exceeds 1E-6, and (2) collective casualty expectation or provisional catastrophic risk criteria would be exceeded given a conservative estimate of typical ship traffic.

The requirement for surveillance of all areas where the presence of an individual would exceed 1E-6 probability of casualty applies to people on land, on ships, or on aircraft. The "one-in-a-million" individual public risk criterion is a foundational requirement adopted by all US agencies that oversee launch and reentry operations. The FAA has never waived this individual risk criterion, even though the old collective risk criterion (30E-6 EC) was waived in several cases, including the first flight test of NASA's Orion capsule.[11,12] There is broad acceptance of the individual risk criterion in many countries for space operations and other industries.[13]

Note that the individual and collective public risk criteria given in the RCC 321 Standard are intended to apply to all members of the public, including people on land, on ships, and on aircraft. Thus, the Standard requires people on non-mission ships observed during the countdown to be accounted for in the determination of compliance with collective risk criteria and provisional catastrophic criteria prior to launch.

The following description of potentially "suitable" ship surveillance techniques are based on a documented RCC study.[7] An important element of modern ship surveillance techniques is the Automatic Identification System (AIS), a ship and shore-based broadcast system. All vessels over 300 tonne on an international voyage, all domestic vessels over 500 tonne, and all passenger carriers are required to operate "Class A" AIS transponders, broadcasting continually updated data, such as identity, position, course, speed, ship characteristics, cargo and voyage information to other vessels and the shore. Class A vessels under way send out updates in 2–10 second intervals, while Class A ships at mooring update every 3 minutes. AIS uses VHF radio signals with a horizontal range of 20 to 40 nm (37-74 km), so it provides reliable information on ship locations within coastal zones and between closely situated ships. Space-based AIS signal reception is so effective that satellites reach the limit of the number of ships that can be tracked at once and are augmented by terrestrial receivers in congested regions. There are several public sources of near real-time data on ship traffic available through the internet, which aggregate AIS data from satellites and ground stations.

Even so, AIS data monitoring certainly does not guarantee ship detection. The requirement to transmit AIS data does not include small non-passenger vessels, which include a significant percentage of near shore fishing and pleasure craft that are not mandated to use AIS. Military ships do not commonly broadcast AIS. In addition, vessels often turn off AIS in areas where piracy is a concern. Detection of vessels without AIS can be accomplished by fixed ground based and ship board radar, as well as surveillance aircraft, all with line-of-sight detection. Range safety organizations with access to classified information can get comprehensive real-time vessel tracking data from the US Navy. The upcoming version of the RCC 321 Supplement will provide more detailed guidelines on "suitable" ship surveillance techniques.

4. SHIP PROTECTION RATIONALE

The rationale for the original standards on ship protection, including the 1E-5 probability of impact, was published in the RCC 321-07 Supplement. The general principle adopted by the RCC was that acceptable risk levels should be less than background risks. The RCC specifically endorsed the following qualitative standard as a guideline for the development and implementation of range safety requirements: "*for any individual uninvolved in the mission, but participating in a voluntary activity that increases their background risk (such as traveling in an aircraft or waterborne vessel), the chances of casualty resulting from the mission should be less than the background risk associated with the voluntary activity.*" Thus, the Risk Committee performed and documented an analysis to compare the tolerable risk criteria due to launch and reentry to actual accident experience in other categories that have a much larger statistical base, to ensure that the proposed risk criteria tolerates only risks that are a reasonable fraction of normal background risks, which the FAA equated with those risks voluntarily accepted in the course of normal day-to-day activities.[14]

The original RCC rationale for ship protection standards cited an International Maritime Organization (IMO) proposal[15] to have a more demanding target for new ship designs and that individual annual fatality risks should be below 1E-4 for crew, and below 1E-5 for passengers and public ashore.

Since 2007, the IMO published revised guidelines for formal safety assessment, including "risk evaluation criteria," defined as criteria used to evaluate the acceptability/tolerability of risk."[16] Similar to the risk tolerability criteria in the RCC 321 Standard, the IMO "risk acceptance criteria" for ship safety are "recommended risk evaluation criteria;" member states and non-governmental organizations are "invited to apply the revised guidelines," but these have NOT been universally accepted by IMO member states.

The IMO guidelines [16] identify a lower bound of one in a million (1E-6) probability of fatality per year for individual crewmembers, passengers, and members of the public ashore. The IMO guidelines state that risks at this lower bound are considered broadly acceptable for maritime activities and should be the target value for new ships. Thus, *the RCC 321 requirement to ensure no ships are present in areas where the individual risk exceeds 1E-6 probability of casualty is consistent, but a bit more conservative than the IMO recommended risk evaluation criteria.* The conservative aspect is that casualties are defined to include serious injuries or worse, including fatalities. There is a chance that some of that conservatism could be offset by the fact that the RCC criteria is per launch or reentry operation; whereas the IMO guidelines are per year. Although it seems unlikely that the same individuals on waterborne vessels would be exposed to risk from multiple launch or reentry operations over a given year, a particular barge has been observed in ship warning areas near Cape Canaveral during multiple launches in a given year. The United Kingdom, which uses risks in their regulation of many industries, regard an individual annual risk of fatality below 1E-6 as "*so low that they merge into the background risks of life, and they require no action.*"[17]

The IMO guidelines identify the following upper bounds as the maximum tolerable individual risks for maritime activities:

1. 1E-3 probability of fatality per year for crewmembers, and
2. 1E-4 probability of fatality per year for passengers, and members of the public ashore, and

3. Values one order of magnitude lower than those listed above for people on new ships.

Note that the originally proposed IMO standards for individual annual fatality risks (below 1E-4 for crew, and below 1E-5 for passengers and public ashore), which were appealed to as part of the rationale in the 2007 version of RCC 321, are at the median between the lower and upper bounds now adopted by the IMO as "recommended risk evaluation criteria." Thus the original rationale in RCC 321-07 remains valid.

An analysis of worldwide and national accident database performed by the International Atomic Energy Agency (IAEA) in 2001, concluded that, "ship collisions frequencies range from about 4E-2 for collisions of any severity to about 4E-4 for collisions that lead to total loss of the ship" per ship-year.[18] In addition, the IAEA analysis found that ship fire frequencies that lead to the total loss of the ship is about 8E-4 per ship-year. A more recent analysis found that the probability of damage sufficient to cause sinking of a passenger vessel was about 5E-4 per ship year due to grounding and 6E-4 due to collision.[19] The analysis in [19] estimated that the probability of a collision for a passenger ship that leads to >5% fatalities is near 6E-4 and about 3E-4 for >90% fatalities per year (see Table 2 in [19]). Therefore, it appears that people on ships are exposed to background risks of more than 1E-4 fatalities per year. The RCC 321 Standard sets limits for warning ships when the cumulative probability of impact with debris capable of causing a casualty onboard exceeds 1E-5. Of course, only a fraction of the impacts "capable of causing a casualty" are expected to produce a casualty and even fewer would produce a fatality. Therefore, *the 1E-5 PI standard appears to be reasonably conservative relative to the IMO criteria for the maximum tolerable individual risks for maritime activities and a reasonable fraction of the background risk from collision and fire hazards.*

The IMO guidelines also introduced a method "*where the Societal Risk acceptance criteria reflect the importance of the activity to the society.*" The RCC 321 document does not address that issue. However, in the preamble to the regulatory collective risk criteria update for commercial space operations, the FAA stated that it "*plans to periodically review and revise this public risk standard, if warranted, based upon factors such as the quantity of launch and reentry activities, demonstrated reliability and safety record and benefits provided, technological capabilities, and maturity of the industry.*" Thus, a conceptual connection between collective public risk acceptability and "the importance of the activity to the society" now exists for both transportation modes (maritime and commercial space). The lack of an explicit limit on societal (i.e. collective) risk in the IMO guidelines complicates the comparison with the part of the RCC 321-16 ship protection criteria that uses collective risks, and that subject is addressed in a subsequent section of this paper.

5. COMPARISION TO IMO CRITERIA

This section discusses several general principles and specific features included in the IMO guidelines [16] that were also incorporated in RCC 321 and other US launch and reentry safety standards.

The IMO states that "*individual risk criteria for hazardous activities are often set using risk levels that have already been accepted from other industrial activities.*" The RCC 321-07 rationale stated that "*the criteria consider similar regulatory experience of local, state, federal, and international organizations.*" A previous paper[20] on the rationale for tolerable risks for commercial space transportation noted that an examination of the risk tolerability limits set for various industries shows (a) the importance of public risk perceptions, and (b) that specific risk tolerability values used by the FAA for launch and reentry fall within the span of values used by government organizations at various levels that govern risks posed by various industries.

The IMO guidelines [16] state, "*if a person is voluntarily exposing himself to a risk and/or has some control over it, then the risk level that is accepted is higher as if this person was exposed involuntarily to that risk or had no control over it.*" The RCC 321 risk acceptability criteria for voluntarily exposed personnel, whom receive direct compensation for their activities, are generally ten times higher than the risk acceptability criteria for involuntarily exposed people. Using a factor of ten was found to be consistent with current practice at the US national ranges[21], past RCC standards (e.g. 321-02), NASA's range safety program requirements [2], and the policy of foreign governments such as the United Kingdom.[22] The IMO criterion for individual risk acceptability also allows ten times higher risks for crew members compared to passengers or members of the public ashore. The RCC 321 Standard sets tolerable individual risks for "mission essential personnel"[2] compared to the general

[2] RCC 321 defines "mission essential personnel" as 'those persons and assets necessary to safely and successfully complete a specific hazardous operation or launch. The ME individuals may include persons in training to perform the specific mission currently being conducted, but excludes those in training for other critical tasks. ME personnel are informed of the hazards associated with the operation and trained in mitigation techniques appropriate to the hazard level. The range commander or mission director (or their

public. Note that the RCC 321 Standard allows "persons in training to perform the specific mission currently being conducted" because training in the real operational environment is deemed critical to the safety of future operations.

The IMO's recommended risk evaluation criteria for individual risks are expressed in terms of annual fatality probability. In order to address not only fatalities, but also disabilities and injuries, the IMO guidelines invoke an "equivalent fatality concept," and suggests that one fatality equals ten severe injuries and one severe injury equals ten minor injuries for maritime risks. RCC 321 includes separate criteria for individual and collective risks in terms of casualties and fatalities. The RCC 321 Standard sets tolerable individual risks in terms of fatalities that are ten times lower than those for individual probability of casualty (e.g. 1E-6 probability of casualty and 1E-7 probability of fatality for member of the public.) Note that the RCC defines a "casualty" as "a serious injury or worse, including death, for a human." Thus, the ratio of acceptable individual risks for fatality and casualty used in RCC 321 appear to be entirely consistent with the IMO guidelines.

The IMO guidelines [16] state that *"an appropriate level for the risk acceptance criteria would be substantially below the total accident risks experienced in daily life, but might be similar to risks that are accepted from other involuntary sources."* The policy objectives in the RCC 321 Standard include that "the general public should not be exposed, individually or collectively, to a risk level greater than the background risk in comparable involuntary activities." In this context, the RCC considers "comparable involuntary activities" as those where the risk arises from manmade activities that are:

(a) subject to government regulations or are otherwise controlled by a government agency, and
(b) of vital interest to the U.S.[3], and
(c) impose involuntary risk of serious injury or worse on the public.

The IMO guidelines [16] state that the *"criteria can be used to assess the acceptability of the total risk on being, e.g. on a passenger ship, but not for assessing the specific risk of dying on a passenger ship due to a fire."* The RCC 321 Standard, starting in 2007, endorsed the use of aggregated risk limits, and also recognized that there had been valid reasons given for setting limits on the risks posed by separate hazards.

Thus, the IMO guidelines include several features of the "recommended risk evaluation criteria" that are entirely similar, if not identical with the criteria incorporated in RCC 321 since 2007. These include the ratio of acceptable risks for people that are voluntarily and involuntarily exposed, the ratio of acceptable fatality/casualty risks, the position that risk acceptance criteria should be similar to risks that are accepted from other involuntary sources, and that criteria are best set for the total risks of consequences that account for fatality and serious injury posed by an activity (although there may be value in setting limits on the risks posed by separate hazards).

6. SOCIETAL RISK CRITERIA

This section examines the criteria for tolerable collective (aka societal) risks, including multiple casualty events. The focus is on comparisons between the RCC 321 criteria for ship protection and the "recommended risk evaluation criteria" published by the International Maritime Organization (IMO) in [16].

As previously mentioned, the IMO guidelines lack an explicit limit on societal (i.e. collective) risk, which complicates the comparison with the part of the RCC 321-16 ship protection criteria that uses collective risks. However, the IMO guidelines do include explicit societal risk bounds in the form of "F-N diagrams," which show the relationship between the cumulative frequency of an accident and the number of fatalities. Specifically, Fig. 2 from [16] shows two lines on a F-N diagram that define regions of "negligible" and "intolerable" risks for ship occupants, as well as empirical data on various types of tanker ships from a 20 year period from 1978 to 1998 according to Lloyd's Maritime Information Services Ltd. Fig. 2 shows the same information (also from the same [16]) for passenger ships. The societal risk acceptance criteria shown in these figures were established by the method outlined in reference.[23]

As explained in reference [23], the sample acceptance criteria shown in these Figs. account for "the economic importance of the activity in question, and calibrated against the average fatality rate per unit economic production. The importance of an activity may be measured most adequately in economic terms, assuming that what is paid in an open market represents the importance." The IMO sample risk

designees) should identify the ME personnel in training and justify their designation as ME."
[3] In 2004, Congress identified space transportation as "inherently risky." (see 49 USC Chapter 701, referred to as the Commercial Space Launch Act, §70101 (a)(12), 12/2004). At the same time, Congress found that *"a robust US space transportation industry is vital to the Nation's economic well-being and national security,"* (CSLA §70101, which gives reference Pub. L.106-405, Sec. 2, 11/1/2000, 114 Stat. 1751). The Major Range and Test Facility Bases (MRTFBs) have long been regarded as "national assets," and thus vital to the interests of the US.

acceptance criteria were based on occupational health statistics in the US and Norway, the passenger criteria were derived from comparisons with air-traffic.

A comparison between the IMO sample societal risk acceptance criteria and the RCC ship protection criteria reveals several notable differences involving:

1. The slope of the risk profiles,
2. The maximum societal risk allowed, and
3. The maximum allowable probability of a casualty producing event.

Societies in general often exhibit a strong aversion to accidents involving multiple casualties, which is to say that many people perceive that a single accident involving 100 casualties is more than 100 times worse than 100 accidents involving only one casualty. Thus, risk acceptability criteria sometimes feature some sort of "catastrophe[4] aversion."

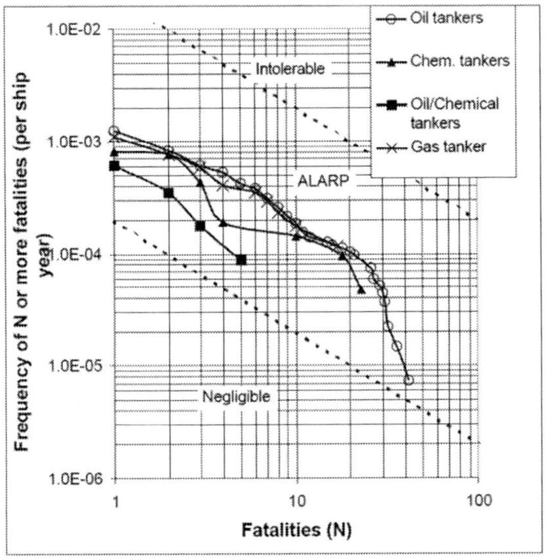

Figure 2: FN Curves for Tanker Ships with Empirical Data from [16]

For example, as shown in Eq. 1, the RCC 321 catastrophic risk criterion uses a slope of -1.5. In general, "catastrophe aversion" comes when the slope of an acceptable risk profile (or F-N diagram) is steeper than one (i.e. < -1 on a log-log plot). The Netherlands[24] has used a slope of -2 as has Santa Barbara County, California[25] (where Vandenberg AFB is located) and some others in the US.

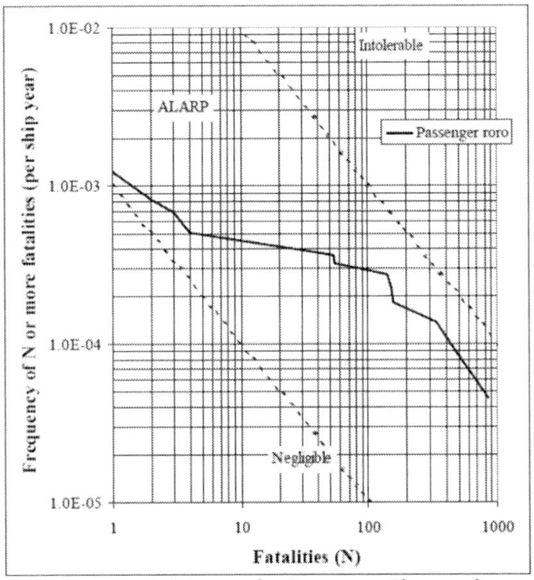

Figure 3: FN Curves for Passenger Ships with Empirical Data from [16]

The slope of -2 has been considered too severe in some countries for certain regulated activities. [24] Specifically, the IMO sample societal risk acceptance criteria (shown in Figs. 2 and 3) do NOT exhibit catastrophe aversion: the acceptable probability of an accident involving, for example, 100 times more fatalities is only 100 times lower. In contrast, the slope of RCC 321-16 acceptable risk profile indicates a catastrophe aversion: for example, the probability of an acceptable event involves 100 times more casualties must be at least 1000 times lower.

Research performed after the publication of RCC 321-07 found that the United Kingdom Health and Safety Executive (HSE) has used a risk aversion exponent of 1.4 (slope of –1.4) for evaluation of societal risk.[26] "The explanation for the choice of this exponent seems to be that it matches the historical data."[24] The HSE has written that "the risk of an accident causing the death of 50 people or more in a single event should be regarded as intolerable if the frequency is estimated to be more than one in five thousand per annum." (See Ref [22] at p. 47)

One of the conveniences of using a discrete representation of the F-N curve is that it is easy to demonstrate that the sum of all of the values of F[≥N] is equal to the number of Expected Fatalities (EF or E_F).[5]

[4] The term catastrophic refers to multiple casualties with a minimum ranging from five to ten depending upon the particular regulation.

[5] F(k) is used instead of P[k] in the formulae because, we are not dealing with probability, but frequency per year.

$$F(\geq 1) = F(1) + F(2) + F(3) + ...$$
$$F(\geq 2) = F(2) + F(3) + ...$$
$$F(\geq 3) = F(3) + ...$$
$$\sum_{k=1}^{n_{max}} F(\geq k) = F(1) + 2F(2) + 3F(3) + ... + n_{max}F(n_{max}) = \sum_{k=1}^{n_{max}} k \times F(k)$$

which is the classical definition of E_F, *i.e.* $E_F = \sum_{k=1}^{n_{max}} k \times F(k)$

The equivalent derivation for equating EF with the area under the continuous distribution form of the F-N curve can be found elsewhere.[27]

Using the area under the F-N curves in the sample acceptance criteria and accounting for the discrete nature of casualties from real events, the maximum EF that is theoretically acceptable can be computed (i.e. all actual acceptable risk profiles that would comply with the F-N criteria would correspond to a lower EF). Thus, the maximum EF that could fit under the "negligible" F-N curve for a tanker (Fig. 2) is 1E-3 per year. Similarly, the maximum EF that could fit under the "negligible" F-N curve for a passenger RoRo (Roll-on/Roll-off for vehicles) is 6E-3 per year.

Whereas the RCC 321 Standard sets limits for warning ships when the collective risks 100E-6 EC (Expected Casualties) or provisional catastrophic risk criteria would be exceeded given a conservative estimate of typical ship traffic. In addition, to comply with the RCC 321 Standard no launch would be allowed to exceed 100E-6 EC. Thus the collective risk limit and surveillance criteria standards in RCC 321 appears to be reasonably conservative relative to the IMO criteria for the "negligible" societal risks for maritime activities (with theoretical maximums of 6E-3 EF for passenger RoRo ships and 1E-3 EF for tankers on an annual basis). Furthermore, the *RCC criteria clearly limit the collective risks to people on ships to levels well below the background risks for tanker and passenger ships based on the empirical data shown in Figs. 2 and 3.*

The IMO guidelines [16] in Figs. 2 and 3 show that the maximum probability of a fatality producing event (another measure of collective risk) corresponding to "negligible" risks for tankers and passenger ships is somewhat larger than the RCC 321 provisional criterion. Specifically, Figs. 2 and 3 show that 2E-4 and 1E-3 are the maximum probabilities of fatality producing events for "negligible" risks to people on tankers and passenger ships respectively, while the RCC 321 Standard allows no more than 1E-4 probability of a casualty producing event. Also, Figs. 2 and 3 indicate that the empirical background risks in terms of probability of fatality producing events are about an order of magnitude higher than the RCC 321 standard limit.

7. FAA WAIVER FOR SHIP PROTECTION

On April 1, 2016, SpaceX submitted a petition to the FAA-AST requesting relief from a regulatory requirement for a launch license for flight of a Falcon 9 launch vehicle carrying the Dragon capsule to deliver cargo to the International Space Station (ISS) as its eighth Commercial Resupply Service mission (CRS-8). SpaceX planned for fly back the first stage to a barge for landing. Specifically, SpaceX requested relief from 14 CFR §417.107(b), which excludes persons in waterborne vessels from the collective risk limit of 30E-6 expected casualties (EC) and limits the Probability of Impact (PI) with waterborne vessels to 1E-5.[6]

During a previous launch of the Falcon 9 from CCAFS to deliver the SES-9 payload to orbit, SpaceX was delayed by the presence of a tug boat towing a large barge inside the ship hazard area in compliance with the FAA's requirement in §417.107(b) to limit the probability of impact for waterborne vessels to 1E-5.

7.1. What was the rationale for the waiver?

The Falcon 9 CRS-8 launch was the ninth launch of an expendable launch vehicle with a Dragon capsule bound for the ISS. SpaceX had attempted three landings of its Falcon 9 first stage on a barge on the ocean off CCAFS. The stages reached their intended landing spot, but did not survive the landings. In no case was public health or safety or safety of third party property jeopardized. The USAF conducted an assessment of the collective risk to people on land due to debris from the CRS-8 launch and determined that the risks are about half the FAA's regulatory limit of 30E-6 EC (at the time the waiver was granted).

On September 25, 2006, the FAA issued part 417 to amend its commercial space transportation regulations governing the launch of expendable launch vehicles. The FAA requirements in part 417 have their genesis in USAF range safety requirements. In addition to the public risk criteria provided in §417.107(b), flight hazard areas were a key element of the performance level requirements in subpart C of 417 to ensure the safety of people on waterborne vessels. Specifically, §417.223(a) states that *"a flight safety analysis must*

[6] In 2016, the FAA clarified the requirements of part 417 concerning hazard areas for ships and aircraft in Ref. [11]. The clarification provided was that "*A launch operator must establish any water borne vessel hazard areas necessary to ensure the probability of impact (Pi) with debris capable of causing a casualty for water borne vessels does not exceed 0.00001* (1×10^{-5})." *Id.* at 47026.

include a flight hazard area analysis that identifies any regions of land, sea, or air that must be surveyed, publicized, controlled, or evacuated in order to control the risk to the public from debris impact hazards. The risk management requirements of §417.205(a) apply."

In addition to the performance level requirements of subpart C of part 417, the FAA included several appendices on flight safety analysis methods. Specifically, §B417.5(a) states that "*a launch operator must perform a launch site hazard area analysis that protects the public, aircraft, and ships from the hazardous activities in the vicinity of the launch site. The launch operator must evacuate and monitor each launch site hazard area to ensure compliance with §§ 417.107(b)(2) and (b)(3).*" The methodology in Appendix B was designed to be consistent with USAF range safety requirements in 2006, and to ensure that the cumulative PI to any ships would not exceed 1E-5 for any debris expected to exceed the kinetic energy or overpressure thresholds established by §417.107(c).

At the time that part 417 was promulgated, safety experts at NASA[7] believed that it would be desirable to apply collective risk[8] management principles to ship safety by including persons in waterborne vessels in the EC calculation. However, the computational tools and input data available at that time made it impractical (and observation techniques posed significant risks to launch operators[9]) to quantify the EC contribution from people in waterborne vessels. In addition, the means to survey ship traffic areas potentially threatened by launch debris were much more limited in the 2006 timeframe. Accordingly, the Federal launch ranges and the FAA adopted the cumulative PI as a surrogate for collective risk and relied on a relatively simplistic approach involving ship hazard areas. Hence, the FAA's current requirements allow launches to proceed with unquantified residual collective risks to people in waterborne vessels.

Since 2006, when the part 417 requirements were promulgated, the capability to compute launch risks to people on waterborne vessels has improved greatly. The Automatic Identification System (AIS) described above, combined with other technological advances, now make real-time ship information readily available, including the position, course, speed, ship size, identity, and cargo data. The real-time data on waterborne vessels provided by AIS and other advanced surveillance techniques, combined with advanced computer models, now enable valid estimates of the individual and collective risks to people on waterborne vessels to be made during a launch countdown. More specifically, advanced computer modeling, including improved software and more computing capacity, now allows reasonably accurate assessments in near real-time of the risks of casualty, fatality, and catastrophe for identified ships.

The FAA assessed the input data and probabilistic casualty models that the USAF at the 45th Space Wing (45th SW) planned to use to quantify individual and collective risks to people on waterborne vessels during the launch countdown for the CRS-8 mission. The FAA found that the 45th SW's public risk analyses use accurate data and scientific methods that are mathematically valid, with reasonably conservative assumptions applied in areas where significant uncertainty exists. For example, the 45th SW uses conservative estimates of the number of occupants on waterborne vessels by assuming that the number of persons on board equals the vessel's maximum capacity and that all occupants are on-deck, and thus exposed to inert debris impacts that might not otherwise pose a threat to people below deck. (This assumption would not be conservative in cases where explosive debris impacts are a significant hazard.)

Additionally, the FAA performed independent analyses using alternative methods to estimate the casualty risks for multiple foreseeable scenarios involving debris impacts on various types of waterborne vessels in the vicinity of Cape Canaveral. The FAA found that large passenger vessels anywhere between the launch point and the first stage disposal zone can contribute significantly to the estimated EC from the CRS-8 launch. The FAA found that small boats (too small to have AIS required) located close to the launch point should not produce significant individual risks, given conditions expected in the vicinity of Cape Canaveral. Specifically, sufficient surveillance with other means

---

[7] For example, the NASA Range Safety Policy requirements (NPR8715.5 dated July 8, 2005 in paragraph 3.2.6.2) stated that "an assessment of risk to the public and workforce due to debris shall account for ...all potential debris, generated intentionally or not, that could cause a casualty, including debris that could affect someone on the ground *or on a waterborne vessel*, or cause an aircraft accident (Requirement)." (emphasis added).

[8] Risk metrics account for both the probability and consequence of foreseeable events. In contrast to the relatively sophisticated casualty consequence models used to compute individual and collective risks according to §417.107(d), the FAA's current requirements limit only the probability of impact on waterborne vessels with only simple threshold values to define what constitutes an "impact."

[9] The only known deaths related to launch operations at Cape Canaveral were five occupants of a helicopter that crashed at sea "shortly after 2 AM...April 7, [1984] while flying surface surveillance for the scheduled launch of a Trident 1 missile from the USS Georgia." *See* Air Force News Print Today (Apr. 8, 2011).

(e.g., radar, and/or using Coast Guard ships or aerial assets) were to be used to ensure individual risks comply with the FAA requirement in §417.107(b)(2). In addition, Notices to Mariners will continue to be issued for the areas where the probability of impact on a ship would exceed 1E-5, which is current practice at the ER, and required by §§ B417.3 and B417.11. Since the FAA's current requirements allow launches to proceed with unquantified residual collective risks to people in waterborne vessels, as long as the collective risk for people on land from each source of hazard (i.e., debris, toxics, or distant focusing overpressure) does not exceed 30E-6 EC, and because the launch will not exceed the 30 E-6 EC with the inclusion of persons on water borne vessels, the FAA found that the Falcon 9 CRS-8 launch would not jeopardize public health and safety or safety of property, and waives 14 CFR §§ 417.107(b)(3) and B417.5(a)'s requirement not to initiate flight absent evacuation of the ship hazard area.

7.2. Public Interest

The USAF conducted an assessment of the risk to property on CCAFS, including assets used for national security space missions, and did not identify national security concerns. The FAA identified no national security or foreign policy implications associated with granting this waiver.

The waiver was deemed consistent with the public interest goals of Chapter 509 and the 2013 National Space Transportation Policy. Three of the public policy goals of Chapter 509 are: (1) to promote economic growth and entrepreneurial activity through use of the space environment; (2) to encourage the United States private sector to provide launch and reentry vehicles and associated services; and (3) to facilitate the strengthening and expansion of the United States space transportation infrastructure to support the full range of United States space-related activities.[28] A successful application of public risk management for the protection of people in waterborne vessels has the potential for reducing launch costs. As it is a major procurer of launch services, reduced launch costs will be of direct benefit to the USG. It will also help to make the US launch industry more competitive internationally.

The 2013 National Space Transportation Policy clearly identified how strengthening U.S. competitiveness in the international launch market and improving the cost effectiveness of U.S. space transportation services are in the public interest: "Maintaining an assured capability to meet United States Government needs, while also taking the necessary steps to strengthen U.S. competitiveness in the international commercial launch market, is important to ensuring that U.S. space transportation capabilities will be reliable, robust, safe, and affordable in the future. Among other steps, improving the cost effectiveness of U.S. space transportation services could help achieve this goal by allowing the United States Government to invest a greater share of its resources in other needs such as facilities modernization, technology advancement, scientific discovery, and national security. Further, a healthier, more competitive U.S. space transportation industry would facilitate new markets, encourage new industries, create high technology jobs, lead to greater economic growth and security, and would further the Nation's leadership role in space." Thus, SpaceX's proposal to apply collective risk management to people in waterborne vessels was found to be in the public interest.

8. SUMMARY

This paper provides an introduction to the RCC as an organization that has been highly successful in the development USG consensus standards. The usefulness and credibility of RCC products are recognized throughout the USG, industry, academia, and internationally. The RCC 321 safety criteria in particular have served as an important benchmark used by domestic and international authorities that govern the safety of launch and reentry operations.

This paper describes key elements of the RCC 321 safety standards for the protection of ships during launch and re-entry. The new ship protection standards address all hazards from planned and unplanned events by specifying risk management criteria that embrace two approaches to define regions where Notices to Mariners shall be issued and where surveillance should be performed. The more traditional approach in the US requires formal warnings to waterborne vessels based on the probability of impact (PI) on a vessel that is "capable of causing a casualty." The second option explicitly ensures that warnings are issued for areas where individual, collective, and catastrophic risks would exceed standard safety limits. RCC 321-16 now specifies that suitable observation techniques (i.e. surveillance) should be used during a launch countdown to locate ships in those region(s) where the: (1) individual probability of casualty exceeds 1E-6, and (2) collective casualty expectation or provisional catastrophic risk criteria would be exceeded given a conservative estimate of typical ship traffic.

A large part of this paper discusses the rationale for the RCC ship protection standards, including recent updates. The 2007 rationale for RCC ship protection standards cited an International Maritime Organization (IMO) proposal to limit individual annual fatality risks crew, passengers and public ashore. More recently, the IMO published revised guidelines for formal safety assessment, including criteria used to evaluate the acceptability [16]. Several features of the IMO's "recommended risk evaluation criteria" are analogous to the criteria incorporated in RCC 321. These include the ratio of acceptable risks for people that are voluntarily and involuntarily exposed, the ratio of acceptable fatality/casualty risks, the position that risk acceptance criteria should be similar to risks that are accepted from other involuntary sources, and that criteria are best set for the total risks of consequences that account for fatality and serious injury posed by an activity (although there may be value in setting limits on the risks posed by separate hazards).

The IMO guidelines introduced a method "where the Societal Risk acceptance criteria reflect the importance of the activity to the society." While RCC 321 does not address societal benefits, the preamble to the regulatory collective risk criteria update for commercial space operations made a conceptual connection between collective public risk acceptability and the benefits provided by a transportation mode. The collective risk limit and surveillance criteria standards in RCC 321 appear to be reasonably conservative relative to the IMO guidelines [16] based on several measures. The maximum EF to qualify as "negligible" societal risks for maritime activities on an annual basis are at least ten times higher than the RCC EC limit per launch mission. Furthermore, the RCC limits the collective risks to people on ships to levels well below the background risks for tanker and passenger ships based on the empirical data shown in Figs. 2 and 3. Overall, the RCC collective risk limits and provisional catastrophe criteria appear reasonably conservative compared to IMO guidelines.

9. REFERENCES

1. Air Force Instruction 91-217, *Space Safety and Mishap Prevention Program*, 18 February, 2010
2. NASA Procedural Requirements, *Range Flight Safety Program*, NPR 8715.5 A, 17 Sept. 2010
3. Range Commanders Council Risk Committee of the Range Safety Group, *Common Risk Criteria for National Test Ranges*, RCC 321-07, White Sands Missile Range, New Mexico, 2007
4. Federal Register, Department of Transportation, Federal Aviation Administration, 14 CFR Parts 417, 420, 431, and 435, *Changing the Collective Risk Limits for Launches and Reentries and Clarifying the Risk Limit Used to Establish Hazard Areas for Ships and Aircraft*; Final Rule, Vol. 81, No 139, July 20, 2016 (see p. 47017)
5 Wilde, P.D., *Range Safety Requirements and Methods for Unguided Suborbital Rocket Launches*, Proceedings of the 2nd IAASS Conference, Chicago, IL, USA, 14-16 May 2007.
6 Wilde P., R. Nyman, C. Draper, and A. Bavandi, *Independent Study of the Risks from the Reentry of the Automated Transfer Vehicle*, 3rd IAASS Conference, Rome, IT, October 2008
7. *Proposed Updates to RCC 321 Regarding Ship Protection*, Larson, E.F., Report No. 14-861-01 Rev A, ACTA, Inc., Torrance, CA, December 2014.
8. *Columbia Accident Investigation Board* (CAIB), CAIB Report, GPO, Washington, DC, August 2003.
9. Wilde P.D. and Murray D., *Catastrophe Risk Management and Evaluation of Maximum Probable Loss for Launch Vehicle Explosions*, 33rd DDESB Seminar, Palm Springs, CA, August 2008.
10.Wilde P.D. and Murray D., *Catastrophe Risk Management and Evaluation of Maximum Probable Loss for Launch Vehicle Explosions*, 33rd DDESB Seminar, Palm Springs, CA, August 2008.
11 *Changing the Collective Risk Limits for Launches and Reentries and Clarifying the Risk Limit Used to Establish Hazard Areas for Ships and Aircraft*, Vol. 81 Federal Register, July 20, 2016, p. 47017.
12 Wilde P, Brinkman C.P., Millard S., *FAA Public Risks and Insurance Requirements for Orion's First Entry Flight Test*, 65th IAC Toronto, Canada, 2014, IAC-14-D6.1.6
13 Health and Safety Executive, *Reducing Risks, Protecting People: HSE's decision-making process*, (R2P2), HSE Books.
14 *Commercial Space Transportation Licensing Regulations*, Vol. 64, Federal Register, No. 76, 21 April 21, 1999 page 19605
15 R. Skjong, *Risk Acceptance Criteria: Current Proposal and IMO Position*, Conference Proceedings from Valencia, Spain 4-6 June 2002
16 International Maritime Organization, *Amendments to the Guidelines for Formal Safety Assessment (FSA) for Use in the IMO Rule-Making Process*, MSC-MEPC.2/Circ.5 16 October 2006
17 Evans et al, *Third Party Risk Near Airport and Public Safety Zone Policy*, Rept. 9636, National Air Traffic Services Limited, London, June 1997
18 International Atomic Energy Agency, *Severity, Probability and Risk of Accidents During Maritime Transport of Radioactive Material*, IAEA-TECDOC-1231, IBSN 1011-4289, July 2001
19 Vanem E., and Skjong R., *Collision and Grounding of Passenger Ships – Risk Assessment and Emergency Evacuations*, ICCGS, 2004
20 Wilde, P.D., *Public Risk Criteria and Rationale for Commercial Launch and Reentry*, 5th IAASS Symposium, Versailles, France, 17-19 Oct 2011.
21 Air Force Space Command Manual 91-710 (AFSPCMAN 91-710), *Range Safety User Requirements*, July 2004
22 UK Health and Safety Executive (HSE), *Reducing Risks, Protecting People: HSE's Decision-Making Process*, ISBN 0 7176 2151 0, 2001, page 44

23 International Maritime Organization, *Formal Safety Assessment Decision parameters including risk acceptance criteria Submitted by Norway*, MSC 72/16 14 February 2000
24 V.M. Trbojevic, *Risk Criteria in the EU*, published by Risk Support Limited, London, 2005.
25 County of Santa Barbara, Planning and Development, *Environmental Thresholds and Guidelines Manual*, October 2008
26 UK Health and Safety Executive (HSE), *Societal Risk: Initial briefing to Societal Risk Technical Advisory Group*, 2009 (See p.20)
27 Vrijling, J.K. and Van Gelder, P.H., *Societal Risk and the Concept of Risk Aversion*, Dept. of Civil Engineering, Delft University of Technology
28 51 U.S.C. 50901(b)(1), (2), (4). *Commercial Space Transportation Licensing Regulations*, Notice of Proposed Rulemaking, Vol. 62 Federal Register, Mar. 19, 1997, p. 13230.

HAZARD AREAS BY AN EXPLOSION OF A LIQUID LAUNCH VEHICLE ON THE LAUNCH PAD

Hyungseok Sim[(1)], Kyusung Choi[(2)], Sangyeon Cho[(3)]

[(1)] *KARI, 169-84 Gwahangno, Yuseong-gu, Daejeon, Korea, Email:aeroshs@kari.re.kr*
[(2)] *KARI, 169-84 Gwahangno, Yuseong-gu, Daejeon, Korea, Email:kschoi@kari.re.kr*
[(3)] *KARI, 169-84 Gwahangno, Yuseong-gu, Daejeon, Korea, Email:chosangy@kari.re.kr*

ABSTRACT

The launch risk should be estimated ahead of the launch of a launch vehicle and appropriate countermeasures should be taken because the failure rate of launch vehicles has been relatively high, and the hazard can be catastrophic in some failure cases. Especially, the explosion of a launch vehicle in early flight phase can pose great risk to the ground due to its high energy of fully loaded fuel. In this research, hazard areas by the explosion of a liquid (RP1/LOX) propelled launch vehicle were analysed and compared, which can help to determine the area to be evacuated during the launch process and to make provisions against some disasters. A hypothetical launch vehicle of the lift-off mass of about 300 ton with about 270 ton of RP1 and LOX was selected to be analysed. First, the overpressure was calculated according to the distance from the explosion source using simple equations. And then, the distance to which the flying debris can reach was computed using ASTOS software. And the hazard areas were analysed where people were subject to the pain or 2nd degree burn by the heat release of the fireballs. Finally, the range in which a forest fire might occur was predicted.

1. INTRODUCTION

Rocket launches can pose hazards in the area near the launch pad and downrange areas along the ground track of the flight in case of a launch failure. Especially, the failures on the launch pad can be catastrophic because the launch vehicles are usually loaded with a large amount of propellant which can explode in some cases. The hazards of the launch vehicle explosion on the pad have been the key restriction for the siting of launch pads and facilities. In addition to this, the analysis of hazard area by explosion may be essential to determine the area to be evacuated during the launch process and to make provisions against explosive failures. The explosions of launch vehicles can be catagorized into two types. For some explosions, considerable amount of liquid propellant of a launch vehicle may enter into detonation type reactions, where the primary hazards are associated with the blast waves and scattering fragments. For other explosions, a significant proportion of the propellants may undergo deflagration type reactions. In these cases, the greatest hazards to facilities and the people are those associated with the fireballs [1].

In this paper, Hazard areas by the explosion of a hypothetical liquid (RP1/LOX) propelled launch vehicle of lift-off mass of 300 ton were analysed and compared. The dry mass to total mass ratio of the vehicle is assumed to be 10 percent, such that the mass of propellant is about 270 ton. For this launch vehicle, the overpressure by the explosion was first calculated according to the distance from the explosion source using simple equations [2]. The distance to which the flying debris can reach was computed using ASTOS software. And then, the hazard areas were analysed where people were subject to the pain or 2nd degree burn by the heat release of the fireballs. Finally, the range in which a forest fire might occur was predicted.

The explosion effects strongly depend on various conditions such as failure mode, propellant mixing, the intrusion of air, and the amount of fuel participating in the reaction, etc. Because the variance resulting from these variables is generally unknown, conservative approaches are usually adopted for those hazard analyses. Conservative values are chosen for this research when the coefficients of equations are not known, and it is almost impossible to get reasonable and accurate values.

2. HAZARD BY THE DETONATION TYPE EXPLOSION

2.1 Overpressure

For such explosions where considerable amount of propellant may enter into detonation type reactions, the principal hazards to adjacent launch facilities and people are those associated with blast overpressure and with dispersing fragments [1]. In those explosions, large energy is released rapidly, generating a pressure wave of finite amplitude. The pressure wave moves forward in air forming shock front and subsequent pressure profile as it progresses further. This shock front proceeds supersonically with discontinuity in pressure and density which may cause some damage to buildings or people.

The main parameters describing blast waves are peak positive overpressure, duration, and impulse etc. Although all those parameters may affect the damage by the blast wave, the peak positive overpressure has been chosen as the indicator for the hazard area in several regulations [3, 5]. It may be because the peak positive overpressure can be predicted simply and can represent the intensity of the blast wave. Figure1 shows the structure of the typical pressure profile for a blast wave in air.

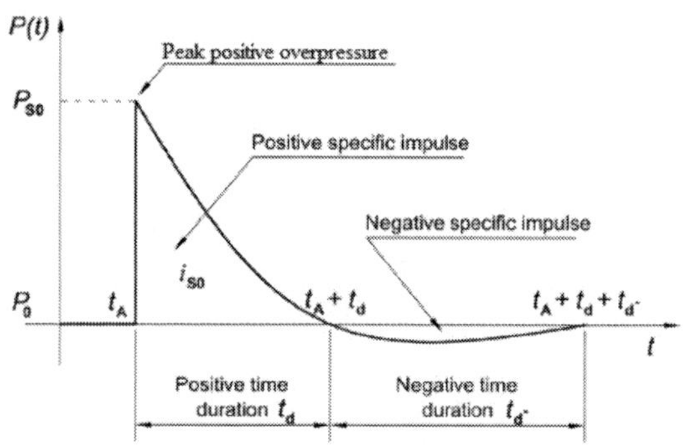

Figure 1. Typical blast wave overpressure resulting from explosion in air

The TNT equivalence of propellants should be determined to predict the explosive strength of a launch vehicle. RP1/LOX launch vehicles can be treated by using a TNT equivalent of 20 percent for quantities up to 500,000 pounds and an equivalent of 10 percent for any excess over that amount [3]. For the vehicle with RP1/LOX of 270 ton, the TNT equivalence can be estimated to be about 50 ton (110,000 lb). While several simple equations that were proposed by many researchers [2] would be applied to predict the overpressure resulting from liquid propellant explosions, Kingery & Pannill method [4] was employed in this paper. They have formulated the relation between the peak overpressure and the distance from the explosion source by compiling overpressure measurements obtained from a series of TNT detonations in Canada, the United Kingdom and the United States. Their method has been widely used as standards in several U. S. and Korean regulations [3, 5].

Table 1. Distant scaling factor vs. overpressure [4]

K-factor	Overpressure (psi)	K-factor	Overpressure (psi)
1	3150	80	0.5
2	315	100	0.35
5	42	110	0.3
10	9.6	125	0.25
20	3	150	0.2
45	1.0	175	0.16

Using Kingery & Pannill relations, the peak overpressure at some distance from the explosion source can be determined as following [3, 4].

$$D = K \times W_{1/3} \qquad (1)$$

Where,
D = distance from the centre of explosion (ft)
K = distant scaling factor (ft/lb$^{1/3}$)
W = TNT equivalent weight (lb)

The distant scaling factors for some peak overpressures are presented in table 1. The distances at which the peak overpressures are 0.5 psi and 1.0 psi are easily determined as about 1170 m and 660 m respectively using equation (1) and table 1.

2.2 Scattering Fragments

Another main hazard of a detonation type explosion of launch vehicles is due to flying debris. The strong expansion of gas might cause the vehicle fragments to fly across the air and to impact the adjacent area. The scattering debris of the hypothetical launch vehicle were simulated using ASTOS (AeroSpace Trajectory Optimization Software) software. Figure 2 shows the trajectories of fragments resulting from the explosion of the launch vehicle including altitude information.

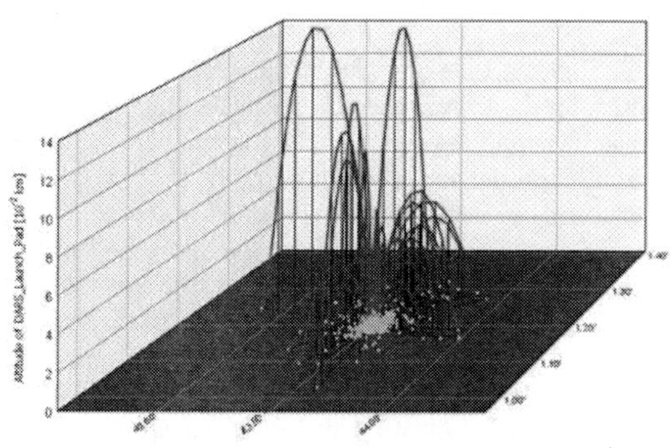

Figure 2. Trajectories of scattering fragments of the hypothetical launch vehicle

The fragments reach an altitude of 170 meter. The impact semi-major axis is predicted to be about 700 m while semi-minor axis to be about 500 m. The hazard area by the scattering debris is almost similar to that by the overpressure of 1.0 psi.

3. HAZARD BY THE DEFLAGRATION TYPE EXPLOSION

3.1 Fireball

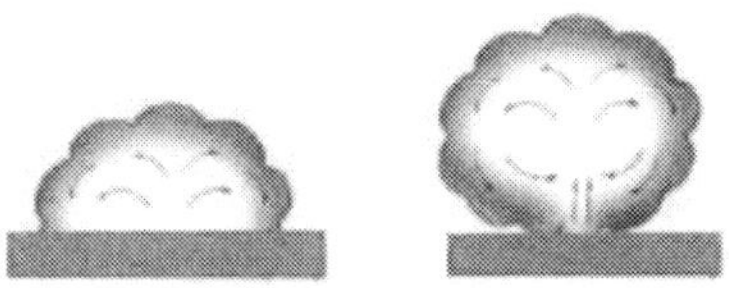

Figure 3. Typical fireball by explosion

For some explosions, the significant proportion of the propellants may enter into deflagration type reactions, where the greatest hazards to facilities and the people are those associated with the fireballs [1]. A typical fireball is shown in figure 3. Fireball size and duration are the most important factors in order to estimate the hazard regarding thermal effects by fireballs. By compiling the data from tests and vehicle incidents, Gayle and Bransford [1] have shown that the fireball diameters and durations were dependent on the cube root of the weight of the combined propellants. Figure 4 and figure5 indicate these relations found by them. They analysed the data for 47 tests and incidents ranging from 10 to 250,000 pounds of combined propellants to get those dependence.

The relations can be expressed as followings;

$$D = 10.05\times W_{0.316} \quad (2)$$

$$d = 0.196\times W_{0.349} \quad (3)$$

Where,
D = fireball diameter (ft)
d = fireball duration (sec)
W = total propellant weight (lb)

For our hypothetical launch vehicle, the diameter and the duration may be predicted as 205 m and 20 second respectively using equations (2) and (3).

3.2 Radiant Heat by Fireball

The greatest hazard of the fireball is caused by the thermal effect. Especially, for the area distant from the fireball, the effect by the radiant heat may be the main concerns to safety

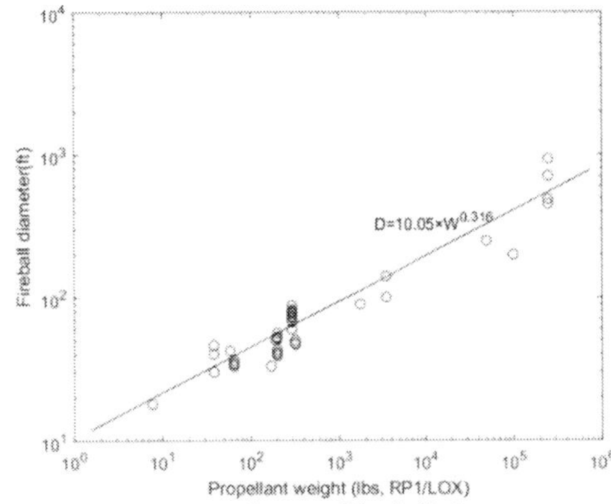

Figure 4. Dependence of fireball diameters on propellant weight for RP1/LOX [1]

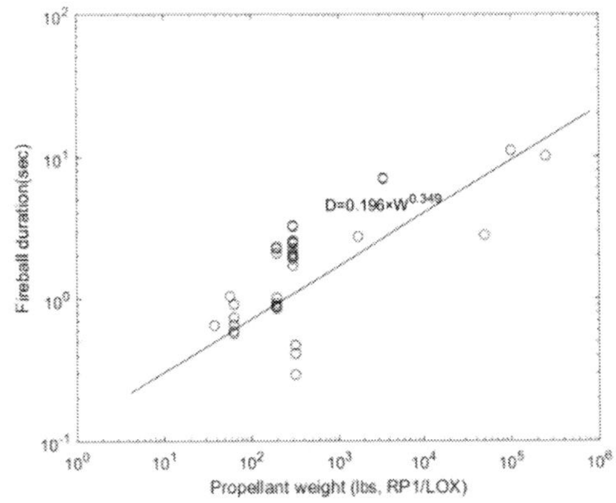

Figure 5. Dependence of fireball durations on propellant weight for RP1/LOX [1]

A mathematical analysis to predict the radiant heat that was proposed by High and Fletcher [6] was employed in this research. Their approach was based on Lambert's Law for thermal relation and the Stephan-Boltzmann law relating radiation intensity to the fourth power of the absolute temperature. The surface temperature of the fireball was estimated to be approximately 2500°F (1644K) for the RP1/LOX fireball [6]. The radiant heat flux at the fireball surface can be expressed as following;

$$I=\varepsilon\sigma T_4 \quad (4)$$

Where,
I = radiant heat flux (W/m^2)
ε = emissivity
σ = Stephan-Boltzmann constant (5.67e-8 W/m^{2}K^4)
T = surface temperature (K) of fireball

The emissivity of a surface relates the emissive power of an actual surface to that of a black body. This value is difficult to accurately determine. The emissivity of 1.0 was used to guarantee conservative result here. The final heat flux at some distance from the explosion center can be expressed as equation (5), which relates the radiant heat flux to the radius of the sphere and the distance between the receiver and the center of the fireball [6].

$$F = \pi I \left[\frac{R^2}{a^2} + \frac{R^4}{a^4} + \cdots \right] \quad (5)$$

Where,
F=heat flux at the distance of a (m) from center of fireball sphere (kW/m^2)
I=radiation heat flux (kW/m^2) at fireball surface
R=radius of fireball (m)
a=distance (m) from center of fireball sphere to receiver

The radiant heat reaching an object in the atmosphere undergoes some attenuation by the thermal absorption by the air. However, because it is thought that the attenuation is relatively small if the distance is not too large and that the area concerned is near to the launch pad, it is assumed here that the radiant heat produced reaches an object without any loss. In addition to this, it is assumed that the heat flux remains at the constant level rather than varies with time forming a curve. It is thought that this assumption also made the analysis of this research conservative.

Using equation (5), radiant heat flux at some distance from the center of the fireball can be easily predicted. Figure 6 shows the radiant heat flux at various distances from the fireball center for our hypothetical launch vehicle which is assumed to explode on the launch pad.

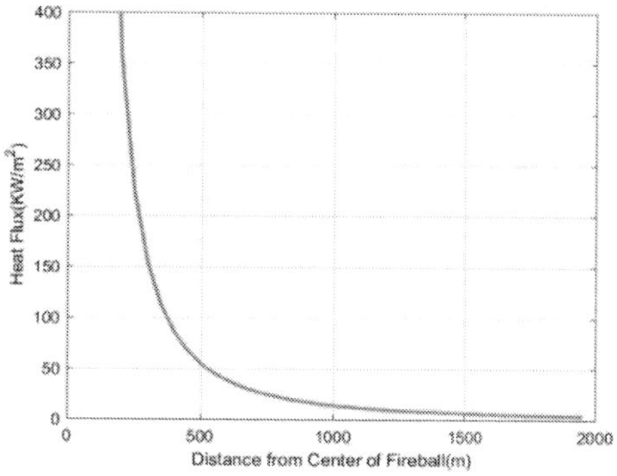

Figure 6. Heat flux vs. distance from center of fireball for the explosion of the hypothetical vehicle

3.3 Hazard by Thermal Effect of Fireball

If we assume that the heat flux at each position lasts for the duration time of a fireball, the hazardous radii regarding skin burn, pain, and fire etc. can be determined based on some information. Figure 7 indicates the pain and 2nd degree burn threshold regarding heat flux. Using these relations, we can easily know that a man will get a 2nd degree burn when he is exposed to the heat flux of 4 (kW/m^2) over approximately 60 seconds.

Because the duration time of the fireball for our hypothetical launch vehicle is about 20 second, it can be seen from the figure that the heat flux above 8 (kW/m^2) would cause 2nd degree burn, and that the heat flux above 4 (kW/m^2) would cause pain. By comparing the above results with the analysis of figure 6, the region where a person may subject to 2nd degree burn (heat flux of 8 (kW/m^2)) may be within 1310 m radius from the center of the fireball. The distance for pain (heat flux of 4 (kW/m^2)) can be estimated as approximately 1870 m. The hazard radius for 2nd degree burn is slightly greater than that for 0.5 psi overpressure (1170 m).

Just like in the procedure for the radius of 2nd degree burn, the hazardous area for forest fire can be predicted using some useful test results. Figure 8 shows the ignition time of dried fallen leaves of pine trees and oriental oak trees which are the most common trees in Korea [8].

These relations are obtained from the ignition test of fallen leaves which are dried and meshed using electric mill. We can tell, by using figure 8, that the heat flux of about 22 (kW/m^2) would be expected to cause dried fallen leaves to ignite in case of the explosion of the hypothetical vehicle whose fireball duration time is about 20 seconds.

Referencing figure 6 again, the distance within which dried fallen leaves would ignite (the heat flux of 22 (kW/m^2)) may be about 790 m, which is slightly larger than the predicted falling radius of scattering vehicle fragments.

The area in which fallen leaves can ignite by an explosion of a launch vehicle may not be small, so special care needs to be taken regarding the fire if there is a forest with fallen leaves near the launch pad and the weather is dry.

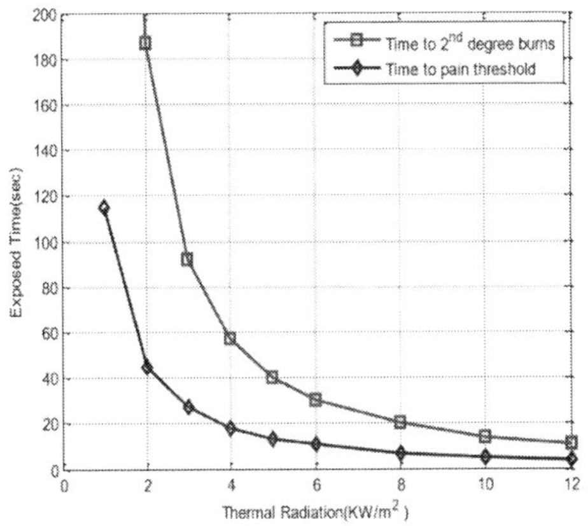

Figure 7. Time threshold for 2nd degree burn and pain with heat flux [7]

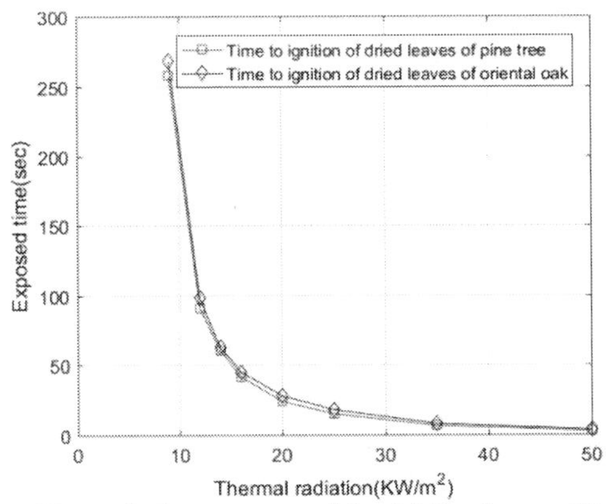

Figure 8. Ignition time of dried fallen leaves [8]

In summary, table 2 compares the hazard radii from the explosion or fireball of the hypothetical launch vehicle whose lift-off mass is about 300 ton and total propellant mass is about 270 ton. The largest distance for an injury might be by 2nd degree burn and is predicted to be about 1310 m. People need to be evacuated outside this distance during launch process of the hypothetical vehicle.

Table 2. Analysis results for the hazard radii from the explosion or fireball of a hypothetical liquid fueled launch vehicle

Hazard	Radii
Overpressure of 1.0 psi	660 m
Overpressure of 0.5 psi	1170 m
Scattering debris	700 m
Fireball radius	103 m
2nd degree burn	1310 m
Pain threshold	1870 m
Ignition of dried fallen leaves	790 m

4. CONCLUSION

Hazard areas by the explosion of a hypothetical liquid (RP1/LOX) propelled launch vehicle of lift-off mass of 300 ton were analysed and compared. The hazard area by the blast wave overpressure, the distance to which the scattering debris can reach, the region in which people were subject to the pain or 2nd degree burn, and the area where a forest fire may occur were analysed for the two type explosions; detonation and deflagration. In order to reduce or avoid the risk regarding the explosion of a launch vehicle on the pad, people within hazard radii should be evacuated before feeding fuel into the vehicle. And fire engines or fire helicopters need to be ready near the launch pad during launch process in order to prepare for the explosion and the subsequent fire.

5. REFERENCES

1. Gayle, J.B. and Bransford, J.W. (1965). Size and Duration of Fireballs from Propellant Explosions. NASA TM X-53314

2. Goel, M.D., Matsagar, V.A., Gupta, A.K., and Marburg, S. (2012). An Abridged Review of Blast Wave Parameters. Defence Sci J. 62(5), pp 300-306.

3. Code of Federal Registration Parts 401, 406, 413, et al. Licensing and Safety Requirements for Launch: final Rule, FAA, US Department of Transportation, USA, Aug. 25, 2006.

4. Kingery, C.N. and Pannill, B.F. (1964). Peak Overpressure vs Scaled Distance for TNT Surface Bursts (Hemispherical Charges). Ballistic Research Laboratories Memorandum Report No. 1518.

5. Safety Management Standards for the Ammunition and Explosives. Korean Ministry of National Defence.

6. High, R.W. and Fletcher, R.F. (1965). Estimation of Fireball from Saturn Vehicles Following Failure on Launch Pad. NASA Program Apollo Working Paper No. 1181, NASA Manned Spacecraft Centre, Aug. 1965.

7. Handbook of Chemical Hazard Analysis Procedures. (1989). Federal Emergency Management Agency, U.S. Department of Transportation, U.S. Environmental Protection Agency.

8. 8 Park, H.J., Kim, E.S., Kim, J.H., and Kim, D.H (2007). A Combustion Characteristic Analysis of Quercus Variabilis and Pinus Densiflora Fallen Leaves Using Radiation Heat Flux. Transactions of Korean institute of fire science and engineering 21(3), pp. 41-

THE ADEQUATE BALANCE BETWEEN AUTOMATION AND HUMAN DECISION

Gérald Grucker

Centre National d'Etudes Spatiales, Guiana Space Center, BP 726 - 97387 Kourou, French Guiana, France
Email: gerald.grucker@cnes.fr

ABSTRACT

Within CSG (Centre Spatial Guyanais – Guiana Space Center), the main goal of Flight Safety is to monitor an in-flight launcher's trajectory so as to be ready to neutralize it, if needed, in order to ensure the protection of people, environment and facilities. This sovereign mission is very demanding and requires that operators are precise, reactive and have composure. In spite of training, human error is always to be feared and it's important to seek how to prevent it. An easy solution would consist in automatizing operations completely in order to avoid human intervention. Unfortunately, this solution would only allow the covering of expected scenarios, a machine would be completely lost dealing with an original situation.
A human operator presence being necessary, in spite of any disadvantages it may present, how could we deal with Flight Safety's mission in the safest possible way? What should be automated? Where does a human operator have an added value?

1 FLIGHT SAFETY'S TEAM CURRENT METHODOLOGY

1.1 Flight Safety team composition

At the European spaceport CSG, safety decisions and actions are made in real time by a team of four safety officers, located in a dedicated small safety room, fully independent from the main control center. The team is placed under the authority of the Chief Safety Officer (RSV). The members have the following roles:

- Telemetry Officer (ACVI) monitors telemetry data received on ground and provides early warnings of any launcher failure (propulsion, occurrence of separation events, on board computer and navigation status, status of the onboard flight safety algorithm, when applicable). He also monitors the quality of the remote control link.

- Deputy Safety Officer (RSVA) manages the localization (tracking) facilities and monitors the trajectory of the launcher. He warns about any deviation of the launcher from the nominal trajectory, and supports the RSV in his decisions.

- Chief Safety Officer (RSV) manages remote control facilities, confirms the trajectory deviation, compares it to the flight corridor and sends the flight termination order if needed.

- Back-up Operator (REMPLAÇANT) is in standby in the safety room to replace any of the three other operators, should the need arise.

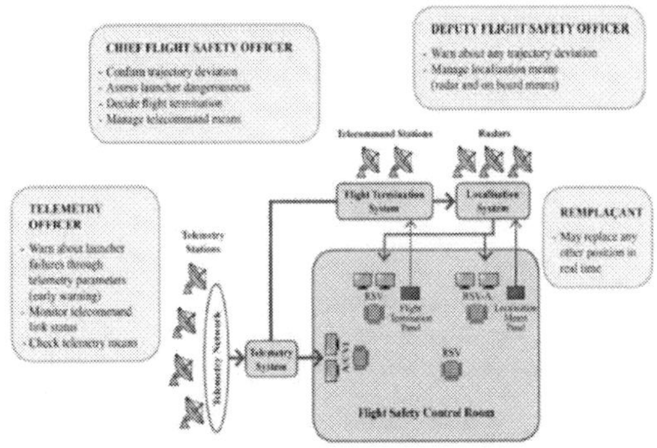

Figure 1: CSG Flight Safety Control Room

1.2 Operational loop

Among the four operators previously introduced, RSV and RSVA are the only ones to deal with localization information in order to analyze the launcher's trajectory.

This analysis requires several steps to ensure its integrity, that is to say its conformity to the real nature of the launcher's position and behavior. The following steps form what is called an « operational loop »:

Step 1 — Thanks to the most « efficient » localization means, RSVA describes his or her displayed trajectory

Step 2 — Thanks to the 2nd most « efficient » localization means, RSV determines if his or her displayed trajectory is coherent with the one stated by RSVA

Step 3 — (optional) In case of an incoherence between trajectory screens, RSV can ask to start a new operational loop, that is going back to step 1 if the incoherence still remains, RSV can rely on « safety principles » in order to exclude the tracking source which is likely at the origin of the incoherence

Step 4) — After having reviewed the situation with ACVI, RSV uses predetermined decision criteria to assess the launcher's dangerousness.

Figure 2 The operational loop

Thanks to this operational loop, RSV has a guarantee that his or her assessment is based on two coherent trajectories that were built with independent tracking sources.

1.3 Visualization and decision

In order to analyze the trajectory and establish a diagnosis, RSV and RSVA rely on two screens:

- A "verticality" screen: this screen displays the altitude of the launcher as a function of its curvilinear distance from the launchpad

- An "impact" screen: this screen displays the instantaneous impact point, defined by where the intact vehicle would impact if thrust were totally and suddenly terminated (taking atmospheric friction into account)

RSVA trajectory screens are fed in real time by data coming from the theoretical best tracking source, while RSV screens are fed by data coming from the second best tracking source.

This is a simplified example of how the trajectory is viewed by operators in Flight Safety Control Room:

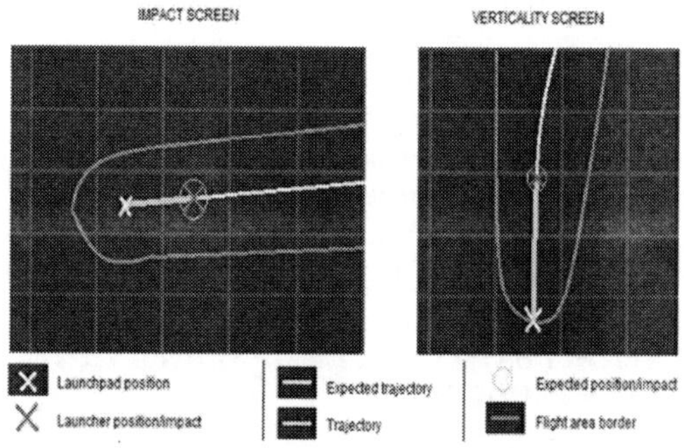

Figure 3 Trajectory screens

Each one of these screens contains information about planned nominal trajectory, actual trajectory, nominal flight area border, topographical elements and safety limits (not displayed in the example above).

Based on the information displayed on these screens and running through the operational loop, RSV applies decision criteria which help to characterize the launcher's dangerousness. Once a diagnosis is announced by RSV, RSVA must tell if he or she approves it or, on the opposite, if RSV has missed something. If RSV and RSVA disagree on the diagnosis, RSV can ask to start a new operational loop. Once the diagnosis is approved by RSVA, the launcher's characterization may bring RSV into getting ready to send a flight termination order then, if necessary, into sending the order effectively.

2. METHOD IMPROVEMENT

Even if the current method responds to Flight Safety mission needs, several details could be improved by making the best use of operators' and automated processes' capacities. The main idea is not to rethink the method from scratch, it would rather consist in optimizing a system that works well by identifying areas for improvement that would help to consolidate current method strong points.

2.1 Areas for improvement

Human factors

Some areas for improvement of the current Flight Safety method are directly linked with human factors.

A human operator has physical limits that, once they are crossed, raise dramatically the risk of an operational error:

- Even if a well-trained operator may be able to manage several tasks at once, a human being can in fact only pay attention to a sole problem at a time; with stress, a "tunnelization" phenomenon may bring the operator into being captivated by a single matter and neglecting to check other potential issues (e.g.: focusing on the verticality screen and neglecting the impact screen).

- An operator may tire himself or herself out and then make careless mistakes (e.g., wrong dialogue).

- An operator has a limited short-term memory, the more parameters he or she has to check the more likely it is for a mistake to happen (e.g.: ACVI makes an incomplete dialogue because too many alarms showed up at once).

- Cognitive bias can distort operators' judgment (e.g.: a novice operator will likely comply with the opinion of an experienced one; e.g.: an operator has an opinion on the current situation but doesn't challenge it once new elements occur).

These phenomena, clearly identified, are extremely rare because training raises awareness among operators and teaches them how to deal with them. However, even if the errors linked with these phenomena are often negligible and corrected right away by the operators themselves, it would be safer to improve the method in order to reduce their occurrence likelihood.

A first area for improvement would be the number of operators. We can observe it during trainings, the error of one operator raises the likelihood of an error made by another operator. This "domino" effect may lead to loss of time which is also a source of error. Indeed, when operators make successive wrong dialogues, correction dialogues may cause loss of precious time, especially if an urgent matter comes out suddenly. Reducing the number of operators would mechanically reduce the risk of an operational error. A back-up operator being necessary to overcome a potential faint of RSV, the minimum number of operators required to ensure Flight Safety mission is two.

A second area for improvement is dialogue restriction (especially the dialogues about trajectory characterization). Indeed, in spite of a very high degree of calibration to optimize concision and avoid any misunderstanding, dialogues can be time consuming and a source of confusion, therefore a source of error.

Another major area for improvement is the limitation of the number of parameters that operators have to monitor. Indeed, high level synthetic information is sufficient for RSV to make a diagnosis on the launcher's behavior. A keen understanding of the origin of the observed anomaly is not necessary to determine if the launcher follows a potentially dangerous path. The less parameters to monitor, the less the operator is likely to make a mistake. This only requires identifying which parameters are truly needed to ensure Flight Safety mission.

Method

Other areas for improvement are linked to the method itself.

During the safety-active phase, for each operational loop, RSV confirms if the trajectory that he or she observes is coherent with the one previously stated by RSVA. The main advantage of this practice is to guard against a false diagnosis, based on a single tracking source which would have an undetected deficiency. Indeed, by viewing two trajectories that were elaborated from two independent tracking sources, it's very unlikely that the two sources have a deficiency at once. If one of them has an undetected deficiency, we would probably observe incoherent trajectories and that would remind RSV that a reconfiguration is needed.

However, this method is perfectible:

- Dialogue loop between RSV and RSVA is really time consuming: in case of incoherent trajectories, reconfiguration requires time to think about it and to identify which tracking source should be excluded.

- Trajectory statements have "blind spots": for concision issues, RSVA makes a brief description of the trajectory; in some anomaly cases, dialogues could let operators think they have coherent trajectories while they do not.

- Dialogues latency between operators can lead to false incoherence that would require restarting an operational loop, which is even more time consuming.

- Only the two best tracking sources are used: each operator observes a single tracking source, the other sources are simply not taken into account.

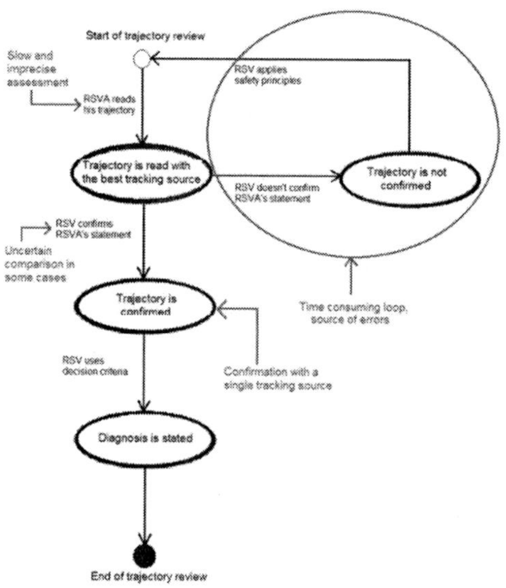

Figure 4 Operational loop imperfections

3. TECHNICAL SOLUTIONS

3.1 Automated tracking confirmation

The automated tracking confirmation is a concept that would let RSV have the guarantee that the only displayed trajectory is reliable and coherent with other independent tracking sources without another human operator's intervention.

Its basic principle is quite simple. A coherence area is calculated using launcher's position and speed vector as they are rendered by a reference tracking source (currently the most efficient source that is displayed on RSV's screens), taking into account its localization errors and delays. We then compare this coherence area with launcher's positions and speed vectors as rendered by the other tracking sources, each time taking into account their own localization errors and delays. As a result of these comparisons, we know if the trajectory rendered by the reference tracking source is confirmed by one or several other tracking sources.

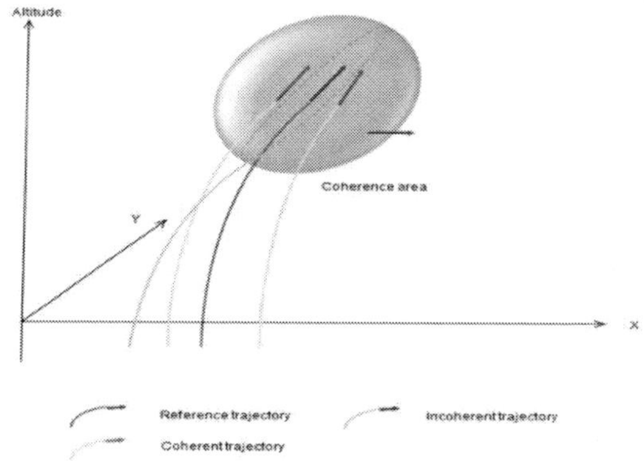

Figure 5 Coherence area

This method has many advantages:

- It's no longer necessary to have two operators comparing their own trajectories.

- Every tracking source contributes to consolidate RSV's diagnosis.

- Tracking sources reconfiguration (e.g.: exclusion of a tracking source) can be dealt with automatically.

- Automatic trajectory confirmation is more reliable because it doesn't have "blind spots".

3.2 Computer aided decision

This concept complements automated tracking confirmation. Indeed, if automated tracking confirmation allows avoiding a second operator's intervention to characterize the launcher's behavior, an operator is still needed to confirm RSV's launcher dangerousness assessment (That is to say: we agree on the launcher's behavior but, is it really dangerous?).

A computer aided decision system, based on pattern recognition algorithms, would allow automatic characterization of the launcher's trajectory. It could identify deviations, fallouts, instabilities, or any behavior that could be seen as dangerous. This automated diagnosis could be directly displayed as superimposed symbols on trajectory screens.

Of course, this system would only be efficient on planned scenarios. It would not prevent us from requiring a second human advice for unprecedented events; but it would still be helpful for trivial cases, where the anomaly is obvious and unequivocal. In these cases, a machine will likely be more reactive and reliable than a human operator will ever be.

This method offers several advantages:

- RSV's launcher dangerousness assessment doesn't require the confirmation of another operator anymore (at least for basic scenarios)

- Decisions making is less time consuming therefore the risk of sending a termination signal too late is reduced

- Careless mistakes are less likely because, for most deviation scenarios, RSV will always be warned by the system

3.3) « Dark-cockpit » telemetry display system

RSV mostly uses telemetry information to anticipate trajectory degradation in case of an on-board alert. The objective consists mainly in raising the state of alert of RSV who doesn't need to have a sharp understanding of the problem's origin.

RSV only requires 4 types of telemetry information:

- Current state of overall propulsion

- Current state of guidance system

- Sequence of nominal events (e.g.: stage jettisons)

- Current state of remote control link

By displaying this kind of information in a synthetic way on trajectory screens, ACVI role becomes obsolete. Moreover, by displaying these parameters in accordance with the « dark cockpit » philosophy, we would avoid overloading trajectory screens. In a "dark cockpit" telemetry display system, telemetry parameters would only be displayed if an alarm occurs. In nominal situation, no telemetry is displayed; it is assumed that if no parameter is displayed it means that everything is fine.

4. IMPLEMENTATION AND ITS ORGANIZATIONAL IMPACTS

Automated tracking confirmation and computer aided decision implementation would make RSVA role almost obsolete. A new operational role could be created in order to merge RSVA and back-up operator roles. This role would consist in replacing RSV in case of a faint and in confirming/disconfirming RSV's decisions in case of an unprecedented situation that would not be covered by procedures.
ACVI role would disappear due to the dark-cockpit's telemetry display system. Indeed, if telemetry information became available on trajectory screens, there would be no further need to have an operator dedicated to tracking on-board anomalies.
From a team of four operators, we could get to a team with only two operators. This reduction would also allow to suppress all dialogues between RSV and ACVI and to drastically reduce those between RSV and the new operator (merging RSVA and back-up roles).

Trajectory screens of RSV and the new back-up operator would display exactly the same information and would be a duplication of one another. Flight Safety team wouldn't have to mind anymore what tracking sources the trajectories they are viewing are a product of. System design would guarantee that operators would always be viewing the available most reliable trajectory, which would enable them to focus on trajectory analysis. This cognitive load decrease would contribute to operational risk reduction.

« Dark-cockpit's » telemetry display system implementation would also reduce operators' cognitive load by displaying the strict minimum of information needed by operators to adapt their state of alert.

A goal that is not sought but that would still be reached is time reduction of training and operators' certification. Indeed, after having suppressed ACVI role and merging RSVA and back-up operator roles, new operators would train to be RSV straight away. Nowadays, it takes almost 30 months to train a certified RSV. It is hard to tell how much time we would save if we implemented all the considered technical solutions. It seems however reasonable to believe that we could at least save the ACVI training part (approximately 6 months).

5. CONCLUSION

Human beings' ability to adapt in front of unexpected situations is their main asset compared to machines. It is important to give operators the means to picture an adequate mental image of the situation in order to be able to intervene when machines are helpless. For the rest, everything that can be automated should be. Complexity of real time calculation doesn't put a brake on innovation anymore, any task that has finite input and output should be automated.

A "semi-automatic" system could even be considered in order to go a step further and let the machine send a termination order in basic and unequivocal scenarios. This last point is not yet on CSG's agenda, but the debate remains open. For the time being, implementation of the previous technical solutions would help Flight Safety team to carry out its mission in a more reliable and, above all, a safer way.

PROCEEDINGS OF THE 9TH INTERNATIONAL SPACE SAFETY CONFERENCE

SESSION 11
SPACE TRAFFIC CONTROL – I

INTEGRATING FORESIGHT ACTIVITIES INTO SPACE SITUATIONAL AWARENESS CAPABILITY DEVELOPMENT AND OPERATIONS: APPROACHES FROM HIGH RELIABILITY ORGANIZATIONS

Regina Peldszus

DLR Space Administration, Department of Space Situational Awareness, SSA Research Section
Königswintererstr. 522-524, 53227 Bonn, Germany
Email: regina.peldszus@dlr.de

ABSTRACT

In view of inherent complexity, uncertainty, and ambiguity, High Reliability Organizations (HROs) in safety-critical or high-risk sectors in and beyond the space domain routinely engage in strategic foresight activities that allow them to parse for and understand incremental and disruptive change. This paper focuses on qualitative foresight efforts in the context of Space Situational Awareness (SSA) as an HRO, with its operational, technological, and policy environment. It discusses contemporary approaches employed in foresight that lend themselves particularly to institutional, governmental, and intergovernmental contexts of SSA. It highlights three methods – horizon scanning, scenario building, and live simulation – that aim at the identification and exploration of weak signals, emerging trends, wild cards, and low probability high-impact events. In offering considerations on practical merits and challenges, it makes the case for tailored foresight efforts for multilateral intergovernmental actors as an indispensable contribution to resilience and a forum for reflection and exchange.

1. INTRODUCTION

The operational, technological, and policy environment of Space Situational Awareness (SSA) is shaped by high degrees of complexity, uncertainty, and dynamics, and is subject to both incremental evolution and radical change – from the shifting ecologies of actors accessing space, to advances in algorithms that enable larger-scale diagnostics and processing, to novel removal technologies, next generation systems enabling more precise means of surveillance and tracking, evolving regulatory agreements that open unprecedented avenues of international collaboration for debris mitigation, or disruptive events such as incidents that change the population of objects in different orbital regimes [1] [2][3].

Although SSA per se is already an effort of foresight – of inherently looking forward and outward in pursuit of safety, resilience, and security – the long-term development and operation of a SSA capability itself relies on foreseeing, understanding, and responding to emerging developments, opportunities, and risks. This collective 'anticipation of the potential' is a key feature of resilient organizations in safety-critical domains [20]. Exploring uncertainties, mapping and evaluating future issues, and translating these into actionable insight is a key prerequisite for devising adaptation strategies. To this end, organizations in high-risk sectors systematically engage in strategic foresight [4] [5] [6]. Particularly in the context of space security and space sustainability, the term foresight is often used synonymously with concepts such as due diligence or care [9], vision or prediction [10], or strategic awareness [11]. These notions allude to the practical definition of foresight or strategic intelligence in an organizational strategy context as the systematic participatory monitoring and sense-making of potential issues and long-term developments [28].

In and beyond the aerospace sector, foresight is increasingly understood from the security, resilience, crisis management, risk management, and business continuity perspective [12]. Foresight is employed – or its dedicated employment advocated – to identify, explore, and understand different futures, for instance in light of debating emerging and potential security challenges, to inform complex organizational decision making, reconcile divergent interests, underpin assumptions for systems requirements, and guide public policy formulation in view of great uncertainty [7] [8].The following considerations outline approaches to integrating foresight activities into SSA as a capability, i.e. as organization, entity, collaborative undertaking, system, and infrastructure that is engaged in operations and development.[1] This requires tailoring not only to

[1] A 'capability' here is understood to consist of a dedicated operational element (e.g. an SSA center that maintains a catalogue, synthesizes a recognized space picture, and provides conjunction alerts), dedicated

the main system characteristics of the domain as such, but also to the organizational context of SSA. From the repertoire available to the foresight analyst, specific qualitative methods are then outlined and appraised which are particularly applicable to s multilateral institutional or intergovernmental context.

2. SSA & HIGH RELIABILTY ORGANISING

2.1 SSA as an HRO Domain

SSA can be framed in the context of High Reliability Organising. High Reliability Organizations (HROs) routinely operate and manage inherently complex or high-risk technologies and processes without major disruptions by ensuring consistently safe operations [13] [14]. The purpose and remit of SSA already involves containing disruptions on a routine basis, predominantly in the form of anticipating events (e.g. conjunctions). It represents a sub-set of the high-level domain of space – at the intersection to other sub-fields such as space operations – which is classified as a high reliability field in the pertinent taxonomies since the advent of thinking on HRO [17].[1] In the classification of sociotechnical systems, an SSA capability would be considered a 'very highly complex' organization containing a set of major and critical technological (sub-) systems (cf. [33] [24]). The concept of HRO fundamentally resonates with other contemporary safety approaches such as Resilience Engineering (RE), which describes an organization's ability and efforts to anticipate, withstand, learn from, adapt and evolve in response to novelty, changes, and disruption within and outside the boundaries of the organization [15] [16].

2.1. Comparable Domains & Salient System Characteristics

HROs comparable to SSA include other large-scale sociotechnical systems that represent critical infrastructure, such as energy grids, air traffic management, nuclear power, chemical processing plants, health care, early warning systems, and others [18].

These fields share a number of system characteristics that not only necessitate safe operations, but coincidentally also render safe operations difficult [17]:

- High or very high complexity;
- Dynamic, tightly coupled, interdependent, nonlinear interactions;
- Intractability, limited comprehensibility (i.e. principle of functions are only partially known, or only understood by fraction of actors);
- Limited observability and accessibility of operational environment (challenge of verification of actuation of e.g. commands);
- Distributed actors, reliance on team work and collaborative practices in operations;
- High degree of causality-versus-intentionality in operations.

These fundamental features are compounded by additional characteristics specific to the domain of space operations, respectively SSA, including [32]:

- Remote interaction with assets, delayed feedback/propagation, reliance on derivation;
- Heterogeneous and distributed assets, infrastructure involving both new and legacy systems that are subject to various life-cycles (incl. long lead-times in hardware procurement versus comparably agile patching of software);
- Heterogeneous and distributed actors on national and international level, across organization, and across functional elements;
- Distributed decision-making (decentralized vs. centralized decision making);
- Great asymmetry between degree of causality in operational domain (e.g. orbital environment subject to laws of physics, rigid proceduralization) versus intentionality in the policy domain (e.g. governing collaboration, regulation).

3. CONTEMPORARY APPROACHES TO FORESIGHT FOR SSA

In the past 15 years, the latest generations of foresight approaches increasingly emphasized high-level systems and cross-actor perspectives while acknowledging conceptual challenges of understanding the unprecedented [19]. These include (i) adaptive planning and strategies addressing different future scenarios, simultaneous exploration and experiment with emerging developments; (ii) combining collective anticipation processes with intra-organizational

sensor elements affording surveillance and tracking, and related elements that afford research and development (R&D), internal governance, and external policy interactions that drive evolution.

[1] Here in the context of Normal Accident Theory (NAT), cf. [17].

strategic foresight; and (iii) horizon-scanning and identification of weak signals.

3.1. Requirements in the Institutional & Multilateral Context for SSA

When undertaken by an entity operating a SSA capability (rather than, e.g. actors purely situated in the policy domain such as a think tank or research institute), foresight will need to address or be embedded in various levels of the organization, as routine operations are likely to performed while maintaining legacy and designing next generation systems.

When this entity with a capability is situated in the public rather than private sector, involves stakeholders from various national institutions, interagency or other multilateral governance models, or operates quasiglobally, foresight efforts will both be shaped by multiple and possibly divergent interest, involve distributed discussions, and – unless outsourced and commissioned externally from dedicated consultants – may not be conducted by a dedicated sub-entity such as a foresight department or unit of a commercial organization.

In scope, granularity, and products, foresight for multilateral SSA needs to be tailored to the specific needs of its end-users. Since the various practitioners involved in the SSA capability – i.e. from flight dynamics, computer engineering, security policy or others – will already be experts in their field, efforts should rely on the collaborative collation, synthesis, and sharing of their existing expert knowledge and awareness in the field, rather than pure fundamental discovery.

In terms of audience, the primary end-users of SSA foresight will be senior-level governance stakeholders in strategy and R&D, who may, however, also participate directly in the foresight process themselves. The type, format, depth and availability of product these users are willing and able to ingest must be decided in the specific context, with the main aim of making results available internally to ultimately facilitate action. In that context, foresight results may complement – and should not necessarily compete – with products from other efforts in internal strategy crafting, co-formulation of policy with external institutional partners, and intelligence products of the individual stake-holding entities in the capability.

In the following, three approaches are outlined that vary in terms of workload, complexity, and depth.

3.2. Method: Horizon Scanning for Comprehensive Understanding of Weak Signals, Wild Cards & Emerging Developments

One particular approach lends itself to these needs of constitutional context due to its flexibility, scalability, and ability to yield relatively exhaustive macro overviews is horizon scanning. Horizon scanning represents a systematic and rigorous outlook for early or weak signals, future developments, a so-called 'big picture', opportunities and risks. Horizon scanning can be conducted either as stand-alone activity or as the initial phase of a larger foresight exercise, on demand or continuously [21] [22].

On a high-level, horizon scanning involves both monitoring of evidence of future trends, as well as identification, review, synthesis, assessment, prioritization of these issues and their translation into insights for a strategic view [22]. Various approaches may be chosen or combined (in parallel or sequentially), including desk research, expert solicitation, interviews, workshops, automated content mining of a wide range of sources (including conference scanning, automated search of publications etc). Approaches are flexible, but main activities include [22]:

- *Signal collection*, including definition and limitation of scan field (e.g. geographic, issue based, sector-based), characterization of scan field (e.g. assumptions, context);
- *Sense-making*, including description of system relationships, description of change drivers; - Preparation of products and reporting.

Methods for the scanning process as such include the monitoring of sources that analysts within SSA will already have access to (e.g. subscription to subject matter related information services or databases, membership in relevant communities of practice, participation in conferences). The key here is to make best use of the insights derived from these forums through capturing and sharing knowledge across stakeholders and enable participants to channel their expertise periodically or ad-hoc.

Due to its malleability, it can be implemented continuously or discontinuously, to resolve ad-hoc questions or feed regular insights on risks and opportunities to decision-makers. An important component of this process is information sharing between various internal actors, developing networks and drawing attention of emerging issues for senior level audiences [22].

Advantages of this approach are its flexibility, possibility to be conducted through desk research of open source material (incl. conference scanning and grey literature), and low intensity of personnel (primarily analysts and subject-matter aspects skilled in

processing and effectively communicating to various stakeholders).
Due to its focus on the external environment, for instance regulatory or technology developments, this approach is suited to inform development processes of a capability.

3.3. Method: Scenarios for In-Depth Exploration of Overall Trends and Low-Probability-High Impact Events

To complement macro-overviews of horizon scanning, scenarios can delve deeper into specific issues. They can reveal context, relationships, and impact of future settings, including those that may appear tangential and unlikely at the outset but are nevertheless entail serious consequences, and whose anticipation warrants awareness, preparation, and response.
Fundamentally narrative in nature, scenarios act as a predictive, explorative, or normative tool to structure thought [23]. As a foresight method, scenario-building has become increasingly valued in providing a platform to recognize, consider, and reflect deep uncertainties in complex technology settings. Scenarios contain goals, agents, and settings [27]; their crafting and discussion allows to address fuzzy problems characterized by trade-off and interdependency and to expose structure and dynamics of problem domains, and affords different perspectives (e.g. use cases, various actors) [25]. Scenarios can highlight not just the most likely, but plausible and important futures [7]. In opposition to simplified, specification-driven design or policy making, a key emphasis here is placed on contextual richness and precision rather than likelihood, while at the same time accepting ambiguity as opportunity [26].
Advantages of this method include that key issues of interest identified in the Horizon Scanning phase of a project can be probed in greater depth, and with more illustrative means, thus aiding communication to various stakeholders. In terms of resources, scenario exercises involve additional personnel, and require advanced skills of the analysts or facilitators including storytelling in workshop settings [22].

3.4. Method: Simulation & Table-Top Exercises for Anticipative Problem-Solving in Policy and Operations

Scenarios may form the basis and context of further understanding of futures through more detailed and fine-grained enactment of a foreseeable setting, e.g. through the unfolding of a discrete situation or event.
Live simulation and tabletop exercises vary in fidelity and scale and enable the link between a future setting and the active response to it. Frequently employed in a training and readiness context for collaborative decision-making, rehearsal of formal processes or protocols, and resolution of anomaly situations (e.g. for contingency simulation of Launch and Early Orbit Operations, cf. [30]), such exercises afford structured, realistic co-discovery across diverse actors, and serve to identify – in cases even tentatively resolve – foreseeable issues, bottlenecks, or problems [31].
Depending on fidelity and scale, preparing and conducting simulations and table top exercise can be complex, time-consuming, and resource-intensive, particularly when conducted across various hierarchies of a system or capability (e.g. involving both operational and policy decisions), or when relying on operational resources (i.e. large simulation sites, computation of results, several moves that involve participants to travel or dedicate considerable time). However, even low-fidelity exercises can be highly evocative and instructive; even convening a half-day meeting with the most basic instructional media (e.g. written scenario outline, conference room, writing material) can yield important first insights into otherwise unforeseen problem cases. Increasingly, simulation exercises are also conducted remotely, particularly when involving distributed teams.
In all cases, for these efforts to be valuable for foresight, they must be accompanied by careful debriefing and discussion, which is performed most effectively by employing observers, instructors, or other types of facilitators.

4. MERITS & CHALLENGES OF FORESIGHT FOR SSA

4.1. Foresight as TCBM and Opportunity for Collective Reflection

Foresight is a intrinsically participatory process. Many participants in activities based on the methods outlined above – such as practitioners from 'shop-floor' operations or those in high-level policy discourse and negotiation – will already be deeply aware of emerging developments in their respective subject matter niches. When intra- and cross-organizational experts channel and contextualize their subject matter insights and merge their thinking and sources on potential futures, a collective synthesis can be an important forum for mutual exchange and act as vehicle for challenging assumptions. Allowing for the inclusion of various cross-capability perspectives through gatekeepers of expertise potentially renders foresight exercises in multilateral settings inherently more fruitful in ways other than those available to globally acting but unilateral commercial actors.
Crucially, foresight offers a window for explorative, solution- and non-solution-oriented thinking in an otherwise highly regimented and proceduralized

domain, and one where there are specific policy decisions to be made across a heterogeneous group of partners and collaborators. Engaging in foresight activities may, in many operations-related settings, be the only opportunity and forum for distributed teams to engage in structured but open-ended reflection and evaluation that is neither time-critical (as in operations) or agenda- and interest-driven (as in policy). Topics may be actively selected that otherwise do not surface on the agenda as they are not immediately urgent but must nevertheless be addressed in the long-term and hence need to be understood, framed, discussed, and reflected on.

If implemented in a meaningful and relevant manner that tailors scope, issues, and products to its specific audiences, foresight may play a critical role in consolidating, synchronizing, and expanding a shared knowledge base between stakeholders of a highly heterogeneous entity, and act as a vehicle for transparency and confidence building both within the organization and towards its partners (cf. [34] [35]).

4.2. Challenges: Tailoring & Coupling to Strategic Activities

Potential options for tailoring of foresight approaches to the specific SSA domain were presented above. Yet, crucial issues remain to be determined when devising a foresight exercise or unit for a multilateral SSA entity. Should foresight be one-off advisory commissioned by external expert or a continuous in-house effort? A response was proposed above, in favor of in-house, continuous engagement to reap the benefits of sharing knowledge and hence evolving jointly. Yet, it is feasible and – in some contexts where methodological expertise is lacking – wise to outsource facilitation to external experts. Large organizations in the commercial sector or industry usually integrate foresight in the context of business intelligence or strategic intelligence in a dedicated unit; where and how this can be sustainably positioned in a multilateral, distributed entity in the long-term is worth discussion; but it is also worth finding a robust solution and persuade stakeholders to entertain even a small-scale continuous effort.

An additional challenge is posed by the identification of effective channels and meaningful dissemination formats for insights to reach the intended audience in the strategy area. Here, the inherent interdisciplinarity of foresight must be leveraged, and technical and nontechnical content and communication styles fused, respectively levelled, in order to render insights palpable to various recipients inside the organization.

This issue is closely linked to the need to actually *act upon foresight*, lest it be simply a generally educational and worth-while but ultimately not impactful pursuit. The type of analysts and participants in the foresight approaches outlined above, bar those involved in high level aspects of table-top exercises in a political or operational context, are not directly related to decision making. They do partake to a certain degree of priority setting, but decisions are made by strategy stakeholders. This connection must be forged.

Finally, the intersection of foresight to other areas that address the external or future environment, whether in scope, depth, or method, needs to be considered. Strategic intelligence activities will eventually intersect with, or derive impetus from, results from numerical modelling and simulation, as well as from other forms of intelligence. Particularly the latter applies when crossing from the safety to the security dimension of SSA, and – regardless of using open sources – constitutes a blurred boundary that requires careful management, but also offers opportunity for further dovetailing and richer insights.

5. CONCLUSION

Engaging in foresight constitutes an indispensable activity to understand emerging opportunities and threats in a dynamically changing environment [29]. In the domain of SSA operations and policy, and specifically in a multilateral, intergovernmental, or institutional context, foresight is a prerequisite for ensuring resilient design and operations.

5.1. Tailored foresight as key contribution to resilience

In order to be effective, relevant, and meaningful, foresight efforts must be actively matched to the remit of the operating entity. When tailored in scope, method, and product to internal audiences in strategy, both the process and output of foresight activities promotes a shared knowledge base among diverse and distributed stakeholders, facilitates structured organizational learning, and supports evidence-based, robust strategy formulation. It provides a forum for open, explorative, structured exchange between different partners in a domain that is otherwise driven by reconciliation of different policy interests and operational constraints. Three top-down approaches – horizon scanning, scenario exploration and table top exercises – lend themselves to the specific domain of SSA, can be scaled to the continuous and ad-hoc needs of various stakeholders, and need not be resource-intensive. Such qualitative methods should complement, inform, and provide impulse and a heuristic base to the employment of, or utilization of results from, numerical modelling approaches, e.g. those that address orbital populations. Frequently, foresight will be the primary or only opportunity for explorative, non-agenda driven exchange between stakeholders. Yet, SSA is such a specialized domain that, in some cases, actors can actually exert real leverage and influence on some of the issues they identify in foresight, due to the double-hat nature of many and reach of its stakeholders, and the strong interdependencies in the field.

5.2. Further R&D Needs to Integrate Additional Methods of Probing and Critique

Specific aspects that remain to be further understood and developed from a methodological and practical perspective include for instance boundaries of foresight as strategic intelligence to other forms of intelligence activities performed as part of the capability and the associated constraints and synergies; the integration and exploitation of insight products derived from live simulation and table-top exercises; and the relationship to other critical thinking or probing efforts for specific problem scenarios such as red-teaming.

ACKNOWLEDGEMENTS

The author thanks her colleagues in the SSA Research Section at DLR, particularly Piera Padula, and in the European Space Surveillance & Tracking Support Framework for interesting exchanges.

9. REFERENCES

1. Ailor, William (2015). 'Space Traffic Management', *Handbook of Space Security: Policies, Applications, Programs*, eds. Schrogl, Kai-Uwe, Hays, Peter L., Robinson, Jana, Moura, Denis & Giannopapa, Christina, Vol. 1, New York: Springer, 231-255.
2. Al-Rhodan, Nayef (2012). *Meta-Geopolitics of Outer Space: An Analysis of Space Power, Security and Governance*, London: Palgrave Macmillan, here 70-74.
3. Pelton, Joe et al. (2015). 'Space Safety', *Handbook of Space Security: Policies, Applications, Programs*, eds. Kai-Uwe Schrogl, Peter L. Hays, Jana Robinson, Denis Moura, Christina Giannopapa, Vol. 1, New York: Springer, 203-230, here 210-214
4. Weick, Karl E., Sutcliffe, Kathleen M., & Obstfeld, David (1999). 'Organizing for High Reliability: Processes of Collective Mindfulness', *Research in Organizational Behavior*, eds. Sutton, R.S. & Staw, B.M. (eds), Vol. 1, Stanford: Jai Press, 81–123.
5. Healey, Mark P. & Hodgkinson, Gerard P. (2008). 'Troubling Futures: Scenarios and Scenario Planning for Organizational Decision Making', *The Oxford Handbook of Organizational Decision Making*, eds. Hodgkinson, Gerard P. & Starbuck, William H., Oxford: Oxford University Press, 565-585, here 566.
6. Catino, Maurizio (2014). *Organizational Myopia: Problems of Rationality and Foresight in Organizations*, Cambridge: Cambridge University Press, here 77
7. Huntley, Wade L., Bock, Joseph G. & Weingartner, Miranda (2009). 'Planning the unplannable: Scenarios on the future of space', Space Policy 2009, 1-14.
8. Axel Volkery, Teresa Ribeiro (2009). 'Scenario planning in public policy: Understanding use, impacts and the role of institutional context factors, *Technological Forecasting & Social Change*, 76 , 1198-1207.
9. McCormick, Patricia K. (2013). 'Space debris: Conjunction opportunities and opportunities for international cooperation', *Science and Public Policy*, 40(6), 801–813.
10. Johnson, Nicholas (2012). 'Cleaning Up Space', *Harvard International Review*, 30 March 2012.
11. Cremins, Thomas E. (2015). 'A New Space Age: Maximizing Global Benefits', *Strategic Foresight: Perspectives on Global Shifts*. Geneva: World Economic Forum, available online 26 June 2017 [http://reports.weforum.org/global-strategicforesight/thomas-e-cremins-nasa-a-new-spaceage/]
12. Elliot, Dominic (2014). 'Disaster and Crisis Management', *The Handbook of Security* (2nd Ed.), ed. Gill, Martin, London: Palgrave MacMillan, 813-836, here 820.
13. Tamuz, Michael & Lewis, Eleanor T. (2008). 'Facing the Threat of Disaster: Decision Making when the Stakes are High', *The Oxford Handbook of Organizational Decision Making*, eds. Hodgkinson, Gerard P. & Starbuck, William H., Oxford: Oxford University Press, 155-173, here 166.
14. Dekker, Sidney W. A. & Woods, David D. (2010). 'The High Reliability Organization Perspective', *Human Factors in Aviation*, 2nd ed., 123-144, here 123.
15. Haavik, Torgeir K., Antonsen, Stian, Rosness, Ragnar & Hale, Andrew (2016). 'HRO and RE: A pragmatic perspective', *Safety Science*, (available online 23 August 2016, in press).
16. Reason, J. (2008). *The Human Contribution: Unsafe Acts, Accidents and Heroic Recoveries*. Farnham: Ashgate, here 8.
17. Perrow, C. (1999). *Normal Accidents: Living with High-Risk Technologies*. Princeton: Princeton University Press.

18. Hesse, Markus & Hornung, Markus (2015). 'Space as Critical Infrastructure', *Handbook of Space Security: Policies, Applications, Programs*, Vol. 1, eds. Schrogl, Kai-Uwe, Hays, Peter L., Robinson, Jana, Moura, Denis & Giannopapa, Christina. New York: Springer, 187-201.
19. Köhler, Jonathan et al. (2015). *Concurrent Design Foresight: Report to the European Commission of the Expert Group on Foresight Modelling*, Brussels: European Commission, DG Research & Innovation, (EUR 26865 EN), here 20.
20. Hollnagel, E., Woods, D. D., Leveson, N. (2006), *Resilience Engineering: Concepts and Precepts*. Burlington: Ashgate.
21. Axelsson, L. (2006). 'Structure for Management of Weak and Diffuse Signals', *Resilience Engineering: Concepts and Precepts*, eds. Hollnagel, E. et al., Burlington: Ashgate, 152-154.
22. Cuhls, Kerstin, van der Giessen, Annelieke, and Toivanen, Hannes (2015). *Models of Horizon Scanning: How to integrate Horizon Scanning into European Research and Innovation Policies*. Brussels: European Commission.
23. Borjeson, L., Hojer, M., Dreborg, K.-H., Ekvall, T., & Finnveden, G. (2006). 'Scenario types and techniques: Towards a user's guide', *Futures*, 38, 723-739.
24. Hobbs, A. N., Adelstein, B. D., O'Hara, J., & Null, C. H., (2008). 'Three principles of human-system integration', *Proceedings of the 8th Australian Aviation Psychology Symposium*, Sydney, Australia, 8-11 April 2008.
25. Varum, C. A., & Mello, C. (2010). 'Directions in scenario planning literature – A review of the past decades', *Futures*, 42, 355-369.
26. Carroll, J. M. (2000). 'Five reasons for scenario-based design', *Interacting with Computers*, 13, 43-60.
27. Rizzo, A., & Bacigalupo, M. (2004). 'Scenarios: Heuristics for action', *Proceedings of XII European Conference on Cognitive Ergonomics*, eds. Reed, D. J., Baxter, G. & Blythe, M., York: EACE.
28. Miles, Ian (2010). 'The Development of Technology Foresight: A Review', *Technological Forecasting and Social Change*, 77(9), 1448-1456, 1448.
29. Weick, Karl E. & Sutcliffe, Kathleen M. (2001). *Managing the Unexpected: Assuring High Performance in an Age of Complexity*. New York: Wiley.
30. Buckle, S., Peldszus, R. & Bessone, L. (2015). Adaptation of the ISS Human Behaviour & Performance Competency Model as Observation & Debriefing Tool for Mission Control Teams During Simulations', *Space Safety is No Accident: The 7th IAASS Conference*, eds. Sgobba, Tommaso & Rongier, Isabelle, New York: Springer, 303-312.
31. Hettinger, L. J. (2003). 'Integrating Training into the Design and Operation of Complex Systems', *Handbook of Human Systems Integration*, ed. Booher, H. R. (Ed.) (2003), New York: Wiley, 433-461.
32. Peldszus, R. (2014). *The Human Element & System Resilience at ESOC: Practical Approaches to Organizational Learning from External High-Reliability and Safety-Critical Domains* (ESA-AMCO-TN-0011 and ESAAMCO-EX-0002). Darmstadt: European Space Agency.
33. Hollnagel, E. (2010). 'Extending the Scope of the Human Factor', *Safer Complex Industrial Environments: A Human Factors Approach* , ed. Hollnagel, E., Boca Raton: CRC Press, 3759, here 52-53.
34. Robinson, Jana (2010). 'The Role of Transparency and Confidence-Building Measures in Advancing Space Security', *ESPI Report* 28, September 2010, Vienna: European Space Policy Institute.
35. Mineiro, Michael C. (2012). *Space Technology Export Controls and International Cooperation in Outer Space*, New York: Springer, 2012, here 212.

THE NETWORK OF PASSIVE CORRELATION RANGING FOR GEOSTATIONARY SATELLITES

Felix Bushuev(1), Mykola Kaliuzhnyi(1), Oleksandr Shulga(1), Leonid Shakun(2), Vladislavs Bezrukovs(3), Oleksandr Reznichenko(4), Sergiy Moskalenko(5), Yevgen Malynovskyi(6)

(1) Research Institute «Mykolaiv Astronomical Observatory», 1, Observatorna St., Mykolaiv, 54032, Ukraine, Email: dir@mao.nikolaev.ua

(2) Research Institute «Astronomical Observatory» of the Mechnikov Odesa National University, 1V, Marazliyivska St., Odesa, 65014, Ukraine, Email: astronomical_observatory@onu.edu.ua

(3) Ventspils University College, 101, Inzenieru St., Ventspils, LV-3601, Latvia, Email: virac@venta.lv

(4) Institute of Radio Astronomy of the NASU, 4, Mystetstv St., Kharkiv, 61002, Ukraine, Email: rai@ri.kharkov.ua

(5) Western Center of Radio Engineering Surveillance, Kosmonavtov St., Mukacheve, 89612, Ukraine, Email: ckako@i.ua

(6) Rivne Minor Academy of Sciences of School Age Youth, 17, Symona Petlyury St., Rivne, 33013, Ukraine, Email: man.rivne@ukr.net

ABSTRACT

An experimental network of passive correlation ranging (NPCR) for geostationary satellites consists of five spatially separated stations located in Ukraine (Kharkiv, Mukacheve, Rivne, Mykolaiv) and Latvia (Ventspils). The measured parameter of the NPCR is a time difference of arrival (TDOA) of the signal to the network stations. Values of the TDOA are computed by cross-correlation of the signal received by the stations, and the satellite's orbit is determined using known coordinates of the stations and the measured values of the TDOA. The GPS is used to synchronize the network stations.

Results of orbit determination of the Eutelsat-13B telecommunication satellite, obtained during more than two years, are presented in the article. The estimate of the error of coordinate determination of the tracked satellite is obtained in the local orbital frame. It does not exceed 225 m at the epoch of the orbital elements. The Eutelsat-13B coordinates, obtained by the NPCR and Eutelsat from August 16 to 31, 2016, are compared.

The presented network can be a prototype of an independent traffic monitoring system at geostationary altitudes.

1. INTRODUCTION

A system of passive correlation ranging (PaCoRa) of geostationary satellites is considered by the European Space Agency (ESA) as a highly accurate alternative to conventional radar used two-way ranging method [1]. The Société Européenne des Satellites (SES) together with the Fraunhofer Institute for Integrated Circuits (IIC) performed the corresponding ESA project during 2010-2013 years. In September 2015, it was reported that this group was implementing the PaCoRa in the USA [2].

A network of spatially separated terminals (stations) is used by the PaCoRa to monitor the position of a geostationary satellite. The stations are located within a satellite's footprint and synchronously receive a signal transmitted by the satellite. The measured parameter is a time difference of arrival (TDOA) of the signal to the network stations. Values of the TDOA are computed by cross-correlation of the signal received by the stations, and the satellite's orbit is determined using known coordinates of the stations and the measured values of the TDOA. It was reported that PaCoRa used 4 main and 1 back-up receiving stations in its first implementation [3].

According to the ESA and SES there are some following advantages of the PaCoRa compare to the classical method of ranging:

- Rx-only: small antennas and standard RF front-end hardware required;
- no uplink for ranging, therefore no impact on the satellite;
- can range all operational geostationary satellites;
- orbit determination of own satellites and 3rd party satellites;
- multi-satellite ranging is possible with a single system, along with multiple orbital positions and bandwidths;
- minimizing the cost of ranging operations by automating measurements.

The above features of the PaCoRa method may be used to create a cost-effective system of independent estimations of traffic in a given area of outer space on geostationary altitudes. This statement is supported by the experience of using the method in the Research Institute "Mykolaiv Astronomical Observatory" (RI "MAO").

The method of passive correlation ranging of geostationary satellites has been employed by the RI "MAO" since the first experiment in August 2011 with two stations spatially separated on 150 km [4]. It has been considered as an independent method for tracking the future Ukrainian geostationary satellite "Lybid'. Since December 2014, regular tracking of geostationary telecommunication satellites had been performed by experimental network of passive correlation ranging (NPCR) consisted of two stations. At the initiative of the RI "MAO" an NPCR with four stations has operated in Ukraine starting from February 2015. The fifth station, located in Ventspils (Latvia), has been included in the network in September 2015. Now the NPCR stations are located in Kharkiv, Mukacheve, Rivne, Mykolaiv and Ventspils. As of the beginning of 2017, an array of measured values of the TDOA lasted more than 800 days has been obtained by the network with two, four and five stations. Three satellites (HotBird-9, Eutelsat-25C, and Eutelsat-13B) also have been tracked by the NPCR in different period of time since 2011. The hardware and software of the network, despite their simplicity and cheapness, allow tracking an arbitrary geostationary satellite emitted DVB-S (Digital Video Broadcasting-Satellite) signal in the direction of network stations.

2. NETWORK HARDWARE AND SOFTWARE

The NPCR consists of spatially separated stations receiving digital satellite television and a data processing center.

The stations have identical hardware and software. Each station consists of [5, 6]:

- the standard antenna-feeder system for the reception of DVB-S signals, with antennas of 0.9 m (Rivne, Mukacheve, Kharkiv and Ventspils) and 1.9 m (Mykolaiv) in diameter;
- DVB-S receiver (SkyStar1 or Skystar2) performed as a PCI-card and upgraded in terms of outputting of in-phase (In) and quadrature (Qu) signals from the receiver's quadrature detector;
- single-frequency ThunderBolt-E GPS receiver;
- digital USB-oscilloscope DSO5200A with 200 MHz passband and 9-bit ADC (Analog Digital Converter);
- personal computer (PC) with USB and RS-232 ports (1 GHz CPU clock rate, 1 Gb RAM, and 100 Gb hard drive capacity);
- Internet connection at data rate of at least 80 Kb per second.

For an example equipment appearance of the Ventspils station is shown in Fig. 1.

Every second the hardware-software complex of the network stations allows recording fragments of the complex DVB-S signal synchronized by the PPS-signal (Pulse-Per-Second), which is generated by the GPS receiver and fed to the external trigger input of the oscilloscope. The fragment duration (T_s) is 200 μs [6].

(a)

(b)

Figure 1. Equipment appearance of the Ventspils station: (a) – antenna of satellite television (in the foreground) and GPS antenna; (b) – PC, digital USB-oscilloscope DSO5200A and ThunderBolt-E GPS receiver

The station software includes programs and drivers of DVB-S receiver, digital USB-oscilloscope, as well as programs developed in the RI "MAO" for every second reading the samples of In and Qu signals from the oscilloscope, and for their initial processing, archiving and saving to the hard disk of the computer [7]. Operation of the stations is monitored over the Internet using the remote desktop application AnyDesk. The FTP-client FileZilla is used to transfer observation data to the data processing center. The daily size of observation data obtained by a station does not exceed 1 Gb.

The data processing center is located in Mykolaiv and designed to calculate orbit elements of tracked satellite by measured values of TDOA. The data processing center consists of PC with an Internet connection of 7.5 Mbps. The PC has a 64-bit CPU with 4 cores and 3 GHz clock rate, and a hard drive of 3 Tb capacities. Note that the daily array of observations of a network of 5 stations is processed for approximately 0.7 days. Herewith, in the case of 5 stations, the values of TDOA are calculated for 10 pairs of stations corresponding to different baselines of the network. Software of the data processing center includes a program of correlation processing of observation data and TDOA evaluation, a program of calculating orbit parameters of a tracked satellite using TDOA measurements [8], as well as auxiliary programs of post-correlation processing and visualization of obtained values of the TDOA and orbital position of the satellite. The programs are written in the Python programming language.

3. THE TDOA AND ORBIT DETERMINATION TECHNIQUE

Let N_s denotes the sample size of the DVB-S complex signal of T_s duration, and A_j and ψ_j be the sampling values of its amplitude and phase at discrete instants of time t_j, where index j changes from 0 to $(N_s - 1)$. Note that the parameter $N_s = 10240$ is a constant for the used oscilloscope type, and T_s depends on the set oscilloscope scale and determines the nominal value of ADC sampling rate: $f_n = \dfrac{N_s}{T_s}$. The peculiarity of our approach is that the synchronous digitization of the DVB-S signal is performed by an external ADC outside the phase locked loop and asynchronously to operating of internal ADC of the receivers. This leads to the fact that at the output of the external ADC, the values of ψ_j contain additional phase shifts, individual for each station. However, considering the modulation type of the DVB-S signal, we can expect that the phase differences $(\psi_j - \psi_{j-1})$ calculated for the same fragment of the signal, that is broadcasted, may be the same or close regardless of the station. Accordingly, the values of the function $F(t_j)$ may be the same or close:

$$F(t_j) = A_j \cdot \sin(\psi_j - \psi_{j-1}) \qquad (1)$$

By applying the transformation (1) we have been able to obtain a convolution of signals at the Nyquist sample rate whose signal-to-noise ratio at the correlator output was about 5 through 10, while the original complex signals did not convolute [5]. Since the normalized values of the function F ($\hat{F}$) are directly used to calculate the correlation function and determine the TDOA, samples of $\hat{F}$ are uploaded from the network stations to the data processing center:

$$\hat{F}(t_j) = \frac{F(t_j) - \bar{F}}{\sigma_F},$$

where $\bar{F}$ and σ_F are the mean and standard deviation (SD) of F.

Let i and k denote the sequence numbers of the pair of stations for which it is necessary to define the TDOA ($\Delta\tau_{ik}$). When calculating the correlation function, the samples of the transformation F obtained by one station, for example, the i-th, are used completely (the first sample), and the samples obtained by the k-th station are used in part, only the middle part of the specified length N_c (the second sample). Therefore, the size of the first sample is fixed and equal to N_s. The size of the second sample N_c is always less than N_s and is equal to the sample size of the correlator. Let n_k denote the specified offset of the second sample. It is measured in counts of the sampling rate. The following formula can be used now to calculate $\Delta\tau_{ik}$ [7, 9]:

$$\Delta\tau_{ik} = \delta_{xi} - \delta_k - \Delta\tau_{hik},$$

where $\delta_{xi} = \dfrac{n_{xi}}{k_{sr} \cdot f_n} + \tau_{PPSi}$ and $\delta_k = \dfrac{n_k}{k_{sr} \cdot f_n} + \tau_{PPSk}$.

Here n_{xi} is the offset of correlation function maximum from the beginning of the i-th station sample measured in counts of the sampling rate;

τ_{PPSi} and τ_{PPSk} are the given initial delays of PPS signals of GPS receivers of the stations;

$\Delta\tau_{hi}$ is the measured value of the difference of hardware delays of the stations;

k_{sr} is the measured coefficient of proportionality between valid (f_v) and nominal sampling rates: $f_v = k_{sr} \cdot f_n$.

The position of the maximum n_{xi} is clarified by using a) spectra interpolation and b) Hilbert's transform of the correlation function [10].

The use of individual delays τ_{PPS} (for each station its own value) allows locating the stations of the network at an arbitrary distance from each other, regardless of the duration of the sample T_s.

It is assumed that $\Delta\tau_{hi}$ depends on equipment characteristics of the stations (e.g., on the difference of the electrical lengths of cables between antennas and receivers and so forth), and k_{sr} is the constant for the given type of USB-oscilloscope. The parameters $\Delta\tau_{hi}$ and k_{sr} are determined during calibration before sending stations to their places of observation.

Let K denote the number of the network stations. The values of $\Delta\tau_{ik}$ are calculated for $K \cdot (K-1)/2$ pairs of the stations corresponding to different baselines of the network.

The orbit of tracked satellite is determined by daily arrays of measured values of the TDOA obtained for all operating baselines of the network. Two models of satellite motion are employed to determine the orbit: the SGP4/SDP4 (Simplified General Perturbation/Simplified Deep Space Perturbation) analytical model and a numerical model of integrating the equations of satellite motion [11]. The following perturbations are taken into account under using the numerical model to determine satellite orbit:

- three-body gravitational attractions of the Sun and Moon (in accordance with JPL DE 405/DE 406 model);
- gravitational attraction of the non-spherical Earth (Earth gravity model EIGEN-6S, truncated to the 9th degree and order).

The satellite position is modeled using the free space dynamics library OREKIT (ORbit Extrapolation KIT) [12].

The orbital parameters are calculated using the least squares method (LS), namely, by minimizing the sum of squared differences of the observed and model values of TDOA. The LS equations are solved using the Levenberg-Marquardt method.

The SGP4/SDP4 orbit elements are given in the so-called TLE format, where TLE is an abbreviation for Two Line Elements. A detailed description of the format can be found at https://www.space-track.org. The orbital elements obtained using the numerical model are given as a state vector consisted of the satellite position and velocity coordinates in the EME2000 coordinate system at the epoch of orbit determination.

4. SATELLITE ORBIT DETERMINATION RESULTS

A catalog of orbital elements of tracked satellite has been created using measured data of TDOA obtained by the NPCR. The catalog contains orbital elements for 441 days within the time interval from February 2015 to February 2017. The epoch of the orbital elements is 12:00:00 UTC of every day. The Eutelsat-13B telecommunication satellite (13° East) has been tracked by the NPCR during this time. Note that it is co-located in the same orbital slot with two other satellites, Eutelsat-13C and Eutelsat-13D. The DVB-S signal at the frequency of 11541 MHz with vertical polarization and at the symbol rate of 22000 kHz is received by the stations during this time. As the duration of sample of recorded signals of quadrature detector is set to T_s = 200 µs then the nominal sampling rate $f_n = \frac{N_s}{T_s} = 51.2$ MHz that is close to the Nyquist rate for the received signal.

The orbital elements are determined using the average values of TDOA ($\overline{\Delta\tau}$) at 60-second intervals of time. Orbital elements are not determined if during a day an interval of observations of the satellite by not less than three stations is shorter than 18 hours or the number of samples of $\overline{\Delta\tau}$ obtained on the interval is less than 720. The following criteria are used in order to reject the failed values of TDOA:

- each second values of TDOA is rejected if the signal-to-noise ratio in the maximum of the correlation function is less than 3.0;
- the value of $\overline{\Delta\tau}$ is rejected if the sample size is less than 10;
- the value of $\overline{\Delta\tau}$ is rejected if the SD of TDOA exceeds 10 ns.

The resulting orbital elements were used to calculate the model values of TDOA. Residuals Δ between model and measured values of TDOA were computed and an average value ($\overline{\Delta}$) and standard deviation (σ_Δ) of the residuals Δ were calculated on fitting intervals. Using obtained arrays of $\overline{\Delta}$ and σ_Δ, their medians $\hat{\Delta}$ and $\hat{\sigma}_\Delta$ were found separately for each model and each baseline. The random component is 5.4 time less on average for the numerical model than for the SGP4/SDP4 model and does not exceed 13.2 ns, while the regular components practically coincide for both model and range from -59.3 ns to +44.8 ns, depending on the network baseline. It follows from the above data that the numerical model is more accurate than the SGP4/SDP4 model.

Two arrays of coordinates were formed on a two-day time interval to estimate an error of coordinate determination of the tracked satellite. The first array of coordinates was obtained using orbital elements for epoch t_0 and the second - for epoch t_1 ($t_0 < t_1$). The time difference between t_0 and t_1 was equal to 24h. The t_0 and t_1 epochs corresponded to the middle of fitting intervals and each interval was equal to 3 days in both cases. The resulting two arrays of coordinates were compared, coordinate residuals were calculated. Mean and standard deviation of the residuals were considered as the unknown estimates of the regular and random errors of coordinate determination of the satellite. The numerical model was used to estimate satellite coordinates in the interval from June 11 to June 14, 2016. TDOA values changed smoothly within this interval so the probability of orbit correction was minimal. The estimates were carried out using a local orbital frame (LOF). The radial (R), tangential (T) and normal (N) satellite coordinates were determined for a point on the Earth's surface corresponding to the location of the Rivne station. Results of estimates of the coordinates R, T and N determination errors are given in Tab. 1 for two (Ventspils, Mykolaiv) and four (Kharkiv, Mukacheve, Ventspils, Mykolaiv) network stations. Note that estimates of the errors of coordinate determination were also obtained for three (Kharkiv, Mukacheve, Rivne) and five (Kharkiv, Mukacheve, Rivne, Ventspils, Mykolaiv) stations which practically coincided with the estimates for four stations, and for this reason here are not given. For comparison, the errors of the PaCoRa system [13] are given in the rightmost column of the table.

Table 1. The results of the estimation of the Eutelsat-13B coordinate determination errors by NPCR *and the PaCoRa system errors*

LOF	2 stations		4 stations		PaCoRa, m
	Mean, m	SD, m	Mean, m	SD, m	
R	+27	±7530	+3	±223	±4
T	–21	±7020	0	±203	±119
N	–7	±1830	0	±2	±10

From the data given in Tab. 1 it follows:

- the regular (mean) and random (SD) errors of coordinate determination of tracked satellite by the NPCR with a minimum number of stations (two) are in the range from 1830 m to 7530 m that is much greater than the errors obtained by the network with four stations;
- the maximum estimate of the errors of coordinate determination of the satellite by the network with four stations does not exceed 225 m that is 1.9 times greater than the PaCoRa system error which is equal 119 m and was obtained by simulation, herewith, the simulation parameters, except the locations of terminals, are not given in [13];
- the regular error in determining the coordinates by the network of four stations does not exceed ± 3 m.

Cartesian coordinates X_N, Y_N and Z_N of Eutelsat-13B in the ITRF2008 coordinate system, calculated using orbital elements from the catalog and the numerical model, were compared with X_E, Y_E and Z_E coordinates which were provided for by the satellite owner Eutelsat. With the company's permission, a table of X_E, Y_E and Z_E values in the time interval from August 16 to August 31, 2016 was downloaded from the archive belonged to the category of the most accurate coordinates. The X_E, Y_E и Z_E values in the table are given in 30-minute increments. The table also contains the message that a correction of the satellite orbit (maneuver) was performed on August 28 at 06:55 UTC.

The graphs of the time variation of the satellite coordinates (X_E, Y_E, Z_E) and (X_N, Y_N, Z_N) and their residuals $dX = X_N - X_E$, $dY = Y_N - Y_E$ and $dZ = Z_N - Z_E$ are given in Fig. 2 (a), (b) and (c), respectively. Note that for each day its own set of orbital elements from the catalog was used to calculate X_N, Y_N and Z_N. For the numerical model, a daily set includes coordinates and velocities of the satellite in the EME2000 coordinate system at epoch of 12:00:00 UTC.

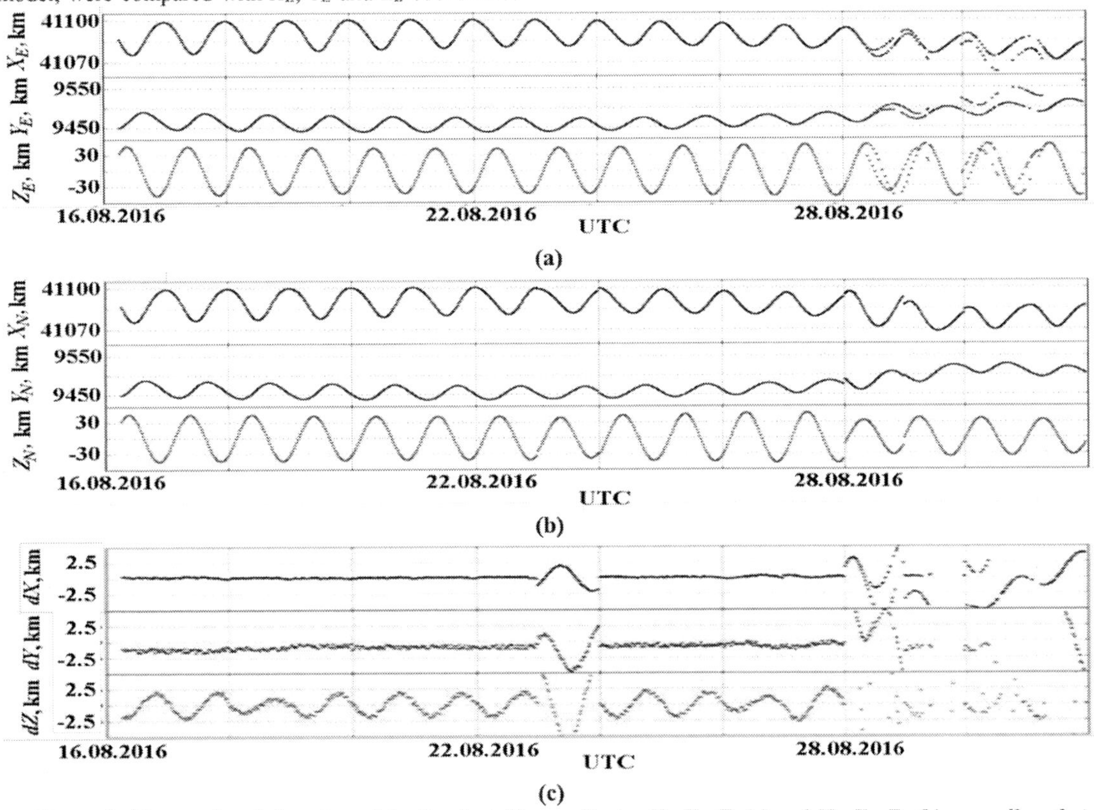

Figure 2. The graphs of changing of the Eutelsat-13 coordinates X_E, Y_E, Z_E (a) and X_N, Y_N, Z_N (b), as well as their residuals dX, dY, dZ (c) in the time interval from August 16 to August 31, 2016

From the graphs, shown in Fig. 2, follows the next conclusions. The X_E, Y_E and Z_E coordinates change smoothly until the moment of the maneuver. After the maneuver, there are jumps between the neighboring values of X_E, Y_E, Z_E. It looks as if there are two separate trajectories of the satellite movement. Herewith the distance between the trajectories exceeds 40 km along Y_E at 19:30 in August 30, for example.

The X_N, Y_N and Z_N coordinates also change smoothly without jumping during the transition from one day to another, except of three days. First of all, this is on August 28, when the satellite's orbit correction was performed, and also on August 23 and 29. The maximum coordinate's jump was 30 km on August 28, whereas it was about 2 km on August 23 and 29. In general, it may be concluded that the X_N, Y_N and Z_N coordinates change much more smoothly than X_E, Y_E and Z_E after the orbit correction, Further studies are needed to determine the reason for the appearance of the jumps of the X_N, Y_N, Z_N coordinates on August 23 and 29. Possible causes include:

- unauthorized changes in the satellite's orbit, due, for example, to collisions with space debris or changes in the satellite cross section in the direction of the solar wind;
- errors of TDOA measurements due to the influence of tropospheric and ionospheric radio signal delays and synchronization errors of the network by the GPS [14];
- inaccuracy of the used model of satellite motion due to not taking into account the pressure of the solar wind, whose influence on the orbit eccentricity can exceed the influence of the Moon [15].

The following could be noted. The ionosphere influence on the operation of the NPCR is much less than on the GPS. The ionospheric delay of the NPCR radio signal is less than tropospheric and does not exceed a meter in unperturbed conditions, when the total electronic content of the ionosphere is 10^{16}-10^{17} m^{-2} [15], since the NPCR operating frequency is more than 10 GHz, and the elevation angle in the direction to the satellite is practically constant and is more than 20°.

From the graphs, shown in Fig. 2 (c), it is following that the coordinate residuals are in the range from -3 km to +2 km for the most part of the analyzed time interval, which does not include August 23 and time after August 27. For this time interval, the average values of dX, dY and dZ are 170 m, -860 m and 0 m, respectively, and standard deviations are ±80 m, ±230 m and ±1280 m. Herewith, unlike dX and dY, the dZ changes are of an oscillating character with an amplitude of approximately 2.5 km. Such character of the dZ changes may be due to the difference of satellite motion models used by Eutelsat and ourselves to calculate coordinates.

5. SUMMARY

1. An experimental network of passive correlation ranging of geostationary satellites, consisted of five stations located in Kharkov, Mukachevo, Rivne, Nikolaev (Ukraine) and Ventspils (Latvia), has been developed and implemented. Continuous operation of the network is ensured. The total duration of arrays of measured values of TDOA, obtained by the network with two, four and five stations, is more than 800 days at the beginning of 2017.

2. The catalog of the orbital elements of the tracked satellite Eutelsat-13B has been created using the measured values of the TDOA. The orbital elements have been determined by daily in the TLE format (the SGP4/SDP4 model of satellite motion) and as the coordinates and velocities of the satellite in the EME2000 coordinate system (the numerical model of integration of equations of satellite motion). The orbital elements have been determined for 441 days at the beginning of 2017.

3. The estimate of the error of coordinate determination of the tracked satellite is obtained in the local orbital frame. It does not exceed 225 m at the epoch of the orbital elements.

4. The Eutelsat-13B coordinates, obtained by the NPCR and Eutelsat from August 16 to 31, 2016 are compared. There is a more stable behavior of the NPCR coordinates obtained after correction of the satellite orbit, in comparison with the Eutelsat data.

5. The presented network can be a prototype of an independent traffic monitoring system at geostationary altitudes.

REFERENCES

1. Passive correlation ranging (PaCoRa) [online]. Available at https://artes.esa.int/projects/passive-correlation-ranging-pacora.
2. Delivery of HMAM antennas for PaCoRa project in USA at the final stage. HILTRON Communications. September 2015. Newsletter. [online] Available at http://www.hiltron.de/media/hiltron/HiltronNewsSeptember2015.pdf.
3. A new way to track geostationary orbiting satellites [online]. Available at https://artes.esa.int/news/new-way-track-geostationary-orbiting-satellites.
4. Bushuev F.I., Kalyuzhny N.A, Shulga A.V., Slivinsky A.P. (2011) Using signals of satellite television to determine the distance to telecommunications geostationary satellites. *The Abstract Book of the International Conference "Astronomical Research: From Near-Earth Space to the Galaxy"*. Mykolaiv, Ukraine, 13-14. http://www.nao.nikolaev.ua/articles/nao2011/2011_Bushuev_a_05.pdf.
5. Bushuev, F.I., Kaliuzhnyi, N.A., Slivinsky, A.P. & Shulga A.V. (2012). Determination of the range to telecommunications geostationary satellites using the signals of satellite television. *Radiofizika i Radioastronomia*. 17 (3), 281–290 (in Russian).
6. Bushuev, F.I., Kaliuzhnyi, M.P., Sybiryakova, Ye. S., Shulga, O.V., Moskalenko, S. S.,

Balagura, O. A. & Kulishenko, V. F. (2016). The radio-engineering complex for determining coordinates of a geostationary telecommunication satellite. *Kosmichna nauka i tehnologia*. 22 (3), 50-59 (in Russian).

7. Bushuev, F.I., Kaliuzhnyi, M.P., Sybiryakova, Ye. S., Shulga O.V. & Gorbanev Yu. M. (2015). Method and first results of calculation of slant-range differences to active geostationary telecommunication satellite. *Radiofizika i Radioastronomia*. 20 (3), 238–246 (in Russian).
8. Shakun, L.S. (2016). The precision orbit determination of geostationary satellite «Eutelsat 13B» by using observations of the signal's time difference of arrival. *The Abstract Book of the 16-th Odessa International Astronomical Gamow Conference-School "Astronomy and beyond: Astrophysics, Cosmology and Gravitation, Cosmomicrophysics, Radio-astronomy and Astrobiology"*. Odessa, Ukraine, p. 45.
9. Bushuev, F., Kaliuzhnyi, M., Sybiryakova, Ye., Shulga, O., Moskalenko, S., Balagura, O. & Kulishenko V. (2016). Results of the ongoing monitoring of the position of a geostationary telecommunication satellite by the method of spatially separated basis receiving of digital satellite television signals. *Latvian Journal of Physics and Technical Sciences*. 53(5), 5–16.
10. Cabot Richard (1981). A Note on the Application of the Hilbert Transform to Time Delay Estimation. *IEEE Transactions on Acoustics, Speech and Signal Processing*. ASSP-29(3), 601-609.
11. Kaliuzhnyi, M., Bushuev, F., Shulga, O., Sybiryakova, Ye., Shakun, L., Bezrukovs, V., Moskalenko, S., Kulishenko, V., Malynovskyi, Ye. (2016). International network of passive correlation ranging for orbit determination of a geostationary satellite. *Odessa astronomical publications*. 29, 203–206.
12. ORbit Extrapolation KIT 8.0 API [online]. Available at https://www.orekit.org/static/apidocs/overview-summary.html.
13. Rodriguez, L., Krier, G., Thill, M. & de Vicente, J. (2014) Passive Ranging for Geostationary Satellites: On a Novel System and Operational Benefits. *SpaceOps 2014 Conferences*. 5-9 May 2014, Pasadena, CA. http://arc.aiaa.org/doi/pdf/10.2514/6.2014-1857.
14. Vandenberg, N. (1999) International VLBI Service for Geodesy and Astrometry: 1999 Annual Report. *NASA/TP-1999-209243*. 309 p. https://ntrs.nasa.gov/archive/nasa/casi.ntrs.nasa.gov/20000011312.pdf.
15. Soop E.M. (1994) *Handbook of geostationary orbits*. Springer-Science+Business Media, B.V.. 309 p.

PROCEEDINGS OF THE 9TH INTERNATIONAL SPACE SAFETY CONFERENCE

SESSION 13
RE-ENTRY SAFETY – II

CASUALTY RISK REDUCTION BY SEMI-CONTROLLED RE-ENTRY

T. Lips and P. Kärräng

HTG – Hypersonic Technology Göttingen GmbH, Germany, Email: t.lips@htg-hst.de

ABSTRACT

At present, there are two classical re-entry scenarios for spacecraft: controlled and uncontrolled re-entry. In a controlled re-entry scenario, also known as targeted re-entry, a final de-orbit thrust maneuver gets the spacecraft onto a well-selected re-entry trajectory directing any surviving fragments into an uninhabited target area. In contrast, if a spacecraft is entering the atmosphere just because of its natural decay without any deliberate actions, surviving fragments could impact anywhere within the latitude band defined by the orbit inclination. This is called uncontrolled re-entry.

The relatively new idea of semi-controlled re-entry is currently defined by ESA as phasing the location of the impact track on the ground. The probable debris fallout zone can extend for a length that can be smaller than one orbit ground-track, but that can also reach several orbit ground-tracks.

This paper describes a method how to find a ground impact arc with minimum population density for a given combination of inclination and arc length. Numerical re-entry simulations have been conducted in order to confirm the general feasibility of drag-assisted removal concepts resulting in a semi-controlled re-entry. This paper will address also some considerations of the ethical problems that come along with semi-controlled re-entry – this type of re-entry reduces the risk for the majority of the Earth's population, but for a small minority the risk might be increased.

1. INTRODUCTION

Widely applicable international rules require a limitation of the on-ground casualty risk to be less than 10^{-4} for re-entries. The mean probability for human casualty P_c resulting from a re-entry event can be calculated as follows:

$$P_c = A_c \cdot \rho_p \tag{1}$$

A person located within the casualty area A_c would be hit by one of the surviving fragments with a probability of P_c. The casualty area of a re-entry object is one of the main results provided by re-entry analysis tools. The applicable population density on the ground ρ_p is a function of the potentially affected impact zone.

For uncontrolled re-entries, the casualty area A_c, defined by the number, size, and kinetic energy of surviving fragments, can be limited, e.g. by design measures such as "design-for-demise". The potentially affected impact zone can not be influenced, as the term "uncontrolled re-entry" implies.

For controlled re-entries, risk limitation is ensured by targeting a relatively small and uninhabited area, typically the South Pacific Ocean, where ρ_p is effectively zero.

For semi-controlled re-entries, compliance with the 10^{-4} limit can be achieved by selecting a ground impact arc, larger than for controlled re-entry, but with lower population density than for uncontrolled re-entry.

Semi-controlled re-entries might be of interest for satellites whose casualty risk would exceed the 10^{-4} in an uncontrolled re-entry, but which do not have the thrust capabilities for a controlled re-entry, e.g. by means of electric propulsion or by drag sail and tether assisted removal concepts. Therefore, the primary objective of semi-controlled re-entry is to minimize the on-ground casualty risk compared to an uncontrolled re-entry. It is assumed that the surviving fragments (number, size, kinetic energy) are not affected by the semi-controlled re-entry. Thus, the only option to reduce the casualty risk is to "reduce" the population density.

2. POPULATION DENSITY

Population distribution data are usually available as 2-dimensional, gridded data files. Longitude λ and latitude ϕ are specified for the center of each grid cell. The size of the grid cells are specified by the cell widths $\Delta\lambda$ and $\Delta\phi$, which are usually identical. The number of cells in longitude direction is $N_\lambda = 2\pi/\Delta\lambda$. The number of cells in latitude direction can be up to $N_\phi = 2\pi/\Delta\phi$ if the complete latitude range is covered by the data file. In order to save hard disk space only the relevantly populated regions are covered, i.e. between -60 and 80 deg of latitude. The total number of cells is $N_c = N_\lambda \cdot N_\phi$. Population distribution data are specified as single, uniform values for each cell, e.g. population density $\rho_p(\lambda, \Delta\lambda, \phi, \Delta\phi)$ or population counts $N_p(\lambda, \Delta\lambda, \phi, \Delta\phi)$.

The basic population database which is referred to in this paper is the *Gridded Population of the World, Version 3* (GPWv3) [1].

The GPW databases mainly consist of global data for population counts and population densities for the several reference years (mid-year estimates). For all versions of the GPW, population densities were calculated as follows:

$$\rho_{p,GPW}(\lambda, \Delta\lambda, \phi, \Delta\phi) = \frac{N_p(\lambda, \Delta\lambda, \phi, \Delta\phi)}{A_l(\Delta\lambda, \phi, \Delta\phi)} \quad (2)$$

where N_p is the population number in a grid cell, and A_l is the actual land area of a grid cell.

For our purpose, calculating the risk for human casualty due to impact within a geodetic grid cell, this type of population density calculation would be inappropriate as also the probability of land impact P_l within a cell has to be considered, which equals the ratio of land area A_l to surface area A_s. Thus, a better approach is to calculate the mean population density for each cell based on its total surface area, including both land and water fraction:

$$\rho_p(\lambda, \Delta\lambda, \phi, \Delta\phi) = \frac{N_p(\lambda, \Delta\lambda, \phi, \Delta\phi)}{A_s(\Delta\lambda, \phi, \Delta\phi)} = \rho_{p,GPW} \cdot P_l \quad (3)$$

Therefore, only the population counts of the GPW databases are needed.

The GPWv3 database provides ASCII data files in the desired ground resolution of $15' \times 15'$ ($\lambda \in (-180\,\text{deg}, +180\,\text{deg})$, $\phi \in (-58\,\text{deg}, +85\,\text{deg})$, 572×1440 grid cells). Historic data files are available for the years 1990, 1995, and 2000. Future estimations are available for the years 2005, 2010, and 2015.

All gridded data files have been expanded to cover the complete latitude range $\phi \in (-90\,\text{deg}, +90\,\text{deg})$. Additional grid cells that were not included in the original GPW file have been set to zero population count values. Therefore the resolution of all GPW grid files was 720×1440 grid cells for the following analyses. The data for the year 2000 have been used as reference. Fig. 1 shows the 2D distribution of the population density for the whole globe.

Averaging the complete 2-dimensional population density distribution $\rho_p(\lambda, \Delta\lambda, \phi, \Delta\phi)$ over the complete longitude range provides longitudinal averaged values of the population density, $\bar{\rho}_p(\phi, \Delta\phi)$:

$$\begin{aligned}\bar{\rho}_p(\phi, \Delta\phi) &= \frac{1}{N_\lambda} \sum_{\lambda=-\pi}^{+\pi} \rho_p(\lambda, \Delta\lambda, \phi, \Delta\phi) \\ &= \frac{1}{N_\lambda} \sum_{\lambda=-\pi}^{+\pi} \frac{N_p(\lambda, \Delta\lambda, \phi, \Delta\phi)}{A_s(\Delta\lambda, \phi, \Delta\phi)}\end{aligned} \quad (4)$$

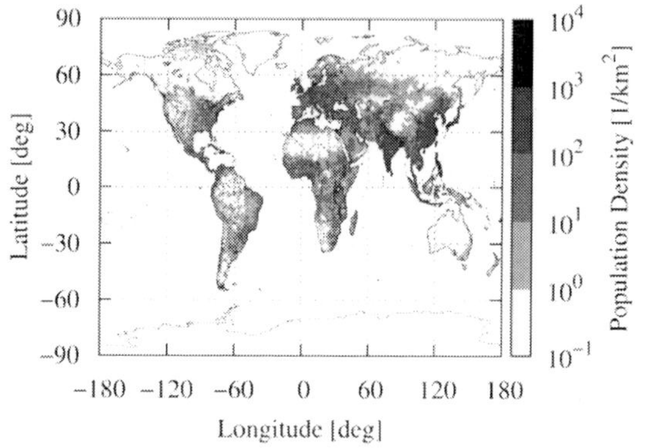

Figure 1. GPWv3, reference year 2000

The result of this averaging process is shown in Fig. 2. This figure illustrates nicely that the mean population densities on the Northern hemisphere are much higher than those on the Southern hemisphere. Re-entries with a potentially affected impact zone on the Northern hemisphere are thus accompanied with a higher casualty risk. Directing re-entry impact zones towards the Southern hemisphere only would be an effective method for risk reduction. However, this is mostly only possible for high eccentric re-entries.

The average population density as function of orbit inclination can be derived by calculating the weighted sum of the longitudinal averaged population density distribution over the whole latitude range. As weighting factors, the impact probability distribution $P_i(\phi, \Delta\phi; i)$ will be used:

$$\begin{aligned}\bar{\bar{\rho}}_p(i) &= \sum_{\phi=-\frac{\pi}{2}}^{+\frac{\pi}{2}} \bar{\rho}_p(\phi, \Delta\phi) \cdot P_i(\phi, \Delta\phi; i) \\ \text{with:} &\quad \sum_{\phi=-\frac{\pi}{2}}^{+\frac{\pi}{2}} P_i(\phi, \Delta\phi; i) = 1\end{aligned} \quad (5)$$

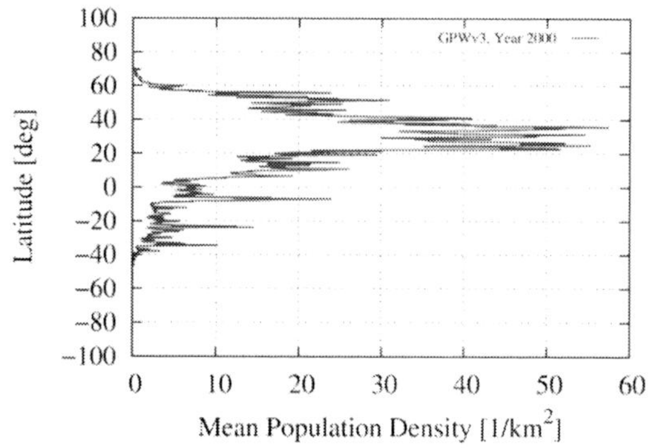

Figure 2. Mean population density as function of latitude

The probability that an uncontrolled re-entry from a near circular orbit of inclination $i \in (0, \pi)$ occurs in a latitude band of width $\Delta\phi$, centered at latitude $\phi \in (-\frac{\pi}{2}, +\frac{\pi}{2})$, can be determined by Eq. 6. Alternatively, impact probabilities P_i as function of latitude ϕ (with orbit inclination i as parameter) can be calculated by Eq. 7.

Fig. 3 shows the impact probabilities as function of latitude for 4 selected orbit inclinations (28.5, 51.6, 90.0, and 98.5 deg). The shown curves have been calculated with Eq. 6, latitude step size $\Delta\phi = 0.25$ deg, and latitude values $\phi_k = -90\text{ deg} + \Delta\phi(k + 0.5)$ with $k \in \{0, 1, 2, \ldots, 719\}$. The sum of all 720 $P_{i,k}$ values equals unity. All curves are symmetric about the equator at $\phi = 0$. Except for an inclination of exactly 90 deg, all curves show significantly higher probabilities for impacts near the minima and maxima of the latitude regimes. As an explanation for these impact probability peaks, Fig. 4 illustrates the dwell time within certain latitude bands. The figure shows the ground track of a near circular orbit of inclination 51.6 deg, with time ticks every 30 seconds of flight. The dwell time within latitude bands, with a width of $\Delta\phi = 2$ deg, near the minimum/maximum of latitude, centered at $\phi = \pm 51$ deg, is much longer than near the equator. Thus also the impact probabilities for these latitude bands are higher.

Fig. 5 shows the average population density for the year 2000 as function of inclination. It provides an indication for orbit inclinations with higher casualty risks due to the higher average population densities within the overflown latitude band. For example, medium orbit inclinations between 20 and 60 deg correspond to the highest average population densities, and thus also the casualty risks for these orbits would be highest. On the other hand, near equatorial or polar orbits correspond to much lower average population densities. Hence, the selection of such orbits would lead to lower casualty risks. However, an active change of the orbit inclination prior to re-entry in order to reduce the casualty risk is unrealistic due to too high delta velocity costs for such a maneuver. Furthermore, most spacecraft for which risk reduction would be needed are already operating in sun-synchronous orbits for which the mean population density is already almost minimal.

3. SEMI-CONTROLLED RE-ENTRY

As explained already in the abstract, semi-controlled re-entry is currently defined by ESA as phasing the location of the impact track on the ground. The probable debris fallout zone can extend for a length that can be smaller than one orbit ground-track, but that can also reach several orbit ground-tracks. Therefore, two principle questions arise:

1. How long is the actual debris fallout zone?
2. Where to direct this fallout zone in order to minimize the on-ground casualty risk?

The first question is actually an uncertainty quantification problem. Semi-controlled re-entry concepts do not have the thrust capabilities needed for a controlled re-entry, i.e. they will apply means of electric propulsion or by drag sail and tether assisted removal concepts. Long duration and low thrust maneuver are more prone to uncertainties. The general assumption for this is that the overall uncertainty is related to the residual lifetime from the initiation of such a maneuver until impact. For example: if the predicted time until impact t_i after the deployment of a drag sail is 7,000 s[1], the width of uncertainty or impact window t_w is of the same size; $t_w = t_i$. This $\pm 50\% \cdot t_i$ approach, considered to be a conservative approach, includes all uncertainties such as maneuver/deployment accuracy, atmosphere/aerodynamics and sail failure.

The second question addresses the optimization problem how to find the optimal location for a ground-track arc with a given length and inclination. Fig. 6 illustrates two impact arcs of 7,000 s length with inclination 98.5 deg. Obviously, the red arc is a better choice than the blue one with respect to affected ground population.

In order to find the optimal arc with minimum population density affected, a numerical tool called `minpopa` (Minimum Populated Arc) has been developed. This tool generates random arc samples of a given length and inclination and extracts the affected mean population density from the GPWv3 database.

Figs. 7 and 8 show the optimum solution for an inclination of 98.5 deg and an arc length of 7,000 s. The mean population density within the affected impact zone is only 0.4% of the value for uncontrolled re-entry. This corresponds to a risk reduction of 99.6%, almost as effective as controlled re-entry.

Figs. 9 and 10 show the optimum solution for an inclination of 51.6 deg and an arc length of 7,000 s. For this ISS orbit, the mean population density within the affected impact zone is still only 3.5% of the value for uncontrolled re-entry. This corresponds to a risk reduction of 96.5%.

This demonstrates that semi-controlled re-entry would be an extremely effective way to reduce the on-ground casualty risk, although very high technical and operational challenges come along with this type of re-entry which have not been addressed in this paper.

4. ETHICAL CONSIDERATIONS AND CONCLUSIONS

The previous sections have shown clearly that – from the pure mathematical point of view – semi-controlled re-entry is an effective method to reduce on-ground casualty risk. The approach is fully in line with the IARA concept in safety analysis (identify, assess, reduce, accept). Under

[1] Computed for an 0.1 m^2/kg object after drag sail deployment at 150 km altitude. If the same deployment occurs already at 200 km altitude, the residual lifetime would be 25,000 s.

$$
\begin{aligned}
i^* &= \begin{cases} i & \text{if} \quad 0 < i \le \frac{\pi}{2} \\ \pi - i & \text{if} \quad \frac{\pi}{2} < i < \pi \end{cases} \\
P_i(\phi, \Delta\phi; i^*) &= \begin{cases} 0 & \text{if} \quad |\phi| > i^* + \frac{\Delta\phi}{2} \\ \Phi(\phi, \Delta\phi; i^*) - \dfrac{1}{\pi} \arcsin\left(\dfrac{\sin(|\phi| - \Delta\phi/2)}{\sin i^*}\right) & \text{if} \quad |\phi| \le i^* + \frac{\Delta\phi}{2} \end{cases} \\
\Phi(\phi, \Delta\phi; i^*) &= \begin{cases} \dfrac{1}{\pi} \arcsin\left(\dfrac{\sin(|\phi| + \Delta\phi/2)}{\sin i^*}\right) & \text{if} \quad |\phi| \le i^* - \frac{\Delta\phi}{2} \\ \dfrac{1}{2} & \text{if} \quad i^* - \frac{\Delta\phi}{2} < |\phi| \le i^* + \frac{\Delta\phi}{2} \end{cases}
\end{aligned}
\tag{6}
$$

$$
\left.
\begin{aligned}
P_i(\phi; i) d\phi &= \frac{\cos\phi}{\pi\sqrt{\sin^2 i - \sin^2\phi}} d\phi \\
P_i(\phi, d\phi; i) &= \int \frac{\cos\phi}{\pi\sqrt{\sin^2 i - \sin^2\phi}} d\phi = \frac{1}{\pi} \arcsin\left(\frac{\sin\phi}{\sin i}\right) \\
P_i(\phi, \Delta\phi; i) &= \frac{1}{\pi} \arcsin\left(\frac{\sin\phi}{\sin i}\right)\Bigg|_{\phi - \Delta\phi/2}^{\phi + \Delta\phi/2} \\
&= \frac{1}{\pi}\left[\arcsin\left(\frac{\sin(\phi + \Delta\phi/2)}{\sin i}\right) - \arcsin\left(\frac{\sin(\phi - \Delta\phi/2)}{\sin i}\right)\right]
\end{aligned}
\right\} \quad (\phi \le i; 0 < i \le \pi/2)
\tag{7}
$$

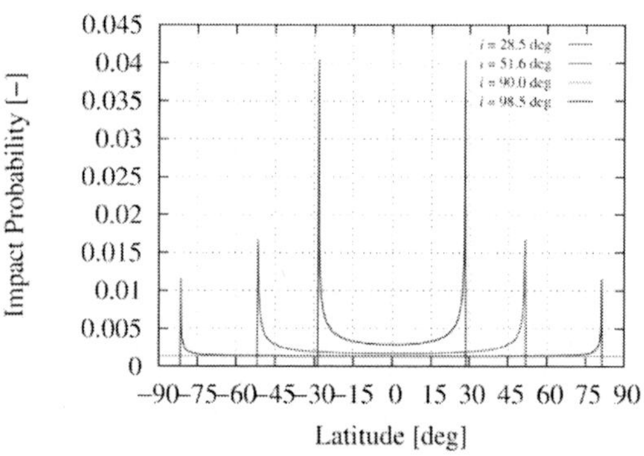

Figure 3. Impact probability as function of latitude

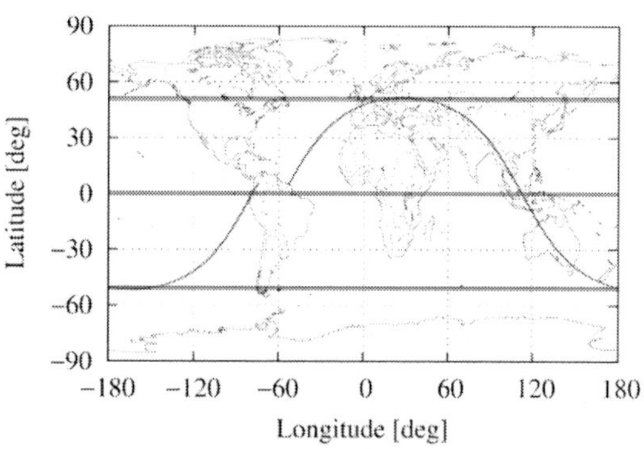

Figure 4. Illustration of dwell times within latitude bands

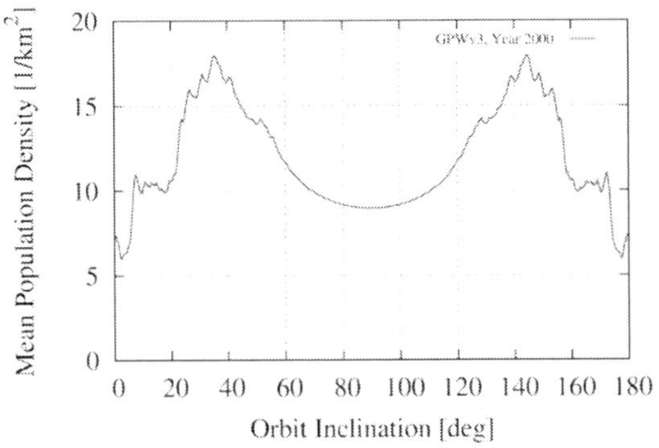

Figure 5. Mean population density as function of orbit inclination

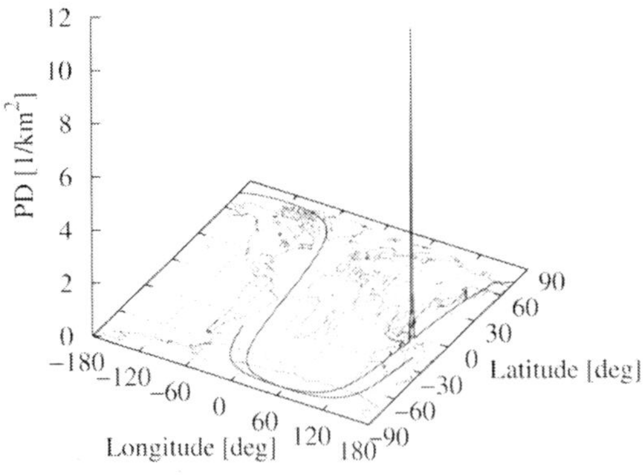

Figure 8. Optimum impact arc of 7,000 s length with inclination 98.5 deg

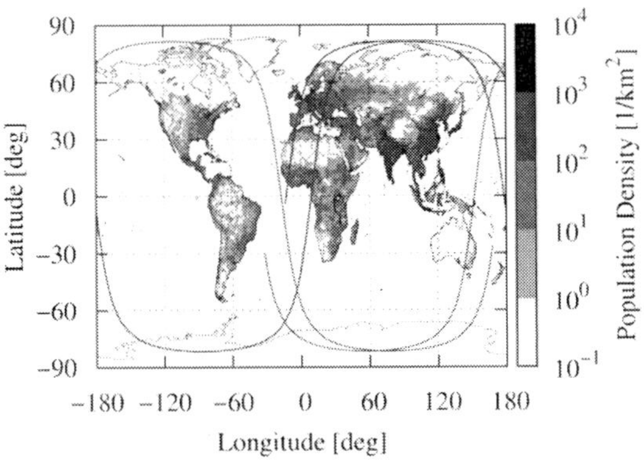

Figure 6. Two impact arcs of 7,000 s length with inclination 98.5 deg

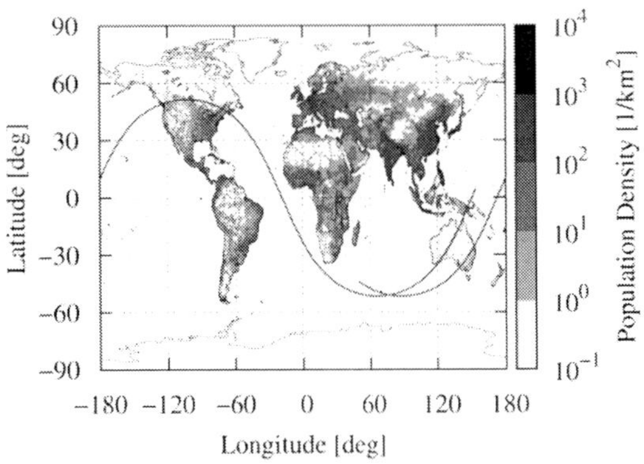

Figure 9. Optimum impact arc of 7,000 s length with inclination 51.6 deg

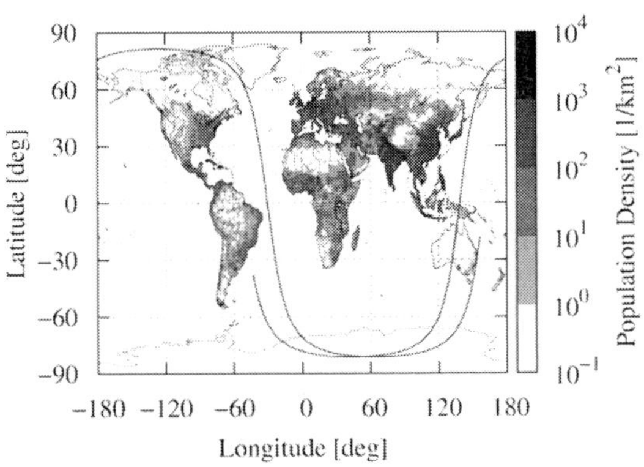

Figure 7. Optimum impact arc of 7,000 s length with inclination 98.5 deg

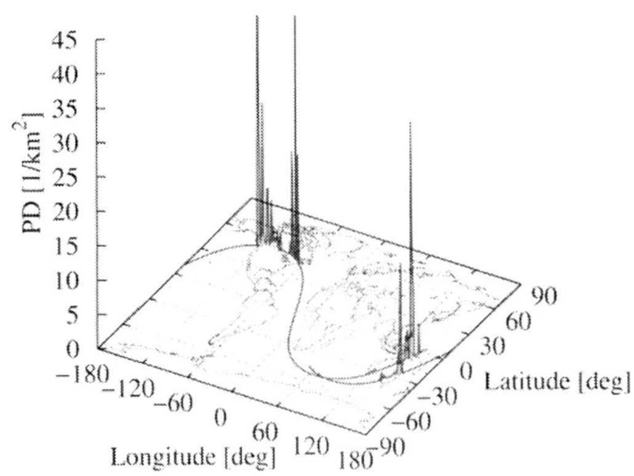

Figure 10. Optimum impact arc of 7,000 s length with inclination 51.6 deg

the clearly defined provisions of re-entry safety standards and guidelines – re-entry casualty risk shall be lower than 1:10,000 – acceptability should not be an issue.

However, objections have been raised because this type of re-entry reduces the risk for the majority of the Earth's population, but for a small minority the risk might be increased. In order to start this discussion, it might be worth to elaborate again on the differences between controlled, semi-controlled, and uncontrolled re-entry.

In the uncontrolled case, basically anybody on Earth is exposed to the same level of risk (minor latitude biases neglected). Casualties can be considered as purely unintended – "act of God". Anybody would benefit to the same extend from technical risk reduction such as design-for-demise.

In the controlled case, the risk is basically zero – as long as no technical failure occurs. Technical failures can again be considered as non-intentional. The persons which might have to suffer from such a mishap are not pre-defined (although the risk is not equally distributed all over the planet).

In contrast, semi-controlled re-entries put people at risk – intentionally and selectively (as shown in Figs. 8 and 10).

Another comparison could be taken into account: launch failures. Also in this scenario, a selected/limited fraction of the Earth's population is exposed to increased risk. However, the root cause in such an event is again a non-intentional failure, and actions are taken just to reduce further/worse consequences. This is an emergency scenario.

Thus, the ethical dilemma seems to stem from the following terms: intention and selection, without emergency justification. The IARA safety concept becomes difficult in its last step: Is it really acceptable to select a certain fraction of people on the ground and to put them at risk intentionally? Although the total risk level is in compliance with the applicable standards and guidelines.

Another point of criticism raised in this context: Why should we consider a questionable approach such as semi-controlled re-entry if there are accepted solutions such as controlled re-entry or uncontrolled re-entry in combination with design-for-demise? This type of criticism insinuates that semi-controlled re-entry is just considered in order to save money.

The authors of this paper are not able to solve the inherent ethical dilemma. Any feedback would be very much appreciated.

ACKNOWLEDGEMENTS

The work presented in this paper has received funding from the European Union's Horizon 2020 research and innovation programme under grant agreement No. 687295.

REFERENCES

1. Center for International Earth Science Information Network (CIESIN), Columbia University; United Nations Food and Agriculture Programme (FAO); and Centro Internacional de Agricultura Tropical (CIAT). *Gridded Population of the World, Version 3 (GPWv3)*. Socioeconomic Data and Applications Center (SEDAC), Columbia University, Pallisades, NY, USA, 2005

ASSISTED NATURAL REENTRY WITH LOW THRUST PROPULSION

Elisabet Cid Borobia[1], Claire Frémeaux[2], Kristen Lagadec[3]

[1] *CNES, 18 avenue Edouard Belin 31401 Toulouse Cedex 9, France, elisabet.cidborobia@cnes.fr,*
[2] *CNES, 18 avenue Edouard Belin 31401 Toulouse Cedex 9, France, claire.fremeauxc@cnes.fr,*
[3] *Airbus, 31 rue des Cosmonautes 31400 Toulouse, kristen.lagadec@airbus.com*

ABSTRACT

Low Earth orbit region is crowded and, in the following years, the number of incoming objects is expected to increase. Deorbiting of satellites at end of mission is, thus, vital to reduce the risks in orbit. Indeed, casualty risk towards the human population should also be limited and several national or agency regulatory texts require the human casualty risk to be lower than 10^{-4}.

Assisted natural re-entry is an attractive alternative to direct re-entry and natural decay, which are the two current options. This deorbiting consists in performing multiple manoeuvers to lower the orbital altitudes as much as possible so that the natural re-entry will be very short, allowing to target a limited impact zone on ground and thus limit the human casualty risk. Maneuvers for an assisted natural re-entry can be performed with classical chemical or even low thrust propulsion, contrary to direct re-entry that needs specific propulsion subsystems.

The casualty risk is only reduced if certain specific constraints, like the impact zone length and its relative position to the Earth, are met. To ensure compliance, a precise final re-entry orbit must be reached on a specific date with a good accuracy. With electrical propulsion, the active deorbiting phase lasts several months, implying a not negligible difference between the predicted conditions and the real conditions experienced in orbit. These differences may impact strongly on the final orbit, preventing to reach the desired target; being necessary to adapt continuously the maneuvers in order to reach the targeted conditions.

This paper presents the principles of an efficient and adaptive strategy for deorbiting maneuvers. It also presents the main challenges, and preliminary performances which appear encouraging as to feasibility.

1 CONTEXT AND MOTIVATIONS

Assisted natural re-entry strategies are motivated by the need for future missions to comply with end-of-life disposal regulations to mitigate the accumulation of space debris. Over the years, these regulatory requirements have become stricter, with a formal introduction into French law in 2008 (*Loi sur les Opérations Spatiales*).

For missions in low-earth orbit, end-of-life disposal requires de-orbiting, by provoking or guaranteeing atmospheric re-entry:

- Either a natural re-entry (or uncontrolled re-entry), which takes advantage of long-term orbital decay from atmospheric drag, such that the spacecraft re-enters the atmosphere in less than 25 years. This option is acceptable only if demise analyses prove that the spacecraft is entirely destroyed in the upper atmosphere or that the fragments that make it to the ground cannot result in a casualty risk above 10^{-4}.
- Or a controlled re-entry, when the surviving fragments are too large or too numerous and the resulting casualty risk would exceed the 10^{-4} limit in case of an uncontrolled re-entry. Controlled re-entry generally requires substantial alterations of the propulsion system, not only in terms of ΔV (propellant) budget but also in terms of thruster size, as it requires lowering the perigee down to 40 km in one last big apogee maneuver.

Assisted natural re-entry (formerly designated as "Semi-controlled" re-entry) is intended as an intermediate solution, applicable when the casualty risk estimated from breakup analyses is moderately above the 10^{-4} limit.

2 PRINCIPLES OF ASSISTED NATURAL REENTRY

The central principle is the following: if the orbital energy can be decreased progressively low enough before attitude control or orbit control is lost, then the uncontrolled part of the descent can then be short enough to predict the impact with an error margin of only few revolutions. Providing that the orbital plane can be properly phased with the Earth's rotation beforehand, it is therefore possible to avoid the more densely populated areas of the globe: this can significantly reduce the casualty risk and thus ensure compliance with the 10^{-4} limit. The ΔV budget is similar to what is required for controlled re-entry, but a large final perigee lowering maneuver is no more needed.

Assisted natural re-entry thus avoids having to

accommodate a high thrust capability, and this is of particular interest for full-electric LEO architectures.

2.1 Risk maps and risk reduction

2.1.1 Casualty area and risk reduction ratio

For a given mission, the casualty area A_c can be estimated from the total area of the surviving fragments, as identified by a breakup analysis.

This casualty area Ac is compared to the maximum casualty area Au allowed for uncontrolled re-entry. We denote $G_{rr} = A_c/A_u$ the risk-reduction ratio that assisted natural re-entry needs to achieve for the mission to comply with the 10^{-4} limit.

For instance, if A_u = 8 m² (typical value for uncontrolled re-entry in 2030), and if the mission's breakup analysis reveals A_c to be ~ 24 m², then assisted natural re-entry has to ensure a risk-reduction ratio G_{rr} = 3: each fragment should be three times less likely to hit a person, compared to uncontrolled re-entry.

2.1.2 Risk maps and maximum dispersion

Risk maps based on population density projections allow to determine how accurate the prediction of the impact point needs to be for reaching a target risk reduction ratio for a given orbit inclination. With a simple Gaussian model for the along-track dispersion in the prediction of the impact, it is possible to compute a casualty risk for every possible latitude/longitude location of the nominal impact point. The resulting risk map shown on Graph 1 highlights the longitude bands which have to be avoided and those which minimize the casualty risk. The risk is minimum when targeting locations that are neither downstream nor upstream from densely populated areas.

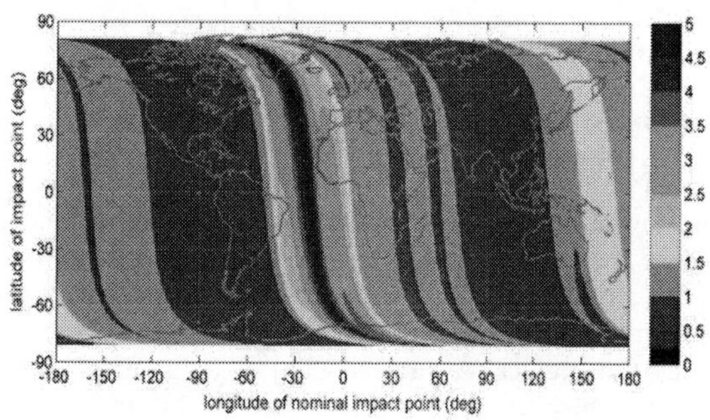

Graph 1. Risk reduction ratio map (wrt. uncontrolled re-entry) for a longitudinal dispersion of 0.75 rev (1-sigma), and an inclination of 98 deg.

By repeating the process we can determine an optimum longitude and a maximum risk reduction ratio for every value of along-track dispersion σ_N.

Graph 2 shows the maximum achievable risk reduction depending on the along-track accuracy of impact prediction. The chart shows that the casualty risk can be divided by almost 3 if the standard deviation of the downrange dispersion is less than 1 (i.e. the impact can be predicted with a +/- 3 orbit accuracy (3-sigma)).

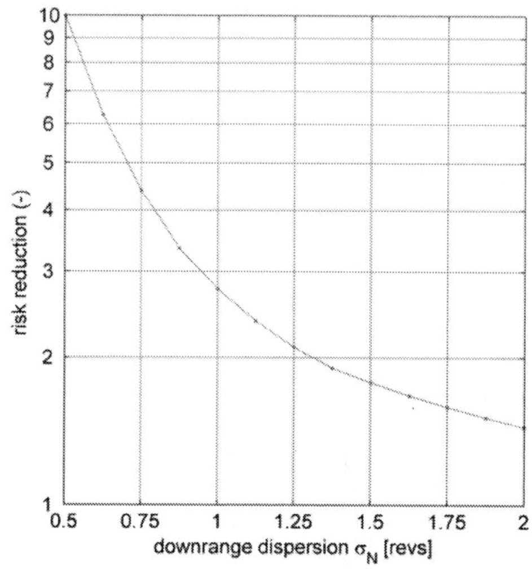

Graph 2. Maximum achievable risk reduction.

2.2 Uncertainty contributors and maximum duration of uncontrolled descent

The factors that may contribute to the uncertainty on the final impact location are mainly related to the drag force uncertainty or variability:

- Atmospheric density : time-variability (real vs. assumed for predicting nominal impact and phasing), regional variability and modelling uncertainty;
- Drag coefficient : attitude-related variability during the uncontrolled descent and modelling uncertainty;
- Accuracy of the controlled descent (time, altitudes)

Thanks to the control strategy, most of these factors can be controlled (individually or in a combined manner) during most of the descent, such that the global uncertainty is essentially determined by the accuracy of the control strategy. This is why we can consider a typical accuracy of 5% (15% 3-sigma) for the prediction of the uncontrolled downrange.

With a relative uncertainty of 5%, the final uncontrolled descent should therefore not exceed 20 revolutions in order to achieve a dispersion of 1 revolution.

2.3 Constraints, scenario and challenges

As the uncontrolled part of the descent must last less than 20 revolutions to achieve the required risk reduction, this means that the orbit and attitude have to remain controlled for nearly the whole descent, including altitudes that are much lower than the mission altitude for which the Attitude and Orbit Control System (AOCS) is designed: the aerodynamic effects become gradually predominant and determine the descent scenario.

2.3.1 Spiraling descent vs. elliptical descent

In a spiraling descent, the orbit remains always quasi-circular and thrust is used to reduce the semi-major axis and adjust the rate of descent depending on atmospheric variations. To do this, the thrust vector has to stay substantially aligned with the velocity vector. However, below a certain altitude, the aerodynamic disturbance torque in this "*thrust attitude*" exceeds the capacity of the AOCS for most spacecraft.

Therefore, except for very specific "GOCE-like" geometries for which the thrust attitude corresponds to an aerodynamic equilibrium, attitude control would be lost prematurely, interrupting orbit control and essentially resulting in an uncontrolled re-entry.

As the majority of satellites are generally not designed with a specific aerodynamic equilibrium, it is necessary to have an elliptical orbit in the descent scenario:

- on the lower portion of the orbit, near the perigee, the satellite flies in an aerodynamic equilibrium attitude to avoid saturating the AOCS actuators;
- on the upper portion of the orbit, the atmospheric density is low enough to resume normal operations, allowing sun pointing to recharge the batteries, and thrust alignment around the apogee to perform perigee lowering maneuvers.

This typically results in a last controlled perigee in the vicinity of 150 km and a last controlled apogee near 300-350 km, to ensure that the nominal duration of the subsequent uncontrolled descent does not exceed the required 20 revolutions.

2.3.2 Phases in the descent scenario

The scenario we suggest for assisted natural re-entry in the general (elliptical) case can be decomposed into three distinct phases, delimited by specific breakpoints in terms of aerodynamic disturbances : initial ascent / descent and phasing when everything is controllable, controlled descent when only apogee maneuvers are still possible, and finally uncontrolled re-entry (see Graph 3).

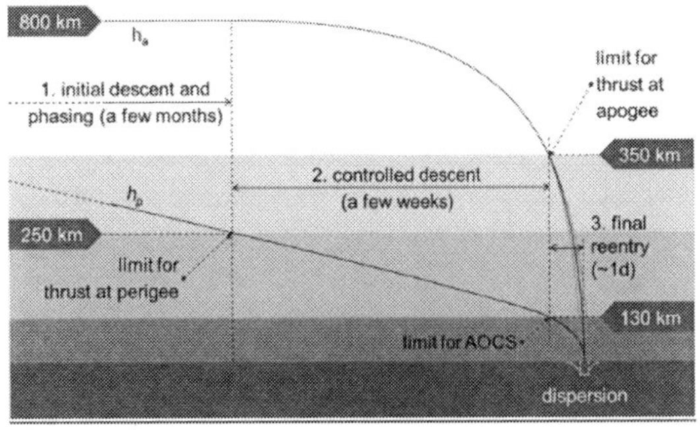

Graph 3. Three phases of assisted natural re-entry. The chart represents the evolution of apogee and perigee altitudes as a function of time.

2.3.2.1 *Phase 1: initial ascent / descent and phasing*

The first phase starts from the mission altitude and its purpose is to prepare orbital phasing (in particular pre-positioning the apogee and the argument of perigee) and to initiate the descent.

It is characterized by full controllability of the attitude and orbit:

- aerodynamic torques are low enough to allow thrust attitude near the perigee;
- aerodynamic drag is lower than the thrust capacity.

In order to match the targeted fallout zone, a certain perigee argument should be reached. The particularity of this orbital element is that, due to perturbation, it undergoes a permanent drift (typically around 1 rev/100 days for a sun-synchronous mission). Correct targeting for perigee argument has thus to be prepared at the beginning of this phase, taking the global drift into account, because for cost reasons it will be possible neither to maintain the perigee argument in a constant position during operations nor to move the line of apsides to a desired value at the end of operations when eccentricity is much higher.

If the initial mission orbit is too low, bringing the satellite to a 130x350 re-entry orbit via apogee maneuvers only may be impossible. It is then necessary to maintain or increase the apogee altitude, which will have a cost in terms or propellant consumption and operations duration. This cost is minimized if an optimal apogee altitude is reached first, its value depending on the satellite characteristics, maneuver capacity and strategy, and mean estimated atmospheric density.

When the optimal apogee altitude is reached and correct eccentricity vector pointing is achieved, perigee descent can begin with apogee maneuvers. In this phase however, it remains possible to execute a perigee maneuver at any time if needed, contrary to Phase 2.

2.3.2.2 *Phase 2: controlled descent*

The second phase starts when it becomes impossible to maintain the apogee altitude:

- either because the drag exceeds the thrust capacity;
- or because maintaining a thrust attitude around the perigee would saturate the AOCS actuators.

From then on, the apogee can only decrease, and the descent is inexorable. However, it is still possible to control the rate of descent by adjusting the thrust profile around the apogee.

To avoid saturating the AOCS actuators because of aerodynamic torques around perigee, the commanded attitude in the lower part of the orbit needs to correspond to an aerodynamic equilibrium attitude.

2.3.2.3 *Phase 3: final uncontrolled descent*

This last phase starts when the orbit control capability is lost:

- either because there is not enough time in the upper part of the orbit to recharge the batteries and achieve a sufficient ΔV;
- or because the aerodynamic disturbances near the perigee exceed the capability of the attitude control system;
- or when aerothermal effects near the perigee exceed the qualification temperatures for exposed organs.

There are no more apogee maneuvers and the orbit starts to circularize under the effect of perigee drag, then the increasing drag at apogee precipitates perigee decay, until final breakup occurs near 80km.

2.3.3 Major technical challenges

There are two major technical challenges to making assisted natural re-entry a valid solution for electric propulsion architectures.

2.3.3.1 *Attitude control around the perigee*

The first major challenge is to avoid losing control of the attitude around the perigee despite aerodynamic effects many orders of magnitude higher than what the AOCS is designed for, without introducing design constraints for the AOCS or the geometry of the spacecraft.

A 2015 Airbus/CNES R&D [7] has allowed to design two promising solutions which are robust to large modelling uncertainty in the aerodynamic coefficients of the spacecraft. The study has showed it was possible to reach perigee altitudes as low as 130 km, even for non-symmetrical spacecraft geometry.

2.3.3.2 *Descent control with low control authority*

The second major challenge consists in controlling the descent in terms of apogee/perigee altitudes as well as in terms of absolute time in order to reach the starting conditions of phase 3 in a deterministic way, despite atmospheric density variations due to the unpredictable nature of solar activity. Parallel advances at Airbus and CNES have determined that this can be done even with low thrust systems. The results are described in more detail in part 3.

2.3.4 Main impacting elements for strategy

2.3.4.1 *Satellite characteristics and solar activity*

The ballistic coefficient depends on both the mass and the surface of the satellite. However, two satellites with exactly the same ballistic coefficient but with different surfaces and masses would have different deorbiting trajectories during the maneuvering part.

The mass has a direct impact in the effectiveness of the maneuver. A smaller mass implies a faster deorbiting strategy than the same satellite with a heavier mass.

The surface as well as the solar activity have an impact on the drag force encountered by the satellite. A larger surface or higher solar activity imply a higher optimal apogee altitude and can lead to longer and more consuming operations even though we benefit from higher drag. The total duration of the deorbiting strategy is more complex to determine than the optimal apogee altitude. It depends on the mass, surface, solar activity and departing orbit. The durations are longer for orbits higher than the optimal apogee altitude, due to the initial perigee and apogee descent. For orbits lower than the optimal apogee altitude, the advantage of starting lower is counterbalanced by the need to increase the apogee altitude first, and the total operation durations are quite similar.

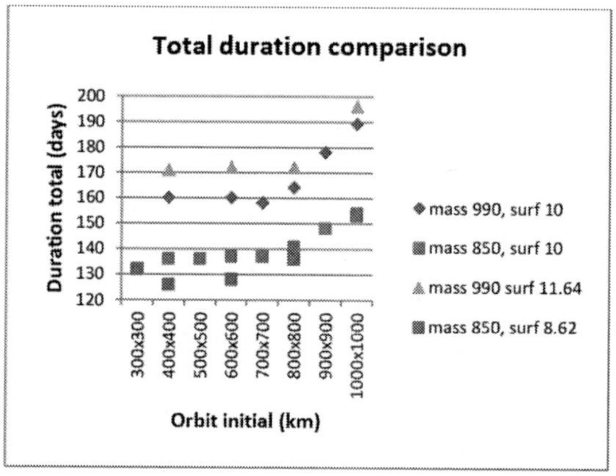

Graph 4. Deorbiting strategy duration comparison.

2.3.4.2 *Inclination*

The mission orbit has a certain inclination that cannot be changed during the deorbiting strategy because it would be too expensive. So, the deorbiting strategy should be performed at the same plane as the original orbit. Unfortunately, the inclination has an impact in the drag force encountered during the descending trajectory, implying a different strategy depending on the drag value. The Graph 5 shows the value that the apogee has to be incremented depending on the inclination and the solar activity. The orbits close to the Equator are the ones with higher drag forces and, thus, their optimal apogee altitudes are higher.

Thus, inclinations close to the Equator add complexity to the strategy and could lead to no realizable strategies due to the high eccentricities needed. On the other hand, if the departing orbit is very high, it would be better to use smaller inclinations. For example, it is preferably to have an Equatorial departing orbit of 1000 km than 700 km. However, for the same satellite in a Polar orbit, it would be better to depart from 700 km.

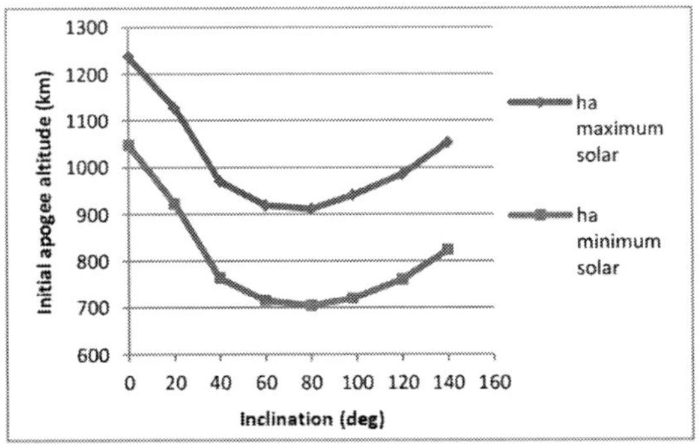

Graph 5. Optimal apogee altitude depending on the inclination for cases of low solar activity (F10.7=65) and high solar activity (F10.7=200).

2.3.4.3 Local time of ascending node

Depending on the position between the orbit and the Sun, the drag force around the orbit is different, especially at very low altitude, as it can be seen in Graph 6, being the perigee the main parameter to suffer from this drag difference. Independently from the season, there are two minima close to 3h and 17h. Then, the orbits with a local time at these points and an argument of perigee at 0º have a faster maneuvering deorbiting strategy. Nevertheless, the variation between the case with more drag and the case with less drag in apogee altitude, and using constant maneuver duration, is approximately of 15 km and in duration about one week. Thus, the impact remains small.

However, if the maneuver duration is variable and dependent on the eclipses, the variations are much more severe. These cases have variations about 90 km in maximum apogee altitude and about one month in total duration.

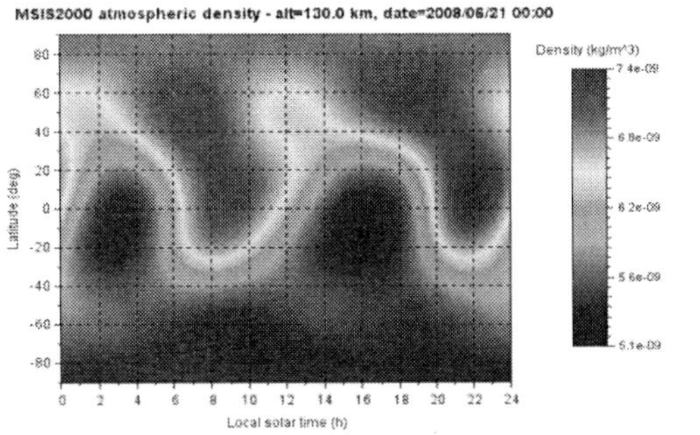

Graph 6. Example of the atmospheric density around the globe at 130 km in a particular date. Higher density in the Northern hemisphere as it is Summer.

3 ADAPTATIVE MANEUVER CONTROL

The nominal deorbiting strategy can be determined and depends on the satellite characteristics, atmosphere and initial mission orbit. However, it cannot be assured that the real conditions will match the nominal ones. Thus, it is necessary to build a control system that follows the nominal trajectory even when some scenario parameters are different.

The nominal strategy (which is the reference case the control will be based on) has the following characteristics:

- Maneuvers have two different durations: a lower one when the maneuver takes place inside the eclipse and a higher one when the maneuver takes place outside the eclipse.
- As the real solar conditions are not known, it uses predicted constant solar activity level, as estimated one year before the reentry date of the simulation. Example in Graph 7.
- It uses a constant ΔV, different from the maximum in order to leave margins for control.

This nominal simulation is stored and then used to compare it with the real trajectory.

Afterward, a propagation simulating the control is launched. The characteristics of the control simulation are the following:

- Adaptive duration of the maneuvers to follow the nominal trajectory.
- Possibility to use both perigee and apogee maneuvers during all the deorbiting simulation (respecting the physical thresholds). Moreover, it also includes the possibility to use accelerating and decelerating maneuvers.
- Use of real solar activity.

Only adapting the maneuvers duration, the satellite should be capable of reaching the correct target date and orbit at the moment of the re-entry.

Monte Carlo simulations are realized. Their number (several hundreds) is determined when the mean and standard deviation errors remain stable.

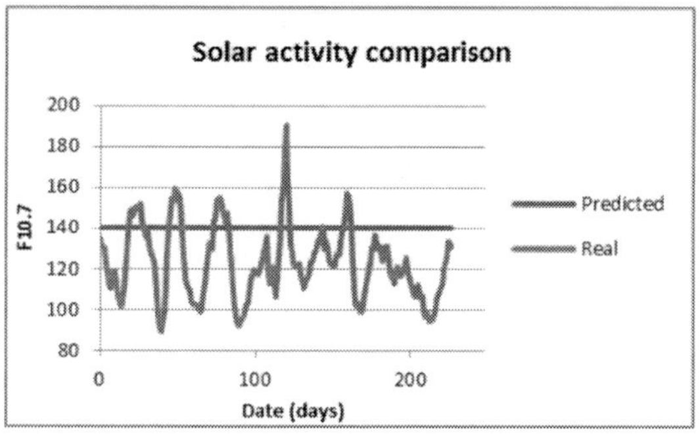

Graph 7. Solar activity differences between the predicted solar activity and the real solar activity.

Table 1 gives the hypothesis of the studied case:

Characteristic	Value
Re-entry orbit	350x130 km (ω=0)
Local time of ascending node	22:30 h
Inclination	98.2 deg
Maneuver duration outside eclipse : nominal / maximum	15 min / 25 min
Maneuver duration inside eclipse: nominal / maximum	10 min / 20 mn
Minimum maneuver duration	1 min
Engine	T=0.082N Isp=1400s
Mass	1 T
Surface	10 m^2
cd	2.25
Earth Potential	8x8
Atmosphere	MSISE2000

Table 1. Simulation characteristics

3.1 Acceptable and unacceptable differences

It is not always possible to reach exactly the same nominal re-entry orbit. Thus, it is necessary to evaluate which are the implications of the differences in the re-entry orbit, in order to evaluate which errors are acceptable and which are not. The inclination and ascending node are not evaluated as these two parameters cannot be controlled and their evolution remains close to the nominal case.

The impact dispersion zone for a natural re-entry from a given orbit is difficult to estimate: it must take into account several uncertain elements such as the atmospheric density model and prediction, the satellite attitude or ballistic coefficients. For observed natural re-entries, it is usually estimated using a dispersion of 20% in the ballistic coefficient which represents the whole drag force uncertainty.

We have considered as acceptable errors on initial re-entry orbit parameters for which the mean re-entry occurs near the nominal mean re-entry point, not exceeding an equivalent of 5% dispersion on drag force.

It has to be assured that statistically the error is acceptable. Thus, the following results should assure that the perigee and apogee altitude differences are compliant with the maximum error, proving that the control is possible.

3.1.1 Perigee altitude

The perigee altitude difference should be very small since at very low altitude, the atmospheric density is very sensitive to altitude. A difference of 1.3 km (1% on 130 km perigee altitude) give equivalent results as 5% drag force error. The simulations will be considered as correct if statistically the perigee altitude difference remains below 1.3 km.

3.1.2 Apogee altitude

The achieved apogee altitude could be less accurate because the drag force at apogee is smaller than near the perigee. A 8 km error on apogee altitude give a similar difference on re-entry point than 5% error on drag force. In the results presented in the following sections, the apogee altitude difference will be considered as correct if statistically it remains below 8 km.

3.1.3 Perigee argument

The case of perigee argument errors is less strict. Indeed, even an error of 5 deg is acceptable. Thus, usually all the cases will be compliant regarding this parameter.

3.1.4 Duration

There is not any error in the deorbiting duration because the two simulations are stopped at the same time.

3.2 Control of the whole strategy

The control should be assured during the whole phases of the strategy: pre-adjusting of the argument of perigee, apogee ascent and perigee descent, as shown in Graph 8.

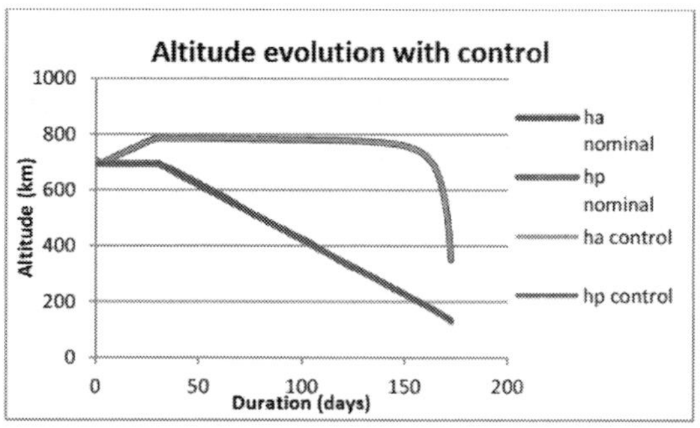

Graph 8. Perigee and apogee altitude during the whole maneuvering part.

Each of the simulation was run at a random date in order to test robustness of the strategy. In a normal distribution of the perigee errors, the errors are inside an interval of 0.7 km, thus, smaller than 1 km. On the other hand, in the case of the same distribution for the apogee errors, the interval remains inside 8 km of errors. Thus, the two cases are compliant with the requirements shown in 3.1.1 and 3.1.2.

3.2.1 Without perigee argument initial control

The control of the perigee argument adjustment is not problematic because it takes place at higher altitudes where the drag force is not very important. Thus, this part is not simulated in the following tests.

3.2.2 Solar activity prediction

The atmosphere plays a central role in the control accuracy. Obviously, the results are better when the predicted solar activity is similar to the real solar activity. The behavior of the solar activity is somehow different between high and low solar activities.

Low solar activities are more stable and thus easier to predict. Indeed, during the years of low solar activity the F10.7 factor varies from 60 to 100, usually without abrupt changes, improving the control. On the other hand, when there is high solar activity, the activity is much more variable (sometimes changing by more than 150 F10.7 units in less than 3 months). Moreover, the maneuvers are less effective during higher solar activity as the drag force is more important. Thus, the control performance is better during low solar activity than during high solar activity.

The Graph 9 and Graph 10 show how, as a matter of fact, the results are degraded during the high solar activity period. Indeed, having an abrupt change in solar activity when the satellite is in very low altitudes can lead to excessive errors.

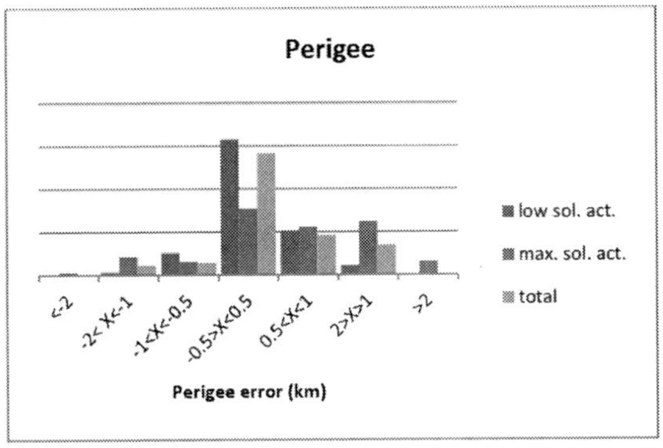

Graph 9. Perigee errors distribution before the natural re-entry. The error standard deviation for: all the cases<1km, limited to low solar activity cases<0.7km and high solar activity cases<1.5km.

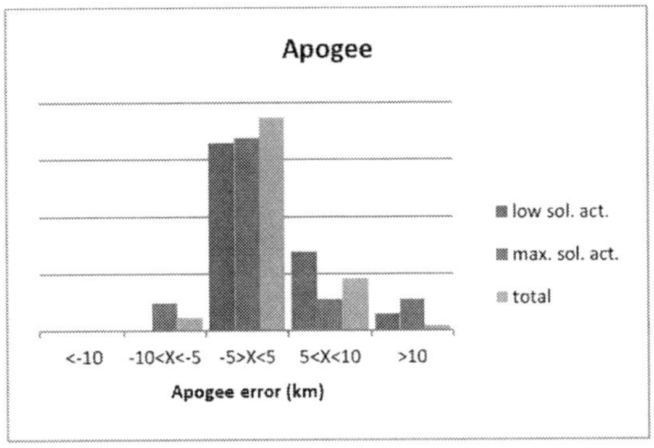

Graph 10. Apogee errors distribution before the natural re-entry. The error standard deviation for: all the cases<6km, limited to low solar activity cases<7km, high solar activity cases<8km.

3.3 On-board vs. Ground control

The computation of the maneuvers could be done on-board or on the ground. The advantages of an on-board computation is that the altitude information is known continuously, allowing a faster adaptability between the real situation and the nominal situation. However, this means cost and complexity to implement the on-board software.

In order to add robustness, it is possible to use a ground control loop, with a less frequent and differed computation of the duration of the next maneuvers, which must be then uploaded. The main problem here is the longer delay needed for the loop, and the lower control frequency.

The on-ground calculations deteriorate the results due to the lack of continuous update of maneuver duration. A demonstration is shown in Graph 11. The previous on-board case is compared to two different on-ground calculation cases. The worst case is the one with a longer period between maneuver updates (12h). Indeed, in this case, the error standard deviation is above 1.3 km, far too high. If the updating period is reduced to a 6h, the results are improved and the error is below to 1 km. Thus, in this example, it is acceptable to use an on-ground system with a minimum update frequency of 4 times a day; however, the on-board calculation remains better.

The tests were performed using a constant update frequency. However, in a real case, it could be wise to use different periods depending on the altitudes: at high altitudes use longer periods and at low altitudes use shorter periods. Indeed, the control is easier when the drag force is less important, thus, at higher altitudes.

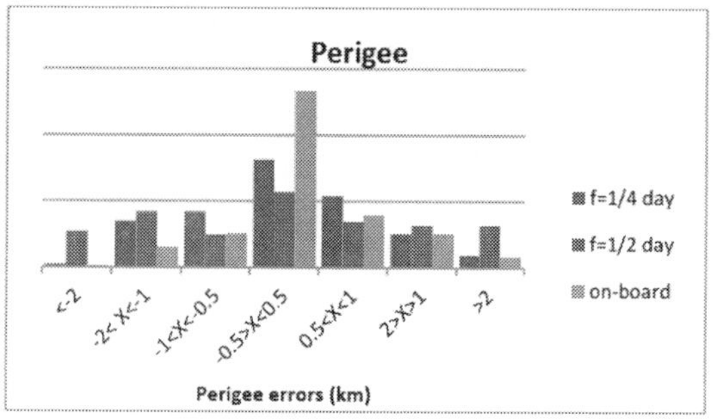

Graph 11. Comparison of the perigee altitude errors in the re-entry orbit depending on the frequency of the on-ground calculations.

3.4 Anomaly cases

It is possible that the satellite suffers some problems that lead to the cancelling of maneuvers during some contingency period. The strategy control should be capable of rallying the re-entry orbit, even with several days of cancelled maneuvers.

There are two major parameters that impact the strategy:

- **Contingency duration**. The impact in the control will be higher if the period is longer. Indeed, when the contingency duration is higher, the gap between the nominal and the real trajectory is larger and it is more difficult to recover the nominal strategy.
- **Altitude where the maneuvers are cancelled**. When the failure takes place at higher altitudes, it is easier to readapt the maneuvers and to rally the nominal trajectory. However, at low altitudes (when the drag force is more important), it will be more difficult to recover from an interruption.

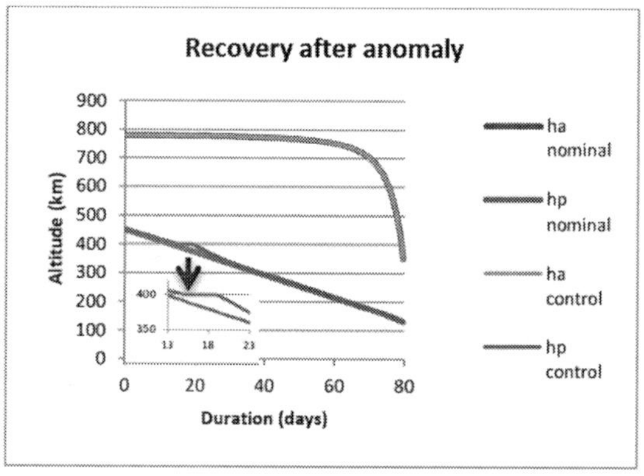

Graph 12. Example of a descending control strategy with 4 days without maneuvers at 400 km (zoomed in the graph). There are also initial perigee and apogee errors at the beginning of the simulation.

Several simulations were performed to test the control capacity about such situations. Graph 13 shows the impact of a 4-day interruption. It is acceptable to cancel the maneuvers 4 days long if the interruption occurs above 300 km. Conversely, below 250 km, it is more difficult to recover from this situation. Indeed, the results show how a non-negligible number of tests with a 4-day interruption at 250 km reach the perigee altitude with an error higher than 1 km and the apogee altitude with one higher than 10 km, which is in excess of the required 8 km.

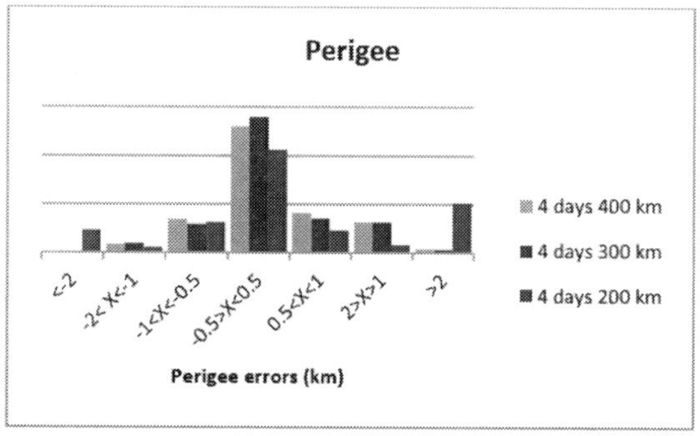

Graph 13. Error in the perigee re-entry orbit altitude depending on the altitude where the maneuvers are cancelled (standard deviation error with an interruption at 400 km <1 km, at 300 km <1.5km and at 200 km <6 km).

On the other hand, if the interruption is shortened to 1 day, the errors become acceptable. Graph 14 shows how it is possible to have a 1-day interruption at 200 km.

Below 200 km, as perigee drag increases exponentially towards the end of the descent, recovery from a late interruption becomes more unlikely.

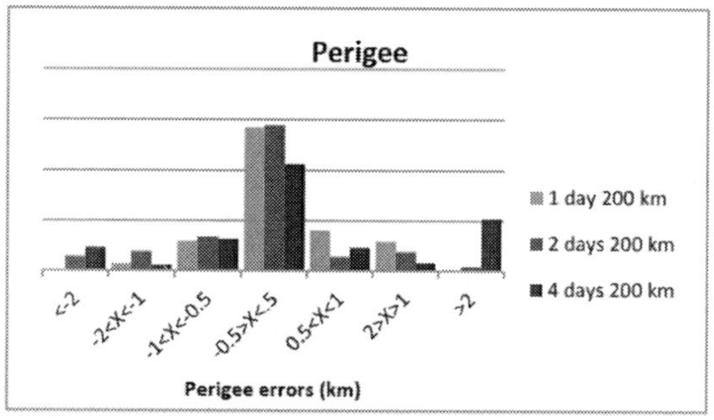

Graph 14. Error in the perigee re-entry orbit altitude depending on the number of days that the maneuvers are stopped (standard deviation error with an interruption of 1 day km <1 km, 2 days <2km and 4 days <6 km).

Yet, as in controlled re-entries, contingency cases are taken into account in the casualty risk computation. Thus, these non-compliant cases involving late interruptions are already taken into account in the global probability.

4 CONCLUSION

The nominal trajectory of the deorbiting strategy is determined by the satellite characteristics and the atmosphere. This strategy imposes the altitude of the perigee and apogee needed at the re-entry orbit. Depending on the initial mission orbit, the first phases of the deorbiting strategy consist, firstly, in ranging the argument of perigee and, secondly, in bringing the satellite to an intermediate orbit: in particular, increasing the apogee if the mission orbit is too low.

By adapting the duration of apogee maneuvers, the timing of the descent trajectory can be controlled despite atmospheric variations in order to reach the target altitudes for apogee and perigee. However, the effectiveness of the control strategy depends on the external factors and in particular on the prediction accuracy for solar activity. Thanks to the lower variability of solar activity, the control performance is best during the more quiet years of the solar cycle and, conversely, less efficient during the years of peak solar activity.

Further studies are under way at CNES and Airbus in order to refine the control strategy and increase the global robustness, but the results of the tests performed so far show how the control can be assured in the whole strategy. Thus, the control is viable and could be used in future missions. Moreover, it exhibits robustness to contingency situations resulting in interruptions, providing that the interruptions are not too long or too late in the scenario.

The results also show that the performance is optimal when the control strategy is implemented on board, allowing the refresh frequency of the maneuver profile to be the highest. Still, we have shown that on-ground implementations with a delay of 6 hours near the end of the descent also yield acceptable performance, thus offering flexibility in the overall design.

5 REFERENCES

1. E. Cid Borobia, C. Frémeaux, J.F. Goester (2017). Fast reentry deorbitation with acceptable risk level. *7th European conference on Space Debris April 18-21, 2017, Darmstadt, Germany.*
2. A. Gaudel, C.Hourtolle, J.F. Goester, M. Ottavian (2015). De-orbit strategies with low-thrust propulsion. *Springer International.*
3. K. Lagadec, R. Pierre (2017). Rapport de l'étude de rentrée semi-contrôlée. *Technical report.*
4. M. Ottaviani, J.F. Goester, A. Gaudel-Vacaresse (2013). Stratégies de désorbitation à poussées faibles. *CNES internship report.*
5. B. Hassan, C. Hourtolle (2015). Maneuvering strategies for semi-controlled reentries applied to low Earth orbits. *CNES internship report.*
6. G. Colas, A. Gaudel-Vacaresse (2014). Computation of risks during uncontrolled atmospheric re-entries on final orbit. *CNES internship report.*
7. K. Lagadec & al., Controle D'attitude Des Satellites A Tres Faibles Altitudes, AIRBUS doc. Ref. GNC_T.TCN.789617.AIRB, October 2016

JELECTRA: NEW FEATURES OF THE CNES LAUNCH AND RE-ENTRY RISK ANALYSIS TOOL

J.F. Goester(1), A. Bellucci(2)

(1)(2)*CNES, 18 avenue E. Belin 31401 Toulouse Cedex 9, FRANCE,*
(1)*+33561281678, jean-francois.goester@cnes.fr*
(2)*+33561287506, aurelie.bellucci@cnes.fr*
(1)(2)*Fax : (+33)561273540*

ABSTRACT

As French National Space Agency, CNES has the prerogative for certain missions, one of which is to control and support French operators for computing risks related to space operations. Then, CNES developed efficient state-of-the-art tools for such evaluation. In this context, ELECTRA development, undertaken in 2007, meets the requirement for precise quantification of the risks involved in the launch and the re-entry of a spacecraft.

Since 2015, thanks to internal CNES decision to use new technologies based on Java™ language for its flight dynamics (FD) tools and libraries, a new development started for ELECTRA leading to a new version V4.0 at the end of 2016 then a V4.1 this year.

In this paper, we will underline an overview of the main functionalities (risk computation after long or short term random re-entry, launch failure, controlled re-entry) then we will focus on some new features. A dedicated topic will also concern the way this new version was validated thanks to the previous one and to the PATRIUS [3] basic reference.

1. BACKGROUND

1.1. Previous versions

Started ten years ago, the development of the ELECTRA tool was originally based on Fortran 95 with GUI (Graphical User Interface) in GENESIS (a language developed by CNES for the last 25 years). In parallel, a viewer called ORESTE, based on multiple layers MapFish library technologies has been developed [1] &[2].

At the beginning, ELECTRA was implemented for internal CNES safety needs and was running on few Operating Sytems (OS). Thus, since December 2010, ELECTRA has been deployed and used operationally to monitor the risk associated to each launch from Guiana Space Centre. The tool was also used to estimate uncontrolled re-entry risk of all CNES LEO missions as well as support to ATV-CC for risk analysis during the controlled re-entry.

But soon ELECTRA has also been provided to space operators, in the frame of the French Space Operations Act. The used technologies became a drawback when deploying it on Operating systems (OS) and platforms not mandatory compatible with CNES ones.

1.2. The SIRIUS project

Some years ago, CNES decided to redevelop its FD heritage mainly based on Fortran language choosing the Java™ object oriented language for reason of portability and sustainability over the next 25 years. Part of the SIRIUS project, the PATRIUS library, originally based on Apache Orekit and Commons Math open source libraries, provides features ranging from orbit extrapolation (with many different force models) to attitude management for any type of orbits. Its exhaustiveness is intended for it to be the building block of both operational software and tools dedicated to studies and mission analyses. Since 2017, the PATRIUS library is also available as an open source project on the CNES Freeware server (https://logiciels.cnes.fr/en).

2. JELECTRA

So, following the development of this new Java™ FD library, it has been decided to redevelop ELECTRA too [4]. It started with two prototype phases in 2014 and 2015 then the development itself started in 2015 leading to a first operational version (V4.0.1) on December 2016. The next version (V4.1) should be available end 2017.

2.1. Portability

These new versions of ELECTRA are then based on Java™ language with its GUI based on GENIUS & GENOPUS (a Java™ framework recently developed by CNES and also available as free open source project). Thanks to Java™ and Commons Math library, same executable files run on a great amount of OS and platforms as Linux and Windows. Moreover, user will obtain exactly the same results independently of the kind of machine used.

2.2. Flight dynamics

The FD functionalities are, of course, based on the PATRIUS library. It allows getting some very interesting new functionality (as orbital events management). On the other side, some changes occurred in the trajectory propagation as force models or numerical integrator were not exactly the same ones. It will be explained more precisely in paragraph 5.2.

2.3. Installer

A real improvement is coming from the installer, which is actually more robust to any encountered configurations. A single jar executable file is needed and no manual manipulations are requested.

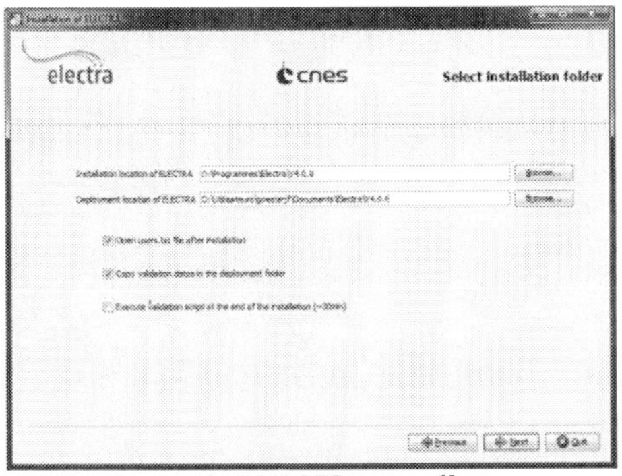

Figure 1: ELECTRA installation

2.4. Main frame

Now, when launching ELECTRA a main frame appears that proposes to select between the different available computation modes as well as a list of additional tools.

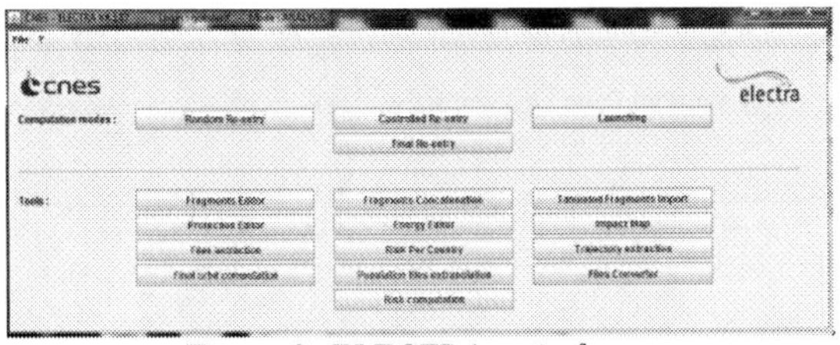

Figure 2: ELECTRA main frame

2.5. Intervals of validity

The management of the validity of number entries is fully improved by the use of the GENIUS library that proposes a more sophisticated way to take it into account. Then, the display will be a more classical look with a red logo if the number leads to a potential error or an orange one if only a warning.

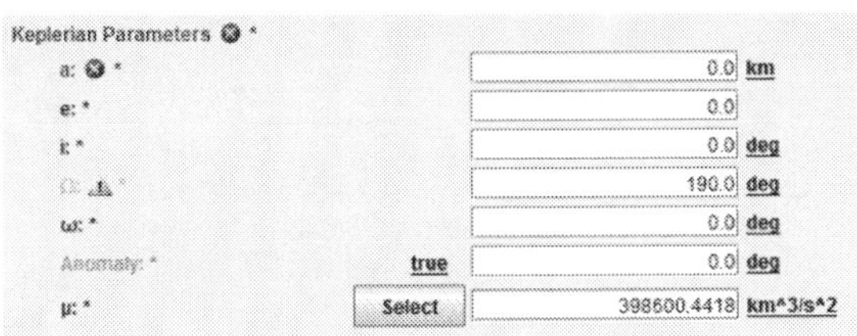

Figure 3intervals of validity

2.6. Parallel computation

Parallel computation is now managed thanks to Java™ multithreading capacity. If this change does not directly appear to the user, the simplicity and the quality of the code is then greatly improved.

2.7. Displays

ORESTE does not exist anymore because part of its functionalities has been included in the ELECTRA tool. And for more sophisticated displays, an evolution of the CNES VTS tool [5] is available (http://timeloop.fr/vts).

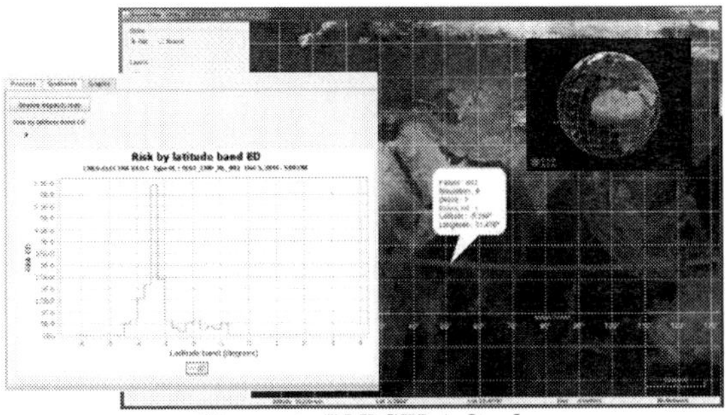

Figure 4: ELECTRA display

3. RISK COMPUTATION

ELECTRA calculates the probability of incurring at least one victim, and the expected value of the number of victims by taking into account the density and the vulnerability of impacted populations.

To do it, it uses as input a population file in GPW format [5]. This file describes the number of inhabitants per unit of a global grid of quadrilateral latitude-longitude cells at a chosen resolution (0.25°, 0.5° or 1°) for a given year. For vulnerability, ELECTRA uses coefficients that represent the level of protection (by a housing structure or others). These coefficients depend on the latitude, the time of day (day/night), the season (summer/winter) and the impact energy of debris. Thus, for each computed impact, ELECTRA knows the density and the level of protection of the population as well as the casualty area of the fragment and can therefore calculate the associated probability and expected number of risk.

4. UNCONTROLLED RE-ENTRY (RA) MODE

4.1. Overview

Assessment of the impact risk during an uncontrolled re-entry is perhaps the simplest one by comparison to the following modes. As it is impossible to predict the "exact" location of the debris impacts, the potential fallout zone corresponds to the area of the Earth's surface between latitudes +/-i (i being the inclination of the orbit). Anyway, to achieve more precise risk, ELECTRA tool discretizes the [+i, -i] latitude range into N latitude bands correlated with the population grid and taking the following elements into account:

- The population density variable according to the latitude band
- The probability of falling in a latitude band depending of the latitude of the band and of the orbit inclination

4.2. Changes

For this mode, there are not great changes (except the GUI look) and the fact that plots are directly available without using a post processing viewer.

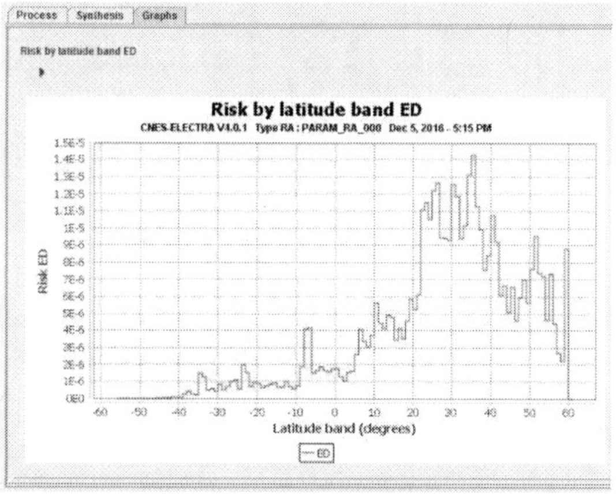

Figure 5: RA mode display

5. LAUNCH FAILURE (RL) MODE

5.1. Overview

For the risk due to a rocket launching, ELECTRA considers failures related to the rocket propulsion premature stop. The boosted phase trajectory is given in an ephemeris file which contains position and velocity of the launcher sampled at a given frequency. ELECTRA considers a stop of the boost at each point of the ephemeris. The probability of occurrence of such a failure can evolve during the boosted phase.

Then, to obtain the impact coordinates (latitude and longitude) and the energy of impact of each fragment, ELECTRA first calculates the trajectory of the intact vehicle from the failure instant until its fragmentation and then the trajectories of each fragment to the ground. This computation is done for each considered propulsion failure.

Several sources of dispersion and/or poor knowledge of some parameters can influence directly on the impact point of debris:

- The initial state vector at the moment of the failure,
- The intrinsic characteristics of the vehicle,
- The altitude of the feared event (rupture or explosion) leading to the spacecraft fragmentation,
- The fragmentation characteristics (including the ejection velocity at the loss of integrity),
- The external environment

Since the effect of these uncertainties on the risk is difficult to predict, the ELECTRA method uses a Monte Carlo process to compute the risk for controlled re-entry and launching phase, taking into account all these dispersion sources.

The forces taken into account in the computation of the trajectory, as well both for the spacecraft and fragments, are:

- The gravitational attraction of the Earth
- The atmospheric forces (drag and lift)
- The forces due to solar radiation pressure
- The gravitational attraction of the Sun and the Moon

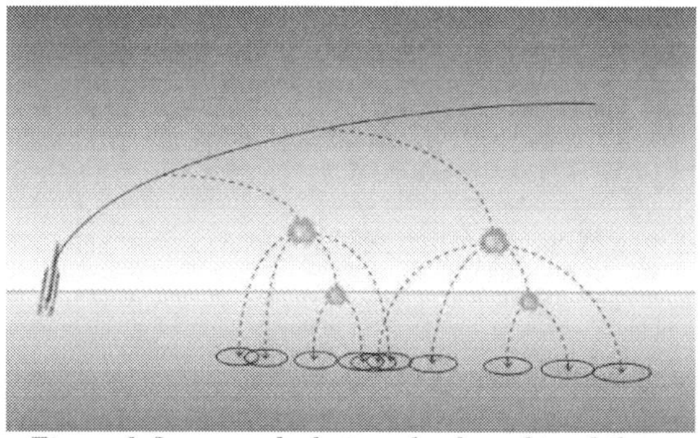

Figure 6: Impact calculation after launching failures

5.2. Changes

As for any other modes, the look & feel has changed but, in terms of functionalities, we may notice:

- The way to tune the numerical integrators
- The available force models

5.2.1. Numerical integrators

The main difference is that, the Cowell numerical integrator previously used for the orbital phases is no more available in the PATRIUS library, replaced by the 8th order Dormand-Prince one which is now proposed as well as for the orbital phase as for the debris phase. It is also possible to choose the propagation inertial frame between three items. At last, the possibility to switch to a Runge-Kutta numerical integrator for specific ballistic coefficients is now proposed as entry parameters (so, on the GUI) as it was hardly coded in the previous versions.

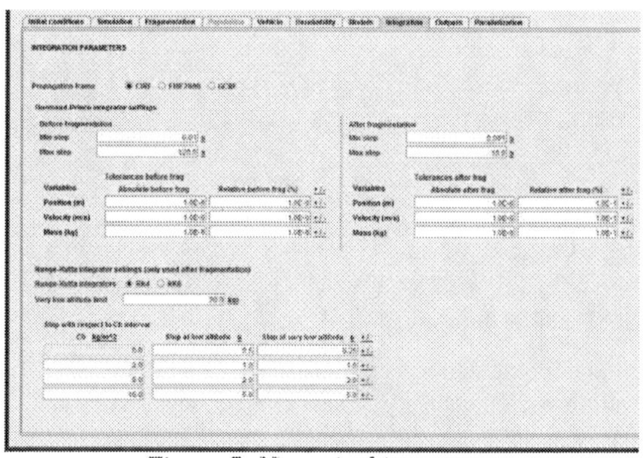

Figure 7: Numerical integrator

5.2.2. Force models

As ELECTRA uses now PATRIUS library, the available force models are slightly different of the previous ones:

- For gravitational attraction, the list of available models is different (more up to date) and we have also the possibility to choose the type of equations between Balmino, Cunningham and Droziner ones (event if it not strictly necessary).
- For gravitational attraction of the Sun and the Moon as well as forces due to solar radiation pressure, constants are not exactly the same and ephemeris too.
- At last, the atmosphere models are limited to the MSISE2000 and the US76 ones.

We can see on the next figure how it is presented via the GUI.

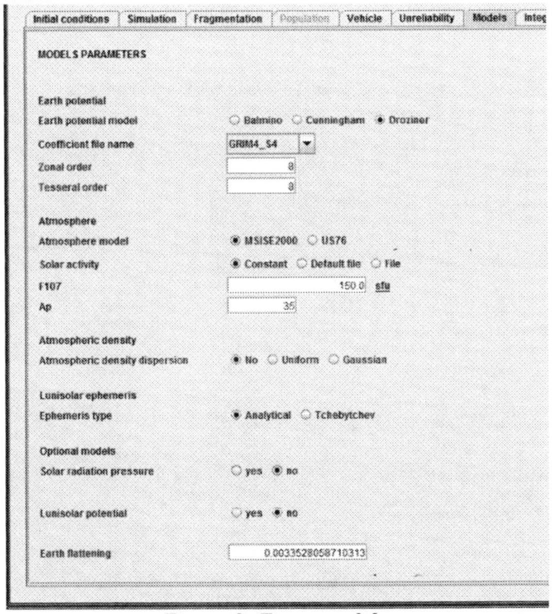

Figure 8: Force models

6. CONTROLLED RE-ENTRY (RC) MODE

6.1. Overview

The principle of this mode is more or less the same as the one used for the RL mode. In fact, the only difference is the way to consider the initial conditions as the trajectory will be fully propagated and not taken from a reference file.

Maneuvers are simulated as continuous constant thrust and ELECTRA allows defining the characteristics of the engine (specific impulse and thrust level), the implementation type on the satellite (maneuver commanded in duration or delta-V), the maneuver characteristics (date, expected direction). Each of these inputs can be scattered by a Gaussian or uniform law.

Under/Over boosts are also well considered (as it has sometimes not to be considered as a failure but a nominal dispersed case).

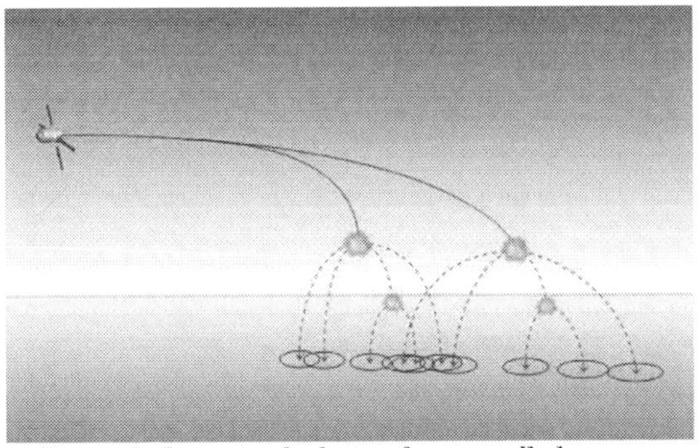

Figure 9: Impact calculation for controlled re-entry

6.2. Changes

In addition to the RL mode evolutions, the way the maneuvers and the attitude are considered has changed.

6.2.1. Maneuvers

In the previous versions, the maneuvers definition, the way they were disturbed and the failure logic were merged in a single tab. Now, it has been preferred to split this information in two separate tabs.

First, the maneuvers and their potential dispersions are defined in a list with all classical information for such continuous constant thrust maneuvers.

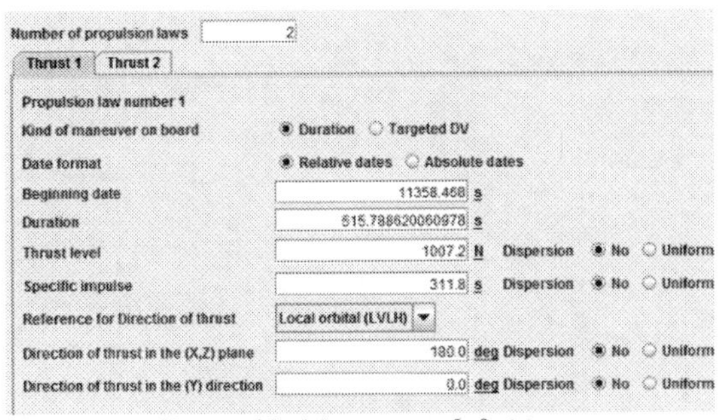

Figure 10: Maneuvers definition

Then, the unreliability parameters are managed in a specific tab as in the following figure.

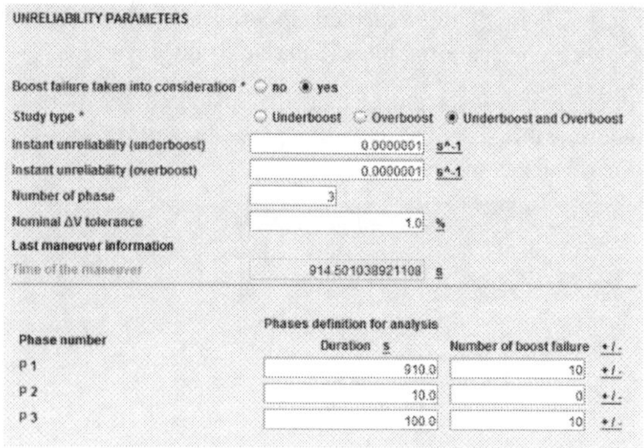

Figure 11: Maneuvers unreliability parameters

6.2.2. Attitude

In the previous versions, the attitude laws management was only depending on the time. A main drawback was the fact that the user has to be very careful to synchronize attitude laws dates to maneuver ones. So, it has been decided to change this management thanks to new PATRIUS possibilities allowing going from a law to another thanks to some orbital events.

Now, the user first defines a list of available laws included in the following list:

- EME200 inertial law
- Local orbital frames laws: QSW, TNW or LVLH
- Solar pointing law
- Yaw steering law
- Aerodynamic law

Except for the last aerodynamic law, it is also possible to add a spinned motion.

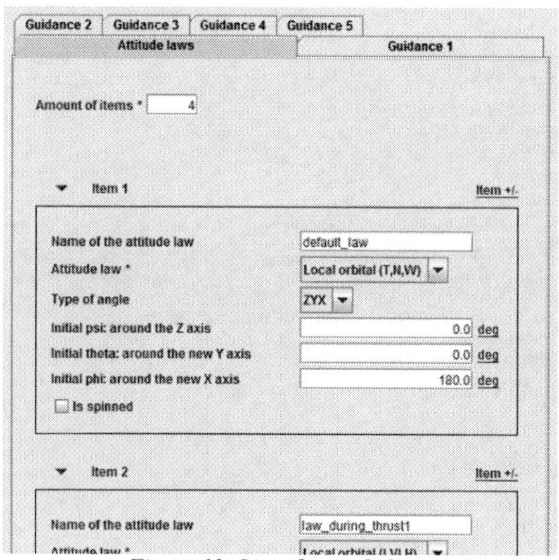

Figure 12: List of attitude laws

Then, depending of the amount of maneuvers, the user defines the guidance principles during and between maneuvers (GUI displays automatically the right amount of guidance tabs):

- During a maneuver, only one law is selected (taken in the previous list)
- Between maneuvers (as well as before the first one and after the last one), it is possible to define a list of "switches" corresponding to the following orbital events and allowing defining which attitude law will be considered before and after this event:
  - Absolute or relative date
  - Eclipse
  - Altitude

As a consequence, with a few switches, it is possible to manage a relatively elaborated guidance between maneuvers rather than to use only date events. The example below shows how to switch between two attitude laws on eclipse criteria during as many orbits as desired.

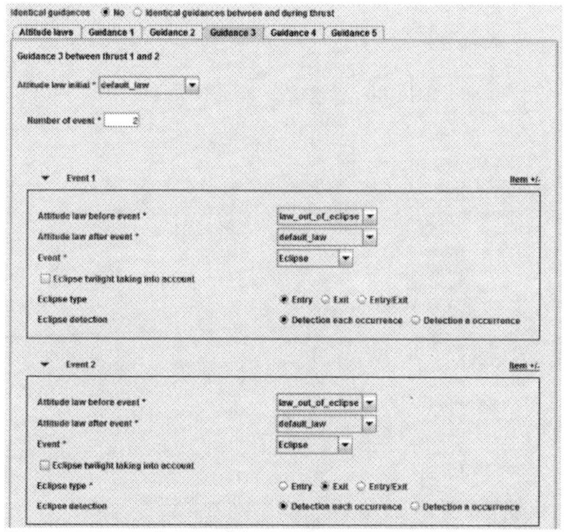
Figure 13: Guidance laws

7. FINAL ORBIT (RF) MODE

7.1. Overview

This new "short-term" re-entry context aims to estimate the risk a few hours or days before the uncontrolled re-entry of a space object. It has been introduced since the V3.x versions ([7]) but not redeveloped for the V4.0.1 one and will appear again in the V4.1.

The principle of this calculus is that, a few days before the possible re-entry, conditions are distributed on a limited amount of orbits and can be characterized by a set of possible state vectors at a given altitude. Then, every point is considered as a probable point of re-entry and is processed in a similar way as what is done for each failure case for launching risk computation.

7.1.1. Determination of probable points of re-entry

A specific tool is introduced in ELECTRA to generate a set of possible state vectors at a given altitude, taking into account uncertainties on the space object itself and on environmental conditions. These probable re-entry points are deduced from initial conditions representing last orbits of the space object few hours or days before its uncontrolled re-entry. These points are obtained by a variation of the ballistic coefficient between a maximal and a minimal dispersed value. A probability of occurrence is joined to each of these points. Notice that this set of points does not represent an orbit itself even if it looks like...

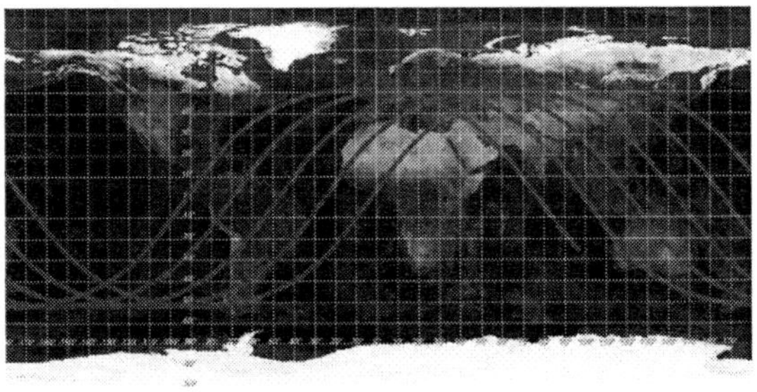

Figure 14: Set of entry points

The method used is to generate circle-equidistant probable re-entry points (characterized by date, position and velocity, and ballistic coefficient), situated at an identical geodetic altitude of re-entry (85 km for example), and associated to an occurrence probability assuming a uniform or a Gaussian repartition of the ballistic coefficient dispersion. In that way, it will be sure not to forget some intermediate points leading to a potential impact on a heavy inhabited area as a full random distribution could lead to consecutive points separated by several hundreds of kilometers.

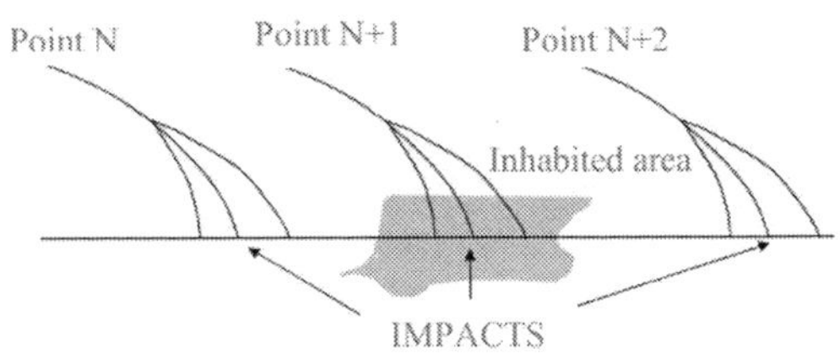

Figure 15: Probable risk between two re-entry points.

As this way of determining this set of entry points may be time consuming, it is possible to do it in two steps: first with numerical propagations, then using interpolation method.

7.1.2. Computation of risk

Once impact points have been computed with a sufficient number of simulations with respect to the level of confidence objective, ELECTRA calculates the probability of incurring at least one victim, and the expected value of the number of victims exactly in the same manner as for the RL and RC modes, i.e. using a Monte Carlo method

So, in order to minimize the CPU time, ELECTRA defines only a sample of re-entry points for the ones a complete Monte Carlo analysis is achieved. Those points are called "anchor points". Anchor points are separated by a defined angular gap. This angular gap value mainly depends on the orbit type.

Then, for the other re-entry points, impacts are evaluated with the help of a "shifting method" of computed "anchor" impacts points.

For each anchor point P_0, impact points are located in a Cartesian form in a R_0 frame. This R_0 frame center is the ground impact point of the un-fragmented vehicle for the re-entry anchor point P_0. Coordinates are defined from Greenwich by rotations in longitude, inclination and argument of latitude of P_0 point.

For re-entry points P_i attached to the P_0 anchor point, impacts points are then computed in relationship with results obtained with the complete Monte Carlo analysis on P_0 in the R_i frame linked to P_i point. R_i frame is centered this time on the impact point of the un-fragmented vehicle for the Pi re-entry point. P_i impact points coordinates are then located in the R_i frame at the same place than the P_0 impacts points in the R_0 frame (see Figure 16).

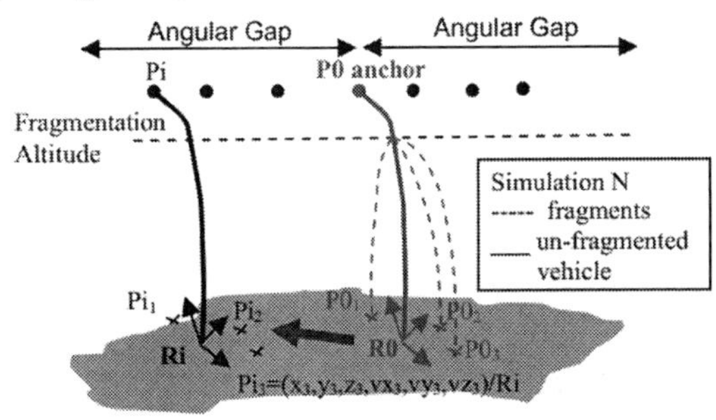

Figure 16: Optimization of impacts zone determination.

7.2. Changes

Three main changes occurred in the tool used to compute the entry points:

- Possibility to customize the force models (as it was hardly coded before)
- Numerical integrator tuning in consistency with the one used in other modes and using feedback of the validation steps described in paragraph 9.3.
- Parallelization of the computation considering a set of ballistic coefficient intervals

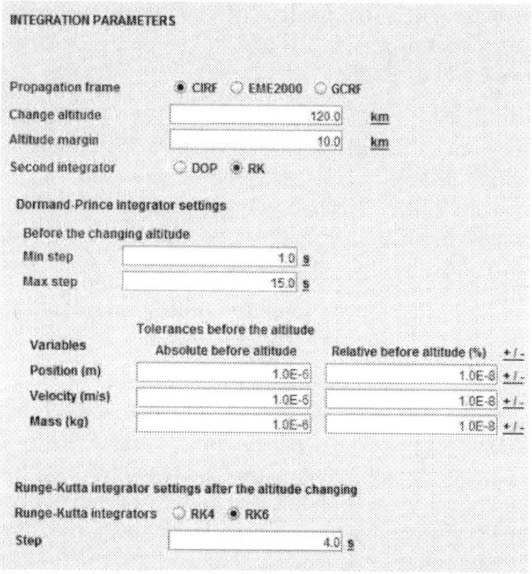

Figure 17: Integrator tuning for the entry points tool

For the RF mode itself, we will find the same modifications as for the RL mode with force models and numerical integrator definitions.

8. ADDITIONAL TOOLS

8.1. Overview

The list of available tools is the following one. Some existed in the previous version but some are new:

- Fragments editor (new look)
- Fragments concatenation
- Tabulated fragments import
- Protection editor
- Energy editor
- Impact Map (new)
- Files extraction (new)
- Risk per country
- Trajectory extraction
- Final orbit computation
- Population Files extrapolation (V4.1 only)
- Files converter (new)
- Risk computation (new in V4.1)

8.2. Changes

8.2.1. Impact Map

This tool is the same as the one directly called via the RL/RC/RF modes but allows using it on previously computed results.

8.2.2. Files extraction

Previously integrated in the ORESTE viewer, this functionality allows extracting files for transferring now to VTS tool in the right format (KML) but also to store in a single zip file a full configuration (data and results) for archive purpose.

8.2.3. Fragments editor

The fragments editor still existed in the previous versions but has been update for better ergonomics especially some possibilities to compute automatically some values.

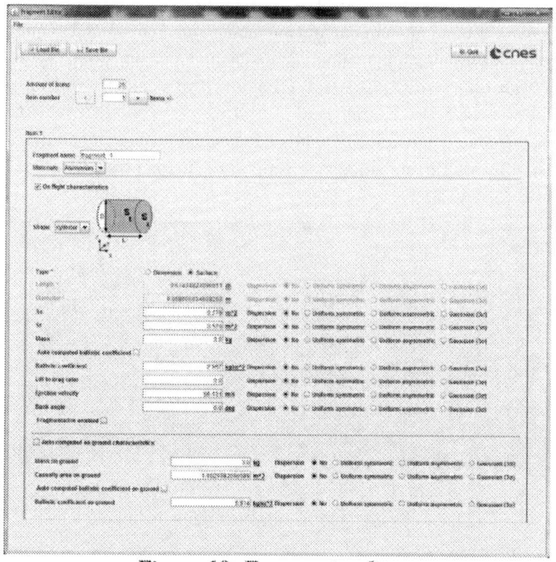

Figure 18: Fragments editor

8.2.4. Files converter

This tool did not exist before for the simple reason that it permits converting old versions context files in new format (XML) but also to adapt the content of this file to new ELECTRA functionalities as, for example, force models or numerical integrator tuning.

8.2.5. A posteriori risk computation

This tool is very interesting as it allows computing risk quickly. Indeed, the CPU time for RL/RC/RF modes are essentially due to Monte Carlo simulations with thousands of trajectory propagations. And previously, if we wanted to compute the risk based on the same trajectories but with different inputs as the population density or protection coefficients, users had to regenerate all these trajectories. Up to now, it is possible to start from a previous computed impact file and to compute again the risk by changing these inputs. Some seconds are necessary to get the results.

Moreover, using the same principle, it is also possible to compute the risk by shifting the impact location longitudes. Strictly speaking, if we compute trajectories with such a shift on initial conditions, the trajectories should be a bit different (due for example to tesseral terms of the Earth potential or longitude effect of some atmospheric models) but the shift on longitude (not on latitude!) for each fragment can be considered as a very good approximation.

This functionality is particularly interesting for post processing RF computation. At last, it is now possible to extrapolate the risk for different years (corresponding to different population densities).

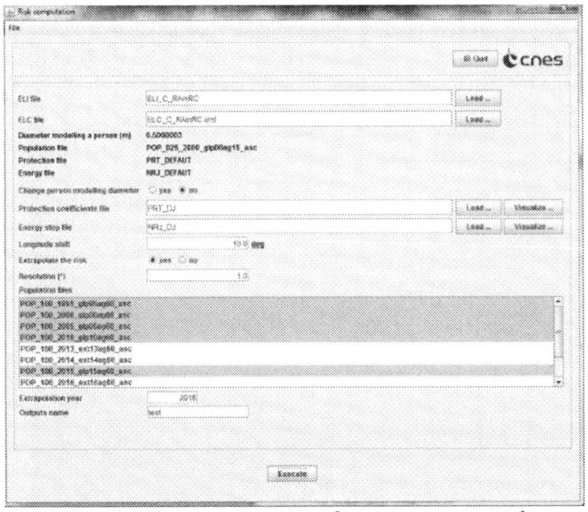

Figure 19: A posteriori risk computation tool

9. VALIDATION

We will not detail all the validations done for these new ELECTRA versions but we will focus on three items:

- Unavailability of some models
- Change in the dispersion of variables management
- Entry points computation

9.1. Unavailability of models

The Fortran version of ELECTRA was based on set of old Fortran libraries (BIBMS) whether the Java™ version uses now the much more recent PATRIUS tool. Therefore, most of the models, such as atmosphere models, are only available in one version of ELECTRA.

In order to validate all the possibilities of the newer version, one must begin with validating cases with the common models. In effect, considering the atmosphere models, only US76 is available in both BIBMS and PATRIUS and is consequently chosen for the validation case. Moreover, what happens to the atmosphere models also occurs for the Earth potential models, the solar system ephemeris, or even the type of integrator.

Fortunately, the use of an object-oriented language, and a carefully designed library, meets this problem perfectly. The solution lies in the usage of interfaces. For instance, all the atmosphere models in Patrius implement the interface Atmosphere, which defines what each model can compute, for instance the atmospheric density at a given point and date. In ELECTRA, the computations only work with this interface, with absolutely no knowledge on the explicit type of model.

As a consequence, each model can be substituted with another, and this has no impact on the ELECTRA algorithm. With this crucial point in mind, we can conclude that if ELECTRA is validated with one model, then the main algorithm is valid for any other model.

9.2. Changes in the dispersion of variables

The dispersion of the variables is done by using the Mersenne-Twister algorithm. This method generates random series of numbers, initialized with a seed. With this seed, it is possible to reproduce random drawings, and thus validate specific results, i.e. for each Monte-Carlo simulation, based on dispersed values. The issue here is that old and new versions of ELECTRA use the random series differently.

In Fortran, one series was used for each failure or re-entry point, and the variables were drawn by using successive values of this series. In Java™ version, this was changed to use one random series per variable. So, for the same configuration, the drawn values are different, and it is no longer possible to validate specific results based on dispersed values.

In order to bypass this issue, large amount of drawings have been considered and the result have been analyzed from a statistical point of view. This principle was applied for the random re-entry mode, with a configuration using a dispersed fragment. The dispersions affect the resulting risk, which means that the risk computed on both versions of ELECTRA should converge to each other for a large number of simulations. The result of such an analysis is shown Figure 20.

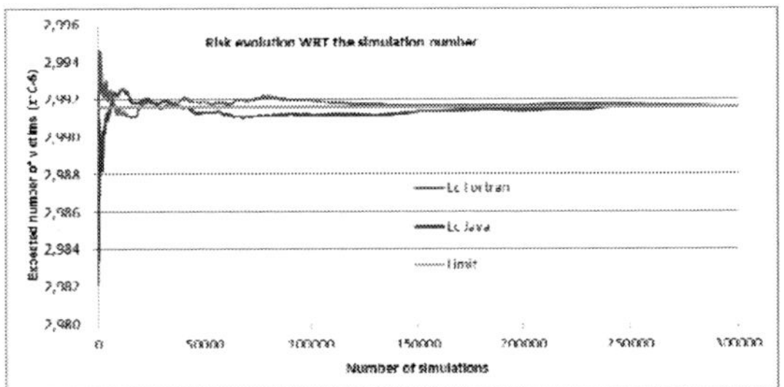

Figure 20: Risk evolution for 1 to 300000 simulations

Note that with even more simulations, one curious behavior can be observed. One would expect the risk to converge with a very high number of simulations, but due to the order of magnitude of the risk (generally about 10^{-4} / 10^{-6}), the machine precision is reached and small variations of the risk are still observed. Figure Figure 21 shows what happens for more than 300000 simulations, on the same configuration as above.

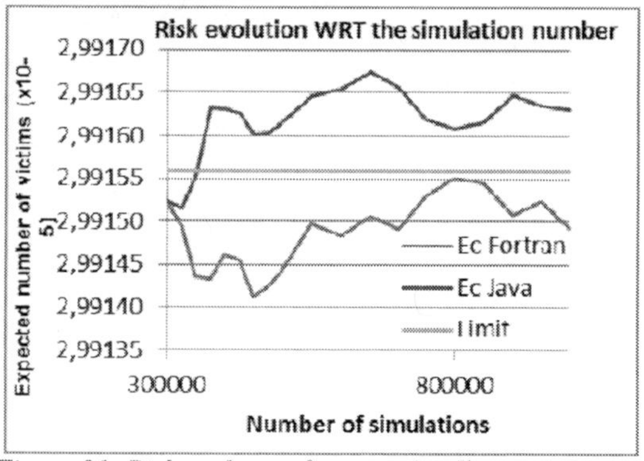

Figure 21: Risk evolution for up to 1 million simulations

9.3. Entry points computation validation

When developing the entry points computation tool, a problem of time computation occured very soon. By comparison with the previous old Fortran versions, the CPU time was much greater than foreseen (as most of the time, we consider that Java code is two times slower than Fortran one). Even if it is now possible to use parallelization to do it, it was not normal as the algorithm was exactly the same one (based on an adapted Newton method [7]). However, after some investigations, it appears that the problem was essentially coming from the iterative searching algorithm with sometimes a higher amount of iterations (up to 20/30 when 3 or 4 was only needed in the Fortran version).

Remember that the basic idea of such algorithm is to iterate on the ballistic coefficient value in order that the entry point will be spaced from the previous one at a constant distance (for example 100 km but with some meters of precision).

The solution came from a better tuning of the parameters of the algorithm but above all from a better tuning of the numerical integrator. Actually, for a normal impact point computation, it is not necessary to get a high precision as, if you drift you impact point of some hundreds of meters by comparison to the population grid scale (~100 km) or the atmospheric dispersions, it is completely negligible. And less precision leads to lower CPU time which is so important for Monte Carlo computation! But using the same tuning as for RC/RL modes for the research of entry points location with a 10 meters precision was no more adapted!

First we tried to test the algorithm on a quasi-equatorial orbit using US76 atmospheric model (to avoid latitude effects) changing the maximum time step of the Dormand-Prince integrator at a given altitude in order to be more precise at low altitudes. We see on Table 1 that, considering only no exit on error (maximum amount of iteration reached), we decrease a lot the CPU time with the right tuning! So, very strict tolerances on the numerical integrator (meaning higher CPU time for one trajectory) lead to more precise trajectories and thus, less iteration.

Switch altitude [km]	High altitudes max step [s]	Low altitudes max step [s]	Relative tolerance	CPU time	Exit on error
km	*s*	*s*	*%*	*min:sec*	-
150	10	2,5	1.10-8	7:17:00	2
150	5	2,5	1.10-8	13:04:00	1
150	10	1	1.10-8	9:40:00	3
150	5	1	1.10-8	16:10:00	5
120	5	1	1.10-8	13:35:00	1
120	5	1	1.10-9	13:13:00	1
120	5	1	1.10-10	13:36:00	0
120	10	1	1.10-10	6:52:00	1
120	15	1	1.10-10	4:37:00	0
120	15	2	1.10-10	4:26:00	0
120	15	2	1.10-11	4:27:00	0
120	20	2	1.10-11	3:31:00	0

Table 1 : inc=0.1deg / US76 / 10m precision / 103pts

Note that the following results have been made on a classical office PC using only one thread (so, with no parallelization).

With 98 degrees of orbit inclination, results were of the same order of magnitude ... But after substituting US76 by MSIS200 atmospheric model, some longitude effects as local hour leads to some new difficulties on the iteration and we decided to save CPU time by:

- Authorizing a precision of 100 m (rather than 10 m) on the entry point location , which is quite sufficient
- Rather than a Dormand-Prince, using a Runge-Kutta integrator for low altitudes as it is more robust and avoiding to use too low time steps.

With such changes, and some new slight changes in the tuning, we obtain the results listed in Table 2. Note that, exit on errors never occurs (reason why it is not mentioned in the table). Moreover, we can see that CPU time significantly changes due to selected models (US76 vs MSISE2000) but also inclination value (more points to compute).

Inc [deg]	Atm. model	Switch altitude [km]	High altitudes max step [s]	Relative tolerance	RK6 step [s]	CPU time	Amount of points
0,1	US76	120	15	1.10-11	4	3:13:00	103
98	US76	120	15	1.10-11	4	12:11:00	176
0,1	MSIS2000	120	15	1.10-11	4	11:22:00	88
98	MSIS2000	120	15	1.10-11	4	37:40:00	162

Table 2 : Final tuning for entry points search

Of course, all these parameters are modifiable via the GUI in case of a very specific initial orbit that will necessitate some customization. And anyway, to decrease the CPU time, we have also the possibility to use the parallelization functionality!

10. CONCLUSION

Now, since end of 2016, ELECTRA is available in its Java™ version allowing to be installed and used on a lot of OS and platforms when previously, it was only usable on an old obsolete Suse™ configuration or eventually on some RedHat™ OS. And more than this usability, we are sure that the numerical results will be strictly the same from a platform to another one.

Taking benefit of this new development, some new features in the four computation modes (RA, RC, RL, and RF) are proposed as:

- More robust and user-friendly installer
- Integrated displays to get a better comprehension of the results
- Easier way to define maneuvers in the RC mode
- More sophisticated management of the attitude laws during and between maneuvers

Moreover, some new tools have been added specially the one allowing to (re)compute very quickly the risk without propagating trajectories again. Some other tools have also been improved as the one use for entry points generation.

At last, these new versions are fully based on CNES FD libraries (PATRIUS) and their associated GUI libraries (GENIUS/GENOPUS) which bring a great level of validation particularly on the trajectory propagations. All these tools are directly available via the CNES Freeware server (https://logiciels.cnes.fr/en). But if PATRIUS, GENIUS and GENOPUS are proposed in an Apache Open Source distribution, ELECTRA (as DEBRISK software), needs a specific approbation by CNES hierarchy even it proposes now a generic license.

Figure 22: CNES Freeware server/ FD page

11. REFERENCES

1. F. Chemama, B. Lazare, C. Aussilhou. ELECTRA tool: Launch and Re-entry safety analysis 4th ICATT (2010), International Conference on Astrodynamics Tools and Techniques, 3-6 May 2010, ESA/ESAC, Madrid, Spain

2. Catherine Hourtolle, Angélique Gaudel-Vacaresse, Alejandro Blazquez. ELECTRA: Launch and Re-entry risk analysis tool. 5th ICATT (2012), International Conference on Astrodynamics Tools and Techniques, 29 May – 1 June 2012, ESA/ESTEC, Noordwijk, the Netherlands

3. R. Houdroge, D. Claude, J. Anton, T. Sabatini, P. Cardoso, G. Mercadier, T. Trapier, Y. Tanguy: the SIRIUS Flight Dynamics Library for the next 25 years. 5th ICATT (2012), International Conference on Astrodynamics Tools and Techniques, 29 May – 1 June 2012, ESA/ESTEC, Noordwijk, the Netherland

4. Philippe Pavero, Guillaume Rochais, Mirentxu Beuvelot. Space Dynamics software ELECTRA. 6th ICATT (2016), International Conference on Astrodynamics Tools and Techniques, 14-17 March 2016, ESA/ESOC, Darmstadt, Germany

5. T. Crosnier, M. Joubert, Q. Minster: VTS, a long-term approach for synchronization of all visualization software. 13th SESP (2015) Simulation for European Space Programmes, 24-26 March 2015, ESA/ESTEC, Noordwijk, the Netherland

6. A. Bellucci, N. Tholey, M. Studer, JF Goester, N. Fuentes: Extrapolation of population grids for risk analysis. 9th IAASS International Association for the Advancement of Space Safety, 18 – 20 October 2017, Toulouse - France

7. A. Gaudel, C. Hourtolle, JF Goester, N. Fuentes: Risk assessment during the final phase of an uncontrolled re-entry. 6th IAASS International Association for the Advancement of Space Safety, 21 – 23 May 2013, Montreal – Canada

INTERNATIONAL SPACE STATION AEROTHERMAL BREAK-UP ANALYSIS USING SCARAB

Patrik Kärräng[1], **Bent Fritsche**[1], **Fabian Zander**[2], **Stefan Löhle**[2], **Tobias Lips**[1], and **Holger Krag**[3]

[1]*Hyperschall Technologie Göttingen, Am Handweisergraben 13, 37120 Bovenden, Germany*
[2]*Institut für Raumfahrtsysteme, Universität Stuttgart, Pfaffenwaldring 29, 70569 Stuttgart, Germany*
[3]*ESA/ESOC, Robert-Bosch-Str. 5, 64293 Darmstadt, Germany*

ABSTRACT

Since November 2000 the International Space Station (ISS) has enabled continuous human presence in orbit. The now aging space station's end-of-life is approaching. Currently the station is expected to be de-orbited in 2024, with its decay ending in a controlled re-entry. Under a European Space Agency contract the Institut für Raumfahrtsysteme at the University of Stuttgart and Hyperschall Technologie Göttingen investigated the aerothermodynamics of the ISS re-entry and its effects on the break-up. In order to assess the effects of such an event the re-entry was simulated using the SCARAB software. We performed numerical flow analysis at different altitudes with ISS geometry variations to find heating amplification caused by shock impingement. The analysis led to adaptations to SCARAB which better predict the break-up of large structures. This paper discusses the ISS re-entry analysis, presenting the a first estimate of the debris footprint, break-up altitudes and mass of the surviving fragments.

1. INTRODUCTION

The International Space Station (ISS) has since its first modules were launched in 1998 served as an in-orbit research laboratory and home to its many crews. The ISS operates in a near-circular Low-Earth-Orbit (LEO) with an altitude of approximately $400\,km$ and inclination of $51.6°$. After many years of successful operation the space station is now approaching its End-Of-Life (EOL). The station is expected to be de-orbited in 2024, although operation could be extended until 2028.

The ISS will be the largest single structure to re-enter Earth's atmosphere. The only event comparable to this is the MIR space station, which in 2001 performed a controlled re-entry over the south pacific ocean. The analysis of the MIR re-entry [1, 2] has given experience and insights regarding debris footprint and fragmentation of large structures. This can be applied to the ISS. There have been early studies regarding the re-entry of the ISS investigating the expected debris footprint and on-ground casualty risk [3]. Also, in order to improve the re-entry environment and fragmentation modeling ESA and NASA had planned that the final ATV mission, ATV-5 "Georges Lemaitre", would perform a shallow re-entry. Previous ATV missions had performed steep re-entry trajectories with perigee at $-70\,km$. The shallow re-entry trajectory was designed to resemble the ISS re-entry conditions, unfortunately this was canceled due to an on board failure [4].

When studying the re-entry of large structures it is important to also consider the flow-structure interaction. For compact satellites one bow shock will envelope the whole structure, but for large branched structures, such as the ISS, the bow shock can hit another part of the structure which might lead to increased heat load at shock impingement. To find what influence the shock has on the aerothermodynamic heating a Computational Fluid Dynamics (CFD) computation is required.

To our knowledge there is no publicly available detailed disposal plan. For this study we have made several assumptions regarding the de-orbit trajectory. The ISS EOL disposal should be consistent with international safety guidelines (i.e. less than 1 in 10,000 casualty risk) therefore the disposal has to be a controlled, destructive re-entry into an unpopulated area on the planet. An uncontrolled re-entry would surely exceed this guideline. The disposal of the ISS will begin at its nominal orbit at $400\,km$ altitude. From there the $360\,ton$ space station experiences natural decay of the orbit by atmospheric drag. The decay rate is difficult to predict, it depends on the drag configuration (shape, frontal area, mass) of the body and atmospheric density. Assuming a low Ballistic Coefficient (BC) the decay lasts around 9-16 months, Fig. 1 shows a possible decay scenario of the ISS. Over several months the natural decay will be used to bring the station orbit between $241\,km$ to $185\,km$ altitude. At this altitude the aerodynamic torques will be significant enough to require excessive propellant for attitude control [3]. Consequently, in our scenario there is a final de-orbit burn sequence at this altitude. The burn from the Progress-M on the Russian segment will be designed to lower the perigee to 75 km. This will be deep enough in the atmosphere so that the entire ISS will re-enter.

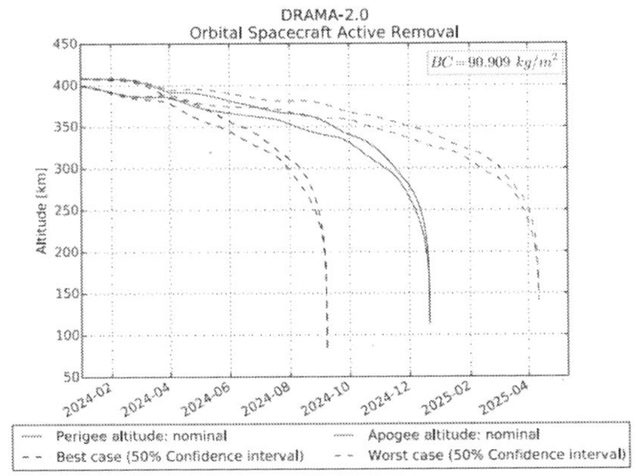

Figure 1: Natural orbital decay estimate for the ISS using the OSCAR tool from ESA's DRAMA 2.0 software.

The de-orbit burn must be timed so that the ISS re-entry debris footprint occurs over an unpopulated area. A suitable region to use as re-entry area is the South Pacific Ocean Uninhabited Area (SPOUA). This region has been used for the re-entry of several satellites, the MIR space station and the ATV missions. The exact dimensions of SPOUA is not well defined but for the ATV re-entry it was defined by the longitude range from 185° East to 275° East (i.e. $-175°$ to $-85°$) and latitude range from 29° South to 60° South (i.e. $-29°$ to $-60°$) [5]. See Fig. 2 for an outline of the region.

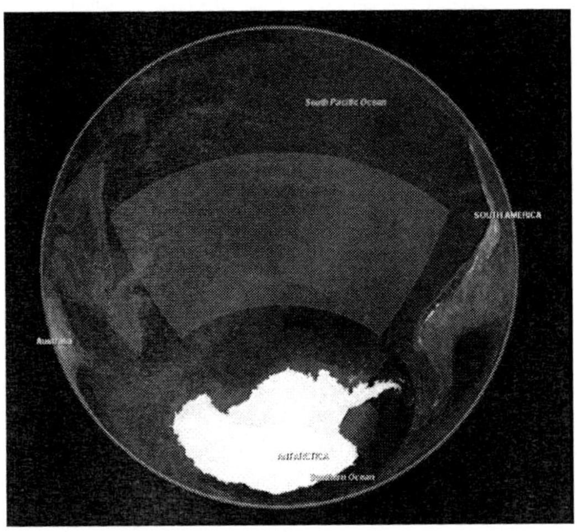

Figure 2: South Pacific Ocean Uninhabited Area as defined for the ATV re-entry. Visualized using NASA World-Wind.

In this paper we present the re-entry analysis performed as part of the work conducted in the 'ISS Atmospheric Break-up Analysis' (ISSABA) project. Within the project the Institut für Raumfahrtsysteme (IRS) at the University of Stuttgart performed flow field analysis at various altitudes, using a numerical flow solver to investigate the aerothermodynamics of the ISS re-entry. The results from the analysis served to improve the break-up modeling of large structures in the re-entry simulation software SCARAB (SpaceCraft Atmospheric Re-entry and Aerothermal Break-up) [6]. The re-entry simulation part of the project was conducted by Hyperschall Technologie Göttingen (HTG).

2. AEROTHERMODYNAMIC ANALYSIS

One of the purposes for investigating the aerothermodynamics is to improve the break-up modeling in SCARAB for regions with shock-structure interaction. The objective is to produce heat flux profiles with altitude and position from the primary shock. This can provide a basis for the adaptation of SCARAB to include localized heating caused by shock impingements. For the computation of the flow field that the ISS will encounter during the re-entry, we used the CFD software EILMER [7]. The simulation assumes viscous flow of a chemically reacting, thermally perfect gas mixture of nitrogen and oxygen [8]. We evaluated three flow conditions, each representative for the re-entry occurring at 80, 70 and 60 km altitude.

Table 1: Aerothermodynamic simulation input paramters [9].

Altitude	80 km	70 km	60 km
Velocity	7668 m/s	7095 m/s	5980 m/s
Temperature	185.6 K	209.9 K	244.4 K
Pressure	1.0 Pa	5.5 Pa	24.0 Pa
Mass fraction, O_2	0.22	0.22	0.22
Mass fraction, N_2	0.78	0.78	0.78

The simulation with the reacting gas is conducted using a two-stage process, an axisymmetric 2D model of the Russian segment of the ISS and a 3D cylinder arm onto which the primary shock impinges. See Fig. 3 for an example visualization. We also investigated the relation between heat flux and the distance from the primary shock for 5 m, 16 m and 25 m.

The calculated heat flux profiles show that the shock impingement causes a heat flux amplification seven times larger than the free-stream heat flux. At the same time the structure inside the shock experience a decrease in heat flux compared with the free-stream.

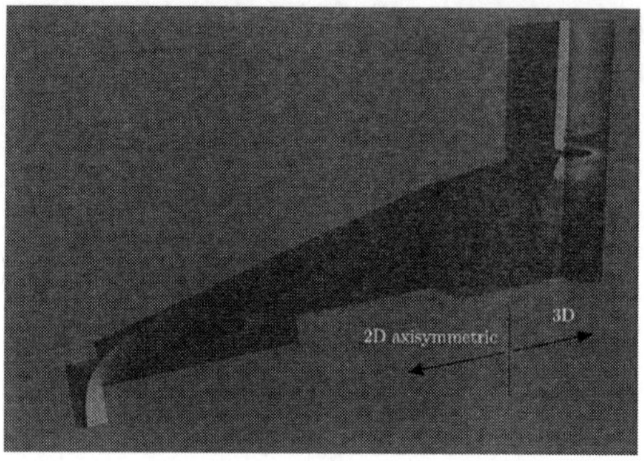

Figure 3: CFD simulation example using both 2d- and 3d- numerical simulations [9].

We also found that for the given flow conditions the peak heating location along the cylindrical arm does not change very much between the different flow conditions ($\pm 0.5\,m$). The shock impingement heat flux amplification does not show a strong dependency on the distance from the primary shock. We used these results and insights to adapt the thermal modeling in SCARAB.

3. ADAPTATION OF SCARAB SOFTWARE

SCARAB is a spacecraft-oriented software tool allowing the analysis of controlled or uncontrolled destructive re-entries (mechanical and thermal destruction). It is a multi-disciplinary analysis tool using six degrees-of-freedom flight dynamics, aerodynamics, aerothermodynamics, thermal- and structural analysis.

3.1. Adaptations for large structures

The standard break-up analysis in SCARAB is for re-entry simulation of compact objects, assuming a global bow shock enveloping the whole geometry, it does not consider large structures where shock-interaction might have a large influence.

The aerothermal analysis in SCARAB is using a global stagnation point which scales the overall heating. With a large branched structure like the ISS this assumption is not valid. We introduced simplified shock-structure interaction handling in SCARAB to solve this issue. The ratios of the heat flux between the undisturbed free-stream and inside the heading shock as well as within the shock-impingement region is calculated in the aerothermodynamic analysis. We modified the standard thermal analysis heat flux from SCARAB with a heating factor coefficient. Tab. 2 shows the bridging formula for the heating factor. For regions on shock-impingement (on-shock) the heating factor is multiplied with the local heat flux and inside shock layer (in-shock) the heating factor is the factor which the heat flux is reduced compared to the free-stream.

Table 2: Bridging formula for shock layers heating factor

Knudsen number	on-shock	in-shock
>1	1	1
$10^{-2}-1$	$1-4.5\log(\mathrm{Kn})$	$1-0.5\log(\mathrm{Kn})$
$10^{-3}-10^{-2}$	$24+7\log(\mathrm{Kn})$	$-\log(\mathrm{Kn})$
$<10^{-3}$	2.5	3

We decided to use one shock originating from one global stagnation point. To describe the shape of the shock we implemented the blast-wave analogy [10]. See Eq. 1 for the equation describing the 3-dimensional shock shape.

$$\frac{r}{d} = 0.795 {C_D}^{1/4} \left(\frac{x}{d}\right)^{1/2} \tag{1}$$

where r is the radial position of the shock, d is the diameter of the body, C_D is the drag coefficient and x is the distance from the nose along the flow direction. Fig. 4 shows the shape of the 3d shock-wave.

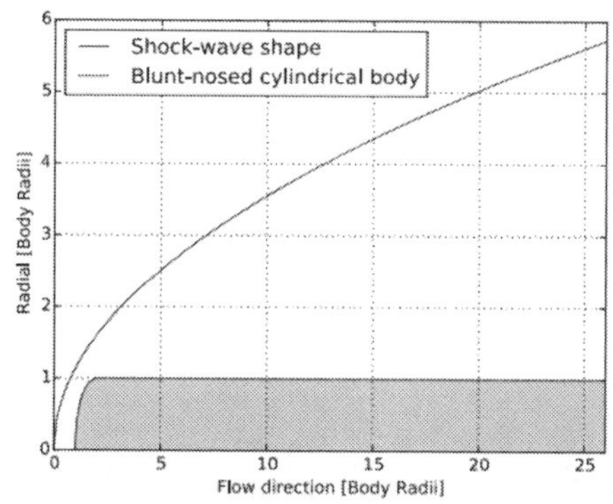

Figure 4: Shock wave shape in relation to the shape of the body used for the blast-wave analogy. Units are the radii of the cylindric body.

4. MODELING OF THE ISS RE-ENTRY

4.1. Spacecraft model

A SCARAB model of a re-entry object consists of a detailed 3D geometric model of the spacecraft with external and internal subsystem compounds represented by simple panelized primitives (spheres, cylinders, cones, circular tori, boxes, flat polygon plates). Each geometric primitive has a material assigned to it, thus the model provides a mass model and the physical basis for all destruction processes to be computed during re-entry.

Table 3: Mass, Center of Mass and Moments of Inertia

	Mass	Center of Mass			Moments of Inertia		
	m [kg]	x [m]	y [m]	z [m]	Ixx	Iyy	Izz
ISS	361108.0	-4.11	-0.16	3.41	121367545.	66054554.	175445177.
SCARAB	357491.5	-5.36	-0.43	2.57	121625456.	58563364.	174058336.

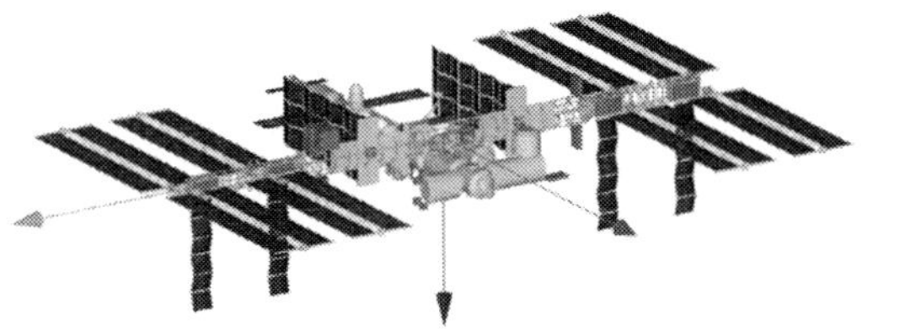

Figure 5: SCARAB model of the ISS

Fig. 5 shows the SCARAB model of the ISS. The geometry consists of a panel grid which is primarily used for calculation of aerodynamic forces and torques as well as the aerothermodynamic heat loads. The panels are combined into enclosed elements creating a volume grid. This is used for the calculation of the total mass, center of mass, moments of inertia (MCI), prior and during the destructive re-entry.

To simplify the modeling and to keep track of the mass budget the ISS has been divided into three main sections: Integrated Truss Structure (ITS), United States Orbital Segment (USOS) and the Russian Orbital Segment (ROS). Tab. 3 shows a comparison between the reported MCI of the ISS [11] and what the SCARAB model resulted in. The origin of the model is located in the center of the ITS. In the model the positive X-axis points in the nominal flight direction, the positive Y-axis is parallel with the starboard truss and the Z-axis completes the triad, pointing to the nadir.

The ITS consist of several truss segments and subsystems such as Heat Rejection System (HRS), Photovoltaic Arrays (PVA) and external instruments. See Tab. 4 for a complete list of the modeled ITS components and their mass in SCARAB.

The pressurized components on the USOS is modeled with a thin outer aluminum layer, a Nextel-Kevlar layer and an inner pressure layer of aluminum. The pressurized components interior is modeled as racks with boxes containing electronic material. The instruments that are attached to the exterior of the USOS are modeled as simplified electronic boxes. See Tab. 5 for a component and mass list of the USOS.

Table 4: ITS components

Name	Mass [kg]
S0	9985.906
S1	6687.801
S3	4508.581
S4	10192.997
S5	1886.397
S6	12903.713
P1	6687.796
P3	4508.580
P4	10192.994
P5	1886.390
P6	12903.693
HRS-Starboard	5567.132
HRS-Port	5567.132
PVA (x8)	9266.547
ELC-1	6158.971
ELC-2	5610.741
ELC-3	5892.608
ELC-4	5892.610
AMS (on S3)	8827.456
Z1 Assembly	8839.519
MSS	5647.306
ITS total mass	**149614.869**

Table 5: USOS components

Name	Mass [kg]
CUPOLA	1943.720
QUEST	9137.329
PMA-1	1318.387
PMA-2	1318.387
PMA-3	1318.387
Unity	14583.684
Harmony	14374.572
Tranquility	14583.694
Destiny	25291.736
Columbus	12917.448
Columbus-EPF	1522.015
JEM-PM	14768.452
JEM-ELM	7566.41
JEM-EF	3892.936
Leonardo	9994.654
USOS total mass	**134531.811**

Table 6: ROS components

Name	Mass [kg]
Pirs (DC-1)	3861.068
Poisk (MRM-2)	3861.069
Rassvet (MRM-1)	7720.367
Zarya (FGB)	25730.467
Zvezda (SM)	25463.666
Progress	6708.196
ROS total mass	**73344.833**

Due to insufficient data the ROS modules are modeled similarly to the USOS. The exterior consists of an outer aluminum layer and a Nextel/Kevlar layer (Note: No inner pressure layer has been modeled for ROS). The interior is also modeled based on the data available about the USOS interior. For the component and mass list see Tab. 6.

4.2. Mechanical break-up modeling

For the structural analysis SCARAB uses an engineering method that is restricted to a group of panels (also called joints) defined by the user via a cut plane. Mechanical fragmentation of the connections in the joint occur when the stress level exceeds the strength of the joint material. The material strength is temperature dependent. There is also the option for the user to define the maximum stress level before breaking. A total of 23 cut planes have been inserted to the ISS geometry.

On the main truss there are two cuts: one between truss segment S3/S4 and another between P3/P4 (one cut on starboard side and one cut on port side). There are cuts at each foot of the Photovoltaic arrays (four cuts, each cut contains two feet) as well as one cut in the Z1 Module, which is part of the ITS but is mounted ontop of USOS.

There are six cuts in the ROS: forward and rear facing side of Zarya, one cut in Pirs, one cut in Poisk and one in Rassvet, finally there is a cut located in the forward part of the Progress module.

In the USOS there are ten cuts: there is one in each of the one hatched modules (Columbus, JEM-PM, JEM-ELM, Leonardo) as well as the airlock module QUEST, the two-hatched Destiny module has one cut on each side, and PMA-1,2 and 3 has each one cut.

Due to insufficient detailed description on module level and limitations of the modeling software, the connections between modules were modeled as solid rings of stainless steel. The strength of such a connection is much higher than what it would be in reality. Therefore we define the maximum stress level when the cut should trigger instead of using the engineering approximation. The maximum strength between modules is determined by using results from the ATV shock analysis [12]. For the truss and solar panels the maximum strength is determined via trial and error, finding the stress at the altitude where the structure is expected to break.

4.3. Initial conditions

The state vector used for this simulation is the one used for the ATV-5 shallow re-entry simulation with SCARAB [13]. The attitude of ISS is assumed to have the rear of the Progress module (attached on the ROS) towards the flow. See Tab. 7 for the used initial state vector, attitude, motion and epoch. The initial temperature for the entire structure was set constant to 300 K.

Table 7: Initial conditions for re-entry

Epoch	
Date	2024-01-01
Time	12:00:00.000 UTC
Cartesian Earth Fixed State Vector	
x-Coordinate	x = 309.827 km
y-Coordinate	y = 6212.735 km
z-Coordinate	z = 2055.842 km
vx-Velocity	v_x = -4.535 km/s
vy-Velocity	v_y = 2.050 km/s
vz-Velocity	v_z = -5.588 km/s
Attitude and Motion	
Roll angle	ϕ = 180.000 deg
Pitch angle	θ = 0.000 deg
Yaw angle	ψ = 180.000 deg
Roll rate	$\dot{\phi}$ = 0.000 deg/s
Pitch rate	$\dot{\theta}$ = 0.068 deg/s
Yaw rate	$\dot{\psi}$ = 0.000 deg/s

The set of initial data in Tab. 7 corresponds to the re-entry conditions in Tab. 8.

Table 8: Derived re-entry conditions

Geodetic altitude, h	175.355 km
Longitude, λ	87.145 deg
Latitude, β	18.400 deg
Aerodynamic velocity, v	7.483 km/s
Flight path angle, γ	-0.181 deg
Azimuth angle, χ	-51.759 deg

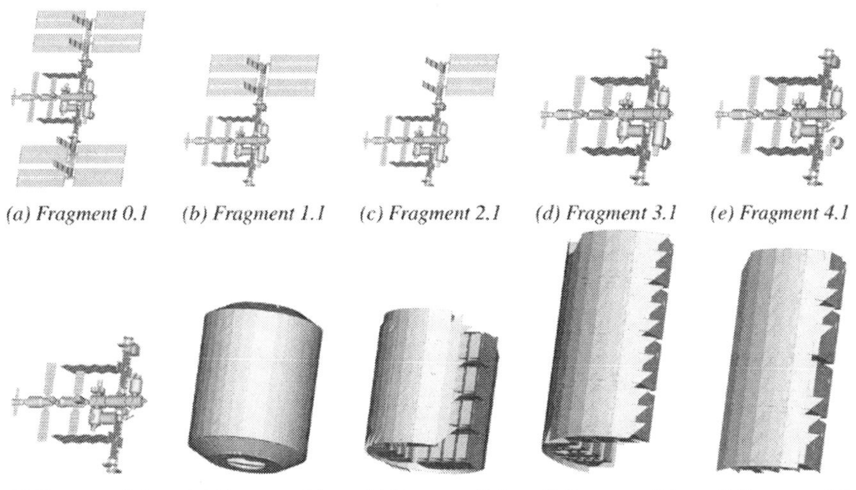

(a) Fragment 0.1 *(b) Fragment 1.1* *(c) Fragment 2.1* *(d) Fragment 3.1* *(e) Fragment 4.1*

(f) Fragment 5.1 *(g) Fragment 6.2* *(h) Fragment 7.4* *(i) Fragment 8.3* *(j) Fragment 9.7*

Figure 6: Columbus module main fragment history

5. RE-ENTRY ANALYSIS

The following section presents the re-entry history for the ISS using the modified thermal model in SCARAB. Sections 5.1. and 5.2. treat all fragments generated during the computation while section 5.3. and 5.4. follow the main ground fragment of the Columbus module from start to the hypersonic limit (Mach 6). See Fig. 6 for the fragment history of the main columbus fragment. The figure shows the main Columbus fragment in different break-up states, from when it was attached to the main structure (Fragment 0.1) to the break-off from the USOS (Fragment 6.2) all the way to the lower mach limit (Fragment 9.7).

5.1. Fragment history

There are three major events during the re-entry of any spacecraft: fragmentation, demise, ground impact. Fragmentation events occur when the structural integrity of the fragment is lost (there are panels or primitives with no connection to the rest of the structure). This either occurs by melting or breaking (triggering of cut-planes). The computation of a fragment ends either when it has demised completely or when it impacts the ground.

See Fig. 7 and Fig. 8 for all events of the re-entry. The plot shows events at the altitude of occurrence against distance traveled and time. During the early stages the altitude is decreasing almost linearly with time until the first break-up occurs after $1181.0\,sec$ at an altitude of $116.9\,km$. This is the outer arm of the truss, holding the PVA, that breaks off due to mechanical stress triggering the cut-plane. After this point more of the cut-planes are triggered one by one, the final cut plane is triggered after $1816.0\,sec$ at $84.3\,km$ altitude.

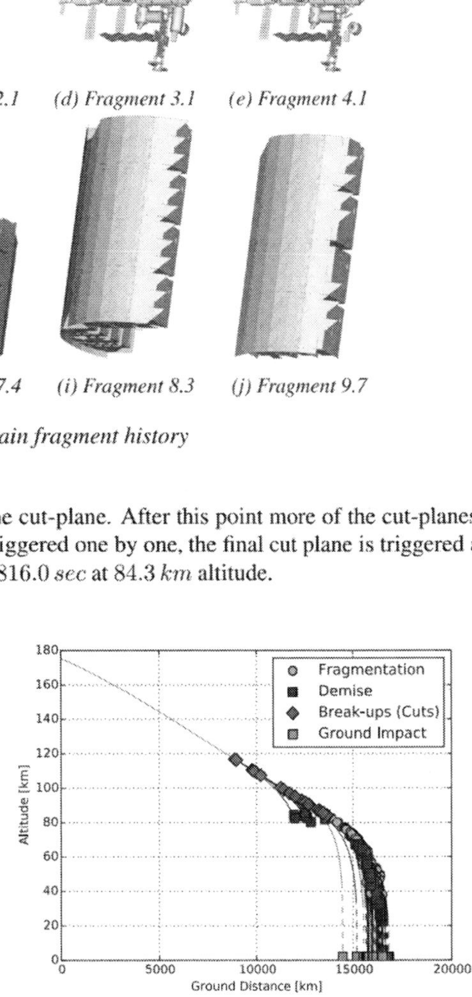

Figure 7: Altitude vs. Ground Distance including Event distribution.

Table 9: Summary of events

Event type	No. of events
Fragmentation	183
Demise	45
Break-ups (Cuts)	22
Ground Impact	150

The break-up of the larger structure exposes more area and smaller features on each module that is normally in shadow of the aerodynamic flow. These parts soon start melting causing further fragmentation. The average altitude where fragmentation occurs is 68.74 km, including all fragmentation events (both melting and breaking). Some of the fragments melt on their trajectory until the melting result in complete demise thus never reaching the ground. Other fragments have enough thermal resistance to reach the ground. Tab. 9 summarizes the number of events of each type. In total 150 fragments reach the ground with a total mass of 241382.058 kg which corresponds to 68.48% of the initial mass (352466.600 kg).

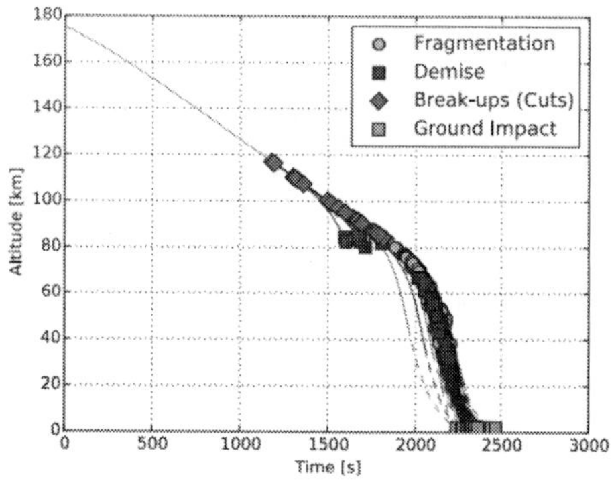

Figure 8: Altitude vs. Time including Event distribution.

5.2. Trajectory and Ground footprint

One of the main interests for a re-entry analysis is the location of ground impacts and the footprint generated by them. See Fig. 9 for the ground track and fragment dispersion of the ISS. The computation starts by the east coast of India and the flight path continues south east passing over Indonesia, Australia, south of New Zealand and ends in the pacific ocean. All impacts are located inside the target area, defined as the SPOUA.

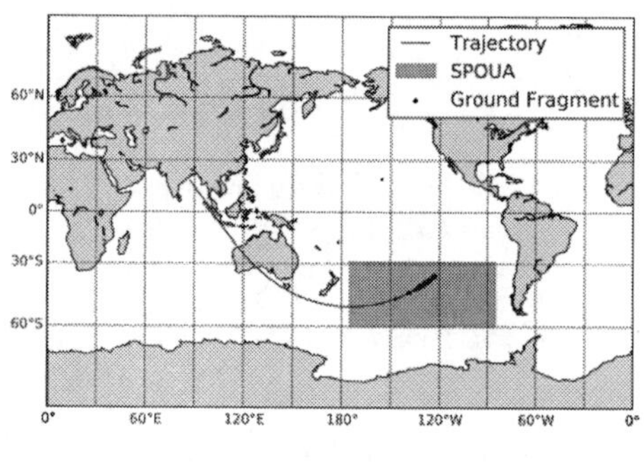

Figure 9: Ground track of all ISS fragments including impact location and SPOUA.

Fig. 10 shows a closer view of the ground impact zone with ground fragments. The fragments related to the Columbus module are highlighted. The footprint is 2360 km along-track and about 51 km cross-track. Most fragments are located near the toe of the footprint while only a few fragments are located close to the heel. All Columbus fragments are located about 2/3rd of the footprint length (heel-to-toe distance). This is in the highest impact density region while the Columbus External Payload Facility (EPF) impacts closer to the heel (located around 1/3rd of the heel-to-toe distance).

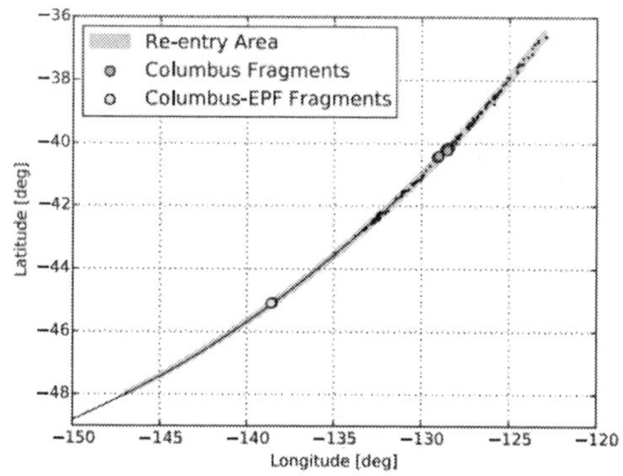

Figure 10: Ground track and fragment dispersion on ground.

5.3. Temperature and Heat Flux

The temperature of the ISS is initialized to a uniform 300 K. Fig. 11 shows the maximum temperature of fragments that include the Columbus module. From its initial temperature, the maximum temperature is increasing until break-up occurs.

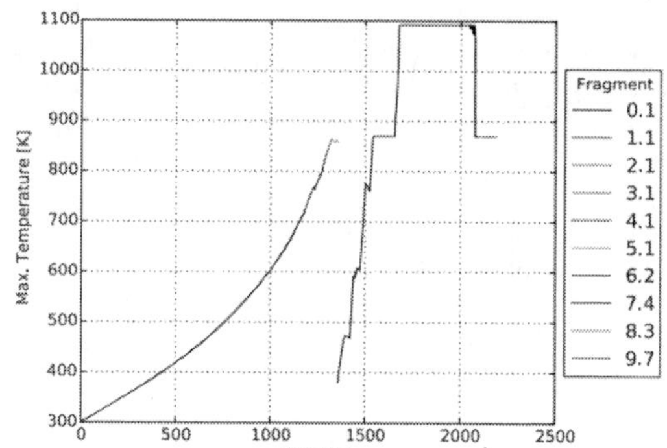

Figure 11: Maximum Temperature vs. Time on the main Columbus fragment.

When the Columbus module breaks off from the main structure the maximum temperature decreases from $860\,K$ to $380\,K$. From this cut it rapidly increases until $880\,K$ (melting temperature of Aluminum) and then $1092\,K$ (melting temperature of Nextel-Kevlar) and finally back to $880\,K$ before the end of the re-entry. Fig. 12 shows the maximum heat flux for the main Columbus fragment (the fragment carrying the Columbus module) against its altitude. The maximum heat flux increases more or less continuously until first break-up. Each fragmentation leads to a new geometry being used, which causes discontinuities in the heat flux. The peak heat flux is $2.451\,MW/m^2$, reached at an altitude of $51\,km$.

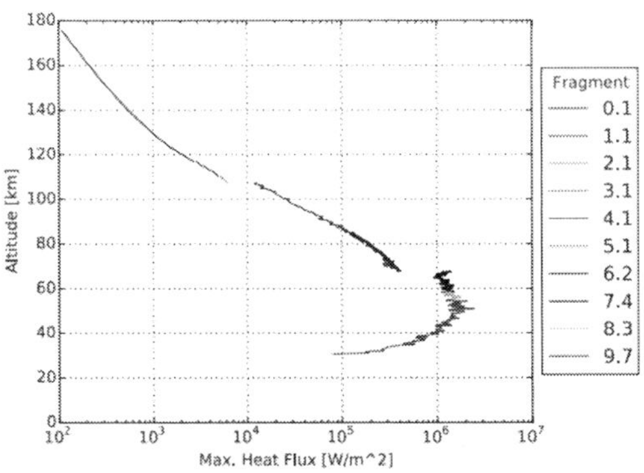

Figure 12: Maximum heat flux on the main Columbus fragment.

5.4. Aerodynamic stress

The aerodynamic influence is proportional to the atmospheric density. As the fragment altitude decreases the forces and torques acting on its body increase. Fig. 13 shows the linear and angular accelerations acting on the

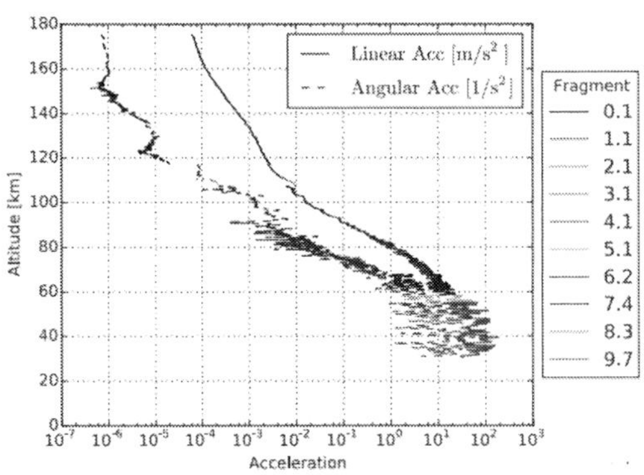

Figure 13: Altitude vs. Aerodynamic Acceleration acting on the main Columbus fragment

main Columbus fragment. See Fig. 14 for the stress in the joint between the Columbus module and the rest of the USOS. At higher altitudes the mechanical loads are small due to the low atmospheric density, while for lower altitudes the aerodynamic moments become important. At around $110\,km$ there is a dramatic increase in mechanical stress and once the maximum stress level of the plane is reached the modules separate. The separation between Columbus module and the USOS occurs at $107.434\,km$ altitude.

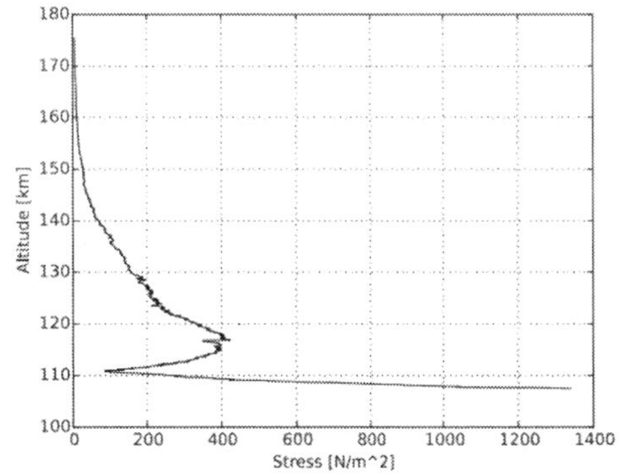

Figure 14: Altitude vs. stress in the joint connecting the Columbus module to the Harmony node

6. EVALUATION OF THE THERMAL MODEL

To evaluate the effect of the adaptation made to SCARAB we compare the modified thermal model to the standard thermal model by performing two re-entry simulations of the ISS model with the same initial conditions.

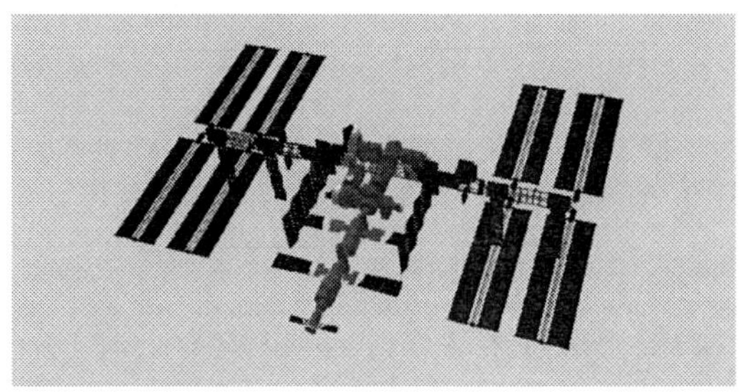

Figure 15: Vizualization of the shock regions on the ISS with the Progress-M in front [9].

As mentioned in section 3.1. the shock shape originates from one global stagnation point. Fig. 15 shows an example of the shock shape as computed from a stagnation point with Progress traveling ahead towards the flow. Red-colored parts are marked as inside the shock (in-shock), white parts are on shock-impingement (on-shock), and black parts are outside of the shock region where free-stream conditions apply.

The largest difference between the results from the two thermals models is in heat flux. The modified thermal model generates a maximum heat flux peak of $1400kW/m^2$, while results from the standard thermal model only produce a peak of $450kW/m^2$. See Fig. 16 for comparison of the heat flux on the main simulation fragment, which in our simulation is the central part of the ITS (S0,S1,P1 and MSS). The heat flux starts to differ between the simulations at $106\,km$ altitude. This is the region where the Knudsen number is 1, which explains the increase in the heat flux (see the bridging formula in Tab. 2).

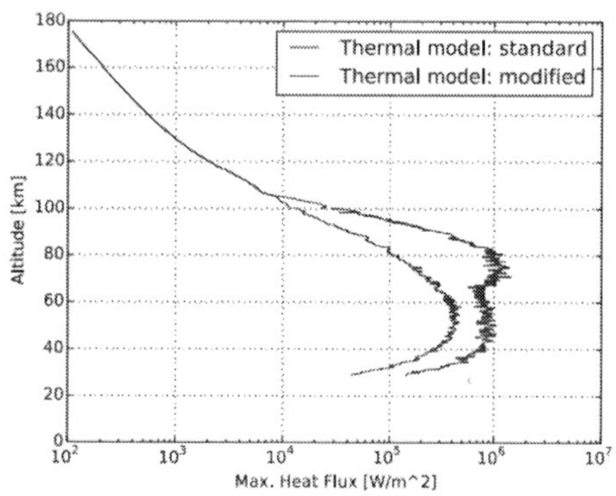

Figure 16: Comparison of maximum heat flux on the main simulation fragment.

The heat flux increases steadily until $82\,km$ where an almost constant heat flux level is computed before it slowly decreases. At this point the Knudsen number passes 10^{-2}. At $67\,km$ the Knudsen number is 10^{-3} and the heat flux reaches a constant offset of the amplification between the thermal models, which again is expected from the formula in Tab. 2.

Despite of this large difference in heat flux the on-ground mass and the number of fragments remain similar. The shock-structure interaction generates a few early fragments which increase the length of the footprint on-ground. The on-ground data is summarized in Tab. 10. See Fig. 17 for the ground footprint generated by the impacting fragments from both thermal models. The trajectory of the fragments in both cases are very similar. The simulation using the standard thermal model results in a footprint with a length of $1335\,km$ and a width of $48\,km$. For the simulation with the modified thermal model the footprint is significantly longer, $2360\,km$, while its width remains similar, $51\,km$.

Table 10: Ground impact summary

	standard	modified
Surviving mass [ton]	248	241
Number of Fragments	135	150
Footprint length [km]	1335	2360
Footprint width [km]	48	51

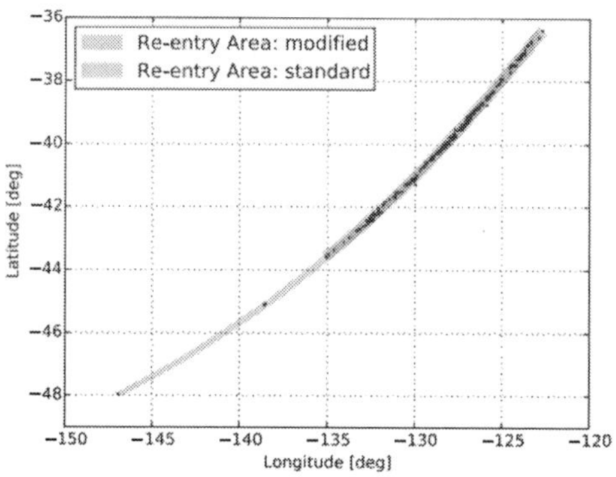

Figure 17: Comparison of footprints.

7. CONCLUSION

In this paper we presented the ISS re-entry results, simulated with SCARAB. We also discussed the modification to the thermal modeling in SCARAB and the aerothermodynamic analysis which we based the modifications on.

In summary the aerothermodynamic analysis found: (1) The heat flux amplification at shock-impingement and heat flux reduction inside the shock-layer compared to the free-stream heating. (2) That between the different flow conditions the location of the shock-impingement does not change much. (3) The heat flux amplification at shock-impingement does not strongly depend on the distance from the shock.

We used these results to modify the thermal modeling in SCARAB. The local heat flux was multiplied with a heating factor coefficient which depends on the local regions relation to the shock (on-shock or inside-shock). In the re-entry analysis the length of the debris footprint was found to be $2360\,km$ along-track and about $51\,km$ cross-track. This footprint is significantly shorter than the suggested footprint of $5370\,km$ [3]. One possible reason for this might be that the simulation drops newly created fragments lighter than approx. $200\,kg$, done to reduce computational time. Including these fragments might have lead to a longer footprint. The first mechanical break-up (cut) occurs at an altitude of $116.9\,km$ and the last is at $84.3\,km$ altitude. The mean fragmentation altitude (both melting and breaking) is $68.74\,km$. Total mass on ground was found to be 68.48% of the orbital mass which also is much higher than the standard re-entry assumption of 10% to 40% of the orbital mass reaching ground.

The present study was the first attempt by the ESA/ESOC Space Debris Office to predict the details of the ISS re-entry. It was based on the currently available data, there are many improvements that would complement the results. The main improvement to be made is to refine the ISS model, refining the assignment of material to the different parts of the structure. Finding the correct specific Nextel-Kevlar material properties (i.e. density, melting temperature, specific heat and latent heat). Add additional details to the geometry (e.g. more detailed internal parts). Finding the correct structural strength between the module compartments on the ISS. Another improvement is to enhance the heating factor coefficient by performing CFD analysis at higher altitudes. Allowing multiple locations for shock creation would grant a more realistic shock-structure behavior. A Monte Carlo simulation using an object-oriented re-entry tool (e.g. ESA's DRAMA/SESAM) could establish a better estimation of the ground footprint size. Finally to get a better understanding of the break-up behavior, due to the shock interaction, more simulations with SCARAB should be done with varied input for uncertainty quantification.

When the space station re-enters it will be the largest single structure to have done so. This event marks the end of an era of Human-spaceflight. The ISS has presented numerous engineering challenges throughout its lifetime, its EOL will not be an exception.

ACKNOWLEDGEMENT

The work presented in this paper was funded by the ESA project 'ISS Atmospheric Break-up Analysis', Contract No. 4000116300/15/D/SR.

REFERENCES

[1] Richard G Stern. Analysis of Mir Reentry Breakup (Aerospace Report TR-2003(8506)-1). Technical report, The Aerospace Corporation, El Segundo CA, Engineering and Technology Group, 2003.

[2] Richard B Mrozinski. Nasa pre-event debris footprint estimates for the deorbit of space station mir. In *MIR Deorbiting*, volume 498, pages 45–55, 2002.

[3] National Research Council et al. *Engineering Challenges to the Long-Term Operation of the International Space Station*. National Academies Press, 2000.

[4] Catherine Hourtolle, Sylvain Delattre, Jean-François Goester, and Emilio De Pasquale. Studies about a shallow re-entry for ATV-5. In *25th International Symposium on Space Flight Dynamics*, pages 19–23, 2015.

[5] ESA Space Debris Mitigation Working Group. ESA Space Debris Mitigation Compliance Verification Guidelines, Issue 1, 2015.

[6] T Lips, B Fritsche, M Homeister, et al. Re-entry risk assessment for launchers - Development of the new SCARAB 3.1L. In *Proceedings of the 2nd IAASS Conference. ESA Communication Production office*, 2007.

[7] R.J. Gollan and P.A. Jacobs. About the formulation, verification and validation of the hypersonic flow solver eilmer. *International Journal for Numerical Methods in Fluids*, 73(1):19–57, 2013.

[8] F. Zander, S. Löhle, H. Krag, S. Lemmens, R.J. Gollan, and P.A. Jacobs. Numerical Flow Analysis of the ISS Re-Entry. In *Proc. of the 7th European Conference on Space Debris, Darmstadt*, 2017.

[9] S. Löhle et al. International Space Station Aerothermal Break-up Analysis Final Report (IRS-17-P11). Technical report, Institut für Raumfahrtsysteme, Universität Stuttgart, 2017.

[10] John David Anderson. *Hypersonic and high temperature gas dynamics*. AIAA, 2000.

[11] NASA Johnson Space Center. On-Orbit Assembly Modeling and Mass Properties Data Book, JSC 26557 REVISION AB VOLUME I, 2008.

[12] ATV Technical Note. ATV-AS-TN-1137 Issue 1 Rev. A.

[13] Tobias Lips. personal communication, February 2017.

PROCEEDINGS OF THE 9TH INTERNATIONAL SPACE SAFETY CONFERENCE

SESSION 14
SPACE TRAFFIC CONTROL – II

TOWARDS A EUROPEAN SPACE TRAFFIC MANAGEMENT SYSTEM

R. Tüllmann(1), C. Arbinger(1), S. Baskcomb(2), J. Berdermann(3), H. Fiedler(4), E. Klock(5), T. Schildknecht(6)

(1)DLR GfR mbH, Galileo Control Center, Oberpfaffenhofen, Germany, email to: ralph.tuellmann@dlr-gfr.de
(2) ROSAS Center Fribourg, Switzerland, Email: stuart.baskcomb@delta-system-solutions.com
(3) DLR, Institute for Communication and Navigation, Neustrelitz, Germany, Email: jens.berdermann@dlr.de
(4) DLR, GSOC, Oberpfaffenhofen, Germany, Email: hauke.fiedler@dlr.de
(5) Austro Control, Vienna, Austria, Email: erich.klock@austrocontrol.at
(6) Astronomical Institute, University of Bern, Switzerland, Email: thomas.schildknecht@aiub.unibe.ch

ABSTRACT

We summarize the main results from an evaluation study which we conducted on behalf of ESA with the aim to identify key pillars for a European Space Traffic Management (STM) system. Our primary focus is on suborbital space transportation, specifically on point-to-point (p2p) connections for human space travel/cargo transportation (including vertical ballistic joyrides), on satellite deployment via air launch systems and on the safe integration of spacecraft with an evolving Air Traffic Management (ATM) system.

A Reference Operations Scenario (ROS) is discussed against which the potential European STM system is designed, playing a fundamental role in the ATM and STM integration process. Initial results from simulations of a hypothetical suborbital p2p flight are presented. They indicate that those flights are generally feasible regarding collision risks with *traceable* space debris/objects. Considering, however, *non-traceable* space debris/objects, a clear gap between the minimum size of detectable objects and the maximum size of objects against which spacecraft can be shielded becomes obvious. If not properly mitigated, the risk caused by this gap could become a show stopper for commercial suborbital space flights. Finally, the Top 10 list of issues is presented that need to be tackled before a European STM system is ready for implementation.

1. INTRODUCTION

There seems to be general consensus that a commercial multi-billion-Euro Space Traffic market is going to develop globally in the next two decades. This market will likely have suborbital point-to-point (p2p) space travel, suborbital cargo transportation and satellite deployment via air launches as primary use cases. Provided suborbital space traffic increases as predicted by several market evaluation studies conducted by the European Commission, NASA, the FAA or the private sector, a safe, efficient and globally (co)operating Space Traffic Management (STM) system is mandatory. Such a system should be designed according to the highest applicable safety & reliability (S&R) standards, e.g., in analogy to those defined for Air Traffic Management (ATM). In this regard, STM shall be responsible for the execution of all necessary managing and monitoring & control operations to ensure safe ballistic travel of manned and unmanned suborbital spacecraft through Near-Earth space and airspace under consideration of an evolving ATM system and infrastructure.

This STM definition is deliberately different from that given by [4] in order to cope with the new suborbital and commercial activities which could, like satellite deployment, Space Weather or Space Debris before, evolve into stand-alone branches of greater STM operations. One key aspect to warrant safe STM operations is the seamless integration of space vessels into the global ATM system. From a European perspective, the required legislation, regulations, concepts and requirements as well as the necessary infrastructure (like spaceports, spaceplanes or control centers) is either missing or not ready to serve the developing STM market in the near future. Therefore, DLR GfR mbH and its partners conducted on behalf of ESA a roadmap study on how to implement a European STM system by 2035 and how to integrate it with the existing ATM system (see [1], [2], [3] for all details).

Our study connects technical, conceptual and organisational aspects related to Space Traffic Control (STC), Space Debris, Space Weather, Clean Space and Safety & Reliability and added a variety of new aspects to the European STM landscape. In this paper, however, we focus on presenting the underlying Reference Operations Scenario (ROS), outlining a possible ATM and STM integration approach and performing an initial feasibility assessment on whether the risk of a space vessel colliding with space debris could actually prevent suborbital p2p travel by comparing it with the risk figure adopted for spaceplane occupants. The paper is concluded by identifying the Top 10 list of STM issues that need to be solved prior to implementing a STM system in Europe.

2. A REFERENCE OPERATIONS SCENARIO

The European STM concept has been developed against a ROS (presented in Fig. 1) comprising 24/7h operations of typical routine and contingency scenarios. Currently, the ROS considers suborbital p2p travel from

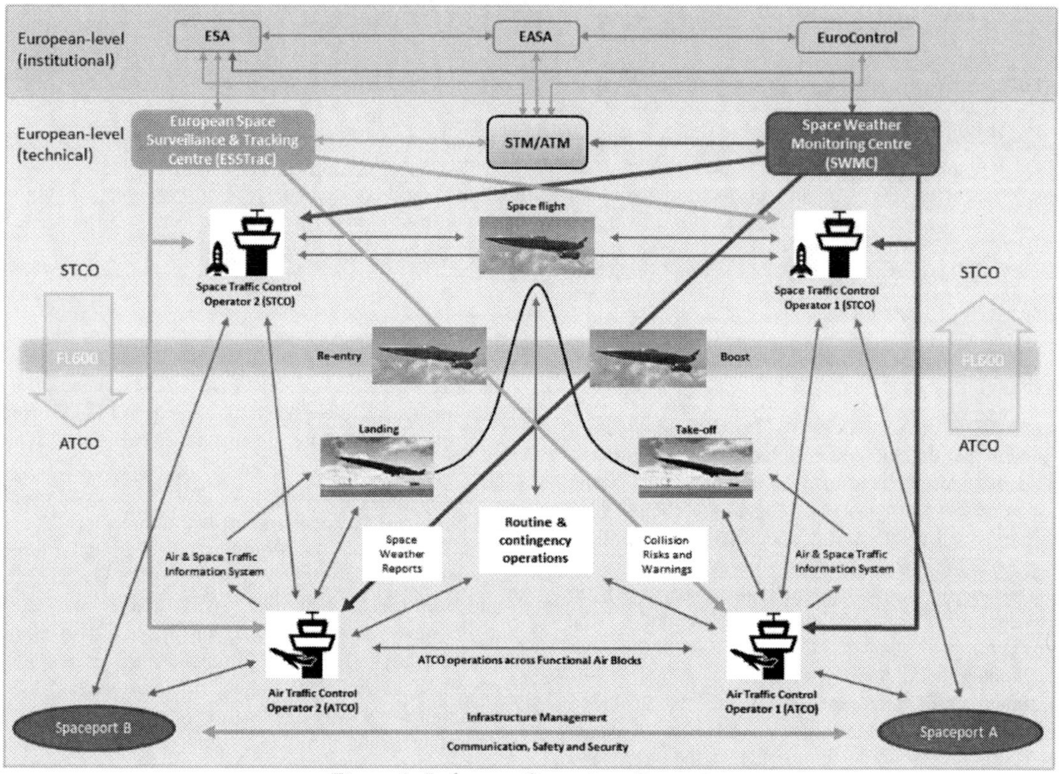

Figure 1: Reference Operations Scenario

Spaceport A to Spaceport B with return to Spaceport A and possible contingency landings at backup sites. The spaceplanes are envisaged to move on ballistic trajectories at maximum altitudes between 100 km and 500 km with the time spent near apogee being much less than 1 hour.

At Spaceport A the spaceplane is cleared for take-off and guided through airspace by the local Air Traffic Control Operator (ATCO). Shortly before the spaceplane leaves controlled airspace (FL600), handover operations are required between the ATCO and the responsible Space Traffic Control Operator (STCO), the latter being in charge of leading the space vessel safely through suborbital space. Before entering controlled airspace, there is a handback between the STCO and the ATCO who will now guide the vessel to its destination (Spaceport B).

Note, however, that internal handover operations between Air Traffic Service providers are needed in case the vessel leaves the respective control area. The same might be required for Space Traffic Service providers.

In case of contingencies which require to return to the departure spaceport or to land at an alternative site, e.g., if Space Weather conditions pose unacceptable risks to crew and passengers aboard the vessel or if the spacecraft got hit by a piece of non-traceable Space Debris (see Sect. 3.2), adequate emergency procedures need to be coordinated among ATCOs, STCOs and flight crew.

It is therefore imperative to warrant the availability of certain service products prior to launch and throughout the flight. Such products include, e.g., Space Weather reports and alerts, Collision Risk Analyses (CRAs), flight trajectories and updates as well as backup trajectories for predefined contingency cases. The service data need to be directly accessible by the controllers and the pilots aboard the air/spacecraft. Ideally, those products are part of a harmonized STM and ATM system in which the service data are circulated via a system similar to the System Wide Information Management (SWIM, [5], [6]) system which is an integral part of the SESAR joint undertaking [7].

Two new STM entities should be mandated with the generation and dissemination of these products, namely the Space Weather Monitoring Center (SWMC) and the European Space Surveillance and Tracking Center (ESSTraC, see also Sects. 3.4 and 3.6 in [1]).

Unfortunately, the operations and activities outlined in the ROS are far from being realized in Europe any time soon. This is because fundamental legislation, critical infrastructure and adequate operations concepts are missing. A significant progress in this field is further hampered by the fact that the safety-critical aspect of integrating spacecraft into today's air traffic flow is

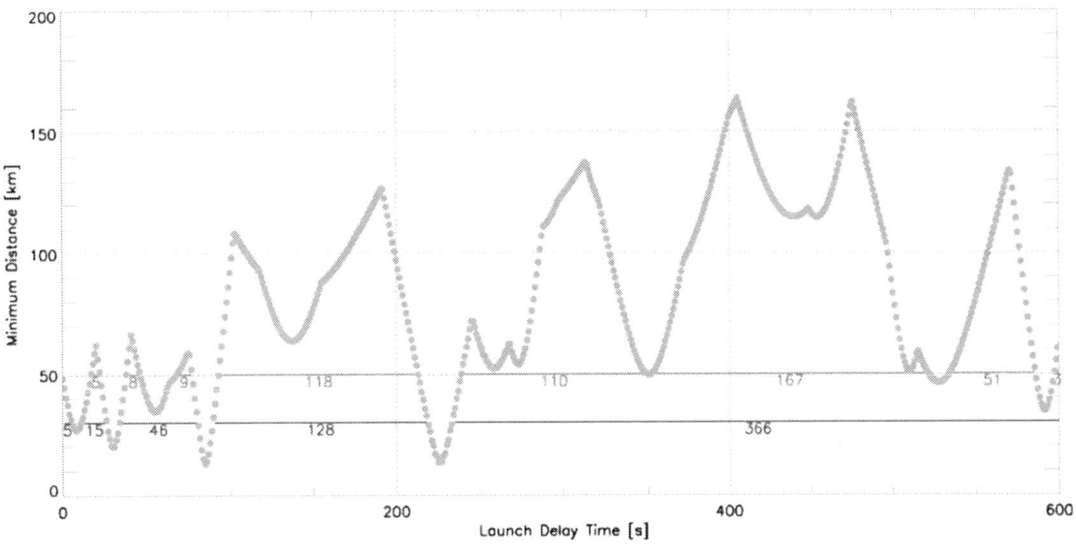

Figure 2: Minimum distance between the space vessel and all TLE-objects for launch delays of up to 600s

currently ignored in Europe's strategic ATM planning (cf. [7]). Therefore, to get Europe ready for commercial space transportation a joint effort is needed by all stakeholders, starting with the acknowledgement that suborbital flights are soon becoming reality and that the time to act is now.

3. COLLISIONS IN SUBORBITAL SPACE

Collisions with space debris or other space objects constitute a non-negligible safety risk for suborbital spacecraft and their occupants/cargo. With advanced space surveillance and tracking networks space debris and space objects (simply called traceable objects hereafter) of sizes larger than about 3cm could become trackable in the coming years. On the other hand, simple shielding technologies can protect a spaceplane only from collisions with ~1mm-sized particles. Hence, this gap requires a careful analysis to tell whether it might become a real show stopper for commercial suborbital space transportation. In the following, we try to give a first answer to this question by evaluating the feasibility of suborbital flights with respect to collisions with traceable objects, by quantifying the risk posed by non-traceable objects and comparing it with the adopted probability for a catastrophic collision.

3.1. Traceable objects

We start by considering a hypothetical ballistic trajectory from Sydney (Australia) to Oberpfaffenhofen (Germany) with a maximum altitude of ~500km. The launch time of the vessel is arbitrarily set to midnight on the 24th of August 2016. For each second of the flight, the distance from the spaceplane to all traceable objects in the NORAD Track Two-Line Element (TLE) database getting closer than 500km is determined (cf. [1] for details) and the minimum distance to the closest object is stored. Next, launch is delayed by 1s and the distance to the closest TLE object is recorded again.

In Fig. 2 minimum distances are plotted as a function of launch delay times of up to 600s. The cyan-colored data points represent the approaching TLE objects and the relative minima indicate their closest distance to the spaceplane. Adopting for simplicity reasons a minimum safety distance of 30km to the TLE objects (red line), there would be two time windows of 128s and 366s each, which the vessel would need to hit to avoid any collisions with traceable objects. If a more conservative safety limit of 50km is applied (orange line), there are three time slots, each of about 120s, for which a collision-free flight would be possible.

Ideally, this iteration is performed for launch delay times covering a period of 24 hours to see whether or not there are sufficiently frequent and long time slots. However, due to computation power constraints, a coarser estimation was performed with a time step size of 120s for the launch delay (assuming this being a representative time slot that can be hit by the vessel) and a time coverage of 12 hours. Again, the minimum distance between the space vehicle and the TLE-objects is calculated, but now every 120s. This results in a statistical distribution of minimum distances, indicating that there could be multiple launch windows per day for which a safe flight path could be calculated in advance (see Fig. 8 in [1]).

In summary, suborbital p2p space flights appear feasible with respect to the risk of colliding with traceable objects. However, we also like to point out that additional work is needed to further substantiate our findings, e.g., by covering launch delays of 24 hours, by considering objects smaller than those listed in the TLE database and by including alternative flight destinations.

3.2. Non-traceable objects

In order to evaluate the collision risk posed by non-traceable objects, i.e. particles <3cm, the object densities provided by ESA's MASTER 2009 model [8] are analysed. As shielding capabilities of potential spaceplanes remain currently unknown, shields with three different shield strengths are considered. First, simple shields capable of withstanding collisions of particles ≤1mm in diameter (e.g., as the nose shield of the Space Shuttle). Second, shields protecting against ≤5mm-sized objects (e.g., as the shield of the Space Shuttle's cabin) and, finally, shields preventing penetrations of ≤10mm-sized objects. Although the latter are successfully deployed aboard the ISS, those Whipple-type shields are not applicable to spaceplanes due to the totally different physical conditions at re-entry. At present, adequate spaceplane shielding capable of blocking hypervelocity impacts of 1cm-sized particles does not exist and represents a severe technological challenge.

To put the assumed shielding capabilities in relation to the particle sizes and spatial densities of the ESA model, the MASTER 2009 population was filtered for a maximum particle size of 10cm (upper bound of the non-traceable object population) and altitudes ranging from 186km (the minimum altitude for which MASTER data is available) to 500km (the maximum altitude of the hypothetical trajectory).

In Fig. 3 spatial densities of the non-traceable object population are plotted as a function of object size. The distribution clearly peaks at 1–2mm-sized particles and exponentially drops towards larger object sizes. The particles are dominated by solid rocket motor slag (Al_2O_3, SRMS) and to a lesser extent by particles produced in collisions (COLL) or by explosions (EXPL). Unfortunately, the MASTER model only covers objects beyond altitudes of 186km. This means, however, that the spatial density distribution for a significant range of altitudes (from 100-186km) remains unknown and represents a gap whose risk potential remains still to be investigated.

In order to estimate the number of impacts received by the spaceplane per unit cross section a and per unit travel distance l, we assume that the vessel travels with a velocity v_{sv} in an environment where the particles move with velocities v_p. While the space vehicle crosses the unit volume element dV, the total number of particles in this volume element is $(1 + v_p/v_{sv})$ times the unit spatial density ρ. The average number of impacts can thus be calculated as:

$$n = \frac{a \cdot l}{dV}(1 + v_p/v_{sv})\rho\, dV. \qquad (1)$$

The factor in brackets is of the order of 2, assuming the velocity of the spaceplane is comparable to that of the

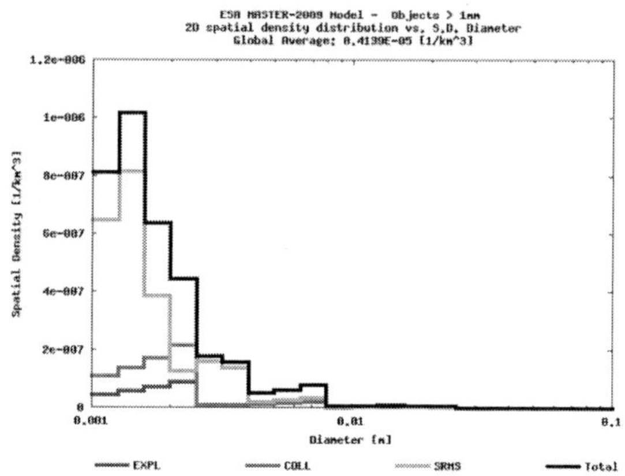

Figure 3: Spatial densities of non-traceable objects as a function of object size

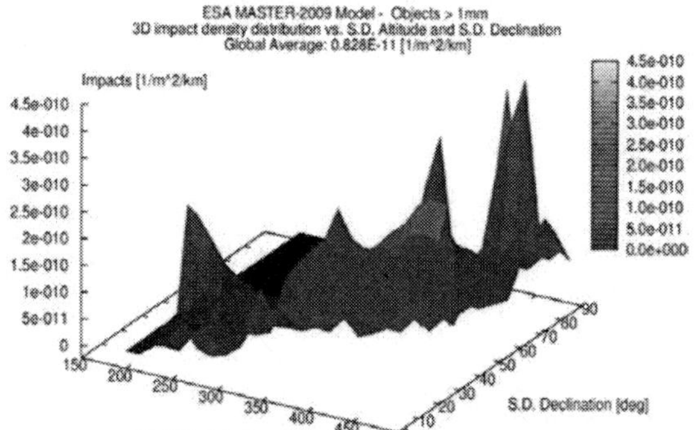

Figure 4: Number of impacts per m^2 cross section and km travel distance for objects >1 mm as a function of altitude and declination.

average velocity of the particles (about 7.8km s^{-1}, cf. [1]). The number of impacts per m^2 and km travel distance can now be computed as a function of altitude and declination. The result is shown in Fig. 4 for the worst case of a minimum particle size of 1mm.

In regions of highest spatial densities, e.g., at altitudes around 270 and 500km, respectively, ~$3x10^{-10}$ impacts m^{-2} km^{-1} are expected. Considering again our hypothetical flight from Sydney to Oberpfaffenhofen, taking into account that the total flight path above an altitude of 200km (limitation of the MASTER 2009 model) is about 13000km and assuming that the entire trajectory would continuously lead through the highest density region (worst case), about $3.9x10^{-6}$ impacts m^{-2} from objects larger than 1mm would be expected. This value would drop to $1.1x10^{-7}$ collisions m^{-2} for the entire flight if an average impact rate of $8.2x10^{-12}$ m^{-2} km^{-1} is adopted (see Tab. 1).

In Tab. 1 global average spatial densities and impact rates for the non-traceable object population are listed for different minimum object sizes (valid for altitudes between 186–500km and for declinations 0°≤ δ ≤ 90).

Table 1: Global average spatial densities and impact rates for the non-traceable object population calculated for different particle sizes.

Particle size	> 1mm	> 5mm	> 10mm
Spatial Density [km^{-3}]	$4.1\cdot10^{-6}$	$2.4\cdot10^{-7}$	$5.0\cdot10^{-8}$
Impact Rate [m^{-2} km^{-1}]	$8.2\cdot10^{-12}$	$4.8\cdot10^{-13}$	$1.0\cdot10^{-13}$
Impacts [m^{-2} $15000km^{-1}$]	$1.2\cdot10^{-7}$	$7.2\cdot10^{-9}$	$1.5\cdot10^{-9}$

Table 2: Shielding capability versus maximum cross section of the spacecraft

Shielding capability (object size in mm)	Maximum cross section (m^2)
1	10
5	100
10	1000

The final row lists the number of impacts per m^{-2} cross section, using a value of 15000km for the full travel distance of the suborbital flight.

Based on these numbers, we can now put the impact risks for spaceplanes with different cross sections in relation to an acceptable level of safety (ALoS) for spaceplane occupants in order to determine the required shielding capability of the vessels (see also [2] for further details).

Due to the lack of historical data and operational experience, an initial ALoS of $1x10^{-4}$ $flight^{-1}$ is adopted for a 1h-flight which is in line with ESA's standard crew safety risk ([9], see also [10]). Because spaceplane passengers are not treated as 'general public', at least not in the early days of suborbital space flights when the industry is still maturing, they will have to accept a significantly higher personal safety risk than today's airplane passenger ($1x10^{-6}$ flight $hour^{-1}$, [11]). On the other hand, to stimulate demand for suborbital flights, the probability of catastrophic accidents has to be significantly below $1x10^{-2}$ flight $hour^{-1}$, i.e. lower than for Space Shuttle missions. Here we adopt a value of $1x10^{-4}$ $flight^{-1}$ which is placed right in the middle between risks for orbital flights and civil aviation.

If we now assume that there are 100 potentially catastrophic scenarios (with equal probability) for the spaceplane and its occupants, than the maximum probability per flight for a catastrophic collision with non-traceable objects amounts to $1x10^{-6}$. This value needs to be compared with the expected impacts m^{-2} for the full space flight (see last row in Tab. 1) to obtain the maximum acceptable cross section of the spaceplane with respect to its needed shielding capability. The results are summarized in Tab. 2.

On this basis, spaceplanes with cross sections $>10m^2$ and shields that protect against collisions with objects $\leq$1mm would not be permitted to launch. This restriction would primarily apply to spaceplane concepts

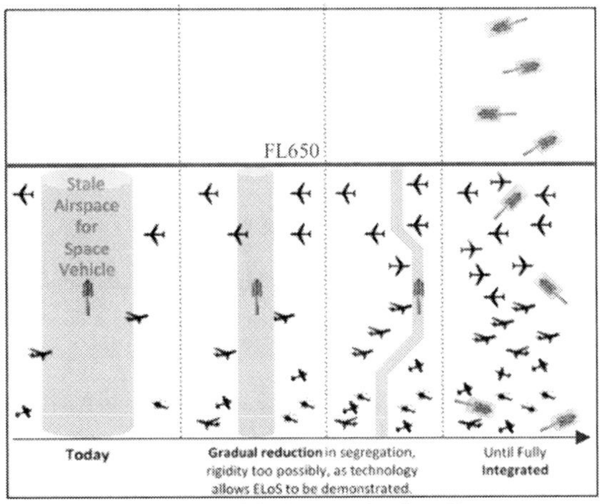

Figure 5: Evolution of flight corridor handling

like the SpaceLiner or Skylon rather than to spacecraft designed for vertical ballistic flights, as the former also operate at higher altitudes with increased collision risks. Currently it appears, as if the gap between today's shielding capabilities (0.1cm) and the expected traceable object size (3cm) could turn out to be a severe show stopper for commercial suborbital flights.

It is important to note, that the above estimates shall only give a first indication of the overall feasibility. The validation of the underlying assumptions, the determination of more accurate impact rates and the in-depth analysis of the risk associated with the gap is beyond the scope of this work and requires a dedicated follow-up study.

In our view it is most important to advance heat and collision shielding technologies and to focus on closing the gap at the lower particle size range, as the smallest objects pose the highest risk (cf. Fig. 3).

4. INTEGRATING STM AND ATM

4.1. Flight Corridor Handling

The approach of safely integrating spaceplanes into the civil air traffic flow calls for new routing and separation standards. One aspect that strongly influences the definition of these standards is the spaceplane's flight performance characteristics, such as its overall maneuverability, turn radius or climb rate.

In Fig. 5 we present an evolutionary high-level flight corridor handling concept which considers four types of separation approaches which in turn require gradual improvements of the spaceplane's flight performance. The first separation approach is already part of today's ATM and demands the closure of large portions of airspace around the vessel (similar to airspace blocking for military purposes). It may turn out that this approach is only feasible in the early days of suborbital flights when there are still infrequent launches, simply because it is the most costly and capacity-constraining one. The

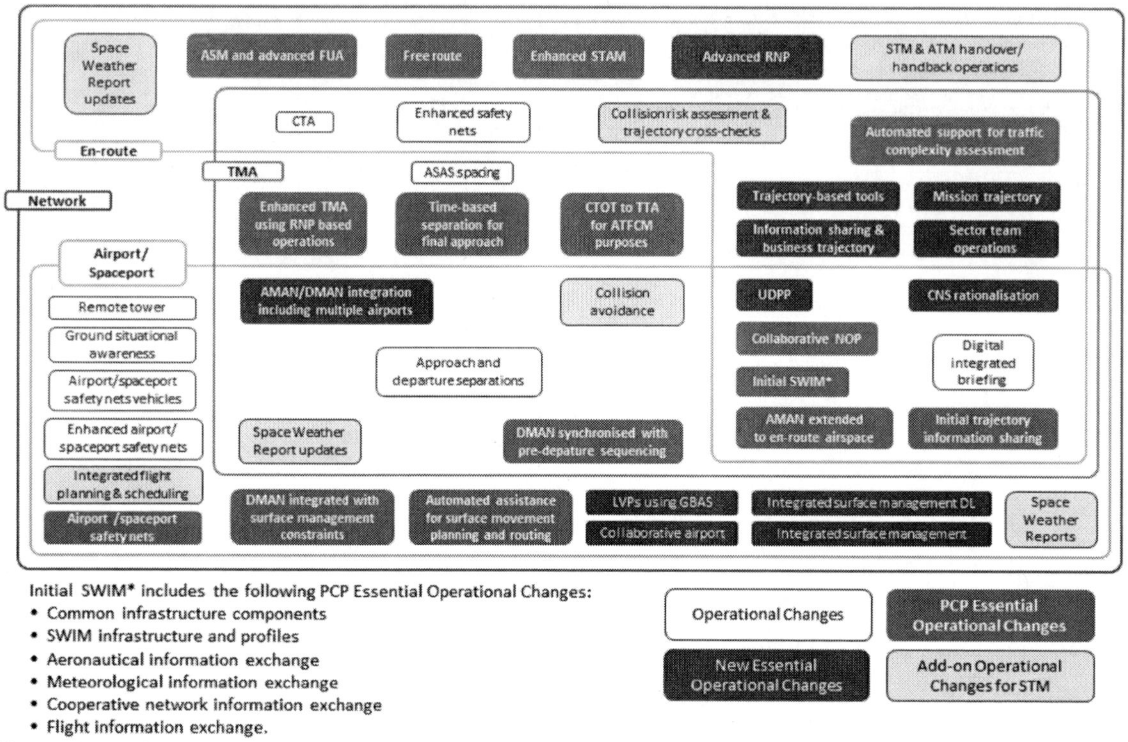

Figure 6: Supplemental STM add-ons (yellow boxes) required to integrate STM with the evolving ATM system (original figure taken from the ATM Master Plan [7] and modified accordingly).

second stage builds up on the first approach and assumes significant advancements on the vertical and lateral navigation performance during launch and re-entry of the spacecraft. This would allow for a significant reduction of the overall containment area and result, compared to the first stage, in a considerable capacity increase due to a better predictability of spacecraft operations in airspace.
The third approach requires an even tighter containment area based on the in-flight NAV performance of the vessel. However, this is currently uncharted territory as no relevant standards yet exist for suborbital spacecraft.
Finally, the fourth approach envisages bubble-like areas around the spacecraft which reflect their individual positioning accuracy and define a three-dimensional space around the vehicle as a non-penetration zone. This concept is very similar to today's separation concept between conventional aircraft in controlled airspace. However, it certainly requires the highest level of accuracy both in position and time and is far from being realized for commercial space operations.
In order to apply ATM concepts for route and radar separation and airspace design principles, e.g., with respect to navigation accuracy and cross-track deviation probabilities, the spacecraft would require a navigation specification that needs to be compatible to aviation standards. In other words, the more airplane-like the in-flight performance of the spaceplane, the easier its integration into routine air traffic flow.
For integrating spaceplanes safely into the air traffic, all spaceplanes and aircraft should be equipped with appropriate technology that allows the determination and dissemination of the vessel's current position and speed (e.g., via GNSS, ADS-B/C, etc.).
Continuous tracking and sharing of positional data with other spaceplanes, airplanes, ATCOs and STCOs as well as 4D positional spaceplane monitoring above controlled airspace should be key pillars in a joint ATM and STM strategy.

4.2. Suggested STM Add-ons to the SES Concept

At present, the topic of suborbital space transportation is largely ignored in the Single European Sky (SES) initiative of the EU [7]. In order to visualise Europe's main deficiencies in the STM and especially in the SES context, Fig. 6 was taken from [7] (their Figure 8) and was adapted to also include the supplemental add-on components and services which we identified missing from the STM point of view (see boxes highlighted in yellow). The following high-level add-ons to the SES ATM Master Plan from 2015 appear mandatory:

1) Space Weather services and products
2) Collision risk assessments and trajectory calculations

3) Integrated (ATM & STM) Flight Planning and Scheduling
4) Handover and handback operations between ATCOs & STCOs.

Add-ons 1) and 2) would require the establishment of ESSTraC and the proposed SWMC together with appropriate operations concepts and globally operating sensor networks. Regarding add-ons 3) and 4), a close collaboration between the ATM sector and future STM key players would be needed to develop and implement suitable flight planning and scheduling concepts and adequate operations procedures. Further details on ATM and STM integration are provided in [1], [2], and [3].

5. TOP 10 STM ISSUES

Based on the analyses performed in our study, the most important issues that have to be tackled for realizing a European STM system are summarized in Tab.3.

Table 3: Top 10 STM issues listed in priority order

Priority	Issue	To be managed by
1	Define and implement binding regulations, procedures and legislation to ensure safe STM operations in Europe	EU/EC
2	UN to coordinate an international consensus on STM implementation (e.g., regarding safety requirements and standards for spaceplane operations in aerospace)	UN
3	Define a European overall safety concept regarding human factors, flight, occupant and on-ground safety (with international interfaces)	EASA/Eurocontrol
4	Development of needed technologies and infrastructure (e.g., for spaceplane engines & shielding, debris tracking, spaceports)	ESA
5	Design and implementation of a European STM/STC operations concept that seamlessly integrates into the global context	ESA
6	Design and implementation of a European Surveillance and Tracking Centre (e.g., including sensor networks, operations, products and services)	ESA
7	Design and implementation of a Space Weather Monitoring Centre (e.g., including sensor networks, operations, products and services)	ESA
8	Design and implementation of an aerospace handling concept for spaceplanes	ESA
9	Design and implementation of a Flight Planning and Scheduling Facility (including international interfaces)	EASA/FAA/Others
10	Design and implementation of Clean Space Concepts (e.g., related to noise reduction, environmental regulations fuel efficiency, or space debris removal)	EU/EASA/ESA

Clearly, there is a long way to go for Europe before first commercial suborbital transportation activities can commence from European territory. For Europe to get its own share of the market, it is vital to conclude on a consolidated STM implementation approach with input from all stakeholders of the aviation and space domain.

6. SUMMARY

The most important results of this work are as follows:

- Europe is not well prepared to serve the evolving space transportation market in the near future.
- A ROS was defined, reflecting typical day-to-day operations for suborbital space flights.
- We identified a gap between current spacecraft shielding capabilities (<1mm) and the minimum traceable object size (>3cm) in future catalogues of traceable objects.
- This gap could turn out to be a real show stopper if it cannot be reduced by applying new heat and collision shielding technologies (focus on small-sized particles as they pose the higher risk).

Acknowledgements: This work was funded by ESA through Contract No. 4000117403/16/F/MOS. The view expressed in this publication can in no way be taken to reflect the official opinion of ESA.

7. REFERENCES

1. Tüllmann, R., Arbinger, C., Baskcomb, S., Berdermann, J., Fiedler, H., Klock, E. & Schildknecht, T., 2017a, On the Implementation of a European Space Traffic Management System – I. The White Paper (Paper I), http://elib.dlr.de/112148
2. Tüllmann, R., Arbinger, C., Baskcomb, S., Berdermann, J., Fiedler, H., Klock, E. & Schildknecht, T., 2017b, On the Implementation of a European Space Traffic Management System – II. The Safety and Reliability Strategy (Paper II), http://elib.dlr.de/112163
3. Tüllmann, R., Arbinger, C., Baskcomb, S., Berdermann, J., Fiedler, H., Klock, E. & Schildknecht, T., 2017c, On the Implementation of a European Space Traffic Management System – III. Technical Requirements (Paper III), http://elib.dlr.de/112204
4. Contant-Jorgenson, C., Lála, P., Schrogl, K.-U. et. al., 2006, Cosmic Study on Space Traffic Management, International Academy of Astronautics (IAA), ISBN 2-9516787-5-4
5. Eurocontrol, 2016, SESAR SWIM Factsheet, System Wide Information Management, N° 01/2016
6. Drescher, J., Morlang, F., Hampe, J., Kaltenhäuser, S., Jakobi, J. & Schmitt, D.-R., 2016, Commercial Space Transportation and Air Traffic Insertion – SESAR Requirements and the European Perspective, 32. Space Symposium, Technical Track, Colorado Springs, Colorado, USA
7. EU & Eurocontrol, 2015, The Roadmap for Delivering High Performing for Aviation for Europe, European ATM Master Plan, https://www.atmmasterplan.eu/downloads/202
8. Wiedemann, C., Flegel, S., Gelhaus, J., Möckel, M., Klinkrad, H., Krag, H. & Vörsmann, P., 2011, The space debris environment model MASTER-2009, 28th International Symposium on Space Technology and Science (ISTS), 5-12 June 2011, Okinawa, paper 2011-r-22
9. ESA 2012, ESSB-ST-Q-003-Issue 1, September 2012, System Safety Engineering, Safety Technical Requirements for Human Rated Space Systems, Sect. 5.2.1
10. IAASS, 2010, Safety Design and Operation of Suborbital Vehicles Guidelines, Suborbital Safety Working Group Manual (October 2010)
11. EASA: Certification Specifications and Acceptable Means of Compliance for Large Aeroplanes. CS-25 Amendment 15. (25.1309 & AMC 25.1309 System Design and Analysis)

EVOLVING SPACE SITUATIONAL AWARENESS

Ruth E. Stilwell, DPA

Aerospace Policy Solutions, LLC, 1835 East Hallandale Beach Blvd., Suite 265, Hallandale Beach, FL 33009 USA
Email: office@aerospacepolicysolutions.com

ABSTRACT

The provision of space situational awareness (SSA) and conjunction assessment is necessarily a global service. However, providers of this service are limited to small number of actors in a few States. Using the model of Copernicus, where raw data is made widely available at no cost, and entrepreneurs develop value added applications, it is possible to develop a more diverse, robust and stable network of space situational awareness service providers. By commercializing the delivery of SSA services, it may be possible to develop products targeted to operators' needs and constraints and more quickly take advantage of innovation. In addition, the encouragement of private sector development can serve to uncouple SSA and conjunction assessment from government agencies and their political and budgetary considerations. The paper will offer a case study in the US deliberations to transfer SSA from a military to civil agency.

1. INTRODUCTION

Congestion in orbit, particularly in low earth orbit has reached a level where systems to prevent collision between satellites or between satellites and debris is necessary. Services are primarily available from the US Strategic Command (US STRATCOMM), and through the European Space Agency and a small number of commercial providers. The services provided by the US STRATCOMM, by the Joint Space Operations Center (JSpOC) include conjunction summary messages to commercial operators free of charge as part of its goal to improve information collection and data sharing for space object collision avoidance [1]. There is considerable debate whether a military organization is the most appropriate structure for providing these services to the expanding commercial space market.

Discussions of the provision of SSA must consider the legal, policy and regulatory frameworks and barriers. Like much of the space industry, systems evolved from government provision to the commercial sector and the policy considerations have lagged behind the technical innovation. This paper focuses on the complex policy elements of the topic, including the questions of funding, authority and international agreements.

2. TERMINOLOGY

The discussion of the need for Space Situational Awareness can be clouded by a complimentary discussion of Space Traffic Management. While these two concepts focus on the same problem, risks from orbital congestion, they are very distinct approaches and should not be conflated. Much of the academic literature distinguishes between these concepts, however some overlap is recognized.

The definition of space situational awareness provided by the Space Foundation is generally accepted [2], "the ability to view, understand and predict the physical location of natural and manmade objects in orbit around the Earth, with the objective of avoiding collisions." In contrast, space traffic management (STM) is still an emerging area and definitions vary [3]. These definitions can range from the provision of a conjunction alerts to a regulatory function that mandates the execution of avoidance maneuvers by space operators, similar to the manner in which air traffic management is applied to aviation operators. In the various discussions of potential space traffic management, there is a common presumption that active intervention by an appropriate authority would exist.

The primary contrast between space situation awareness and space traffic management is one of authority. Using the model of aviation, clear distinctions are made between systems for situational awareness and those used for air traffic control. For both pilots and controllers, situational awareness tools are used to support a decision making process but do not dictate specific action. For example, there is a distinction between Traffic Information Services (TIS), which provide a pilot with situational awareness, while Traffic Collision Avoidance Systems (TCAS) include a resolution advisory to be executed by the pilot. While both provide safety information regarding a potential hazard, the latter requires a higher certification standard and requires an appropriate regulatory framework.

The distinction between SSA and STM is further clouded by the inclusion of conjunction assessment and alerts as part of SSA. In order to clarify the distinguish between the services provided to facilitate the policy discussion, these three services should be discussed separately. In addition, it is important to discuss these as services, rather than concepts when attempting to define how the services should be provided. This paper provides these definitions for the distinct services:

SSA – the detection, collection and dissemination of information on the location and trajectory of natural and manmade objects in orbit around the Earth.

Conjunction Assessment and Alerting (CAA) – the evaluation of natural and manmade objects in Earth's orbit to identify potential collisions and notification of operators to determine if avoidance maneuvers are necessary.

STM – the control of the orbital environment by an appropriate authority responsible for the prevention of collisions between operational satellites and natural or manmade objects.

Discussions of STM look toward far term solutions that face a number of international barriers. The foundational concept of the Outer Space Treaty precludes any claim of sovereignty in space, this presumes that no state could claim the regulatory authority required by the STM concept. This paper focuses on the near term opportunities available for SSA and CAA and does not extend to the discussion of STM.

3. PROPOSED CHANGES TO SSA IN US

In the United States, there is discussion of moving from military to civil provision of the existing SAA services. This would require the transition from US STRATCOMM to a civil agency, presumably to the US Federal Aviation Administration (FAA). While the US has a space agency, NASA, it is a research and development organization, not a regulator or service provider. In contrast, the FAA includes the Office of Commercial Space Transportation (AST) as the regulator for space launch activities and is the current US air navigation service provider. The provision of SSA services by STRATCOMM emerged from the military mission to protect US space assets as a matter of national defense. In 2003, the Department of Defense conducted a pilot program to provide space surveillance data to non-government entities. The US Air Force provided SSA support through a web based service at www.space-track.org and mission support for commercial and foreign entities was transferred to STRATCOMM in 2009 [4]. A transfer of SSA services to a civil agency, would not eliminate the need for the US Department of Defense to continue its program for military purposes including the agreements to provide SSA information to US Allies.

The proposal is supported by both the US Air Force and the FAA, but remains controversial, particularly with regard to funding. The regulatory functions of FAA are funded through the general treasury, while licenses and permits are generally funded through direct fees. The provision of air navigation services are funded through a combination of dedicated excise taxes from aviation system users and contributions from the general treasury. As the military need for SSA services would continue, the transfer of the civil portion of the services would not necessarily include a transfer of funding.

The US proposal recognizes that the Department of Defense maintains and would continue to maintain the space surveillance network. As a result, a civil agency would be a pass through for SSA data and would only assume responsibility for the processing and analysis to provide collision risk models to provide the CAA portion of the services. Representative Bridenstine, sponsor of the Space Renaissance Act describes the proposal as, "A commercialized conjunction analysis and warning center, overseen by a civil agency, should fuse unclassified Defense data with data from international and commercial sensors to provide a single integrated space picture. Eventually, this information should be used by a civilian agency with authority to take action to minimize collisions in space. This will preserve space access for generations to come." Assuming the identified civil agency is FAA, considerable amendment to statutory authority and indemnity would be needed before such responsibilities could be assumed.

Within the proposal is the recognition that the collection and aggregation of data, SSA, is a distinct activity from the action to minimize collisions, CAA.

3.1 Policy and Political Considerations

This proposal raises a number of policy considerations. Overriding many of these policy issues is the absence of sovereignty claims in space. A civil agency is designed to serve the citizens of that nation, and in some cases, like air traffic service provision in ICAO delegated airspace, meet international obligations. This limitation is not generally applied to military activities as narrowly as US military obligations extend well beyond sovereign borders.

As the Outer Space Treaty precludes claims of sovereignty, it is unclear how the US would establish the civil authority to provide services to on orbit operations. As STRATCOMM currently provides those services free of charge, it can be argued that it is free data sharing rather than a particular service provision. As the data and conjunction assessment is performed to meet a national defense mission, the practice of making that data available for other purposes represents added value rather than a service funded for non-US customers. However, if the CAA services are transferred to FAA, a civil agency, for the purpose of preventing collisions for non-aviation, including non-US users in space for which the US does not have an international obligation to provide services, the basis for the free service does not transfer.

The question of funding cannot be overlooked when examining the proposal. The funding philosophy for FAA, specifically for the non-regulatory functions, has evolved over time. Prior to 1970, the FAA was funded through allocations from the general treasury (although aviation taxes were collected, they were also deposited into the general treasury). The Airport and Airway Development Act of 1970 segregated the aviation excise taxes to provide funding for capital improvements to the aviation infrastructure. The legal structure of the aviation trust fund, originally designated for capital investment, has been amended several times and now provides funding for all budget categories of the FAA, including more that 80% of the Operations and Maintenance accounts [5]. Currently, the general treasury contribution for operations is intended to compensate for government use of air navigation services, including services to State and military aircraft and emergency services.

In 1996, there was a temporary expiration of the trust fund authority that coincided with a global push toward the corporatization of air navigation service provision based on the direct collection of user fees as a funding mechanism. This trend supported a philosophical belief among some US entities that air navigation services, including air traffic control, were services provided to specific customers rather than a public good. This philosophical shift in the FAA funding debate is relevant in the provision of conjunction assessment and alerting services to prevent collisions on orbit. It is likely it would be seen in the same way as air traffic services to prevent collisions between aircraft in US controlled airspace, with the notable exception that the US has no claim of ownership of space, even that portion of outer space that overlies its territory.

Putting the funding for orbital CAA in the FAA budget process must consider the manner in which the agency is funded. When examining public policy proposals, a primary step in the evaluation is the assessment is the question of who benefits and who pays. For FAA to continue to provide CAA services on a free basis to commercial and foreign entities, it is necessary to make a convincing and enduring argument that the service is in the public interest, as it is the general treasury that will fund those services. While within the space community a clear case can be made that the avoidance of collisions on orbit are necessary for the preservation of the orbital environment for the public good, it is important to recognize that a similar argument, that the prevention of collision between aircraft is a public good, has not endured.

If the funding for the service is aggregated within operations, instead of a separate funding line, the question becomes even more controversial. Funding from the aviation trust fund has fluctuated between 70% and 100% of the operations account over the last 20 years [6]. As a result, this aggregation would represent the use of excise taxes collected by aviation users, dedicated for aviation purposes, to fund services provided to space operators.

The third option, to create a fee-based service brings us back to the issue of sovereignty. The US does not have the authority to levy taxes on a foreign entity operating outside its borders. A fee-based service, while feasible, would have international implications. The interest of preventing collisions between any objects on orbit, may require the provision of data to States under embargo or other restrictions or States my have restrictions that would preclude their operators from doing business with the US government, resulting in the refusal of services and degrading on orbit safety.

Finally, the political considerations should not be ignored. Unlike aviation, where in the event of a suspension of government services, operators have the option of suspending operations, satellites on orbit are continuous operators. While one would like to assume that CAA would be considered an essential service, FAA is not exempt from budget shortfalls restricting services.

In addition to the proposal in the Space Renaissance Act, other proposals are being evaluated for the provision of SSA [7]. These include:

- Continued Service Provision by DoD
- Provision by a Civil (US) Government Organization
- Provision by NGE (Industry Self-Provision)
- Provision by International Organization

Each of these proposals considers that a single entity is appropriate for the provision of CAA. It appears that this approach may be designed to support the eventual evolution to STM. However, unlike ATM or STM where a volume of airspace is controlled by a single entity, there is no technical reason to restrict SSA or CAA to a single service provider for a given volume of outer space. The single provider analysis appears to be used in the evaluation to preclude duplicative bureaucracies.

4. ANOTHER PATH FORWARD

There is nothing inherently governmental in the development of applications and algorithms to predict the location and trajectory of space objects. Commercial providers of these services already exist. Within the STRATCOMM, commercial providers are under contract to provide these analyses. The overriding argument for maintaining SSA as a government service in the US is the desire to restrict access to information related to space assets engaged in national security activities. However, this presupposes that only the US national security interests are protected and that other State actors are not similarly concealing information on the position of sensitive State space assets.

It is important to recognize the legitimate interests of States in limiting public access to information related to the position and trajectory of certain State satellites. In developing a safety system dependent on voluntary participation, the framework should consider alternate means of compliance. For State actors who have an interest in concealing position and trajectory information, the risk can be mitigated by placing a higher burden for such actors in a collision avoidance scenario. For example, if in the event of a conjunction alert, the State operator assumes responsibility to execute the avoidance maneuver, it is not necessary to include the unaffected satellite operator in the mitigation decision.

Space itself has evolved from the domain of governments to the private sector and SSA and CAA should evolve with it. We have seen this evolution in Earth sensing, first with the US Geological Service decision to provide free public access to data derived from the Landsat and expanded by the European Commission decision to provide users with free, full and open access to Copernicus data derived from the Sentinel satellites [8].

Following these models of free and open access, and applying them to information from the various space surveillance networks could spark innovation and rapid improvements in the provision of commercial CAA services. High quality sensing data from multiple sources, coupled with operational data from satellite operators could lead to better quality conjunction analysis and alerting services. These services could be provided by multiple commercial providers, creating market competition. By decoupling CAA from SAA, we eliminate many of the policy barriers that emerge from a single State or single entity provider.

Since the US is already willing to provide this data free of charge, and the European Space Agency (ESA) has accepted and promoted the concept of free and open access to Earth sensing data, the barriers to providing orbital sensing data to commercial enterprises to develop CAA should be minimal. The benefits of creating opportunities for commercial development of CAA services can provide both safety and economic benefits.

Transitioning from a free government provided service to a commercial and privately funded service requires that the users perceive additional value to support an additional cost. While the STRATCOMM services are provided without fee, that does not mean they are without cost. False alerts or alerts requiring excessive maneuvers are costly both in terms of fuel expended and in lost service life of the satellite. A commercial provider for CAA that can take advantage of rapid industry innovation has the potential to reduce false alerts and unnecessary maneuvers. This added value supports the business case for a paid service.

A government service, provided at no cost, does not have the same incentives to invest in continual upgrades as a commercial provider in a competitive environment. The provider with the best algorithms and most precise analysis will capture the market, there is a financial incentive to continually upgrade. This incentive is not available in the government model. In addition, if the SSA data is available on a free and open basis from the agency responsible for collecting it, it is not necessary to have an additional bureaucracy for the processing of the data.

In addition to potential value added in the commercial provision of CAA, the insurance markets may further drive a requirement for a satellite operator to purchase CAA services. It is reasonable to assume that the near term may see a change in the liability convention, however even if it id unchanged, States can impose a maximum loss value for which operators are required to be insured. Reducing the risk of on orbit collision, like any safety equipment in an insurable activity, reduces the insurance risk costs. Conversely, insurance providers may consider a provider who is unwilling to subscribe to CAA services as an unacceptable risk.

5. CONCLUSION

If Conjunction Assessment and Alerting services are considered separate and distinct from Space Situational Awareness, opportunities emerge for the development of a robust commercial sector to provide these services. The provision of CAA by multiple actors allows the industry to move away from reliance on one or two States, and overcome certain political and policy barriers. The concept of free and open access to data from the space surveillance network, in the same manner that as Earth sensing data, can provide the impetus to create a robust and competitive commercial industry for the provision of CAA.

6. REFERENCES

1. Leeder, Guinevere. Lieutenant Colonel United States Strategic Command Chief, Space Policy (2011). Sharing Space Situational Awareness. Presented to UNOOSA, Vienna.
2. Space Foundation Public Policy and Government Affairs. Space Situational Awareness. www.spacefoundation.org
3. Office of Commercial Space Transportation, Federal Aviation Administration, U.S. Department of Transportation. Towards a Civil Space Traffic Management System (undated). https://www.faa.gov/about/office_org/headquarters_offices/ast/media/6_space_traffic_management_plans.pdf
4. US Department of Transportation (April 2016). Report on Processing and Releasing Safety-Related Space Situational Awareness Data. Submitted to United States Congress, Washington, DC.
5. Fischer, John W. & Kirk, Robert S. (1999). Aviation: Direct Federal Spending 1918-1998. Congressional Research Service, Washington, DC.
6. Marlin, Ruth E. (2005). Understanding Air Traffic Control Financing. National Air Traffic Controllers Association, Washington, DC.
7. Lal, Bhavya. Picard, Alyssa & Weeden, Bryan (2016). Approaches to Civil Space Situational Awareness. IDA Science and Technology Policy Institute, Washington, DC.
8. Martin, Jolyon (2015). The Sentinels Operational Concept – Data Access. European Space Agency, France.

NEWS FROM SWIM IN SPACE

Frank Morlang

German Aerospace Center DLR, Lilienthalplatz 7, 38108 Braunschweig, Germany, frank.morlang@dlr.de

ABSTRACT

All the future air traffic participants are requested to act as system wide information management (SWIM) communicating sub-systems by the future Single European Sky Air Traffic Management Research (SESAR) SWIM "Intranet for ATM" concept. Against the background of the global character of future commercial space transportation (CST) operations and the associated SWIM harmonization need referring the U.S. Next Generation Air Transportation System (NextGen) and SESAR, a first solution based on the already harmonized data format standards Aeronautical Information Exchange Model (AIXM) and Flight Information Exchange Model (FIXM) had been realised in the Tool Command Language (Tcl). The new version's improved performance by the usage of a C code runtime embedding module, a data representation proposal for vehicle specific hazard area information as well as the way for flexible extensions and enhancements in the future by a fundament for mixed language programming additions are presented.

1. INTRODUCTION

The SWIM ideology can be abstracted as the enabler of data delivery at the right time to the right people in terms of quality information that is commonly understood, achieved with the help of open standards based on service oriented architecture (SOA). Facing the integration challenge of space traffic in the current Air Traffic Management (ATM) needs SWIM compliancy of future commercial space transportation (CST) vehicles having "landing like an aircraft at an airport" characteristics.

2. THE SOLUTION ENHANCEMENT

2.1. Performance Improvement

A greater than 30% performance (Fig. 1 and Fig. 2) improvement was successfully realised by replacing the CalculateHeading and CalculateHazardZone Tcl procedures by C code, benefiting from the "Compiled Runtime In Tcl" (CriTcl) package [1][2]. It is a self-contained package to build C code into an extension on the fly, making it possible to wrap C code into cached chunks which compile into a Tcl extension. This allows the usage of embedded C code like callable Tcl procedures.

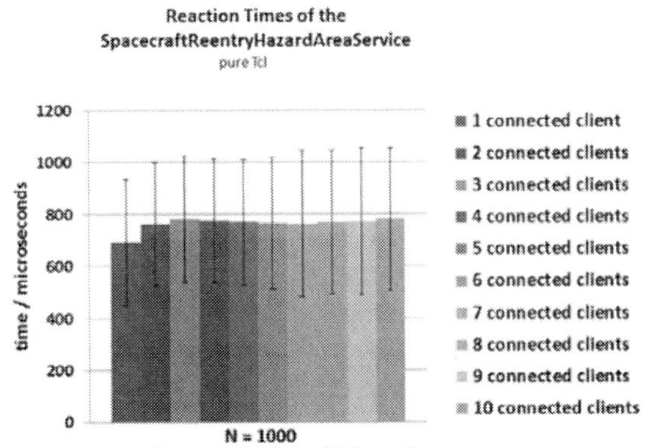

Figure 1. Pure Tcl performance

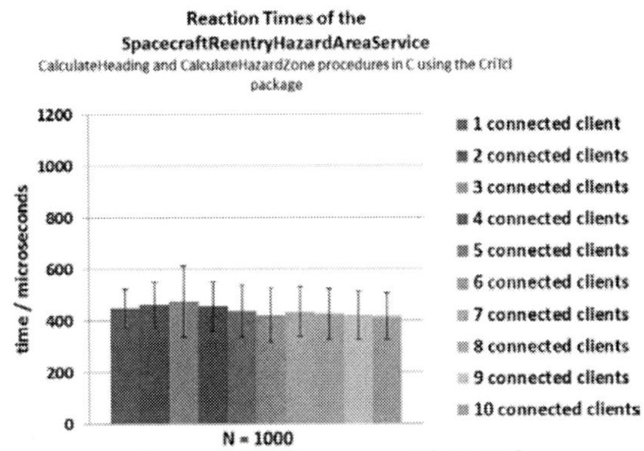

Figure 2. Performance with CriTcl

The principle is exemplarily shown in Fig. 3 to Fig. 5 by means of the CalculateHeading procedure. An equivalent in C only needs integration in the Tcl source code and can be easily switched to this alternative by commenting out the invocation of the original Tcl procedure. While the C function compiling needs a few seconds when running the first time, starting without code modifications does not need that time, just loading of the corresponding shared library, which was generated and cached during the compilation process, is requested. An also cached MD5 checksum of the source code is used to identify if recompiling is needed, which results in an almost instantaneous start behaviour of subsequent runs.

```
# ------------------------------------------------------------------
# CalculateHeading --
#     Calculates heading based on two points lat and lon.
#
# Arguments:
#     lat and lon of the two points
#
# Returns:
#     calculated heading
proc CalculateHeading {lat1in lon1in lat2in lon2in} {
   variable Pi
   set LocalOperator1 [expr {cos(($lat2in*$Pi)/180) * sin((($lon2in*$Pi)/180) - (($lon1in*$Pi)/180))}]
   set LocalOperator2 [expr {cos(($lat1in*$Pi)/180) * sin(($lat2in*$Pi)/180) - sin(($lat1in*$Pi)/180) \
          * cos(($lat2in*$Pi)/180) * cos((($lon2in*$Pi)/180) - (($lon1in*$Pi)/180))}]
   set LocalHeading [expr {atan2($LocalOperator1, $LocalOperator2) * (180 / $Pi)}]
   if {$LocalHeading < 0} {
      set LocalHeadingBuffer $LocalHeading
      set LocalHeading [expr { 360.0 + $LocalHeadingBuffer }]
   }
   return $LocalHeading
}
```

Figure 3. CalculateHeading procedure in Tcl

```
critcl::cproc c_calcheading {double lat1in double lon1in double lat2in double lon2in} double {
   /* this is C code */
   double localoperator1;
   double localoperator2;
   double localheading;
   localoperator1 = cos((lat2in * pi) / 180) * sin(((lon2in * pi) / 180) -  ((lon1in * pi)/180));
   localoperator2 = cos((lat1in*pi)/180) * sin((lat2in*pi)/180) - sin((lat1in*pi)/180) * \
        cos((lat2in*pi)/180) * cos(((lon2in*pi)/180) - ((lon1in*pi)/180));
   localheading = atan2(localoperator1, localoperator2) * (180 / pi);
   if (localheading < 0)
      localheading += 360.0;
   return localheading;
}
```

Figure 4. CalculateHeading procedure in C using CriTcl

```
# set CurrentHeading [CalculateHeading $HistoryLat $HistoryLon $CurrentLat $CurrentLon]
set CurrentHeading [c_calcheading $HistoryLat $HistoryLon $CurrentLat $CurrentLon]
```

Figure 5. Simple switching from Tcl to C in the Tcl main file

Although the Simplified Wrapper and Interface Generator (SWIG), driven by its high level of automation, was taken into consideration as a tool to glue Tcl and C [3], CriTcl was chosen, because it is a loadable package for Tcl itself, thus fitting better in the Tcl extension structure and making the whole solution easier to deploy.

2.2. Data Representation

The paper proposes vehicle specific hazard area information to be stored and managed as a representation of data with the extensible markup language (XML) according to Fig. 6 and Fig. 7. Although this representation has a data-centric character, which is usually associated with XML-enabled databases [4], a native XML database (NXD) called BaseX[1], using XML documents as the fundamental unit of storage, was chosen because of the following reasons:

- Planned service output enrichment of the SpacecraftReentryHazardAreaServer might turn the data representation to a document-centric character, usually associated with NXD [4].
- It provides handling of multi user write and simultaneous read operations based on a client/server architecture with future usage potential towards database splits, where a server dedicated to a subset of clients will only store data related to these clients.
- It offers a wide range of interfaces.


```xml
<?xml version="1.0" encoding="UTF-8"?>
- <hazardareas>
  - <spacecraft status="test">
      <id>00001</id>
      <name>test</name>
    - <hazardarea>
        <footprintrefalt unit="FL" relative="agl">0</footprintrefalt>
      - <altitude unit="FL" relative="msl" value="0">
          <arealength unit="km">0</arealength>
          <areawidth unit="km">0</areawidth>
        </altitude>
      - <altitude unit="FL" relative="msl" value="2000">
          <arealength unit="km">200</arealength>
          <areawidth unit="km">25.0</areawidth>
        </altitude>
      - <altitude unit="FL" relative="msl" value="2050">
          <arealength unit="km">205</arealength>
          <areawidth unit="km">25.6</areawidth>
        </altitude>
      - <altitude unit="FL" relative="msl" value="2100">
          <arealength unit="km">210</arealength>
          <areawidth unit="km">26.2</areawidth>
        </altitude>
      - <altitude unit="FL" relative="msl" value="2160">
          <arealength unit="km">216</arealength>
          <areawidth unit="km">27.0</areawidth>
        </altitude>
      - <altitude unit="FL" relative="msl" value="2210">
          <arealength unit="km">221</arealength>
          <areawidth unit="km">27.6</areawidth>
        </altitude>
      - <altitude unit="FL" relative="msl" value="2270">
          <arealength unit="km">227</arealength>
          <areawidth unit="km">28.4</areawidth>
        </altitude>
      - <altitude unit="FL" relative="msl" value="2310">
          <arealength unit="km">231</arealength>
          <areawidth unit="km">28.9</areawidth>
        </altitude>
      </hazardarea>
    </spacecraft>
</hazardareas>
```


Figure 6. Hazard area description example in xml


```xml
<?xml version="1.0" encoding="UTF-8"?>
<hazardareas>
   <spacecraft status="test">
      <id>00001</id>
      <name>test</name>
      <hazardarea>
      </hazardarea>
   </spacecraft>
   <spacecraft status="test">
      <id>00002</id>
      <name>test2</name>
      <hazardarea>
      </hazardarea>
   </spacecraft>
</hazardareas>
```


Figure 7. Hazard areas' representation of two spacecrafts

A first database connection test setup was successfully realised, benefiting from the "BaseXClient-Tcl" package of Danilo Chang[2], which uses the BaseX server protocol to communicate with the database server.

[1] www.basex.org

[2] https://github.com/ray2501/BaseXClient-Tcl

2.3. Flexibility For The Future

Scalability

Although the prototype's single system performance is sufficient for most small and medium sized cases (Fig. 8), considerations were taken how to make the solution scale better. A first version of a SWIM asset connector proxy was developed that works on top of the prototype solution servers (Fig. 9).

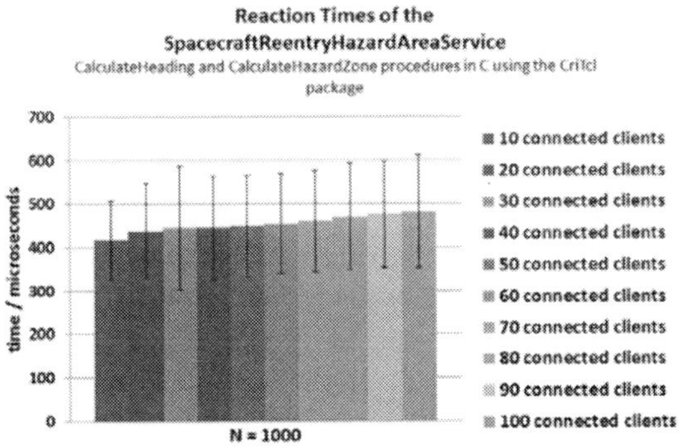

Figure 8. Performance with CriTcl for larger numbers of connected clients

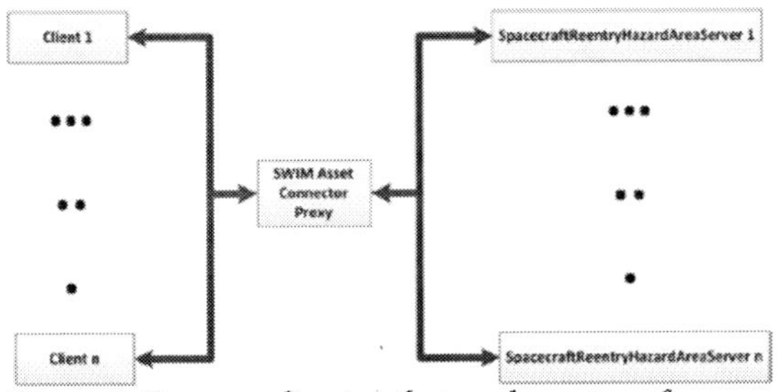

Figure 9. Proxy application that works on top of servers

A single interface is exposed and points clients to the appropriate servers by sending the requests to target servers and passing the results back to the callers. The realisation benefits from a remote communication facility application programming interface (API) for Tcl called "comm", part of the standard collection of utility modules for Tcl called "Tcllib". This API provides an inter-interpreter remote execution facility with control over the remote execution path. If needed, the SWIM asset connector proxy is able to remotely start a new server instance as well as to automatically actualise a list of channel connections. Asynchronous result generation is implemented by the technique of having a remotely invoked command indicate that it will not deliver an immediate, synchronous result. Thus the proxy can continue processing further requests with no blocking and no nesting of event loops. The code realisation consists of the two procedures shown in Fig. 10. The forward procedure takes the next target server as the first argument and the other arguments are processed as commands to be passed to that target server, setting up the asynchronous return command to be used in the forwardReturner procedure, where the result is just passed back to the caller of the forward procedure.

```
proc forward {target args} {
  comm::comm send -command [list forwardReturner [comm::comm return_async]] \
      $target {*}$args
}

proc forwardReturner {todo args} {
  $todo return -code [dict get $args -code] \
      [dict get $args -result]
}
```

Figure 10. Realisation of asynchronous operations invocation

Mixed language programming

The way for flexible extensions and enhancements in the future is paved by the fact that the prototype solution bases on a solid, mature and evolving cross-platform fundament for mixed language programming additions because of the following reasons:

- Tcl's cross-platform high-level API permits written code working on a wide range of Unix / Linux platforms, Macintosh and Windows.
- According to its roots in 1988 (version 1.0) and the fact that new features are still being added under active development (latest release: version 8.6.6 as of July 27th, 2016), Tcl/Tk benefits from both, being mature and evolving.
- Interfacing with other programming languages is possible by taking advantage of the fact that Tcl interfaces natively with the C language on the one hand and has associated bridging extension modules on the other hand, e.g. loading a Java interpreter into an existing Tcl process for using functionality implemented in Java [5].

3. OUTLOOK

Currently, the SpacecraftReentryHazardAreaServer solution consumes the aircraft state data within the Flight Object FIXM message and publishes the hazard area as temporary flight restriction (TFR) airspace output using AIXM version 5.1 for consumption and display by other interested parties. Enrichment of this output is planned in terms of delivering air traffic information about flights that will interfere with the hazard area and might need special attention. Realisation is foreseen to benefit from using the FlightXML 2.0 API for fetching relevant information from Flight Aware's hyperfeed [6].

4. SUMMARY

The paper presents the improved performance by the usage of a C code runtime embedding module of the SpacecraftReentryHazardAreaServer's new version. A data representation proposal of vehicle specific hazard area information based on a native XML database is introduced and justified. Flexibility for the future is addressed by a SWIM asset connector proxy realisation working on top of servers for facing possible future scalability needs. An outlook is given to enrich the service output with hazard area surrounding air traffic information.

9. REFERENCES

1. Landers, S. & Wippler, J-C. (2002). CriTcl - Beyond Stubs and Compilers. In Proc. 9th. Annual Tcl/Tk Conference, Tcl Community Association, Vancouver, Canada.

2. Kupries, A. (2016). C Runtime In Tcl. In Proc.23rd. Annual Tcl/Tk Conference, Tcl Community Association, Texas, USA.

3. Beazley, David M. (1998). Tcl and SWIG as a C/C++ Development Tool, http://www.swig.org/doc.html

4. Pavlovic-Lazetic, G. (2007). Native XML databases vs. relational databases in dealing with XML documents. Kragujevac Journal of Mathematics. 30(2007), 181-199.

5. B. Johnson, T. Pointdexter and D. Bodoh. 2011. JTcl and Swank: What's new with Tcl and Tk on the JVM. Proc. of 18th Annual Tcl/Tk Conference. (Oct. 2011), http://www.tclcommunityassociation.org/wub/proceedings/Proceedings-2011.html

6. Conn, Z. (2016). Hyperfeed: FlightAware's parallel flight tracking engine. In Proc.23rd. Annual Tcl/Tk Conference, Tcl Community Association, Texas, USA.

THE IMPACT OF SECURITY AND DEFENSE POLICIES ON THE ESTABLISHMENT OF A SPACE TRAFFIC MANAGEMENT REGIME

Ms. Ntorina Antoni(1), Ms. Angeliki Papadimitriou(2), Dr. Christina Giannopapa (3)

(1) *Eindhoven University of Technology, 5612 AZ Eindhoven, The Netherlands, Email: n.antoni@tue.nl*

(2) *European Space Agency (ESA), 8-10 rue Mario Nikis, 75738 Paris Cedex 15, France, Email: angeliki.papadimitriou@esa.int*

(3) *European Space Agency (ESA), 8-10 rue Mario Nikis, 75738 Paris Cedex 15, France, Email: christina.giannopapa@esa.int*

ABSTRACT

The objective of this paper is to examine and provide a holistic assessment on the establishment and operational effectiveness of a comprehensive Space Traffic Management (STM) regime in Europe, as seen in the context of European space and security/defense policies, programs and governance. The paper focuses on the current space and security/defense activities and programs of selected European States and the European Space Agency (ESA). It particularly analyses the STM related aspects of existing policies and programs. Recent STM related policy developments in Europe include the European Commission's 'Space Strategy for Europe' in October 2016 calling for a stronger role in space traffic management. Also, the ESA Council at Ministerial level in December 2016 announced its intention to further engage in space safety and STM. Separate civilian and military requirements of security/defense and space programs call for a regime that satisfies the interests of all user communities. The study addresses the way the European space and security/defense policies could affect the potential establishment of space traffic management regime in Europe.

1. INTRODUCTION

In the past decade Europe has been subjected to crises detrimental to the European security. Such crises include security threats, ranging from climate change, migration, asylum flows and neighboring regional conflicts, to terrorism. The demand to tackle these unprecedented crises and multidimensional challenges is increasing, in particular as regards the security dimension. 'Space' is already contributing and could further contribute with existing systems and future ones to reinforcing security. Society relies more and more on space applications and technology, the safety and sustainability of space activities is threatened by various risks deriving from the changing space environment. [1] Challenges include the globalization of space activities, the growing number of governmental and private operators offering new space services and products, as well as the increasing number of satellites in orbit. Space assets located in outer space have a fundamental role in providing "security from space" by means of accurate positioning, navigation and timing services, telecommunication services and Earth observation services. However, it is also critical to safeguard the integrity of the space environment itself for safeguarding existing and future space-based infrastructure, in line with sustainable development principles. ("security in space") It is an essential and fundamental task to ensure the safe conduct of space activities so that the exploration and utilization of space, including space applications and space-based services, can meet social demands. Space Traffic Management (STM) could provide for a systematic and coordinated approach capturing all relevant technical and regulatory aspects of space activities in a coherent manner. STM is a concept to provide a framework for the safety, stability and sustainability of space activities in the future. It can include technical as well as regulatory elements. [2]

In this context, the International Academy of Astronautics (IAA) has revisited the concept of Space Traffic Management (STM) in the IAA Cosmic Study of 2017 where Space Traffic Management has been defined as a concept encompassing 'the set of technical and regulatory provisions for promoting safe access into outer space, operations in outer space and return from outer space to Earth free from physical or radiofrequency interference.' Situational awareness is a key element for any traffic management systems, and so it constitutes a basis for the establishment of STM. The Space Security Index (SSI) [3] describes SSA as the "technical ability of different spacefaring actors, [...] to detect, track, identify, and catalogue objects in outer space, such as space debris and active or defunct satellites, as well as observe space weather and monitor spacecraft and payloads for maneuvers and other events." The quality of a STM system is inherently linked to how accurate, comprehensive and up-to-date the databases are it depends upon. Therefore, the availability of accurate means of Space Situational Awareness (SSA) is a condition *sine qua non* for a comprehensive STM system. [4]

The paper examines the security and defense policies in Europe, taking into consideration ESA, EU and national strategies, programs and governance. It addresses STM elements in the space and security/defense strategies, policies and programs of the European Union (EU), European Space Agency (ESA) and other institutions, and its respective Member States. It further identifies the impact of these policies and programs on the establishment of a comprehensive space traffic management regime in Europe.

2. SPACE TRAFIC MANAGEMENT DIMENSIONS OF EUROPEAN POLICIES AND PROGRAMS

The increased need for security in Europe and for Europe's space activities had led to a number of policies and programs developed by the space actors, towards the strengthening of European security and defense. In this context, ESA, the EU and their respective Member States have strengthened their existing partnerships and developed new ones with security stakeholders in Europe. Figure 1 shows the concerned European States, within the frame of the European Union as well as the European Space Agency.

2.1 European Space Agency (ESA)

ESA has established close collaboration with European institutions dealing with space security/defense, namely the European Commission (EC), the European External Action Service (EEAS), the European Defense Agency (EDA), the European Union Satellite Centre (SatCen) and the European GNSS Agency (GSA). [5] ESA has been more and more involved with security-related projects and programs and addresses the security threats to its own activities, in order to secure the investments of the Member States. [6] Over the past years, the Agency has set up a comprehensive regulatory framework in order to be able to cope with security-related requirements. The main features include the

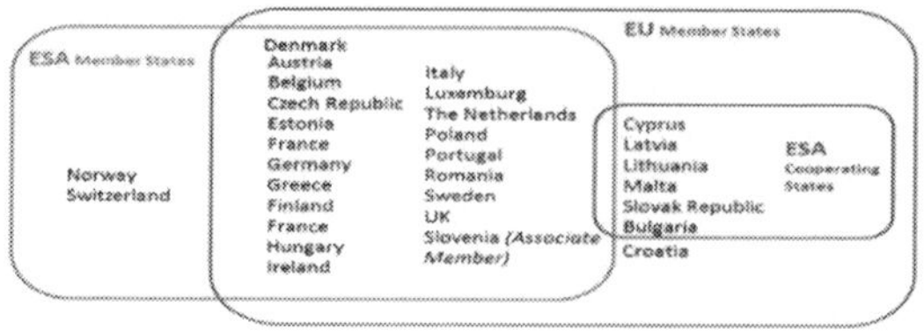

Figure 1 2017 ESA and EU Member States

Security Agreement and Security Regulations and implementing procedures and facilities. [7] This entails the capability to receive, store, and produce ESA classified information and exchange classified information with third parties such as the EU Council (supported by dedicated Security Agreements).

A document entitled "Elements of ESA's Policy on Space and Security", [8] was presented in October 2016 to ESA council and subsequently in December to the ESA Security Committee. The document contained ESA's involvement in space and security activities and provided guiding principles for ESA's future involvement in the domain of space and security in view of strengthening the dialogue with Member States and supporting the current and future activities of the agency. This document introduced the current European context for space and security, the European goals in this domain and the specific objectives to which the Agency intends to contribute. It also presented the draft guiding principles for ESA's contribution to space and security in Europe. The document defines Space and Security as a two-dimensional term which includes security from space and security in space. In particular:

- "Security from space" can be defined as the contribution of space systems in achieving security and defense purposes. this term is also used to imply the protection of human life and environment in earth against natural threats from outer space;
- "Security in space" can be defined as the protection of space assets in the outer space environment against natural and human threats and establishing a sustainable development of space activities.

In addition, space and security are affected by *inter alia*: the space environment, space situational awareness (SSA), policies and laws, commercial and civil space and global utilities, space support for terrestrial military operations and space systems protection and negation. [9]

In the Joint Statement on Shared Vision and Goals for the Future of Europe in Space signed on 26 October 2016, [10] the EU and ESA re-affirm Europe's ambitions in space and agreed on the following overarching goals:

- maximize the integration of space into European society and economy, by extending the use of space technologies and applications to support public policies, providing effective solutions to the big societal challenges faced by Europe and the world;
- foster a globally competitive European space sector, by supporting research, innovation, entrepreneurship for growth and jobs, seizing larger shares of global markets;
- ensure European autonomy in accessing and using space in a safe and secure environment, by consolidating and protecting its infrastructures.

The shared vision and goals have been the high-level policy guiding element for the Commission with its Communication for a Space Strategy for Europe and ESA with its strategy on Space 4.0 for a United Space in Europe, in order to set the way forward for Europe.

The ESA Director General Jan Woerner announced in its proposal of November 2016 that ESA will invest in a safe and secure space environment by investing into Space Traffic Management consisting in: [11]

- Space Situational Awareness (SSA) activities;
- augmenting capabilities for the monitoring of space weather with the development of a Space Weather (SWE) mission to a Lagrange point (L1 or L5) including relevant pre-development activities;
- preparing the first steps towards protection from NEOs by providing a companion observatory and micro-lander mission to the NASA DART impactor, which will strike the small moon orbiting the Didymos binary asteroid with the Asteroid Impact Mission (AIM) mission;
- preparing the necessary technologies for the removal of debris through the preparatory activities for e.Deorbit/Tug; and
- continuing to invest efforts in other debris mitigation and remediation measures, including a system in analogy to TCAS (Traffic collision avoidance system in use for aircraft): spacecraft collision avoidance System, SCAS.

The main projects and programs of ESA with a space and security dimension are, among others, GNSS, Satellite Communications, Earth Observation, Integrated Applications, Critical Space Technologies, Cyber Security and SSA. The ESA SSA programs aims to develop a European system in response to the recognized need for a SSA system at European level, and the recognized need for ESA and other European stakeholders to have access to independent SSA data and services. This system aims at providing timely and reliable data, information and services regarding the space environment and the risks that is poses to infrastructure in orbit and on the ground and ultimately to the economies and communities that depend on them. ESA's SSA programs is focusing on three main areas: [12]

- Space Weather (SWE): monitoring conditions at the Sun and in the solar wind, and in Earth's magnetosphere, ionosphere and thermosphere, that can affect space borne and ground-based infrastructure or endanger human life or health;
- Near-Earth Objects (NEO): detecting natural objects that can potentially impact Earth and cause damage;
- Space Surveillance and Tracking (SST): watching for active and inactive satellites, discarded launch stages and fragmentation debris orbiting Earth. The SSA system will eventually form the basis for the development of an ESA Space Safety initiative which will link SSA with other relevant activities maintaining security and safety of operations in outer space.

Each of these activities is being developed as a 'segment' (analogous to a satellite's traditional 'ground segment') in parallel sets of capabilities and services, supported by data centers and a Tasking Centre with contributions by existing European infrastructure. The current Phase II activities place increased emphasis on developing space weather and NEO services, while research, development and validation activities continue in the space surveillance and tracking domain. [13]

In implementing activities and programs for Europe's independent access to and secure use of space, ESA aims to build upon its experience in space activities and engage in a dialogue with the Member States, EU and international institutions for shaping a coherent and functioning legal and regulatory framework for space activities. Such a comprehensive approach includes efforts towards an international space traffic management system, meaning the management of suborbital space traffic as well as space debris mitigation and removal. [14] This will facilitate the protection of ESA's and the Member States' space assets from risks and will ensure the continuation of services and adequacy and suitability of infrastructures. In addition, it will help enhance the resilience of critical space-based assets and essential services from threats such as space debris and space security.

Furthermore, the ESA Council at Ministerial Level in December 2016 has welcomed the efforts of the Agency to 'provide Europe with the technology needed to monitor, predict and mitigate the risks associated with the expanding commercial use of space, the growing number of small satellites, and the potential arrival of mega-constellations as well as the proliferation of orbital debris and the space weather events that are bound to increase safety issues'. [15] In this regard, the Council has highlighted the need to safeguard in the short-medium term the operational satellites, primarily the Agency's spacecraft, and in the long term to aim for a safe and sustainable space environment. In order to achieve this goal, it is suggested that all ESA missions follow:

a. the United Nations Committee on the Peaceful Uses of Outer Space (UN COPUOS) Draft Guidelines of 2007 for the Long-Term Sustainability of Outer Space Activities, [16] which address, inter alia, the sustainable utilization of outer space, space debris and space situational awareness, space weather and regulatory regimes, and
b. the Inter Agency Space Debris Coordination Committee (IADC) Space Debris Mitigation Guidelines of 2002, as amended, [17] which served as a basis for the development by UNCOPUOS of the UN Guidelines.

In this regard, the Agency aims at broadening the knowledge of the space environment so as to better forecast risks and continue to develop the required technologies. ESA has attained a certain degree of maturity towards forming an Agency devoted to space safety as seen in the framework of the SSA programs. [18] The Agency's contribution to the establishment of STM can be further elaborated and implemented, through exchanges with its Member States as well as national, European, international organizations and institutions.

2.2 Member States

A number of the ESA and EU Member States respectively have developed a dedicated space and security/defense strategy or policy. Such strategy or policy takes into consideration the specificities of each Member State as depicted in Table 1. [19] The table incudes the space and security or defense strategies, policies and programs of selected Member States, as well as the STM dimension. According to Euroconsult, France, United Kingdom, Italy, Germany, Spain, Poland, Norway, Greece, Luxembourg, The Netherlands, Belgium and Denmark are the ESA and EU Member States with a dedicated government budget for defense space programs. [20]

France

The 2015 French space strategy stresses for defense purposes the relevance of very high resolution optical observation, electronic intelligence, ultra-secure telecommunications and SSA. [21] The top priorities of the French military space policy 2025 are: - to ensure the security of all space systems

- to keep up the current standard of service regarding support for operations
- to develop space capabilities not held so far. [22]

SSA missions include space situation collection, knowledge of opposing space capabilities and protection of defense interest space capabilities. SSA capability objectives consist in: [23]

a) Exploiting at best available data, specifically the American «Space Track» catalogue (complementary access);
b) Sustaining LEO surveillance capability (post-GRAVES) and develop an initial GEO surveillance capability (sovereign access)
c) Getting access to global GEO surveillance data and to LEO trajectography information (guaranteed access) via European SST and OOC from CNES.

France operates three SST radar systems. The GRAVES-Wide Network Suitable for Space Surveillance), [24] owned by the French Ministry of Defense and operated by the CDAOA (Air Defense and Air Operations Command), can track about 2,000 objects larger than 10 cm in low earth orbit. The SATAM tracking radars are operated by the French Air Force and their main role is to track debris for management of collision risks and to predict atmospheric re-entry. [25] The Missile Range Instrumentation Ship, named BEM Monge, equipped with three tracking radars and operated by the System Evaluator and Test Directorate of the French Ministry of Defense. [26] They are both used to monitor objects in LEO. In addition, CNES operates TAROT, an optical telescope to track objects in GEO. [27] One telescope is installed on the Calern plateau in France and the other in the Silla observatory in Chile.

Germany

The 2010 German Space Strategy encompasses space-based systems for Earth observation, communication, navigation and the observation of the Sun and near-Earth asteroids as aspects of national security. The strategy recognizes the need for creating synergies between civil and military space research and stresses the need to building up a national competence for space situational awareness through the use of existing resources. [28]

Germany developed its space surveillance radar called Tracking and Imaging Radar (TIRA) and operated by the Research Establishment for Applied Science. TIRA is a unique system that offers space agencies the possibility to measure the orbit with high precision or produce a high-resolution image of objects such as satellites. The system is therefore used to gain precise measurements of space debris, prevent evasive maneuvers for operative satellites or create an image of an object that has gone out of control. [29]

United Kingdom

The United Kingdom has three space surveillance systems, two radar systems (Fylingdales and Chilbolton) and one optical system (Starbrook). The primary role of RAF Fylingdales is to maintain an uninterrupted missile warning service, thereby demonstrating that a surprise missile attack could not succeed. The Unit's secondary role is space surveillance, which forms an essential element of the primary task. [30] Chilbolton is the largest steerable meteorological radar in the world. It is able to remotely determine the shape and orientation of cloud and precipitation particles in the atmosphere. [31] Starbrook, located in Cyprus, is a ground-based optical sensor system for the surveillance of space. Designed and implemented in 2006, Starbrook has an added experimental survey sensor and is one of a new class of wide field of view telescopes that scan near-Earth space to discover and catalogue objects. The first is part of the US Space Surveillance Network (SSN) and is operated by British armed forces. The second is managed by the Rutherford Appleton Laboratory and is mainly used for atmospheric and ionospheric research. The radar has been upgraded to track down to 10 centimeters objects at an altitude of 600 kilometers. [32]

Italy

The 2010 Italian Space Strategy recognizes the relevance of space in support of national security objectives. [33] The 2015 Italian White paper for International Security and Defense puts forward the relevance of ability to access and use of outer space. Space is considered a strategic asset that contributes to the safety of the "national system" and increase the solidity of the political, economic and social structures. [34]

Spain

Spain houses two major European sites relevant for SSA/SST: The Space Surveillance Test and Validation Centre (SSTC) located at ESA's European Space Astronomy Centre (ESAC) and the EU Satellite Centre (EU Satcen). As reflected in the Spanish strategy for ESA and EU space programs released in 2011, [35] both entities have grown through their role as reference centers for SSA among other reasons. The finalization of SSA as a main running programs is included into its priorities for the preparation of the Spanish space sector for the new challenges in times of budgetary restrictions.

Additionally, Spain, represented by CDTI, is a Member of EU SST Consortium which implements EU SST Support Framework. [36] The ESA SST precursor services are taking advantage of the numerous resources that exist across Europe. For data acquisition, data from radar and optical systems are used to obtain both routine and ad hoc observations of both satellites and debris. These sensors are provided by national space agencies, European defense ministries, academia and private entities. SSTC also works closely with international partners; the availability of international channels – such as the data provided by the US Air Force – greatly adds to SSTC capabilities. [37]

In addition to the leading role of Spain in ESA SSA programs, Spain boasts national assets in the frame of its Earth Observation Satellite Programs, PNOTS, notably the development of Paz (radar) and Ingenio (optical). Operated by Hisdesat, the satellites will operate in a complementary way to contribute, for example, to increase the safety of Spanish troops destined abroad and to improve civil protection, monitor forest fires and floods, border control and the detection of discharges into the sea.[38] The Spanish MoD procures a relevant part of its secure military communication services from a partly commercial system, Secomsat. Secomsat is composed by two satellites: SpainSat, which services are provided by HisdeSAT - a government satellite services operator - for the MoD, and XTAR-EUR, a spacecraft operated by XTAR LLC, a US-based joint venture between HisdeSAT (44 %) and Loral (56 %).[39] In the frame of EDA, Spain has taken the leadership of the Member States involved in the GOVSATCOM Pooling and Sharing Demonstrator Project.[40] Finally, from a policy perspective, the 2012 National Defense Directive explicitly mentions the breakdown of space security as a risk to be accounted for. [41]

Italy and Germany have followed France in the development of their own satellite systems for dual (Italy) and military (Germany) applications. Both nations concentrated their efforts on the development of SAR capabilities, resulted in the COSMO-SkyMed constellation, and the German SAR Lupe series, followed by two COSMO-SkyMed Second Generation (CSG) satellites. Germany launched TerraSAR-X and TanDEM-X. Spanish efforts are focused on the PAZ military SAR and the INGENIO dual use optical satellite. Italy and France are also parties to the bilateral agreement Optical and Radar Federated Earth

2.3 The European Union (EU)

The main EU stakeholders in space and security matters include the European Commission, the Council of the EU, the European Council, the European Parliament, the European Defense Agency (EDA), the EU Satellite Centre (SatCen), the European GNSS Agency (GSA), and the European External Action Service (EEAS). Recent security and defense related policy developments in Europe carried out by the said stakeholders demonstrate the increasing importance of space and security.

Based on the implementation of the Conclusions of the European Council of December 2013 [56] and subsequent Defense Foreign Affairs Councils of the EU, the EEAS and EDA have made efforts to strengthen the Common Foreign and Defense Policy (CSDP) and the European defense capabilities. In June 2016 the High Representative for Foreign Affairs and Security Policy, Vice President (HR/VP), Federica Mogherini, presented the "EU global strategy on foreign and security policy (EUGS)", [57] which highlights the autonomy and security of European space-based services, the need for responsible behaviour principles and the necessary investments in space systems for security. It calls for "improv[ing] the monitoring and control of flows which have security implications. The objective of responsible behaviour is in line with the EU-sponsored multilateral initiative in 2015 towards an International Code of Conduct for Space Activities seeking to achieve enhanced safety, security, and sustainability in outer space. [58]

In addition, in the same period the European Commission (EC) released two documents of direct relevance to space, security and defense: The Communication on a Space Strategy for Europe and a European Defense Action Plan that stress the need for Europe's non-dependence. The Communication on Space Strategy was released in October 2016 [59] reiterating the strategic importance of space for Europe in reinforcing its role as a stronger global player and as an asset for its security and defense.

Space's crucial role is highlighted with regard to enabling reinforced synergies between civil and space security activities. Particular emphasis is put on the *'growing threats also emerging in space: from space debris to cyber threats or the impact of space weather. These changes make greater synergies between civil and defense aspects increasingly relevant. Europe must draw on its assets and use space capacities to meet the security and safety needs of the Member States and the EU'*. [60] Reference is made to STM in the Space Strategy, wherein the Commission draws attention to the fact that: *'Increased human activity in space and the rapid growth of new entrants is testing the UN conventions on outer space to the limit, including on issues of space traffic management and mining'.* Furthermore, *'Europe should be among the leaders in navigating global challenges such as climate change or disaster risk reduction, while promoting international cooperation and building the global governance or appropriate legal frameworks for space'*.

The European Commission has expressed the intention to work closely with the EDA and ESA to assess the

Table 1 Space Implementing Entities and Space and Defense Strategies/Policies in Selected ESA Member States

Member State	Space Implementing Entity	National Space Strategy/Policy	Space and/or Security/Defense Strategy/Policy
France	CNES	Ambition 2020 – 2015, space for the climate [43] Stratégie Spatiale Française[44]	2013 French White Paper on defense and national security [45]
Germany	DLR	Making Germany's space sector fit for the future - The space strategy of the German Federal Government [46]	2016 White Paper on German security Policy and the Future of the Bundeswehr [47]
Italy	ASI	Strategic Vision 2010-2020 [48]	2015 White Paper on International Security and Defense [49]
Spain	CDTI	Spanish Strategy for ESA and EU Space Programs 2007-2011[50]	The National Security Strategy [51]
United Kingdom	UKSA	National Space Policy [52] Space Innovation and Growth Strategy 2014-2030 - Space Growth Action Plan [53]	National Space Security Policy [54] National Security Strategy and Strategic Defense and Security Review 2015 [55]

demand for and feasibility of initiatives to provide next generation of resilient satellite communication services for governmental and institutional security users under the GOVSATCOM initiative. [61] ESA has been working closely with EDA for GOVSATCOM preparations. [62] Additional space and security/defense activities include the two European flagship programs – Galileo and Copernicus – the research ones under Horizon 2020 [63] and the Space Surveillance and Tracking (SST) framework. [64] The latter is regarded one of the three main areas of SSA, the other two being Space Weather and Near-Earth Objects. The SST support framework seeks to contribute '*to ensuring the long-term availability of European and national space infrastructure facilities and services which are essential for the safety and security of the economies, societies and citizens in Europe*'. The objective is to create a partnership where MS will contribute with their existing and future assets to the SST capability at European level.

The SatCen is a significant asset for the SST Consortium (currently composed of five national designated entities: ASI, CDTI, CNES, DLR and UKSA) through the provision of SST services (geospatial imagery information services and products with various levels of classification to civil and military users) to SST users. [65]

In November 2016 the Commission released the European Defense Action Plan (EDAP) [66] which contributes to ensuring that the European defense industrial base is able to meet Europe's current and future security needs and, in that respect, enhances the EU's strategic autonomy, strengthening its ability to act with partners. The EDAP promotes civil/military synergies within EU policies including the space sector as highlighted in the Space Strategy for Europe.

In April 2016, the European Parliament released a report on "Space capabilities for European security and defense". The report among others underlines the important role of space-based capabilities and services in, among other areas, the context of European security and defense. [67]

In July 2017 a report of the European Parliament was released on the motion for a Resolution on "a Space Strategy for Europe" proposed by the Commission. [68] The report has stressed the vulnerability of space assets threats ranging from interference or attack from state and non-state actors to collisions with space debris or other satellites. It has highlighted the need for ensuring security of critical infrastructure and communications as well as the development of resilient technologies, in particular because of the dual-use implementation of space and space-based technologies. The Parliament calls for further investments to boost the development of enhanced space capabilities and technology in order to facilitate Europe's response to the challenges in the space sector and changing markets. Especially, the risks posed by space debris to European space infrastructure call for mitigation measures that reinforce current space surveillance and tracking (SST) services in order to avoid collisions and, in the longer term, to remove orbital debris.

The report has acknowledged the importance of the EU SST to further include space-based weather forecasts and near-earth objects to address the potential risk of collisions with Earth. The European Space Agency capabilities and expertise in this domain could be made available for this purpose and further provide a platform for the conception and the development of future capabilities. Crucial to such platform is the data sharing policy. In addition, enhancing Europe's global presence and furthering international cooperation, requires '*ensuring a peaceful and safe space environment will require engagement with international partners to promote norms of responsible behaviour and sustainability*'. The need for international coordination on space traffic and debris management is considered highly important to tackle the planned mega-constellations and the congestion of near-earth orbits that are expected to increase because of the low cost of satellite launch.

3. CONCLUDING REMARKS

Space actors in Europe recognize that an STM regime is essential for the security and sustainability of European space activities. Accordingly, they have stepwise developed a number of activities, strategies and policies which among others put the basis for STM development and reflect the important role of STM in space and security.

In the frame of those space strategies and policies, STM is included in the broader frame of sustainable use of Outer Space. The concept of sustainability has steadily risen in importance since its introduction by the 1983 Brundtland Commission, where sustainable development has been defined as "development that meets the needs of the present without compromising the ability of future generations to meet their own needs".[69] The principles underlying sustainable development [70] are increasingly applied to space activities and therefore reflected in relevant space strategies and policies.

Although STM is not explicitly reflected in all space and security strategies of all the European players at national and supranational, the relevance of sustainable use of Outer Space and "security in space" is universally recognized. The application of the principles of sustainable development in space strategies and policies also addressed the contribution of space assets to sustainable development on Earth. This can for instance be seen in the French Space Strategy, entitled "Ambition 2020 - 2015 Space for the Climate". [71] The increasing use of space assets for climate change (and sustainable development more broadly) in the frame of "security from space" and how this is reflected in the strategies and policies of European and global players offers potential for future research. Although fragmentation of strategies, policies and programs persists, – both at European and Global level – STM is emerging as a major concern and priority for space faring nations.

This paper has looked into the various STM elements of space and security/defense policies programs and governance in Europe. The main space actors in Europe including the European Space Agency (ESA), the European Union (EU) and their respective Member States have been examined. Member States constitute the main policymakers for both space and security matters in Europe. This refers not only to their action in national level but also in multilateral, intergovernmental and European level in the frame of their membership at ESA and EU. The majority of these Member States have adopted a space strategy and/or policy with some of them, such as the United Kingdom, having put in place a dedicated space security policy. Space traffic management is more and more becoming a priority for the Member States and as part of engagement in space activities.

The establishment of an STM regime is a complex procedure depending on various factors. Those could include a comprehensive and interdisciplinary approach, technical, legal and institutional framework as well as international cooperation. However, the political support constitutes the prerequisite which enables all the others.

4. REFERENCES

1. Schrogl K-U., Hays P., Robinson J., Moura D., Giannopapa C. (eds.) Handbook of Space Security, Springer, Vienna, 2015
2. IAA, Cosmic Study, Space Traffic Management, Towards a roadmap for implementation, 2017 p.15-16
3. IAA, Cosmic Study, Space Traffic Management, Towards a roadmap for implementation, 2017, p.78
4. IAA, Cosmic Study, Space Traffic Management, Towards a roadmap for implementation, 2017
5. ESA, Status Report on ESA's security-related developments and activities, ESA/C (2017)62, 1 June 2017.
6. Giannopapa C., Adriaensen M., Antoni N., Schrogl K-U., Elements of ESA's Policy on Space and Security, 68th International Astronautical Congress (IAC), IAC-17,E3,4,5,x41447, 2017
7. ESA, Highlights of ESA Rules and Regulations, http://www.esa.int/About_Us/Law_at_ESA/Highl i ghts_of_ESA_rules_and_regulations (last accessed on 6 September 2017).
8. ESA Council, ESA's elements of space and security policy, ESA/C(2016)106, October 2016 9. Schrogl K-U., Hays P., Robinson J., Moura D., Giannopapa C. (eds.) Handbook of Space Security, Springer, Vienna, 2015
10. Joint Statement on Shared Vision and Goals for the Future of Europe in Space by the European Union and the European Space Agency, 26.10.2016. http://m.esa.int/About_Us/Welcome_to_ESA/Shar ed_vision_and_goals_for_the_future_of_Europe_i n_space (last accessed on 6 September 2017)
11. ESA, The Director General's Proposal 'Towards Space 4.0 for a United Space in Europe' ESA/C-M (2016)1, Paris, 9.11.2016
12. ESA SSA Programs Overview, http://m.esa.int/Our_Activities/Operations/Space_ Situational_Awareness/SSA_Programs_overvie w (last accessed 25 September 2017)
13. ESA, The Director General's Proposal 'Towards Space 4.0 for a United Space in Europe' ESA/C-M (2016)1, Paris, 9.11.2016
14. ESA, Resolution "Towards Space 4.0 for a United Space in Europe" ESA/C-M(2016)2, 9.11.2016
15. The Guidelines were adopted by the STSC at its 44th session in 2007 (A/AC.105/890, para.99), endorsed by the Committee at its 50th session in 2007 (A/62/20, paras. 118 and 119) and subsequently submitted to the General Assembly at its session in 2007. Endorsed by the General Assembly through Resolution 62/217 of 22.12. 2007
16. IADC Document Registration List, http://www.iadconline.org/index.cgi?item=docs_p ub (last accessed 25 September 2017)
17. ESA, Resolution "Towards Space 4.0 for a United Space in Europe" ESA/C-M(2016)2, 9.11.2016
18. M. Adriaensen, C. Giannopapa, D. Sagath and A. Papastefanou, Prioirities in National Space Strategies and Governance of the Member States of the European Space Agency, Acta Astronautica, Volume 117, December 2015, p. 356-367,
19. Profiles of Governments Space Programs, Euroconsult, 2015
20. CNES, Ambition 2020 – 2015, space for climate, http://www.cnes-csg.fr/automne_modules_files/standard/public/p11 636_99f9990e67f7bbf293ff282b37d9bc1eCNES_ plk_instit_2015_171214_GB.pdf (last accessed on 6 September 2017)
21. European Parliament, Space, Sovereignty and European Security – Building European Capabilities in an advanced institutional framework, 2014 http://www.europarl.europa.eu/RegData/etudes/etu des/join/2014/433750/EXPOSEDE_ET(2014)433 750_EN.pdf (last visited 25.09.2017)
22. A GRAVES Sourcebook https://fas.org/spp/military/program/track/graves.p df (last accessed 25 September 2017)
23. La Division Surveillance de l'Espace du Commandement de la Défense Aérienne et des Opérations Aériennes. http://www.defense.gouv.fr/actualites/dossiers/les pace-au-profit-des-operations-militaires/fichestechniques/dse-cdaoa (last accessed 25.09.2017)
24. Monge (A 601) http://www.defense.gouv.fr/marine/equipements/b atiments-de-soutien/batiments-specialises/dexperimentations/monge-a-601 , (last accessed on 25.09.2017)
25. CNES, Ground Based Assets https://debrisspatiaux.cnes.fr/en/node/187 (last accessed 25.09.2017)
26. BMWI, Making Germany's space sector fit for the future - The space strategy of the German Federal Government, http://www.dlr.de/rd/en/Portaldata/28/Resources/d okumente/Raumfahrtstrategie_en.pdf (last accessed on 6 September 2017).
27. Space observation radar TIRA https://www.fhr.fraunhofer.de/en/theinstitute/techn ical-equipment/Space-observationradar-TIRA.html (last accessed 25.09.2017)

28. RAF Fylingdales https://www.raf.mod.uk/raffylingdales/aboutus/whywearehere.cfm (last accessed 25 September 2017) 31. Chilbolton Facility for Atmospheric and Radio Research (CFARR) Chilbolton Advanced Meteorological Radar (CAMRa) 3 GHz S-Band Radar http://catalogue.ceda.ac.uk/uuid/364bb67ddd8a439eaf648bc9d183c906 (last accessed 25.09.2017)
32. An introduction to Space Insight's Starbrook systems for the surveillance of space, http://www.spaceinsight.co.uk/assets/uploads/SpaceInsight_Starbrook.pdf , (last accessed 25.09.2017)
33. ASI, Strategic Vision 2010-2020, http://www.asi.it/sites/default/files/ASI_DVS_2010_2020_ENG_0_0.pdf (last accessed on 6.09.2017).
34. Italian Ministry of Defense, White Paper for International security and defense, July 2015, https://www.difesa.it/Primo_Piano/Documents/2015/07_Luglio/White%20book.pdf (last accessed 6.09.2017).
35. The national Security Strategy of Spain, http://www.lamoncloa.gob.es/documents/estrategia_seguridad_baja_julio.pdf (last accessed on 06.09 2017).
36. EU SST Governance, http://www.eusst.eu/project/who-we-are/ (last accessed 18.10.2017)
37. ESA, Space Surveillance and tracking – SST Segment, http://www.esa.int/Our_Activities/Operations/Space_Situational_Awareness/Space_Surveillance_and_Tracking_-_SST_Segment, (last accessed 17.10.2017)
38. Marca Espana, The news Spanish satellites, http://marcaespana.es/en/news/society/newspanish-satellites, (last accessed 17 October 2017) 39. Quoting: European Parliament, Space and Sovereignty and European Security – Building European Capabilities in an advanced institutional framework, Brussels, 2014, http://www.europarl.europa.eu/RegData/etudes/etudes/join/2014/433750/EXPO-SEDE_ET(2014)433750_EN.pdf, (last accessed 17.10.2017)
40. EDA, 14 EDA Member States to pool & share GOVSATCOM capabilities, Brussels, 16 June 2017, https://www.eda.europa.eu/info-hub/presscentre/latest-news/2017/06/16/14-eda-memberstates-to-pool-share-govsatcom-capabilities, (last accessed 17.10.2017)
41. Presidency of the Spanish Government, National Defense Directive 2012 – For a necessary and responsible defense, Madrid, July 2012, https://www.eda.europa.eu/docs/defaultsource/documents/dgl-national-defense-directive-2012.pdf,(last accessed 17.10.2017)
42. European Parliament, Space, Sovereignty and European Security – Building European Capabilities in an advanced institutional framework, 2014 http://www.europarl.europa.eu/RegData/etudes/etudes/join/2014/433750/EXPOSEDE_ET(2014)433750_EN.pdf (last visited 25 September 2017)
43. CNES, Ambition 2020 – 2015, space for climate, http://www.cnes-csg.fr/automne_modules_files/standard/public/p11636_99f9990e67f7bbf293ff282b37d9bc1eCNES_plk_instit_2015_171214_GB.pdf (last accessed on 06.09.2017)
44. Stratégie Spatiale Française, https://cache.media.enseignementsuprecherche.gouv.fr/file/Politique_spatiale_francaise/09/8/Strategie_spatiale_francaise-marsBD_211098.pdf (last accessed on 06.09. 2017)
45. Presidency of the French Republic, The French White Paper on defense and national security, http://responsibilitytoprotect.org/Livre_blanc_Press_kit_english_version.pdf (last accessed 6 September 2017)
46. BMWI, Making Germany's space sector fit for the future - The space strategy of the German Federal Government, http://www.dlr.de/rd/en/Portaldata/28/Resources/dokumente/Raumfahrtstrategie_en.pdf (last accessed on 06.09.2017)
47. German Federal government, 2016 White Paper on German Security Policy and the Future of the Bundeswehr
48. ASI, Strategic Vision 2010-2020, http://www.asi.it/sites/default/files/ASI_DVS_2010_2020_ENG_0_0.pdf (last accessed on 6 September 2017).
49. Italian Ministry of Defense, White Paper for International security and defense, July 2015, https://www.difesa.it/Primo_Piano/Documents/2015/07_Luglio/White%20book.pdf (last accessed 6 September 2017).
50. Spanish Strategy for ESA and EU Space Programs 2007-2011, https://www.cdti.es/recursos/doc/Programas/Aeronautica_espacio_retornos_industriales/Espacio/2943_7_101110112011135137.pdf (last accessed on 06.09.2017).
51. The national Security Strategy of Spain, http://www.lamoncloa.gob.es/documents/estrategia_seguridad_baja_julio.pdf (last accessed on 06.09.2017).

52. HM Government, National Space Policy, https://www.gov.uk/government/uploads/system/uploads/attachment_data/file/484865/NSP_-_Final.pdf (last accessed on 06.09.2017).
53. IGS Steering Board and UKSA, Space Innovation and Growth Strategy 2014-2030 - Space Growth Action Plan, https://www.gov.uk/government/uploads/system/uploads/attachment_data/file/298362/igs-actionplan.pdf (last accessed on 06.09.2017)
54. HM National Space Security Policy, https://www.gov.uk/government/uploads/system/uploads/attachment_data/file/307648/National_Spac e_Security_Policy.pdf (last accessed on 06.09.2017)
55. HM National Security Strategy and Strategic Defense and Security Review 2015, https://www.gov.uk/government/uploads/system/uploads/attachment_data/file/478933/52309_Cm_9161_NSS_SD_Review_web_only.pdf (last accessed on 06.09.2017).
56. European Council, Conclusions, EUCO 217/13, 20 December 2013.
57. European External Action Service, Shared Vision, Common Action, A Stronger Europe – A Global Strategy for the European Union's Foreign and Security Policy, Brussels, 28 June 2016
58. EEAS, Security and Sustainability in Outer Space, https://eeas.europa.eu/headquarters/headquartershomepage/8466/security-and-sustainabilityouterspace_e
59. European Commission, Communication from the Commission to the European Parliament, the Council, the European Economic and Social Committee and the Committee of the regions on 'Space Strategy for Europe', COM(2016) 705, Brussels, 26 October 2016
60. European Commission, Communication from the Commission to the European Parliament, the Council, the European Economic and Social Committee and the Committee of the regions on 'Space Strategy for Europe', COM(2016) 705, Brussels, 26 October 2016
61. European Council, Conclusions, 19/20 December 2013; Council of the European Union, High Level Civil Military User Needs for Governmental Satellite Communications (GOVSATCOM), 7550/17, 22.03.17.
62. ESA, Status Report on ESA's security-related developments and activities, ESA/C(2017)62, 1 June 2017
63. European Commission, Space research – Horizon 2020, https://ec.europa.eu/growth/sectors/space/research/horizon-2020_en (last accessed on 29 September 2017)
64. European Parliament and the Council of the European Union, Decision No 541/2014/EU of 16 April 2014 establishing a Framework for Space Surveillance and Tracking Support, OJ L 158/227.
65. SatCen SSA, https://www.satcen.europa.eu/services/ssa (last accessed on 25.09.2017)
66. European defense action plan, http://europa.eu/rapid/press-release_IP-16-4088_en.htm (last accessed 25.09.)
67. European Parliament, Report on space capabilities for European security and defense (2015/2276(INI), http://www.europarl.europa.eu/sides/getDoc.do?pubRef=-//EP//TEXT+REPORT+A8-20160151+0+DOC+XML+V0//EN (last accessed 18.10.2017)
68. European Parliament, Report on a Space Strategy for Europe, (2016/2325(INI)), PE 602.728v03-00, A8-0250/2017, Committee on Industry, Research and Energy, 05.07.2017.
69. United Nations, Report of the World Commission on Environment and Development: Our Common Future, 20 March 1987, http://www.undocuments.net/our-common-future.pdf , (last accessed 06.10.2017)
70. See for instance: M. Uchitomi, Sustainable Development in Outer Space – Applicability of the concept of sustainable development to space debris problems, IAF/IISL, Proceedings of the forty-third colloquium on the law of outer space: 2-6 October 2000 (Rio de Janeiro, Brazil), Reston (Va.), American Institute of Aeronautics and Astronautics, 2001, 71-80, p. 75.
71. CNES, Ambition 2020 – 2015, space for climate, http://corporate.cnes.fr/plk_instit_2015_171214_GB.pdf, last visited 6 October 2017.

PROCEEDINGS OF THE 9TH INTERNATIONAL SPACE SAFETY CONFERENCE

SESSION 15
COMMERCIAL SPACEFLIGHT – II

AUTONOMOUS NAVIGATION USING GRAVITY GRADIENT MEASUREMENTS

Rachit Bhatia(1), David K. Geller(2)

(1)Graduate Student, 4130 Old Main Hill, Utah State University, Logan-84322, Utah (USA), Email: rachitbhatia@aggiemail.usu.edu
(2)Associate Professor, 4130 Old Main Hill, Utah State University, Logan-84322, Utah (USA), Email: david.geller@usu.edu

ABSTRACT

The future of deep space exploration depends upon technological advancement towards improving spacecraft's autonomy and versatility. This study examines the feasibility of orbit determination using gravity gradient measurements, and investigate its potential as a prospective solution for futuristic autonomous orbital navigation.

The long term objective of this research is to ascertain specific sensor requirements to meet pre-defined mission navigation error budgets. For this paper, a Monte Carlo simulation based on a simple three degree of freedom environment model and an extended Kalman filter are developed. While the results are low fidelity, they can be used as a guide for more detailed and complete analysis.

Traditional inertial navigation (dead reckoning and external aiding) is not considered. Instead, measurements from pairs of advanced, highly sensitive accelerometers (e.g. cold atom accelerometers), on three mutually perpendicular baselines, are used to determine gravity field gradients which are then correlated to onboard gravity maps and used to determine orbital information. This study is repeated for a variety of orbital regimes and initial conditions.

1. INTRODUCTION

The objective of this research is to determine the feasibility and requirements for an autonomous navigation that can potentially apply to all flight regimes. For any space mission, navigation relies primarily on external aids such as the Global Positioning Systems (GPS), the Tracking and Data Relay Satellite (TDRSS), or the Deep Space Network (DSN). This limits the range of our exploration capability, and requires specialized communication and ground-based navigation systems to achieve acceptable level of spaceflight safety. These additional systems not only require precious onboard resources, but are also subject to failures that can result in the Loss of Crew/Loss of Vehicle condition. For next generation space navigation, there is a need to relieve the traditional navigation techniques by implementing autonomous navigation system onboard and thus reduce the risk level of Loss of Crew/Loss of Vehicle condition.

Autonomous navigation with reduced dependence on ground-based infrastructure can provide the capability to better optimize and control mission trajectories. Enhanced automation and versatility can be beneficial for both crewed and robotic space missions, and may result in increased scientific productivity over mission timeline.[1]

This study evaluates the role of advanced accelerometers, used in recent gravity-mapping missions like GRACE-2 and GOCE, in developing and executing autonomous orbital navigation for different mission requirements. Future autonomous orbital navigation architectures need to be versatile and reliable for varying space environments. The navigation approach addressed in this study has the potential to satisfy these requirements and, considering the universal nature of gravity, this approach provides a generic solution for autonomous navigation in almost all types of space environment, thus giving it an edge over other potential autonomous orbital navigation techniques.

Gravity gradiometry has been in use since mid 20th century and has find applications in wide ranging fields like mineral exploration[2], field survey, submarine navigation[3]and gravitational mapping[4]. The technology has been used for many airborne and terrestrial surveys[3], predominantly to image subsurface geology to aid hydrocarbon and mineral exploration. Over 2.5-million-line km has been surveyed using the technique.[2] During the Cold War, US Navy submarines used gravity gradiometry for covert navigation.[3] In recent years, the technology has matured and requisite instruments have evolved and been upgraded. Because of this, there is renewed interest in space applications for this technique. Recently, engineers and scientists have used various measurement principles based on electrostatics, superconductivity, and cold atom interferometry, to considerably advance the measurement sensitivity and precision of accelerometers.[3]

This research explores the viability of using pairs of advanced accelerometers and onboard gravity field maps to autonomously determine orbital position and velocity for LEO, GEO, and other orbital regimes. The results presented in this paper have been generated for nearly-ideal conditions (not possible for real mission scenario), nevertheless the analysis of these results provide an important glimpse into the technological requirements for autonomous orbital navigation, and help contemplate a plan for future work required to develop the technology.

Development of autonomous orbital navigation can help break new ground in improving safety for human exploration, reducing complexity and cost of spacecraft navigation systems, and reducing costly efforts of traditional ground-based navigation. This study presents a step in this direction.

Section 2 provides an overview and history of the gravity gradiometry, Section 3 shows the derivation of the gravity gradient measurement model, Section 4 outlines the basic simulation, and Section 5 provide an overview of the extended Kalman filter. Simulation results and analysis are presented in Section 6. Conclusions and required future work are listed in Section 7 and Section 8.

Table 1: Gravity Gradiometer Instruments(GGI)[3]

Gradiometer	**Developer**	**Noise, $1-\sigma$ Eö**	**Data Rate, sec**
Rotating Accelerometer GGI	Bell Aerospace/Textron	2 (Lab.), 10 (Air)	10
Rotating Torque GGI	Hughes Research Lab	0.5 (Goal)	10
Floated GGI	Draper Lab	1 (Lab.)	10
Falcon AGG	Lockheed Martin(LM)/BHP Billiton	3	Post Survey
ACVGG	Lockheed Martin(LM)	1	1
3D FTG	LM/Bell Geospace	5	Post Survey
FTGeX	LM/ARKeX	10 (Goal)	1
UMD SGG (Space)	University of Maryland(UMD)	0.02 (Lab.)	1
UMD SAA (Air)	University of Maryland(UMD)	0.3 (Lab.)	1
UWA OQR	University of Western Australia(UWA)	1 (Lab.)	1
Exploration GGI	ARKeX	1 (Goal)	1
HD-AGG	Gedex/UMD/UWA	1 (Goal)	1
Electrostatic GGI	European Space Agency(ESA)	0.001 (Goal)	10
Cold Atom Interferometer	Stanford University/Jet Propulsion Laboratory(JPL)	30 (Lab.)	1

2. OVERVIEW OF GRAVITY GRADIOMETRY

The study and measurement of the changes in the gravitational acceleration, at a given position, requires the measurement of the gravity gradient tensor at that position. Gravity Gradiometer Instrument (GGI) is used to measure the gravity gradient tensor.

Hungarian physicist Loránd (Roland von) Eötvös is credited for inventing the first gravity gradiometer instrument, in the late 1880s.[5] While working on series of experiments on the proportionality of inertial and gravitational masses, Eötvös specialized torsion balance to measure gravitational gradient.[5] To recognize this ingenious invention, the unit of the gravitational gradient has been named after him.[5] One Eötvös is equal to $10^{-9}s^{-2}$.[6]

The gravity gradient tensor (GGT) is the 3x3 matrix, consisting of 9 components of the derivative of the gravitational vector with respect to position vector.

$$\nabla g = \begin{bmatrix} \nabla g_{XX} & \nabla g_{XY} & \nabla g_{XZ} \\ \nabla g_{XY} & \nabla g_{YY} & \nabla g_{YZ} \\ \nabla g_{XZ} & \nabla g_{YZ} & \nabla g_{ZZ} \end{bmatrix} \tag{1}$$

$$\nabla g_{ij} = \frac{\partial^2 U}{\partial \mathbf{r}_i \partial \mathbf{r}_j}, \; i,j = X,Y,Z \tag{2}$$

where U is the gravitational potential at the given position vector $\mathbf{r}$.

The gravity gradient matrix is symmetric ($\nabla g_{ij} = \nabla g_{ji}$) with zero trace ($\sum_i \nabla g_{ii} = 0$). Thus, only five out of nine components are independent.[4]

Table 1 summarizes the chronological evolution of the gravity gradiometer instruments and parallel improvement in their sensitivity, respectively.[3] This study is based upon the measurement model for electrostatic gravity gradiometer instrument, used onboard European Space Agency's GOCE (Gravity field and steady-state Ocean Circulation Explorer) mission.[4]

3. MEASUREMENT MODEL

Each gradiometer instrument comprises a configuration of 3 pairs of accelerometer, arranged on three mutually perpendicular baselines. And each accelerometer is designed to detect the difference between the acceleration of the spacecraft/spacecraft-frame and that of the proof-mass, by measuring the electrostatic force required to keep the proof-mass in the center of the accelerometer-frame.

An analysis of the forces acting on the the proof-mass helps to model the accelerometer measurement as a function of forces acting on the spacecraft. Acceleration of the proof-mass ($\bar{a}_{proof}$) as observed in inertial frame, can be given as

$$\mathbf{a}^I_{proof} = \mathbf{a}^I_{veh} + \omega^I \times \left(\omega^I \times \mathbf{R}^I_{com_i}\right) + \omega^I \times \mathbf{R}^I_{com_i} + 2\omega^I \times \dot{\mathbf{R}}^I_{rel_i} + \ddot{\mathbf{R}}^I_{rel_i} \quad (3)$$

where $\mathbf{a}^I_{proof}$ is the acceleration of the COM of the proof-mass, $\mathbf{a}^I_{veh}$ is the acceleration of the COM of the spacecraft, $\mathbf{R}^I_{com_i}$ is the position of the COM of the proof-mass of the accelerometer A_i relative to the COM of the spacecraft, ω^I is the angular velocity of the spacecraft, $\dot{\mathbf{R}}^I_{rel_i}$ is the velocity of the COM of the proof-mass as viewed relative to the rotating spacecraft-fixed frame, and $\ddot{\mathbf{R}}^I_{rel_i}$ is the acceleration of the COM of the proof-mass as viewed relative to the rotating spacecraft-fixed frame. All vectors with superscript 'I' are coordinatized in an inertial frame.

Using Newton's second law

$$\mathbf{F}^I_{veh} = m_{veh}\mathbf{a}^I_{veh} \quad (4)$$

$$\mathbf{F}^I_{proof} = m_{proof}\mathbf{a}^I_{proof} \quad (5)$$

where $\mathbf{F}^I_{veh}$ is the total force acting on the spacecraft, m_{veh} is mass of the spacecraft, $\mathbf{F}^I_{proof}$ is the total force acting on the proof-mass, m_{proof} is mass of the proof-mass within the cage,

Force analysis of the vehicle, yields

$$\mathbf{F}_{veh} = \mathbf{F}_{gravity_{veh}} + \mathbf{N} \quad (6)$$

where $\mathbf{F}_{gravity_{veh}}$ is the force acting on the spacecraft due to gravitation, and $\mathbf{N}$ is the force acting on the spacecraft due to non-gravitational forces, like drag.

And, force analysis of the proof-mass, yields

$$\mathbf{F}_{proof} = \mathbf{F}_{gravity_{proof}} + \mathbf{F}_{emf} \quad (7)$$

where $\mathbf{F}_{gravity_{proof}}$ is the force acting on the proof-mass due to gravitation, and $\mathbf{F}_{emf}$ is the electro-motive force acting on the proof-mass, to keep the proof-mass at the center of the cage.

From the free body analysis of the vehicle and the proof-mass, the forces acting on the vehicle and the proof-mass can be expressed as below

$$\mathbf{a}^I_{veh} = \mathbf{g}^I\left(\mathbf{R}^I_{veh}\right) + \frac{\mathbf{N}^I\left(\mathbf{R}^I_{veh}, \mathbf{V}^I_{veh}\right)}{m_{veh}} \quad (8)$$

$$\mathbf{a}^I_{proof} = \mathbf{g}^I\left(\mathbf{R}^I_{veh} + \mathbf{R}^I_{com_i}\right) + \left(\frac{\mathbf{F}^I_{emf}}{m_{proof}}\right) \quad (9)$$

where $\mathbf{R}^I_{veh}$ is the position of the COM of the spacecraft, $\mathbf{V}^I_{veh}$ is the velocity of the COM of the spacecraft, $\mathbf{g}^I\left(\mathbf{R}^I_{veh}\right)$ is the gravitational acceleration at the position $\mathbf{R}^I_{veh}$, and $\mathbf{g}^I\left(\mathbf{R}^I_{veh} + \mathbf{R}^I_{com_i}\right)$ is the gravitational acceleration at the position$\left(\mathbf{R}^I_{veh} + \mathbf{R}^I_{com_i}\right)$.

Since, the detected acceleration is the difference between the acceleration of the spacecraft/spacecraft-frame and the acceleration of the proof-mass. This means that by taking the difference between Eq. 8 and Eq. 9, and noting that the accelerometer measurement is proportional to $\left(\frac{\mathbf{F}^I_{emf}}{m_{proof}}\right)$, the detected acceleration is given as

$$\mathbf{a}^I_{detected} = \frac{\mathbf{F}^I_{emf}}{m_{proof}} = \mathbf{g}^I\left(\mathbf{R}^I_{veh}\right) - \mathbf{g}^I\left(\mathbf{R}^I_{veh} + \mathbf{R}^I_{com_i}\right) + \frac{\mathbf{N}^I\left(\mathbf{R}^I_{veh}, \mathbf{V}^I_{veh}\right)}{m_{veh}} + \omega^I \times \left(\omega^I \times \mathbf{R}^I_{com_i}\right) + \dot{\omega}^I \times \mathbf{R}^I_{com_i} + 2\omega^I \times \left(\dot{\mathbf{R}}\right)^I_{rel_i} + \ddot{\mathbf{R}}^I_{rel_i} \quad (10)$$

The difference of the terms $\mathbf{g}^I\left(\mathbf{R}^I_{veh}\right)$ and $\mathbf{g}^I\left(\mathbf{R}^I_{veh} + \mathbf{R}^I_{com_i}\right)$ is approximately (to first order) equal to $-\left(\nabla g^I\right)\mathbf{R}^I_{com_i}$, where ∇g^I is the gravity gradient at the position $\mathbf{R}^I_{com_i}$. Thus further simplification yields

$$\mathbf{a}^I_{detected} = -\left(\left[\nabla g^I - \left(\bar{\Omega}^2\right)^I - \dot{\bar{\Omega}}^I\right]\right)\mathbf{R}^I_{com_i} + \frac{\mathbf{N}^I\left(\mathbf{R}^I_{veh}, \mathbf{V}^I_{veh}\right)}{m_{veh}} + 2\bar{\Omega}^I\left(\dot{\mathbf{R}}\right)^I_{rel_i} + \ddot{\mathbf{R}}^I_{rel_i} \quad (11)$$

where ∇g^I is a matrix consisting of the components of the GGT, and Ω^I, $\dot{\Omega}^I$, and $\left(\Omega^2\right)^I$ are matrices, defined as

$$\Omega^I = \begin{bmatrix} 0 & -\omega_z & \omega_y \\ \omega_z & 0 & -\omega_x \\ -\omega_y & \omega_x & 0 \end{bmatrix}, \dot{\Omega}^I = \begin{bmatrix} 0 & -\dot{\omega}_z & \dot{\omega}_y \\ \dot{\omega}_z & 0 & -\dot{\omega}_x \\ -\dot{\omega}_y & \dot{\omega}_x & 0 \end{bmatrix},$$

$$\left(\Omega^2\right)^I = \begin{bmatrix} -\omega_z^2 - \omega_y^2 & \omega_x\omega_y & \omega_x\omega_z \\ \omega_x\omega_y & -\omega_z^2 - \omega_x^2 & \omega_y\omega_z \\ \omega_x\omega_z & \omega_y\omega_z & -\omega_x^2 - \omega_y^2 \end{bmatrix}$$

where ω_x, ω_y, ω_z are the components of the angular velocity of the spacecraft, relative to inertial frame, and $\dot{\omega}_x$, $\dot{\omega}_y$, $\dot{\omega}_z$ are their time derivatives (angular accelerations).

For this study, several important assumptions have been made. These assumptions have been used to create an hypothetical situation, so as to help generate a simplistic result and thereby, determine the feasibility of this concept. It is understood that such a hypothetical situation may never exist in reality, and thus for any practical use, further evaluation (with more realistic models) is required.

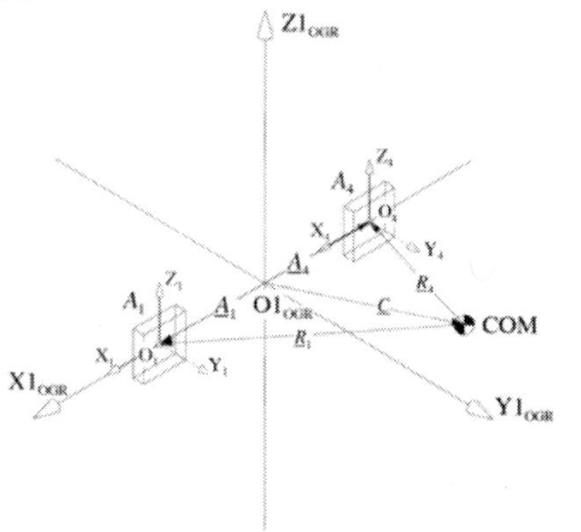

Figure 1: Schematic Model of a Gravity Gradiometer[4]

This study assumes a conceptual scenario with a non-rotating spacecraft. An ideal gradiometer is assumed, except for the measurement noise. The measurement noise has been assumed to be zero mean Gaussian, and a constant and uncorrelated noise has been applied to all the components of gravitational gradient matrix (∇g). However, based on the analysis of the problem, it is suspected that in reality, the noise on different components of gravitational gradient matrix (∇g)is correlated and should be taken into account, for more accurate modeling.

For this study, there exist no accelerometer bias, scale factor, and misalignment. It is assumed there is no center of mass uncertainty, and uncertainty in the accelerometer location. Lastly, the environmental model has no drag, solar radiation pressure or any other non-gravitational acceleration, like acceleration due to external magnetic field or self-gravity. Using the aforementioned assumptions and Eq. 11, to isolate the components of $\mathbf{N}$ and ∇g, common-mode acceleration and differential-mode acceleration, for each pair of accelerometer, are computed as[4]

$$\mathbf{a}^I_{c,ij} = \frac{1}{2}\left(\mathbf{a}^I_i + \mathbf{a}^I_j\right) \tag{12}$$

$$\mathbf{a}^I_{d,ij} = \frac{1}{2}\left(\mathbf{a}^I_i - \mathbf{a}^I_j\right) \tag{13}$$

where $\mathbf{a}^I_{c,ij}$ is common-mode acceleration measured by the accelerometer A_i and A_j, and $\mathbf{a}^I_{d,ij}$ is differential-mode acceleration measured by the accelerometer A_i and A_j, such that accelerometer A_i and A_j are the accelerometers on the same axis of the gradiometer, as shown in Fig. 1

On expanding and simplifying expressions for common-mode acceleration and differential-mode acceleration, important equations defining the measurement model are computed, and listed in Table 3

In Table 3, equations for the elements of gravitational gradient matrix (∇g) and components of angular acceleration have been approximated to first order.

Table 3: Measurement Model Equations[4]

	Measurement Model Equations
Components of Non-gravitational forces	$a_{c,14,X} = a_{c,25,X} = a_{c,36,X} = \mathbf{N}_X$ (14)
	$a_{c,14,Y} = a_{c,25,Y} = a_{c,36,Y} = \mathbf{N}_Y$ (15)
	$a_{c,14,Z} = a_{c,25,Z} = a_{c,36,Z} = \mathbf{N}_Z$ (16)
Elements of Gravitational Gradient Matrix (∇g)	$\nabla g_{XX} = -\frac{a_{d,14,X}}{A_{1,X}} - \omega_y^2 - \omega_z^2$ (17)
	$\nabla g_{YY} = -\frac{a_{d,25,Y}}{A_{2,Y}} - \omega_x^2 - \omega_z^2$ (18)
	$\nabla g_{ZZ} = -\frac{a_{d,36,Z}}{A_{3,Z}} - \omega_x^2 - \omega_y^2$ (19)
	$\nabla g_{XY} = \frac{1}{2}\left(-\frac{a_{d,14,Y}}{A_{1,X}} - \frac{a_{d,25,X}}{A_{2,Y}}\right) + \omega_x\omega_y$ (20)
	$\nabla g_{XZ} = \frac{1}{2}\left(-\frac{a_{d,14,Z}}{A_{1,X}} - \frac{a_{d,36,X}}{A_{3,Z}}\right) + \omega_x\omega_z$ (21)
	$\nabla g_{YZ} = \frac{1}{2}\left(-\frac{a_{d,25,Z}}{A_{2,Y}} - \frac{a_{d,36,Y}}{A_{3,Z}}\right) + \omega_y\omega_z$ (22)
Components of Angular Acceleration	$\dot{\omega}_x = \frac{1}{2}\left(\frac{a_{d,25,Z}}{A_{2,Y}} - \frac{a_{d,36,Y}}{A_{3,Z}}\right)$ (23)
	$\dot{\omega}_y = \frac{1}{2}\left(-\frac{a_{d,14,Z}}{A_{1,X}} + \frac{a_{d,36,X}}{A_{3,Z}}\right)$ (24)
	$\dot{\omega}_z = \frac{1}{2}\left(-\frac{a_{d,25,X}}{A_{2,Y}} + \frac{a_{d,14,Y}}{A_{1,X}}\right)$ (25)

4. SIMULATION MODELING

In this section, the simulation dynamics and measurement models are presented.

4.1 Dynamics

The true state vector is defined as

$$x^I = \begin{bmatrix} \mathbf{r}^I \\ \mathbf{v}^I \end{bmatrix} \tag{26}$$

where $\mathbf{r}^I$ and $\mathbf{v}^I$ are the true position and velocity of the spacecraft, respectively. Correspondingly, the dynamics for the truth model are given as

$$\dot{x}^I = f(x^I, t) = \begin{bmatrix} \dot{\mathbf{r}}^I \\ \dot{\mathbf{v}}^I \end{bmatrix} = \begin{bmatrix} \dot{\mathbf{v}}^I \\ \mathbf{g}(\mathbf{r}^I) \end{bmatrix} \tag{27}$$

where $\mathbf{g}(\mathbf{r}^I)$ is 18x18 spherical harmonics gravity model, and is computed analytically by evaluating the second derivative of gravitational potential U with respect to the position vector $\mathbf{r}$, expressed as an infinite series of spherical harmonics [7,8]

$$U = \frac{\mu}{r}\left[1 + \sum_{n=1}^{\infty}\sum_{m=0}^{n}\left(\frac{r_E}{r}\right)^n P_{n,m}(s_\lambda)\,(C_{n,m}cos\,(m\phi) + S_{n,m}sin\,(m\phi))\right] \tag{28}$$

where μ is the universal gravitational parameter, $\mathbf{r}$ is the position vector from a point O fixed in body E (say Earth) to a generic point Q, r denotes the magnitude of $\mathbf{r}$, r_E is a scaling radius for body E. $P_{n,m}$ is the associated Legendre function of the first kind, of degree n and order m, and has as its argument s_λ, the sine of λ, the latitude of Q. [7,8]The longitude of Q is denoted by ϕ. $C_{n,m}$ and $S_{n,m}$ are unnormalized gravitational coefficients of degree n and order m. [7,8]If point O is coincident with the mass center of E, then $C_{1,0}$, $C_{1,1}$, and $S_{1,1}$ all become zero. [7–9]

4.2 Measurements

For simulation modeling, the measurement model is the same as presented in Section 3, in this paper. Although, for easier computation, the measurement vector is rearranged using the elements of gravity gradient matrix, given as

$$\tilde{\mathbf{z}} = \nabla\mathbf{g}_{measured} + \eta \tag{29}$$

$$\tilde{\mathbf{z}} = \begin{bmatrix} \nabla g_{XX} & \nabla g_{YY} & \nabla g_{ZZ} & \nabla g_{XY} & \nabla g_{XZ} & \nabla g_{YZ} \end{bmatrix}^T + \eta \tag{30}$$

where $\tilde{\mathbf{z}}$ is the true measurement vector, η is the white gaussian noise (with standard deviation as $\sigma_\eta = 1E\ddot{o}$) added to the measurements, and R denotes the corresponding measurement covariance, respectively.

5. EXTENDED KALMAN FILTER FORMULATION

The Extended Kalman Filter for orbital navigation using gravity gradient measurements, is outlined below

5.1 Defining State Vector

The state vector is defined as

$$\hat{x}^I = \begin{bmatrix} \hat{\mathbf{r}}^I \\ \hat{\mathbf{v}}^I \end{bmatrix} \tag{31}$$

where $\hat{\mathbf{r}}^I$ and $\hat{\mathbf{v}}^I$ are the position and velocity of the spacecraft, respectively.

5.2 Initial State and State Covariance

The initial state is defined as the position and velocity of the spacecraft at the given epoch. The initial velocity error is computed using the following thumb rule

$$\sigma_v = \frac{2\pi}{T} * \sigma_r$$

where T is the orbital period of the spacecraft's orbit, σ_r and σ_v denote initial standard deviation of the position and velocity, respectively. Thus, the initial filter state vector ($\hat{x}_0$) and initial state covariance P_0 are defined as

$$\hat{x}_0^I = \begin{bmatrix} \hat{\mathbf{r}}_0^I & \hat{\mathbf{v}}_0^I \end{bmatrix}^T,\; P_0 = \begin{bmatrix} (P_0)_r & Z_{3\times3} \\ Z_{3\times3} & (P_0)_v \end{bmatrix} \tag{32}$$

where $(P_0)_r = \sigma_r^2 (I_{3x3})$ is initial position covariance matrix, and similarly, $(P_0)_v = \sigma_v^2 (I_{3\times3})$ is initial velocity covariance matrix, while I_{3x3} is a square identity matrix of size 3x3, and $Z_{3\times3}$ is a square matrix with all elements as zero, and has size of 3x3

5.3 Dynamics, State and Covariance Propagation Equations

The dynamics are non-linear and continuous, given as

$$\dot{x}_{DM}^I = \begin{bmatrix} \dot{\mathbf{r}}_{DM}^I \\ \dot{\mathbf{v}}_{DM}^I \end{bmatrix} = \begin{bmatrix} \mathbf{v}_{DM}^I \\ \mathbf{g}(\mathbf{r}_{DM}^I) \end{bmatrix} + \begin{bmatrix} \mathbf{0}_{3\times1} \\ \mathbf{w} \end{bmatrix} \tag{33}$$

where $\mathbf{g}(\mathbf{r}_{DM}^I)$ is 8x8 spherical harmonics gravity model, and the subscript DM denotes the design model.

Now, linearizing the dynamics as

$$\delta\dot{x}_{DM}^I = F\delta x_{DM}^I + G\mathbf{w} \tag{34}$$

$$G = \begin{bmatrix} 0_{3\times3} \\ I_{3\times3} \end{bmatrix} \tag{35}$$

$$F = \left.\frac{\partial f}{\partial \bar{x}}\right|_{\hat{x}_k^I} = \begin{bmatrix} 0_{3\times3} & I_{3\times3} \\ \left(\frac{\partial f}{\partial \bar{r}}\right)_{3\times3} & 0_{3\times3} \end{bmatrix} = \begin{bmatrix} 0_{3\times3} & I_{3\times3} \\ \nabla g_{3\times3} & 0_{3\times3} \end{bmatrix} \tag{36}$$

where $\hat{x}^I$ is the defined filter state vector, $\mathbf{w}$ is the process noise, and $\nabla g_{3\times3}$ is the gravity gradient at the given position.

The process noise is considered as a zero-mean stationary white noise, constituting the unmodeled perturbation accelerations.[10]

Once the state and state covariance has been initialized, the recursive propagation steps of the filter are defined and executed at each time step, by first computing the state estimation vector ($\hat{x}_k$), using the above defined dynamical equations, and then the state transition matrix is computed at the midpoint by

$$\phi = I_{6\times6} + F dt + F^2 \frac{(dt)^2}{2} \tag{37}$$

such that

$$F = \left.\frac{\partial f}{\partial \bar{x}}\right|_{\hat{x}_k^I} = \begin{bmatrix} 0_{3\times3} & I_{3\times3} \\ \nabla g_{3\times3} & 0_{3\times3} \end{bmatrix} \tag{38}$$

Next, the state covariance matrix is propagated using the Discrete Lyapunov Equation as

$$P_{k+1} = \phi_k P_k \phi_k^T + Q_{d,k} \tag{39}$$

where k is the time index in the recursive estimation process by the filter, $Q_{d,k}$ is the strength of the discrete white noise.

5.4 Measurement and update equations

The estimated measurement vector ($\hat{\mathbf{z}}$) is defined and linearized as

$$\hat{\mathbf{z}}_k = h(\hat{x}_k^I, t_k) + \boldsymbol{\eta}_k \tag{40}$$

$$H_k = \left.\frac{\partial \mathbf{h}}{\partial \mathbf{x}}\right|_{\hat{x}_k^I} = \begin{bmatrix} \frac{\partial \mathbf{h}}{\partial \mathbf{r}}_{6\times3} & \frac{\partial \mathbf{h}}{\partial \mathbf{v}}_{6\times3} \end{bmatrix} = \begin{bmatrix} \frac{\partial \mathbf{h}}{\partial \mathbf{r}}_{6\times3} & 0_{6\times3} \end{bmatrix} \tag{41}$$

where H_k is the measurement partial.

The measurement partials are computed analytically, using Eq. 42 and the measurement model (as presented in Section 3, in this paper). The complete derivation of the measurement partial is given in the Appendix. The measurement partial has been derived as

$$H_k = \begin{bmatrix} -\frac{2}{L_X}\left(\frac{\partial(A_{D,14,X})}{\partial \mathbf{r}}\right)_{1\times3} & \mathbf{0}_{1\times3} \\ -\frac{2}{L_Y}\left(\frac{\partial(A_{D,25,Y})}{\partial \mathbf{r}}\right)_{1\times3} & \mathbf{0}_{1\times3} \\ -\frac{2}{L_Y}\left(\frac{\partial(A_{D,36,Z})}{\partial \mathbf{r}}\right)_{1\times3} & \mathbf{0}_{1\times3} \\ \left(-\frac{1}{L_Y}\left(\frac{\partial(A_{D,25,X})}{\partial \mathbf{r}}\right)_{1\times3} - \frac{1}{L_X}\left(\frac{\partial(A_{D,14,Y})}{\partial \mathbf{r}}\right)_{1\times3}\right) & \mathbf{0}_{1\times3} \\ \left(-\frac{1}{L_X}\left(\frac{\partial(A_{D,14,Z})}{\partial \mathbf{r}}\right)_{1\times3} - \frac{1}{L_Z}\left(\frac{\partial(A_{D,36,X})}{\partial \mathbf{r}}\right)_{1\times3}\right) & \mathbf{0}_{1\times3} \\ \left(-\frac{1}{L_Z}\left(\frac{\partial(A_{D,36,Y})}{\partial \mathbf{r}}\right)_{1\times3} - \frac{1}{L_Y}\left(\frac{\partial(A_{D,25,Z})}{\partial \mathbf{r}}\right)_{1\times3}\right) & \mathbf{0}_{1\times3} \end{bmatrix} \tag{42}$$

The measurement residual is defined as

$$\text{residual} = \tilde{\mathbf{z}}_k - \hat{\mathbf{z}}_k \tag{43}$$

Next, residual covariance matrix V_k is computed using the measurement partial, the measurement covariance and the state covariance, as [11]

$$V_k = H_k P_k^- H_k^T + R_k \tag{44}$$

where R_k is the measurement covariance (as defined in Section 4.2)

Further, the Kalman gain K_k is given as [11]

$$K_k = P_k^- H_k^T (V_k)^{-1} \tag{45}$$

Using the Kalman gain and measurement residual, the state vector is updated as [11]

$$\hat{x}_k^+ = \hat{x}_k^- + K_k (\tilde{\mathbf{z}}_k - \hat{\mathbf{z}}_k)^T \tag{46}$$

And the state covariance matrix (P_k) is computed using Joseph formulation, as [11]

$$P_k^+ = (I_{6\times6} - K_k H_k) P_k^- (I_{6\times6} - K_k H_k)^T + K_k R_k K_k^T \tag{47}$$

To ensure that the state covariance matrix remains symmetric

$$P_k^+ = \frac{P_k^+ + P_k^{+T}}{2} \tag{48}$$

Finally, the new measurement residual and residual covariance matrix are computed and the procedure is repeated for the next time step. [11]

6. SIMULATION RESULTS AND ANALYSIS

Simulation results have been generated for varying important modeling parameters like filter gravity model, measurement noise, measurement frequency, initial position error, gradiometer baseline length, and orbital regime. This is to achieve the objective of determining the role and importance of each modeling parameter in designing an autonomous orbital navigation system. This analysis will help to decide modeling parameters and sensitivity requirements for achieving stipulated mission objectives. Nominal parameters have been listed in Table 5

6.1 Effect of varying Filter Gravity Model

In this section, the effect of filter gravity model, varied from a simple two body model to 18x18 spherical harmonics gravity model, is investigated. The results have been generated using nominal modeling parameters (as given in Table 5), while varying the degree and order of the spherical harmonics gravity model used in the filter.

It was determined that there is negligible change in 3σ error with respect to the change in the filter gravity model, except for the case when the filter gravity model is set to two body. And, the filter performance is nearly the same for any filter gravity model, above two body model.

In Fig. 2, the filter seems to be tuned for filter gravity model set at two-body plus J2, with 18x18 true gravity model, and noise level of 1Eö $1-\sigma$. This result can help conserve onboard resources, because this means that under specific conditions, a low resolution filter gravity model can be used onboard while ensuring reasonable estimation of filter states. These results have been verified using the Monte Carlo Analysis, and the 3σ errors have been found to be within 95% confidence interval.

Table 5: Nominal Parameters

Parameter	Nominal Value
Altitude	400,000 m
Eccentricity	0.01
Inclination	30 deg
Right Ascension of the Ascending Node	30 deg
Argument of Perigee	90 deg
True Anomaly	0
Integration step-size	30 sec
Number of Orbits	2
Measurement Frequency	2 per min
Truth Gravity Model	18x18
Filter Gravity Model	8x8
1σ Initial Position uncertainty	10,000 m
1σ Initial Velocity uncertainty	10 m/s
1σ Measurement Noise	1Eö
Gradiometer Baseline Length	1 m
Process Noise	0

6.2 Effect of varying Measurement Noise

The noise levels were varied from 0.0001Eö 1-σ to 100Eö 1-σ, at a factor 10, to record the effects of varying noise levels on the filter performance. The results have been generated using nominal modeling parameters (as given in Table 5), while measurement noise level is varied. Results have been generated over five orbital periods. It was observed that the 3σ error is directly proportional to the measurement noise levels.

After 5 orbital periods, 3σ position error for 0.0001Eö 1-σ was observed to be approximately 4 cm, while the corresponding 3σ velocity error being 0.004 cm/s. This contrasts to 3σ position error of 36.7 km and 3σ velocity error of 38.9 m/s for 100Eö 1-σ, while for 1Eö 1-σ, 3σ position error of 396 m and 3σ velocity error of 0.437 m/s is recorded.

These results have been verified using Monte Carlo analysis, and the 3σ errors have been found to be within 95% confidence interval. In Fig. 3, the LVLH True Navigation error for 100 Monte Carlo runs for 0.1Eö 1-σ measurement noise, with measurement frequency of one measurement every 10 seconds, using 8x8 truth gravity model and 8x8 filter gravity model, and other nominal parameters (as given in Table 5) are shown. Similar performance has been recorded, when Monte Carlo results have been analyzed for other measurement noise levels.

6.3 Effect of varying Measurement Frequency

In this case, the effect of measurement frequency, varying from one measurement every 2s to one measurement every 300s, are studied. The results have been generated using nominal modeling parameters (as given in Table 5), while measurement frequency is varied. For all cases, a fixed integration time-step equal to 2 sec is used.

After 2 orbital periods, 3σ position error for no measurements has been noted to be about 1550.9 km, while corresponding 3σ velocity error being 1.77 km/s. This contrasts to 3σ position error of 1.933 km and 3σ velocity error of 2.1375 m/s for one measurement every 5 minutes, while for one measurement every 2 seconds (time step), 3σ position error of 162.3 m and 3σ velocity error of 0.179 m/s is recorded.

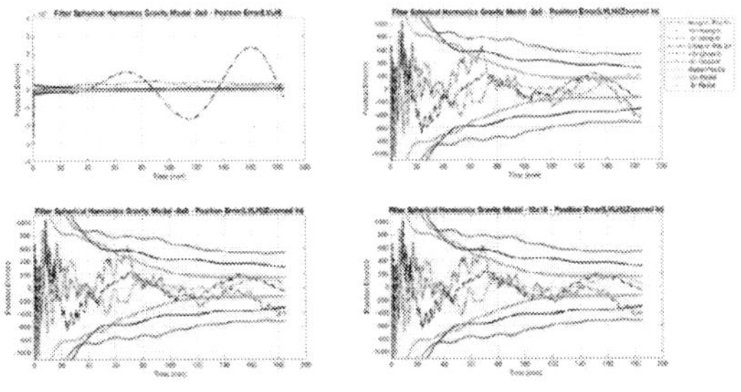

Figure 2: Actual and 3σ LVLH position error as a function of time, for filter gravity model with 2 X 0 spherical harmonics

As expected, an increase in 3σ error is seen, as the measurement update frequency is increased or as the measurements are delayed, such that the total 3σ position error for 300 second update frequency is almost 5.5 times the error for 10 second update frequency. This is because, with the increase in update frequency, the measurement information is delayed and dispersed over a larger time span, thus delaying the filter update and worsening the estimate over time. However, the convergence of the filter in every case, verifies and justifies the satisfactory performance of the filter.

6.4 Effect of varying Gradiometer Baseline Length

In this case, the change in filter performance is recorded after varying the gradiometer baseline length. The results have been generated using nominal modeling parameters (as given in Table 5), while gradiometer baseline length is varied.

From this study, it is noted that the 3σ error changes are absolutely negligible with respect to the change in the baseline length. This result is helpful to identify that for any practical application, and having a constraint on physical length, gradiometer baseline length cannot be

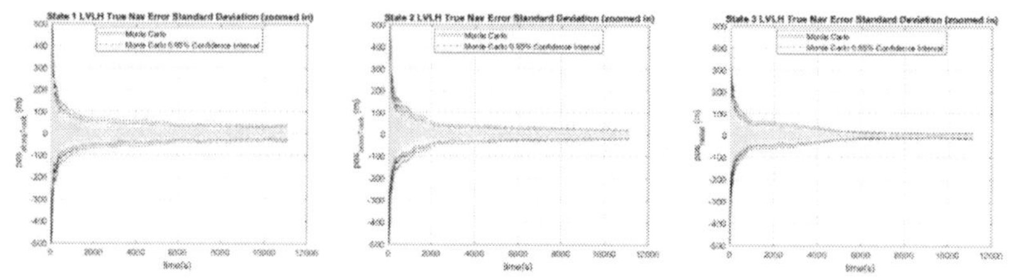

(a) LVLH true navigation position error (100 samples) and standard deviation as a function of time, for measurement noise: 0.1Eö

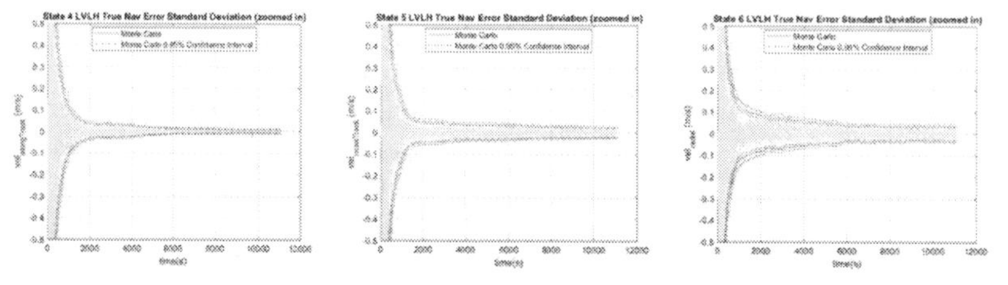

(b) LVLH true navigation velocity error (100 samples) and standard deviation as a function of time, for measurement noise: 0.1Eö

Figure 3: LVLH True Navigation error for 100 Monte Carlo runs, using 0.1Eö 1-σ measurement noise, 8x8 truth gravity model and measurement frequency of 10 sec (with other nominal parameters)

considered a significant modeling parameter for designing an autonomous orbital navigation (using gravity gradient measurements). However, it is acknowledged that for high precision in estimation, the measured baseline length must be known very accurately.

6.5 Effect of varying Orbital Regime

The results have been generated for different orbital regime, using the nominal parameters (as given in Table 5). Fixed-time step of integration has been varied for different orbits (as given in Table 6).

Results have been generated over two orbital periods, using 1500 measurements, and for a measurement frequency of one measurement every time-step. The orbital specification for different orbital regimes, implemented in the simulation are given in Table 6

6.5.1 For Measurement Noise of 1 Eö $1-\sigma$

In this case, the measurements have been processed for measurement noise of 1Eö $1-\sigma$, and it has been found that the performance of the proposed concept has been nominal for low earth orbit, such as GOCE orbit, however, the 3σ error increases for higher altitude orbits. With measurement noise of 1Eö $1-\sigma$, and after 2 orbital periods, 3σ position error for GOCE orbit has been noted to be about 283.24 m, while corresponding 3σ velocity error being 0.3129 m/s. This contrasts to 3σ position error of 614.32 m and 3σ velocity error of 0.63438 m/s for Vanguard-1 rocket. Similarly, for Ariane 5 orbit, 3σ position error of 2229.1 m and 3σ velocity error of 1.4872 m/s is recorded, while for SES-16 GTO orbit, 3σ position error is 49641 m and 3σ velocity error is 9.6563 m/s.

6.5.2 For Measurement Noise of 0.1 Eö $1-\sigma$

In this case, the measurements have been processed for measurement noise of 0.1Eö $1-\sigma$. The 3σ error has been noted to vary by about a magnitude, when measurement noise is set to 0.1 Eö $1-\sigma$, instead of 1 Eö $1-\sigma$.

Param.	SES-15	Vanguard1	Ariane-5	GOCE
a*	34,874,137 meters	8,817,137 meters	24,247,137 meters	6,494,637 meters
e*	0.4399832	0.2024625	0.7256533	0.000983
i*	2.3115°	34.2687°	3.8115°	96.5717°
Ω*	217.0247°	177.2534°	218.7161°	344.5256°
ω*	214.6288°	319.0597°	331.6670°	296.2811°
ν*	0	0	0	0
dt*	86.4758 s	10.9934 s	50.1337 s	6.9498 s
Total Steps	1500	1500	1500	1500

* a - Semi-major axis, e - Eccentricity, i - Inclination,
Ω - Ascending node, ω - Argument of perigee,
ν - True Anomaly, dt - Integration time-step

Table 6: Orbital Specifications

With measurement noise of 0.1Eö $1-\sigma$, and after 2 orbital periods, 3σ position error for GOCE orbit has been noted to be about 28.325 m, while corresponding 3σ velocity error being 0.03129 m/s. This contrasts to 3σ position error of 61.443 m and 3σ velocity error of 0.0635 m/s for Vanguard-1 rocket. Similarly, for Ariane 5 orbit, 3σ position error of 230.84 m and 3σ velocity error of

0.1523 m/s is recorded, while for SES-16 GTO orbit, 3σ position error is 5757.7 m and 3σ velocity error is 1.431 m/s. These results have been verified using the Monte Carlo analysis, and the 3σ errors have been found to be within 95% confidence interval.

In Fig. 4, LVLH True Navigation error for 50 Monte Carlo runs for SES-15 orbit, using 0.1Eö 1-σ measurement noise, 15 seconds of integration step-size, and measurement frequency of 60 sec, with 8x8 Filter Gravity Model, and other nominal parameters (as given in Table 5) have been shown. Similar performance has been recorded, when Monte Carlo results have been analyzed for other orbital regimes.

7. CONCLUSION

The objective of this study was to provide a starting point for more rigorous analysis of the complex problem to design an autonomous orbital navigation system. The results, presented in this paper, have been helpful in recognizing the important modeling parameters and acknowledging the advantage, disadvantage and corresponding influence on the filter performance.

The characteristics of some of the important modeling parameters and their inter-relationship have been attempted and recorded, during this study. Though the results, for this study, are only conceptual and based on a very simple environmental model, they do provide an insight into the complexity of designing an autonomous orbital navigation system.

8. FUTURE WORK

This study has been important in recognizing and validating the efficacy and the potential of the proposed concept. However, the model presented here is very basic and low fidelity. Thus, further study and analysis of the proposed concept is required.

There is a need to recognize and include inherent uncertainties in the system and environment, and test the measurement model with prominent sources of error. This help to identify the sensitivity requirements and optimal sensor configurations, for autonomous orbital navigation.

9. ACKNOWLEDGEMENT

This is to acknowledge that partial financial support has been provided by the International Space Safety Foundation.

10. APPENDIX - MEASUREMENT PARTIAL DERIVATION

From the measurement model (as presented in Section 3, in this paper), it is known that

$$\nabla g_{XX} = -2\frac{A_{D,14,X}}{L_X} - \omega_y^2 - \omega_z^2 \tag{49}$$

Here, Eq. 49 is a first order approximation. On differentiating both sides of Eq. 49 with respect to position vector, yields

$$\left(\frac{\partial \nabla g_{XX}}{\partial \mathbf{r}}\right)_{1\times3} = -\frac{2}{L_X}\left(\frac{\partial (A_{D,14,X})}{\partial \mathbf{r}}\right)_{1\times3} - \frac{\partial\left(\omega_y^2\right)}{\partial \mathbf{r}} - \frac{\partial\left(\omega_z^2\right)}{\partial \mathbf{r}} \tag{50}$$

Since $\omega^I = \begin{bmatrix}\omega_x & \omega_y & \omega_z\end{bmatrix}^T = \begin{bmatrix}0 & 0 & 0\end{bmatrix}^T rad/sec$, this implies

$$\left(\frac{\partial \nabla g_{XX}}{\partial \mathbf{r}}\right)_{1\times3} = -\frac{2}{L_X}\left(\frac{\partial (A_{D,14,X})}{\partial \mathbf{r}}\right)_{1\times3} \tag{51}$$

From the definition of the differential-mode accelerations

$$\mathbf{A}_{D,14} = \frac{(\mathbf{A}_1 - \mathbf{A}_4)}{2} \tag{52}$$

In Eq. 52, differentiating both sides with respect to position vector, yields

$$\left(\frac{\partial (\mathbf{A}_{D,14})}{\partial \mathbf{r}}\right)_{3\times3} = \left(\frac{\partial \left(\frac{(\mathbf{A}_1-\mathbf{A}_4)}{2}\right)}{\partial \mathbf{r}}\right)_{3\times3} = \frac{1}{2}\left(\left(\frac{\partial \mathbf{A}_1}{\partial \mathbf{r}}\right)_{3\times3} - \left(\frac{\partial \mathbf{A}_4}{\partial \mathbf{r}}\right)_{3\times3}\right) \tag{53}$$

where $\left(\frac{\partial \mathbf{A}_1}{\partial \mathbf{r}}\right)_{3\times3}$ and $\left(\frac{\partial \mathbf{A}_4}{\partial \mathbf{r}}\right)_{3\times3}$ are the gravity gradient tensors at the position of Accelerometer A_1 and A_4, respectively. And, thus $\left(\frac{\partial (A_{D,14,X})}{\partial \mathbf{r}}\right)_{1\times3}$ is the first row of $\left(\frac{\partial (\mathbf{A}_{D,14})}{\partial \mathbf{r}}\right)_{3\times3}$. Similarly, partial of the differential-mode 25 and 36 are given as

$$\left(\frac{\partial (\mathbf{A}_{D,25})}{\partial \mathbf{r}}\right)_{3\times3} = \frac{1}{2}\left(\left(\frac{\partial \mathbf{A}_2}{\partial \mathbf{r}}\right)_{3\times3} - \left(\frac{\partial \mathbf{A}_5}{\partial \mathbf{r}}\right)_{3\times3}\right) \tag{54}$$

$$\left(\frac{\partial (\mathbf{A}_{D,36})}{\partial \mathbf{r}}\right)_{3\times3} = \frac{1}{2}\left(\left(\frac{\partial \mathbf{A}_3}{\partial \mathbf{r}}\right)_{3\times3} - \left(\frac{\partial \mathbf{A}_6}{\partial \mathbf{r}}\right)_{3\times3}\right) \tag{55}$$

where $\left(\frac{\partial \mathbf{A}_2}{\partial \mathbf{r}}\right)_{3\times3}$ and $\left(\frac{\partial \mathbf{A}_5}{\partial \mathbf{r}}\right)_{3\times3}$ are the gravity gradient tensors at the position of Accelerometer A_2 and A_5, and $\left(\frac{\partial \mathbf{A}_3}{\partial \mathbf{r}}\right)_{3\times3}$ and $\left(\frac{\partial \mathbf{A}_6}{\partial \mathbf{r}}\right)_{3\times3}$ are the gravity gradient tensors at the position of Accelerometer A_3 and A_6, respectively. Using the formulation as given in Eq. 49-55, the measurement partial is given as

$$H_k = \begin{bmatrix} \left(\frac{\partial \nabla g_{XX}}{\partial \mathbf{r}}\right)_{1\times3} & \mathbf{0}_{1\times3} \\ \left(\frac{\partial \nabla g_{YY}}{\partial \mathbf{r}}\right)_{1\times3} & \mathbf{0}_{1\times3} \\ \left(\frac{\partial \nabla g_{ZZ}}{\partial \mathbf{r}}\right)_{1\times3} & \mathbf{0}_{1\times3} \\ \left(\frac{\partial \nabla g_{XY}}{\partial \mathbf{r}}\right)_{1\times3} & \mathbf{0}_{1\times3} \\ \left(\frac{\partial \nabla g_{XZ}}{\partial \mathbf{r}}\right)_{1\times3} & \mathbf{0}_{1\times3} \\ \left(\frac{\partial \nabla g_{YZ}}{\partial \mathbf{r}}\right)_{1\times3} & \mathbf{0}_{1\times3} \end{bmatrix} \tag{56}$$

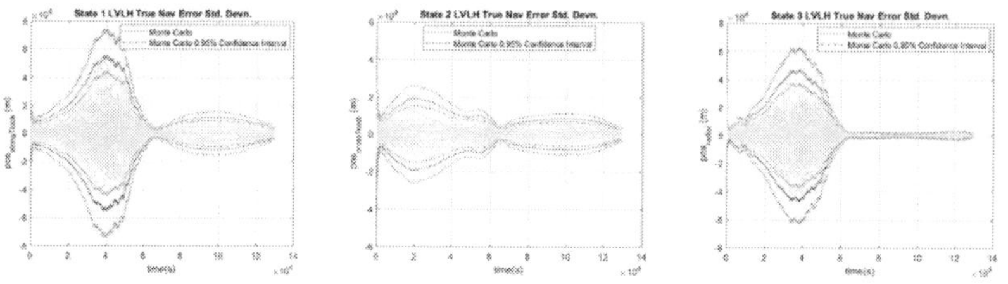

(a) LVLH true navigation position error (50 samples) and standard deviation as a function of time, for measurement noise: 0.1Eö

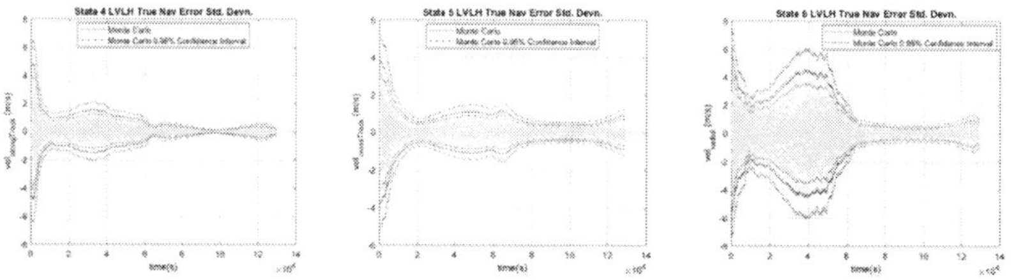

(b) LVLH true navigation velocity error (50 samples) and standard deviation as a function of time, for measurement noise: 0.1Eö

Figure 4: LVLH True Navigation error for 50 Monte Carlo runs for SES-15 orbit, using 0.1Eö 1-σ measurement noise, 15 seconds of integration step-size, and measurement frequency of 60 sec (with other nominal parameters)

$$H_k = \begin{bmatrix} -\frac{2}{L_X}\left(\frac{\partial(A_{D,14,X})}{\partial \mathbf{r}}\right)_{1\times 3} & \mathbf{0}_{1\times 3} \\ -\frac{2}{L_Y}\left(\frac{\partial(A_{D,25,Y})}{\partial \mathbf{r}}\right)_{1\times 3} & \mathbf{0}_{1\times 3} \\ -\frac{2}{L_Y}\left(\frac{\partial(A_{D,36,Z})}{\partial \mathbf{r}}\right)_{1\times 3} & \mathbf{0}_{1\times 3} \\ \left(-\frac{1}{L_Y}\left(\frac{\partial(A_{D,25,X})}{\partial \mathbf{r}}\right)_{1\times 3} - \frac{1}{L_X}\left(\frac{\partial(A_{D,14,Y})}{\partial \mathbf{r}}\right)_{1\times 3}\right) & \mathbf{0}_{1\times 3} \\ \left(-\frac{1}{L_X}\left(\frac{\partial(A_{D,14,Z})}{\partial \mathbf{r}}\right)_{1\times 3} - \frac{1}{L_Z}\left(\frac{\partial(A_{D,36,X})}{\partial \mathbf{r}}\right)_{1\times 3}\right) & \mathbf{0}_{1\times 3} \\ \left(-\frac{1}{L_Z}\left(\frac{\partial(A_{D,36,Y})}{\partial \mathbf{r}}\right)_{1\times 3} - \frac{1}{L_Y}\left(\frac{\partial(A_{D,25,Z})}{\partial \mathbf{r}}\right)_{1\times 3}\right) & \mathbf{0}_{1\times 3} \end{bmatrix} \tag{57}$$

REFERENCES

1. Board, S. S., Council, N. R., et al., *Vision and voyages for planetary science in the decade 2013-2022*, National Academies Press, 2012.
2. DiFrancesco, D., Meyer, T., Christensen, A., and FitzGerald, D., "Gravity gradiometry–today and tomorrow," *11th SAGA Biennial technical meeting and exhibition*, 2009.
3. Richeson, J. A., *Gravity gradiometer aided inertial navigation within non-GNSS environments*, phdthesis, University of Maryland, College Park, 2008, Copyright - Database copyright ProQuest LLC; Last updated - 2016-05-27.
4. Cesare, S., "Performance requirements and budgets for the gradiometric mission. Technical Note," Tech. rep., GOC-TN-AI-0027, Alenia Spazio, Turin, Italy, 2002.
5. Király, P., "Eötvös and STEP," *Proceedings of the Symposium STEP (esa WPP-115, Noordwijk, The Netherlands)*, July July 1996, pp. pp 399–406.
6. Gray, S. D., Parmentola, J. A., and LeSchack, R., "Estimating the weight of very heavy objects with a gravity gradiometer," *Journal of Physics D: Applied Physics*, Vol. 28, No. 11, 1995, pp. 2378.
7. Roithmayr, C. M., "Contribution of zonal harmonics to gravitational moment," 1990.
8. Roithmayr, C. M., "Contributions of spherical harmonics to magnetic and gravitational fields," 2004.
9. Takahashi, Y., Busch, M. W., and Scheeres, D., "Spin state and moment of inertia characterization of 4179 Toutatis," *The Astronomical Journal*, Vol. 146, No. 4, 2013, pp. 95.
10. Sun, X., Chen, P., Macabiau, C., and Han, C., "Autonomous orbit determination via kalman filtering of gravity gradients," *IEEE Transactions on Aerospace and Electronic Systems*, Vol. 52, No. 5, October 2016, pp. 2436–2451.
11. Crassidis, J. L. and Junkins, J. L., *Optimal Estimation of Dynamic Systems, Second Edition (Chapman & Hall/CRC Applied Mathematics & Nonlinear Science)*, Chapman & Hall/CRC, 2nd ed., 2011.

CHALLENGES OF DETERMINING "SAFE ENOUGH" IN HUMAN SPACE FLIGHT

Robert P. Ocampo[(1)] and David M. Klaus[(2)]

[(1)]*University of Colorado Boulder, 429 UCB Boulder Colorado 80309 , robert.ocampo@colorado.edu*
[(2)] *University of Colorado Boulder, 429 UCB Boulder Colorado 80309 , klaus@colorado.edu*

ABSTRACT

No spacecraft will ever be perfectly safe. Consequently, engineers must strive to design, develop, and operate spacecraft that are safe *enough*. But how safe is safe enough? The process of answering this question can be difficult and contentious, as the inherent uncertainties associated with defining and measuring "safe" are complicated by the subjective challenges of determining "enough." These complications, which include *uncertainty in terminology, subjectivity in the choice of metrics, uncertainty in the measurement itself,* and *subjectivity in the acceptance of the measurement*, must be eliminated, circumvented, or accounted for in order for "safe enough" to be objectively evaluated. This article describes these complications and presents recommendations in the hope that such discussion may help to facilitate the evaluation of "safe enough" in future spacecraft.

1. INTRODUCTION

Weeks before he would die in the tragic Apollo 1 fire, astronaut Gus Grissom told a reporter:

"If we die, we want people to accept it. We're in a risky business, and we hope if anything happens to us, it will not delay the program. The conquest of space is worth the risk of life."

To this day, space flight remains a "risky business". Recent accidents, including the loss of Orbital ATK's Cygnus spacecraft, the destruction of Space X's Dragon capsule, and the in-flight death of a Virgin Galactic SpaceShipTwo test pilot serve to underscore this point. But how risky is *too* risky? Conversely, how safe is safe enough?

These questions—and others like it—have surrounded the U.S. space program since its inception [1-10]. They remain relevant today because no comprehensive "right" answer exists: Different programs, flying different missions of various durations, may be willing to accept varying definitions and levels of risk, as well as varying degrees of uncertainty within the risk assessment [11]. The original Atlas booster was test flown 73 times before it was considered "safe enough" for crewed orbital flight; conversely, the Saturn V was test flown just *twice* before it sent humans to the moon. Consequently, the question of "how safe is safe enough" must be continuously addressed, on both a program-by-program and flight-by-flight basis. This can be a difficult and contentious process, as the inherent uncertainties associated with defining and measuring "safe" are often complicated by the subjective challenges of establishing "enough." These complications—*uncertainty in terminology, subjectivity in the choice of metrics, uncertainty in the measurement itself,* and *subjectivity in the acceptance of the measurement*—are described below in the hope that such discussion may help to facilitate the evaluation of "safe enough" in future spacecraft.

2. COMPLICATIONS

2.1. Uncertainty in Terminology

The Aerospace Safety Advisory Panel (ASAP)—the independent group tasked with evaluating NASA's safety performance—has stated that the one of the primary obstacles to determining "safe enough" stems from the ambiguous use of the term "safe" in the English lexicon. In their 1978 annual report, they wrote:

"The very nature of safety determinations and the wide-spread confusion about the nature of safety decisions would be dispelled if the very meaning of the term were clarified" [4].

This task of clarifying terminology is more complicated than a cursory dictionary definition would suggest, as the terms "safe" and "unsafe"—though *linguistically* antithetical—are often readily (if unknowingly) applied to the *same* spacecraft. Consider the case of two theoretical spacecraft: *Spacecraft A* and *Spacecraft B*. *Spacecraft A* has flown for 40 years without a single accident, while *Spacecraft B* suffered a catastrophic accident in its first flight. When asked to label these vehicles as "safe" or "unsafe", engineers readily characterize *Spacecraft A* as "safe" and *Spacecraft B* as "unsafe". However, both descriptions are derived from the same vehicle—in this case, the Soyuz spacecraft [12]. Such paradoxical labeling also appears to be present in descriptions of the Space Shuttle [11], suggesting the terms "safe" and "unsafe" are actually context dependent, rather than (entirely) vehicle-specific.

These examples highlight just how difficult it can be to properly characterize crewed space systems. Labeling a spacecraft as "safe" is technically inaccurate, as no spacecraft can ever be free of catastrophic hazards (9, 11, 13-14]. Conversely, designating a spacecraft as "unsafe" implies that the vehicle cannot or should not be flown. These difficulties extend beyond a simple linguistic challenge, as using one of these terms over the other can potentially mean the difference between program viability and cancellation.

To resolve this dilemma, we have taken to describing "unsafe" in terms of *degrees*—articulated here and elsewhere as "risk" [11, 15]. Under this framework, if risk is determined to be sufficiently low, the spacecraft can be considered "safe enough." Conversely, if risk is determined to be unacceptably high, it can be rejected as "not safe enough"[1].

2.2. Subjectivity in the Choice of Metrics

Given these definitions, a metric for measuring spacecraft risk can now be selected. This metric serves to define the method of measurement, as well as the reference units for risk (e.g. number of successful launches, levels of failure tolerance, or performance in a flight readiness review [e.g. pass vs. fail]).

Selecting a metric to measure risk can be a difficult process. Unlike other physical variables, such as mass or length, risk cannot be measured empirically; it must be abstracted from the spacecraft and its interaction with the environment [16]. How this abstraction should proceed is ultimately a subjective choice (albeit one based on objective benefits). Should it based on the rate of successful launches? Or should it be quantified using probabilistic calculations? Or should some other metric be applied?

Apollo engineers were "deep[ly] and irreconciab[ly]" divided as to whether risk should be measured statistically or qualitatively [3]. Later Space Shuttle engineers faced a similar disagreement regarding the use of Probabilistic Risk Analysis (PRA) (*statistical*) vs. Failure Modes and Effects Analysis (FMEA) (*qualitative*) as the primary means of measuring risk post-*Challenger* [13, 17-19].

[1] Although there remains a certain degree of subjectivity to this choice of definitions (despite its heavy reliance on NASA terminology), they are employed here to ensure the discussion can proceed; without a set of firm definitions, the remaining complications (though largely independent of the definitions) cannot be effectively described.

These disputes were and remain prevalent because no metric can ever serve as a perfect proxy for risk. In order for statistical metrics to be reliable, the spacecraft must be tested thousands (if not millions) of times at both the component and system level. As some opponents have argued, this time may be better spent "searching for design flaws than mindlessly running tests" [3]. On the other hand, measuring risk based on qualitative engineering judgments tends to produce widely divergent assessments of risk. Prior to the *Challenger* accident, experts qualitatively estimated the probability of a catastrophic Space Shuttle accident to range between 1 in 100 and 1 in 100,000—*a range of three orders of magnitude* [17].

Even relatively simple metrics, such as dividing the number of successful launches by the number of total launches, cannot serve as perfect proxies for risk. Consider a spacecraft that has been successfully launched one time (and one time only). This spacecraft would have a mathematically perfect safety record, but could not in good conscience be described as perfectly "safe" [20].

To date, a consensus method for evaluating risk remains undefined in the United States. NASA currently places special emphasis on prospective metrics, such as Probabilistic Risk Assessment (PRA), to measure risk in its Commercial Crew Program (CCP) [21]. This contrasts with FAA regulations, which specify the use of retrospective, actuarial metrics to define vehicle safety records [22].

2.3. Uncertainty in the Measurement Itself

The choice of a metric is further complicated by the fact that few (if any metrics) can quantify risk with perfect precision. Even something as simple as counting redundant parts within a system—an approach loosely (and more qualitatively) employed by the Space Shuttle Program in its early years under the auspices of the Critical Items List (CIL) [13]—can generate uncertainty. The Space Shuttle Solid Rocket Booster (SRB) O-rings were originally classified as "criticality 1R"—meaning they were considered redundant to catastrophic failure. However, this classification was later changed to "criticality 1" (e.g. *not* redundant to catastrophic failure) in 1982, when engineers realized that leakage of the primary O-ring during certain phases of launch was actually a single-point failure [23]. Notably, this classification change occurred in the absence of any modifications to the SRB design, suggesting that even simple metrics for measuring risk can have uncertainty associated with their measurements.

This uncertainty must be accounted for to ensure evaluations of "safe enough" are statistically appropriate. Consequently, certain metrics explicitly list an uncertainty component in their measured values. Probabilistic Risk Assessment (PRA), for example, generates both a mean estimate of risk *and* an estimate of risk uncertainty.

2.4. Subjectivity in the Acceptance of the Measurement

Once a method for measuring risk has been selected, a threshold value (either quantitative or dichotomous) can be assigned to "safe enough" (e.g. ***18*** successful launches, ***3*** levels of failure tolerance, Probability of Loss of Crew less than ***1/200***, ***successful completion*** of a flight readiness review). If a spacecraft meets (or in certain cases, exceeds) the threshold value, it can be considered "safe enough"; if it does not meet this threshold, it can be rejected as "not safe enough" [11].

This threshold value must balance what is *achievable* given the program's budget, schedule, and engineering capabilities with what is desirable (or *acceptable*) from a programmatic or personal standpoint [7, 11]. Identifying such a value is no simple or arbitrary task. Predicting *achievable* risk requires the use of expert opinion during the initial stages of program development [9]. As described in the Space Shuttle example above, such opinions can vary widely, and even then may not encompass the "true" risk of the system.

Determining what is *acceptable* in terms of risk is also "far from straightforward", as anticipated mission benefits must be shown to demonstrably outweigh the potential for mishap [24]. Given that the "weight" of each benefit naturally varies from individual to individual (and program to program), this process is largely subjective. As such, it is amenable to rejection *ex post facto*. Consider a theoretical spacecraft that meets its assigned risk threshold with perfect certainty. Even though such a spacecraft is "safe enough" by the letter of the law, it would be hard to argue as such in the event of a catastrophic accident.

3. RECOMMENDATIONS

Developing techniques to eliminate, account for, or circumvent these complications is critical to the consistent and unequivocal evaluation of "safe enough". Although a detailed description of these techniques is beyond the scope of this article, several general recommendations are described below:

- Define all relevant terminology (e.g. "safe", "unsafe", "safe enough", and "risk") during program startup, and strictly adhere to these definitions throughout the program's development and operations. This may serve to minimize confusion as to what is "safe" versus what is "safe *enough*."

- Select a risk metric that either minimizes or accounts for measurement uncertainty. This helps to ensure that "safe enough" can be reliably evaluated (to a pre-determined level of statistical certainty) despite any potential unknowns in the vehicle's performance or design.

- Establish an acceptable risk threshold based on what is realistically achievable given technical, budget, and schedule constraints. Expert opinion and risk heuristics should be used early in the design process to coarsely predict achievable risk. If risk is determined to be acceptably low, the program can continue forward; conversely, if risk is determined to be unacceptably high, then the program can be restructured (or cancelled) before significant resources are committed to a specific design or operational paradigm [25].

4. CONCLUSION

Answering the question "how safe is safe enough?" is a difficult and contentious process, impeded by elements of subjectivity and uncertainty. Despite these complications, the question of "safe enough" should be placed at the forefront of all human space flight programs, as the answer will determine the level of risk that is considered acceptable by the stakeholders and also serve as a benchmark for assessing whether or not it is achieved.

To Gus Grissom and his fellow astronauts, the "conquest of space [was] worth the risk of life." We must continuously ask: "is it still?"

5. ACKNOWLEDGMENTS

The FAA has sponsored this project in part through the Center of Excellence for Commercial Space Transportation. However, the agency neither endorses nor rejects the findings of this research. The presentation of this information is in the interest of invoking technical community comment on the results and conclusions of the research.

6. REFERENCES

1. Swenson Jr, L. S., Grimwood, J. M., & Alexander, C. C. (1966). This New Ocean: A History of Project Mercury. NASA SP-4201. *NASA Special Publication, 4201.*

2. Hacker, B. C., & Grimwood, J. M. (1977). On the Shoulders of Titans: A History of Project Gemini:'Spirit of'76. *NASA, Rept. SP-4203*, 265-298.

3. Murray, C. A., & Cox, C. B. (1989). *Apollo, the Race to the Moon*. Simon & Schuster.

4. Aerospace Safety Advisory Panel. (1979). Aerospace Safety Advisory Panel Annual Report for 1978. *Washington, DC.*

5. Aerospace Safety Advisory Panel. (2009). Aerospace Safety Advisory Panel Annual Report for 2008. *Washington, DC.*

6. Aerospace Safety Advisory Panel. (2010). Aerospace Safety Advisory Panel Annual Report for 2009. *Washington, DC.*

7. Aerospace Safety Advisory Panel. (2011). Aerospace Safety Advisory Panel Annual Report for 2010. *Washington, DC.*

8. Aerospace Safety Advisory Panel. (2012). Aerospace Safety Advisory Panel Annual Report for 2011. *Washington, DC.*

9. Aerospace Safety Advisory Panel. (2014). Aerospace Safety Advisory Panel Annual Report for 2013. *Washington, DC.*

10. Aerospace Safety Advisory Panel. (2015). Aerospace Safety Advisory Panel Annual Report for 2014. *Washington, DC.*

11. Ocampo, R. P., & Klaus, D. M. (2016). A Quantitative Framework for Defining "How Safe is Safe Enough" in Crewed Spacecraft. *New Space*. 4(2): 75-82

12. Ocampo, R.P. (2016). *Unpublished data*.

13. Slay, A. (1988). *Post-Challenger Evaluation of Space Shuttle Risk Assessment and Management*. National Academies.

14. U.S. House of Representatives (1986). Investigation of the Challenger Accident. *Washington DC.*

15. NASA. (2008). NASA NPR 8715.3 NASA General Safety Program Requirements (w/Change 9 dated 2/08/13). *Washington, D.C.*

16. Aerospace Safety Advisory Panel. (2002). Aerospace Safety Advisory Panel Annual Report for 2001. *Washington, DC.*

17. Feynman, R. P. (1986). Personal observations on the reliability of the shuttle. *Report of the Presidential Commission on the Space Shuttle Challenger Accident, 2*, 1-5.

18. Fragola, J. R. (1996, May). Risk management in US manned spacecraft: From Apollo to Alpha and beyond. In *Product Assurance Symposium and Software Product Assurance Workshop* (Vol. 377, p. 83).

19. Vaughan, D. (1997). *The Challenger launch decision: Risky technology, culture, and deviance at NASA*. University of Chicago Press.

20. Ocampo, R. P., & Klaus, D. M. (*2016*). Comparing the Relative Risk of Space Flight to Terrestrial Modes of Transportation and Adventure Sport Activities. *New Space*. 4(3): 190-197.

21. NASA. (2015). NASA CCT-REQ-1130 Rev D-1, ISS Crew Transportation and Services Requirements Document. *Washington, D.C.*

22. Operator informing space flight participant of risk, 14 C.F.R. § 460.45 (2016).

23. Presidential Commission On Space Shuttle Challenger, & Rogers, W. P. (1986). Report of the Presidential Commission on the Space Shuttle Challenger Accident.

24. Aerospace Safety Advisory Panel. (2013). Aerospace Safety Advisory Panel Annual Report for 2012. *Washington, DC.*

25. Ocampo, R. P., & Klaus, D. M. (2016). *A Heuristic Method for Predicting Achievable Risk in Human Space Flight*. Manuscript submitted for publication.

PROCEEDINGS OF THE 9TH INTERNATIONAL SPACE SAFETY CONFERENCE

SESSION 16 LAUNCH SAFETY

RAFAEL'S TEST RANGE SAFETY ANALYSIS TOOL

Ronen Ingbir [(1)], Nave Ben-Yakov [(2)], Dima Kanevsky [(3)]
Meir Cohen [(4)], Shay Acco [(5)], Yeshayahu Atzmon [(6)]

RAFAEL Advanced Defence Systems LTD., RAFAEL/WTEC P.O.B. 2250, Haifa 31021, Israel
Email: (1)ronenin@rafael.co.il, (2)naveb@rafael.co.il,(3)dimaka@rafael.co.il,
(4)meircohe@rafael.co.il, (5)shayac@rafael.co.il, (6)yeshaya@rafael.co.il

ABSTRACT

RAFAEL is developing a Test Range Safety Analysis Tool (TRSAT) to be used by Rafael Weapon Test and Evaluation Centre (WTEC). A new approach that combines comprehensive safety analysis capabilities into one framework is implemented. TRSAT is intended to support future challenges of modern systems under test and complex flight test missions.

1. Rafael WTEC

The Israeli Ministry of Defense (IMOD), Directorate of Research and Development, has nominated RAFAEL-Advanced Defense Systems Ltd, to develop, operate and maintain its national test range facilities.
The unit in RAFAEL dedicated to this assignment is the Weapon Test and Evaluation Center (WTEC).
WTEC runs advanced test ranges, to meet the high demands of the state-of-the-art surface to surface, surface to air, air to surface and air to air tactical weapon systems. The end-to-end tests are conducted under realistic operational scenarios, and are fully monitored by tracking and measuring instruments.

2. WTEC Safety

Prior to each mission, the test range safety officer studies the structure and capabilities of the system under test. The mission scenario is planned, taking into account the test range boundaries, the trajectory of the tested body and its safety envelope. Risk assessment is performed, to ensure that safety criteria are met.
During the mission, the test range instrumentation provides real-time safety data which is presented both graphically and numerically on the monitors of the safety officer for real-time control. The central computer calculates Instantaneous Impact Point (IIP) estimations as a decision making aid for the safety officer. The safety officer is the highest authority concerning safety decisions prior, during and after the mission. His decisions are independent of any hierarchical intervention.

3. TRSAT and System Architecture

TRSAT is a new tool (software package) under development that incorporates numerous capabilities in one framework:

- Risk analysis,
- Safety and operations concept planning,
- Mission plan optimization while minimizing risks,
- Updated physical models,
- Sensors deployment planning and optimization taking into account nominal and malfunction trajectories.

The system is modular and can be operated fully or partially (selected modules only). Fig.1 illustrates full use sequence of the system in the case of malfunction trajectories (solid arrow) and partial use for the case of nominal trajectories (dashed arrow).

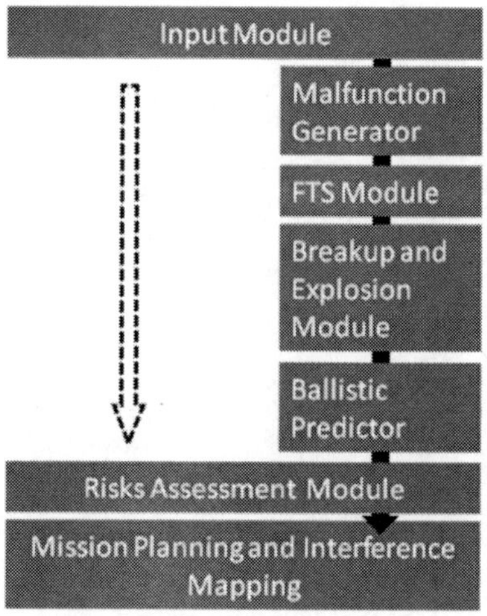

Figure 1. TRSAT System Architecture

The Malfunction module generates realistic initial physical conditions of malfunction trajectories. The Flight Termination System (FTS) module simulates the outcome of the safety officer acts or that of the Autonomous Flight Termination System (AFTS). The Breakup and Explosion module sets the initial dynamics and properties of all the objects including fragments of the system after its flight was terminated. A ballistic predictor is used to propagate the trajectories to the altitude of interest (ground/sea impact points and/or aviation typical cruising altitudes) taking into account atmosphere and wind. After the full set of required data including impact properties is generated, a comprehensive risk assessment is performed. The Mission planning and interference mapping module is usually the last to be operated, completing the test range planning. In the following paragraphs the functionality of the various modules is demonstrated using synthetic study cases.

4. Input Module

The input module Graphical User Interface (GUI) is utilized to feed the Flight safety data package (FSDP) into the system. System specifications including aerodynamic coefficients, nominal and Monte-Carlo dispersions (MC) trajectories are requirements. If needed and in order to improve statistics, additional trajectories are generated by interpolating the given FSDP data.

5. Malfunction Generator

The Malfunction Generator is the core module of TRSAT. This module generates malfunction trajectories. One possibility is to systematically scan initial physical conditions originating from every possible point/time of the nominal trajectories. In this approach all trajectories are assigned with equal weights. Another mode of operation is to generate specific initial conditions, assigning them weight according to the probability of occurrence. An example of the first approach is demonstrated in Fig.2

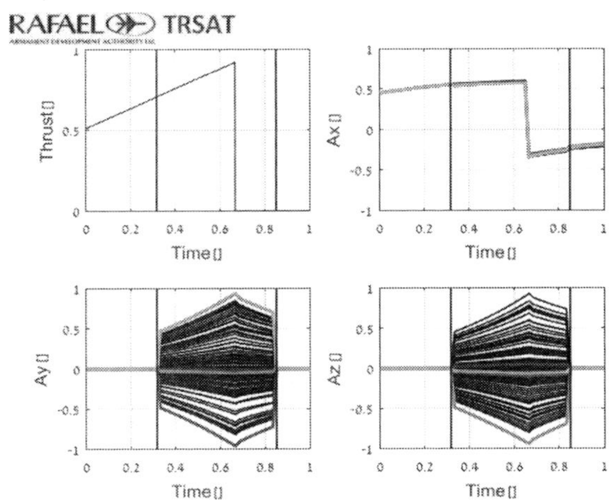

Figure 2. Nominal thrust and malfunctions accelerations

In the upper left figure the nominal normalized thrust is presented as a function of the normalized time. The resulted **Ax** accelerations are presented in the upper right figure. Malfunctions accelerations **Ay** and **Az** are presented in the bottom figures. Geographic extreme trajectories accelerations are marked with non-blue colours.

The TRSAT malfunction generator integrates numerically body accelerations in time and calculates the velocities and positions of the System Under Test (SUT). Different SUT differs by the values of the acceleration and overall flight time.

An example of the malfunction velocities and spatial body angles is presented in Fig.3. In the upper left figure the normalized velocities are presented as a function of the normalized time. The body angles (role, yaw, pitch) are also presented. Geographic extreme trajectories velocities and angles are marked with non-blue colours.

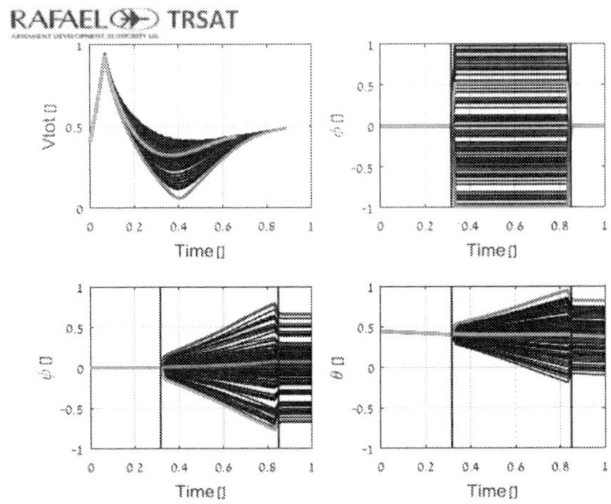

Figure 3. Malfunction MC: Normalized Velocities and Spatial body angles

Fig.4 demonstrates the malfunction trajectories and ground impact points. In the upper figure the normalized ground impact points are presented. In the lower figure the altitude as a function of position is presented. Geographic extreme trajectories are marked with non-blue colours, impact points are marked in black.

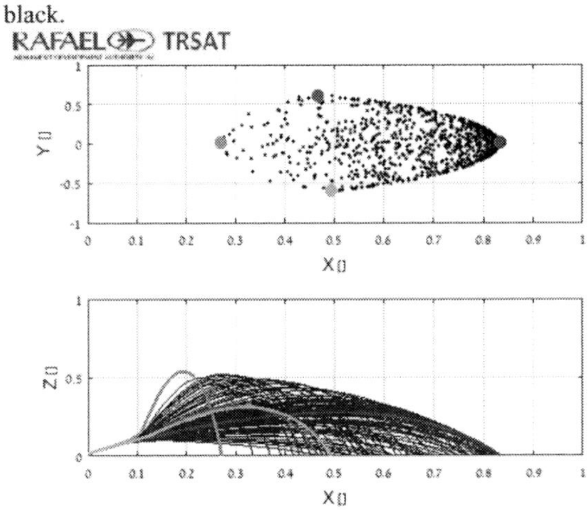

Figure 4. Ground impact points and trajectories MC

The TRSAT malfunction generator can also produce extreme trajectories by optimization. For example, a simple propelled rocket that lost its stability in the pitch plane can be found at any given angle of attack. Maximum range is achieved by selecting the optimal angles of attack needed at any given time. This process is done taking into account the thrust, mass profile, aerodynamic lift and drag coefficient with respect to different Mach numbers and angles of attack.

6. FTS Module

High energetic systems under test are compelled to include a Flight Termination System (FTS) in order to ensure that they, or their debris, will not exit predefined safety boundaries. The TRSAT FTS module simulates the safety officer or the AFTS (Autonomous Flight termination System) role. Termination polygons are constructed by taking into account the test range boundaries and typical human reaction time (or machine decision logic/process for the case of AFTS). This module enables the safety officer to plan the mission safety concept and analyse the expected outcome if the FTS (or AFTS) is operated.

7. Breakup and Explosion Module

The breakup and explosion module includes physical models that simulate debris generation in case of a planned break-up/explosion or unexpected malfunction. The motivation for including this module is to realistically simulate all the potential risk elements. One example is the flight termination process that often generates numerous fragments with unique properties and associated risks. Another example is systems under test that include live warheads or other components that can explode in a malfunction. The explosion model performance is demonstrated in the following figures.
In Fig 5 the normalized velocity magnitude as a function of the polar angle for scattered fragments is presented.

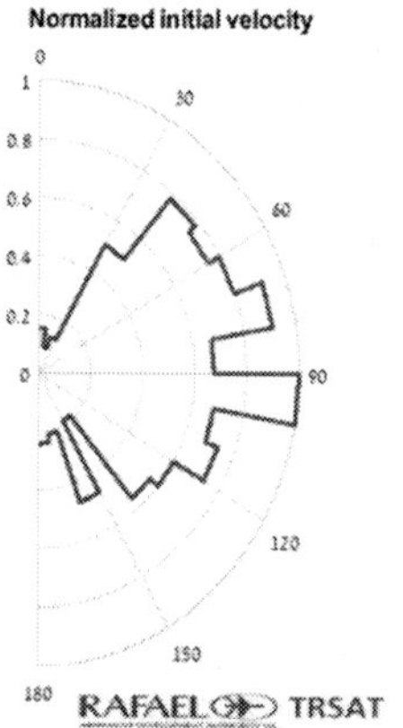

Figure 5. Normalized average velocities as a function of the polar angle

In Fig 6 the normalized number of fragments and mass (in colors. Higher number – greater mass) as a function of the polar angle is presented.
In Fig 7 the normalized number of fragments as a function of the polar angle is presented.

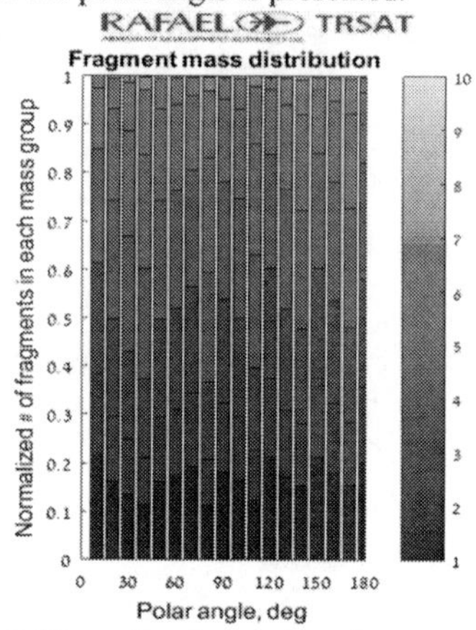

Figure 6. Normalized mass distribution polar angle dependence

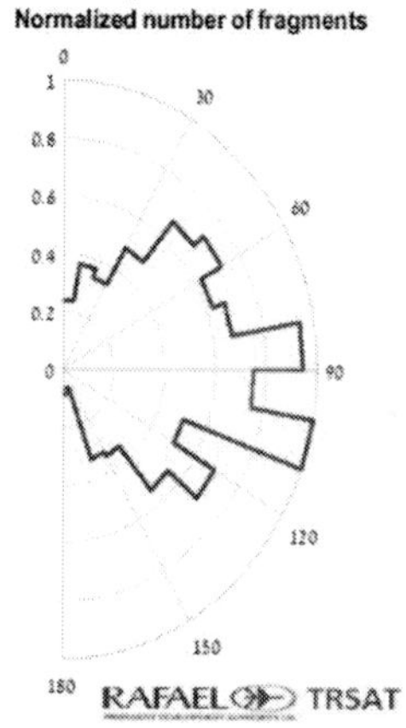

Figure 7. Normalized number of fragments polar angle dependence

The results of this module are a set of initial condition (dynamics and properties) for every object and fragment/debris.

8. Ballistic Predictor

The TRSAT uses a ballistic trajectory model to construct trajectories when an object has only drag. This applies mainly to debris/fragments. The predictor uses the test range atmosphere, wind model including dispersions and Mach number dependent ballistic coefficient. Earth rotating ellipsoid model is used, in order to get accurate results. This module propagates the trajectories to the ground impact point or to any other altitude of interest (for example: aircrafts cruising altitude).

9. Risk Assessment Module

The TRSAT comprehensive risk assessment module takes into account the nominal scenarios and malfunctions. This module uses the mass, material type, dimensions and velocity (impact energy) of every object/fragment/debris that was generated in the previous modules.
Risk assessment analysis example is demonstrated on a single case study (not a full MC) and presented in the following figures.
Fig 8 illustrates a malfunction that causes an explosion. The tested object breaks and the explosion debris are propagated to the ground.

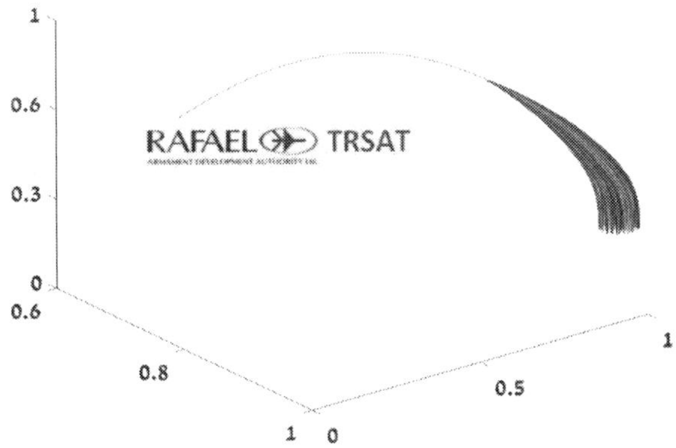

Figure 8. Fragments trajectories

In Fig 9 the histogram of the individual probability of fatality is presented as a function of the normalised impact point.

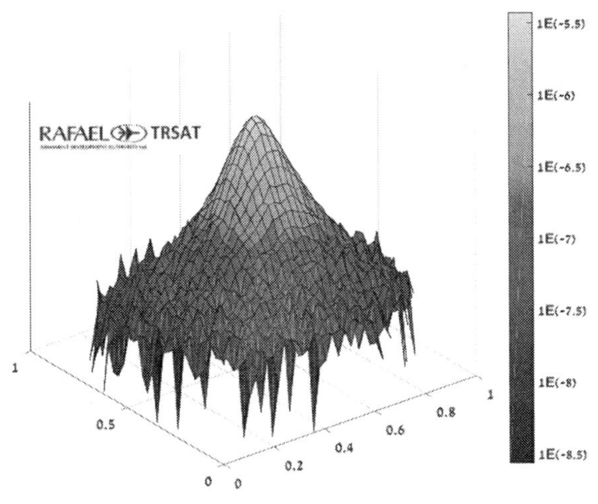

Figure 9. Risk assessments

In Fig 10 on top of the objects/fragments/debris normalised impact points (blue dots), the individual probability of fatality contours are drawn.

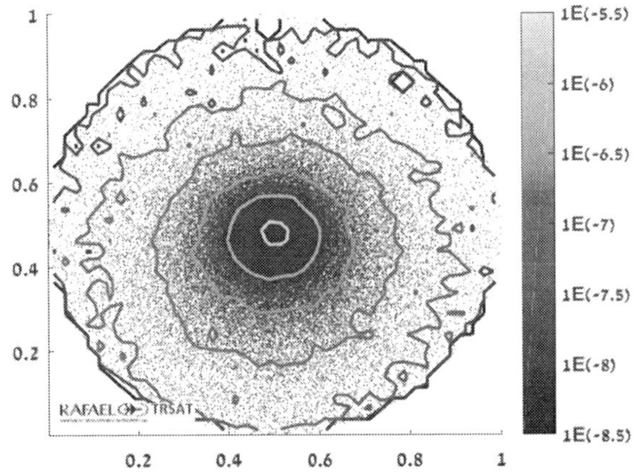

Figure 10. Risk assessments

The TRSAT supports risk assessments to personal, infrastructure, ships, aircrafts and spacecrafts.
Risk to aircrafts is calculated taking into account the mass of the objects while assuming typical aircrafts cruising velocities. In this approach the risk is mass and material dependent and varies with aircrafts size/type.
For the other categories the impact energy is calculated directly and a set of energy based criteria are used to evaluate the risk.

TRSAT risk assessment module also enables cumulative risk calculation using the individual risks probabilities and the population/traffic (aerial, maritime...) data bases.

One of the parameters to be newly tested using TRSAT is the statistics significance and physical model uncertainties effect on the calculated results. Mission go ahead approval is given only when all the individual and cumulative safety criteria are met and the overall uncertainty is at least one order of magnitude lower than the mean or most probable risk value.

10. Mission Planning and Interference Mapping

TRSAT, as a safety tool, incorporate mission planning capabilities while taking into account not only the nominal scenarios but also possible malfunction trajectories. Using this module the test range performance is evaluated while optimizing sensors array configuration, deployment and mutual interference avoidance.

An example of the test range tracking accuracy for various sensors configuration is presented in Fig 11.

The normalised distance relative to the actual tracked body is shown as a function of normalised time for different sensors configurations as depicted in the figure.

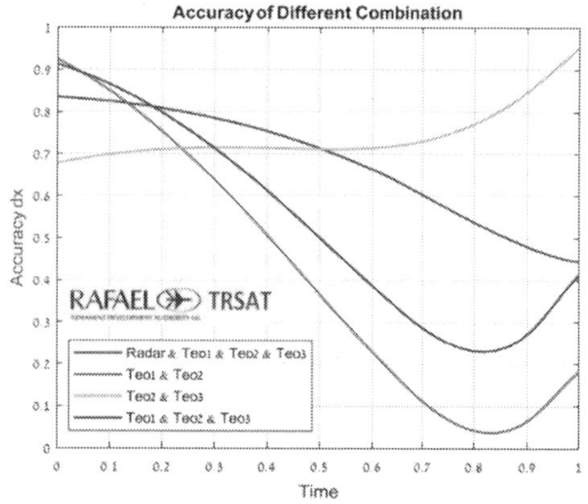

Figure 11. Performance comparison of various sensor configuration as depicted in figure

An example of the test range average tracking accuracy is presented in Fig 12. The normalised, time averaged, distance relative to the actual tracked body is shown for the same array of sensors deployed at alternating positions.

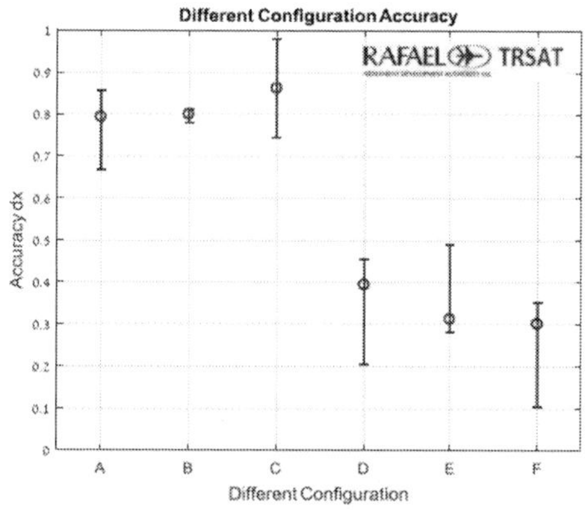

Figure 12. Performance comparison of the same sensors array and alternating positions (A..B......F)

Using TRSAT, it is possible to optimise sensors' configuration and position to get the best performance under budget constraints.
Further performance improvement can be achieved by incorporating specific mission tailored parameters to the BET (Best Estimated Trajectory) algorithm.

In the following figures, an example of performance improvement is presented for the case of same sensors configuration and deployment.

In Fig 13, three sets of time based weights were given to the sensors in the tracking algorithm.

Fig 14 presents the tracking performance after using Fig 13 weights relative to using uniform weighting. Performance improvement of an order of magnitude is achieved by using a time based sensors weighting, tailored to this specific mission.

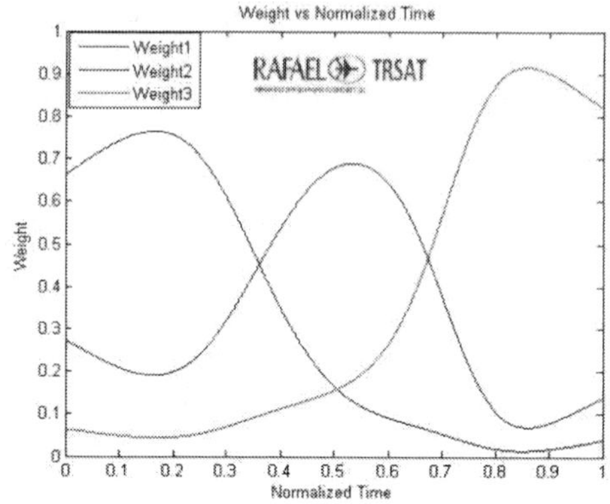

Figure 13. sensors weighting Vs. normalized time

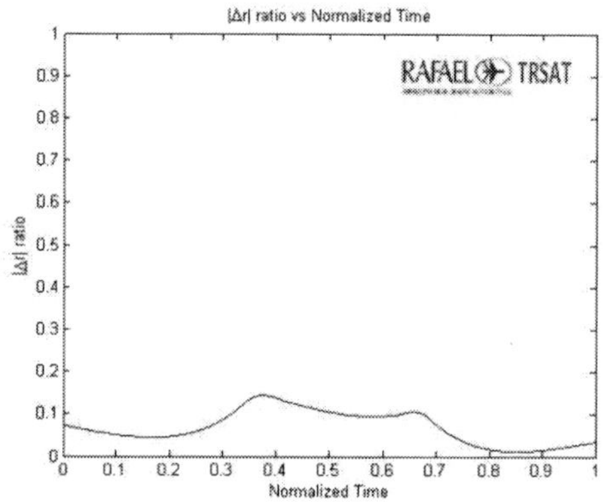

Figure 14. Performance improvement after applying optimal sensors weighting (fig.12)

TRSAT interference mapping capabilities are especially relevant to complex missions with multi object scenarios. Using the nominal and malfunction trajectories the line of sight, tracking performance, telemetry and FTS link budget is evaluated (including aerial/maritime sensors). Using this module, the test range apparatus deployment or flight tests scenarios are modified to minimize interference.

11. Closure

Rafael's TRSAT is a new tool under development that incorporates numerous capabilities in one software framework. Updated physical models and fast simulation capabilities will be incorporated to reach desired level of confidence of the risk calculations. This tool is intended to support future challenges of modern systems under test and complex flight test missions at Rafael WTEC.

DESIGN-TO-SAFETY: ANALYSIS OF THE EXPLOSION AND FRAGMENTATION INFLUENCE ON INERT DEBRIS IMPACT FOOTPRINTS AND MITIGATION SOLUTIONS FOR INNOVATIVE LAUNCHER CONCEPTS

A. Martinez Torio(1), D. Delorme(1), L. Rozenberg(1), V. Guenard(1), F. Bignalet-Cazalet(2)

(1)CNES Launchers Directorate, 52, rue Jacques Hillairet, 75012 Paris (France),
Email: alexandra.martineztorio@cnes.fr, david.delorme@cnes.fr, laurence.rozenberg@cnes.fr, vanessa.guenard@cnes.fr

(2)CNES Guiana Space Center, Centre Spatial Guyanais, 97387 Kourou (French Guiana),
Email: francois.bignalet-cazalet@cnes.fr

ABSTRACT

The ever increasing price pressure on the worldwide launch market has shaped the strategy for the next generation of European launchers preparation. In addition to the development of low-cost propulsion technologies based on LOX/CH4, CNES Launchers Directorate, investigate the introduction of affordable partly-reusable launchers with a main stage return phase performed by a Toss-Back maneuver.

These launcher concepts, which have never been operated from French Guiana, give rise to new safety pitfalls different from those already handled for the launches operated from GSC, challenging the flight safety principles.

This article provides an overview of the most critical scenarios, following an accidental explosion, in a Toss-Back maneuver, with an analysis of the effects of the resulting explosion (in terms of energy and fragmentation), and its impact on ground. The main goal is to identify the most critical parameters driving the scenarios' criticality and to determine the possible risk mitigation measures such as "design-to-safety" solutions.

1. INTRODUCTION

The ever increasing price pressure on the worldwide launch market has shaped the strategy for the next generation of European launchers preparation. In addition to the development of low-cost propulsion technologies based on LOX/CH4, CNES Launchers Directorate, investigate different launcher conception approaches, following two main guidelines:

- Launcher intrinsic simplification: aiming at reducing the number of engine and propellant types used and at maximizing synergies between boosters and central body and between lower and upper stages.

The introduction of hydrocarbon-based propellants and low-cost propulsion technologies allows a simpler engine conception and thus, lower development and production costs.

Taking into account the higher technological maturity level of Methane with respect to Kerosene in Europe, LOX/CH4 seems to be a promising propellant for a reusable stage. In this frame, a technological demonstrator, called PROMETHEE (Precursor Reusable Oxygen METHane cost Effective Engine), aims at consolidating the future LOX/CH4 low-cost reusable engine [1].

- Launcher reusability: limited to lower stages (boosters or central body) due to accessibility constraints of an upper stage.

Among these launcher concepts, this paper deals in particular with a LOX/CH4 two-stage partially reusable vehicle, with a main stage return phase back to the launch pad, performed by a Toss-Back maneuver.

Lower stage (reusable): C400 composed of 7 engines. Thrust level 1000kN approx.

Dimensions: 30m height, 5.4m of diameter

Upper stage: C91 composed of 1 engine derived from the lower stage.

Figure 1 LOX/CH4 two-stage partially reusable launcher configuration and staging

This maneuver is performed, after the booster/1st stage jettisoning, by the main propulsion system of the stage, which must be reignitable and throttable. This maneuver enables to obtain a return ballistic trajectory with an impact point near the landing site, to brake and thus reduce dynamic pressure and to land vertically on the landing site.

The resulting return trajectory consists of the following maneuvers:

- After the stage separation, a tilting maneuver to reorient the vehicle, by means of the attitude control system, before the boost-back burn
- a first boost (i.e. boost-back burn) to reverse the velocity vector and to point it to the landing site
- a ballistic reentry phase, and
- a second boost (i.e. landing burn) to control the final vertical landing.

For a hard-landing strategy, the vehicle goes through a deceleration all along the final phase of the descent, reaching a zero velocity at the precise instant of touch-down.

On the other hand, a soft-landing strategy, the vehicle, with a capability of achieving a thrust-to-weight ratio of 1, slowly lands with an acceleration close to zero, maintaining a targeted vertical velocity during its descent. Compared to a hard-landing strategy, this landing strategy can be more fuel consuming, and it may require a more vertical ascent phase.

Eventually, an additional intermediate boost (i.e. re-entry burn) could be foreseen to brake so as to reduce the impact of general loads and dynamic pressure during the fallout.

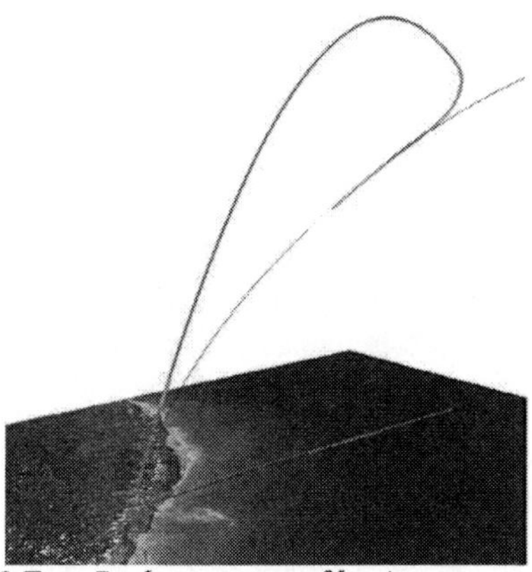

Figure 2 Toss-Back mission profile. Ascent trajectory is depicted blue (boosted phases) and in yellow (ballistic phases), return trajectory is depicted in green (boosted phases) ant pink (ballistic phases)

These new launcher concepts, which have never been operated from French Guiana, give rise to new safety pitfalls (i.e. steep vertical ascent flight, natural overflight of land areas for return maneuvers, high maneuverability and deviation capacity, etc.), different from those already known and handled for the launches operated from GSC, challenging the flight safety principles.

In parallel to the study of innovative flight safety approaches and strategies, the understanding of the physics involved in an accidental explosion or a flight termination is deemed essential.

Hence, this paper is focused on the identification of the most critical events leading to an explosion that can happen at each flight phase of a Toss-Back maneuver, and the analysis of the causes and consequences, in terms of energy of explosion and the resulting fragmentation effects.

Depending on the different specific features, inherent of each flight phase (in terms of trajectory and vehicle dynamics), the effects of such accidental scenarios on the debris impact footprint are then estimated.

The aim is to identify the most sizing parameters (i.e. debris geometry, ejection speed, aerodynamics, wind sensitivity, etc.) driving the violence of the explosion and the criticality of the debris ballistic fallout, and thus, the dangerousness of the generated footprints, at each phase.

Finally, following a design-to-safety approach, a list of possible palliative measures aiming at mitigating the effects on ground will be proposed. These measures can consist of:

- design recommendations for the vehicle conception (e.g. specific materials and technologies) so as to master the fragmentation and energy effects of an accidental explosion in order to obtain less critical footprints,
- the modification of the vehicle architecture by the introduction of safety aerodynamic devices so as to reduce the aerodynamic capacity or induce instability of upstanding bodies falling back to ground,
- the re-optimization of the vehicle ascent profile and the return strategy,
- the introduction of innovative safety logics and architectures.

2. SAFETY ISSUES OF A TOSS-BACK MANEUVER

Compared to a standard expendable mission, a reusable mission implies some inherent safety issues.

The trajectory optimization process maximizes the targeted performance while minimizing its cost in terms of propellant.

A return phase requires the enchainment of ballistic phases and boosted maneuvers in order to reverse the velocity vector of the separated stage, to approach its impact point to the final landing site, to reduce the dynamic pressure and to finally brake in order to vertically land at almost zero velocity.

So as to minimize the propellant cost of the return maneuver, the distance of the impact point and the landing site at stage separation should be reduced.

The resulting optimized ascent trajectory is characterized by a steep verticality. As a consequence, compared to a standard expendable mission, the launcher reaches higher altitudes before its impact point clears off the launch pad area. In case of accidental explosion or neutralization, the impact footprint will be larger and will remain closer to the launch site, and thus, the protected areas, becoming a potential hazard for property and population.

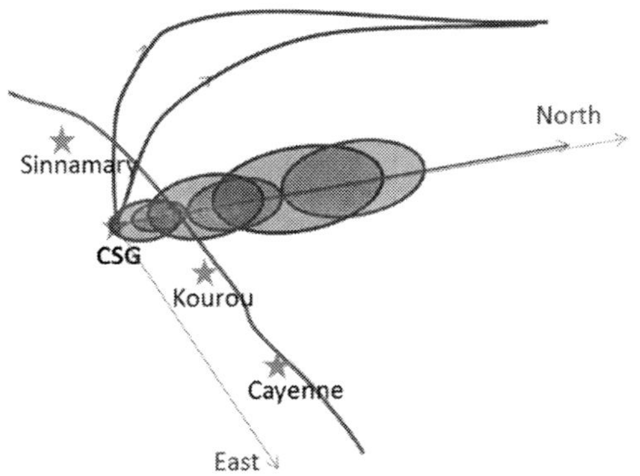

Figure 3 Qualitative representation of debris impact footprints during the ascent trajectory of a Toss-Back mission (blue) compared to a standard trajectory (red)

Moreover, a nominal return trajectory naturally overflies land areas during its descent to the landing site. Without any specific constraint, in case of accidental explosion or neutralization, the stage debris may fall over protected areas. Especially at the beginning of the return maneuver (end of first return boost), where the launcher is at higher altitudes, debris impact footprint extents will be larger.

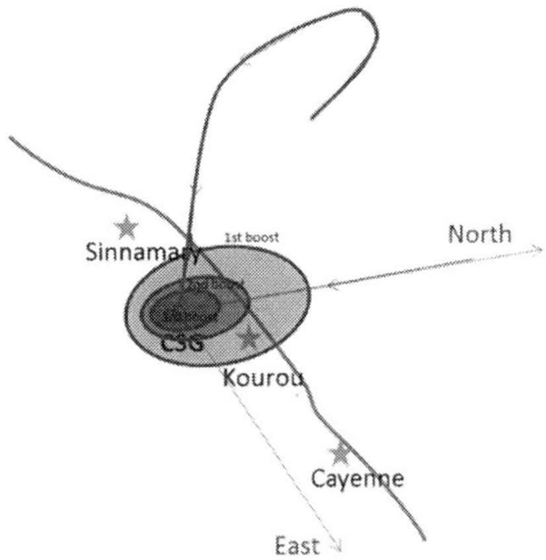

Figure 4 Qualitative representation of debris impact footprints at return boosts extinctions

By controlling the ascent trajectory vertical profile, the launch azimuth and the strategy used for the return maneuver (number and duration of the boosts needed, utilization of the aerodynamic capacity of the vehicle, landing strategy), these safety issues can be mitigated, while respecting the system constraints in terms of performance.

3. FEARED EVENTS ANALYSIS

With the objective of analyzing the criticality of this kind of missions by assessing the eventual debris fallout footprint extents that could be produced in case of any accidental scenario, the worst case scenarios have to be determined. The final purpose is to determine the critical parameters driving the size and the dangerousness of the produced debris impact footprints.

Consequently, an exhaustive hazard assessment has to be conducted. This analysis consists of a description of the possible hazards involved during each phase of the ascent and return trajectories, in both nominal and degraded cases, regardless of the origin of the failure. Special attention will be paid to those feared event leading to a ballistic reentry of debris or upstanding bodies after a vehicle or engine explosion.

3.1. Feared Events

A top-down analysis of the different feared events on such a vehicle has been conducted in order to identify the worst case scenarios for each flight phase. These worst case scenarios are considered to be generic, regardless of the vehicle's fuel. This analysis has concluded that at least the following scenarios should be taken into account on the flight safety studies due to their potential criticality:

FE1: External mixture of fuel and air
FE2: Internal mixture of fuel and oxidizer in the ITS/ISS
FE3: External mixture of fuel and oxidizer
FE4: Explosion in the thrust frame (i.e. main engine)

<u>FE1: External mixture of fuel and air</u>

A breach on the fuel tank or on the feeding lines of the fuel circuit, a tank rupture due to an exceed of the general loads or the thermal flux encountered during the atmospheric phase (in a deviated trajectory or in a nominal with winds out of specification), an airtightness loss, an impact of an external debris (landing legs, fins, etc.) causing the tanks ruin or a line rupture, can lead to an external mixture of fuel and air. The final consequence of such scenario will be an unconfined explosion of the vehicle.

<u>FE2: Internal mixture of fuel and oxidizer in the ITS/ISS</u>

An internal mixture of fuel and oxidizer in a confined structure like the tanks, the inter-tank or the inter-stage structures, following a breach on either both the fuel and oxidizer tanks or on the lines, can lead to a violent confined explosion.

<u>FE3: External mixture of fuel and oxidizer</u>

The rupture of the inter-tank skirt (in case the vehicle does not have common bulkhead between tanks), due to an exceed of the general loads or the thermal flux encountered during the atmospheric phase (in a deviated trajectory or in a nominal with winds out of specification) can lead to the collision of both tanks and the external mixing of the fuel and oxidizer, causing a unconfined explosion.

This scenario can also be originated by the ruin of both tanks by an external body, by the buckling of the fuel tank and the rupture of the oxidizer feeding line, or vice-versa, or by the rupture of feeding or pressurization lines. Also the pneumatic burst of a tank at maximum rupture pressure, causing the rupture of the other tank at nominal pressure, can lead to the external mixture of both propellants, and a subsequent unconfined explosion.

At high altitudes, the violence of an external explosion is reduced as the resulting overpressure is proportional to the atmospheric pressure. Consequently, we can consider that at high altitudes, the worst case scenario, in terms of debris ejection speed, would be the event of a pneumatic burst. We can consider that the debris caused and ejected by the pneumatic burst would be sufficiently far from the eventual center of explosion at the moment of the subsequent chemical detonation, so as to consider that such explosion would have a negligible effect on the final ejection speed of the different fragments.

The energy of the pneumatic burst would depend on the maximum burst pressure of the tanks, which also depends on the technology and the design of the vehicle.

<u>FE4: Explosion in the thrust frame</u>
An explosion of the main engine can occur, during boosted phases, due to different reasons: a detonation or an overpressure in the combustion chamber, or at the igniter level, an over-speed of the disks of the turbines of the turbo-pump system, etc. This can lead to the ruin of the tanks and infer a pneumatic burst (which can be followed by a chemical explosion in a non-confined space). Nevertheless, an eventual debris ejected from the engine local explosion could pierce the bulkheads of a tanks (going through the lower tank up to the upper one), leading to the mixing of fuel and oxidizer inside the inter-tank structure, and causing a confined chemical explosion.

The term fragmentation refers to the different debris generated by a very fast rupture of a structure subjected to a physical (pneumatic) or chemical explosion.

For our safety studies, in order to be as much robust as possible to the uncertainties and the random nature of the explosion and fragmentation phenomena, a large population of fragments with different sizes, m/S, mass, topology and ballistic coefficients has to be defined.

Today, the fragmentation scheme following the accidental explosion of a launch vehicle is the result of the structural engineer expertise and the lessons learned from previous accidents or experiments. The materials, the technologies, the vehicle architecture and conception (including the use of neutralization lines) and the understanding of the structural behavior under static and dynamic loads, are key parameters for the engineer to determine – in consistency with the observations - the most probable rupture lines of the vehicle. The solicitation (pressure time rise and amplitude) - linked to the violence of the explosion (by instance confined or not) - is an additional driver of the degree of fragmentation.

We can consider that the closer a structure is to the center of explosion, the smaller the fragments generated will be, as the generated blast wave will have more devastating effects on the structure. Given the fact that the propellant is initially contained in the tanks, we can make the hypothesis that the smallest fragments will be those of external walls of the tank or the adjoining structures.

These fragments will be characterized by an important ejection speed, a light mass and a quite low ballistic coefficient (depending on the technology and the thickness).

In the meantime, structures situated further from the explosion center are less ruined by the blast wave generated by the explosion. As a consequence, these fragments are heavier and experience a lower ejection speed. Some structures, depending on the material and technology, can even withstand the effects of the explosion and remain upstanding with a high ballistic coefficient.

Another type of fragments is those with a good tradeoff between ballistic coefficient and ejection velocity. Light and close enough from the explosion center, and with a relatively reduce aerodynamic surface so as to compensate its mass and induce a high a ballistic coefficient, or heavy and really close to the explosion center, with an important m/S. An example of this kind can be the propellant feeding lines, or the bigger fragments of the tank walls, or the aft bay.

3.2. Criticality assessment

Considering that the vehicle dynamics and the altitude, at which the failure arises, can have an important impact on the criticality of the feared events and their consequences on ground, the ascent and return trajectories have been split up in the following phases:

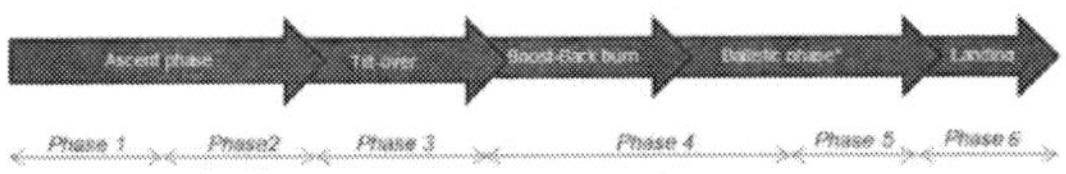

- Phase 1: This phase corresponds to the first seconds of vertical flight. Atmospheric pressure is high as the altitude is still relatively low. Along this phase, the experienced dynamic pressure (as well as thermal fluxes) increases.

  In case of vehicle explosion during this purely vertical flight, the scattering of debris is driven by the ejection velocity. Nevertheless, fragments with a high ejection speed but low ballistic coefficient brake sufficiently and produce reduced impact footprints due to their high drag. Therefore, the global impact footprint would be determined by those fragments or ejected equipments with the best tradeoff between ejection speed and ballistic coefficient, like the oxidizer feeding line, or the heavier fragments of the aft bay.

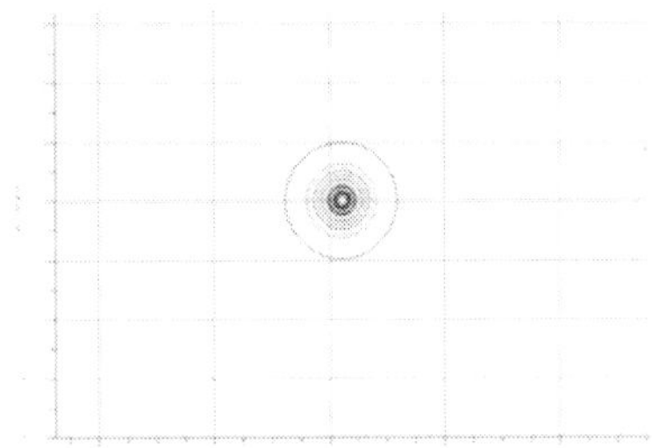

Figure 5 Debris impact footprints at the vertical ascent flight. The most sizing fragment is the heavier fragment of the aft bay

The criticality of the verticality of the ascent phase can be observed on the proximity of the impact footprints to the protected areas. In fact the more vertical the trajectory is, the higher the altitudes reached (and thus, the larger the footprints), while the impact point remains closer to the launch pad area.

- Phase 2: At the beginning of this phase, the vehicle starts slowly tilting and increasing its horizontal velocity. Altitude increases fast while its impact point starts slowly distancing from the protected areas. Along this phase, the experienced dynamic pressure (and thermal fluxes) is at its highest values. Atmospheric pressure decreases and reaches insignificant values.

  As the horizontal velocity of the vehicle increases, fragments with extreme values of ballistic coefficient stand out from the global impact footprint, sizing the front and the back of the global footprint.

  Impact footprints caused by the light fragments ejected at the highest speeds begin to enlarge as a consequence of the increase of altitude, compensating their low ballistic coefficient.

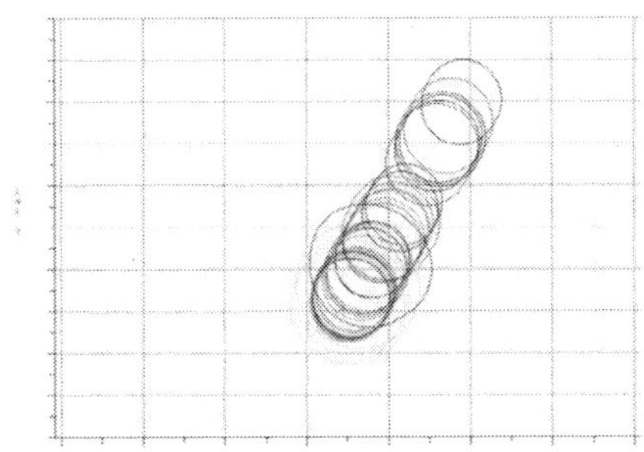

Figure 6 Debris impact footprints at the end of

the vertical ascent flight. The global impact footprint is determined by the light fragments from the fuel tank walls (at the back in yellow) and by those with a high ballistic coefficient (at the front in red and black)

- Phase 3: This phase includes the end of the ascent boost and the tilt-over maneuver (ballistic maneuver).

As a consequence of their high ejection velocities, light fragments are sensitive to the increase of altitude. Indeed, an increase of several hundreds of meters can enlarge the final impact footprint by several kilometers. Their impact footprints on ground now cover all the other impact footprints.

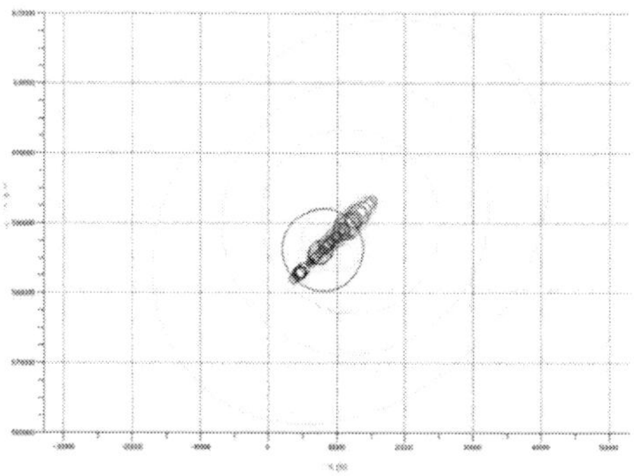

Figure 7 Debris impact footprints at the end of the ascent flight. The global impact footprint is determined by the light fragments from the fuel tank walls

The criticality of this phase can be reduced by adapting the mission profile : reducing the verticality profile of the ascent phase, in order to distance the impact footprint from the protected areas. However, this solution would imply a reduction of performance. In order to determine a compromise between performance and safety compliance, iterative loops of launcher concept, staging and trajectory have to be performed.

This can also be treated by design in order to obtain a more convenient fragmentation or by preventing the potential generated fragments to experience high ejection velocities by rendering the architecture more robust to explosion.

- Phase 4: This phase comprises the boost-back burn and the first seconds of the reentry ballistic phase (with apogee overflight).

This phase is the most critical one in terms of size of debris impact footprints and proximity to the protected areas. This dangerousness is justified by the high altitude and by energy inferred to the vehicle at the end of the boost-back burn. Indeed, at the end of the boost the impact point of the vehicle is already near the landing site and thus the protected areas. The horizontal velocity entailed by the burn can have an impact on the dangerousness of footprints generated by fragments with high ballistic coefficients.

As a consequence of their high ejection velocities, light fragments impact footprints are still the most critical ones in terms of size, still covering most of the other debris impact footprints.

The criticality of this phase can be reduced by adapting the mission profile : by limitting the maximum altitude (at apogee) in order to reduce impact footprint sizes, or by modifying the boost-back burn in order to target a different impact point at the end of the boost, distancing it from the protected areas. This can be done by shortening the first boost duration, preventing the debris fallout footprint from approaching the protected areas.

Such strategy will imply a reduction of performance and the need for trajectory re-optimization of the rest of the return phase. The final approach of the impact point to the landing site can be performed by an intermediate boost or using the aerodynamic lift capacity of the vehicle.

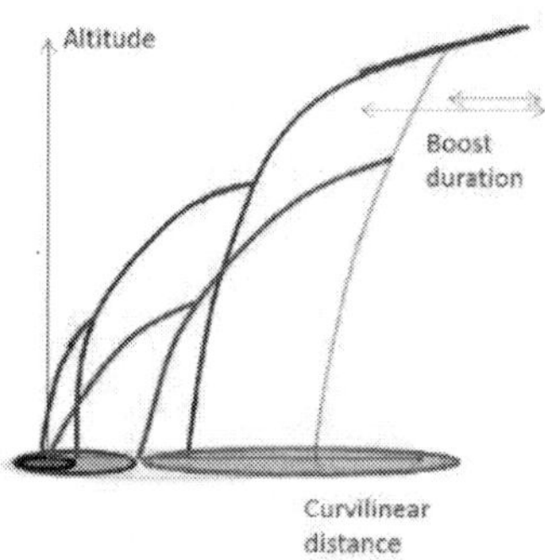

Figure 8 First boost duration shortening distances the impact footprint at extinction from the protected areas. Nevertheless, the re-optimization of boosts 2 and 3 implies a reduction of performance

- Phase 5: During this phase the vehicle performs its atmospheric reentry. Along this phase, an intermediate boost can be foreseen in order to reduce the impact of general loads and dynamic pressure during the reentry. The impact point of the vehicle can also be approached closer to the landing site by means of this intermediate or using the natural aerodynamic lift capacity of the vehicle. The criticality of such phase will depend on the strategy employed, the altitude and the duration of the boost, if applicable.

- Phase 6: This last phase includes the landing boost.

  Due to the low altitudes and the low quantity of fuel remaining in the vehicle, this maneuver is not critical in terms of the impact footprint sizes.

  Since the vehicle dynamics along this phase is similar to the phase 1, the critical fragments driving the size and the dangerousness of the impact footprints will also be those with the best tradeoff between ejection speed and ballistic coefficient.

4. SOLUTIONS – Design-to-safety

The results of the conducted safety analysis show the potential criticality of the impact footprints in case of vehicle accidental explosion for the different flight phases of a Toss-Back maneuver (especially during high altitude phases).

This criticality can be reduced by introducing palliative measures aiming at mitigating the effects on ground of the critical fragments, following a design-to-safety approach.

4.1. Design-to-fragmentation

These measures can consist of recommendations for the vehicle conception and architecture design (i.e. the use of specific materials and technologies) so as to avoid the most sizing explosion scenarios and to master the fragmentation and energy effects of an accidental explosion.

<u>Palliative measures for explosion effects mitigation</u>

- <u>Confined explosion</u>:

The most sizing scenario in terms of explosion energy is the event of an internal mixture of propellants in a confined structure, leading to a violent explosion.

In order to exclude this scenario of confined explosion, the following risk reduction measures can be considered on the design of the vehicle:

- Separated bulkheads on the propellant tanks.
- No equipments in the inter-tank structure that could be detached or explode, causing the ruin of the tanks bulkheads.
- No internal propellant feeding lines or pressurization vessels, contained within the tanks.
- Bulkheads and welded joints should not be the weakest points of the tanks (tank rupture should initiate at the external surface level).
- Pressurization vessels should have a "Leak Before Burst" design, qualified to the different flight phases.

Equipping inter-tank structures with a venting system ensures the depressurization of the space as the altitude increases. Thus, at high altitudes, it can take a substantial quantity of propellant to ensure that the internal pressure of the structure is greater than the minimum combustion pressure needed for the reaction to initiate. Therefore, the risk of confined explosion, as a consequence of a small breach on the tanks, at high altitudes, can be excluded.

Considering these risk reduction measures, the only risk of having a fuel and oxidizer mixture in a confined space can be originated by a projected debris coming from another local explosion (e.g. attitude control system or main engine PROMETHEE). This debris, ejected longitudinally like a projectile, could pierce the aft and forward bulkhead of a tank and the aft bulkhead of the other tank, if it comes from the rear bay (and the other way round in case of a debris coming from the upper part), leading to a mixing of fuel and oxidizer in a confined space (as the inter-tank structure), and originate a violent vehicle explosion.

The credibility of such scenario will depend on the size, mass and energy of the ejected debris. This debris should be small enough not to cause a pneumatic burst of the tanks before, and big enough to ensure the enough propellant flow rate for the initiation of the explosion for the given pressure and temperature conditions.

Different solutions to avoid this scenario can be considered:

- Introducing a ballistic protection (i.e. a shield or a protective screen using multilayer materials) to protect the tanks bulkheads from the eventual debris.
- Introducing confinement solutions around the possible projectile sources (e.g. attitude control system or main engine).

- Using lattice external structures to ensure deconfinement of the inter-tank volume.

In case of introducing such kind of solutions to avoid such scenario, an eventual engine explosion would only lead to the ejection of engine debris (or debris coming from the local explosion), and the ballistic fallout of the upstanding body, which can also be very critical. The consequences of such scenario depend on the neutralization system. Upstanding body fallout can be avoided by commanding a destruction of the rest of the vehicle by activation of pyrotechnical cords. In case of neutralization by propulsion cutoff, the safety issue related to an upstanding body with an aerodynamic lift capacity should be treated by the introduction of specific aerodynamic devices to maximize its drag, reduce its lift and induce instability of the body.

- Unconfined explosion:

An unconfined explosion due to the rupture of the inter-tank skirt (in case the vehicle does not have common bulkhead between tanks), due to an exceed of the general loads or the thermal flux encountered during the atmospheric phase, can be excluded for certain flight phases, for which the dynamic pressure encountered respects a certain margin with respect to the maximum dynamic pressure encountered during atmospheric reentry (value used for the sizing of the structural design of the vehicle).

However, the risk of an unconfined explosion with a different origin, like a breach on the fuel tank or on the feeding lines of the fuel circuit, an airtightness loss, or an impact of external debris, cannot be so easily excluded.

Palliative measures for fragmentation effects mitigation

Among the most sizing fragments in terms of impact footprint size are light fragments coming from tank wall external structures. These fragments, which are very sensitive to the increase of altitude due to their high ejection velocities, despite the fact of having a relatively low ballistic coefficient, drive the dangerousness of the most critical phases (high altitude phases 3 and 4).

This effect can be mitigated by design in order to obtain a more convenient fragmentation from an aerodynamic point of view, or by preventing the potential generated fragments to experience high ejection velocities by rendering the architecture more robust to explosion.

- Impact of debris mass

Indeed, the criticality of these light fragments can be reduced by reducing their ejection speed. This can be done by applying a strengthening of the structure external walls, increasing their thickness and thus their mass, making them more resistant to the effects of the blast wave.

However, this mass gain will induce a higher ballistic coefficient (assuming a similar rupture scheme). The higher the ballistic coefficient the better the fragment will preserve its ejection velocity throughout its ballistic fallout down to ground, producing larger impact footprints.

At high altitudes, the effects on ground of the reduction of the ejection velocity are more important than the effects of the increment of ballistic coefficient. Then, the reinforcement of the structure could be a potential solution to mitigate the criticality of vehicle at high altitudes.

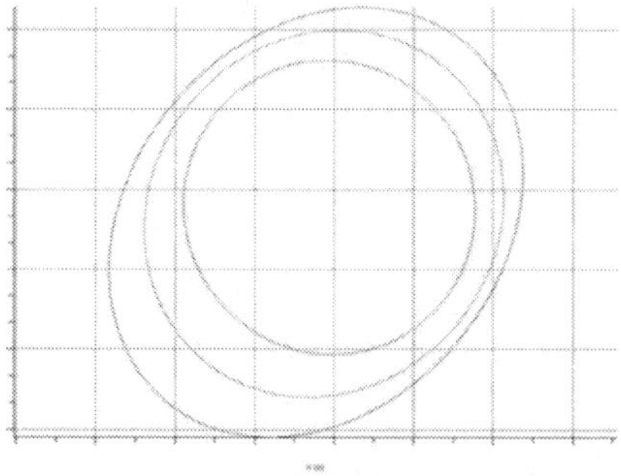

Figure 9 Debris impact footprints at the end of the ascent flight for a debris of x kg (dark blue), 2x kg (purple) and 4x kg (light blue)

Nevertheless, this solution can turn out to be problemmatic with respect to the mass budget of the vehicle, implying a performance cost for the vehicle.

- Impact of debris surface (debris mass considered constant)

Another parameter driving the sensitivity of the fragment to the gain in ejection velocity and the aerodynamic behavior is the surface of the fragment. Controlling its aerodynamic surface, the aerodynamic behavior of the fragment throughout the fallout can be optimized.

A sensitivity analysis of the impact of modifying just the surface of a fragment, considering a constant mass (the thickness of the fragment is adapted to ensure a constant mass, in order to be able to analyze exclusively the impact of the surface), showed the influence of the surface on the ejection velocity and the ballistic coefficient. For large fragments, the force that the explosion blast wave induces on the fragment is higher than for small fragments. However, a larger surface implies a lower ballistic coefficient, and thus a more important drag throughout the fallout to the ground, inducing mitigation of the effects of the initial ejection velocity on the ground.

On the contrary, small surface fragments are ejected at lower speed at the explosion but due to theirs higher ballistic coefficient, the effects of the initial ejection velocity are more visible on the impact footprint on ground.

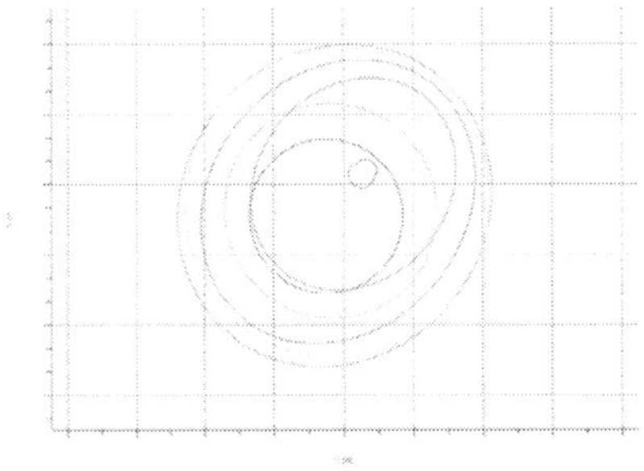

Figure 10 Sensibility analysis of fragment surface with respect to impact footprints. In dark blue, surface of 64x m^2; in light blue, surface of 32x m^2; in green, 16x m^2; in purple, surface of 8x m^2; in yellow, surface of 4x m^2; in orange, surface of 2x m^2 and in red, surface of x m^2

The results of such sensitivity analysis showed that there is an optimal surface for which the impact footprints can be considerably reduced. Indeed, for large surfaces, large impact footprints are generated as a consequence of the ejection speed. Reducing its surface decreases this effect, reducing also the impact footprints. Nevertheless, over a certain value, the effect of the ballistic coefficient takes over and the impact footprints are re-enlarged. There is then an optimal value for which both effects can be minimized.

- Impact of debris shape

It has also been proven that fragments coming from the cylindrical tank walls, for the same unfolded surface and mass, <u>longitudinal debris</u> (characterized by a large height and a small arc section) present a larger aerodynamic surface, compared to circumferentially-sliced fragments (characterized by a small height and a great arc section), having a lower ballistic coefficient. As the unfolded surface facing the explosion center is the same, the ejection speed will also be similar for both fragments.

Yet, longitudinal fragments fallout velocity is considerably more reduced at the impact on ground, producing less sizing impact footprints on ground.
In conclusion, mastering the fragmentation scheme so as to obtain longitudinal fragments is a promising solution for the criticality of the vehicle at high altitudes.

- Impact of debris surface (and mass)

For this sensitivity analysis, mass is coherent to the surface of the fragment (and will depend on the density of the material and the technology).

In general, the force applied by a blast wave on a given fragment, and thus its quantity of motion, depends on its reference surface (effective surface seen by the pressure wave). A fragment of large surface will experience greater quantity of motion compared to a smaller one. However, the increment of mass would make it more robust to the gain of ejection velocity. As the gain in mass will be greater (which depends on its real surface) than the gain in quantity of motion (which depends on its reference surface), it can be considered that increasing a fragment surface implies a reduction of the ejection velocity.

As the impact on the fragment mass is greater compared to the increment on the aerodynamic reference surface, a larger surface implies an increment on the ballistic coefficient. This fact has an opposite effect to the reduction of the ejection velocity on ground footprints.

On the other hand, reducing the surface of a fragment will then imply an increment on its ejection velocity and a reduction of its ballistic coefficient.

As for the previous sensitivity, the results showed that there is an optimal fragmentation scheme for which the impact footprints criticality can be considerably reduced. Indeed, for fragments with reduced surfaces, impact footprints are large as a consequence of their light mass. Increasing their surface will increase their robustness to the ejection velocity, reducing the effects on ground. Nevertheless, over a certain value, the effect of the ballistic coefficient takes over and the impact footprints are re-enlarged (as it is depicted in Figure 10). There is then an optimal value for which both effects can be minimized.

This strategy would require the characterization of the structural response of a fuel tank (which is the most sizing for the impact footprint) subjected to a fast dynamic solicitation. The experimental campaign would investigate the rupture scheme of a smooth wall structure and the influence of the presence of singularities, rigidity discontinuities or additional functionalized layers. The prediction and calibration of the structural response with appropriate fast dynamic simulation tools should come along with a experimental campaign, in order to support the identification of the most promising design-to-fragmentation solutions. A dedicated metallic materials characterization at high deformation velocities would be required.

4.2. Trajectory optimization

<u>Vertical ascent profile</u>

The verticality of the trajectory can be a major issue entailing more critical impact footprints as the vehicle would reach higher altitudes (producing larger impact footprints) while remaining close to the protected areas.

Reducing the verticality profile of the ascent phase, in order to distance the impact footprint from the protected areas, has an impact on the performance. The propellant cost of the return maneuver will be increased, implying a reduction of performance. In order to determine a compromise between performance and safety compliance, iterative loops of launcher concept, staging and trajectory have to be performed.

Trajectories with a hard landing strategy, as the propellant cost of the landing burn is reduced compared to a soft landing strategy, ascent trajectories have a less steep vertical profile.

Maximum apogee altitude

The criticality of phases 3 and 4 are driven by the maximum apogee altitude. Limitting the maximum altitude of the trajectory mitigates the effect on ground of the critical fragments, which are ejected at high ejection velocities.

Indeed a reduction of several hundreds of meters entails an important reduction of the final impact footprint on ground.

This can be ensured by controlling the duration and the direction of the boost-back burn, tightening the reentry trajectory profile.

Translation of the impact point at the end of the boost-back burn

In order to distance the impact footprints at the end of the boost-back burn from the protected areas, a different impact point, further from the protected limits, should be targeted at the boost extinction.

This can be done by controlling the duration and the direction of the boost-back burn. The rest of the return maneuver should be re-optimized so as to ensure the shift of the impact point towards the final landing site.

In case the natural lift capacity of the vehicle allows covering such distance (by means of aerodynamic devices), an intermediate burn should not be needed.

Nevertheless, when safety needs impose a greater distance from protected areas, an intermediate boost should be needed to perform the impact point shift to the landing site. However, this solution would imply a reduction of performance.

The criticality of an explosion arising during the intermediate burn will be driven by the altitude of the burn and the horizontal velocity induced to the vehicle. In general, the greater the altitude at which the burn is performed, the lower the need of horizontal velocity to shift the impact point a certain distance. In the meantime, the greater the distance between the initial impact point and the landing site, the greater the inferred horizontal velocity needed.

So for an intermediate burn performed at low altitude, the important horizontal velocity inferred can reveal the criticality of fragments with extreme values of ballistic coefficient, which will determine elongated footprints.

The higher the altitude of the intermediate boost, the larger the impact footprints will be (determined by the light fragments with high ejection velocities). So an intermediate boost should be performed at an instant with the best tradeoff between altitude and horizontal velocity needed.

Nevertheless, this solution implies a reduction of performance. In order to determine a compromise between performance and safety compliance, iterative loops of launcher concept, staging and trajectory have to be performed.

4.3. Safety logic

In parallel to the investigation of the design-to-safety solutions in order to reduce the impact footprints in case of accidental explosion, the study of innovative safety architectures is a promising lead so as to ensure the near range protection from deviated trajectories.

This safety logic could be defined in order to minimize the delay of the safety chain to ensure a safe intervention. This can be done by means of a localization system based on non-biased measurements or automated safety systems using reactive detection criteria.

5. REFERENCES

1. Bonhomme, C., Palerm, S., Petitot, S., Louaas, E., Tchou-kien, D., Chopinet, J.N., Danous, P., Dreyer, S. (2015). The Future of Liquid Propulsion: French Activities'. IAC-15-C4.1.4
2. Baiocco, P., Bonnal, C. (2015). The Future for Technology Demonstration for Reusable Launchers. Planet. AA-5620, Acta Astronautica 120 (2016) 43–58
3. Martinez Torio, A., Bahu, J.M., Delorme, D., Guenard, V., Poussin, H. (2016). Near Range Safety Analysis for a Reusable Launcher Concept based on Toss-Back

FIELD PROGRAMMABLE GATE ARRAY FAILURE RATE ESTIMATION GUIDELINES FOR LAUNCH VEHICLE FAULT TREE MODELS

Mohammad Al Hassan (1), **Paul Britton**(1), **Glen Spencer Hatfield**(2), **Steven D. Novack**(2)

(1) *National Aeronautics and Space Administration (NASA);* mohammad.i.alhassan@nasa.gov
(2) *Bastion Technologies Incorporated, 17625 El Camino Real #330, TX 77058, USA* steven.d.novack@nasa.gov

ABSTRACT

Today's launch vehicles complex electronic and avionics systems heavily utilize Field Programmable Gate Array (FPGA) integrated circuits (IC) for their superb speed and reconfiguration capabilities. Consequently, FPGAs are prevalent ICs in communication protocols such as MILSTD-1553B and in control signal commands such as in solenoid valve actuations.

This paper will identify reliability concerns and high level guidelines to estimate FPGA total failure rates in a launch vehicle application. The paper will discuss hardware, hardware description language, and radiation induced failures. The hardware contribution of the approach accounts for physical failures of the IC. The hardware description language portion will discuss the high level FPGA programming languages and software/code reliability growth. The radiation portion will discuss FPGA susceptibility to space environment radiation.

1. INTRODUCTION

The digital integrated circuit that makes up the FPGA is based on Complementary Metal-Oxide-Semiconductor (CMOS) technology. This integrated circuit is designed to be configured by the end user or customer after manufacturing. Unlike Application Specific Integrated Circuit (ASIC), FPGAs are designed with the capability to be configured and reconfigured, hence the name "Field programmable". As shown in Figure 1, the internals of the FPGA IC consist of programmable logic blocks and a hierarchy of reconfigurable interconnects that can be inter-wired in different configurations. Those interconnects are made possible through the CMOS-based IC transistors. The user gets to program the hardware of the FPGA by programing the logic structure of the device: logic blocks and interconnects.

In complex electronics, such as those used in the spacecraft, FPGAs are generally used to perform command, control and communication signal functions. The FPGA is used as the interfacing device between the controlling/commanding device (e.g., flight computer) and the commanded component, such as solenoid valves controlling flow from fuel tanks or thrust vector controllers.

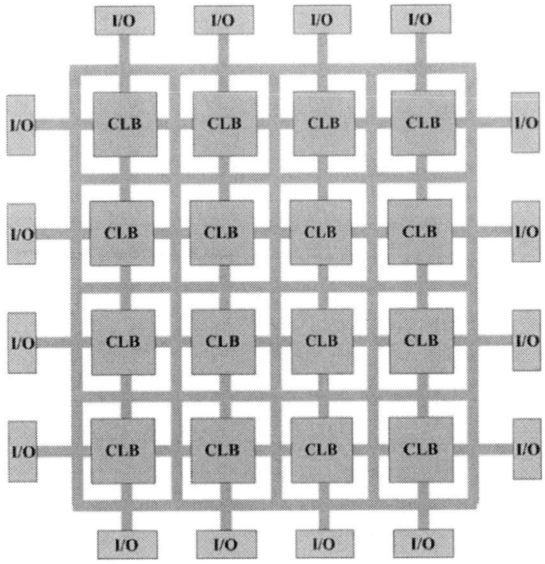

Figure 1. SRAM-Based FPGA Logic Blocks and Interconnects [1]

Herein lies the ability of FPGAs to introduce catastrophic failures for launch vehicles, such as loss of mission, vehicle, or loss of crew. FPGA hardware has the potential to experience different failure modes, such as fail-in-place or fail high/low. Likewise, Hardware Description Language (HDL) coding errors and radiation induced failures have the potential to drive the FPGA to initiate erroneous actuation of the FPGA-controlled components.

2. GUIDELINES TO ESTIMATE FPGA FAILURE RATE

The approach described below aims to provide guidelines to consistently estimate FPGA failure rates across generic spacecraft subsystems. The discussion of this approach will be divided into three sections, hardware, hardware description language code, and radiation effects.

It is important to note that Bayesian updates apply to all three risk contributors discussed in this paper to incorporate data that becomes available from testing and flight operations.

2.1 Hardware Contributions

The bathtub curve, shown in Figure 2, characterizes the hazard function and comprises three parts, infant mortality, useful life, and wear out. The "Infant mortality" steep slope of the curve represents initially high failure rates that decrease with time as defective parts are identified and discarded. The curve then flattens as the failure rate becomes more constant and the curve is referred to as constant failure rate region or useful life region. Eventually, the failure rate increases in the wear out region as age and wear induce failures,

In the Useful Life region, the time between random failures, is a reliability figure of merit known as Mean Time Between Failures (MTBF), MTBF is the inverse of the component's failure rate (λ=1/MTBF).

Hardware failure rate data sources for an FPGA include historical data, similar component/model demonstrated reliability data, testing, prediction as in MIL-HDBK217FN2, or expert elicitation.

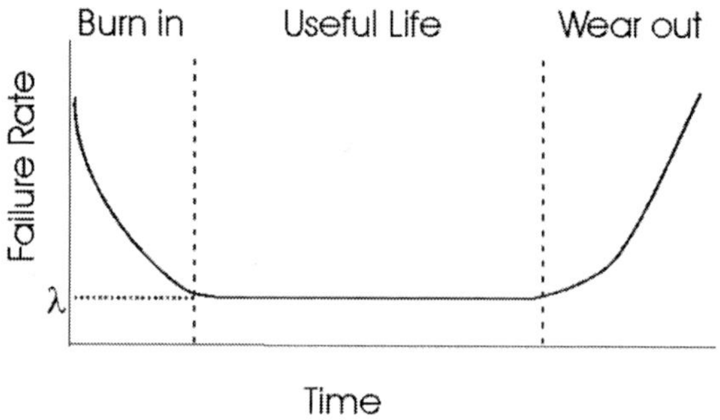

Figure 2. Bathtub Curve Hazard Function for Hardware Failure Characterization

2.2 Hardware Description Language Contributions

The goal of this Section is to provide guidelines to account for failures arising from programming languages used to program FPGAs.

The logic blocks and interconnects of an FPGA are considered hardware, and are programmed/synthesized by programming software such as Very High Speed Integrated Circuit Hardware Description Language (VHDL) or Verilog where the code is subject to software "failure" causes such as bad requirements, programming errors (coding bugs), latent errors, etc. According to NASA Primary Avionics Software System (PASS) report by Johnson Space Center, latent error is defined as "A segment of code that fulfills its requirements except under certain off-nominal, and probably unanticipated conditions" [2]. Latent errors make it past testing and onto operational flights before they are discovered.

It is necessary in this Section to make a distinction between hardware and the software used to program the hardware in terms of failure rate/reliability. This is due to the fact that software and hardware are dissimilar in many aspects. The PASS report [2] points out that software does not wear out over time as hardware does. Software is not susceptible to fatigue or to environmental stressors such as temperature, pressure, shock, vibration and radiation. Therefore, the software hazard function cannot be characterized by the bathtub curve, but is rather modeled with the software reliability curve, as shown in Figure 3.

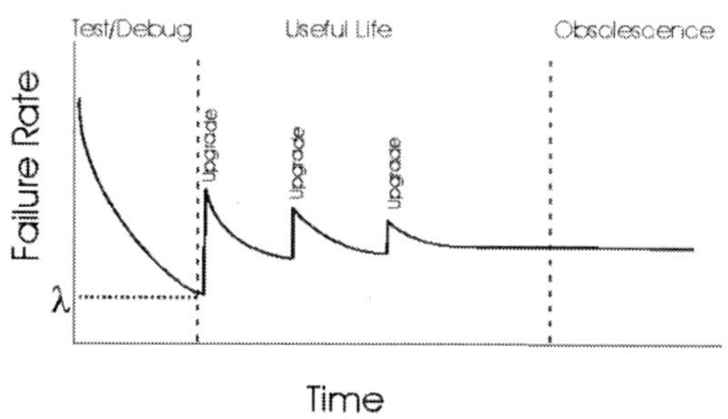

Figure 3. Software Reliability Curve

The Test/Debug region of the curve represents discovery and correction of code faults prior to or during operational use. In the Useful Life region, upgrades introduce new code faults and are evident by the spikes in failure rates. However, the maturity of the code (early mature, mid-mature, and late mature [2]) during Useful Life must be factored in estimating the code's probability of failure. Late-matured code is expected to be the most robust of the three maturity levels. Software risk assessment is often considered relatively more difficult than hardware risk assessments, and every spaceflight program with an interest in quantifying FPGA HDL risks would need to leverage historical data, test data, and prediction data when possible. Finally, in the Obsolescence region, no more upgrades to the code are conducted and the failure rate in this region becomes entirely driven by latent errors.

2.3 Space Radiation and FPGAs

Space environment is characterized by different sources of radiation that exist within the various space environments (e.g., South Atlantic Anomaly, or Van Allen Belt). Ionizing radiation, has the potential to strip off electrons from the molecules they interact with, hence the name "ionizing radiation". Listed below are the most common types of radiation found in space [3].

I. Galactic Cosmic Radiation (Cosmic Rays) This type of high energy ionizing radiation comes from exploding stars (Supernovae) and has strong potential to strip-off electrons or leave ionic tracks in the insulation layer of the gates and is considered the most damaging. It is very difficult to shield spacecraft components from this type of radiation.

II. Trapped Radiation
Trapped radiation is comprised of highly energetic charged particles trapped in the Earth's magnetic field, also known as the Van Allen Belt. The threat associated with this type of radiation is eliminated once the space vehicle is travelling outside of the Van Allen Belt.

III. Solar Energetic Particles
The source of these particles is the sun and they appear in high intensity. Protection from these high-energy particles is easier than cosmic rays and trapped radiation.

2.3.1 FPGA Hardware and Space Radiation

As mentioned above, ionizing radiation deposits energy onto the molecules or atoms it interacts with, and is capable of stripping off their electrons. These high energy particles can interact with the CMOS semiconductor doping of the FPGA, causing erroneous FPGA operation, which poses a threat to the spacecraft reliability.

In general, ionizing radiation effects on integrated circuits such as the FPGA, are classified into two categories: Total Ionizing Dose (TID) and Single Event Effects (SEE). TID is defined as the radiation accumulation thresholds before a transistor starts to experience variation in voltage thresholds and its junctions start to leak currents, leading to functional failure of the transistor. The significant sources of radiation in this case varies from trapped electrons, trapped protons, and solar protons. Fortunately, TIDs do not pose a threat to modern spacecraft as their FPGAs may come equipped with radiation hardened technologies that can withstand long years of radiation accumulation.

On the other hand, SEEs are a serious concern to spacecraft and must be accounted for in the fault tree analysis. They are capable of interrupting a data path and/or causing loss of key spacecraft control function (e.g., loss of communication with flight computers, loss of propulsion control or erroneous valve actuation) leading to loss of mission/crew. A SEE occurs when an energetic particle, such as a cosmic ray's heavy ion or a heavy proton in the Van Allen belt strikes the FPGA integrated circuit leading to disruptive effects. SEE comprises two main categories: soft SEEs and hard SEEs. A soft SEE is referred to as Single Event Upset (SEU) and includes data upsets like bit flips to memory cells or transient pulses in the logic circuitry. Hard SEEs are Single Event Functional Interrupts (SEFI) and Single Event Latch-up. (SEL). SEL is considered the most severe case of SEE that leads to physical destruction of the IC. Fortunately, modern designs and technologies of the spacecraft FPGAs have rendered SELs unlikely to occur.

2.3.2 FPGA Programming Technology and Space Radiation

Space-flight FPGAs come in different memory/programming technologies such as flash-based, Static Random Access Memory (SRAM) based or antifuse-based. Flash-based FPGAs and SRAM cells are more vulnerable to TID and SEU, respectively. A penetrating cosmic ray heavy ion has the capability, depending on the material density and shielding thickness, to penetrate and change logic gates voltage thresholds which can lead to changes in the logic structure. However, antifuse based FPGAs are not reprogrammable and are significantly less sensitive to data upsets or damaged by heavy ions at the energy levels found in space [4].

Some modern spacecraft technologies are inclined toward lowering costs by reducing requirements for components physical parameters such as weight, size, and power consumption, without compromising performance. In order to accomplish this objective, ICs like SRAM utilize new technologies including high speed and lower power CMOS and fiber optics, which are very vulnerable to SEEs [5].

2.4 Failure Rates and the Fault Tree

Table 1 below provides the most common data sources to each failure category of the FPGA along with examples to illustrate the expected format of the failure rate or probability of failure (Pf). A typical spacecraft FPGA high level fault tree should conform to the fault tree shown in Figure 4, which illustrates FPGA high level fault tree logic.

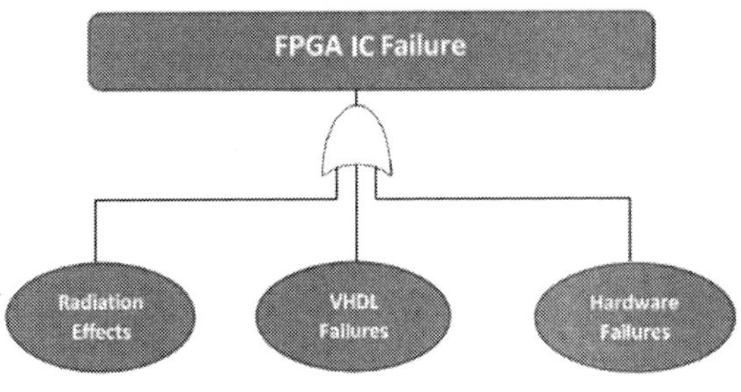

Figure 4. FPGA Fault Tree High Level Logic (OR)

Table 1. Data Sources and Example Failure Rates

Failure Category	***Data Sources***	***Notes***	***Arbitrary Example Failure Rate/Pf***
Hardware	Historical data, prediction methods, and demonstrated reliability data from reliability databases such as EPRD	Modern technology and robust manufacturing techniques have rendered the hardware risk category to be of low-impact, relative to the other two failure categories	1.45 FPMH (68,965 MTBF) *FPMH = Failure per Million Hour
VHDL	Historical data, demonstrated data, and software prediction programs data	Software reliability growth should be factored in (early mature, mid-mature, and late mature). Failure rate/failure probability is expected to progressively improve with each growth category. The fault tree should account for the most current growth category only.	Pf per KSLOC: Early-Mature 7E-06 Mid-Mature 4E-06 Late-Mature 1E-06 *Pf = Probability of Failure *kSLOC = 1,000 SLOC
Radiation	Historical data, demonstrated data, and SEE prediction programs such as CREME96	The predominant contributor to the SEE prediction is the soft and transient errors (SEU)	500 FPMH (2,000 MTBF)

3. CONCLUSION

FPGAs speed, configuration flexibility, and cost effectiveness have made the ICs highly sought after in space mission programs to implement high-speed signal processing in spacecraft. However, the FPGAs reliability have been rendered vulnerable to three failure categories: physical hardware, programming-induced failures, and radiation-induced failures. FPGA hardware is an integrated circuit of components with proven reliability track record such as transistors and multiplexors, therefore, it is safe to assume that FPGAs hardware reliability estimates are more reliable than the hardware programming languages and radiation effects by a significant margin. Programming of the hardware logic blocks and interconnects are susceptible to failures introduced to the code including wrong requirements, coding errors, and latent errors. Radiation effects pose a substantial threat to the reliability of the FPGAs and are the predominant risk contributor to FPGA failures [5] in space environment. The ionizing radiation of the space environment interact with the CMOS technology of the semiconductors of the FPGAs. Depending on the energy level of these radiations, the effects could slowly accumulate over the years until a functional failure occurs (TID), or the functional failure could be instant (SEE). In general, an FPGA fault tree should conform to Figure 4 and account for the three failure categories as independent failures (OR logic).

4. REFERENCES

[1] Farooq, Marrakchi, Mehrez, Tree-Based Heterogeneous FPGA Architecture, Application Specific Exploration and Optimization, 2012, Springer.

[2] Russell Robin, Thompson Nelson, Zhu Shangyi, NASA Primary Avionics Software System (PASS) Probabilistic Risk Assessment, SSMA-08-011 Rev. B, August 27, 2010.

[3] NP-2014-03-001-JSC, Types of Radiation in Space, NASA.

[4] Kevin Morris, FPGA Reliability in Space-Flight and Automotive Applications, Electronic Engineering Journal, September 6, 2005.

[5] NASA, Goddard Space Flight Center, Radiations Effects & Analysis: https://radhome.gsfc.nasa.gov/radhome/see.htm

PROCEEDINGS OF THE 9TH INTERNATIONAL SPACE SAFETY CONFERENCE

SESSION 18 REGULATIONS & STANDARDS – II

SPACE SAFETY AND GLOBAL SPACE GOVERNANCE

Ram S. Jakhu[(1)] and Joseph N. Pelton[(2)]

[(1)]*Institute of Air and Space Law, McGill University, 3690 Peel Street, Montreal (Quebec) H3A 1W9, Canada*
Email: ram.jakhu@mcgill.ca
[(2)]*International Association for the Advancement of Space Safety, Kapteynstraat , 2201BB Noordwijk*
The Netherlands, Email: joepelton@verizon.net

1. INTRODUCTION

Space safety means the protection of human life, property and/or spacecraft during all phases of a space mission. In other words, the term refers to the protection against technical risks, which could be avoided and/or mitigated through certain technical means and procedures that ought to be implemented through appropriate regulatory regime(s), both at international and national levels. The international regulatory regime that applies to space safety that can be generally characterized as global space governance, is today quite limited. There is the Outer Space Treaty of 1967 and the four follow-on international agreements that relate to Astronauts, Registration, Liability and the exploration of the Moon.[1] These were all negotiated in the 1960s and 1970s. There has been little new added since. After years of negotiation there are now voluntary guidelines to limit the generation of orbital space debris and enhance safety with regard to nuclear power source materials in space. There are also national legislative acts and regulations adopted by a number of countries around the world. The newly developed "Space Legal Tech platform" developed by Prof. Lucian Rapp in France now provides information on such national legislative acts and regulations. Overall, global space governance related to space safety is today limited and in need of strengthening in the years ahead. The United Nations Committee on the Peaceful Uses of Outer Space (UNCOPUOS), through the UNISPACE+50 process is currently moving to define its next 12 years of work and the International Association for the Advancement of Space Safety (IAASS) needs to be a part of this.

At McGill University Institute of Air and Space Law, in Montreal, a recently concluded three-year study of global space governance addresses some 20 different topics where improvements in global space governance might be sought. Many of these involve issues directly or indirectly related to space safety.[2] Indeed the IAASS and the Secure World Foundation (SWF) have been formally requested to present to the UNCOPUOS and other interested bodies the findings and recommendations from this effort known as Global Space Governance: An International Study. (hereinafter referred to as the International Study).

This paper outlines briefly five areas related to space safety that merit priority attention and urgent international actions to specify means, methods and processes to improve space safety through treaties, laws, regulations, guidelines or codes of conduct, transparency and confidence building measures. This paper specifically outlines possible improvements to global space governance related to space safety in the following areas:

1. Space Traffic Management (including the *Protozone* (20 km to 160 km) and Space Situational Awareness
2. Orbital Space Debris and the related issue of Large Scale Deployment of Satellite Constellations in Space
3. Cosmic Hazards and Planetary Defense
4. Space Security and Non-Deployment of Weapons in Space
5. The Mission and Functioning of the UNCOPUOS.

2. SPACE TRAFFIC MANAGEMENT

Space Traffic Management, including the *Protozone* (20km to 160km) and Space Situational Awareness, is clearly a key issue of increasing concern to those who address space and near space safety. There are many matters of concern here, including the safety of passengers on space plane flights, high altitude dirigible and balloon ascensions, and future flights on hypersonic transport. Further, this is a key issue for those concerned with space situational awareness, increasing threats from orbital space debris, monitoring the launch of missile systems, and possible military actions in space or near space. This is a difficult and complex issue, particularly as space systems and capabilities emerge, not only from the perspective of international law and regulation but also in terms of the technical, operational, and safety systems needed to improve space situational awareness and safe traffic control and management.

It might assist the UNCOPUOS in its consideration of this difficult issue if additional global expertise were invoked to examine this issue with a view to identifying possible new solutions and/or new technical competencies that need to be introduced. Several groups with special expertise on this subject thus might usefully be asked to consider the need and methods for implementing a comprehensive space traffic management system. The groups might include: (i) the Working Group on the Long-Term Sustainability of Outer Space Activities (LTSOSA), (ii) the Inter-Agency Space Debris Committee (IADC), (iii) the International Academy of Astronautics (IAA), (iv) the International Law Association, and (v) the IAASS and (vi) the SWF.

This analytical effort on space traffic management might specifically be mandated to study the issues raised in the book on International Study. This should be done on a priority basis with a view to rendering a brief report within three to six months' time. This solicited report should seek specific recommendations on the following questions:

(i) Which international organization should be explicitly asked to undertake the future conduct of Space Traffic Management and Control, including related activities involving international operations in "near space" (or the so-called *Protozone*, i.e. the area between 20Km and 160Km)?

(ii) Are there new, upgraded or expanded technical systems or competencies needed to implement necessary systems for improved safety of operations? and

(iii) Are there specific actions needed to advance international safety systems associated with implementing space traffic management and control, perhaps in terms of amending the constitutive document(s) of specialized agencies of the UN, increased coordination between or among national or regional aerospace safety regulatory bodies, or funding actions to allow progress in this key area moving forward?

In order to achieve effective and efficient space traffic management, it imperative to have effective space situational awareness (SSA). The objective of SSA is to provide timely and extensive information about conjunctions, on global and nondiscriminatory basis. Significant new capabilities are being added to aid in global situational awareness such as the new S-band space fence that is being deployed in the Micronesia region of the Pacific Ocean. These and other new capabilities are now key.[3] This is particularly so given the plans to deploy large scale small satellite constellations—some that would envision deployment of thousands of satellites and real concerns about the effective de-orbit of such a large number of satellites and the prospect of more orbital collisions. The UNCOPUOS Working Group on the Long Term Sustainability of Outer Space Activities (LTSOSA) and other interest groups should focus on this issue with renewed emphasis.

3. ORBITAL SPACE DEBRIS AND LARGE SCALE DEPLOYMENT OF SATELLITES IN SPACE

Since the International Study was completed, additional filings have been made to deploy low Earth orbit small satellite constellations.
Some of the systems even proposed deploying over 10,000 small satellites in a single communications satellite constellation. This development raises two concerns; i.e. the first one relates to the increase in space debris and the second one related to radio frequency interference and jamming.

The latest estimates as to the likelihood of a significant new collision from the space agencies—even without the deployment of these so-called Mega-LEO constellations– is between every five years and every ten years. This possibility of the so-called Kessler Syndrome ultimately leading to an ever-increasing cascade of deadly space debris has been strongly indicated in the International Study. The issue of how many new constellations of small satellites can be safely deployed and not threaten access to space by all nations is a serious concern.

It is widely believed that this subject needs urgent attention. Before very large small satellite constellations are deployed, there should perhaps be a moratorium (perhaps of a two-year duration) on deploying such Mega-LEO constellations pending international agreement on new regulations pertaining to such networks. This would presumably entail some additional form of controls and limitations on what seems to be potential dangers to the long-term sustainability of space. These new "regulations" might be in the form of an international code of conduct or internationally agreed guideline as to the reasonable limits on the number of mega constellations and especially whether there should some explicit safety precautions against space collisions or radio-frequency interference. A prime objective might be to ensure that there is a nearly "fail-safe" process to de-orbit satellites in these large constellations at the end of life. This is because recent studies support that this represents the prime space debris problem.[4] These new regulations would presumably need to be overseen by an international agency with some oversight powers and procedures. This process would need to be monitored and enforced by either the International Telecommunication Union (ITU) or some other agreed international body. This review might also consider guidelines for national model laws as to enforcement procedures for the 25 year deorbit guideline, such as has been implemented in the French Operations Space Act. In addition to the two-year moratorium on system deployment pending new regulatory action, there also might a moratorium on new ITU filings for such systems until the ITU Secretariat carries out a study as to the impact such large-scale systems pose to the future long-term use of space due to the expanded threat of excessive orbital space debris and the potential Kessler Syndrome. This study should also consider the additional technical issue of potential interference between these systems and protected geostationary orbit (GEO) satellite systems.

One of the recurring international space related issues that has now become a chronic problem is that of radio frequency interference and jamming involving spacecraft in orbit. The deployment of high altitude platform systems (HAPS) in the stratosphere, and the proposed positioning of electronic networking devices in the stratosphere, could further increase the intensity and frequency of this problem. Issues of interference now increasingly involve not only governmental spacecraft, security and defense-related spacecraft, but also affect all types of commercial spacecraft that are involved in the provision of a variety of space services. As large-scale constellations and HAPS are deployed, the potential for greater interference will only increase. Currently, other than bringing the issue to the attention of national administrations of where the interference has originated, the ITU has limited capabilities to address this problem.

It is recommended that the UN Office of Outer Space Affairs (UNOOSA) and officials of the ITU consult regarding this problem. A report should be prepared that highlights the frequency, nature and magnitude of this problem and suggests possible actions that might be taken within the UN system to ameliorate this problem.

4. COSMIC HAZARDS AND PLANETARY DEFENSE

The recent measurements carried out by the Swarm satellites of the European Space Agency (ESA) Swarm and NASA's Magnetospheric Multiscale (MMS) Satellite mission have confirmed significant changes in the Earth's magnetosphere. These measurements indicate a shift of the magnetic poles and the possibility that the protective magnetic and ionic shielding represented by the shaping of the Van Allen belts that affords protection to vital planetary infrastructure could be at increasing significant risk as these trends continue. A Lloyds of London study by technical consultants have indicated the possibility of a "3 trillion-dollar black swan event" from a massive solar storm.

Various studies have indicated that threats from asteroid impacts, coronal mass ejections, solar flares, etc. are truly increasing as global populations, rapid urbanization and dependence on electronic power systems, continental pipelines, information technology networks and telecommunications grow in importance.

Action Team 14 from UNISPACE III, which led to the implementation of the International Asteroid Warning Network (IAWN) and the Space Mission Planning Advisory Group (SMPAG), played a key role. Perhaps a similar process might be envisioned at the UNISPACE+50 event that could gather and process the latest information from spacecraft monitoring the Sun, changes to the Earth's magnetosphere, threats of asteroid and comet impacts, etc.

This would be with a view to developing new ideas on planetary defense, safeguard research initiatives, and even the possible feasibility of creating an internationally-deployed and operated solar shield system at L-1.[5]

5. SPACE SECURITY AND NON-DEPLOYMENT OF WEAPONS IN SPACE

The Space Security Index that is prepared each year by a number of respected organizations, including McGill University, from around the world notes that the future prospect of conflicts in space seems to be increasing each year.[6] Currently, there are some efforts being made by certain nations to stem this tide, including a code of conduct, but none has been successfully adopted. There is clearly a need to seek new means to mitigate space-based hostilities (or hostilities involving space technology and applications) and attacks on space infrastructure. Codes of conduct, standards of practice, or even transparency and confidence-building measures could all assist to mitigate possible attacks in space, from space, or from Earth to space. Any serious conflict in space would have a devastating effect not only on assets in space but also on international relations on the Earth and, in addition to the humanitarian crisis as a result of an armed conflict, could potentially result in major global economic consequences.

It is recommended that the Union of Concerned Scientists, in collaboration with the International Academy of Astronautics and other appropriate organizations, conduct an objective study on the reasons, nature and scope of the emerging conflicts in space as well as their implications for space sustainability and international relations. If this study indicates serious negative implications, the UN General Assembly should adopt a declaration of legal principles designed to avoid conflicts in space. In parallel to this study, additional efforts by an international panel of qualified space experts should be taken to draft a uniform code of conduct governing all types of space activities so as to minimize the potential destruction of or harm caused to any space assets—public or private. Importantly, procedures for redress and enforcement for violations of the proposed code of conduct should be agreed upon. Further, States and other stakeholders in the space domain should take note of and contribute to the efforts to draft *McGill Manual on International Law Applicable to Military Uses of Outer Space*. When completed, the Manual will go a long way to clarify existing international law in relation to strategic and military actions that are prohibited or permissible in times of peace and in times of armed conflict, thereby strengthen international security and the rule of law in outer space.

6. THE MISSION AND FUNCTIONING OF THE COPUOS

As mentioned in the Introduction, to practically ensure the achievement of space safety to a maximum possible extent, it is imperative to appropriate international institution that could adopt and implement safety standards and procedures.

The swift changes to the world's economy and labour structure, to the natural environment, and governance structures since the start of the 20th Century have not been clearly understood. Since 1900 to today, the world's population grew exponentially from 1.8 billion to 7.3 billion and is projected to exceed well over 10 billion by 2100. Space and space-related activities, issues, and problems will become much more important in the years ahead for both developed and developing nations. As a result, the regulatory problems, disputes, and yes, even conflicts, could all become more challenging and complex to resolve in the years ahead. Governmental programs and regulatory oversight have historically been reactive in nature, thus responding to problems rather than proactively anticipating concerns and creating rules and processes before the worst happens. This is, in a world of rapid technological growth and social, economic and cultural change as well as faced with a myriad of environmental issues is simply not viable, particularly in the face of the sustainability and survival of humankind.

Global space governance refers not only to international, regional, or national laws, policies and non-binding guidelines but also imply the existence of regulatory institutions mandated with actions/mechanisms for (a) governing or regulating space-related affairs or activities and (b) for the compliance monitoring and enforcement processes. As Jean Monnet, one of the founders of modern day European Union, correctly asserted, "Nothing is possible without men, but nothing is sustainable without institutions[8]; [and] Institutions may, if they are well constructed, accumulate and transmit the wisdom of successive generations"[9]. Undoubtedly, it would have been extremely difficult, if not impossible, to the see the orderly and highly progressive development of global telecommunication and aviation activities without the active and systematic roles played by the ITU and the International Civil Aviation Organization (ICAO). The upsurge in the number of space players with very divergent political agendas and economic interests, the exponential growth of space activities, and the rapid increase in the variety and complexity of space operations underline the necessity and urgency of one central international institution with the sufficient capacity and capability for the effective, efficient and orderly international coordination and global governance of space related activities. It is therefore important for the UN General Assembly to seek ways to think ahead, to proactively diagnose problems as they are emerging and to develop processes or regulations to address the next frontier of issues and problems arising from humankind's increased reliance on and venture into outer space.

It is recommended that the UN General Assembly take action to strengthen the role of UNCOPUOS. This international body with the mandate to promote and ensure the peaceful use of outer space must be endowed with the necessary financial resources, sufficient and competent staff, and supported by teams of experts from a variety of spacefaring and non-spacefaring nations so that it can continue to identify rapidly emerging issues and to study space-related problems and concerns as they occur. UNCOPUOS, as the central international body for space affairs, ought to actively pursue the progressive development of global space governance but also to coordinate the actions of all the concerned international institutions in order to avoid the fragmentation or duplication of the space governance system. It should be recognized that specific issues will require specific solutions and specific expertise, and, hence, the challenge for the international community
(as embodied by UNCOPUOS) is to coordinate the many layers of governance to ensure the global public interest remains at the core of all forms of regulatory and oversight actions. In particular, this might involve identifying how space activities and applications can be used to assist in meeting all the space-related governance needs, as well as in meeting the U.N. Sustainable Development Goals for 2030.[7] In addressing priority issues and strengthening global space governance, the proposed actions would not necessarily involve the costly creation of new international projects and institutions. Rather, in addressing issues and concerns of global consequence and impact, the starting point should be identifying ways to incorporate national or bi-lateral space initiatives, efforts by existing nongovernmental institutions or stakeholders and incorporate both the public interests and private space initiatives.

7. CONCLUSION

One of the key focus areas of the UNISPACE+50 event is improved global space governance that will be an area of UNCOPUOS concern in setting its action agenda for the next 12 years through 2030. The IAASS needs to work with UNCOPUOS and its LTSOSA Working Group to address all of the above seven priority areas to enhance space safety going forward. In this respect it needs to create new partnerships with other interested groups and organizations such as the Secure World Foundation, the International Academy of Astronautics, the Association of Space Explorers, the International Space University, the McGill University Institute of Air and Space Law and others that share the concerns and interests addressed in this paper.

8. REFERENCES

1. U.N. Treaties and Principles on Outer Space (2017). http://www.unoosa.org/pdf/publications/STSPACE11E.pdf

2. Ram Jakhu and Joseph Pelton (editors), Global Space Governance: An International Study (2017) Springer Press, N.Y.

3. Joseph Pelton, "Upgrading Debris Tracking Capabilities" Space Debris and Other Threats from Outer Space (2013) Springer Press, N.Y. pp 23-25

4. H. Klinkrad, "Large satellite constellations and related challenges for space debris mitigation" Journal of Space Safety Engineering, Vol. 4, No. 2, June 2017 p. 59.

5. Joseph N. Pelton and Firooz Allahdadi (editors), Handbook of Cosmic Hazards and Planetary Defense (2014)

6. Space Security Index (2017) http://spacesecurityindex.org/2016/11/space-security-2016/

7. U.N. Sustainable Development Goals for 2030, (2017). http://www.un.org/sustainabledevelopment/sustainable-development-goals/

8. http://www.azquotes.com/quote/652155

9. http://www.newworldstandardcritique.com/25279-2/

A MODEL FOR SETTING A REGULATORY FRAMEWORK FOR THE DEVELOPMENT OF SUBORBITAL OPERATIONS IN ITALY

Giovanni Di Antonio(1), Marco Sandrucci(2), Alessandro Cardi(3), Francesco Santoro(4), Alberto Del Bianco(5), Cristoforo Romanelli(6),

(1)*ENAC – Italian Civil Aviation Authority, Airworthiness Regulation Department, Via Gaeta 3, 00185 Rome – Italy,* g.diantonio@enac.gov.it
(2)*ENAC – Italian Civil Aviation Authority, Turin Operations Office, Via R. Montecuccoli 2, 10121 Turin – Italy,* m.sandrucci@enac.gov.it
(3)*ENAC – Italian Civil Aviation Authority, Deputy Director General, Via Gaeta 8, 00185 Rome – Italy,* a.cardi@enac.gov.it
(4)*ALTEC S.p.A., Corso Marche 79, 10146 Turin – Italy,* francesco.santoro@altecspace.it
(5)*ALTEC S.p.A., Corso Marche 79, 10146 Turin – Italy,* alberto.delbianco@altecspace.it
(6)*ALTEC S.p.A., Corso Marche 79, 10146 Turin – Italy,* cristoforo.romanelli@altecspace.it

ABSTRACT

This paper describes the model currently under discussion in Italy which has been developing for setting up a regulatory framework for commercial suborbital operations. This model is expected to accommodate in the near-midterm HOTOL (Horizontal Take-Off and Horizontal Landing) suborbital operations within the national aviation system, the national airspace and using existing facilities, taking into account the relevant international regulatory experience in the field. An adequate level of safety for uninvolved people on the ground, in the air, at sea and for critical infrastructures will be primarily sought along with due considerations for the safety of involved people onboard, in line with the state of the art and the technological development of the sector.

1. INTRODUCTION

The growing interest and development of novel technologies in commercial suborbital operations of a number of consolidated and innovative firms and stakeholders pushed ENAC, the Italian Civil Aviation Authority, since 2014, to study, evaluate and analyze this emerging field [3] with the aim to develop an effective and, as much as possible, flexible regulatory framework which can accommodate sub-orbital operations within the existing aviation and airspace system. Possible applications of suborbital operations include tourism, business, microgravity experimentations and astronauts' training. The goal is to enable the development of HOTOL suborbital operations from existing airports within the Italian territory, primarily assuring an adequate level of safety for third parties on ground, namely uninvolved people and properties, and other airspace and sea users. As far as commercial operations are concerned, the mandatory control of the hazards for involved people onboard shall be required as well to an extent in line with the state of the art and with the development of the sector.

2. LEGAL FRAMEWORK

Whether or not suborbital operations has to be regulated under the air law or under the space law is currently debatable, hence each State has to clarify which is the entity in charge to regulate and oversight the nascent sector of the commercial suborbital flights. This aspect will be even more relevant on the long term, when suborbital flights will connect two different points of the Earth setting forth for the future generation transportation. From a functionalist point of view [19] it could be reasonably agreed that the Aviation Authority could regulate a suborbital operation, though by setting an *ad hoc* regulatory regime and this for a few reasons. Firstly, HOTOL vehicle (at least some of its stages) is, in most cases, a winged vehicle capable to derive (part of) its support by the atmosphere from the reaction of the air, at least in some phases of the trajectory, namely during the launch and the re-entry. In this regard the spaceplane could meet in some way the ICAO (International Civil Aviation Organization) definition of aircraft as any machine that can derive support in the atmosphere from the reactions of the air [15]. Secondly, any suborbital trajectory will intersect and share the airspace below FL650 (Flight Level 650) during the launch and re-entry phase, hopefully by means of a dynamic 4D segregation. One more important factor to be considered is the fact that, at least in Italy, the sites likely to be used as a spaceport for commercial launch and re-entry operations will be selected among existing civil aerodrome approved under the EU (European Union) Regulation 139/2014. Most of these sites are State airports assigned to ENAC by law according to the Art. 693 of the Italian CoN (Code of Navigation) [1], in turn assigned by ENAC to the aerodrome operator through a specific agreement. Finally, the Article 687 of the CoN recognizes ENAC as the unique national regulatory Authority for civil aviation and the law n. 265/2004 recognizes ENAC as the unique authority for Air Navigation Services regulation. From an international perspective, in 2013 Italy and the US (United States of America) signed the Framework Agreement for Cooperation and Use of Outer Space for Peaceful Purposes [2] further transposed into law n. 197/2015. On the basis of this agreement, in 2014 ENAC signed with the US Office for Commercial Space Transportation of the Federal Aviation Administration (the FAA-AST Office) a Memorandum of Cooperation in the Development of Commercial Space Transportation [4] later renewed and extended to the Italian Space Agency (ASI) on 30 July 2016. Taking into account the above considerations, the Italian Ministry of Infrastructure and Transportation (MIT) issued its Decree n. 354 of 10 July 2017 [5] which identifies ENAC has the national entity in charge to develop the regulatory framework for the commercial suborbital transportation in Italy. The priorities highlighted by the MIT are: (i) to develop a three-year plan to define a regulatory framework for HOTOL suborbital flights; (ii) to identify the issues that can be directly regulated by ENAC and those which may need to be addressed by the primary legislation or other government acts; (iii) the regulatory framework shall be flexible enough to support the development of the sector, provided that the safety of third parties on ground, in the air and at sea must be primarily pursued along with the safety of occupants; (iv) the level of safety for the system and operations shall be established by ENAC and shall be updated, as necessary, according to the state of the art and the advancement of the technology; (v) suborbital operations shall take place within the national border and shall not have a detrimental impact on the civil aviation development; (vi) security and cyber-security issues shall be taken into account; (vii) the criteria for the selection of candidate spaceport site(s) (subsequently described at Para. 5.) shall be defined giving priority to the existing infrastructures; and finally (viii) ENAC will coordinate with the MIT and will cooperate with the military and ASI.

3. SAFETY CONSIDERATIONS

In a very general form *safety* can be defined as the minimum level of risk deemed acceptable by the society for a specific situation. This minimum level of risk can be specified either in a qualitative or quantitative manner. The risk associated to a specific event that can lead to unwanted consequences can be expressed as the product (or the combination) of the probability that the event occurs and the measure (or severity) of its consequences. Two types of risk can be considered [21]: the collective (or societal) risks expressed in terms of costs or sacrifice suffered by the society as an all (such as the *number* of casualties or fatalities) and the individual risks expressed in terms of *probability* that a person suffer a casualty or a fatality. These two types of risk lead to the definition of two different types of safety objectives: a collective safety objective (expressed in terms of maximum admissible societal cost), and an individual safety objective (expressed e.g. as the maximum probability of having a casualty for a person exposed to the hazard). Another distinction has to be done in terms of the people exposed to the hazard. In that respect for suborbital operations we will distinguish two groups:

- The uninvolved people on ground, in the air and at sea, along with critical infrastructures (which are referred to as *third parties*), and
- The involved people onboard the spaceplane vehicle, either the crew or the other (possibly paying) participants (which are referred to as *occupants*).

For suborbital flights, the US regulation [8] addresses third parties' safety by the collective safety objective *Ec*, expressed as the maximum average number of casualties[1] per launch or re-entry mission and by an individual safety objective expressed as the maximum allowed probability for an individual of suffering a casualty during a mission. Three types of hazards are supposed to produce third parties casualties [8]: (i) the direct impact with a vehicle debris, (ii) the overpressure blast due to the explosion of a falling debris, and (iii) toxic release. In addition to that, the regulator should also pursue the minimization of the risk for the occupants (which is referred to as *occupant safety*) as an essential pillar to allow – even from a legal point of view – manned suborbital flights. The safety for the third parties in the air, i.e. the other airspace or outer space users, can be addressed in the near-midterm by segregation of the airspace below Fight Level 650 (FL650) and by limiting the maximum apogee altitude (e.g. well below the parking circular orbit of 186 km altitude). In the future this hazard is expected to be controlled by an STM (Traffic Management) system fully integrated with the existing ATM (Air Traffic Management) system [20].

3.1. Third parties safety

Let P_C be the probability that a debris from a suborbital vehicle crashes on a populated area on the ground during a launch or re-entry mission, and let N_C be the number of people that would be directly hit by a debris or by a harmful blast overpressure caused by the explosion of the debris. Defining the debris casualty area A_C as the harmful ground footprint of the debris [10], the number N_C of people exposed to the risk can be worked out as the product of the casualty area A_C times the average population density D of the region where the debris is expected to fall, i.e. $N_C = A_C \cdot D$. Assuming that everyone who is hit suffers a casualty (i.e. the conditional probability of suffering a casualty, given a person is struck, is assumed equal to 1), then the mean number of casualties per mission (or the risk R_C) can be worked out

[1] According to the US regulation [8], Part 401, Para. 401.5, a *casualty* is defined as a fatality (i.e. a death) or a serious injury.

and it may not be greater than a pre-established value E_C:

$$R_C = P_C \cdot N_C \leq E_C \qquad (1)$$

Hence, it is necessary to define the above acceptable upper limit E_C for the risk R_C. This upper limit is the *safety objective* that has to be met for assuring the safety of third parties on the ground. A possible value for E_C was originally set by the US regulation equal to 30 casualties per million of missions, i.e.: $E_C = 30 \cdot 10^{-6} = 3 \cdot 10^{-5}$ casualties per mission (see [8] Requirement 431.35). Recently this requirement has been relaxed based on a more accurate review carried out by NASA (National Aviation and Space Administration) and USAF (US Air Force) on the launch data available in the US since 1989 [9]. As a result of this review, the E_C value was increased up to 10^{-4} casualties per each launch and re-entry phase of a RLV (Reusable Launch Vehicle). In conjunction with the collective safety objective $E_C = 10^{-4}$ casualties per each launch or re-entry, the US regulation also mandates an individual safety objective in terms of a maximum probability $P_I = 10^{-6}$ for an individual of suffering a casualty during a mission[2]. If N is the total number of people within the overflown region then, besides the (1), we should also impose the following additional risk condition[3]:

$$R_C/N \leq P_I \qquad (2)$$

The According to [10], the former safety objective $E_C = 3 \cdot 10^{-5}$ was deemed consistent with the risk level attained by the US Government launch activities at the Federal Ranges. In [10] it is also shown that, once expressed in terms of individual risk out of a population of 100000 people, the safety objective E_C becomes orders of magnitude less than the individual risk per year for a US citizen, due to non-working activities. For example, assuming that 100 missions are carried out per year over a population of 100000 inhabitants, the individual probability to have a casualty during a mission would be equal to the mean number of casualties per mission out of the total population, i.e.: $3 \cdot 10^{-5}/100000 = 3 \cdot 10^{-10}$; then considering 100 missions per year the individual probability to have a casualty in one year would be equal to $3 \cdot 10^{-10} \cdot 100 = 3 \cdot 10^{-8}$. This figure must be compared with the yearly individual probability of fatality of $2 \cdot 10^{-4}$ due to non-occupational accidents in the US and with the yearly individual probability of fatality of 10^{-6} in Western Countries due to every cause not depending on personal willing [21]. Moreover it can be observed that a collective safety objective E_C in the order of 10^{-5} casualties per mission (or per phase of mission) appears to be consistent with the current commercial civil aviation fatalities rate per departure (or take-off). According to the Boeing Statistical Summary of Commercial Jet Airplane Accident [17] the worldwide operations from 1959 to 2016 resulted in a total of 29784 onboard fatalities out of 1389 million of FH (Flight Hours), corresponding to a fatality rate of $2.1 \cdot 10^{-5}$ fatalities per FH. The same order of magnitude results from the ICAO statistics for the Scheduled Commercial Flights of aircraft with a MTOM (Maximum Take-Off Mass) greater than 5700 kg during the period from 2012 to 2016[4]. According to [16] in this five-year period it is possible to count 2126 fatalities[5] out of 163.9 million departures, which results in a fatality rate of $1.3 \cdot 10^{-5}$ fatalities per departure (or take-off)[6]. Moreover, as far as the GA (General Aviation) is concerned, based on the GAMA (General Aviation Manufacturers Association) statistics for the period 1985-2016 [18] in US, it is possible to derive a rate of $1.5 \cdot 10^{-5}$ fatal accidents per FH (Flight Hour) and an average number of 2 fatalities per fatal accident. Considering that for the GA the average duration of one single flight can be typically assumed of 1 hour, it is possible to work out a rate of $3 \cdot 10^{-5}$ fatalities per departure, again in the same order of magnitude. We have to highlight now that, even if the above noted civil aviation fatalities rates refer to onboard fatalities, from a societal point of view it possible consider them as the sacrifice in terms of fatalities that the today society accepts to take benefit from the possibility of a freely movement of individuals (actually a causality is always a cost for the society whether it happens onboard or on the ground). It can thus argue [21] that in the today Western society travelling by air is more and more becoming a normal daily life necessity, and it could not be considered as an "avoidable" choice any more, should people want to live in and take benefit from the modern society. All the above considerations give us a clear indication that, at least in the Western countries, an acceptable safety objective for uninvolved third parties on the ground during suborbital operations could be reasonably set in the order of 10^{-5} or 10^{-4} casualties (or fatalities) per mission or launch and re-entry phase.

[2] A recent French low for launch activities [14] requires a "collective" (sic) safety objectives of $2 \cdot 10^{-5}$ expressed in terms of maximum admissible probability to make at least one victim, which can be increased up to 10^{-4} in case of impossibility to carry out a controlled atmospheric re-entry.

[3] In fact the probability that an individual within a group of N people suffers a casualty during a mission can be estimated as the ratio R_C/N between the average number of people that would suffer a casualty during a mission out of the total number of people exposed to the hazard.

[4] For the year 2012 the ICAO statistics take into account the aircraft with a MTOM greater than 2250 kg [16]

[5] Of which 683 are due to security reasons.

[6] Considering only the fatalities due to safety reasons a fatality rate per departure of $8.8 \cdot 10^{-6}$ would be obtained, which is very near to the 10^{-5} order of magnitude.

3.2. Occupants safety

The aviation sector probably most comparable with suborbital flights is the highest segment of the GA. The level of safety required by the current EU and US aviation regulations for a GA aircraft (expressed in terms of maximum allowed probability per FH for a technical failure of a system that would lead to a catastrophic event) ranges from a failure rate of 10^{-4}/FH for Class I aircraft (typically SRE, Single Reciprocating Engine aircraft with a MTOM of 2722 kg or less) to a failure rate of 10^{-9}/FH for Class IV aircraft (typically commuter category aircraft) which is the same required for the FAR/CS 25 large transport aircraft (ref. Tab. 1); moreover no single failure should result in a catastrophic failure condition [11].

Table 1 – GA aircraft system safety targets (maximum failure rate, or probability per FH) for catastrophic failure of a system (source [11])

Aircraft Class	Typical Description (*)	System safety target (probability / FH)
I	SRE with MTOM of 2722 kg or less	10^{-6}
II	MRE, STE, MTE with MTOM of 2722 kg or less	10^{-7}
III	SRE, STE, MRE,MTE with MTOM greater than 2722 kg	10^{-8}
IV	Commuter Category	10^{-9}
(*) SRE=Single Reciprocating Engine aircraft; MRE=Multiple Reciprocating Engine aircraft; STE=Single Turbine Engine aircraft; MTE=Multiple Turbine Engine aircraft		

Not considering the special Class IV aircraft (commuter category) which is more similar to the large transport aircraft and which is devoted to routinely transportation, a suborbital vehicle could be better compared with a Class III aircraft, with a safety target of 10^{-8}/FH for catastrophic failure. Considering that a suborbital vehicle is by nature complex, it is possible to assume, as it is usually done for large transport aircraft, that 100 independent technical failures can result in a catastrophic event, which could lead to a total probability of catastrophic failure of 10^{-6}/FH. Assuming then, as usual in aviation, that only 10% of the total accidents are due to technical causes related to the systems (the rest 90% accidents being due to operational causes, like human factor, hostile environment, incorrect maintenance, etc.) then the total catastrophic rate becomes 10^{-5}/FH which is equivalent to a rate of $5 \cdot 10^{-6}$ catastrophic events per mission if we assume an average 2-hour suborbital operation. If we now compare this figure with the accident rate of the Space Shuttle, which is in the order of 10^{-2} catastrophic events per mission [22], it is believed that a suborbital vehicle will hardly achieve the GA safety objective in the near-midterm. In this respect, it is also of significance that the safety target requested by the NASA CCP (Commercial Crew Program) for the human space flights is in the order of 10^{-3} accidents per mission [22]. Probably this last safety objective represents a lower limit for the reliability of a space vehicle allowed by the current state of the art. In the future, when routinely commercial suborbital flights will become a reality, it is expected that the suborbital vehicle will be designed, manufactured and certified against a specific-developed flightworthiness standard with a catastrophic safety target for system failures of at least 10^{-4} /FH. In the meantime, imposing this target appears largely impractical due to the current state of the art (especially in the field of rocket propulsion), therefore an alternative approach has to be pursued in order not to block the development of this emerging sector. As far as the occupant safety is concerned, according to the MIT policy [5], ENAC is evaluating the possibility to mandate qualitative performance based requirements for the safety of the occupants, which can assure a minimum level of safety by controlling the hazards and reducing the related risk as low as reasonably practicable. A possible set of occupant safety performance based requirements today available are the FAA-AST Recommended Practices for Human Space Flight Occupant Safety [12] developed on the basis of NASA, ESA (European Space Agency) and IAASS (International Association for the Advancement of Space Safety) requirements for HSF (Human Space Flight) [13]. Three main assumptions are made there, which could make them particularly suitable to implement in this phase. Firstly, as said, the FAA-AST Recommended Practices are performance based in that they state high level objective requirements to be complied with by applying industry standards tailored on the specific architecture of the system, vehicle configuration, design solutions and type of operations to be carried out. Secondly, the FAA-AST Recommended Practices do not impose a quantitative (i.e. numeric) safety objective in term of a specific minimum level or risk to be achieved because of the variety and complexities of system architectures, vehicles and operations that are anticipated to be implemented in the future and for which different levels of risks could be appropriate. The compliance with (and the implementation of) these objective requirements throughout the life-cycle of the operations (including the vehicle design and construction) will assure that potentially catastrophic hazards are sufficiently controlled in order that the occupants are not exposed to avoidable risks. Thirdly, the FAA-AST Recommended Practices may be conveniently implemented and verified by using an operation centric approach (see Para. 4.3). In this regard, the objective requirements are supposed to be complied with not only at the vehicle level but, in principle, at the system level, where in this context the term "system" is intended as "an integrated composite of personnel, product, subsystems, elements and processes that when combined together will safely carry occupants on planned space flight" [12]. This implies e.g. that a certain hazards related to the vehicle design may be controlled (e.g. by adopting adequate operational procedures, conditions and limitations) in a way very similar to that currently used in aviation for issuing PTF (Permit To Flight) to manned aircraft with a non (fully) approved type design or to RPAS (Remotely Piloted Aircraft Systems) operated in the Specific Category B by using the holistic SORA (Specific Operation Risk Assessment) methodology developed by the JARUS (Joint Authorities for Rulemaking on Unmanned Systems), [23]. This approach is well illustrated by the FAA-AST practice 1.5.2 (System Safety Engineering) which essentially recommends to implement an hazard analysis consistent with common aerospace practice, able to: a) identify and describe hazards and the associated causes; b) identify and describe each safety-critical system, its function and the consequences of its failure; c) implement a hazard control strategy that will prevent the occurrence of the hazard or mitigate the risk to an acceptable level[7]; and finally d) demonstrate that the hazard control and risk mitigation measures have been successfully implemented thought objective verification evidence [12]. This type of process is a significant aspect of the regulation because it will allow for a certain degree of flexibility without necessarily asking for the certification of the vehicle. In this regard the vehicle and the vehicle operator, the spaceport and spaceport operator along with the STM/ATM service providers will be considered as an integral part of the "system". The level of verification to be implemented by the competent Authority to gain sufficient confidence that the system complies with the occupant safety objective requirements is a crucial point to be discussed further. One possibility is using the ALARP (As Low As Reasonably Practicable) method developed in UK by the HSE (Health and Safety Executive) [24]. The adapted ALARP approach for occupant safety would be based either on establishing a set of "good practices" (i.e. stakeholders consensus standard, possibly performance based) to be fully complied with, or on the "first principle" that any measure capable to control the risk should be taken unless the sacrifice it poses is grossly disproportionate with respect to the safety gain [24]. Based on this principle the Authority could accept non compliances with the "good practices" standards where implementing them would result impractical or "would not contribute materially to the level of safety"[8]

4. REGULATORY STRUCTURE

To regulate the commercial suborbital transportation sector ENAC envisages to issue two distinct Regulations: one Regulation for issuing the spaceport authorization and one Regulation for issuing operation authorizations for both experimental and commercial activities. In addition to that ENAC intends to set bilateral agreements with foreign Countries (starting with the US government) to ease the international interoperability.

4.1. Scope and Applicability

In order to alleviate the effort and to gather experience on this novel domain, and on the basis of the on-going projects, the relevant mission profiles and in line with the MIT Policy [5], ENAC have identified a number of characteristics for the suborbital flight operations intended to be regulated.

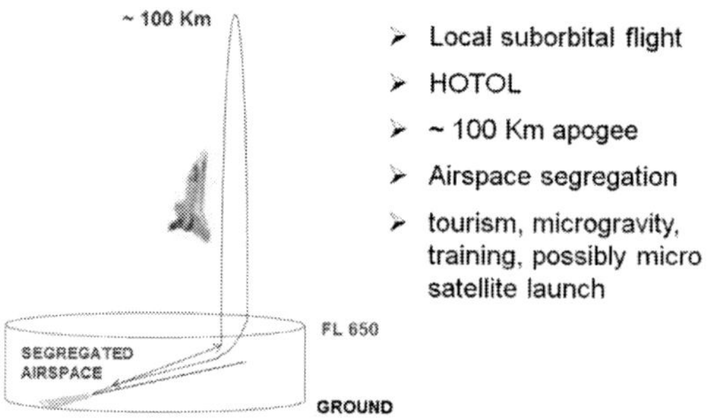

Figure 1 - Italy Initial Reference Operational Scenario

Therefore the initial ROS (Reference Operational Scenario) will be HOTOL experimental and commercial suborbital flights from spaceports located within the Italian territory and engaging the national airspace, only. The scope of operations is anticipated to cover at least the launch of micro and small satellites into orbit, research, astronauts' training and space tourism. Fig.1 provides a simplified example of the ROS for this initial phase.

4.2. Spaceports regulation

Setting up a regulation for issuing an authorization and, eventually in later stage, a Certificate of Approval or a License for a site adequate for hosting suborbital flight operations is a near term target that ENAC and Italy have established in the roadmap. As matter of fact that brand

---

[7] This point could be implemented by using a *bow-tie* methodology as described in [23] and [25].

[8] The ALARP approach appears very similar to the one presently followed by the EU regulation (Part 21, Para. 21.A.101) for establishing the certification basis (i.e. the set of applicable airworthiness requirements) for a changed aircraft design [6]. As a matter of fact the Part 21 requirement n. 21.A.101(b)3. allows the Applicant not to comply with the latest amendment of an airworthiness rule if the effort "would not contribute materially to the level of safety of the changed product or would be impractical" [6].

new spaceports are unlikely to be built in Italy, and probably in the majority of the other European States. All the potential future spaceports will derive from a conversion or upgrading process of existing civil or military aviation airports. A pragmatic approach for Italy is to consider the current European Regulation laying down requirements and administrative procedures related to civil aviation aerodromes EU Regulation 139/2014 [7] as a baseline regulation on which additional specific requirements should be stacked to cope with the peculiar issues of HOTOL spaceplanes. In this scenario, considering the compliance to EU Regulation (EC) 139/2014 as a baseline will not be a burden. Besides, there is a need of harmonize any future spaceport regulation with other international regulations already in place in this sector. This is considered a driving factor in any rulemaking process and, for this specific rulemaking task, ENAC has chosen the US Part 420 [8] as reference on which the harmonization should be based. Hence, the future spaceport regulation will take into consideration both the existing EU Regulation 139/2014 and the US Part 420 and it will include any additional requirements deemed necessary to accommodate HOTOL suborbital flight operations.

4.3. Operations regulation

The operations regulation will have the scope to authorize experimental and commercial suborbital flights by using a comprehensive risk based operation centric approach. In this regard a specific type of operation should be associated to a specific scenario, which in turn should be characterized by a well defined set of (foreseeable) hazards. For the sake of interoperability, ENAC intends to develop the operations regulation starting from the existing US regulation, namely Part 431 [8], Part 437 [8] and Part the 460 [8] adapted as necessary to fit with the National and European legal framework. Different type of authorizations could then be issued for a single operation with a single flight or for a specific type operation (i.e. a specific scenario) with multiple similar flights. A sufficient number of experimental operations should be conducted within a specific operational scenario before any commercial activity is authorized in that scenario. A qualitative and/or quantitative total hazard and risk assessment covering all the hazards for both uninvolved and involved parties during a specific suborbital operation is considered the best suitable way to authorize suborbital operations in the near-midterm. The regulation will have also to define the minimum level of risk for third parties, critical infrastructure and the occupants. As said above, while the level of risk for third parties could be easily quantified in terms of maximum allowed casualties per mission or mission phase, the level of risk for the critical infrastructures and for the occupants will be probably defined in a qualitative way by using an ALARP concept in conjunction with the requirement of some recognized good practice. The areas that should be covered by the operation regulation are the following (the spaceport is not included here because it will be the object of a specific regulation, see the above Para. 4.2):

- Flightworthiness (mainly related to occupant safety);
- Third parties risk assessment;
- Air and Space Traffic Management (STM/ATM) including tracking and segregation, contingency procedures especially in case of falling debris after a mishap in flight (mainly related to third parties' safety);
- Crew license;
- Medical requirements (for both the crew and the other flight participants);
- Authorization procedures (for experimental and commercial operations), including eligibility.

Finally the level of verification of the effectiveness of the controlling measures should be agreed with the Authority unless otherwise specified in the regulation.

4.4. International mutual recognition

Although the initial ROS in Italy (see above Para. 4.1) within the national ground and airspace boundaries, international harmonization and mutual recognition of authorizations, Certificates/Licenses in the sector of suborbital flight operations is considered by ENAC a goal that should be pursued since the beginning of the rulemaking process. A number of elements and logical considerations bring to this necessity, namely:

- present and near-midterm potential suborbital flights operators are exclusively foreigner holding an "authorization" issued by foreign Authorities;
- as soon as both Authorities and stakeholders reach an adequate level of knowledge and experience, the operation scenarios will potentially expand and impact other States (States in the proximity of the launch/take-off sites, A-to-B mission profile, etc.);
- by similarity with the civil aviation sector, a common set of rules will promote international partnerships and commercial arrangements among stakeholders;
- unless in contrast with national laws and regulations, adopting existing regulations or sharing a common approach can only be a benefit for this emerging sector.

In line with the above considerations, ENAC has been promoting and supporting the cooperation in the rulemaking process both among the European involved Countries and at international level (see e.g. the ICAO/UNOOSA Space Learning Group initiative and the cited FAA-ENAC-ASI Memorandum of Cooperation [4]). The issue of the two above mentioned regulations for the spaceport and operations authorization is obviously considered a necessary milestone to be reached before to coordinate, discuss and eventually sign bilateral or multilateral agreements with foreign States and

Authorities.

5. SPACEPORT SELECTION AND PERFORMANCES

Spaceport is the operational base for the execution of suborbital flight operations. As already pointed out, the Italian approach will not be to build a new spaceport from scratch, but rather to evaluate and designate an official spaceport among already existing sites.

5.1. Spaceport selection criteria

Spaceport selection shall be based on several aspects which are important in order to allow the safe operations of a spaceplane. On the basis of a number of technical, administrative and legal considerations ENAC is issuing criteria for existing airports to be eligible to host suborbital operations. The criteria will constitute the basis of the selection procedure that will be adopted by the MIT to eventually designate potential spaceport sites. The criteria have been grouped in the following clusters:

- Airspace & Traffic (ATM impact, route density, airspace congestion …): since airspace segregation from aviation traffic will be unavoidable for the intrinsic characteristics of the current spaceplane projects, parameters as airspace congestion and ATM capacity play a primary role to lessen the impact on aviation operations. Besides, the use of site for suborbital operations shall be verified to avoid any misuse of the existing airport or conflict with national security interests.
- Territory & Surrounding (population density, flight corridors, properties on ground,..): as detailed in Para. 3, the safety on third parties on ground is a requirement strictly related to the elements of this cluster.
- Meteorology (winds, weather conditions & statistics, …): spaceplanes' limitations and safety of operations are affected by weather to the same extent or more than in the aviation domain. Spaceport area weather shall be analyzed in detail to identify possible weather phenomena affecting the spaceport area and region that could limit the operation of a spaceplane (wind, fog, rain, humidity, etc.).
- Environment: EIA (Environmental Impact Assessment) considerations, analysis, mitigation and restrictions applicable to airports need to be updated to the characteristics of this type of operations and the peculiar propellant/fuel used.
- Infrastructures (runway, propellant storage, logistics, …): elements of this cluster are derived directly by the characteristics of the spaceplanes the site intends to be eligible for, by prescriptive requirements set for propellant storage and by economic constraints. As an example, existing airport layout (access, infrastructures, communications, runways, radio and navigation facilities) shall be analyzed to verify the compatibility with the spaceplane operations and requirements and ground and flight safety shall be verified, in particular for what concerns the use and storage of high energy propellants and debris control.
- Spaceport Operator Organization (SMS - Safety Management System, risk management …): in accordance to what explained in Para 4.2, aerodromes certified in accordance with EU Reg. 139/2014 is considered in terms of priority as far as the spaceport operator's organization capability is concerned.

5.2. Spaceport performances

There are different suborbital spaceflight initiatives currently being developed especially in the US, that feature vertical and horizontal takeoff, single or multi stage to orbit, so that in general a spaceport should be able to accommodate pretty much all these capabilities. However, the Italian activities are aimed at supporting the horizontal takeoff on a properly featured runway. It has to be pointed out that a spaceport should not only be outfitted for suborbital flights but it should provide enough flexibility to accommodate other kind of aerospace activities that have promising markets; for instance the air launch of satellites which is a capability that is attracting growing interest among the international communities. Also, since many of the above referred vehicles will in perspective be considered reusable, a spaceport shall feature the capabilities for allowing the refurbishment and maintenance operations on the vehicle and associated hardware. In general a spaceport can also be seen as a hub for executing a wide range of activities, including aerospace experimental operations. In general, the term spaceport designates a site featured with all the technical, infrastructural, operating, safety elements, and the relevant license from the competent Authority, to allow takeoff, landing and ground/flight operations for spaceplanes suitable for suborbital or point to point flight. The Italian approach to the spaceport identification does not include building brand new facility, but is based on the selection of already existing infrastructures with possible infrastructures upgrading work to better match the selected spaceflight system requirements. According to the initial ROS in Italy, the following considerations address some specific operational aspects that have significant relevance for a spaceport.

- The spaceport shall feature runways 3000 m (9800 ft) long in both directions. It is also necessary to consider the direction of the prevailing winds and crosswinds in that specific location. The spaceport shall be easily accessible by public and private transportation and intermodal transportation. Also, the site layout shall allow the intra-site transportation relevant to transfer the vehicle and associated hardware to/from the ground processing infrastructures to the runway.

- The spaceport shall be equipped with all the infrastructures and facilities like hangars, ground support equipment and tools needed to prepare the spaceplane and payload for flight, The pre-launch activities include preparation and execution of the vehicle preflight test and checkouts along with the relevant subsystems and equipment to ensure functional and interface checks and final integration and configuration activities.
- The spaceport shall be equipped with all the needed infrastructures that allow preventive and corrective maintenance on the vehicle and with a proper system to manage the spare parts and relevant procedures.
- The spaceport shall be equipped with all the infrastructures needed to store dangerous materials in accordance to the specific spaceplane. Such infrastructures shall have to be built and handled in accordance with the national regulations. Such infrastructures shall also feature temperature and humidity control capabilities and lightning protection. Specific containment areas shall be foreseen to mitigate risks resulting from emergencies such as explosions, fire, overpressure and toxic materials release. In accordance with the Safety requirements, the maximum amount of dangerous materials is a function of the distance from the operating areas of the spaceport and mostly from populated areas and roads. Basing on this, this may be a major constraint when evaluating already existing sites, because such safety distances may not be compatible with the layout of already existing sites.
- The spaceport shall be equipped with a proper radio communication infrastructure, operating both on aeronautic used frequencies and internal offline ones to allow proper execution of ground and flight operations.
- The spaceport shall include a planning function to support both flight and ground operations.
- The spaceport shall include proper training facilities for flight crew, flight participants and flight passengers for space tourism.
- The spaceport shall feature the proper capabilities aimed at the coordination and support of the launch operations and all the subsequent phases till the landing at the completion of the mission. Specific ground stations have to be put in place for the telemetry acquisition and the vehicle tracking. Different studies approach are currently being evaluated based on specific non cooperative signal filtering techniques, like the extended Kalman filter.
- The spaceport shall feature the proper infrastructures to execute all the planned post landing operations; these include crew, flight participants, passengers disembarking, spaceplane safing, post flight check of the spaceplane system and equipment, vehicle refurbishment and reconfiguration to the next mission. A proper data archiving system maintenance record and configuration control systems shall be in place.
- Since space tourism is a promising market to be captured by a suborbital flight initiative, the spaceport shall include all relevant services to passengers that will enhance their spaceflight experience. These services are included in the package most of the operators are planning to offer to their customers. Such services include training, medical, entertainment, public observation areas.
- A spaceport SMS shall be organized, including a detailed Safety Management Plan identifying all operations, the related hazards control and the hazard mitigation activities.

6. ECONOMIC AND INDUSTRIAL PERSPECTIVES

To analyze the economic development and industrial perspectives for the Country, it is important to understand that the identification of a spaceport and the setup of a regulatory framework are instrumental to creating specific capabilities in the Country and offer unique services to potential customers. The main peculiarity of suborbital flight is to provide some minutes of microgravity through the execution of a parabolic[9] trajectory with apogee of about 100 km of altitude. Typically, a suborbital spaceflight system can carry crew and flight participants, as well as payloads to perform microgravity experimentation. While the suborbital parabolic flights do not allow extended microgravity like the International Space Station, the 4 to 7 minutes achievable microgravity level still offers significant more time than the traditional parabolic flights carried out in the commercial airspace that offer 20 to 40 seconds each parabola. Setting up a suborbital spaceflight initiative in Italy can open different business opportunities, in particular the following markets have been identified: (i) Microgravity experimentation. The suborbital vehicle is equipped with specific cargo capabilities featuring Shuttle-like middeck drawers that can accommodate experimental payloads and allow the execution of the experimental protocol required by the principal investigator. Provisions will be on board for both unattended and man-tended payloads and it is possible that resources like power and data will be provided so that the payload may downlink housekeeping and scientific telemetry to a Control Center on ground. The provided services can represent for the payload an initial

[9] Actually, an elliptical orbit that intersect the Earth's surface, approximated by a parabola around the apogee.

test and validation step in preparation to long duration microgravity exposure in the International Space Station. Part of this market segment also includes testing of new technologies that can take advantage of the frequent rate of suborbital flights to grow in maturity level in preparation for a subsequent experimental phase in long duration missions. (ii) Pilots and astronauts training. During the execution of suborbital flights different levels of accelerations are experienced, ranging from microgravity reached during the flight coasting phase and peaking to 4-6 g's during takeoff and landing. This specific environment can be a tremendous opportunity for pilots and astronauts to perform training on specific tasks. In particular, astronaut can rely upon different minutes of microgravity to train on the execution of operations that they will be carrying out during the actual mission. (iii) Space Tourism. The coasting phase reached by a suborbital spaceflight system is a great opportunity for paying tourists to experiment the thrill and feeling of "weightlessness" and enjoy some minutes of Earth spectacular view, and live a totally space immersive experience. According to previous studies, a market demand of more than 4000 seat-cargo equivalent is expected worldwide for the suborbital reusable flights over ten years. According to studies, there is also possibility to grow up to more than 13.000 in an optimistic economy scenario. A significant share of this market demand can be captured by the European/Italian market taking advantage of a suborbital spaceflight initiative based on an Italian spaceport. Setting up a spaceport in Italy will significantly affect the Italian economy. In general, it will boost the new commercial space economy, catalyze the Italian industry, and attract business development in the area. The Italian Government value is driven by tax revenues from the space industry growth and incremental tourism in Italy. There will also be a significant demand of direct and indirect jobs to support the spaceflight operations services. The return for the Italian economy is strictly related to the business model that will be developed to carry out the suborbital operations. It is likely that a non-Italian suborbital system will fly within the Italian territory. This implies to set up specific agreements with the suborbital system provider and to develop a business model that exploits the relevant market opportunities. The basic idea is to create a suborbital operator that will conduct operations from an authorized spaceport site. Part of the business agreement shall include a significant opportunity of involvement for the Italian industry in the development of the spaceflight system.

7. CONCLUSIONS

The introduction of suborbital flight systems is going to be a real challenge for regulators in the next years. A traditional "certification-like" approach for the vehicle, organizations (design, production, maintenance, operations and training), crew, spaceports and STM/ATM services providers able to assure both the safety of uninvolved third parties and of the occupants onboard appears impractical to be implemented in the near-midterm, though it will certainly be needed in the future to allow national and international routine transportation fully integrated in the existing aviation system. Building a flexible regulatory framework for commercial suborbital operations at national level will play an essential role in allowing the operations to be carried out meanwhile. To be effective and also to back the economic development of the sector, the regulatory framework will have to assure an adequate level of safety primarily to uninvolved third parties, by minimizing at the same time the risk for the occupants onboard the spaceplane. In line with the Italian MIT policy [5] ENAC has been developing a national regulatory framework tailored on the initial ROS, based on the existing foreign regulations, such as the US Part 400-460 [8], and taking in to account possible additional specific national requirements which will be recognized necessary for complying with the current national and EU legal framework; all the above with the prospective to contribute to a future harmonized European system. Priority will be given to the spaceport selection criteria and to the issuance of a spaceport regulation to authorize both the launch and re-entry site and the spaceport operator, who will have to have adequate organizational and SMS (Safety Management System) capabilities. Regarding the operational aspects, a specific performance-based, operation-centric and holistic regulation for the suborbital vehicle operations will be developed as well, based on the actual foreseeable operational scenarios. Eventually, once a consistent and complete national regulatory framework for suborbital operations is in place, it will be possible to start bilateral discussions with other foreign Countries' regulators to allow international interoperability and mutual recognition.

8. REFERENCES

1. Italian Code of Navigation, Decrees n. 96/2005 and n. 151/2006
2. Italy-US Framework Agreement for Cooperation and Use of Outer Space for Peaceful Purposes of 19 March 2013, transposed into law n. 197/2015
3. ENAC, A Regulatory Policy for the Prospective Commercial Space Transportation Certification and Operations in Italy, Edition 1, July 2016, https://www.enac.gov.it/La_Comunicazione/Pubblicazioni/info-84061321.html
4. FAA-ENAC MoC in the Development of Commercial Space Transportation of 12 March 2014, renewed on 30 July 2016
5. Italian Republic, Ministry of Infrastructures and Transportation, Sustainable Development of the Commercial Suborbital Flights Sector, Decree n.

354 of 10 July 2017 (in Italian).
6. EU Regulation 748/2012, Annex I Airworthiness and Environmental Certification (Part 21), http://www.easa.europa.eu/document-library/general-publications/easy-access-rules-initial-airworthiness
7. EU Commission Regulation n. 139/2014 of 12 February 2014 laying down requirements and administrative procedures related to aerodromes.
8. US Code of Federal Regulation (CFR), Title 14 Aeronautics and Space, Volume 4, Chapter III, Parts 400-460, Commercial Space Transportation
9. US Federal Register, Vol. 81, No. 139, July 20, 2016, Amdt. No. 431-4, 81 FR 47027.
10. FAA, Advisory Circular (AC) 431.35-1, Expected Casualty Calculations for Commercial Space Launch and Reentry Missions (superseded by the FAA Flight Safety Analysis Handbook, Version 1.0, 2011)
11. FAA, Advisory Circular (AC) 23.1309-1E, System Safety Analysis and Assessment for Part 23 Airplanes, 17 November 2011
12. FAA AST, Recommended Practices for Human Space Flight Occupant Safety, Version1.0, 2014
13. IAASS, Space Safety Standard, Commercial Human-Rated System, 2010, doc. n. IAASS-ISSB-S-1700-Rev-B
14. French Republic, Ministry of National Education, Higher Education and Research Order of amending the Order of 31 March 2011 on the technical regulation pursuant to Decree No 2009-643 of 9 June 2009 concerning authorizations granted pursuant to Law No 2008-518 of 3 June 2008 on space operations, Article 7
15. International Civil Aviation Organization (ICAO) Annex 8 Amendment 105, Airworthiness of Aircraft, Eleventh Edition, July 2008, Part I, Chapter 1, Definitions
16. ICAO, Safety Report 2017, 2016, 2015, 2014, 2013
17. Boeing (2016), Statistical Summary of Commercial Jet Airplane Accidents Worldwide Operations | 1959 – 2016,
18. GAMA, 2016 General Aviation Statistical Databook & 2017 Industry Outlook
19. R. Abeyratne, Regulation of Commercial Space Transport. The Astrocizing of ICAO, Springer, 2015
20. R. Tüllmann, C. Arbinger, S. Baskcomb, J. Berdermann, H. Fiedler, E. Klock, T. Schildknecht, On the Implementation of a European Space Traffic Management System – I. A White Paper, DLR, 2017
21. J. D. Andrews, T. R. Moss, Reliability and Risk Assessment, Second Edition, Professional Engineering Publishing, 2002.
22. A. Quinn, Regulating Airworthiness/ Spaceworthiness of Future Suborbital Commercial Spaceflight (SCS) Vehicles, Saturn, Presented at the 3rd ICAO/UNOOSA Symposium SPACE 2017, Wien 29-30 August 2017
23. JARUS, Guidelines on Specific Operations Risk Assessment (SORA), doc. n. JAR-DEL-WG6-D.04, Edition 1.0, 26 June 2017, http://jarus-rpas.org/content/jar-doc-06-sora-package
24. Health and Safety Executive (HSE), ALARP "at a glance", http://www.hse.gov.uk/risk/theory/alarpglance.htm
25. A.L.C. Roelen, J.G. Verstraeten, L.J.P. Speijker, S. Bravo Muñoz, J.P. Heckmann, L. Save, T. Longhurst, Risk models and accident scenarios in the total aviation system, ASCOS Safety Certification, 2014

PROCEEDINGS OF THE 9TH INTERNATIONAL SPACE SAFETY CONFERENCE

SESSION 19
SPACE DEBRIS – II

ECONOMIC FUNDAMENTALS OF MITIGATING ORBITAL DEBRIS

Martin K. Zhu, Ph.D.

*Federal Aviation Administration (FAA), 800 Independence Avenue SW (AST)
Washington, DC 20591, United States of America
Email: martin.zhu@faa.gov*

1. INTRODUCTION

The U.S. Government developed a set of Orbital Debris Mitigation Standard Practices based on the NASA guidelines of 1997 [1]. Other countries and organizations, including Japan, France, Russia, and the European Space Agency (ESA), have followed suit with their own orbital debris mitigation guidelines.

The U.S. Government Orbital Debris Mitigation Standard Practices (USGODMSP or the standard practices) remains voluntary for the aerospace industry to minimize the growth of orbital debris. The Standard Practices were approved by all US Government agencies in February 2001 contain guidelines that no debris greater than 5 millimeters remain in orbit more than 25 years, dispose of post mission spacecraft into disposal space orbits and minimize collision [2]. Among the standard practices, the key is the post mission disposal.

Orbital debris comprises fragmented material (resulting from anti-satellite tests, upper stage explosions, etc.), nonfunctional spacecraft, rocket bodies, and mission related items (explosive bolts, vehicle shrouds, etc.), but excludes naturally-occurring debris such as meteoroids. Orbital debris are space hazards to mission spacecraft and astronauts. As more and more space-faring nations launch objects into Earth orbit, space is becoming increasingly crowded with orbital debris. If left unchecked, orbital debris can diminish the usefulness of certain orbits and present a hazard to operations on-orbit.

Conversely, the valuable Earth space is a large volume, touted as the "Big Sky," so large such that for many years there was no limitation on disposing of "space junk" (mostly used rocket bodies and satellite debris) in the same space where mission spacecraft resides. "In the early days of spaceflight, the "Big Sky" theory was the near universally accepted paradigm for dealing with collisions of orbiting objects. This theory was also used during the early years of the aviation industry. Just as it did in aviation, the "Big Sky" theory breaks down as more and more objects accumulate in the environment." [3] [4]. Over time the accumulation of "space junk" induced congestion and increased the hazard to valuable space orbits despite the "Big Sky."

If space operators were asked about containing the growth of orbital debris, they will respond with "yes" because space operators are increasingly aware of the orbital debris and other resident space objects (RSOs) in high traffic space regions can be hazardous to their spacecraft in missions. However, if space operators were checked if they would like to be voluntarily access the protected space regions (LEO and GEO) in compliance with the standard practices, some of them would probably not comply at their cost at this time.

Space operators holding the above two positions actually mirror two perspectives—being aware of that highly utilized Earth orbits become increasingly competitive and congested; at the same time, space operators seem unconcerned about the impact of orbital debris on mission spacecraft. The former position reflects orbital debris awareness and the latter position reflects a negative externality problem. The negative externality could lead to a "tragedy of the commons."

Space junk provides a classic example of a negative externality. A negative externality is an involuntary cost experienced by someone who is not a party to the transaction that produced it. Post mission space objects impose debris spillover costs to satellite owners and other space mission operators. As all other market-failure problems or "tragedy of the commons," satellite owners suffering the cost will not be compensated for their losses. Consequently, the public space resource of space has become no longer a "free" domain for the disposal of orbital debris and space missions.

Orbital debris exposes spacecraft to a risk of being hit and forces them to undertake costly avoidance maneuvers and deploy unnecessary backups. These risks and avoidance costs are not reflected in the private cost of a launch, and as such, a divergence between private and social costs is an externality cost.

International adoption of the standard practices would internalize this externality and assure that the space operators incorporate all the costs of future launches, including debris disposal cost.

This paper provides economic fundamentals to respond to orbital debris mitigation skeptics and for orbital debris mitigation advocates.

2. LIMITED AND VALUABLE RESOURCES

Humankind shares common resources, as we rely on same natural physics and behave similarly. In the space, human beings concentrated in using certain Earth orbits, such as LEO and GEO, based on physics and common needs. These Earth orbits are believed to present high utility value to space-faring nations or individual operators. According to FAA AST [5], every dollar spent in direct space assets and services generated six times the total of the indirect and induced space economy for the last decade. This six-fold multiplier was derived by taking the total space industry impact value divided by the direct impact value. Most of these non-government space assets reside in commonly used space regions, LEO and GEO, and thus these space regions are valuable.

Space-faring economies cannot afford to have highly used orbital orbits being freely disposed and thus let space junks take away protected space orbits from mission spacecraft, because LEO and GEO space resources are relatively limited, compare to the vast space disposal regions (see figure 1).

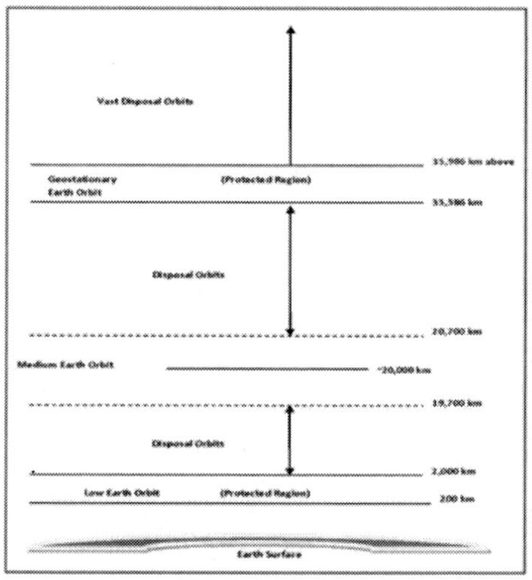

Figure 1 Protected Regions and Disposal Orbits

The valuable Earth orbits show a clear trend towards congestion as Figure 2 illustrates the increasing risks of spacecraft collisions along the timeline [6].

Current international modeling efforts indicate that even if there were no further space launches, collisions between objects already in space will eventually become the major source of debris. NASA predicted there would be another 10 hypervelocity collisions in the next 50 years approximately one incident every five years (see figure 2) [6].

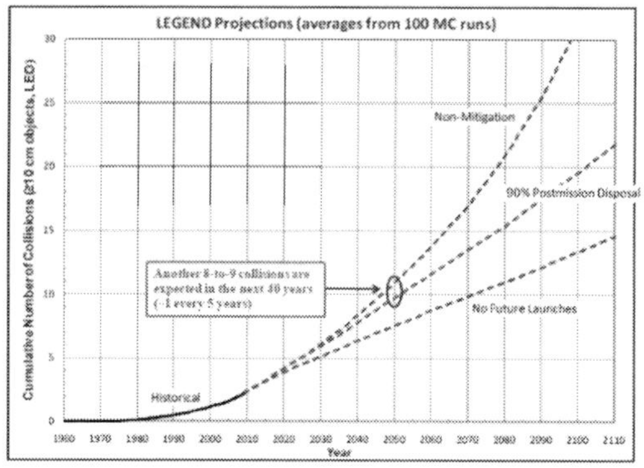

Figure 2: Predicted Accidental Collision Activities in LEO [6]

Due to rising numbers of RSO and Kessler syndrome in the next 50 years, debris would be multiplied through high velocity collisions to reach from 38,000 to 48,000 pieces (≥ 10 cm debris).

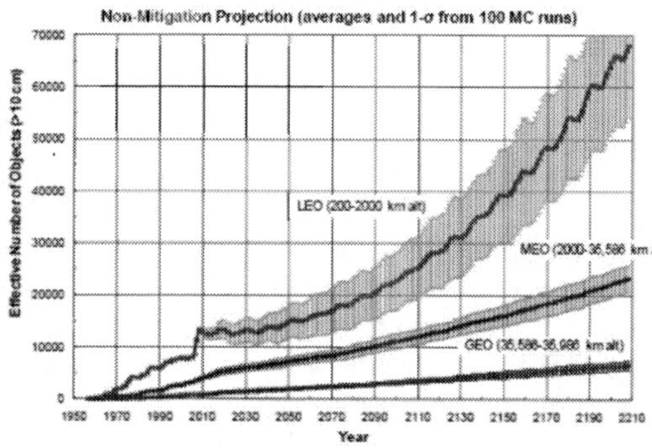

Figure 3: Projected growth of the ≥ 10 cm debris population in LEO, MEO, and GEO for the next 200 years. [7]

LEO and GEO regions are already congested. There were no man-made objects in the near-Earth space prior to 1957 when the USSR launched Sputnik 1. The first non-government space launch was in the early 1960s. Nearly 60 years later, the population of orbital debris above 1 cm in diameter is over 500,000 pieces [8]. Earth orbit already used by "space junk" is a significant amount relative to the available key Earth orbits. This congestion problem has been growing at an exponential rate since mankind put the first manmade object in space. Space history has already recorded four hypervelocity collisions, which induced thousands of orbital debris pieces greater than 10 cm (figure 3). In the near future LEO environment, we will see mega-constellations made of small satellites to be deployed. Some space experts estimated a much higher catastrophic collision number, which would be about 40 catastrophic collisions assuming 95 percent success rate with orbital debris mitigation measures—e.g. post-mission disposal. [9] Some leading space scientists believe that those useful and popular orbits have passed a tipping point by stating "in some of the most congested regions of low Earth orbit, this point was actually passed more than 10 years ago." [10] Moving objects in space at a speed of 11.7 km per second is a speed moving 7 times faster than a bullet. At such speeds, an impact to a typical operational spacecraft by debris 5mm and larger will most likely cause damage that ends the mission of the spacecraft. Orbital debris due to its kinetic energy constitutes a deadly force to be avoided, which makes the threshold of congestion in space much lower than that of commonly defined congestion on the Earth surface.

3. RISING SPACE OPPORTUNITY COSTS

The growth of orbital debris induced congestion results in rising orbital debris awareness costs (ODAC), which is a part of societal opportunity costs. The rising costs are related to increased collision avoidance analyses, shielding, collision related R&D, debris avoidance maneuvering, and back-up satellites, all of which derives directly from orbital debris induced congestion. Though ODAC are business proprietary information, they are certainly significant. The FAA Office of Commercial Space Transportation (FAA/AST) estimates up to a 10% of 20-year operations costs are spent on ODAC, some of which are back up satellite, shielding and maneuvering. [11] At roughly 0.5% of satellite industry sales, debris imposes an annual cost of about $44 million for the U.S. satellite industry. [11] Such costs will rise without any mitigation.

Rising ODCA costs are also attributable to the increasing need for space monitoring and sensing costs as space management requirements. For example, Space-5 (ORS-5) satellite launched in August 2017 from Florida's Cape Canaveral Air Force Station [12] contributes to rising ODCA costs due to increasing numbers of RSOs in the future. Some of the $87.5 million price tag for this launch is the contributes to the societal costs of ODCA.

Human space utilization will increase substantially. The new space economy will be symbolized by low cost and frequent access to Earth orbits. U.S. Contractionary Act -- The Ultra-Low Cost Access to Space (ULCATS) Act (2016)—means more companies entering the space market and more objects being sent into space. According to a SpaceX news release, it planned and filed with the U.S. Federal Communication Commission an initiative to deploy 4,425 satellites operating in 83 orbital planes at altitudes ranging from 1,110 km to 1,325 km in 2019. Considering an annual launch rate of 270 new spacecraft, space experts estimated there would be 50% more orbital debris resulting from sending constellations of satellite into LEO without a proper space management. [9]

ODCA cost is significantly increasing due to orbital debris induced maneuvering. The International Space Station (ISS) frequently conducted debris avoidance maneuvers, with one such move in 2008, two in 2009, three in 2011, four in 2012, and two in 2014 [14]. The ISS again conducted debris avoidance maneuvers twice in the second quarter and twice again in the third quarter of 2015 for a total of 25 times in its 15-year history [14]. In addition, 1 debris avoidance maneuver was failed; and ISS crews were forced to retreat to Soyuz spacecraft in four incidents that is called "Shelter-in-place" when insufficient time permitted a collision avoidance maneuver. Collision avoidance maneuvers place a huge burden on each mission in space orbit. ISS maneuvering cost was estimated to be $1 million or higher per action. [15] The 25 debris avoidance maneuvers had costed about $25 million or higher, without taking astronaut opportunity costs into account.

The largest part of space opportunity costs is attributable to lost valuable space orbit slots to orbital debris occupation. Presume in the next 50-year period that there is the same amount debris created, which equals to about 14,000 counts of rocket bodies and debris (the ≥ 10 cm debris). Based on the current numbers and projections, prohibiting post-mission upper stages from remaining in orbit for more than 25 years would save 14,000 valuable orbital slots from orbital debris occupations, which are about nine times the near-Earth orbital slots currently used by U.S. satellite payloads [16]. In 2013 dollars, such near-Earth orbital slots could provide roughly an additional $195 billion value to the global satellite industry revenue and $320 billion value to the global space industry revenues, and $5 trillion to the global telecommunication industry, assuming other things constant.

The crowded Earth orbits will be required to have some kind of space management to promote investment. This is situation is another part of societal opportunity cost to be taken into account. It will not be cheap, but eventually the space players will have to pay for it, including orbital debris mitigation cost.

Figure 4 shows the trend line (thick) of predicted non-governmental space market value based on $323 billion in the year of 2015. The predicted value was based on the 5.6% growth rate of non-governmental space market value using 2006-2015 series. Summing up all societal opportunity costs, the space industry cost curve (thin line) will move up and squeeze the space industry profits. The upward cost curve will eventually push cost to exceed revenue in the future. The movement of upward space cost curve from point A to B indicates, assuming overall RSOs are doubled in the next 50 years (see Figure 3), and inevitably continued space operation generates "market failure" because no business wants to operate at the point where cost curve is above revenue curve.

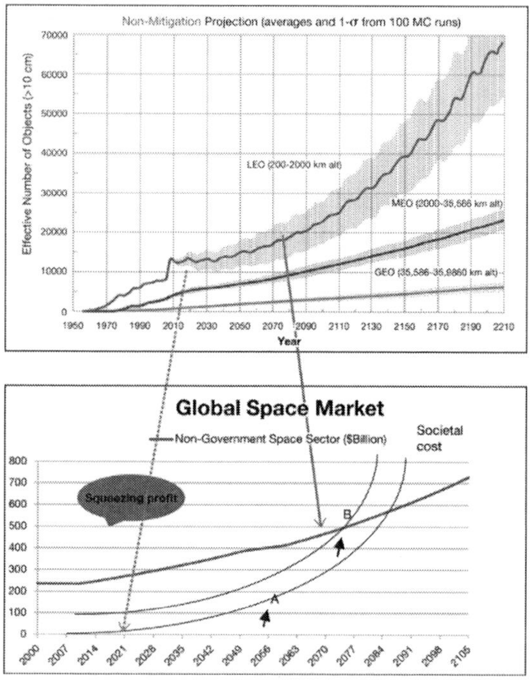

Figure 4

The last category of societal costs involves the mitigation cost with the standard practices, which have not been incurred and contribute a very small part of costs comparing with all above DA costs. If a global adoption of standard practices is in the space industry's future, we should focus intensely on the question of the magnitude of cost saving. The standard practices would result in a significant cost savings, which is a topic in following cost-saving discussion.

4. COST SAVING

To quantify cost savings, two kinds of costs are to be determined. They are mitigation costs with the standard practices (i.e. prior mitigation costs) and the likely mitigation costs passing end of the 25-year requirement (i.e. posterior mitigation costs). The prior mitigation costs are associated with expenditures of loaded extra propellant and mechanism built into spacecraft to conduct post mission debris disposal. The posterior mitigation costs pertain to likely future expenditures of active debris removal. To actively remove a large batch of small size debris is not practical at this time because of the international space treaty implication to preventing anyone to remove other nation spacecraft and orbital debris. Conversely, the active removal of targeted-and-large orbital debris has been tested and is practical, but would incur with significant posterior mitigation cost. The future active debris removal could cost between $600,000 and $3.5 million per post mission upper stage (targeted and large orbital debris).

Costs incurred with the standard practices would be much lower than future removal cost when the industry will have no choice as post mission induced congestion worsens over time. The difference is the cost saving.

4.1 The prior mitigation costs

The standard practices require any space operators under the 25-year post-mission disposal guideline to take one of three approaches to mitigate orbital debris. They are for operators to maneuver post-mission upper stages to storage orbits within the first 30 days after, or to administer an uncontrolled reentry, or to conduct a controlled reentry into the broad ocean area. These mitigation methods would result in additional costs by possibly adding extra propellant and reducing payload. In most cases, a mitigation method might not require adding a rocket motor although it is deemed to be a low probability scenario. These are primary sources of costs associated with the standard practices and we discuss them below in detail.

Data originate from government agencies and space experts' estimates. [19] [20] The Launch vehicle forecasts data used for this analysis are only a part of the global demand for commercial space industry, mainly the demand addressable by the US space industry. The world impact would be the extrapolation of the applied data.

Methodologies used in this paper are the following. First, we estimate the number of launch vehicles maneuvering for natural decay (or uncontrolled reentry), controlled reentry and storage orbit. Second, we comprehensively estimate unit costs associated with each mitigation method. Further, we multiply each unit cost by the number of upper stages times the number of analytical years, and summarize the three categorical mitigation costs. Finally, we calculate present value and annualized cost per upper-stage mitigation.

The three approaches would incur a wide range of mitigation costs, from $20,000 to $900,000 for each post mission disposal supposedly in a 15-year period to reach steady state. We estimate the annualized cost using a 7% discount rate to be around $250,000 per post-mission disposal.

4.2 The posterior mitigation costs

Now we are discussing about the posterior mitigation costs. The posterior mitigation costs do not have to occur because the widely adoption of the standard practices would likely contain the further growth of orbital debris. If the industry has to incur posterior mitigation costs, it is not certain for how long the highly used Earth orbits would remain sustainable without post mission debris disposal, plus the deployment of mega constellations of satellite in the crowded space regions. This situation leaves us with the question of how long we have to conduct active debris removal (ADR), which, in turn, determines the posterior mitigation timeline. As stated above, when the space opportunity costs, orbital debris awareness costs in particular, are high enough to crowd out the industry opportunities and squeeze the industry profits, active debris removal becomes inevitable. In addition, if we consider the widely adopted standard practices as the first step toward future space traffic management, post mission spacecraft have to be mitigated either before or at the posterior end of the 25-year requirement.

Considering ADR is an example of mitigations posterior end of the 25-year requirement, ADR could happen much earlier without post mission disposal. ADR is an effective way of removing targeted and large used upper stages in desirable space orbits, given the fact it is not practical to indiscriminately remove batches of small space objects. The U.S. and European space agencies have tested ADR. However, it is not a common practice now but it is a reasonably realistic scenario in the future if there would be only a low adoption rate of post mission disposal over a long period of time. These factors lead us to believe ADR assumption is reasonably realistic: the finite and highly used Earth orbits, the rivalry nature of utilizing certain Earth orbits, the large size of post mission upper stages, no international space law allowing a nation to randomly remove other nation's space junk, each nation being responsible for its own space objects and liable for damages to others, space industry demands in the future exceeding the capacity of highly utilized Earth space, as well as space traffic management.

Here are further supporting explanations why we think ADR is a baseline for a future mitigation approach under a scenario of the minimal adoption of the standard practices. When a company has a satellite that occupies a useful orbit slot, the space orbit is usually valuable and highly utilized. If the satellite passes its life span, the original satellite owner has an interest in continuing to utilize the same orbit slot and its spectrum with a replacement. When a rocket launcher has a post mission rocket body, a space junk, in a space orbit competing against potentially hundreds of orbital slots due to Kessler syndrome, the satellite owner would lose useful orbits to the "space junk." This limited availability is akin to orbital "real estate," where location is a main reason to be congested and contested. As an analogy, when "real estate" still has unwanted tenants in it, until the tenants are removed desirable tenants cannot occupy the "property" again. Therefore, orbital debris removal is an approach to preserve existing property in the protected space regions

As long as post mission rocket bodies and spent satellite accumulate in orbits, a "tragedy of the commons" caused by "space junk" induced congestion worsens. It is not inconceivable that, eventually, space orbit users will have to pay debris removal costs to deal with the Earth-orbit congestion problem. Thus, we let ADR value be the baseline of posterior costs for post mission upper stage disposal.

As a note, ADR does not apply to satellite debris disposal because, unlike upper stages, a satellite is technically in compliance with the standard practices with different mitigation measures such as design for demise. Therefore, the following posterior costs computation does not factor in any number of used satellite disposal for the calculation of posterior costs.

Data regarding ADR are obtained from public sources, including a French Space Agency (CNES) study and Bonnal, et al estimate. [21] [22] We average ADR costs for the estimate of $31.5 million to represent the expected value of per large post mission unit removal.

Like the methodology used for the prior mitigation costs, we estimate posterior mitigation cost per upper stage and then calculate the number of upper stages that would otherwise remain in orbit. There were 86 orbital launches around the world in 2015. [23] Due to orbital altitude and inclination and solar rays, about 74% or 64 of these 86 post mission upper stages reentered or will reenter Earth atmosphere within 25 years. The remaining 24% of these 86 post mission upper stages would be about 24 averted number of upper stages if operators would choose to be in the compliance with the standard practice guideline. The averted number multiplied by estimated removal cost per unit equals the posterior mitigation costs. We estimate the present value of cost-saving beginning in the 50th year after a rocket launch as forecasted orbital debris counts would be doubled. [24] Assuming benefits after 50 years is conservative, because the later benefits occur, the lower present value benefits will be. The annualized posterior mitigation costs using a 7% discount rate would be around $688,000 per post mission upper stage. [25]

The net cost saving is the difference between the prior mitigation cost and the posterior mitigation cost, which is the present value of $438,000 per post mission upper stage. Using the same 15-year timeline as the calculation of prior mitigation cost, the standard practices would result in approximately $158 million of the present value net cost saving ($438,000 x 24 unit x 15 year) using a 7% discount rate, although we assume present value benefits occur after 50 years of a launch.

5. INTERNATIONAL ADOPTION OF STANDARD PRACTICES

The key question here is whether or not space-faring nations will actually adopt the standard practices. I believe there are a few rationales for most space-faring nations to acquire the standard practices.

5.1 Nation's self-interests: opportunities in the space

Traditionally, a nation's economy is measured by goods and services generated in the conventional economic sectors. However, we are increasingly shifting our attention to goods and services supported by what we have in the space. As economic activities including trade, entertainment, communications, science and research relying on satellite services, the global space impacted economies are growing at a rate that is outpacing other global economic components. Along with the increasing demand for space assets, space governance with more space monitoring and tracking will be unavoidable to target the core of orbital debris problems—future growth of orbital debris. An effective economic solution to the problem is an incentive for nations to agree and act on the mitigation of orbital debris.

5.2 Collision avoidance relying on data

The largest dataset collected from sensing and monitoring RSO in Earth orbits is called the U.S. Strategic Command (USSTRATCOM) data. Benefits of access to the data include for operators to conduct safe launches and collision avoidances. Many space-faring nations and organizations depend on the data on a daily basis. By using USSTRATCOM data, these nations and organizations have been de facto voluntarily participating in a seemingly primary space management system, which is currently maintained and expanded by the United States. With the future deployment of very large constellations, the space-faring nations are aware of the inadequacy of space management, which depends on the high stakes of space infrastructure and data collection. As the world space activity increases, deploying constellations of satellites becomes more competitive and leads to congestion in protected space regions. All space-faring nations would be interested in safe space mission and collision avoidance. The standard practices set the goal of containing orbital debris growth. Organizations and nations using the data would eventually voluntarily comply with the standard practices guideline.

The inadequacy of space management reflects the lack of global cooperation to mitigate growth of orbital debris. Further, space industries need a global space traffic management (GSTM), which is inevitable by looking forward through global space cooperation. At this time, there is no such GSTM. Space faring nations are looking at what major space players will do. The United States is likely to take the lead and other world space players will like to follow by joining a GSTM. To a great extent the adoption of standard practices would be the very first step of GSTM. Post mission debris disposal, as an example, would be a critical component of GSTM.

5.3 Commercial space industry wants to do more

The adoption of the standard practices would provide initial benefits of space management to the industry. Let's consider NASA on-orbit separation rule [26], just like the separation rule for commercial aviation. As space operators become aware of orbital debris induced congestion, the requirement of separation means that the availability of orbital "slots" are important for satellite operations in a crowded space region. Further, comparing sending 20 to 30 satellites into orbit per launch today, a current trend indicates that future launches will send hundreds of mini satellites or CubeSats (e.g. 10 cm. x 10 cm. size satellites) into near-Earth space. There is more than one of such companies planning a large-scale constellation of hundreds of small satellites, by mass production. [27] Desiring for more activities in space than ever, the industry wants to advance their commercial interests in space tourism, commercial spaceflight, commercial satellite imagery based services, satellite communications in commercial aviation, high-throughput satellite communication services, and augmented satellite based navigation systems. Large-scale constellation satellite technologies that are evolving to be practical will require more orbital slots and spectrum than ever before. All these factors demand require more safe and accessible orbits. The standard practices provide minimal requirements for the industry to return useful orbits to the space commence.

5.4 High stakes in the economy

Orbital debris impacts on the direct space industry, which in turn amplify impacts by six times on the indirect and induced space economies according to data published by FAA AST Commercial Space Impact on the US economy, 2009. That is, reducing $1 direct space related spending due to orbital debris would contribute a $6 loss of indirect and induced revenue in the economy. The stake of orbital debris mitigation is so high that space-faring nations are increasingly interested in effective solutions to mitigate orbital debris for their operations.

In conclusion, global adoption of standard practices would result in an annualized cost saving to be approximately $158 million under a conservative assumption. The global adoption of the standard practices will be inevitable, as economic "infestation theory" predicts that one nation's adoption of orbital debris mitigation will eventually influence other nations to follow.

6. REFERENCES

[1] http://www.aerospace.org/crosslinkmag/fall-2015/orbital-debris-mitigation-guidelines/

[2] https://www.orbitaldebris.jsc.nasa.gov/library/USG_OD_Standard_Practices.pdf

[3] Gene Stansbery and Nicholas Johnson, NASA, Orbital Debris Program Office, *Orbital Debris: Past, Present, and Future* http://ntrs.nasa.gov/archive/nasa/casi.ntrs.nasa.gov/20130000305.pdf

[4] Knecht, W.R. (2001). Modeling the Big Sky Theory. Proceedings of the Human Factors and Ergonomics Society 45th annual meeting, 87-91, Santa Monica, CA: Human Factors and Ergonomics Society.

[5] The Economic Impact of Commercial Space Transportation on the U.S. Economy in 2009. https://www.faa.gov/news/updates/media/Economic%20Impact%20Study%20September%202010_20101026_PS.pdf

[6] "An Updated Assessment of the Orbital Debris Environment in LEO", "NASA JSC Orbital Debris Quarterly News", Vol. 14, Issue 1, January 2010, pg 8.

[7] Liou, J-C., "Engineering and Technology Challenges for Active Debris Removal," 4th European Conference for Aerospace Sciences, 2011.

[8] NASA Orbital Debris Office website: http://orbitaldebris.jsc.nasa.gov/faqs.html#3, Frequently Asked Questions, Johnson Space Center, Houston, TX, Updated July 2009.

[9] "Could Cubesats Trigger a Space Junk Apocalypse" Tereza Pultarova, Space.com. https://www.space.com/36506-cubesats-space-junk-apocalypse.html

[10] "The longer we wait, the tougher and more expensive it will be to safeguard satellites" --By Darren McKnight, Donald Kessler, 2012. http://spectrum.ieee.org/aerospace/satellites/weve-already-passed-the-tipping-point-for-orbital-debris

[11] Data source: The FAA/AST internal study and expert judgments.

[12] USAF **ORS-5 satellite successfully launched.** http://www.schriever.af.mil/News/Article-Display/Article/1291775/ors-5-satellite-successfully-launched/

[13] Data source: U.S. Space Surveillance Network

[14] NASA Orbital Debris Quarterly News, Volume 19, Issue 3, July 2015 and Volume 19, Issue 4, October 2015.https://orbitaldebris.jsc.nasa.gov/quarterly-news/pdfs/ odqnv19i4.pdf

[15] This information was based on a conversation with ISS team members in 2011.

[16] Current U.S satellite payloads only use 1,529 near-Earth orbital slots. Page 12, NASA Orbital Debris Quarterly News, Volume 21, Issue 3, August 2017. https://orbitaldebris.jsc.nasa.gov/quarterly-news/pdfs/odqnv21i3.pdf

[17] Space Foundation: "The Space Report 2016, The Authoritative Guide to Global Space Activity"

[18] State of the Satellite Industry Report, published by Satellite Industry Association, prepared by the Tauri Group, September 2014, page 6. http://www.sia.org/wp-content/uploads/2014/05/SIA_2014_SSIR.pdf

[19] Harrison, Hunter, Jonson and Roberts, CSIS "Implications of Ultra-low-cost Access to Space" Figure 3. Page 20.

[20] Launch forecast are from the 2015 Commercial Space Transportation Forecasts report, April 2015 by FAA/AST, and Commercial Space Transportation Advisory Committee (COMSTAC). Cost are estimated based on government space experts at NASA, USAF and FAA provided information.

[21] Centre National d'Etudes Spatiales (CNES) istThe French space policy agency, http://www.cnes.fr/web/CNES-en/7114-home-cnes.php (The HTML link was copied on July 21, 2014)

[22] C. Bonnal and P. Bultelat, "High Level Requirements for an Operational Space Debris Detroiter," NASA/DARPA Orbital Debris Conference on Orbital Debris Removal, Chantilly, VA, 8-10 December 2009.

[23] Space foundation: "The Space Report 2016, The Authoritative Guide to Global Space Activity," page 2.

[24] NASA estimated orbital debris counts would be doubled in 50 years. See Figure 2.

[25] ADR price tag for averting one used rocket in 2015 dollar is an average of $31.5 million.

[26] NASA, Space Debris and Human, September 26, 2013, "The on-orbit separation rule is about a mile deep by 30 miles across by 30 miles long (1.5 x 50 x 50 kilometers), with the vehicle in the center. http://www.nasa.gov/mission_pages/station/news/orbital_debris.html

***Disclaimer**: The views and opinions expressed in this article are those of the author and do not necessarily reflect the official policy or position of any agency of the U.S. government. Examples of analysis performed within this article are only examples. Assumptions made within the analysis do not reflect the position of any U.S. government entity*

UPPER STAGE PASSIVATION AS A MEANS OF PREVENTING SPACE DERBIS APPEARANCE

Roman Mykhalchyshyn

State Enterprise «Yuzhnoye State Design Office», 3, Krivorizka Str., Dnipro, 49008, Ukraine,
E-mail: mykhal89@gmail.com

ABSTRACT

In the context of the rapidly growing activity of space flights, the threat of spacecraft collision with fragments of space debris is increasing, since quantity of space debris grows every year significantly.
The analysis of statistical data on space debris has shown that about half of its elements occur in fragments of destroyed launch vehicles and spacecraft. Therefore, prevention of the launch vehicles stages destruction is the key factor in combating of space debris.
The materials present a set of activities to prevent the destruction of the space stage, ensuring not only the safe removal of propellant components and gases from the stage, but also stage transfer to the disposal orbit.

1. INTRODUCTION

In people's minds, space and near-Earth have always been boundless and at the beginning of space activities, it seemed that to furrow these expanses from the viewpoint of technological hazards is not and will never be dangerous. However, these convictions were wrong. The infinite near-Earth space turned out to be not so infinite, but rather limiting, at least, its separate areas. Unless something is done in the near future, space debris can completely paralyze the further penetration of people into the depths of space in the not so distant future.
Due to the regular launching of various artificial satellites into the orbit (about 4,700 satellites have been launched since 1957) and the destruction of space-rocket technology objects, the saturation of near-Earth space with bodies of technogenic origin is constantly increasing. Among the objects, "populating" near-Earth space, are actively functioning spacecraft and space debris objects. Space debris includes failed spacecraft, launch vehicles upper stages, space tugs, LV and spacecraft destruction fragments in orbit, technological elements related to the spacecraft delivery and functioning. The most significant for the space debris formation are the destruction of spacecraft and launch vehicles last stages (including the space tugs), accompanied by the clouds formation from a large number of fragments. These disruptions are often accompanied by explosions, which leads to uncontrolled self-propagation of space debris. In this case, the process of self-propagation of space debris prevails over the process of self-purification due to braking on the residual atmosphere, so the amount of debris irreversibly increases.

2. SPACE DEBRIS APPEARANCE ANALYSIS

The first two decades after the launch of the first artificial Earth satellite, the problem of space debris had no clear outlines. During this period, the launch vehicles were launched by two states – the USSR and the United States – mainly in the interests of state authorities. Since 1986, the competition on space services markets of the USA and France began. Later on, India, Japan, a number of European countries, Israel, China joined this process. In these years, the peak of the launch activity in the world was reached – 116 launches in 1988. All this, of course, has significantly increased the amount of space debris and the problem has become quite tangible. At the same time, operating spacecraft make up only about 6% of large objects of near-Earth space.
Large objects of space debris include bodies larger than 10 cm. As of the beginning of 2017, the total number of such objects officially cataloged according to the US space object control system as per NASA Debris Quarterly News [1] is about 17 400 (Fig. 1).

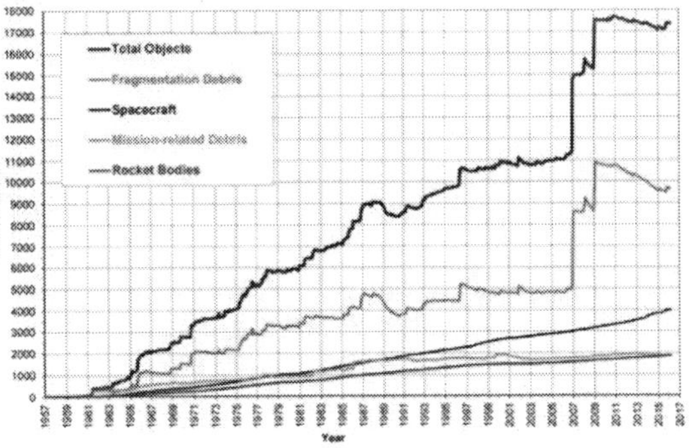

Figure 1. Yearly Number of Objects In Earth Orbit by Object Type

Medium-sized space debris are bodies of 1-10 cm in size. Their total number reaches 200 000 and is constantly growing due to the destruction of spacecraft and explosions of the launch vehicle last stages. This fraction is also unsafe due to their movement with cosmic velocities. Also, there is space debris of small sizes (0.1-1 cm), the number of which reaches several million objects [2].
A feature of technogenic pollution of near-Earth space is that the bodies of space debris most densely

"populate" the areas of working orbits of operating spacecraft from low (300-900 km), including solar-synchronous, to geostationary orbit (36 000 km). A schematic representation based on computer animation of the space debris belt in low orbits (up to 1000 km) and geostationary orbit (36 000 km) is shown in Fig. 2 and 3 respectively [2].

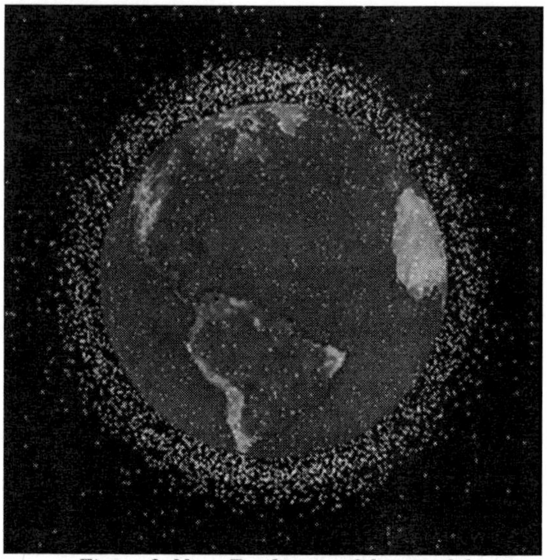

Figure 2. Near-Earth space debris model (Low Earth orbits)

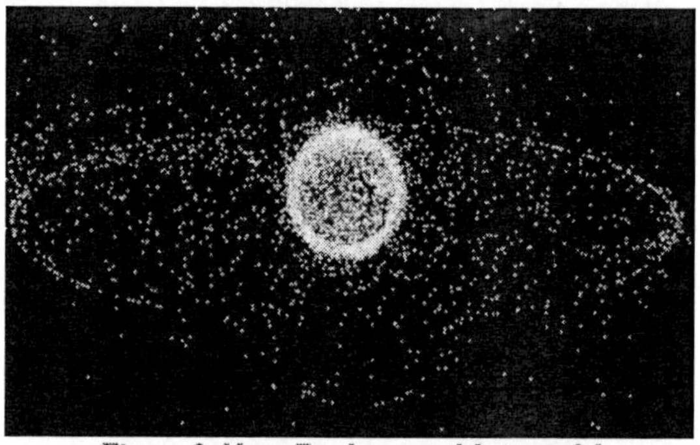

Figure 3. Near-Earth space debris model (Geostationary orbit)

Another problem that arises in the presence of technogenic objects in near-Earth space is the uncontrolled de-orbiting and the fall of large space objects and their unburned fragments to the Earth. Especially dangerous are spacecraft with nuclear power sources falling on the Earth and collapsing at the entrance to the atmosphere, as well as fragments of stages with the remains of toxic rocket fuel components that can cause great damage to human activity and serious environmental consequences. Examples of such events, which caused alarm in governmental circles of different countries and a great resonance of the world community, are the drop of fragments of Soviet satellites Kosmos-954 (1978), Kosmos-1402 (1983) and Kosmos-1900 (1988) with nuclear power installations on board, as well as the termination of existence and uncontrolled de-orbiting of the American orbital station Skylab with its fall into the territory of Australia. Therefore, an independent set of activities is being developed and implemented to reduce the factors that affect pollution of the Earth's surface as a result of space activities [3].

3. PASSIVATION SYSTEM DEVELOPMENT

The passivation system is one of the most effective measures to combat space debris. The principle of operation is based on the use of special devices for the removal of residual energy resources from the stage, after executing of the mission.

Schematic diagram of the system for upper stage is presented in fig. 4 and 5, it consists of:

- pyrovalves 3, 6 for the removal of residual propellant components from the tanks 1, 2 and feeding lines through jets 4, 5, 7, 8;
- pyrovalves 16, 17 for the discharge of gas residues from bottles 9, 10, 12, 18 through jets 13, 14 [4].

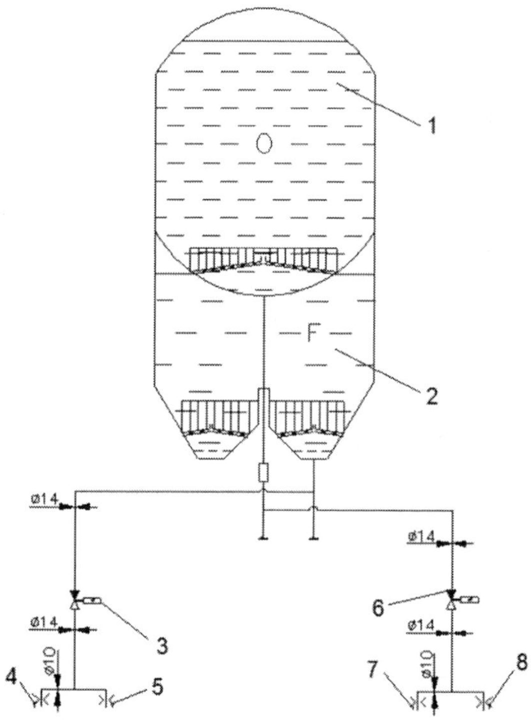

1 – oxidizer tank; 2 – fuel tank; 3, 6 – pyrovalves;
4, 5, 7, 8 – metering jets

Figure 4. Scheme of the passivation system for residuals removal of propellants and gases from tanks

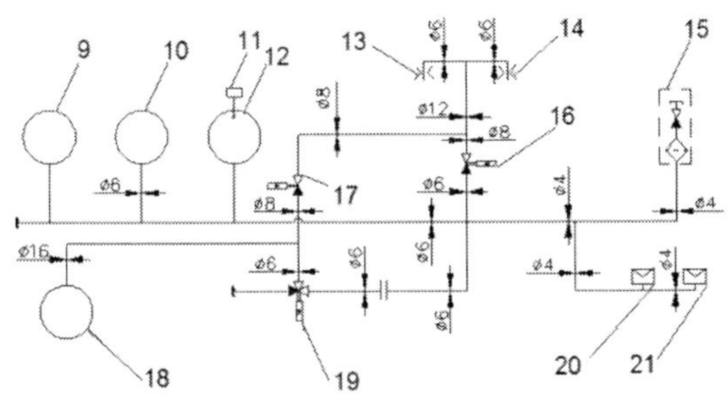

9, 10, 12 –bottles of the pressurization system; 13, 14 – metering jets; 16, 17 – pyrovalves; 18 – bottle of main engine; 19 – pyroswitch; 20, 21 – pressure sensors

Figure 5. Scheme of the passivation system for gas residual removal from bottles

3.1. Gases removal from bottles of pressurization system and main engine

Upper stage receives some impulse at separation of a satellite. Propellant components in tanks move to the upper bottoms under the action of that impulse. In this case, some part of propellant components remains under tank capillary stabilizers and in feeding lines.
Before the beginning of gas removal from bottles in weightlessness conditions, propellant components are located in tanks under capillary stabilizers where they are held by surface tension forces; they also can take an arbitrary position in tanks above capillary stabilizers.
After generation of a command for operation of pyrovalves 16 and 17, the gases remove from the bottles of pressurization system and main engine as a result of connection of the systems collector with the ambient space (Fig. 6) [4].

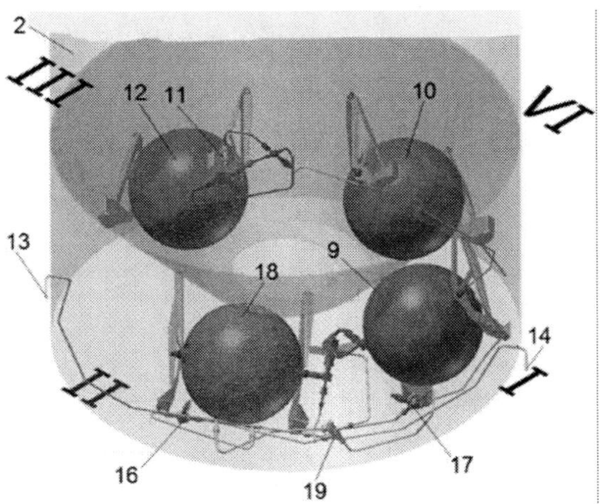

Positions according to fig. 5

Figure 6. The passivation system for gas residual removal from bottles

Removal is performed through the metering jets installed symmetrically at 15° angle to the longitudinal axis of stage. The lines from the collector up to the metering jet have approximately the same configuration and hydraulic resistance; that ensures the same thrust after gas release. In this case, the stage obtains the positive longitudinal acceleration.
Under the action of positive longitudinal acceleration, propellant components (located in tanks out of the limits of capillary means for provision of continuity) move to capillary stabilizers and to main engine inlets.

3.2. Propellant component and gases removal from fuel tank

Line of diameter 14 mm and pyrovalve 3 (Fig. 7) is installed for removal of fuel and gases from stage fuel tank. That line is mated to the on-board part of the filling line at one side; it is divided into two lines 10 mm diameter of about the same resistance at the other side. At the pipeline's outlet, there are metering jets located symmetrically with regard to the longitudinal axis of the stage in its diametrical plane. The plane of the section of metering jets is directed to the side of engine nozzle of the stage at 15° angle to the longitudinal axis with account of intersection of thrust action lines above the stage canter of mass.
After operation of pyrovalve 3, there is liquid fuel outflow under gas pressure. Thrust of fuel (flowing through metering jets) creates longitudinal acceleration. Propellant components in tanks are pressed to intake devices under the action of that acceleration. Gas pressure in tank is reduced because of increase of the gas volume at propellant component removal.
At achievement by fuel level of "critical" height of dip of that level, there is gas breakthrough into the feeding line above the plate of the intake device; propellant component is expelled from the feeding line and after that, gas is removed from the fuel tank. Propellant component located in the portion from the inlet into the charging line up to engine inlet and some part of component (located in the chute of the intake device) are not discharged. When sun heat is supplied, then fuel residual (while melting) completely evaporates with outflow of vapors into the ambient space [4].

3.3. Propellant component and gases removal from oxidizer tank

Line of diameter 14 mm and pyrovalve 6 (Fig. 7) is installed for removal of fuel and gases from stage oxidizer tank. That line is mated to the on-board part of the filling line at one side; it is divided into two lines 10 mm diameter of about the same resistance at the other side. At the pipeline's outlet, there are metering jets located symmetrically with regard to the longitudinal axis of the stage in its diametrical plane. The plane of the section of metering jets is directed to the side of

engine nozzle of the stage at 15° angle to the longitudinal axis with account of intersection of thrust action lines above the stage canter of mass.
After operation of pyrovalve 6, there is liquid oxidizer outflow under gas pressure. Thrust of oxidizer (flowing through metering jets) creates longitudinal acceleration. Propellant component in tank is pressed to intake devices under the action of that acceleration. Gas pressure in tank is reduced because of increase of the gas volume at propellant component removal.
At achievement by oxidizer level of "critical" height of dip of that level, there is gas breakthrough into the feeding line above the plate of the intake device; propellant component is expelled from the feeding line and after that, gas is removed from the oxidizer tank. Propellant component located in the portion from the inlet into the charging line up to engine inlet and some part of component (located in the chute of the intake device) are not discharged. When sun heat is supplied, then oxidizer residual (while melting) completely evaporates with outflow of vapors into the ambient space [4].

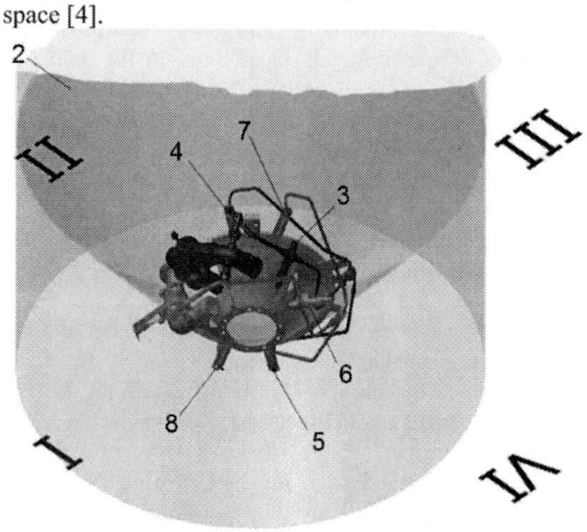

Positions according to fig. 4

Figure 7. The passivation system for residuals removal of propellants and gases from tanks

4. CALCULATION OF PROPELLANT COMPONENTS AND GAS REMOVAL SYSTEM

4.1. Gases removal calculation from pressurization system bottles and main engine bottle

The gas release process was reviewed with the following assumptions:

– thermal and physical characteristics of gas (gas constant, specific heat capacity and adiabatic index) are constant within the reviewed range of temperatures and pressures;

– the following pressure difference was observed in the bottles at the moment of helium release from those bottles (operation of pyrovalves), kgf/cm^2:

	min	max
the bottles of the pressurization system	– 20	– 60
the bottle of main engine	– 60	– 280

Table 1. Gas removal thrust by the time

τ, s	0	5.6	10	20	30	50	70	110	130	160
F_{max}, kgf	5.4	9.3	10.8	8.4	6.5	3.8	2.3	0.8	0.3	0
F_{min}, kgf	1.8	2.7	2.5	1.75	1.5	0.7	0.5	0	0	0

Zero time meets the moment of voltage supply to pyrovalves 16 and 17 after generation of control system command on helium release from pressurization system and main engine bottles.
Thrust difference of the efforts between two release branch pipes does not exceed ~10% with account of errors of installation and production of the metering jets.

4.2. Computation of PC and vapors residual dumping

During the computation, we determine the time of propellant components dumping from tanks **t**, mass flowrate **G**, thrust of fuel and oxidizer flowing through metering jets **F** and pressures in tanks at the moment of completion of propellant components dumping $\mathbf{P_k}$.
Propellant components dumping time is determined by a formula

$$t = \frac{G}{\dot{G}_{av.}}, \quad (1)$$

where G – mass of propellant components residuals;

$\dot{G}_{av.}$ – Propellant components average mass flowrate during the dumping.

Propellant components weight flowrate per second **G** during the dumping is determined by a formula

$$\dot{G} = \sqrt{P_t \frac{2F^2\rho}{\zeta_{tot}}}, \quad (2)$$

wher

ρ – propellant components density;
P_t – pressure in tanks at the dumping moment;
F – area of the pipeline's section;
ζ_{tot} – total coefficient of hydraulic resistance/

$$F = 2\dot{G}_j \, \mathrm{w} \cdot \cos\alpha, \quad (3)$$

where

$\dot{G}_j$ – propellant components flowrate through one metering jet;

α – angle of deviation of the metering jet's axis from the longitudinal axis of the stage.

Pressure value in tanks as of completion of propellant components dumping is determined by a formula

$$P_{fin.} = P_{in.} \cdot \frac{V_{in.}}{V_{fin.}}, \quad (4)$$

where

$P_{in.}$ – pressure in the tanks before the beginning of propellant components dumping;

$V_{in.}$ – volume of gas cushion before the beginning of propellant components dumping;

$V_{fin.}$ – volume of gas cushion at the moment of propellant components dumping completion.

Table 2. Calculation results of propellants removal

Name of characteristic, dimension	Value of characteristic	
	oxidizer system	fuel system
Time of components dumping from tanks, s	from 90 up to 290	from 65 up to 230
Propellant components weight flowrate, kgf/s	from 1.37 up to 1.47	from 0.68 up to 0.75
Trust of propellant components flowing through the metering jets, kgf	from 1.29 up to 1.44	from 0.58 up to 0.69
Value of pressure in tanks as completion of propellant components dumping, kgf/cm^2	from 4.80 up to 6.1	from 2.27 up to 2.95

Gas forces through intake at the end of removal with low longitudinal acceleration ($n_x \approx 10^{-4} - 10^{-3}$). After gas and fluid mix is discharged up to full practically removal from stage At the forcing moment propellant components weigh are

- oxidizer – not more than 20 kg;
- fuel – not more than 10 kg.

Gas removal from tanks is carried out just past propellant components removal approximately after 65-290 s from generation of a command on operation of pyrovalves 3 and 6 of fuel and oxidizer tanks respectively.

Fig. 8 and 9 show the calculated values of thrust changes along the axis of one branch pipe as a result of gas removal from tanks.

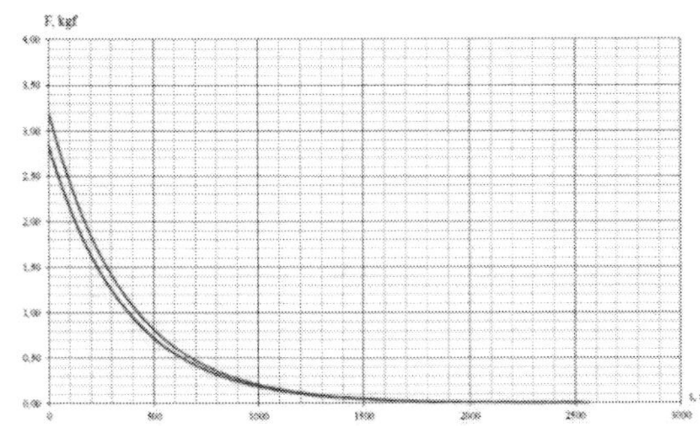

Figure 8. Thrust during gas removal from the oxidizer tank

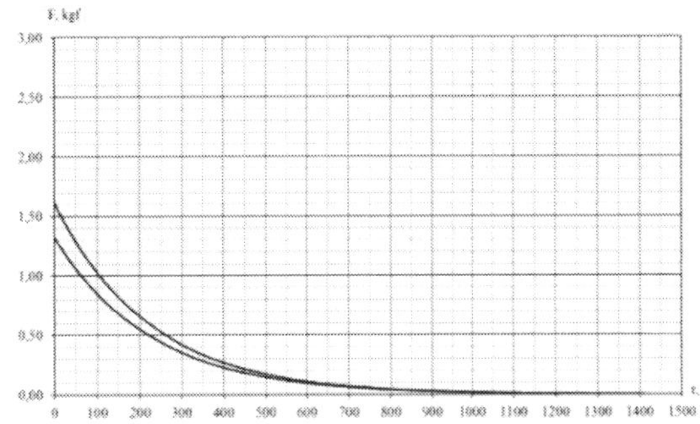

Figure 9. Thrust during gas removal from the fuel tank

As, ambient pressure (where stage locates after mission – more than 200 km) is considerably lower than propellant components saturated vapor pressures, then unremoved residuals will be vaporized and boiled and after that vapor removal through removal lines [4].

Since the residual removal is performed by flight axis of the stage and the thrust during removal is sufficient, at the time of passivation the stage will move from the working orbit to the disposal orbit.

5. CONCLUSION

- The analysis of the composition and formation causes of space debris has shown that the main part of it consists of fragments of the launch vehicle stages and spacecraft.
- It has been identified and substantiated that the using of the passivation system is the most optimal way to prevent the formation of space debris.
- Passivation system structure has been developed that allows removing all residual energy resources from the stage.
- The performed calculations have confirmed operability of the passivation system.
- The calculated values of thrusts during residual removal have confirmed provision of moving the stage to disposal orbit.

9. REFERENCES

1. ESA, Requirements on Space Debris Mitigation for ESA Projects, ESA/ADMIN/IPOL(2017) 1 April, Annex 1.

2. European Code of Conduct for Space Debris Mitigation, Issue 1.0, 28 June 2016, 68–71.

3. ISO 24113:2011, Space Systems – Space Debris Mitigation Requirements, 23–24.

4. Upper stage development for future space programs, Technical Report Yuzhnoye SDO, 2016. – 82–103.

EVALUATING MICROMETEOROID AND ORBITAL DEBRIS RISK ASSESSMENTS USING ANOMALY DATA

Michael Squire[(1)]

[(1)]*National Aeronautics and Space Administration, NASA Langley Research Center, 11 Langley Blvd, Mail Stop 116, Hampton, Virginia, USA, 23681, Michael.D.Squire@nasa.gov*

ABSTRACT

The accuracy of micrometeoroid and orbital debris (MMOD) risk assessments can be difficult to evaluate. A team from the National Aeronautics and Space Administration (NASA) Engineering and Safety Center (NESC) has completed a study that compared MMOD-related failures on operational satellites to predictions of how many of those failures should occur using NASA's MMOD risk assessment methodology and tools. The study team used the Poisson probability to quantify the degree of inconsistency between the predicted and reported numbers of failures for a selected group of robotic satellites. Many elements go into a risk assessment, and each of those elements represent a possible source of uncertainty or bias that will influence the end result. There are also challenges in obtaining accurate and useful data on MMOD-related failures.

1. INTRODUCTION

Micrometeoroid and orbital debris (MMOD) particles traveling at more than 70 km/s (micrometeoroids travel at the highest velocities, orbital debris velocities are up to ~15 km/s) can potentially strike any orbiting spacecraft, and MMOD damage is the highest risk factor for most spaceflight missions. Spacecraft designers and mission planners must satisfy MMOD risk requirements in order to ensure that the vehicle can protect human occupants, satisfy mission objectives, and preserve the minimum capacity needed to perform end-of-mission disposal. MMOD risk assessments (MRAs) quantify that impact potential based on several factors including the spacecraft's configuration, location, construction, operational status, and tolerance to damage. The values provided by the MRAs are used to satisfy MMOD requirements. So the users of MRAs are interested in understanding the accuracy of the MRAs in predicting the probability of a failure for their respective spacecraft.

A team formed by the National Aeronautics and Space Administration (NASA) Engineering and Safety Center (NESC) sought to define how well NASA's MRA methodology quantified risk by comparing the values produced by MRAs to operational spacecraft history. Ideally, a study of this kind would use a large dataset including many different satellites to obtain an average MMOD failure rate and compare this value to an MRA average predicted failure rate for that dataset. However, this kind of bulk comparison was not practical. Where spacecraft failure and anomaly data are available (which was not always the case), the team found a wide disparity in reporting and cause attribution making it inappropriate to combine results. Furthermore, MRAs are time and resource intensive, and performing a large number of them to encompass a variety of spacecraft was not realistic.

So the team decided to take a "micro-analysis" approach and select a few satellites based on the possibility of obtaining the data necessary to both determine MMOD failures and perform detailed MRAs. The study achieved a statistically significant number of data points by using satellite constellations, where the same vehicle design was used in several individual spacecraft. This allowed a common MRA for the spacecraft but applied throughout the constellation. In addition, the MRAs were not performed for the spacecraft as a whole, but for individual components. For example, the MRA for Spacecraft #1 resulted in a predicted number of failures not for the spacecraft but for Spacecraft #1's tanks and batteries. This increased the dataset size and allowed the team to concentrate on those spacecraft components where an MMOD impact failure was relatively easy to diagnose (e.g. an MMOD tank failure results in a catastrophic rupture of the tank). Finally, with the values of predicted number of failures and reported number of failures in place, the team could compare them and judge the level of consistency between the two, which would in turn reflect the relative agreement between the MRA and the reported history.

2. NASA's MRA PROCESS

NASA's Hypervelocity Impact Technology (HVIT) group is responsible for most of the MRAs performed to support NASA's programs. Fig. 1 shows HVIT's MRA process and how spacecraft operators and designers use it to evaluate MMOD risk and design MMOD protection. The MRA tool, Bumper, is at the center of the MRA process. As shown in the diagram, Bumper uses input including orbital debris and micrometeoroid environment models, spacecraft geometry, failure criteria for each component and shield included in the analysis, ballistic limit equations (BLEs), and operating parameters including spacecraft orbit and attitude. Each of these input elements can contribute to uncertainty, bias, or error in the overall assessed risk produced by Bumper.

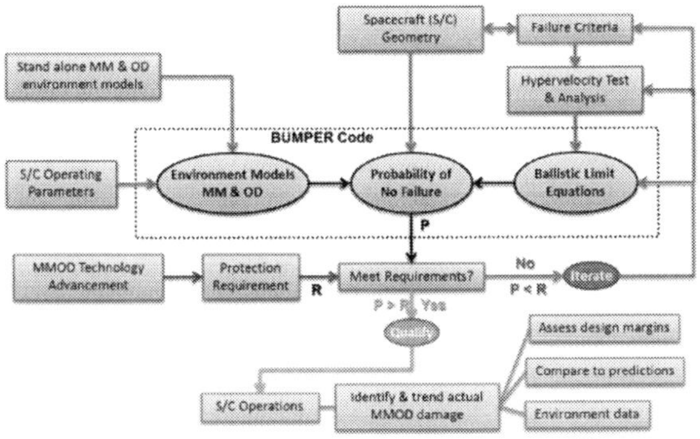

Figure 1. NASA's Bumper MRA Process [2]

The Bumper code uses this information to produce the number of events predicted to occur over a given time. For simplicity, these events are referred to here as "penetrations," but those events can be MMOD penetrations, impacts, or failures, depending on the criteria set for the analysis. This result is typically presented as a probability of no penetration (PNP), determined using Eq. 1:

$$PNP = e^{(-N)} \tag{1}$$

where e = the base of the natural logarithm (~2.718...), and N = number of penetrations calculated. The probability of penetration (PP) is simply 1-PNP. PNP and PP are typically presented as percentages or odds (e.g., 1% or 1 in 100).

The Bumper MRA process is used to iteratively design and improve MMOD protection. During design, if a Bumper MRA results in a risk greater than what can be tolerated, then some of the input parameters can be adjusted to observe what the effect on risk will be. The predicted number of penetrations is what is predicted to occur during a given time period in the future. However, for this study, it was necessary to look backward and use the MRA to predict how many penetrations (or failures) should have occurred starting a given time in the past. One of the ramifications of this was a difficulty in obtaining some of the needed configuration data for spacecraft that have been flying for many years after they were designed.

3. POISSON PROBABILITY

The Bumper MRA provides a single value for the number of predicted failures, and this was compared to the number of failures reported to the team. The team desired a simple way to quantify this comparison and used the Poisson probability for this purpose. The Poisson probability is calculated using the generalized version of Eq. 1:

$$P(k) = \frac{\lambda^k e^{-\lambda}}{k!} \tag{2}$$

where P(k) = the Poisson probability, λ = the predicted number of occurrences, and k = the reported number of occurrences. The value for λ is calculated by the MRA, and the value for k comes from the failure history. The Poisson probability indicates the level of consistency between the reported number of occurrences and what was predicted to occur. Typically a threshold of 5% or below is used to indicate inconsistency. The Poisson probability is interpreted as follows:

> There is a P(k)% chance that k occurrences will occur given a prediction of λ occurring.

If P(k) is low, then that indicates an inconsistency between the reported and predicted, and the cause may be either the reporting or predicting or both. The Poisson probability should not be confused with the PNP. The PNP is an alternative expression of λ, while the Poisson probability quantifies how well the PNP agrees with reported data.

4. ROBOTIC SPACECRAFT FAILURE ANALYSIS

4.1. Baseline Analyses

For this study, the team evaluated three satellite designs comprising two constellations and one single satellite. This represented a total of 73 individual spacecraft. The team performed MRAs for each of the spacecraft's pressurized tanks and batteries; determining a failure for these components was relatively unambiguous compared to other components. The total exposure time for all of the components was 1436.3 years, and the total exposed area-time product (i.e., area x time) was 2847.6 m^2-year. The individual MRA results are shown in Tab. 1.

Table 1. Failure MRA Results for Robotic Assets

Asset	Component	Area-Time Product (m^2-year)	Number of Failures Predicted
Satellite#1	Battery Cell	39.4	0.8
	Tank	12.1	1.1
Satellite#2	Battery	349	0.05
	Tank	2253	5.0
Satellite#3	Battery Cell	65.1	4.0
	Tank	129	0.4
Total		**2847.6**	**11.3**

It is important to note that the MRAs for these assets were performed based on failure, meaning an MMOD impact would have to not only strike the component, but must damage it to a degree that the component fails, which for tanks is a rupture and batteries is the removal of that battery from service (and possibly a rupture also).

The results in Tab. 1 show that there is a wide variation between the number of failures predicted from as low as 0.05 up to a high of 5.0. The predicted failures can be thought of as a failure rate over the respective time frame for each satellite (they are different for each satellite system). The total number of predicted failures was 11.3 over the life of the spacecraft assessed. Tab. 2 shows that a total of two failures were recorded for the same components over the same time period. Tab. 2 also shows the Poisson probabilities for each of the assessed components based on the predicted versus reported failures.

Table 2. Poisson Probabilities for Robotic Assets

Asset	Component	Number of Failures Reported	Poisson Probability
Satellite#1	Battery Cell	0	44%
	GN2 Tank	0	33%
Satellite#2	Battery	0	95%
	Tank	1	4%
Satellite#3	Battery Cell	0	9%
	Tank	1	70%
Total		**2**	**0.096%**

The Poisson probability gives the probability that the number of failures reported is consistent with the number predicted. For example, there is a 44% probability that there would be zero failures of one of Satellite#1's battery cells given the predicted failure rate of 0.8 (from Tab. 1). This indicates that the predicted risk and the reported failures are consistent. Conversely, the Satellite#2 tank reported one failure (Tab. 2), but 5.0 were predicted (Tab. 1), resulting in a 4% probability that if the prediction was correct, there would be 1 failure. It is appropriate to sum the predicted and reported numbers of failures to calculate an overall Poisson probability for these satellites. Tab. 2 shows that this overall probability is only 0.096%, indicating very low consistency.

4.2. Sensitivity Studies—Tank Wall Thickness

As shown in Section 2, NASA's Bumper MRA process has several elements that contribute to the result, and uncertainties and errors in any of them can have a substantial effect on the final result. Elements related to how robust exposed regions of the spacecraft are to MMOD are important to accurately portray in the MRA. For example, because it is integral to the ability of the tank to withstand an MMOD impact, the tank wall thickness has a substantial effect on the MRA risk. To observe the sensitivity of the MRA to tank wall thickness, the team performed MRAs with alternative thicknesses for Satellite#1 and Satellite#2.

The initial MRAs performed for Satellite#1 included a hydrazine tank, but the team later discovered that this tank was only used during ascent. After ascent, the tank was isolated from the rest of the hydrazine system and any liquid in the tank was allowed to freeze. This meant that any MMOD impacts into the tank either may not rupture the tank, or might not be detected because that subsystem was no longer monitored. The MRAs that did include the tank resulted in substantially higher predicted failures than when the tank was removed from the MRA. There were also conflicting data received by the team on the thickness of this tank's wall. The effects including this tank and altering its wall thickness had on the MRA and the Poisson probability can be seen in Tab. 3 (there were zero reported failures for this tank).

Table 3. Effect of Tank Wall Thickness on Satellite#1 Poisson Probability

Tank Wall Thickness (cm)	Predicted Number of Failures	Poisson Probability for Satellite#1 Tank	Total Poisson Probability for all Satellites
Tank Not Included (Baseline)	NA	NA	0.096%
0.178	7.3	0.1%	0.0002%
0.254	3.2	4.1%	0.006%
0.343	1.5	22.3%	0.03%

For the greatest thicknesses (0.343 cm from Tab. 3), the predicted number of failures is relatively consistent with the no failures reported, resulting in a Poisson probability of 22.3%. However, if thinner tank walls are chosen, the predicted numbers of failures are greater by a factor of two or three and practically guarantee that every satellite in the constellation should have experienced a tank penetration and failure. As expected, the Poisson probabilities indicate much higher inconsistency between predicted and zero reported for the tanks with smaller wall thicknesses.

The team found challenges securing consistent configuration tank wall thickness data for Satellite#2 like they did for Satellite#1. Two different sources provided different thicknesses for the Satellite#2 tank, and neither source was judged to be more or less credible than the other. For the baseline MRA, 1.0 mm was used as the tank thickness because it resulted in a more conservative result. A thickness of 1.5 mm was the alternative thickness reported to the team. The effect of the difference was to reduce the predicted number of failures by more than half and reduce the Poisson probability by an order of magnitude as shown in Tab. 4.

Table 4. Effect of Tank Wall Thickness on Poisson Probability for Satellite#2

Tank Wall Thickness (mm)	Predicted Number of Failures	Poisson Probability for Satellite#2 Tank	Total Poisson Probability for all Satellites
1.0 (Baseline)	5.0	4%	0.096%
1.5	1.9	43%	1.15%

Again, the 0.5 mm variation between tank wall thicknesses resulted in a significant change in the resulting risk prediction.

4.3. Sensitivity Study—Failure Reporting

For Satellite#3, the team modified the reported number of failures instead of changing MRA input. Analysis of the calculated size and path (provided by the satellite operator) of the presumed MMOD particle that penetrated the one battery cell that failed revealed that to hit the battery cell that failed would have required an unlikely (but possible) path to negotiate a crowded region in front of the target. Another cause of the battery failure, which would have had to produce not only the loss of performance but also an observed attitude perturbation to the spacecraft, was unlikely but also possible. If this second scenario had occurred, then the reported number of MMOD failures for this component would have been reduced from one to zero. Tab. 5 shows that for the battery alone, the Poisson probability decreases from a marginally inconsistent 9% to a strongly inconsistent 2% when removing the single reported failure. The effect on the total Poisson probability for all of the assets is negligible; the baseline value was already very low (i.e., very low consistency), and the change only reinforced that. When trying to determine the size particle that caused a recorded perturbation, there are numerous assumptions and guesses that have to be made including the velocity, size, composition, direction, and shape of the particle. These parameters are interrelated in how they produce the perturbation witnessed by the spacecraft operators, and changes in one will affect the estimated and calculated values for the others.

Table 5. Effect of Number of Reported Failures on Poisson Probability

Number of Reported Battery Failures Due to MMOD	Predicted Number of Failures	Poisson Probability for Satellite#1 Tank	Total Poisson Probability for all Satellites
1 (baseline)	4.0	9%	0.096%
0	4.0	2%	0.015%

5. ROBOTIC SPACECRAFT IMPACT ANALYSIS

The MRAs described in Section 4 were failure risk assessments, meaning the criteria used to establish the risk (i.e. predicted number of events) were component failures. However, each spacecraft, and each component, is impacted by many more MMOD particles that do not cause a failure than those that do. An *impact* risk assessment, as opposed to a *failure* risk assessment, will predict the number of impacts a component will experience. Impact MRAs do not require as much data concerning the material characteristics and damage modes since what is sought is simply how many times the spacecraft is hit. However, it is much more difficult to determine when a component is impacted if there is no indication (failure is an obvious indication) visible to the spacecraft controllers.

Fortunately, one of the spacecraft operators that provided the team with failure data also provided the team orbit perturbation events that occurred on some of their spacecraft that did not result in a failure. These were uncommanded changes in the spacecraft velocity, which resulted in changes to the vehicle altitude, and pitch, yaw, and roll. The spacecraft operator determined that the probable cause of these events was MMOD impacts. The locations in the orbit of each of the impacts were consistent with orbital debris (as opposed to micrometeoroid or other failure causes). This is seen in Fig. 2, which displays the cumulative distribution of satellite latitude where each anomaly occurred (blue line) against what would be predicted by NASA's orbital debris environment model, the Orbital Debris Engineering Model Version 3.0 (ORDEM3.0) (red line). The black dashed line represents where the data would fall if it were random, as would be expected with micrometeoroids.

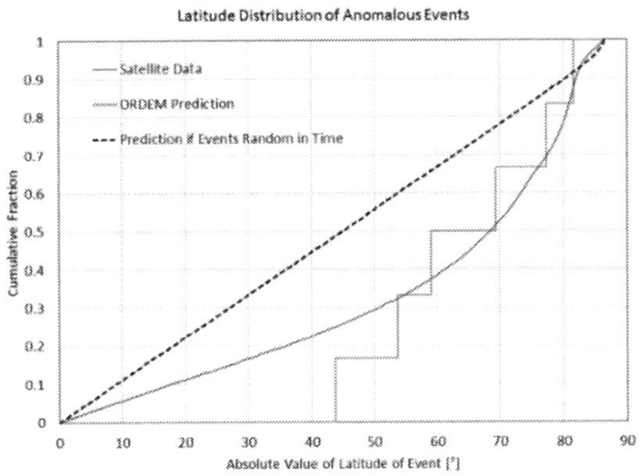

Figure 2. Latitude Distribution of Anomaly Events

The team used these measurements to determine the momentum imparted to the spacecraft and the momentum of the MMOD particle. The momentum could then be converted to a particle size dependent on assumptions of velocity and shape. The average orbital debris particle velocity at that altitude is 13.94 km/s according to ORDEM3.0, so that was taken as the impact velocity. With a velocity, the team could calculate a mass. ORDEM3.0 is expressed in terms of characteristic length (L_C), a measure of the average of the longest dimension of the particle and the two dimensions orthogonal to that longest dimension (e.g., for a sphere, all three of those dimensions are equal). To get the L_C from a mass requires assuming a shape and density. In most MRAs, and all those performed by NASA/HVIT, the shape of each piece of orbital debris is assumed to be spherical. However, a sphere has the highest mass possible for a given L_C, so using this assumption results in the highest particle mass, and greatest assessed risk possible. Looking at it another way, the mass for the sphere will result in the smallest possible L_C. Using the spherical assumption and the density of aluminum, the particle size (i.e., L_C) for each of the impact events was estimated using the calculated momenta. The impact frequency for these size particles was then compared to what would be expected using ORDEM3.0.

An additional consideration is the momentum enhancement factor (MEF). This is the additional momentum that is imparted to the satellite from the impact ejecta. Because this ejecta has mass originating from the spacecraft and is travelling in the direction opposite the direction of the incoming MMOD particle, a force is directed against the spacecraft in the same direction as the simple momentum exchange of the MMOD particle to the spacecraft. This MEF has been experimentally measured for different types of materials [8, 9, 10], and is estimated for this case to be between 1-3 (i.e., an MEF of 1 would mean no additional momentum is exchanged). Assuming an MEF of 1, ORDEM3.0 would predict 24 impacts in the size range observed/calculated. When an MEF of 2 is assumed, the predicted number of impacts increases to 70 because smaller particles are required to produce the same net momentum observed, and there are more smaller particles—thus more impacts. If an MEF of 3 is chosen, there would be 164 impacts. Any of these values is greater than the six impacts registered.

The team then investigated the results if a shape other than a sphere was used for the orbital debris. Different shapes were applied to the results including oblate ellipsoids and octahedrons. A sphere with voids was also explored, which resulted in a reduced net density that varied as a function of critical length ($L_C^{-0.25}$, $L_C^{-0.5}$, $L_C^{-0.75}$, $L_C^{-1.0}$) . This "voided sphere" is analogous to a debris particle that may have material folded over on itself or crumpled like a sheet of paper, resulting in a nonhomogeneous density. When these different shapes were applied to ORDEM3.0, the number of impacts was reduced. The degree of reduction varied depending on the shape assumed. Fig. 3 shows the curves for each of the shape/voided spheres plotted on a log-log scale as a function of momentum versus flux. The blue line on the graph represents the baseline spherical assumption, the red lines represent the impacts (one red line includes a catastrophic impact in addition to the other six), and the other lines are different shapes as indicated in the legend. As can be seen on the figure, the data for the impacts are more consistent with the curves for the voided spheres and the octahedron than for spheres.

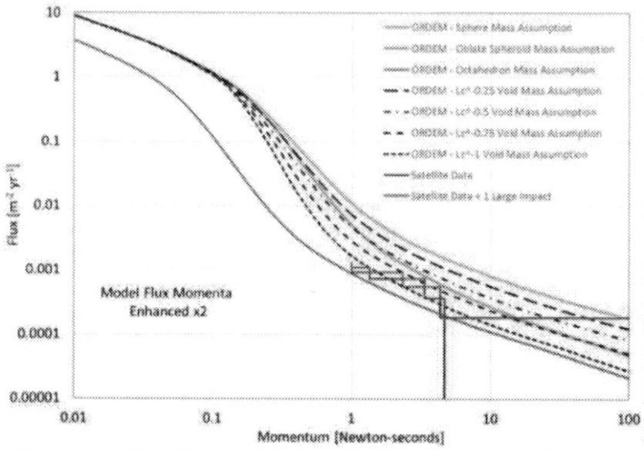

Figure 3. Predicted Orbital Debris Flux as a Function of Impact Momentum (MEF=2)

Any decrease in flux as a function of momentum (i.e., any of the lines in Fig. 3 below the blue one) would represent a reduction in risk when applied to an MRA. So if alternative shapes to spheres are applied to ORDEM3.0 and Bumper, then the end result would be a reduction in risk. However, the shape and orientation of an impacting projectile can influence the penetration characteristics of the projectile. Consider the different damage that would be caused by a rod-shaped particle if that particle impacted on the pointed end versus the flat side. These penetration characteristics would need consideration and may, to some degree, counteract risk improvements made with alternative shapes.

6. CAUSES OF INCONSISTENCIES

The overall results for this very limited number of robotic spacecraft indicate that the predicted number of failures is inconsistent with the reported number. This means that either the failure predictions are overpredicted, failures are under-reported, or a combination of these factors is responsible. Most likely, the inconsistencies are a result of a combination of error or bias that stem from uncertainties present in several sources on the prediction and reporting sides. This section discusses some of these sources of uncertainty that can result in inaccurate and/or inconsistent results.

Some of the causes of uncertainty or bias in the MRAs are:

- Sensitivity to design parameters, such as dimensions and materials (e.g., the ambiguity of the tank wall thicknesses discussed in Section 4). Design and construction details are routinely unavailable, difficult to obtain, or contain unusual construction features and might require assumptions based on engineering judgment. This is especially true for spacecraft already on orbit and far removed from design and construction, as was the case in this study. Even for spacecraft currently in development, there can be issues obtaining accurate input information due to proprietary or classified data or the actual build differing from drawings.

- MMOD shape represented as solid spheres in ORDEM3.0 and Bumper. As discussed in Section 5, spheres represent the maximum possible mass for a given characteristic length and density. The impact assessment showed that using different shapes and voided spheres results in reduced assessed risk. However, the complex interaction of mass, shape, orientation, and size-dependent debris flux affects the overall risk of nonspherical particles compared to that of the solid sphere assumption.
- A lack of applicable test data for most robotic spacecraft MRAs. NASA uses BLEs derived from testing and analysis for crewed vehicles, which differ in materials and layup from typical satellite construction.
- Limitations of HVI testing. BLEs must be extrapolated in order to represent on-orbit collision velocities since most HVI testing is typically only up to 10 km/s. Compare that to orbital debris relative impact velocity up to 15 km/s, and up to 72 km/sec for micrometeoroids.
- Worst-case assumptions for component failure conditions, especially in first order risk assessments. As more information is gathered and a more detailed and accurate spacecraft configuration is applied to the MRA, the assessed risk tends to decrease.
- Limited HVI test data to derive BLEs. Statistically-based uncertainty bounds are not normally derived due to limited HVI data.

Challenges associated with using spacecraft failure and anomaly data include:

- A lack of consistency in categorizing and reporting anomalies across the aerospace community.
- Lack of experience in the process of associating failures to causative mechanisms. Observed failures may be the result of a combination of multiple events occurring on the spacecraft.
- Lack of motivation on the part of spacecraft operators to fully investigate the root cause of an anomaly or failure. Many commercial operators prioritize returning the spacecraft to service rather than spending resources troubleshooting and identifying the cause of an anomaly.
- Reluctance on the part of spacecraft operators to share anomaly data due to privacy fears or a perception that public disclosure of failures will reflect negatively on the company.
- A lack of adequate onboard sensors, and requirements to provide those sensors, to measure physical perturbations to better correlate particulate impacts to failure mechanisms.
- Unavailability of telemetered data, especially after a catastrophic failure.
- Lack of understanding and natural variability in other MMOD failure mechanisms (e.g., plasma) other than physical penetration and ejecta.

7. CONCLUSION

A primary conclusion of this study is that verifying the accuracy of the MRAs can be very challenging. For the limited sample set used in this study, there is an overall inconsistency between the magnitude of predicted risk and the actual number of MMOD-induced failures reported. There are uncertainties for both the prediction and the reported that could be contributing to the inconsistencies. The team used Poisson probabilities as a tool to illustrate the inconsistencies quantitatively. These values are not meant to be applied to other spacecraft.

Much attention recently has been given to ORDEM3.0 as a contributor to uncertainty. MRAs performed using ORDEM3.0 have, in many cases, resulted in higher assessed risk. Indeed, the impact analysis in this study looking at spacecraft perturbations shows that perhaps the number of impacts expected by the model is greater than what is occurring. However, those results are only focused on a narrow range of particle sizes—particles large enough to perturb the orbit of the affected satellites. The study did not gain any insight to how well ORDEM3.0 is representing particles smaller or larger than ~3-5 mm. The team also showed promising data in evaluating a more appropriate shape for orbital debris and how this may improve this element of the risk assessment. At least two tests have been performed where a grounded satellite has been impacted by a hypervelocity projectile under laboratory conditions, and the resulting debris was counted, measured, and catalogued by shape [11, 12]. The data from these experiments might be useful to characterize orbital debris shape for use in MRAs.

The component failure analyses for the robotic spacecraft revealed a wide range of Poisson probabilities for the different components, even on the same spacecraft. This seems to imply that there is more contributing to MRA uncertainty than simply ORDEM3.0 since all of the component MRAs used the same ORDEM3.0, but not all had the same degree of inconsistency. The team showed how important knowing the accurate design parameters of the spacecraft/component is by varying tank wall thicknesses and getting very different predicted failure results. If a tank wall thickness was a little thicker than originally thought, then it would take a larger MMOD particle to penetrate that tank, and the population of MMOD particles decreases as they get larger. Other sources of error and uncertainty as discussed in Section 7, while not targeted in this study, offer contributions that may be in the same order of magnitude as the shape effect and tank wall thicknesses as far as influencing the inconsistencies in this study and in general.

There are opportunities for improvement on the MRA side and the anomaly tracking side. *In situ* measurement of MMOD impacts and continued use of returned impacted surfaces will help refine and upgrade environment models. HVI testing on materials and components specific to the spacecraft being assessed (when possible) will help to improve damage prediction and assignment of failure criteria. Standardization of anomaly reporting and characterization, and a greater willingness to share that information, will help to understand actual failure rates. With robotic spacecraft having to meet end-of-mission disposal requirements, the robotic community is taking a greater interest in MMOD protection than previously—a risk that has been familiar to the crewed missions for a long time.

8. ACKNOWLEDGEMENTS

The author would like to thank the members of the NESC Evaluation of MMOD Risk Prediction with Available On-orbit Assets assessment team for participating in this study and contributing to the results presented here. A list of the team members, plus more details on this study, is available to US citizens in the detailed report [1] published by the NESC.

9. REFERENCES

1. Squire, M., *et al.* (2017). Evaluation of Micrometeoroid and Orbital Debris (MMOD) Risk Predictions with Available On-orbit Assets. NESC-RP-14-01000
2. Christiansen, E., *et al.* (2009). Handbook for Designing MMOD Protection. NASA/TM-2009-214785
3. Squire, M., *et al.* (2014). Joint Polar Satellite System Micrometeoroid and Orbital Debris Assessment. NASA/TM-2015-218780
4. Hyde, J. (2017). ISS Module Leak Risk from MMOD Impacts. NASA JSC-67020
5. Evans, H. and Hyde, J. (2017). ISS Radiator Damage Risk from MMOD Impacts. NASA JSC-67022
6. Hyde, J. (2017). Damage Risk to International Space Station Pressurized Mating Adapter Cover from Micrometeoroid and Orbital Debris Impacts. NASA JSC-67030
7. Squire, M., *et al.* (2009). Independent Review of Constellation (Cx) Orion Vehicle Micrometeoroids and Orbital Debris (MMOD) Risk Analysis. NASA/TM-2009-215912
8. Anderson, C. & Mullin, S. (1989) Hypervelocity Impact Phenomenology: Some Aspects of Debris Cloud Dynamics
9. LeMaster, P., Mount, A. & Zee, R. (1992). Momentum Distribution in Debris Cloud during Hypervelocity Impact. AIAA Space Programs and Technologies Conference, Huntsville, AL, March 24-27, 1992. AIAA-92-1794
10. Williamsen, J. & Evans, S. (2017). Quantifying Orbital Debris Momentum Transfer Coefficients into Satellite Shields Following Hypervelocity Impact, and Its Potential Use in Orbital Debris Flux Validation. Presented at Hypervelocity Impact Symposium, April 24-28, 2017, University of Kent, Canterbury.
11. McKnight D., *et al.* (1995). Satellite Orbital Debris Characterization Impact Test. Kaman Sciences Corporation
12. Polk M., *et al*, (2015). Orbital Debris Assessment Testing in the AEDC Range G. Proceedings from 13th Hypervelocity Impact Symposium. Boulder, CO.

FEASIBILITY STUDY OF THE MODELLING OF HYPERVELOCITY IMPACTS IN ULTRA HIGH MOLECULAR WEIGHT POLYETHYLENE COMPOSITES USING AN SPH DISCRETISATION METHOD

Bob Verheijen(1), Derek I. Gransden(1), Otto K. Bergsma(1), Ulrich Heisserer(2), Harm van der Werff(2), Torsten R. Lässig(3), Werner Riedel(3)

(1) *Faculty of Aerospace Engineering, Delft University of Technology, Kluyverweg 1, 2629 HS Delft, The Netherlands*

(2) *DSM Dyneema, P.O. Box 1163, 6160 BD Geleen, The Netherlands*

(3) *Fraunhofer Institute for High-Speed Dynamics, Ernst-Mach-Institut, Eckerstraße 4, D-79104 Freiburg, Germany*

ABSTRACT

This paper presents and compares two new formulations of a non-linear orthotropic hydrocode model for the simulation of hypervelocity impacts in Ultra High Molecular Weight Polyethylene (UHMWPE) based composites using Smoothed Particle Hydrodynamics (SPH) discretisation. This project was prompted by the observation that combining existing material models for UHMWPE-based ballistic composites, designed for use with Lagrangian elements, with an SPH discretisation, yielded results that did not match validation data. The cause for this discrepancy is identified and remedied in this paper. The two proposed models are compared to each other and to impact experiments on Dyneema® HB26 targets.

1. INTRODUCTION

Due to their high specific strength and stiffness, Ultra High Molecular Weight Polyethylene (UHMWPE) fibres are widely recognised for the ballistic protection they offer when used in ballistic composites for terrestrial applications. Given the threat posed by Micro Meteorite and Orbital Debris (MMOD) impacts to space infrastructure, it is clear the potential benefits of UHMWPE based composites to spacecraft shielding should be investigated.

In the early days of research into the use of ballistic composites in Stuffed Whipple Shield (SWS) protective configurations for spacecraft, both aramid and (U)HMWPE based composites were considered [1] [2]. For use aboard the ISS, Kevlar® based protection was selected. However, with ongoing developments in material science, and improving understanding of the hazards posed by space, continuous re-evaluation of the means available for the protection of spacecraft is required. Amongst these hazards, and of particular importance to astronaut health during long-term or interplanetary missions, are High Atomic number and Energy (HZE) particles that are part of the space radiation spectrum [3] [4] [5] [6]. Against this type of radiation, PE based materials are the best structural material currently available [3] [7].

For development and analysis of UHMWPE based protection systems, numerical simulation tools are required. Towards this end, Lässig et al. [8] [9] performed an extensive material characterisation study of the Dyneema® HB26 ballistic composite, and derived the material parameters to create a continuum, non-linear hydrocode model in ANSYS® AUTODYN®. The model was validated with impacts from spherical aluminium impactors at a velocity range from 2052 km/s to 6591 km/s. This model was then subsequently improved by Nguyen et al. [10] [11] by introducing updated material parameters and implementing an in-plane damage based erosion criterion and sub-laminate discretisation to better capture out-of-plane failure behaviour.

Both these models, though, use Lagrangian elements for discretisation. An inherent limitation of element based formulations for impacts is that they depend on the artificial erosion of excessively deformed elements for their stability, constituting a removal of energy and matter. Moreover, the need for shared nodes between elements prohibits the fragmentation of impactor and target at impacts in the hypervelocity range. Therefore simulation of impacts in spacecraft structures sometimes use Smoothed Particle Hydrodynamics (SPH) discretisation, as seen in the orthotropic material model for Kevlar®-epoxy (KFRP) proposed by Clegg et al. [12] [13] and Carbon Fibre Reinforced Polymer (CFRP) model proposed by Wicklein et al. [14].

This paper covers the initial steps taken towards converting the model presented by Nguyen et al. [11] to a non-linear orthotropic hydrocode model for SPH simulations in ANSYS® AUTODYN® for the analysis and design of UHMWPE based shielding systems. Two material model formulations for the Dyneema® HB26 composite are presented and compared.

2. BOUNDARY CONDITIONS AND TEST CASES

ANSYS® AUTODYN® 2D R16.2 Academic was used with symmetry set to axial. No 3D simulations were performed, because of limitations on available particle numbers.

The validation cases used in this paper are those provided by Lässig et al. [9]. The experimental set-up consisted of Dyneema® HB26 target samples with in-plane dimensions of 200 mm x 200 mm and an Areal Density (AD) of 15 kg/m^2. This equates to a thickness of 15 mm. The impactor consists of a 6 mm diameter spherical aluminium impactor. Impact velocities ranged between 2052 m/s and 6591 m/s.

In this material model, the 11 direction corresponds to the through-thickness direction. The set of orthogonal material directions is completed by the 22 and 33 in-plane material axes. For the 2D impact simulations, the 11 direction corresponds to the x direction.

When making equal AD comparisons with the KFRP model [12], 9 mm thick KFRP samples are modelled. Riedel et al. [13] mention a nominal thickness of 5.7 mm for their composite samples. When investigating thickness effects, a second set of thicknesses was therefore used, consisting of 5.7 mm thick KFRP samples and 9.6 mm, equivalent AD, Dyneema® samples. For the KFRP model, the shear strength in 23 direction was changed from $2.0 \cdot 10^4$ kPa, as mentioned in [12] to $2.0 \cdot 10^5$ kPa to preserve consistency with unit test data presented in the article.

All targets were modelled as 100 mm tall (y direction) rectangles of appropriate width (x direction). The top row of particles was constrained in the x direction, and left unconstrained in the y direction. The impactor was modelled as a semi-sphere with a 6 mm diameter, placed at least 2 particle diameters in front of the target. For all simulations the maximal particle velocity was capped at 1.5 times the impactor velocity.

3. PROBLEM DESCRIPTION

The initial step in converting the existing material model proposed by Nguyen et al. [11] for use in SPH simulation, was to reset the material through-thickness material tensile strength. Nguyen et al. set this value to infinite, effectively disabling through-thickness tensile failure (as intended). In the absence of a published spall strength value for UHMWPE based composites, the choice was made to set the tensile failure stress in 11 direction (note: the 11 direction represents the through-thickness direction) to 45 MPa, which is the published value for KFRP [12]. The damage coupling coefficient C was set to 0.5 in line with the value used by Lässig et al. [9]. This material model will be labelled DM0.

The impactor was modelled using the Al-1100 material model available in AUTODYN®, with a Hydro (Pmin) Failure criterion, with a tensile limit of $-1.0 \cdot 10^{-6}$ kPa [15].

Simulations with an impactor velocity of 3100 m/s and 6591 m/s were used as initial test case. These two cases constitute the lowest experimental velocity yielding penetration and the highest experimental velocity available from the experimental dataset respectively [9]. For these initial test cases a particle diameter of 0.10 mm was used, which is significantly finer than the 0.25 mm reported by Glegg et al [12]. For both cases it was found that during the simulation no penetration occurred, and that the debris cloud caused by spallation achieved velocities far exceeding the experimental values. Systematic variation of values in the Orthotropic Softening criterion, mesh refinement and sample thickness, excluded these factors as likely causes of the observed discrepancy in penetration and achieved velocities.

Changing the strength model from the Orthotropic yield to Von Mises and the Orthotropic failure model to the Material Stress model, whilst keeping strength values equal, did yield penetration and realistic residual velocities. This resulted in the creation of the DM2 material model. This indicated the cause of the discrepancy lays in the parameters used in the orthotropic yield model of the Dyneema® model [11] when combined with an SPH model.

Unit tests were performed on unit cubes of DM0 as detailed in [11]. The results of the through-thickness compressive test are presented in Figure 1 Examination of the through-thickness compressive curve obtained from the model, showed that even though the numerical curve follows the experimental curve on which it was based closely, one important discrepancy occurs. Where the experimental curve stops at a value of approximately 700 MPa [16], the numerical curve does not exhibit failure until a stress of 3.4 GPa. For a material model with a Lagrangian finite element formulation, this is not necessarily a problem. The Orthotropic Softening model can allow for effective failure once in-plane stresses become too large and strain based erosion can remove elements that have become too deformed as a consequence. For the model presented by Nguyen et al. [11] this becomes even less of a problem, since out of plane failure of elements has actively been disabled. However, for an SPH based formulation, this explains the observed problem of stagnation of the impactor during penetration, followed by excessive spallation at the rear of the target plate.

Comparing the compressive response to one from a unit test on the KFRP material model [12], also presented in Figure

1, shows a far lower ultimate stress value in compression for this latter model. A clear distinction between the models proposed by Lässig et al. [9] and Nguyen et al [11], and those presented by Clegg et al. [12] and Wicklein et al. [14], is that the former two base their master effective stress plastic strain curve on in-plane shear test data. This choice allows for effective capturing of the non-linearities in both in-plane and through-thickness shear data. The latter two both use in-plane tensile test data. Therefore it was decided to base the Orthotropic Yield model for DM1 on available in-plane tensile test data, rather than in-plane shear data.

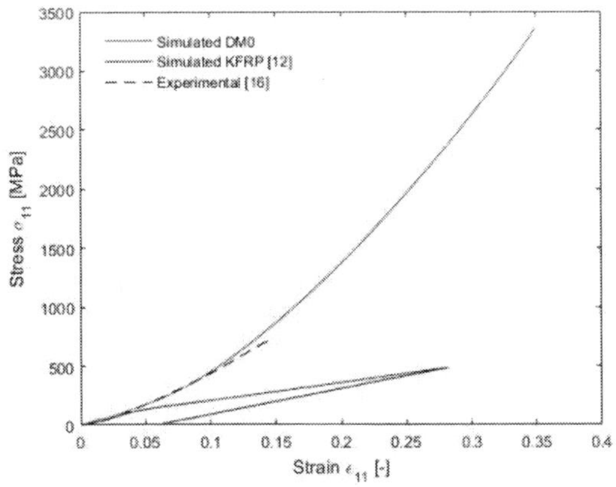

Figure 1: *Comparison of compressive unit tests on the DM0 and KFRP models to experimental input data as used by Nguyen et al. [11].*

Note that an alternative hypothesis for the encountered discrepancy between experimental and simulated data for the DM0 model is the use of the 2D axial symmetry setting. Since strength values for orthotropic composites vary greatly depending on the loading direction, it may be that the axial symmetry assumption overestimates material strength. This should be further investigated by performing 3D simulations on both the DM0 and the proposed material models. For this, either more computational resources, experimental validation data on thinner samples (to reduce the number of particles) or more efficient meshing techniques, such as the use of Lagrangian elements for non-critical material regions, are recommended.

4. PROPOSED MODELS

This paper builds upon the work published in [8] [9] [10] [11]. Where possible, numerical values used in [11] are kept. Both proposed models make use of material model components available in ANSYS® AUTODYN® 2D R16.2. For more details on the evolution of this model and explanation of its various components, and determination of certain parameters, the reader is referred to these publications. For extra background material, [12] and [14] are recommended, as well as the AUTODYN® user's manual [17] and composite material modelling guide [18].

Two material models for the simulation of Dyneema® HB26 under hypervelocity impacts are proposed here. Both use for their Equation of State (EOS) the Ortho type with the Shock subtype. DM1 combines this with an Orthotropic Yield strength and an Orthotropic Softening failure criterion. DM2 uses a Von Mises strength and a Material Stress failure criterion.

4.1. Orthotropic yield criterion

Since the behaviour of the Orthotropic yield criterion, proposed by Chen et al. [19], proved critical to this work, a short explanation of its workings is provided here. This yield criterion is used by Lässig et al. [8] [9] Nguyen et al. [10] [11] Clegg et al. [12] and Wicklein et al. [14] to capture orthotropic hardening. The model consists of a yield surface defined by Equation 1.

$$\begin{aligned} f(\sigma_{ij}) &= a_{11}\sigma_{11}^2 + a_{22}\sigma_{22}^2 + a_{33}\sigma_{33}^2 \\ &+ 2a_{12}\sigma_{11}\sigma_{22} + 2a_{23}\sigma_{22}\sigma_{33} + 2a_{31}\sigma_{33}\sigma_{11} \\ &+ a_{44}\sigma_{23}^2 + a_{55}\sigma_{31}^2 + a_{66}\sigma_{12}^2 = k \end{aligned} \tag{1}$$

Here the σ terms are the stresses in the main material directions, the a terms define the hardening behaviour in the different material directions, and k is the current yield surface. By replacing k with a ten point piecewise linear master effective stress ($\bar{\sigma}$) plastic strain ($\bar{\varepsilon}^p$) curve, hardening can be captured. This master effective stress-plastic strain curve can be defined by mapping ten stress and corresponding plastic strain values of one of the tensile or shear characterisation curves onto the master effective domain. This mapping is performed using Equation 2 for normal tensile test data and Equation 3 for shear test data.

$$\bar{\sigma} = \sigma_{ii}\sqrt{\frac{3a_{ii}}{2}} \quad \bar{\varepsilon}^p = \varepsilon_{ii}^p\sqrt{\frac{2}{3a_{ii}}} \tag{2}$$

$$\bar{\sigma} = \sigma_{ij}\sqrt{3a_{ij}} \quad \bar{\varepsilon}^p = \varepsilon_{ij}^p\sqrt{\frac{1}{3a_{ij}}} \tag{3}$$

The remaining a coefficients are determined by fitting the transformed curves for the remaining stress strain characterisation curves onto this master effective curve using Equation 2 and 3 and manipulation of the a parameters. Requirements on the yield model are further detailed in [18] and [19].

4.2. DM1

First, the Orthotropic Yield model is defined. In line with [12] and [14], the orthotropic yield model was redefined based on in-plane-tensile tests. Parameters a_{22} and a_{33} are subsequently set to 1.0. The result of a verification unit test can be seen in Figure 2.

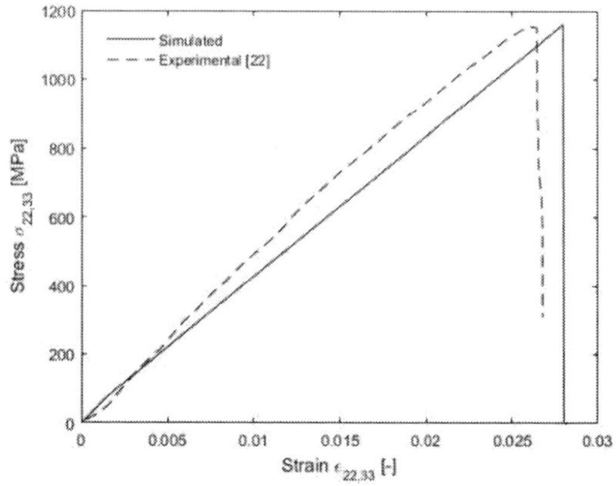

Figure 2: *In-plane-tensile verification unit test.*

The parameter a_{11} was set to match Split Hopkinson Bar data from Shaker et al. [20] as well as possible. Note though, that the peak reported strain rate from these tests lays around $6 \cdot 10^3$ s^{-1}, which is significantly short of the rates on the order of 10^6 s^{-1} [21] generally associated with hypervelocity impacts. Verification unit test data are presented in Figure 3.

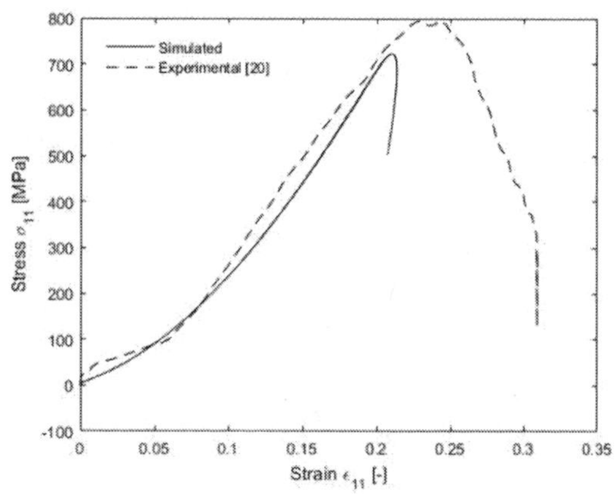

Figure 3: *Through-thickness compressive verification unit test.*

A consequence of the decision to base the master effective stress plastic strain curve on a linear base curve, is that the nonlinearities encountered in the experimental shear data cannot be effectively captured. It has been shown that in-plane shear strength strongly depends on sample width, ranging up to 120 MPa for wide samples [22]. Here the choice was made to fit the in-plane shear response of the material model to the same experimental data used by Nguyen et al. [11], taken from Lässig [23]. The Maximum Shear Stress 23 was set to 55 MPa, and a_{44} was chosen to match the observed failure strain. The verification unit test is presented in Figure 4.

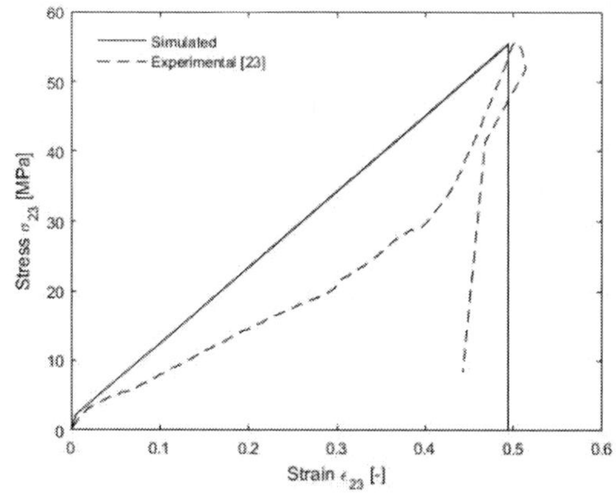

Figure 4: *In-plane shear verification unit test.*

Most up to date dynamic characterisation data for through thickness shearing of Dyneema® HB26 can be obtained from [24]. This source became available relatively late in this work, and was therefore only used to update the through thickness shear strength value in the model to 156 MPa. The yield model was left unaltered to avoid complications due to parameter interactions. The experimental shear test data provided by [9] and used in [11] did not extend up to failure. Ultimate strain for the DM1 model was therefore chosen to match the intersection point with the simulated curve from [11]. The a_{55} and a_{66} parameters where set accordingly, as can be seen in Figure 5. Note that in the unit tests used to determine these simulated material curves, Lagrangian elements were used. The maximum stress reached during these tests is limited by the used spall strength, as noted by Nguyen et al. [10]. By varying through-thickness shear strengths in simulations, it has been observed that this assumption does not fully hold for the SPH discretisation. The effects, appear to be limited, but more investigations on this topic are needed.

Unless mentioned differently, values as used by Nguyen et al. [11] are used. For the EOS, the parameter S1 was modified to match experimental data as well as possible. From Figure 6, it can observed that a more elaborate EOS is required to

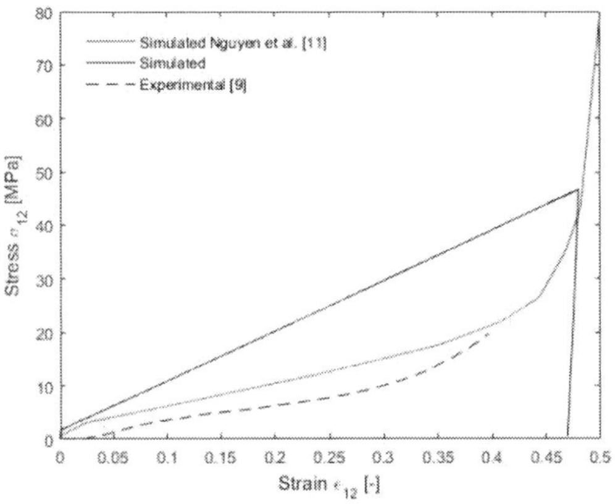

Figure 5: *Through-thickness shear verification unit test.*

obtain a better match. This change was deemed beyond the scope of this work. A summary of all used input parameters for the DM1 model is presented in Table 2.

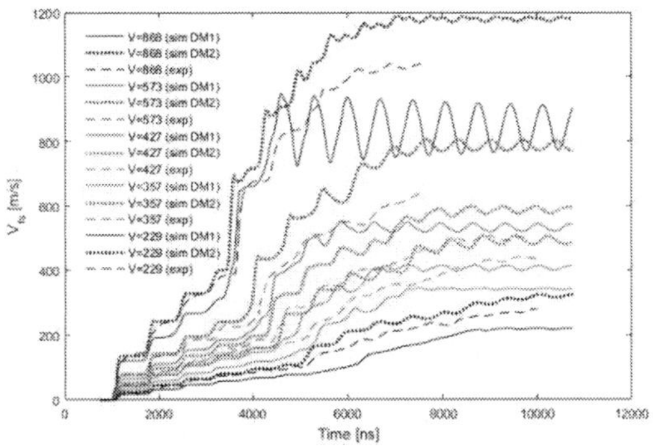

Figure 6: *Comparison between experimental [9] free surface velocity trace from inverse flyer plate test and numerical verification form DM1.*

4.3. DM2

For the DM2 model, the input values of DM0 were used, with the exception that the through-thickness shear strengt was updated to 156 MPa [24]. S1 was set to match experimental flyer plate data, as presented in Figure 6. For the Material Failure strength criterion the Orthotropic Post Failure Option was used. The material model can be found in Table 3.

5. RESULTS

Due to the nature of a discretisation using SHP particles, extensive fragmentation of both target and impactor is observed. As a consequence, identifying a single numerical value for the residual velocity of fragments belonging to either the impactor or the target is difficult. For the validation data available [9], it is often not possible to distinguish between the material components in the debris clouds. Reported velocities extracted from simulations are always taken at 100 microseconds after impact and are rounded to the nearest multiple of 50 m/s. For simulations using DM2, a particle diameter of 0.15 mm was used. For DM1 and the KFRP model a diameter of 0.10 mm was used.

5.1. 15 mm thick Dyneema® target

Figure 7 and 8 present experimental and simulated debris cloud, for consultation when required. For DM1, for the cases with impact velocities of 5370 m/s and 6591 m/s, the shape of the debris cloud was matched quite well, as seen in Figure 7b. Here the choice was made to report best estimate maximum velocities of this main debris cloud using engineering judgement. For DM2 and the other simulations performed on DM1, velocities were taken from the debris cloud within 15 mm of the interface between impactor and target. Higher observed maximal velocities due to over penetration or excessive spallation are denoted between brackets (). Test cases where excessive fragmentation and spallation lead to trouble interpreting the data are denoted by a *.

Table 1: *Residual velocity values for the DM1 and DM2 models with 15 kg/m² AD targets.*

[m/s]	2052	2453	3100	3532	5370	6591
Exp [9]	0	0	679	958	1730	2457
DM1	150 (200)	300 (1600)	1000* (2950)	700* (3500)	1700 (3000)	2400 (3500)
DM2	100 (1650)	0 (2000)	650 (2100)	900 (2200)	1650 (3250)	2300 (3250)

As can be seen from Table 1, DM2 yields a reasonably close prediction of the experimentally determined peak debris velocities. It can also be observed that for the cases with impact velocities of 5370 m/s and 6591 m/s this model struggles to capture lateral dilation of the debris cloud. As can be seen in Figure 7, the simulated cloud fans out significantly less than the experimental cloud. Observed excessive spall cloud velocities were found to be well in excess of the residual velocities of the impactor and main target cloud. This model may be expected to predict the residual velocities of impactors reasonably well, even for thick targets. Prediction of secondary damage using this

Table 2: *Input parameters for the DM1 material model*

Parameter	Value	Unit	Source	Parameter	Value	Unit	Source
EOS: Orthotropic				Strength: Orthotropic yield			
Reference Density	0.98	g/cm^3	[25]	Plasticity constant 11	140	-	Fit to [20]
Young's modulus 11	$3.62 \cdot 10^6$	kPa	[9]	Plasticity constant 22	1.0	-	Based on [22]
Young's modulus 22	$5.11 \cdot 10^7$	kPa	[22]	Plasticity constant 33	1.0	-	Based on [22]
Young's modulus 33	$5.11 \cdot 10^7$	kPa	[22]	Plasticity constant 12	0	-	-
Poisson's Ratio 12	0.013	-	[9]	Plasticity constant 23	0	-	-
Poisson's Ratio 23	0	-	[9]	Plasticity constant 31	0	-	-
Poisson's Ratio 31	0.5	-	[9]	Plasticity constant 44	700	-	Fit to [23]
Shear modulus 12	$2.0 \cdot 10^6$	kPa	[11]	Plasticity constant 55	1100	-	Fit to [11]
Shear modulus 23	$1.92 \cdot 10^5$	kPa	[11]	Plasticity constant 66	1100	-	Fit to [11]
Shear modulus 31	$2.0 \cdot 10^6$	kPa	[11]	Eff. Stress #1	$1.71 \cdot 10^5$	kPa	Based on [22]
				Eff. Stress #2	$3.09 \cdot 10^5$	kPa	Based on [22]
Volumetric Response: Shock				Eff. Stress #3	$4.46 \cdot 10^5$	kPa	Based on [22]
Grüneisen coefficient	1.64	-	[26]	Eff. Stress #4	$5.84 \cdot 10^5$	kPa	Based on [22]
Parameter C1	$3.57 \cdot 10^3$	m/s	-	Eff. Stress #5	$7.21 \cdot 10^5$	kPa	Based on [22]
Parameter S1	2.3	-	Set to match [9]	Eff. Stress #6	$8.59 \cdot 10^5$	kPa	Based on [22]
Reference Temperature	293	K	-	Eff. Stress #7	$9.96 \cdot 10^5$	kPa	Based on [22]
Specific Heat	$1.85 \cdot 10^3$	J/kg K	[25]	Eff. Stress #8	$1.13 \cdot 10^6$	kPa	Based on [22]
				Eff. Stress #9	$1.27 \cdot 10^6$	kPa	Based on [22]
Failure: Orthotropic softening				Eff. Stress #10	$1.41 \cdot 10^6$	kPa	Based on [22]
Tensile failure stress 11	$4.50 \cdot 10^4$	kPa	[12]	Eff. Plastic Strain #1	0.00021	-	Based on [22]
Tensile failure stress 22	$1.15 \cdot 10^6$	kPa	[22]	Eff. Plastic Strain #2	0.00064	-	Based on [22]
Tensile failure stress 33	$1.15 \cdot 10^6$	kPa	[22]	Eff. Plastic Strain #3	0.00107	-	Based on [22]
Maximum shear stress 12	$1.56 \cdot 10^5$	kPa	[24]	Eff. Plastic Strain #4	0.00150	-	Based on [22]
Maximum shear stress 23	$5.50 \cdot 10^4$	kPa	[23]	Eff. Plastic Strain #5	0.00193	-	Based on [22]
Maximum shear stress 31	$1.56 \cdot 10^5$	kPa	[24]	Eff. Plastic Strain #6	0.00236	-	Based on [22]
Fracture energy 11	790	J/m^2	[9]	Eff. Plastic Strain #7	0.00279	-	Based on [22]
Fracture energy 22	30	J/m^2	[12]	Eff. Plastic Strain #8	0.00322	-	Based on [22]
Fracture energy 33	30	J/m^2	[12]	Eff. Plastic Strain #9	0.00365	-	Based on [22]
Fracture energy 12	$1.46 \cdot 10^3$	J/m^2	[12]	Eff. Plastic Strain #10	0.00408	-	Based on [22]
Fracture energy 23	$1.46 \cdot 10^3$	J/m^2	[12]				
Fracture energy 31	$1.46 \cdot 10^3$	J/m^2	[12]				
Damage coupling coefficient	0.50	-	[9]				

model may be feasible, when regions of excessive spallation are removed and reason exists to expect that the axial components of the debris cloud will dominate.

For the DM1 model, it can be seen that for the scenarios with impact velocities of 2052 m/s and 2453 m/s stopping of the projectile was not fully captured. Fragments of Dyneema® were dislodged and perforation was observed. For the impactor velocities of 5370 m/s and 6591 m/s, the former case displayed under-penetration, followed by excessive spalling and the latter case presented over-penetration by the impactor. However, the general shape of the debris cloud seems to match that of the experimental data quite well. For the scenarios with the 3100 m/s and 3532 m/s, excessive fragmentation of both impactor and target, as presented in Figure 8b, made accurate quantification of meaningful peak velocities impossible. It is clear that for these two cases, the simulation does not accurately match the validation data. A potential explanation for this is the use of a faulty spall strength in combination with the observed discrepancy in the shock wave propagation behaviour of the model. It should also be noted that the reported bulk speed of sound in Dyneema® HB26 is 1922 m/s [9], and that the degradation of model performance coincides with the two data points where the experimental residual velocity is well below this threshold, but penetration still occurred.

5.2. Comparison to 9 mm thick KFRP target

Impact simulations on 9.0 mm thick KFRP targets were performed at velocities of 3532 m/s, 5370 m/s and 6591 m/s. All simulations showed a tendency towards excessive spallation comparable to that observed for the 15 mm thick DM1 targets, as illustrated by Figure 7d and 8d.

Table 3: *Input parameters for the DM2 material model*

Parameter	Value	Unit	Source	Parameter	Value	Unit	Source
EOS: Orthotropic				Failure: Material Stress			
Reference Density	0.98	g/cm^3	[25]	Tensile Failure Stress 11	$4.50 \cdot 10^4$	kPa	[12]
Young's modulus 11	$3.62 \cdot 10^6$	kPa	[9]	Tensile Failure Stress 22	$1.15 \cdot 10^6$	kPa	[22]
Young's modulus 22	$5.11 \cdot 10^7$	kPa	[22]	Tensile Failure Stress 33	$1.15 \cdot 10^6$	kPa	[22]
Young's modulus 33	$5.11 \cdot 10^7$	kPa	[22]	Maximum Shear Stress 12	$1.56 \cdot 10^5$	kPa	[24]
Poisson's Ratio 12	0.013	-	[9]	Maximum Shear Stress 23	$1.20 \cdot 10^5$	kPa	[11]
Poisson's Ratio 23	0	-	[9]	Maximum Shear Stress 31	$1.56 \cdot 10^5$	kPa	[24]
Poisson's Ratio 31	0.5	-	[9]				
Shear modulus 12	$2.0 \cdot 10^6$	kPa	[11]	Post Failure Option: Orthotropic			
Shear modulus 23	$1.92 \cdot 10^5$	kPa	[11]	Failed in 11, Failure mode	11 only		
Shear modulus 31	$2.0 \cdot 10^6$	kPa	[11]	Failed in 22, Failure mode	12 & 22 only		
				Failed in 33, Failure mode	31 & 33 only		
Volumetric Response: Shock				Failed in 12, Failure mode	12 & 11 only		
Grüneisen coefficient	1.64	-	[26]	Failed in 23, Failure mode	23 & 33 only		
Parameter C1	$3.57 \cdot 10^3$	m/s	-	Failed in 31, Failure mode	31 & 11 only		
Parameter S1	1.5	-	Set to match [9]				
Reference Temperature	293	K	-	Strength: Von Mises			
Specific Heat	$1.85 \cdot 10^3$	J/kg K	[25]	Shear Modulus	$2.0 \cdot 10^6$	kPa	[11]
				Yield Stress	$1.0 \cdot 10^4$	kPa	[24]

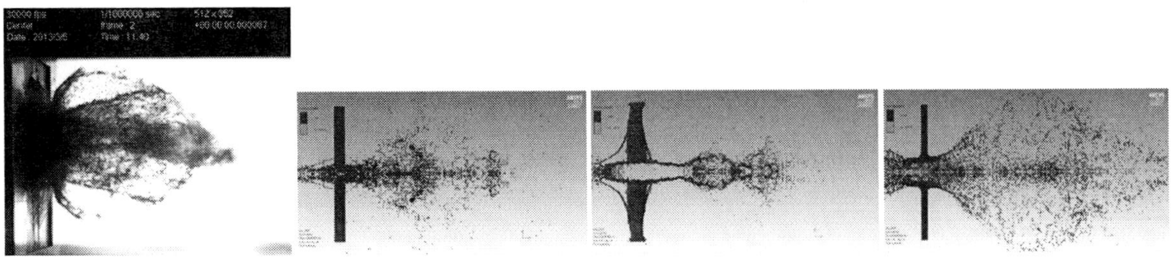

(a) *Experimental HB26 [9]* (b) *DM1* (c) *DM2* (d) *KFRP*

Figure 7: *Comparison of experimental and numerical debris clouds at* 66.7 *microseconds after* 6591 *m/s impact in* 15 *kg/m*^{2} *target.*

In order to make a first-order comparison between the stopping potential predicted by DM1, DM2 and KFRP, Table 4 presents a comparison of the momentum in x direction of the aluminium impactor, 100 microseconds after impact. The parameter (p_{x0}) denotes the momentum of the impactor at the associated velocity before the impact.

Table 4: *Residual x momentum values of the impactor for the DM1, DM2 and KFRP models with 15 kg/m^2 AD targets.*

[m/s]	3532	5370	6591
p_{x0} [kg · m/s]	1.06	1.61	1.97
DM1 [kg · m/s]	0.030	0.150	0.288
DM2 [kg · m/s]	0.081	0.272	0.397
KFRP [kg · m/s]	-0.402	-0.192	-0.074

For all 9.0 mm KFRP simulation, it was observed that significant amounts of the impactor were expelled back in negative x direction. The parts of the impactor that did penetrate, encountered significant stagnation and excessive fragmentation. This corresponds to the negative momentum values encountered for the KFRP model from Table 4.

Two explanations are proposed for this observation. First, it is possible that the KFPR model predicts residual velocities accurately, meaning a 9.0 mm KFRP target would effectively stop a 6.0 mm diameter Aluminium impactor at the considered velocities. Second, it is possible that the complex orthotropic models used in DM1 and KFRP struggle with capturing the response of very thick laminates modelled with SPH. As far as the author is aware, no experimental direct impact data exist for the KFRP model for this material thickness. Therefore no conclusive statements can be made on the topic.

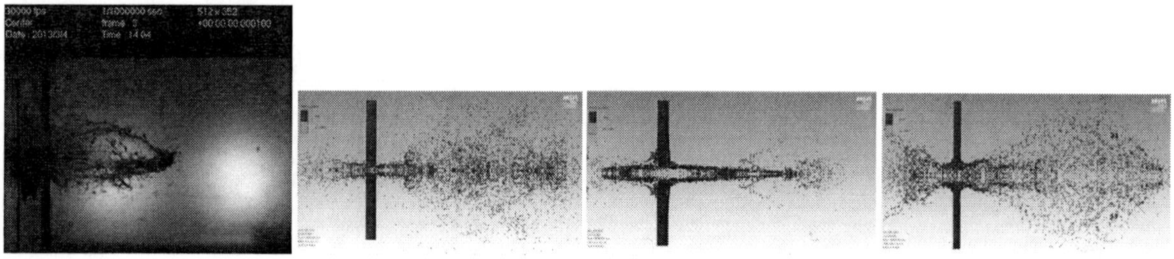

(a) *Experimental HB26 [9]* (b) *DM1* (c) *DM2* (d) *KFRP*

Figure 8: *Comparison of experimental and numerical debris clouds at 66.7 microseconds after 3532 m/s impact in 15 kg/m^2 target.*

5.3. 9.6mm thick Dyneema® target

Since the DM1 model showed difficulties capturing the impact response in thick targets, and the KFRP model shows a comparable response, the effect of thickness was investigated as described in Section 2. The impactor diameter was kept constant at 6 mm. In Table 5, the reported velocities are taken at or around the interface between the bulk of the impactor fragments and the fragments originating from the target plate. The values between brackets represent approximate velocity values encountered in the front of the debris cloud.

Table 5: *Residual velocity values for the DM1 and DM2 models with 9.5 kg/m^2 AD targets.*

[m/s]	3100	3532	4500	5370	6591	8000
DM2	1250	1550	2200	2800	3650	4400
	(2700)	(2400)	(3150)	(3500)	(4500)	(5300)
DM1	1050*	1050*	1850	2350	3750	5350
	(4500)	(4500)	(4400)	(3500)	(4250)	(5350)

As was the case for the 15 kg/m^2 setting, it can be seen that for both models the difference between the residual velocity of debris from the impactor and from the target tends to become larger as the impact velocity becomes smaller. The ballistic limit was not encountered in either the simulations using DM1 or DM2, but it was observed again that DM1 showed increasingly messy results as residual velocities drop well below the speed of sound in the bulk material. It was also observed again that the debris clouds produced using DM1 tend to fan out more than those produced using DM2.

Simulations using these thinner target plates are unsupported by experimental validation at this point in time. Therefore no conclusive comments can be made as to which modelling approach represents reality best.

Simulations using a Lagrangian element-based discretisation, using the model proposed in [9] were attempted for this thickness. However, distorted elements caused these simulations to regularly abort before the 100 microsecond wrap-up time was reached, illustrating one of the limitations of the element based simulation approached for hypervelocity impacts.

5.4. Comparison to 5.7mm thick KFRP target

A summary of the linear momentum in x direction for both 9.6 mm DM1 and DM2 and 5.7 mm KFRP targets is presented in Table 6.

Table 6: *Residual x momentum values of the impactor for the DM1, DM2 and KFRP models with 9.5 kg/m^2 AD targets.*

[m/s]	3532	5370	6591
p_{x0} [kg · m/s]	1.06	1.61	1.97
DM1 [kg · m/s]	0.184	0.423	0.595
DM2 [kg · m/s]	0.244	0.535	0.730
KFRP [kg · m/s]	-0.081	0.291	0.528

In Table 6 it can be seen that for the KFRP model, the values for residual impactor momentum systematically lay below those for the values predicted by both the DM1 and DM2 models. This trend continues to the point where the momentum value for the KFRP model at an impactor velocity of 3532 m/s becomes negative. Absent experimental validation for direct impacts on these configurations, no conclusive statements can be made on the accuracy of these values.

6. CONCLUSION AND DISCUSSION

In this article, the first step was set towards an SHP based hypervelocity impact model for Dyneema® HB26, to be used for the design and analysis of spacecraft protection systems. Both the flexibility, and some shortcomings of the SPH discretisation method have been observed. The computational robustness issues associated with element based simulation methods of hypervelocity impacts are mostly negated. However, tensile instabilities, excessive spallation and sensitivity to through-thickness compressive material data have been encountered.

For thick laminates with an AD of 15 kg/m^2, DM2 has been demonstrated to be able to accurately capture the residual velocities over the whole available velocity range. However, it failed to capture the scatter of the debris cloud. DM1 showed promise in capturing the residual velocities and debris cloud shape for thick laminates over a narrow range of velocities (observed for 5370 and 6591 m/s). However, poor performance and excessive spallation was encountered at impact velocities between 3100 and 5370 m/s. It was, demonstrated that the KFRP model also predicts stagnation of the impactor and excessive spalling for these scenarios.

For thinner laminates, with an AD of approximately 9.5 kg/m^2, it was found that the tendency of DM2 to produce linear debris trajectories, and of DM1 to create debris clouds that fan out, was preserved. No convergence of the predicted residual velocities for the two models was observed for the considered velocity range.

Additional experimental validation, preferably on targets of varying thickness, is required. For DM2 it should be assessed whether its ability to accurately predict residual velocities remains intact for laminates of different thicknesses. For DM1, it should be assessed whether it accurately predicts the size and shape of the debris cloud for other target thicknesses. Being able to distinguish between the part of the debris cloud originating from the impactor and target during experiments would be very valuable. Further experimental characterisation of high strain rate behaviour of UHMWPE based materials should continue. Through-thickness compressive and tensile behaviour require attention in particular.

For DM1, a higher order formulation of either the shock EOS as used now, or of the polynomial EOS used by Lässig et al. [9] will be required to properly match the shock wave response of the more complicated model. Additional work should also be performed on establishing a master effective stress plastic strain curve that allows for the capturing of nonlinear shear behaviour, but that does not introduce un-physical hardening in the through-thickness direction.

Comparison between the post impact x momentum values from the DM1, DM2 and KFRP models indicate that the KFRP model [12] consistently shows a greater loss of momentum from the impactors for equal AD shielding. However these observations should be taken as indicative only. Ultimately, a set of composites with UHMWPE, aramid and other high potential fibre systems (PBO or PIPD based for example) should be created and subjected to side by side experimental comparison. As it stands, different matrix choices, fibre volume fractions and different material philosophies prevent accurate one-to-one comparison between the two protection systems.

In the end, this work has succeeded in identifying several difficulties in creating a reliable set of parameter with which to simulate the hypervelocity impact response of ballistic Dyneema® composites, and overcoming several of them. Further research into the topic should lead to increased understanding of the behaviour of composite materials under hypervelocity impact. Combining potentially equal or superior impact resistance to aramid based shielding ballistic composites, and superior radiation shielding behaviour, the investigation into the potential of UHMWPE based composites for spacecraft structures and shielding should be continued.

ACKNOWLEDGEMENTS

This work was supported by DSM Dyneema®, the Fraunhofer Institute for High-Speed Dynamics and the International Space Safety Foundation. The author would like to thank Dr Long H. Nguyen from the Australian Government Department of Defence Science and Technology Group for the many invaluable discussions on the modelling of impact in general, and in Autodyn® in particular.

REFERENCES

[1] W P Schonberg and E J Walker. Use of composite materials in multi-wall structures to prevent perforation by hypervelocity particle impact. *Composite structures*, 19(1):15–40, 1991.

[2] E L Christiansen and J H Kerr. Mesh double-bumper shield: a low-weight alternative for spacecraft meteoroid and orbital debris protection. *International Journal of Impact Engineering*, 14(1-4):169–180, 1993.

[3] F A Cucinotta, M Y Kim, and L J Chappell. Evaluating shielding approaches to reduce space radiation cancer risks. Report, NASA Johnson Space Center, Houston, TX 77058, United States, May 2012.

[4] G A Nelson. Space radiation and human exposures, a primer. *Radiation research*, 185(4):349–358, 2016.

[5] M Durante and F A Cucinotta. Heavy ion carcinogenesis and human space exploration. *Nature Reviews Cancer*, 8(6):465–472, 2008.

[6] F A Cucinotta and M Durante. Cancer risk from exposure to galactic cosmic rays: implications for space exploration by human beings. *The lancet oncology*, 7(5):431–435, 2006.

[7] S Nambiar and J T W Yeow. Polymer-composite materials for radiation protection. *ACS applied materials & interfaces*, 4(11):5717–5726, 2012.

[8] T R Lässig, F Bagusat, M May, and S Hiermaier. Analysis of the shock response of uhmwpe composites using the inverse planar plate impact test and the shock reverberation technique. *International Journal of Impact Engineering*, 86:240–248, 2015.

[9] T R Lässig, L H Nguyen, M May, W RiedeL, U Heisserer, H van der Werff, and S Hiermaier. A non-linear orthotropic hydrocode model for ultra-high molecular weight polyethylene in impact simulations. *International Journal of Impact Engineering*, 75:110–122, 2014.

[10] L H Nguyen, T R Lässig, S Ryan, W Riedel, A P Mouritz, and A C Orifici. Numerical modelling of ultra-high molecular weight polyethylene composite under impact loading. *Procedia Engineering*, 103:436–443, 2015.

[11] L H Nguyen, T R Läessig, S Ryan, W Riedel, A P Mouritz, and A C Orifici. A methodology for hydrocode analysis of ultra-high molecular weight polyethylene composite under ballistic impact. *Composites Part A: Applied Science and Manufacturing*, 84:224–235, 2016.

[12] R A Clegg, D M White, W Riedel, and W Harwick. Hypervelocity impact damage prediction in composites: Part I-material model and characterisation. *International Journal of Impact Engineering*, 33(1):190–200, 2006.

[13] W Riedel, H Nahme, D M White, and R A Clegg. Hypervelocity impact damage prediction in composites: Part II-experimental investigations and simulations. *International Journal of Impact Engineering*, 33(1):670–680, 2006.

[14] M Wicklein, S Ryan, D M White, and R A Clegg. Hypervelocity impact on cfrp: testing, material modelling, and numerical simulation. *International Journal of Impact Engineering*, 35(12):1861–1869, 2008.

[15] C J Hayhurst, I H Livingstone, R A Clegg, G E Fairlie, S J Hiermaier, and M Lambert. Numerical simulation of hypervelocity impacts on aluminum and nextel/kevlar whipple shields. In *Proceedings of the Hypervelocity Shielding Workshop*, pages 61–72, 1998.

[16] S Chocron, A E Nicholls, A Brill, A Malka, T Namir, D Havazelet, H van der Werff, U Heisserer, and J D Walker. Modeling unidirectional composites by bundling fibers into strips with experimental determination of shear and compression properties at high pressures. *Composites Science and Technology*, 101:32–40, 2014.

[17] ANSYS, Inc. Ansys autodyn user's manual v15.0. Manual, ANSYS, Inc., ANSYS, Inc., Southpointe, 275 Technology Drive, Canonsburg, PA 15317, United States, November 2013.

[18] ANSYS, Inc. Autodyn composite modeling v15.0. Manual, ANSYS, Inc., ANSYS, Inc., Southpointe, 275 Technology Drive, Canonsburg, PA 15317, United States, November 2013.

[19] J K Chen, F A Allahdadi, and C T Sun. A quadratic yield function for fiber-reinforced composites. *Journal of composite materials*, 31(8):788–811, 1997.

[20] K Shaker, A Jabbar, M Karahan, N Karahan, and Y Nawab. Study of dynamic compressive behaviour of aramid and ultrahigh molecular weight polyethylene composites using split hopkinson pressure bar. *Journal of Composite Materials*, 51(1):81–94, 2017.

[21] S Hiermaier. *Structures under crash and impact: continuum mechanics, discretization and experimental characterization*. Springer Science & Business Media, 2007.

[22] U Heisserer. Dyneema® material information for ballistic modeling. Report, DSM, Geleen, The Netherlands, 2013.

[23] T Lässig. Dynamic characterization and modeling of dyneema. Report I-49/2012, Fraunhofer EMI, Freiburg, Germany, 2012.

[24] M Ruiz-Ripoll, O Millon, T Lässig, and W Riedel. Experimental characterization of the out-of-plane shear strength of ultra-high molecular weight polyethylene composite by using the split hopkinson bar device. In *17th European Conference on Composite Materials*, 2016.

[25] DSM Dyneema. Fact sheet: Dyneema high strength, high modulus polyethylene fiber. Report, DSM, Urmond, The Netherlands, 2008.

[26] R G McQueen. Selected hugoniots. Report, Los Alamos National Laboratory, Los Alamos, New Mexico, 1969.

PROCEEDINGS OF THE 9TH INTERNATIONAL SPACE SAFETY CONFERENCE

SESSION 21 DESIGNING SAFETY – II

RADIATION SHIELDING FOR LONG DURATION MANNED SPACE MISSIONS

W. J. Burger[1], F. Ambroglini[2], R. Battiston[3], V. Calvelli[4], R. Musenich[4], R.B. Meinke[5], and M. Giraudo[6]

[1] *FBK, Via Sommarive 18, and TIFPA-INFN, via Sommarive 14, 38123 Trento, Italy, email: william.burger@tifpa.infn.it*
[2] *University of Perugia and INFN-Perugia, Via Pascoli, 06123 Perugia, Italy*
[3] *TIFPA-INFN and University of Trento, via Sommarive 14, 38123 Trento Italy*
[4] *INFN-Genova, Via Dodecaneso 33, 16146 Genova, Italy*
[5] *Advanced Magnet Lab, 1720 Main St, NE Bldg. 4, Palm Bay FL 32905, USA*
[6] *Thales Alenia Space Italy, Strada Antica di Collegno, 10146 Torino, Italy*

ABSTRACT

A manned mission to Mars would present an important health risk to the crew members due to the prolonged exposure to the ionizing radiation of galactic cosmic-rays. The radiation levels would largely exceed those encountered in the Apollo missions. A corresponding increase in the passive shielding provided by the spacecraft in order to maintain an acceptable dose level implies a significant increase of the mass. The advent of superconducting magnets in the early 1960's was considered an attractive alternative. The technology allows to generate magnetic fields capable to deflect the cosmic-rays in a manner analogous to the reduction of the particle fluxes in the upper atmosphere, due to the Earth's dipole magnetic field. A series of three active magnetic shield studies have been made, funded successively by European Space Agency (ESA), the NASA Innovative Advanced Concepts (NIAC) program and the Union European's Seventh Framework Programme (FP7). The performance of the shield configuration designs is evaluated with a 3-dimensional Monte Carlo simulation. The principal results are presented.

Key words: manned space missions; active magnetic shielding; high temperature superconductors; radiation protection; Monte Carlo simulation.

1. INTRODUCTION

The exposure to the ionizing radiation of galactic cosmic-rays (GCR) and solar energetic particles (SEP) represent a health risk for the crew members on a long duration interplanetary mission. Figure 1 shows the proton flux of a 10-day SEP event and the GCR fluxes for protons, carbon and iron nuclei. The SEP events are characterized by the emission of very high fluxes of low energy, relatively high ionizing protons, which may last on the order of hours or days. The GCR fluxes are modulated by the solar cycle. Periods of maximum solar activity result in the decrease of the low energy GCR flux.

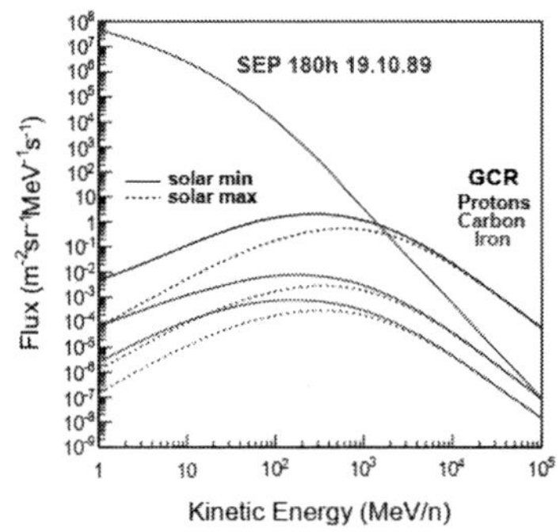

Figure 1. The proton flux of a 10-day SEP event in October 1989, and the GCR proton, carbon and iron nuclei fluxes for solar minimum and maximum periods. [1].

The ionizing radiation of the SEP protons may be considered a short-term risk, which could jeopardize the success of the mission. The GCR fluxes represent a longer term risk affecting the life expectancy of the crew due to radiation induced cancers.

In addition to the ionization losses of the primary GCR, the protons and nuclei are subject to strong interactions with the nuclei in the material of the spacecraft, which produce lower energy secondary charged particles and neutrons. The ionization loss in the materials of the spacecraft provide effective passive shielding if sufficiently thick to contain the primary particles and their secondaries. In the case of SEP events, the required thickness of the shielding material would limit the protection to a small volume of the spacecraft, a shelter which would be occupied during the duration of the event.

The SEP and charged GCR may be deflected by the presence of a magnetic field. A particle with charge q and velocity vector **v** is deflected in the plane perpendicular to the field direction by the Lorentz force,

$$\mathbf{F} = q\mathbf{v} \times \mathbf{B}. \quad (1)$$

The particle moves in a circle in the deflection plane, with an angular deviation θ from the incident direction of

$$sin(\theta) = \frac{BL}{R} \quad (2)$$

where B is the magnetic flux density, L the length of the field region in the deflection plane and R the magnetic rigidity, i.e. the particle's momentum divided by electric charge.

The advent of superconducting magnets in the early 1960s provided the possibility to attain sufficiently high values of B to produce a magnetic shield surrounding a spacecraft [2]. The superconducting materials operate at the temperature of liquid helium (4 K). The discovery of high temperature superconductors (HTSC) [3], which operate at $\sim$ 25 K, represent a significant advancement since the use of liquid helium in space is highly problematic.

2. THE MAGNET SHIELD CONFIGURATIONS

Magnet shield configurations based on the high temperature superconductors yttrium-barium-copper-oxide (YBCO) and magnesium diboride (MgB_2) were evaluated in the ESA [4], NIAC [5] and the Space Radiation Superconducting Shield (SR2S)[1] studies.

YBCO superconductors are manufactured in the form of a 4 cm wide tape with a thickness $\sim 0.2\,\mathrm{mm}$. Multilayer cylindrical coils composed of YBCO tape may be used to form pure multipole fields by modulating the angle of the helical turns in successive tape layers [6]. The magnesium diboride superconductors are produced in cable form. The individual MgB_2 wire elements are brittle and require a rigid support.

The shield configurations are based on realistic extrapolations of the existing technology. The field strengths and dimensions, and the exact material composition of the engineering designs, are described in a Monte Carlo simulation in order to evaluate the performance taking into account the contribution of the passive shielding elements, as well as the magnetic field.

2.1. ESA Study

Two toroidal field configurations were considered in the ESA study. Figure 2 illustrates the double-helix solenoid coil shield concept based on the YBCO superconductor. A dipole field is produced in each solenoid coil by reversing the current direction in the opposite-tilt-direction layers of the helical windings. The superposition of the fields of the 2 m diameter coils results in a nearly homogeneous axial field with a field integral BL $\sim 4\,\mathrm{Tm}$. In the Monte Carlo simulation, the solenoid coils are oriented to produce a toroidal field around a 4 m diameter cylindrical habitat.

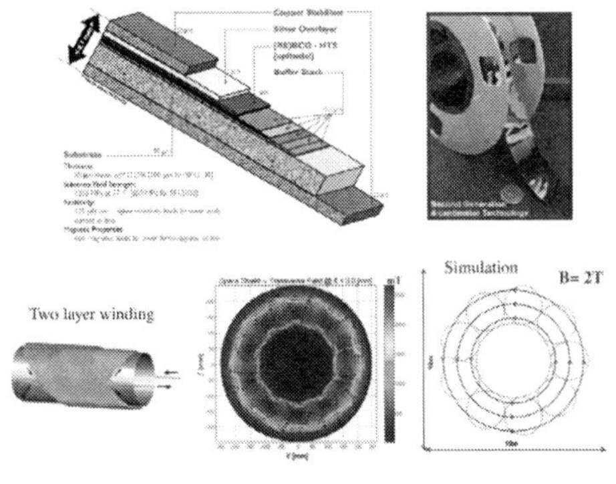

Figure 2. The composition of the YBCO tape from SuperPower (top left) and a photograph of a 0.2 mm thick tape produced by American Superconductor (top right). The two-layer winding used to produce the double-helix solenoid field (bottom left). The field map of the toroidal field [4] around the habitat (bottom center) and the field used in the Monte Carlo simulation (bottom right).

The second configuration of the ESA study was a racetrack coil toroid composed of the MgB_2 superconducting cable. The field integral of the 12 racetrack coil toroid is 4.9 Tm. The racetrack coil geometry is used in high energy physics. The dimensions of the low temperature superconducting racetrack toroid of the Atlas experiment at the Large Hadron Collider (LHC-CERN) are comparable with the dimensions required for a radiation shield in the space application.

2.2. NIAC Study

A modified version of the YBCO coil shield was developed in the NIAC study. The dipole field, obtained with the multiple layer winding in the ESA study, was abandoned in favor of a single orientation winding used to produce a solenoid field. The coil diameter was increased to 8 m and the field flux density reduced to 1 T, yielding a maximum field integral of 8 Tm at the center of the coil, and an average value of 6.3 Tm, taking into account the path length variation across the cylindrical coil.

The reduction of the number of YBCO tape layers increases the flexibility of the coils, and allows a compact storage for launch. After deployment in space, the coils expand to their full diameter when the current is applied, due to the effect of the Lorentz force acting on the current flowing in the flexible coils. The fringe fields of the 20 m

[1] funded by grant agreement no. 313224, FP7 (2007-2013)

long solenoids are compensated by a central solenoid coil concentric with the cylindrical habitat, which reduces the field inside the habitat to an acceptable level. The 6+1 extendable solenoid coil shield is illustrated in Fig. 3.

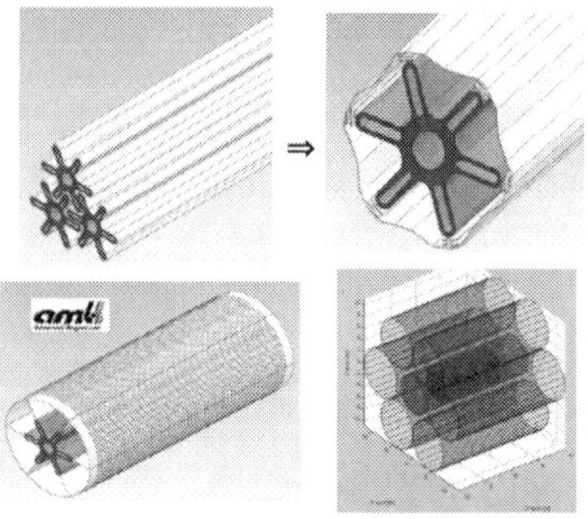

Figure 3. The 6+1 extendable solenoid shield of the NIAC study [5]: coils packed for launch (top left), deployment in space (top right), the fully deployed 8 m diameter, 20 m long solenoid coil (bottom left) and the final 6+1 configuration with the compensation solenoid (red) in the center (bottom right).

2.3. SR2S Study

The SR2S consortium has chosen to pursue the racetrack coil toroid configuration based on the MgB_2 cable. A continuous-coil toroid (CCT), consisting of 120 racetrack coils, with a field integral of 8 Tm, protects a 4.5 m diameter, 6 m long cylindrical habitat based on the dimensions of the ESA Columbus scientific module on the International Space Station (ISS).

The habitat and the position of the coils around the habitat are shown in Fig. 4. Each coil is surrounded by a 0.6 mm thick KEVLAR support sheath. A 8.6-m long support cylinder composed of a metal matrix composite material, aluminum-boron-carbide (Al-B4C), surrounds the habitat to support the Lorentz force acting on the coils in the direction of the habitat.

The multiplication of the number of coils in the continuous-coil design reduces the force acting on a single coil. The result is a reduction of the magnet shield mass and a greater mechanical tolerance for the toroid assembly.

The performance results of the SR2S CCT shield configurations are presented in the next section. The ESA study provided the framework for the NIAC and SR2S studies[2]. The engineering designs did not attain the level of detail of the succeeding studies. The principal results of the ESA and NIAC studies are summarized in the discussion

Figure 4. The 4.5 m diameter, 6 m long Columbus habitat (top), and the CCT shield of the S2RS study. A single racetrack coil (lower left) and the 120 coil toroid (lower right) are shown with the Al-B4C support cylinder (gray) which surrounds the Columbus habitat (blue-green). Each coil (yellow) is supported by a 0.6 mm thick KEVLAR sheath (cyan).

section. The three studies are presented in greater detail in Ref. [7].

3. THE SR2S CCT DESIGN

The HTSC racetrack coils of the CCT consist of a cable core with a density of $3.0\,g/cm^3$, composed of 57.4% aluminum, 8.6% MgB_2, 23% titanium and 11% SiO_2, surrounded by 1.2 cm thick aluminum. Each coil is surrounded by a 0.6 mm thick KEVLAR support sheath (90 kg). The total mass of the 120 coil toroid is 79 t.

The 5.5 m diameter, 4.4 cm thick, 8.6 m long aluminum-boron-carbide cylindrical support shell (60% boron), has a density of $2.6\,g/cm^3$, and a mass of 16.8 t. The total mass of the 8 Tm field configuration is 95.8 t. The individual contributions of the different elements to the total mass are shown in Fig. 5.

The base-line shield design performance for GCR protons and $Z \geq 2$ nuclei was evaluated with a field integral of 8 Tm. Higher field value configurations were evaluated to study the evolution of the proton and He nuclei doses with field strength and shield mass, in particular the relative contributions of the charged secondaries and neutrons. The different toroid fields used in the simulation are shown in Fig. 6. The highest field values, 50 and 100 Tm, represent extrapolations to cable current densities beyond the present-day performance of the HTSC.

The 11.5 Tm shield configuration is obtained by increasing the coil dimensions. With the larger dimension coil, and an increase of the current density to the limiting value

[2] ESA (2011), NIAC (2012-2013) and SR2S (2013-2015)

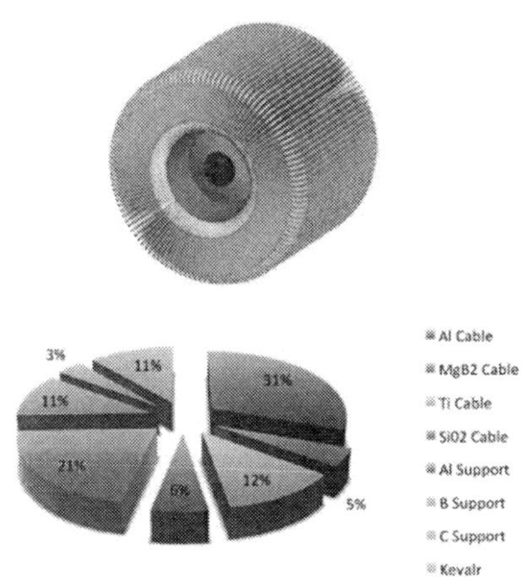

Figure 5. The material composition of the 8 Tm CCT shield, including support structures, in the Monte Carlo simulation. The total mass of the 8 Tm shield is 95.8 t.

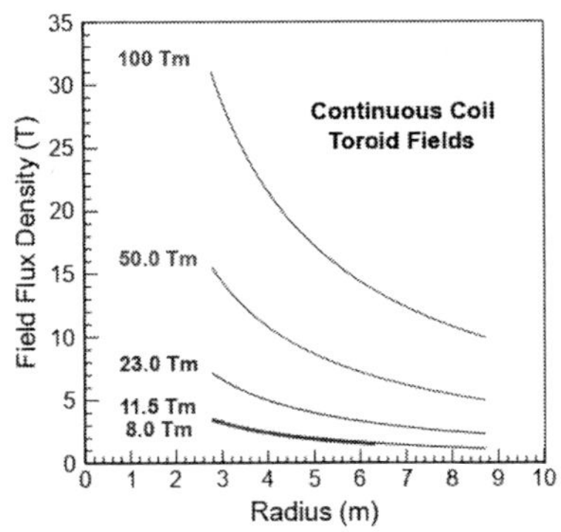

Figure 6. The toroid fields evaluated in the performance study. The field flux densities of 50 and 100 Tm require current densities which exceed the performance of the present-day HTSC.

of the cable dimensions, the integral flux attains a value of 23 Tm. Figure 7 shows the coil dimensions of the 8 Tm, and 11.5 (23) Tm configurations. The total shield mass in the simulation of the larger dimension coil configurations is 137 t.

The 4.5 m diameter, 6 m long Columbus habitat (Fig. 4) is composed of a 1.5 cm thick aluminum cylindrical shell and two 3.0 cm thick aluminum endcaps. The mass of the habitat in the simulation is 4.36 t.

3.1. The Monte Carlo Simulation

The Geant4 Monte Carlo simulation [8] allows a detailed description of the geometry and composition of the

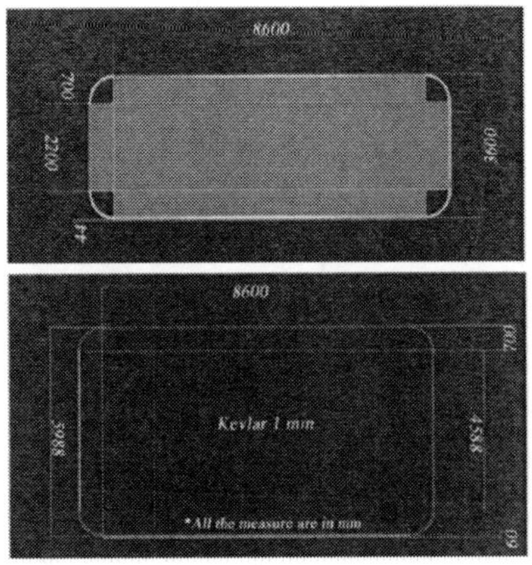

Figure 7. The dimensions (mm) of the racetrack coils of the 8 Tm (top) and 11.5 (23) Tm (bottom) CCT shield configurations.

shield, spacecraft and dose-sampling volumes. Particle propagation in the magnetic field and materials is performed with a detailed treatment of electromagnetic and hadronic interactions, including the generation and propagation of secondary particles.

The GCR radiation risk arises from the eventual damage caused by the energy lost by the charged particles in human tissue. The mean energy loss rate in a material due to ionization is given by the Bethe-Bloch equation,

$$\frac{dE}{dx} = K\frac{z^2}{\beta^2}\frac{Z}{A}\left[\frac{1}{2}\ln\frac{2m_ec^2\beta^2\gamma^2T_m}{I^2} - \beta^2 - \frac{\delta}{2}\right] \quad (3)$$

where z and β are the particle's charge and velocity relative to the speed of light c. Z, A and I are respectively the atomic number, mass and mean excitation energy of the material. The other terms in the equation are K which regroups various physical constants, m_e the mass of the election, the relativistic term γ expressed in terms of β, the maximum kinetic energy lost in a collision with a free electron T_m and a density correction term δ.

The ionization energy losses recorded during the track propagation dE_i are converted to a dose equivalent ϵ_i (Sv) by multiplying the absorbed dose $\frac{dE_i}{m}$ (Gy), where m is the mass of the volume considered, by the quality factor $Q(L)$ defined by the unrestricted linear energy transfer (LET) in water L (keV/μm) [9]:

$$\epsilon_i = Q(L) \cdot \frac{dE_i}{m}, \quad (4)$$

with

$$L = \frac{dE_i}{dx} \quad (5)$$

and

$$Q(L) = \left\{ \begin{array}{ccc} 1 & \text{for} & L \leq 10 \\ 0.32 \cdot L - 2.2 & \text{for} & 10 < L < 100 \\ 300/\sqrt{L} & \text{for} & L \geq 100 \end{array} \right\}$$

The total dose equivalent $d_z(E_j)$ for an exposure time t, due to GCR of charge z and kinetic energy E_j is the sum of the dose equivalents recorded for N_j incident particles generated with the flux $f_z(E_j)$ ($\mathrm{cm^{-2}sr^{-1}s^{-1}MeV^{-1}}$), in the energy interval ΔE_j (MeV), over the acceptance A ($\mathrm{cm^2sr}$):

$$d_z(E_j) = \sum_i \epsilon_i^j \cdot \frac{A}{N_j} \cdot f_z(E_j) \cdot \Delta E_j \cdot t \qquad (6)$$

The total GCR dose D is obtained by extending the generation over suitable ranges in charge and kinetic energy. The contribution from charge z is

$$d_z = \sum_j \left[\sum_i \epsilon_i^j \cdot \frac{f_z(E_j) \cdot \Delta E_j}{N_j} \right] \cdot A \cdot t \qquad (7)$$

and the total dose equivalent including all charges up to Ni, $D = \sum_{z=1}^{z=28} d_z$.

The kinetic energy spectra $f_z(E_j)$ are taken from the CREME 2009 GCR model [1] which extends over the energy range 1 to 10^5 MeV/n. The GCR fluxes for solar minimum and maximum conditions are shown in Fig. 1. The solar minimum fluxes are used for the shield performance evaluations.

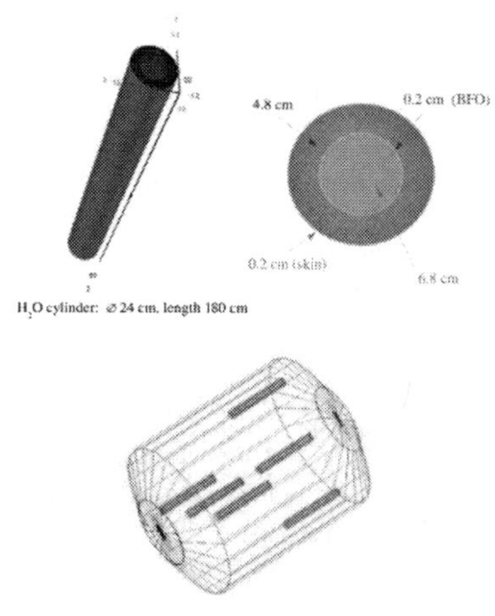

Figure 8. The cylindrical water volume used to compute the dose of the skin and BFO. The total body dose refers to the full volume of the 24 cm diameter, 180 cm long cylinder (81.4 kg). The location of the six water cylinders in the Columbus habitat (bottom).

The human body is represented in the simulation as a 24 cm diameter, 180 cm long water cylinder. The cylinder is subdivided to define the regions used to compute the dose associated with the skin and blood-forming organs (BFO), respectively the first 2 mm at the surface of the cylinder and a 2 mm thick layer located a depth of 5 cm from the cylinder surface. The body dose refers to the ionization losses recorded in the full volume of the cylinder. The reported doses represent the average values recorded in the six water cylinders, which are used to sample the dose levels on and off the axis in each half-volume of the cylindrical habitat (Fig. 8).

3.2. Performance Results

The annual GCR dose equivalents of the CCT shield, with and without the 8 Tm field, are compared in Table 1. The dose levels refer to the dose received from the protons and nuclei, which traverse the magnetic field volume, generated on the four lateral sides (barrel region) of the $30 \times 30 \times 30\,\mathrm{m^3}$ cube, positioned around the center of the Columbus habitat (Fig. 9). The Geant4 QBBC physics list was used for the hadronic and nuclear interactions [10].

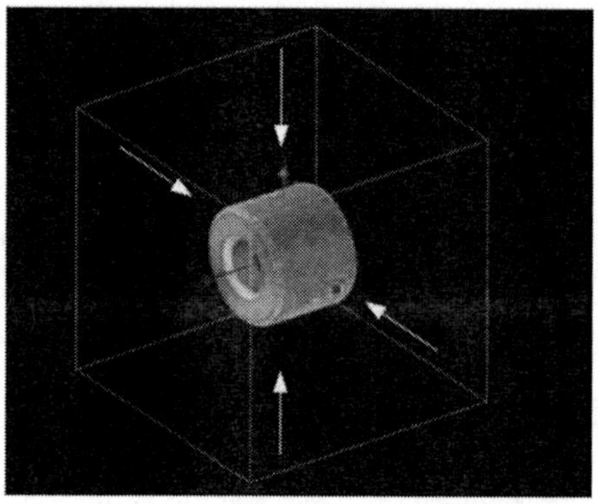

Figure 9. The surface of the $30 \times 30 \times 30\,\mathrm{m^3}$ cube surrounding the cylindrical Columbus habitat. The quoted dose levels refer to the GCR generated on the four lateral sides indicated in the figure by arrows.

Table 1. The Geant4 annual GCR dose equivalents (cSv/y) for the CCT shield configuration, with and without the 8 Tm field, for the acceptance corresponding to the barrel region (Fig. 9). The quoted uncertainties () represent the root-mean-square deviation of the average dose recorded in the six water cylinders (Fig. 8).

Z	CCT no field		CCT field	
	BFO	body	BFO	body
1	15.2 (0.9)	15.1 (0.3)	11.3 (0.4)	11.6 (0.3)
2	3.84 (0.36)	3.88 (0.08)	3.10 (0.27)	3.40 (0.07)
3-10	1.32 (0.12)	1.38 (0.08)	1.03 (0.04)	1.06 (0.04)
11-20	0.80 (0.10)	0.84 (0.09)	0.60 (0.04)	0.65 (0.04)
21-28	0.74 (0.12)	0.84 (0.14)	0.64 (0.11)	0.72 (0.09)
total	21.9 (1.0)	22.1 (0.4)	16.7 (0.5)	17.4 (0.3)
fraction field on/off			0.76 (0.05)	0.79 (0.02)

The magnetic field produces an additional 20-25% reduction of the dose levels provided by the passive shielding of the shield and habitat material. The corresponding annual BFO and body dose equivalents of the Columbus habitat are respectively, 28.1 ± 1.3 and $30.2 \pm 0.4\,\mathrm{cSv/y}$. The passive shielding elements and the magnetic field of the CCT shield reduces the habitat doses by 40-50%.

The evolution of the dose reduction with field strength and shield mass is illustrated in Fig. 10. The reported annual body dose equivalents refer to GCR proton and He nuclei, which represent the dominant contribution ($\sim$ 85%) to the estimated total doses in Table 1. The 50 and 100 Tm fields, which require current densities exceeding the performance of the present-day HTSC, provide an indication of the dose reduction expected for significantly higher field flux densities.

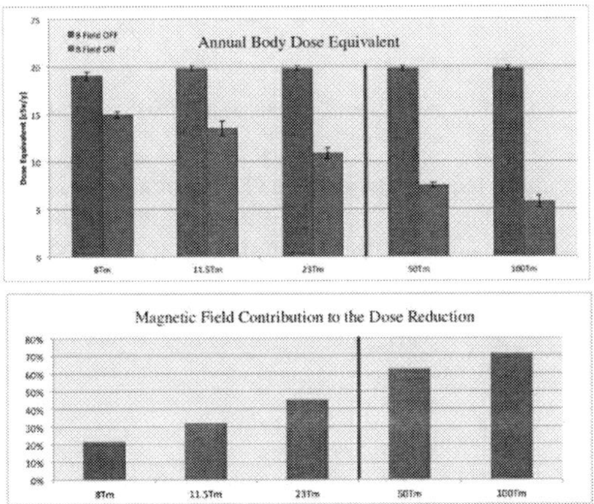

Figure 10. The annual body dose equivalent doses due to GCR protons and He nuclei, with the field off and on, for the 8, 11.5, 23, 50 and 100 Tm CCT (top), and the corresponding contributions of the field to the dose reduction (bottom). The 50 and 100 Tm configurations require field flux densities which exceed the performance of the present-day HTSC. The displayed errors represent the root-mean-square deviation of the average dose recorded in the six water cylinders (Fig. 8). The quoted doses refer to the barrel region acceptance (Fig. 9).

The increase in the screen mass required to attain the larger integral field values results in a higher field off dose level, due to the increase in secondary particle production. The increase is more than compensated by the magnetic field, which results in a steady decrease of the total dose with increasing field strength. The contribution of the field to the overall dose reduction is 45% at 23 Tm, the technological limit of the present-day HTSC. At 100 Tm, the field represents 70% of the observed dose reduction.

3.3. Secondary Particle Production

The effect of the magnetic field on the dose due to the secondary particles was studied in detail. Figure 11 shows the contributions of the primary GCR protons and He nuclei, and their secondaries, for the Columbus habitat alone, and the five CCT field configurations. The presence of the 8 Tm field does not compensate the dose due to the secondaries created in the shield material, consequently the total dose due to protons and He nuclei exceed the level of the Columbus habitat. A decrease of both the primary and secondary doses is observed as the field integral BL increases.

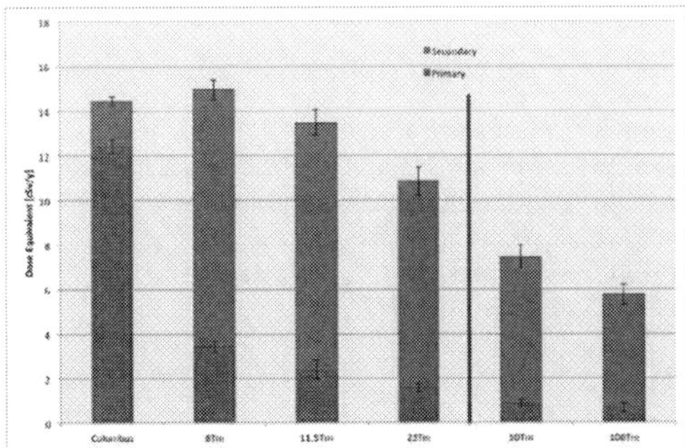

Figure 11. The contributions of the primary GCR proton and He nuclei, and the secondaries produced in the material of the habitat and magnet shield to the annual body dose equivalent. The error bars represent the root-mean-square deviation of the average primary and secondary doses recorded in the six water cylinders (Fig. 8). The quoted doses refer to the barrel region acceptance (Fig. 9). The 50 and 100 Tm configurations require field flux densities which exceed the performance of the present-day HTSC.

The individual contributions of the primary and secondary particles are shown in Fig. 12. The principal contributions to the secondary dose are due to protons and neutrons. The reported neutron (gamma) dose levels refer to the ionization losses recorded in the water cylinders of the charged particles created by the neutrons (gamma) in hadronic (electromagnetic) interactions in the material of the shield and habitat. The individual contributions of the secondaries denoted "others" include the mesons and baryons not explicitly quoted, and light nuclei.

The vertex distributions of the secondaries created by the primary GCR protons and He nuclei in the materials of the magnet shield and habitat, which contribute to the dose in the 8 and 23 Tm CCT shield configurations, are shown in Fig. 13. The particles created in Al-B4C support cylinder surrounding the habitat are the principal source of the secondary dose.

The geometric acceptance limits the contribution of the particles created at a greater distance from the habitat. The contribution of the charged secondaries is further re-

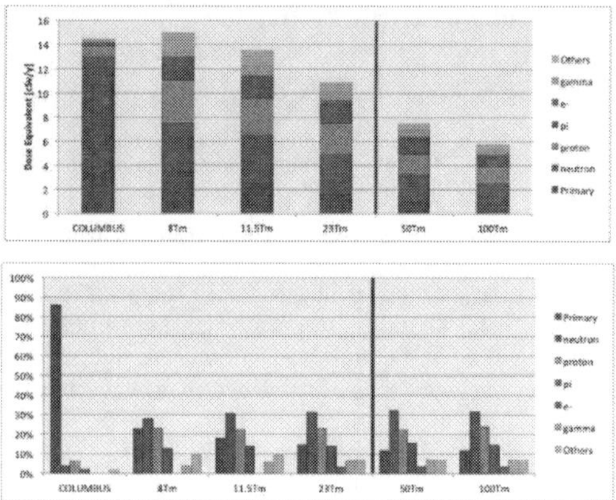

Figure 12. The annual body dose equivalent of the primary GCR proton and He nuclei, and the secondary particles (top), and the the contribution of each category to the total dose (bottom). The 50 and 100 Tm configurations require field flux densities which exceed the performance of the present-day HTSC.

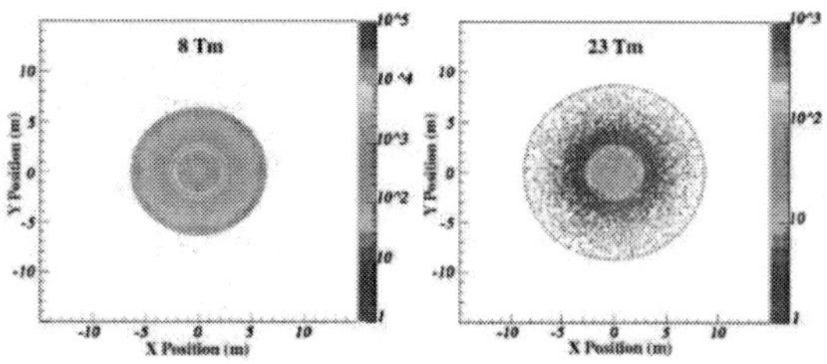

Figure 13. Vertex distributions of the generated secondaries which contribute to the dose in the 8 and 23 Tm CCT shield configurations. The largest contribution to the secondary dose is due to interactions in the Al-B4C support cylinder surrounding the habitat, denoted by the red circular regions located inside the field volume in the xy projections.

duced by the presence of the magnetic field. The differences observed in the two vertex distributions are due to the different coil dimensions and field integrals BL.

The vertex distributions of the secondary neutrons and protons which contribute to the recorded dose levels are presented in Fig. 14. The effect of the field on the secondary protons results in a shorter radial extension of the volume which contributes to the dose, compared to the region corresponding to the neutron contribution, which reflects principally the geometric acceptance.

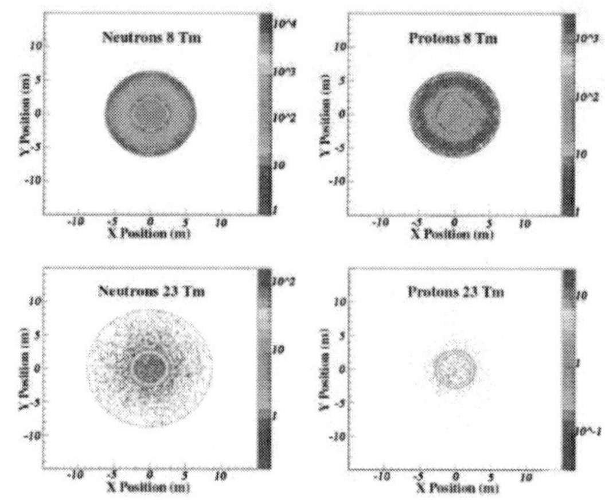

Figure 14. Vertex distributions of the generated secondary neutrons and protons which contribute to the dose in the 8 and 23 Tm CCT shield configurations.

3.4. CCT Current Technological Limit Performance

The body dose equivalent due to GCR protons and He nuclei exceed the value observed for the Columbus habitat for the 8 Tm CCT configuration (Fig. 11). A further improvement in the performance is obtained by increasing the field integral BL. The effect of the increase in BL on the proton and He nuclei annual BFO dose equivalents is summarized in Table 2.

Table 2. The GCR proton and He nuclei dose reductions. The reported dose equivalents correspond to the barrel region acceptance (Fig. 9). The quoted uncertainties () represent the root-mean-square deviation of the average dose recorded in the six water cylinders (Fig. 8). R_{hab} *is dose reduction with respect to the habitat.* R_{field} *is the contribution of the field to the dose reduction.*

BL	Mass	BFO Dose Eq. B Off	BFO Dose Eq. B On	R_{hab}	R_{field}
(Tm)	(t)	(cSv/y)	(cSv/y)	%	%
8	96	19.0 (1.0)	14.4 (0.6)	1 (4)	24 (2)
11.5	137	20.3 (2.1)	12.8 (1.8)	12 (12)	37 (9)
23	137	20.3 (2.1)	10.5 (1.2)	28 (8)	48 (6)

The 137 t, 23 Tm CCT shield configuration would result in a BFO dose equivalent of $\sim 10\,\mathrm{cSv/y}$. The contribution of the $Z > 2$ nuclei, 15% for the 8 Tm toroid shield (Table 1), is further reduced for the higher field, larger mass configurations due to the increase in the material thickness, which increases the efficiency of the passive shielding elements for the higher charge nuclei. The contribution of the GCR $Z > 2$ nuclei is negligible for the 23 Tm configuration

The presence of the 23 Tm field is responsible for 48% of the total shield reduction. The dose level, which corresponds to the part of the acceptance protected by the magnetic shield, ~ 70% of the total, is a factor 5 smaller than the limit for LEO (50 cSv/y) quoted in Ref. [11].

A comparison may be made with the equivalent dose estimate provided by the *in situ* measurement of the Mars Science Laboratory during its voyage to the planet, $67.2 \pm 12.0\,\mathrm{cSv/y}$ [12]. The quoted value is based on the ionization recorded in a thin silicon detector of the Radiation Assessment Detector (RAD), mounted on the top deck of the *Curiosity* rover inside the spacecraft. The measurements were made during a period of low-to-moderate solar activity [13]. The areal material density of the spacecraft was less than $10\,\mathrm{g/cm^2}$ for most of the solid angle acceptance of the RAD.

The annual skin dose equivalents of the Columbus habitat and the 8 Tm CCT are listed in Table 3. The dose of the barrel region (Fig. 9), shielded by $4\,\mathrm{g/cm^2}$ of aluminum, were scaled to account for the full angular acceptance in order to compare with the experimental result.

Table 3. The annual skin dose equivalents of the Columbus habitat and 8 Tm CCT. The quoted uncertainties () represent the root-mean-square deviation of the average dose recorded in the six water cylinders (Fig. 8). R_{hab} *is dose reduction with respect to the habitat.* R_{field} *is the contribution of the field to the dose reduction.*

	Skin Dose Eq. (cSv/y)	R_{hab} %	R_{field} %
Columbus Habitat	51.2 (1.3)	-	-
8 Tm CCT Field Off	34.4 (0.7)	33 (2)	-
8 Tm CCT Field On	27.0 (1.4)	47 (3)	22 (4)

The Columbus habitat dose level is 25% lower than the dose equivalent quoted in Ref. [12]. A lower dose equivalent is quoted in Ref. [13], $63.8 \pm 10.9\,\mathrm{cSv/y}$, after a re-evaluation of the conversion factor used to obtain LET in water from the ionization losses recorded in the silicon of the RAD.

The 20-25% discrepancy between measurement and simulation is comparable to the difference in the proton free space, skin dose equivalents obtained with the water cylinder geometry (Fig. 8) and an ICRU sphere [14] in the Monte Carlo simulation. The former are 30% higher [15], which suggests the cylindrical geometry results in an underestimation of the skin dose equivalent[3]. In this case, the ~ 50% reduction of the habitat dose (Table 3) may be applied to the *in situ* measurement to obtain a dose equivalent estimate for a spacecraft equipped with the 8 Tm CCT, $\sim 30\,\mathrm{cSv/y}$.

[3]The corresponding BFO dose equivalent is 10% lower than the value obtained with the ICRU sphere [15].

4. DISCUSSION

4.1. Physics Models and Dose Equivalent

The Geant3 Monte Carlo simulation [16] was used for the dose evaluations in the ESA study. The free space dose equivalents of the two Geant simulations, computed with the water cylinder (Fig. 8), are compared in Table 4. The Geant4 proton dose equivalents increase by ~ 10% with respect to the values obtained with the Geant3 simulation, while the doses for the $Z \geq 2$ nuclei are lower, with the notable exception of the BFO and body doses of the highest charge nuclei, $21 \leq Z \leq 28$.

Table 4. Annual GCR dose equivalents (cSv/y) in free space at the solar minimum. The effect of the generation statistics on the quoted free space doses is negligible (~ 1%).

	Geant3			Geant4		
Z	skin	BFO	body	skin	BFO	body
1	10.8	11.3	11.1	12.2	12.9	12.8
2	5.3	5.2	5.1	5.8	4.6	4.8
3-10	35.9	22.2	11.8	13.6	6.1	7.4
11-20	38.4	16.6	14.8	19.9	9.2	11.4
21-28	27.3	7.1	8.7	24.7	11.1	14.3
total	117.7	62.4	51.5	76.2	43.9	50.7

The updated nuclear interaction models in Geant4 explain the difference. In the Geant3 simulation, the ionization losses of the GCR $Z > 2$ nuclei dominate, representing ~ 90% of the BFO total dose equivalents. The corresponding contribution of the charged secondaries of the GCR $Z > 2$ nuclei is significantly higher in Geant4, between 30-40%, increasing with the charge of the GCR nuclei[15]. The secondaries created in the water cylinders may explain the higher Geant4 BFO and body equivalent doses reported in Table 3 for the highest charge nuclei group.

The Geant3 free space BFO dose equivalents of the ESA study were compared to previous estimates (58-70 cSV) quoted in Ref. [11] to verify the dose calculation. The Geant4 free space BFO dose equivalent is ~ 30% lower. The reduction of the estimated dose due to the different hadron physics models is significant, comparable to the variation due to solar activity, 40% [7].

The LET is used for the evaluation of radiation damage for electronic devices. The effects of large energy deposition on small size scales is appropriate for microelectronic components. The dose equivalents computed with LET in silicon are compared to failure rates to qualify the radiation hardness of semi-conductor devices.

The corresponding LET in water is employed to estimate the biological damage, which is ultimately the result of the effect of the ionizing radiation on the DNA molecules of the cells in human tissue. The dose equivalents may be correlated with the lifetime cancer risks of long-term clinical studies [17]. The long term health risks due to a prolonged exposure to GCR are not well known. The two atomic bombings, irradiation during nuclear accidents, the lunar manned missions and human activity in LEO do not represent exactly the conditions encountered in interplanetary space. The effects due to the high charge GCR nuclei are an additional concern [18].

4.2. NIAC Study Results

The 6+1 extendable solenoid shield performance of the NIAC study (Fig. 3) was evaluated with the two Geant Monte Carlo simulations. The results corresponding to the barrel region acceptance of the generation cube used in the NIAC study are summarized in Table 5.

Table 5. Annual GCR dose equivalents at solar minimum, and the corresponding field off/on fractions, for the 6+1 extendable solenoid shield of the NIAC study. The quoted uncertainties () represent the root-mean-square deviation of the average dose recorded in the six water cylinders positioned in the same configuration (Fig. 8) in the cylindrical NIAC habitat.

	dose equivalent (cSv/y)		
	skin	BFO	body
G3	36.5 (1.9)	26.1 (1.0)	20.8 (0.7)
G4	**33.2 (1.7)**	**23.6 (1.4)**	**26.2 (0.8)**
	fraction field on/off		
G3	0.97 (0.07)	0.95 (0.06)	0.95 (0.04)
G4	**0.91 (0.06)**	**0.92 (0.07)**	**0.91 (0.04)**

The estimated dose levels are lower in the Geant4 simulation. The shield performance, quoted as the ratio of the recorded dose levels without and with the magnetic field, although relatively modest differ by a factor of two, respectively 5% (Geant3) and 10% (Geant4),

In order to estimate the effect of additional spacecraft structures on the dose levels evaluated over the full acceptance, the 6+1 extendable solenoid shield was incorporated in a preliminary spacecraft design for a near Earth asteroid (NEA) mission, shown in Fig. 15.

A roughly equivalent mass, 19 and 22 t, was added on each side of the 1.8 cm thick cylindrical aluminum habitat (10.1 t). The additional aluminum structures consist of 1.8 cm thick spherical and cylindrical fuel tanks, a combustion chamber, an access tube, a re-entry vehicle and hatches. The total mass present in the simulation was 88 t, including the shield mass (35 t), without the chemical propellants.

The effect of the additional spacecraft structures is negligible. The ratio of the full acceptance, NEA spacecraft skin, BFO and body doses to the corresponding doses of the shield and habitat alone are respectively 0.97 ± 0.22, 1.01 ± 0.19 and 1.01 ± 0.16. The dose reduction observed for the GCR generated on the two surfaces at either end of the cylindrical spacecraft is compensated by an increase of the dose received in the barrel region due to

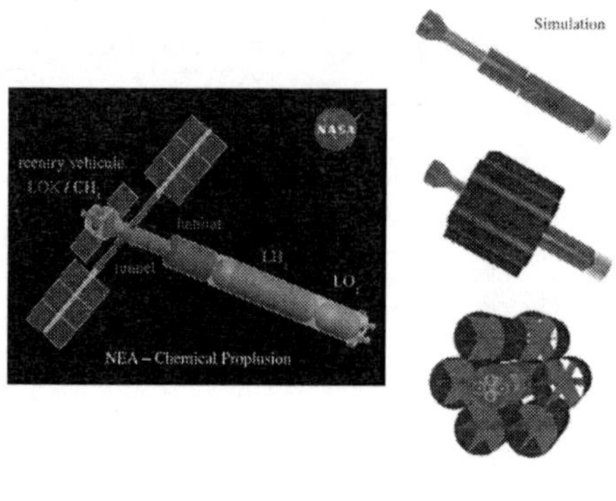

Figure 15. Spacecraft design for a near Earth asteroid (NEA) mission with chemical propulsion and the 6+1 extendable solenoid shield. At right, the elements implemented in the Geant3 Monte Carlo simulation.

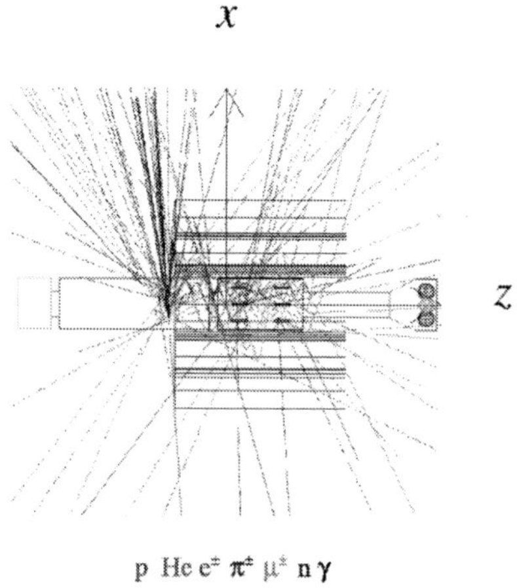

Figure 16. Proton and He nuclei events generated in the barrel which result in an ionization loss in the water cylinders in the Geant3 Monte Carlo simulation. Low energy secondaries generated in the liquid hydrogen container on left are redirected towards the habitat by the magnetic field (red, blue and green tracks).

charged secondaries created in the additional spacecraft structures, illustrated by the events shown in Fig. 16.

A further optimization of the performance requires a sufficient quantity of material to shield the magnetic field volume and limit the entry of particles which would be deflected in the direction of the habitat. The performance could be improved by a reconfiguration of the existing structures of the NEA spacecraft, and if required, additional passive shielding elements. The exercise demonstrates a magnetic shield cannot be treated as add-on feature, but must be considered an integral part of the spacecraft design.

4.3. Performance Comparison of the Advanced Designs

A comparison of the NIAC 6+1 extendable solenoid (ES) and the SR2S CCT shields is presented in Table. 6. The BFO dose equivalents represent respectively 47% and 33% of the quoted LEO limit (50 cSv/y). The reduction of the GCR dose due to magnetic field is marginal for the NIAC 6+1 EC shield. The field integrals of the magnetic shielding configurations considered in the three studies render redundant the presence of a high mass shelter for SEP events.

Table 6. The performance of the NIAC 6+1 extendable solenoid (ES) and the SR2S CCT toroid shields. The annual BFO dose equivalents at solar minimum correspond to the barrel region acceptance (Fig. 9). The quoted uncertainties () represent the root-mean-square deviation of the average dose recorded in the six water cylinders (Fig. 8). R_{hab} *is dose reduction with respect to the habitat.* R_{field} *is the contribution of the field to the dose reduction.*

Shield	BL (Tm)	Mass (t)	BFO Dose Eq. (cSv/y)	R_{hab} (%)	R_{field} (%)
6+1 ES	6.3	35	23.6 (1.4)	12.3 (8.1)	7.8 (7.2)
CCT	8	96	16.7 (0.5)	40.6 (3.3)	23.7 (4.2)

5. CONCLUSION

The same methodology used to design state-of-the-art particle detectors for accelerator and astrophysics experiments was applied to evaluate the space radiation shield configurations. The Monte Carlo simulation reproduces in detail the underlying physics processes with an exact description of the materials and geometry of the detector (dose sampling volumes) and shield configuration. The results obtained with different engineering designs provide a realistic assessment in terms of the trade-off between dose reduction and shield mass.

The field contribution to the overall shield dose reduction attains 37% and 48% for the 11.5 and 23 Tm configurations of the 137 t CCT (Table 2), with BFO dose equivalents a factor 4-5 lower than the quoted LEO limit. The results of the 23 Tm CCT indicates a further significant reduction, for example a factor of 10 with respect to the LEO limit, is not a realistic scenario today. The shield mass is considerable and a technological limit solution is not compatible with safety standards which require important margins of security. The corresponding dose reduction factors of the lower mass shields in Table 6 are 2 and 3. The field contribution represents 25% of the total dose reduction for the 8 Tm, 96 t CCT.

There are no established dose limits for long duration manned space missions. A limit based on the risk of long-term cancers, the ultimate criteria for the acceptable GCR dose, is premature given the complexity of the problem. Risk assessment is affected often by the subjective perception of the danger, an aspect which cannot be neglected in the present context. A manned mission to Mars will likely exceed the world-wide impact of the first landing on the Moon. The challenge is considerable in a world preoccupied by the reduction of risks and costs.

The author wishes to recognize the efforts of Dr. Domenico D'Urso of INFN-Perugia, which made possible the high statistic Monte Carlo studies, the indispensable contribution in matters of administration of the ESA and SR2S studies of dott.sa Marta Perucci, and the TIFPA director Prof. Marco Durante for his pertinent comments and the financial support which allowed the work to be presented at the conference.

REFERENCES

[1] Cosmic Ray Effects on Micro Electronics https://creme.isde.vanderbilt.edu/CREME-MC

[2] W. von Braun, Popular Science, 1969, 98-100

[3] J. G. Bednorz and K. A. Müller, *Zeitschrift für Physik B*, 64 (2), 1986.

[4] R. Battiston *et al.*, *Superconductive Magnet for Radiation Shielding of Human Spacecraft*, Final Report ESTEC Contract N. 4200023087/10/NL/AF, 2011

[5] S. C. Westover *et al.*, *Magnet Architectures and Active Radiation Shield Study*, NASA Institute for Advanced Concepts, Phase 1 Report Final, 2012

[6] Meinke R.B., Ball M.J., Goodzeit C.L., 2003, IEEE Proceedings of PAC 03, Vol. 3, 1996

[7] F. Ambroglini, R. Battiston and W.J. Burger, Front Oncol. 6:97, 2016, doi:10.3389/fonc.2016.00097

[8] J. Allison *et al.*, IEEE Transactions on Nuclear Science, Vol. 53, No. 1, 2006, 270-278

[9] ICRP Publication 60, Ann. ICRP 21 (1-3) 1991

[10] A.V. Ivantchenko *et al.*, Int. J. of Radiation Biology 2012, 88(1-2), doi:10.3109/09553002.2011.610865

[11] Rapp D., *Radiation Effect and Shielding Requirements in Human Missions to the Moon and Mars*, Mars Vol. 2, 2006, 46-71

[12] C. Zeitlin *et al.*, Science, Vol. 340, 2013, 1082

[13] J. Guo *et al.*, A & A, 577, 2015, doi:10.105/0004-6361/201525680

[14] *Radiation Quantities and Units*, Report 33, ICRU, Bethesda MD, 1980 No. 6, 2294-2499

[15] F. Ambroglini *et al.*, proceedings of the 7th IAASS Conference, *Space Safety Is No Accident*, Springer International Publishing Switzerland, 2015, doi:10.1007/978-3-319-15982-9_18

[16] R. Brun *et al.*, CERN Report DD/EE/84-1, 1987

[17] *Health Risks From Exposure to Low Levels of Ionizing Radiation, BEIR VII Phase 2*, National Academies Press, Washington DC., 2006

[18] M. Durante and F. A. Cucinotta *et al.*, Rev. Mod. Phys, 83, 2011, doi:10.1103/RevModPhys.83.1245

FLY SAFELY
WITH EUROPEAN DESIGN

Florian Bittner[(1)]

[(1)]*Airbus Defence and Space, Airbus Allee 1, 281991 Bremen, Germany,*
Florian.Bittner@airbus.com

ABSTRACT

The NASA Orion vehicle that will fly to the moon in the next years is propelled along its mission by the European Service Module (ESM), developed by Airbus Defence and Space on an ESA contract.

This paper is about the design of the Propulsion Drive Electronics (PDE) that provides the interface between the thrusters of the European Service Module and the Orion on board computers.

As this is an astronautical mission, safety is of utmost priority.

1 THE ORION ESM ARCHITECTURE

Airbus Defence and Space is developing the European Service Module (ESM) for the NASA Orion vehicle for human space exploration mission on an ESA contract [2]. The ESM provides to Orion the generation of power through 4 deployable solar arrays, the power distribution to the ESM and Crew Module (CM) users, the passive and active thermal control system for ESM and CM using thermistors and fluid loops, the storage and delivery of water and gas to the CM, and the propulsion system. Fig. 1 shows the modules of the Orion Vehicle, in particular the service module and the crew module.

The hardware in the ESM is controlled by a set of electronic units developed by Airbus Defence and Space and its subcontractors:

- Propulsion Drive Electronics (PDE), controlling the propulsion hardware
- Pressure Regulation Unit (PRU) controlling the regulation of pressure in the ESM propulsion tanks
- Power Control and Distribution Unit (PCDU), providing power at 28V and 120V
- Solar Array Driving Electronics (SADE), controlling the solar arrays
- Thermal Control Unit (TCU), managing the active and passive thermal control systems
- Fluid Control Assembly (FCA) controlling the pumps for the active cooling system.

The mission management is performed by the Orion flight computers located in the CM. These are developed by NASA and subcontractors. The ESM electronic units communicate with the Orion computers via a 1GBit real time Ethernet network called ODN (Orion Onboard Data Network).

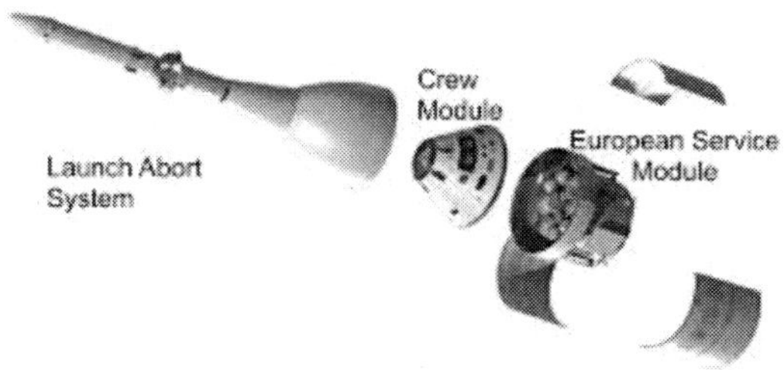

Fig. 1. Modules of Orion Vehicle

Fig. 2 depicts the overall avionics subsystem architecture and interfaces. The interface to the SLS launcher is shown to the left side, to the crew module on the right.

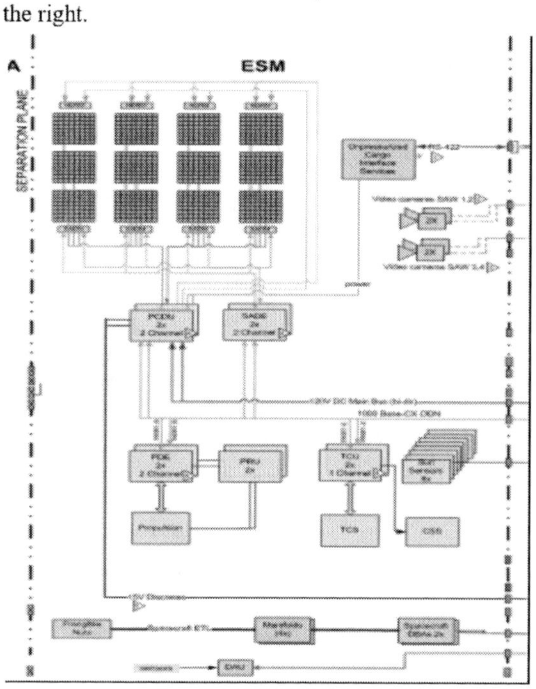

Fig. 2. ESM Avionics architecture

1.1 The ESM propulsion subsystem and the PDE mission

A primary subsystems in the ESM is the Propulsion Sub System (PSS) [3], which is connected to three different types of engines: The 27.7 kN main engine, a reused thrust vector controlled Orbital Maneuvering System thruster from the Space Shuttle program, eight auxiliary engines (AUX) providing a thrust of 490 N each, are used together with the Main Engine for the orbital manoeuvers and 24 Reaction Control System (RCS) thrusters for attitude control.

The ESM propulsion is equipped with a pressure regulation system (PRU), controlled autonomously in the ESM depending on the phases of the mission.

The ESM Propulsion Drive Electronics (PDE) is the interface between the Orion Flight Computers and the ESM Propulsion hardware. Through the ODN interface the PDE receives commands packets to operate the engines on demand of the crew vehicles Flight Software, and returns readings from sensors to the Flight Software, which enables monitoring and failure detection of the propulsion hardware. For a successful mission two aspects are of utmost importance: reliability and time reactivity.

1.2 The ATV heritage

The ESM design is based on the Automated Transfer Vehicle (ATV) heritage. The ATV successfully serviced the ISS between 2008 and 2015 on five different missions.

The ESM PDE's design is an evolution of the ATV PDE. The functions have been expanded to control a broader range of engines and a gimballing mechanism, the command interface changed the Mil-Bus interface to a Time-Triggered Ethernet, thus optimizing its reactivity from the 100ms ATV cycle to the 25ms of Orion and increasing its reliability to allow crewed missions.

Another noteworthy point is critical software. The ATV concentrated the most critical functions in the Monitoring and Safing Unit (MSU). This device was designed to establish and maintain a safe position of ATV away from the ISS by executing the Collision Avoidance Maneuver (CAM) and overriding the ATV Software and the ATV PDE by direct commands to the propulsion hardware. Thus it was possible to reduce the criticality of the ATV and PDE software as only the MSU software had been developed to the highest criticality A grade [1]. This is not the case on ESM. As a result the PDE software had to be developed to meet the highest safety requirements

On ESM an independent safing system such as the MSU is not present, and the reliability requirements are not only associated with collisions but also spread over the entire flight; this forced the development of PDE SW to criticality category A.

Until now the ATV MSU and the ESM PDE software are the only two of Category A software developed for European space projects.

2 PDE ARCHITECTURE

2.1 PDE description

The design of the PDE architecture has the focus on high reliability and fault-tolerance to protect against un-wanted activation of critical actuators.

A full PDE is comprised by two identical PDE boxes (Fig. 3) each consisting of two independent channels providing actuator controllers and sensor interfaces. Redundancy on system level is achieved by connecting nominal and redundant actuators and sensors to different PDE channels. Each PDE channel has its own network interface for communication with the crew modules computers via the ODN.

The required PDE functionality will be provided by sub-functions realized either by dedicated hardware functions implemented in radiation hardened FPGAs or dedicated firmware functions executed on a radiation hardened LEON processor.

In general the communication via SNIC/ODN with the Orion Flight computers will be handled by firmware modules executed on the LEON processor. The firmware will decode and check the correctness of the received ODN command message. Effector IDs will be extracted and checked by the firmware out of the validated ODN package and the effector information will be copied to the hardware based driver or serial interfaces to execute/send the decoded command in the next minor frame.

The timing of the command processing will be based on a synchronized clocking system which synchronizes itself to the ODN Master Frame Interrupt (MFI) and is implemented in hardware.

Fig. 3. ESM PDE box

The generation of Measurement and Status Packets (MSP) is based on the same terminology. Each hardware function (e.g. an analogue signal conditioner) will perform a scan cycle based on synchronized clocks and sends the converted information via serial link to hardware buffer accessible by firmware. The firmware will generate ODN packets based on the information available in the different hardware buffer related to different functions (Measurement Buffer, PRU Buffers, Actuator Status register, PDE channel status register).

2.2 Time-Triggered Ethernet interface

Propulsion related commands in order to activate/deactivate thrusters or scanning of propulsion related sensor data will be issued by the Orion Flight Computer via the ODN. The ODN transports this information to the network interface switches in the CMA and the switches will distribute the information to the PDE boxes. NASA provided network interface cards of type SNIC providing three TT-GbE interfaces are installed on each PDE channel and act as subscriber to the ODN. This triple redundancy adds to the reliability of the system.

2.3 PDE On-board computer

The controller of each PDE channel is based on a LEON2 fault tolerant processor, which is based on a SPARC V8 32-bit RISC architecture and executes at 100 MHz. The CPU is radiation-hardened and applies several single event effect mitigation techniques: the architecture provides full triple module redundancy, transient filtering, error detection and correction (EDAC) for registers and external memories, and parity protection of the caches. In addition a 33 MHz PCI interface, a memory controller, and several peripherals including UARTs, general purpose I/Os plus an embedded Debug Support Unit (DSU) is provided.

The processor bus addresses an external 2 MByte (+EDAC) SRAM, Flash memories containing four independent 4 MByte software images, as well as three high-reliability anti-fuse FPGAs. The SRAM and the Flash memory are protected by LEON-internal error correction and detection (32 data + 7 EDAC bits). FPGA registers and memory blocks are mapped into the processor's I/O space. In addition to using a highly reliable radiation-hard processor device, a processor watchdog is implemented in a dedicated Supervisor FPGA.

2.4 Supervisor FPGA

The major task of the Supervisor FPGA is in general to monitor all the functions of the MPCC. In detail the Supervisor FPGA execute the following tasks:

- Hold the LEON2, the Flash-PROM and the other two FPGA in reset during the power-up sequence and releases them sequential after power-up.
- Monitors the status of the 1.5V rail, the 1.8V rail and the 3.3V rail and in case of an under-voltage event all components will be reset.
- Reset the LEON2 and the Flash-PROM in case of a watchdog event.
- Provide a sync pulse to the other FPGAs (including the FPGAs on the MASC PCBs) and a sync counter for the LEON2.
- Provide a common Alive-Trigger-Pulse for the Clock Monitor.
- Provide communication links for both Pressure Regulation Units via the serial PRU interfaces.
- Flash control functions including:
    - Write protection
    - Partition selection
    - Flash reset
    - Flash ready status
- SRAM control function: decoding SRAM enable signals from Leon-Bus addresses

2.5 Actuator FPGA

The Actuator FPGA controls the different valve driver stages on the three MDSU boards. In detail the Actuator FPGA

- Commands and monitors the MDSU valve driver stages. The task of commanding and monitoring the valve drivers is split to the Actuator FPGA and the Inhibit & Measurement FPGA. The different command and status signals are divided between both FPGAs in a way that if one FPGA drives a command signal the other FPGA reads the corresponding status signal and vice versa. This concept improves the reliability of the PDE architecture.
- Commands and monitors the two PRUs.

2.6 Inhibit and Measurement FPGA

The I&M FPGA commands and monitors the different valve driver stages on the three MDSU boards. In detail the I&M FPGA execute the following tasks:

- Command and monitor the MDSU valve driver stages. Valve driver control is split between the Actuator FPGA and the I&M FPGA. (see above).
- Command and monitor the MASC boards via serial differential links.

2.7 Interface with Pressure Regulation Unit

Since the PRU of the ESM does not provide a direct ODN interface the PDE acts as a router for communication between the flight computers and PRU channels.

An ESM-specific implementation of the High-Level Data Link Control (HDLC) protocol is used for point-to-point serial data links between individual PDE and PRU channels. Data integrity across the entire data chain is guaranteed on several levels: the HDLC interfaces provide CRC protection at the link level; the PDE-internal data buffers implement error detection and correction; the actual data frames exchanged between CM flight computers and PRU are transported transparently by the PDE to provide end-to-end CRC protection.

3 PDE SOFTWARE

3.1 Software criticality

A major step in the design of the PDE SW development has been the determination of its criticality [4]. In a nutshell catastrophic consequences and/or loss of life might be a possibility when a malfunction occurs, thus the highest criticality is applied. In ESA terminology this is category A, NASA calls it class A.

3.2 Software architecture

The software architecture is kept as simple as possible and is implemented as a stand-alone modular system for the LEON processor:

- no operating system,
- no synchronous interrupts
- no 3rd party libraries
- no board support packages or automatic start-up code
- reuse of a well-proven boot loader for basic processor set-up and PDE software loading during start-up
- a simple control main loop, synchronized by events derived from the Time-Triggered Ethernet, with a strict processing schedule and a clearly predictable timing behavior

The result of the software architectural design is a simple main loop with two 25 ms sub-cycles (one 'command', one 'measurement' sub-cycle) and an overall cycle time of 50 ms, visualized in Fig. 4.

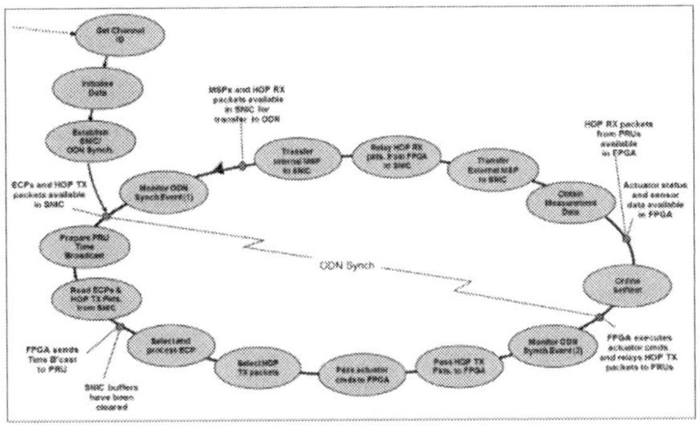

Fig. 4. PDE software main loop

The start-up code or bootloader is written in assembler all other modules of the PDE software are coded in 'C' based on the venerable GNU tool chain used in various space projects.

3.3 ISVV

A result of the Category A classification is the need to execute Independent Software Verification and Validation, ISVV [5].

The PDE SW ISVV has focused on design analysis, code analysis and validation.

Design analysis has been limited to the verification of timing & sizing budget based on the results by the software team. Code Analysis has been performed and yielded 178 potential defects and some recommendations for improvement. The majority of the identified issues were false alarms by Polyspace, the tool used. A review led to just a few recommendations to be followed and implemented.

Finally testing of non-nominal functions, including stress tests, robustness tests, tests at the boundaries, tests with invalid inputs and test extending beyond the domain defined by the requirements were executed in a real environment. No show stopper was identified albeit two issues with distorted timing were detected, analysed and considered to be within limits.

4 ERROR HANDLING

The MPCV-ESM system specification requires that the MPCV-ESM propulsion subsystem shall be single fault tolerant. This means that the occurrence of single fault inside the propulsion subsystem shall not degrade the performance of the propulsion subsystem of the MPCV-ESM.

As already stated each PDE channel will be equipped with a SNIC being the interface to the ODN data communication network. Data transportation errors on the ODN will be handled by the ODN autonomously without interaction with the PDE interface controller.

Data validated by the SNIC will be transferred from the SNIC memory to the PDE interface controller memory area by the PDE interface controller. DMA mechanisms provided by the SNIC will not be used, because the PDE interface controller shall master the data flow inside the PDE channel. Each command package transported in the MAC-Payload area of the ODN datagram will be protected by additional CRC checksum covering the whole MAC-Payload section, because the ODN datagram checksum is not visible to the PDE interface controller.

A missing command in the ODN schedule will *not* be seen as a catastrophic failure requiring an immediate shut-down of the affected PDE channel. Flow control valve actuators will be switched off, Latch valve and Ball valve activations started in previous PDE command cycles and the Thrust vector control output voltage for pitch and yaw will keep the previous commanded output values in case of a missing command. The MSP package send right after the slot of the missing command will reflect which command source has been executed or that no command has been executed in the previous command slot if all source CRCs have failed.

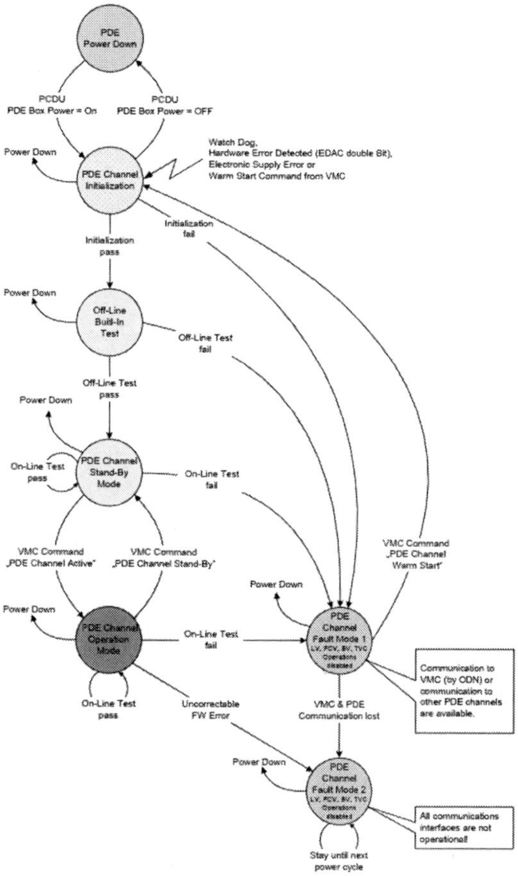

Fig. 5. PDE operating modes

5 NEXT STEPS: QUALIFICATION TESTING

Currently two flight boxes are being assembled and environmentally tested. The FPGA code is already accepted by ESA and NASA. The PDE software is integration readiness tested, a review is pending as of this writing, only formal aspects have to be closed out.

Qualification campaign on PDE level is starting in autumn 2017.

EM PDEs are deployed to the avionics test platforms, both in Europe for test of the integrated functions at ESM level, and in the US to test the end-to-end chain with the ESM and the CM including the Orion Flight Software.

6 FLIGHT

The first uncrewed flight of Orion driven by the ESM is expected in 2018 and will go to the moon and back. A crewed mission is expected two years later.

REFERENCES

[1] O. Boudillet, D. Berthelier and D. Dalemagne, "Category A software development for the ATV", DASIA 2006.

[2] M. Bottacini, J. Grantier, M. Gronowski and Bill Johns, "The European Service Module Contribution to the Orion Program", IAC 2015

[3] Jan-Hendrik Meiss, Markus Jaeger, Matthias Gronowski, Thierry Kachler, and Kevin Dickens. "Evolution and Status of the Orion-ESM Propulsion Subsystem", AIAA SPACE 2016, AIAA SPACE Forum, (AIAA 2016-5622)

[4] Antonio Preden, Jens Kaschner, Felix Rettig, Michael Rodriggs, "Critical Software For Human Spaceflight - The Equipment Software for Orion Propulsion, DASIA 2017

[5] ESA Guide for Independent Software Verification & Validation, Issue 2, 29.12.2008

DESIGN FOR MINIMUM RISK APPROACH FOR ARIANE 6

Thierry Garnier

ArianeGroup, 51/61 route de Verneuil – BP71040, 78131 Les Mureaux Cedex, France
Email: thierry.garnier@ariane.group

ABSTRACT

Safety requirements applicable to Ariane 6 are expressed by French Space Operation Act. There are failure tolerance requirements - catastrophic hazards shall be controlled by two failure tolerance - and probabilistic requirements. Some catastrophic hazards cannot be controlled using failure tolerance. Where failure tolerance is not the appropriate approach, hazards shall be controlled by Design For Minimum Risk (DFMR).
For Ariane 6 Launcher System, ArianeGroup is responsible of the design rule choice, so there is no a priori approved standard by the safety authority. It is needed to define areas for DFMR approach.
This paper presents DFMR approach for Ariane 6 Launcher System in the new context of this program.

1. CONTEXT

Ariane 6 is the new European launcher with two versions A62 and A64. Its performances to geostationary transfer orbit will be 5 tons (A62 with 530 t weight at lift-off) and more than 10,5 tons (A64 with 860 t weight at lift-off).

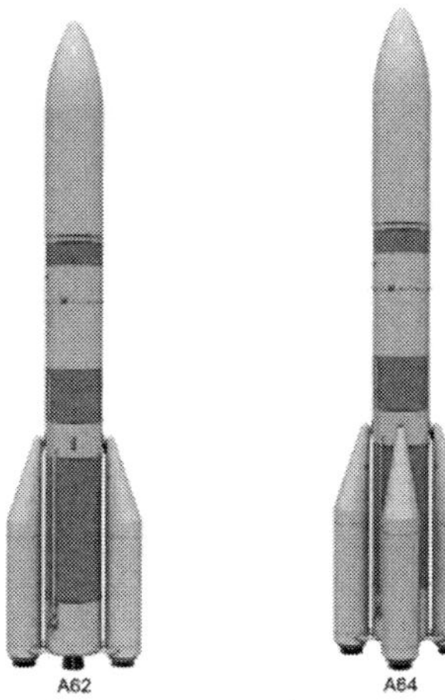

Figure 1. Ariane 6 versions

Ariane 6 development is in progress. First flight is planned in 2020.
European Space Agency (ESA) is in charge of Ariane 6 Launch System (Launcher System and Launch Base). ArianeGroup (former Airbus Safran Launchers) is prime contractor and design authority for Ariane 6 Launcher System. As design authority, ArianeGroup is in particular responsible for requirement derivation, design rules and requirement verification. ESA has technical insight activity.
Main safety requirements are expressed by decrees of French Space Operation Act. Safety requirement verification is submitted to French national space agency CNES.
Successful safety file submission is not the sole need for safety analyses. Safety is part of system engineering as for other performances and there is an ArianeGroup internal need of precise rationales for safety requirement derivation and a need of clear rules for requirement verification. Sound system engineering approach is part of the future competitiveness.

1.1. Safety Requirements

Main safety requirement for ground operations is Failure Tolerance (FT). CNES decree [1] gives failure tolerance requirements. Two severity classes are defined.

- (1) Catastrophic hazards: loss of life or disabling injury.
- (2) Critical hazards: loss of assets, severe environmental damage, or non-disabling injury.

Failure tolerance design requirements are:

- 1FT for activity with potential critical consequences, Fail Safe (FS) criterion,
- 2FT for activity with potential catastrophic consequences, FS/FS criterion.

There is an explicit exception for structural parts provided adapted standards and methods are applied.
In addition to failure tolerance, FSOA contains also probabilistic requirements for debris re-entry, fall-back of jettisoned parts of the launcher, and catastrophic scenarios relative to activities inside the space center.
Failure tolerance is not strictly limited to safety purpose: a fail operational criterion for electrical systems is part of Ariane 6 launcher design rules mainly for reliability reason. It can be mentioned that reliability contributes to launch pad safety but there is no regulation requirement about asset safety after lift-off.

1.2. System Safety Precedence

As listed in ECSS-Q-ST-40C [2], safety design order of precedence in the control of catastrophic and critical hazards is:

- Failure tolerance design;

- Design for minimum risk.

Lowest priority design solutions like automatic safety devices, warning devices, and special procedure are mentioned for information but won't be further discussed.

Failure tolerance is the basic safety requirement to control hazards for Ariane 6, in line with usual space safety policy. Some catastrophic or critical hazards cannot be controlled by using failure tolerance. Usual examples are failures of structures and structural failure of pressure vessels. Where failure tolerance is not the appropriate approach, hazards shall be controlled by Design For Minimum Risk (DFMR).

1.3. Design For Minimum Risk

A Google search shows that, for space systems mater, design for minimum risk or DFMR is a term mainly used by NASA (and IAASS). DFMR exists in ECSS-Q-ST-40 [2] since its first issue in 1996 but it was not developed in Europe.

First, we can remind NASA approach. Safety Design for Space Systems [3] states "Hazards controls based on compliance with specific safety requirements other than those utilizing fault or failure tolerance fall collectively within a category called design for minimum risk. These requirements rely on safety factors and safety margins established by analysis and test, past experience, and international safety standards to ensure an established level of acceptable risk." A DFMR definition is given in the glossary of terms of Constellation Architecture Requirements Document [4], "A process in which risks are reduced/controlled (in lieu of compliance with specific failure tolerance requirements) through specified factors of safety, material properties, or other properties inherent to the design of the part, component, subassembly, or assembly. The process includes design and certification in accordance with approved consensus standards, with design implementation and verification provisions to enhance the reliability of safety critical space systems to the maximum extent practical. Examples include structures, pressure vessels, pressurized lines and fittings, material compatibility, and flammability."

DFMR was explained during the fourth IAASS Conference [5]. IAASS 2010 presentation [5], [3], and [4] emphasize the importance of standards for DFMR. They make a distinction between DFMR when approved standard are applied and zero failure tolerance (0FT) when no previously well documented approach exists.

ECSS Safety standard [2] describes DFMR as "Hazard which cannot be controlled by compliance to failure tolerance shall be reduced to an accepted level by compliance with specific safety related properties and characteristics of the design."

For Ariane 6, ECSS sense for DFMR will be used. It is less restrictive than NASA acceptation. There will be no distinction between DFMR candidates and 0FT candidates.

In the past European launcher programs, some design choices were still based on qualitative requirements different from FT. Verifications could be based on qualitative analyses. Application of DFMR approach is an old practice for space programs even if "Design For Minimum Risk" term was seldom used in Europe.

An incentive to fix rules for DFMR approach is the increasing rigor in system engineering.

2. TECHNICAL FIELDS FOR DFMR

DFMR area is an area where hazard mitigation is not achieved by failure tolerance.

It is possible to mitigate hazards without FT but the question is "Which hazard mitigations are acceptable in a given space program?" Lessons learnt give a starting point to answer.

Well-known DFMR areas are structures, pressure vessels and pressurized lines.

Various approaches were accepted in the past to justify failure mitigation of fittings and seals but now they can be DFMR candidates.

Some areas were implicit: hazard scenarios were little developed in hazard analyses for toxicity (of non-toxic materials in usual conditions), material compatibility, and some aspects of flammability. In practice, hazard mitigation is DFMR

Pyrotechnic items are always listed as hazardous items. Unwanted pyro-ignition in normal environment without ignition signal is often listed in hazard analyses and various mitigation justifications can be found. In fact the underlying approach is DFMR.

Mechanisms and hazard mitigation is a more complex question. It is possible to look at an example of the day to day life: the braking system of our car. The design of this system is not full failure tolerant and failure may have potential catastrophic consequences. This "not simple" mechanism seems to be a DFMR candidate for car equipment and it is commonly accepted. There is no braking circuit in a launcher but an equivalent mechanism is generally not fully mastered in space systems. Currently, it would not be a DFMR candidate for Ariane 6 launcher when catastrophic hazard mitigation is needed.

A real FT is not easily applicable to software; therefore the use of mature software development and design practices reduces the risk of software error contributing to system failure. The ArianeGroup logic relies on software criticality determination, rules for development, verification, and validation.

As a result, the guideline is that DFMR approach can be used when hazard mitigation relies on well-known physical or chemical properties and phenomena and when margins can be defined.
Software is a special area: when FT is not applicable hazard mitigation relies on accepted rules but no margin can be defined with respect to physical characteristics.
The list of potential DFMR candidates remains open to systems like batteries as for JSC 20793 Rev C [6]. When consequences are critical, the list of acceptable mechanisms is also not limited to the simple ones.

3. CHALENGES FOR DFMR APPLICATION

There is a common agreement in space community to apply DFMR approach in structure area. In this situation, a global agreement on applicable rules exists from the beginning of the development.
For other areas, the responsible of the design shall propose the approach for hazard mitigation associated to his design.
Some launcher sub-system designs depend on ArianeGroup sub-contractors (considered as partners) and the prime contractor must avoid interference in partner contractual responsibility
First challenge is the number of entities contributing to safety: partner, Launcher System prime contractor and design authority ArianeGroup, ESA as customer and Launch System coherence, and safety submission to CNES. For DFMR application a common agreement shall be found in the short planning of Ariane 6 development.
Second challenge is the good understanding of "DFMR candidate" by partners. DFMR candidate doesn't means forced solution nor accepted without justification.
Third challenge is to avoid confusion by every people of the program between mitigation of all hazards for an item and mitigation for a hazard scenario. For example solid propellant won't ignite in ambient conditions (single event catastrophic scenario mitigated by DFMR) but if there is a very high temperature it will burn (hazard scenario to be mitigated by FT).
Fourth challenge is to explain that DFMR is not only margin. If an environment condition is missing in the technical specification, DFMR with high margin or FT can't guarantee safety. In that case, margin doesn't mean anything and environment out of specification is a common cause of failure.
Fifth challenge is the limitation of imposed constraints by ESA at the beginning of Ariane 6 development in order to avoid over-cost causes. The consequence is that ArianeGroup is fully in charge of the proposition of DFMR areas and associated standards/rules to ESA for Launch System coherence and to CNES safety department as safety authority.

4. CONCLUSION

It is possible to mitigate hazards in areas where failure tolerant design is not reasonable. DFMR is a way to have a structured approach in that case.
DFMR is part of the engineering and production process which aims at a safe and reliable Launcher System on ground and in flight.
We have to remember that DFMR doesn't simplify the specification process. In particular, the technical specification shall include all the environments and physical constraints the system is likely to meet. On another side, even for the best design, engineering analyses are not bounding in case of excessive process variability and lack of process control.
DFMR is a demanding approach. It encourages to simplify the design and to fix common rules. It is a way to get lower costs due to design simplification. Cost reduction with the help of DFMR doesn't mean lack of tests and verifications.

5. REFERENCES

1. Decree CNES n° 2010-1, Decree regulating the operations of the Guiana Space Center facilities.

2. ECSS-Q-ST-40C Rev.1 (2017) *Space Product Assurance – Safety*.

3. *Safety Design for Space Systems* IAASS Butterworth-Heinemann 2009.

4. CxP 70000 Rev. B, *Constellation Architecture Requirements Document (CARD)*, Appendix B.

5. Wetherholt, J. & Heimann, T.J. (2010). *Design For Minimum Risk*, 4th IAASS Conference.

6. JSC 20793 Rev. C, *Crewed Space Vehicle Battery Safety Requirements*.

PROCEEDINGS OF THE 9TH INTERNATIONAL SPACE SAFETY CONFERENCE

SESSION 23
SPACE TRAFFIC CONTROL

EFFECT OF LARGE CONSTELLATIONS ON LIFETIME OF SATELLITES IN LOW EARTH ORBITS

William Ailor (1), Glenn Peterson (2), James Womack (3), Megan Youngs (4)

(1) The Aerospace Corporation, 2310 East El Segundo Blvd, El Segundo, CA 90274, USA, william.h.ailor@aero.org
(2) The Aerospace Corporation, 2310 East El Segundo Blvd, El Segundo, CA 90274, USA, glenn.e.peterson@aero.org
(3) The Aerospace Corporation, 2310 East El Segundo Blvd, El Segundo, CA 90274, USA, james.m.womack@aero.org
(4) The Aerospace Corporation, 2310 East El Segundo Blvd, El Segundo, CA 90274, USA, megan.a.youngs@aero.org

ABSTRACT

A 2010 study found, assuming business-as-usual satellite operations and replenishment activities, that for large satellites operating in the low Earth orbit regime, a mean lifetime reduction over the next 50 years of ~13% might be expected (i.e., a satellite with a mean lifetime of 10 years in a no-debris environment would see a reduction of 1.3 years in an environment that included space debris), with much of this reduction coming from degradation of solar panels by small, un-trackable debris. Since 2010, proposals have been put forward that would add thousands of new satellites to the LEO environment within the next few years, many more than the number projected in the 2010 study. The current study reexamines the earlier prediction by assuming that over 5,000 new small satellites are operating in high LEO orbits, consistent with proposals announced by commercial companies. As with the 2010 study, this projection includes the effects of collisions of both tracked objects, objects greater than 10 cm, and small debris down to 1 mm—a size that the 2010 study showed could "sandblast" and reduce the power output of solar panels—and adds changes in the debris environment due to the previously un-modeled new satellites colliding in their constellations and with other objects as they undergo constellation replenishment activities. Primary results are that satellites being disposed from or added to the new constellations and debris associated with collisions involving these objects could potentially double the reduction in the mean operational lifetime of satellites operating in LEO. As in the 2010 study, the degradation of solar panels due to small debris impacts is a significant factor in this degradation, with the mean lifetime reduction increasing from the 13% predicted in the earlier study to as much as 60% in some cases (i.e., a large satellite with a mean lifetime of 20 years in a no-debris environment could have its mean lifetime reduced by as much as 12 years). Results suggest that minimizing collisions during replenishment activities and making solar panels and space hardware more resilient to small debris impacts will be increasingly important in the future.

1. BACKGROUND

Fig. 1 shows the evolution of the population of orbiting objects larger than about 10 cm since the beginning of the space age. Objects classified as "debris" represent the largest subset of the total population, with objects classified as "spacecraft" totaling fewer than about 4,000. There are about 1,300 satellites currently operating, with about 900 of those operating in low Earth orbits (LEOs), orbits where the satellites remain lower than 2000 km above the Earth's surface. In addition to the tracked objects, there are estimates that as many as 8000-10000 additional debris objects larger than 10 cm also reside in LEO orbits. These are joined by an estimated 500,000 objects in the range 1 to 10 cm in size and millions of still smaller objects.

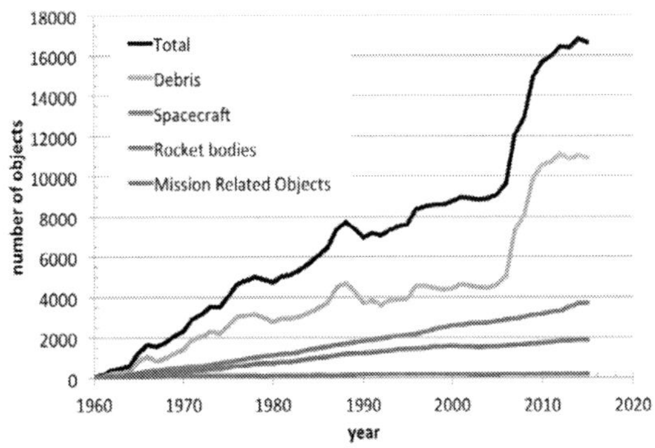

Figure 1. Population of tracked orbiting objects.

2010 Projection and Consequences of Debris Environment

Two 2010 studies [1,2] examined the consequences of operating and maintaining constellations of three different satellite designs at 850 km, near the peak of the 2010 LEO flux environment shown in Fig. 2 for 50 years. Satellites in each constellation were assigned lifetimes typical for LEO satellites and were replaced when missions were ended by normal parts failure (the baseline case) or suffered normal parts failures and were exposed to the debris environment. In the latter case, spacecraft were removed and replaced when debris larger than 10 cm impacted anywhere on the satellite, where an object between 1 and 10 cm impacted a critical component, or when solar panel output fell below a sustaining value as a result of impacts by debris 1 mm and larger. The 850-km altitude is also the location where the probability of collision was the highest, as Fig. 3 shows. The studies assumed satellite launch rates typical for the 10-years preceding the studies and future collision rates between two orbiting objects of roughly two to three collision events every ten years, with each event similar in effect to the 2007 Chinese ASAT test and the 2009 Iridium/Cosmos collision.

Given these assumptions, the flux of objects in the 1mm to 1 cm, 1 cm to 10 cm, and >10 cm size ranges would vary as shown in Fig. 4. The periodicity in the flux shown in the figure resulted from the decay of objects in low orbits being influenced by the changing F10.7 cm flux during the 11-year solar cycle.

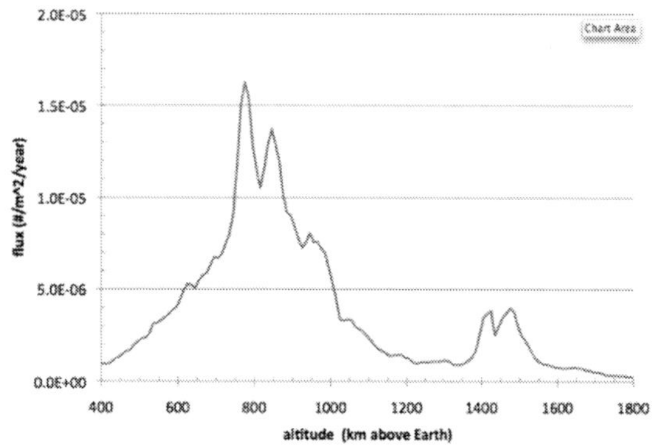

Figure 2. Estimated flux of tracked and untracked objects in LEO 2010.

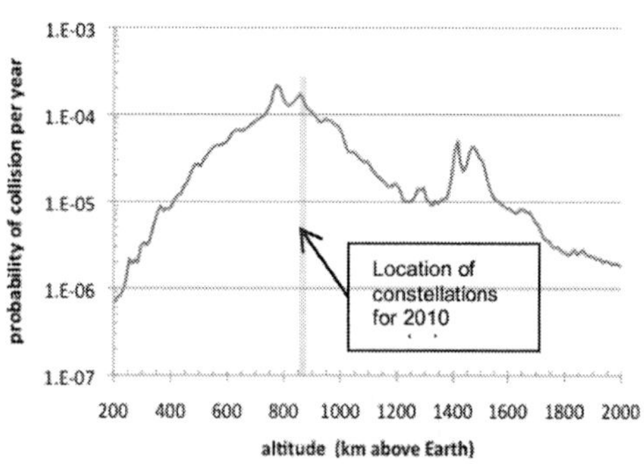

Figure 3. Probability of collision per year in LEO for 2010 studies.

The 2010 study considered the effect of this long-term debris environment on the lifetime of three different satellite design models and estimated the cost of replacing satellites that failed prematurely due to collisions with space debris. The designs, a small government satellite, a medium-sized commercial satellite, and a large commercial satellite, all had extended solar panels, with the government satellite's solar panels having the most strings.

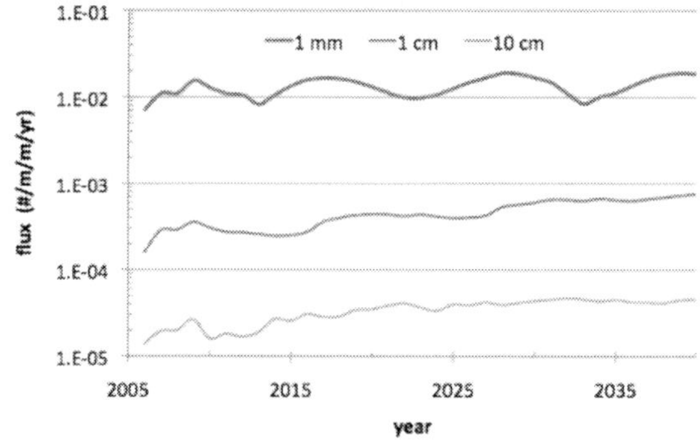

Figure 4. Predicted debris flux at 850 km for next 30 years from 2010 studies.

Table 1 compares the mean lifetime of the three satellite designs in the 2010 study given:

- Impacts by object larger that 10 cm or impact of 1-cm objects on a critical component within the satellite
- Impact by all objects including objects smaller that 1 cm; and
- Satellites experiencing no debris impacts.

Table 1: Mean lifetime for three satellites designs.

Satellite Type	Mean Lifetime (No Debris)	Mean Lifetime With Fatal Impacts Only (% reduction from No Debris)	Mean Lifetime (All Impacts) (% reduction from No Debris)
Small	8.97 years	8.50 years, 5.2%	8.18 years, 8.8%
Medium	12.26 years	11.55 years, 5.9%	10.69 years, 12.8%
Large	5.67 years	5.53 years, 2.5%	5.46 years, 3.7%

The studies concluded there would be a relatively small decrease in the mean satellite lifetime due to operating in the debris environment for the next 30 to 50 years based on the then-projected environment. A primary driver in the lifetime reduction was the solar panel degradation due to impacts by small debris--impacts that gradually lowered the power provided by the solar panels until power dropped below a critical value, ending the satellite's mission. These effects resulted in a 3.7% reduction in the mean lifetime of the large satellite to a 12.8% reduction for the medium satellite. The large satellite was characterized by a solar panel design that was the most robust to impact by small debris. Details on the designs of the satellites, etc., are provided in the references.

1.1 Proposed Commercial Constellations

The evolution of technologies that enable fabrication of smaller, lighter, less expensive, but still very capable satellites portends major changes in the LEO environment not anticipated in the previous long-term studies. Recent proposals would take advantage of this small-satellite revolution by creating large constellations of these satellites to provide new and enhanced services. Tab. 2 gives representations of five new constellations that have been proposed, and Tab. 3 gives notional designs for satellites in each constellation. Details on these assumptions are provided in [3].

Table 2: Conceptual designs for five proposed constellations.

Constellation	Total Satellites	Orbit Altitude (km)	Orbit Inclination (deg)	Number of planes	Satellites per Plane
1	4080	1100	87	48	85
2	720	1200	88	18	40
3	120	1800	89	6	20
4	100	651	97.9	5	20
5	30	576	97.8	5	6

Table 3: Conceptual satellite designs for each constellation.

Constellation	Lifetime (years)	Satellite Area (m^2)	Satellite Diameter (m)	Satellite Mass (kg)
1	6	0.35	0.77	50
2	6	2.0	1.7	150
3	12	10.5	6.9	700
4	5	0.16	0.5	5
5	5	0.28	0.7	100

As is evident, should these constellations be realized, the number of operating satellites in the LEO regime would dramatically increase--from ~900 now to over 6000. Maintenance of these constellations would add large numbers of objects to that population as ageing satellites are replaced and satellites are disposed at their end of life, with some of these operations moving the replacement satellites slowly through orbital altitudes where other operational satellites reside. Satellites that failed in a constellation were not replaced, but were added to the debris background at the constellation's altitude. And of course, any collisions of dead satellites with operating satellites within the constellations would add to the object count.

1.2 New Debris Environment

Reference [3] describes the **Baseline Flux** for this study, which was created by projecting the 2015 flux into the future assuming business as usual launch trends from 2005-2015 and includes the presence and continued replenishment of the then-existing 66-satellite Iridium constellation at ~778 km, the 42-satellite Globalstar constellation at ~1415 km, and the 36-satellite Orbcomm constellation at ~785 km. Each of these constellations is undergoing upgrading to the next generation of satellites; the future replenishment was constructed to reflect those upgrades (i.e., different physical satellites and possibly slightly different orbits).

Reference [3] also describes the Future Constellation Model (FCM). The **FCM Flux** used for the present study was developed assuming the future constellations were in place in 2015, maintained according to the launch schedule given in Tab. 4, and propagated into the future in the similar fashion as the Baseline model.

Fig. 5 repeats the 2010 prediction of the future debris environment at 850 km altitude shown in Fig.4, and adds a figure showing the 2015 Baseline Flux at that altitude for comparison given the additions of the large constellations. Both predictions were made using the Aerospace Debris Environment Projection Tool (ADEPT)[4, 5] (note that the references also developed population projections that included CubeSats, which were not considered here). As seen, if the new constellations were in place today, in 2025 the flux of the smallest debris fragments would be about $1.3 \times 10^{-2}/m^2$/year, very close to the earlier predictions, but in the following ten years would grow to ~$5 \times 10^{-2}/m^2$/year or approximately four times larger, and would remain at that level for the prediction period. The most dramatic growth is in the flux of the small particles and, given the results of the 2010 study, the most pronounced effect on satellite lifetime might be due to the reduced lifetime of solar panels.

Table 4: Conceptual launch schedule

Constellation	Satellites per Launch	Launches per year	Launch Vehicle
1	85	8	Falcon
2	40	3	Falcon
3	10	1	Falcon
4	20	1	Minotaur
5	6	1	Minotaur

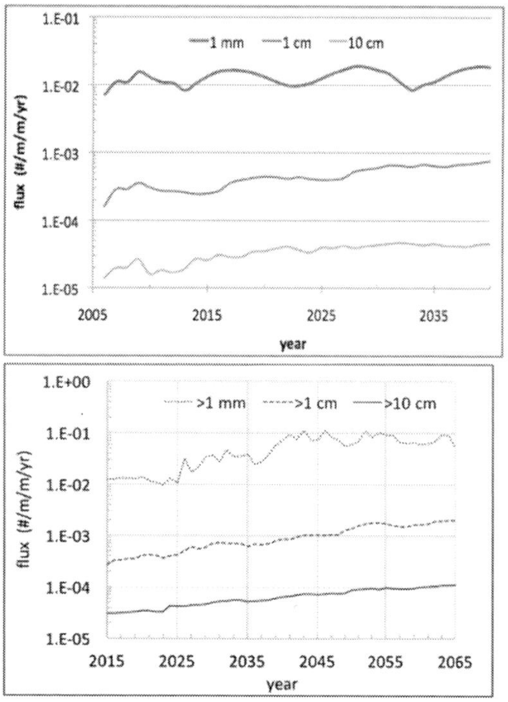

Figure 5. Baseline flux at 850 km (left) from 2010 study [1, 2]; 2015 Baseline Flux at 850km (right).

2.0 DESCRIPTION OF STUDY

2.1 Objectives

The primary objective of this study is to update the 2010 study given the addition of the large constellations and their effects on the overall debris population.

2.2 Target Satellites

Consistent with the 2010 study, models of small, medium, and large spacecraft were used as "targets" for the debris environment, with the minimum size bound of satellites set at 50-kg launch mass.

Fig. 6 shows an example of the generic satellite. The blue panels represent the solar arrays, the red boxes represent the "critical payload components" that exist inside the spacecraft bus. These critical components were defined such that a debris strike by an object 1 cm or larger to any of these payload elements would make the satellite unable to perform its primary mission, and it would be replaced.

In orbit, the spacecraft is oriented such that the Nadir face is –Z and the ram face is +X. Additionally, the solar arrays were assumed to be canted at 45 degrees to account for the variation in their angle throughout a typical orbit.

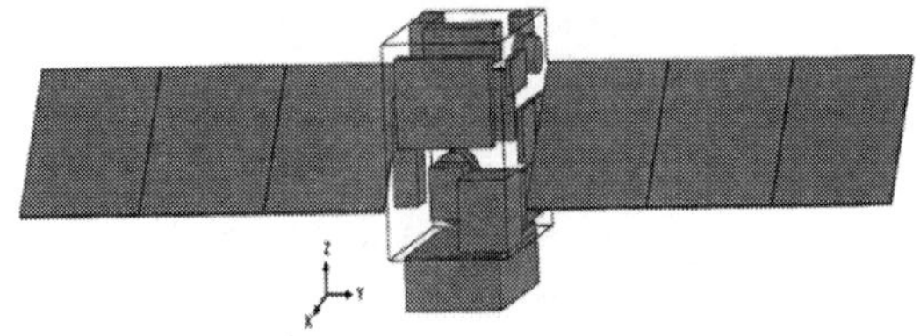

Figure 6. Representative Spacecraft

Front, side, and top views (Fig. 7) for each representative spacecraft were used to calculate the projected areas of the critical components and the total percentage of each side that could be considered critical. The total projected area for each of the spacecraft bodies and their solar arrays are listed in Tab. 5.

The debris field to which the target satellites was exposed was grouped into particle sizes of 1mm to 1cm, 1cm to 5cm, and greater than 5cm. The projected area of criticality was only used for debris of 1cm to 5cm and represents the percentage likelihood that debris in that size range would disable the spacecraft. Smaller debris (<1cm) was assumed to have no consequence should it impact the spacecraft body, and debris larger than 5cm would terminate all satellite operations no matter where it strikes the spacecraft. This can be seen numerically in Tab. 6. The solar arrays had different definitions of strike likelihood and performance injury based on particle size, which can be found in Tab. 7.

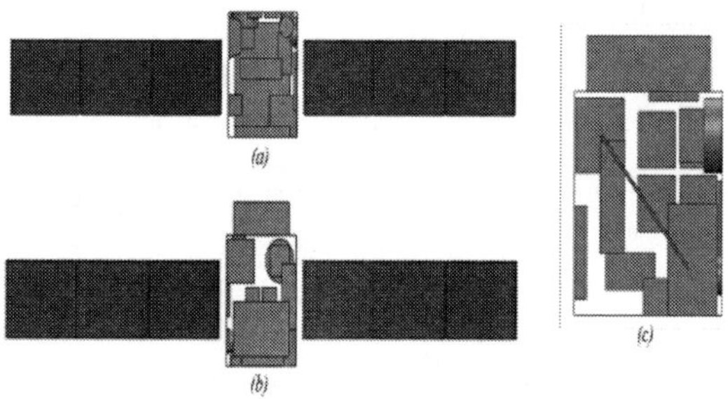

Figure 7. Top (a), Front (b), and Side (c) Views of the Representative Spacecraft.

Table 5. Exposed Areas, Solar Panel Strings, and Mean Lifetime for Each Satellite Type.

Satellite	X Total Area, m^2	Y Total Area, m^2	Z total Area, m^2	Solar Array Area, m^2	Spare Strings	Mean Satellite Lifetime, years
Small	0.44	0.39	0.29	3.04	3	12
Medium	0.74	2.06	4.87	9.6	4	16
Large	4.79	10.44	6.96	13.5	5	20

Table 6. Critical Area Percentage for Three Satellite Designs (without Solar Arrays).

Particle Size	Small, %			Medium, %			Large, %		
	X	Y	Z	X	Y	Z	X	Y	Z
1mm-1cm	0	0	0	0	0	0	0	0	0
1cm-5cm	86	82	94	77	74	87	84	64	80
>5cm	100	100	100	100	100	100	100	100	100

Table 7. Effects of Debris Impact on Solar Array Performance.

Particle Size	Likelihood and Damage Description
>5cm	100% chance of power loss included in fatal impacts
1cm - 5cm	50% chance of negligible power loss
	40% chance loss of 1 string -- 2% power loss
	5% chance loss of 2 strings -- 4% power loss
	4% chance of connector or harness strike -- 25% power loss
	1% chance of root connector or yoke strike -- 100% power loss
1mm - 1cm	88.4% chance of negligible power loss
	10% chance loss of 1 string -- 2% power loss
	1% chance loss of 2 strings -- 4% power loss
	0.5% chance of connector or harness strike -- 25% power loss
	0.1% chance of root connector or yoke strike -- 100% power loss

1.3 Target Spacecraft Lifetime

The satellite reliability model used for this study used a bus/payload reliability model and a solar array model to assess the effects of space debris impacts on the reliability of the spacecraft and solar panels. The annual probability of debris strikes (flux x area) was used to modify the bus/payload model, and a Monte Carlo simulation was used to estimate the number of string failures on solar arrays as a function of time based on the annual probability of debris strikes. The satellite was assumed to end its primary mission and require replacement when the number of spare strings dropped to zero or the panel experienced a total power loss. Modified bus/payload and solar panel models were statistically combined into a single model for spacecraft reliability in the debris environment.

2. RESULTS

The results are presented as probability of impact, probability of survival, and percent reduction of the mean lifetimes of each of the three satellites when exposed to the environment present on their launch date and beyond. For example, Fig. 8 shows the probability of impacts to the solar array (top figure) and the bus/payload (lower figure) of a large satellite placed in an 850-km orbit on a given year up to 2075. Fig. 9 shows the same information for the large satellite exposed to the FCM Flux.

Fig. 10 shows the probability of the large spacecraft with solar panels surviving as a function of its launch date in both debris environments. The large spacecraft launched in 2020 into the Baseline environment at 850 km would have a 0.95 probability of surviving 10 years; if launched in 2040, the probability of survival drops to 0.93. If the large satellite is launched in 2020 into the FCM environment at 850 km, the probability of 10-year survival would be 0.93; if launched in 2040, its survival probability drops to 0.89.

Comparing the results given in Figs 9 and 10 shows the overall effect of the debris environment and inclusion of solar panels on satellite lifetime as a function of launch date for cases using the Baseline Flux and FCM Flux environment, respectively.

Fig. 11 shows that the lifetime reductions for the small satellite design **without solar panels** exposed to the peak FCM Flux remain less than 1% for satellites launched into LEO through 2041, and lifetime of the satellite with the largest bus/payload cross-sectional area would possibly be ~15% less if launched that same year. Note that the maximum lifetime reduction for all three satellite bus sizes is the greatest for satellites in orbit in the 750-850 km range, near the peak of the debris flux shown in Fig. 4, but well below the operational orbits of the proposed large constellations. A large satellite without solar panels launched in 2016 into the peak flux would have a lifetime reduction about 30% lower than that of a similar satellite launched into a baseline flux environment; if launched in 2041 the lifetime would be about 60% lower.

Fig. 12 shows the effect of **adding solar panels** to the satellite designs (note that the solar panel designs are more robust for the largest design, so the effects are not strictly due to the increased exposed area). With the solar arrays included in the analysis, the lifetime reductions remain small for the small satellite launched into either environment; the lifetime reduction of a large spacecraft with solar panels launched into the FCM debris environment at 800 km in 2041 would be ~60%, approximately twice that for the same satellite launched into the Baseline environment at that altitude. For our large satellite with solar panels and a mean lifetime of 20 years in a no-debris environment, its mean lifetime would be reduced to 8 years. Note that the lifetime reduction at the higher altitudes is only slightly higher for FCM environment vs. the Baseline environment, most likely due to the transition over time of debris from collisions at higher altitudes to lower altitude regions.

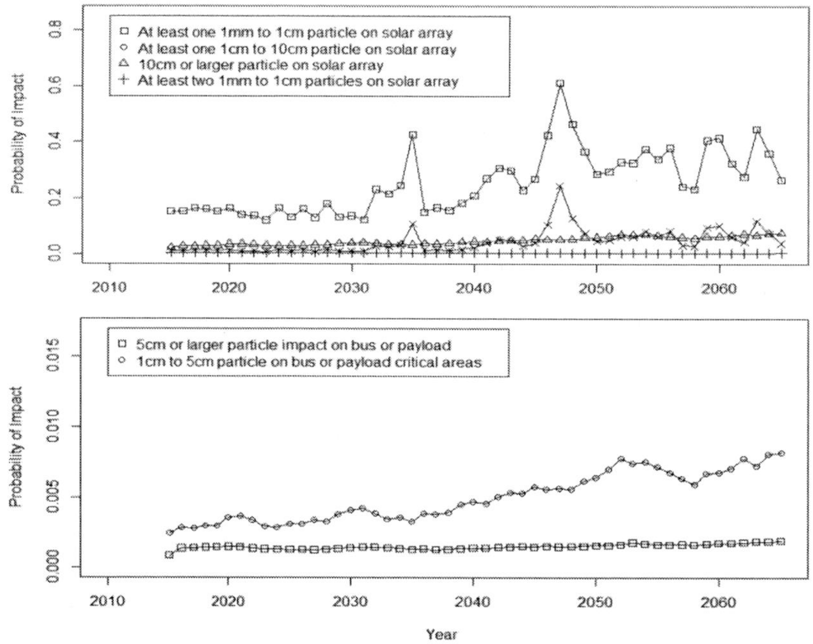

Figure 8. Probability of impact of debris particles on solar array of large satellite launched in given year into an 850-km orbit and exposed to the Baseline Flux (top: Impacts on solar array; bottom: Impacts on bus/payload).

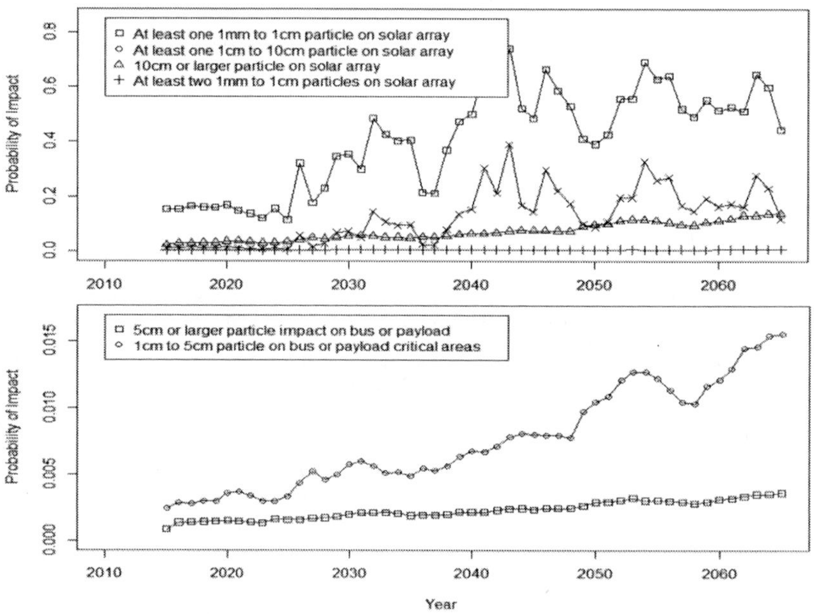

Figure 9. Probability of impact of debris particles on large satellite launched in given year into an 850-km orbit and exposed to the FCM Flux (top: Impacts on solar array; bottom: Impacts on bus/payload).

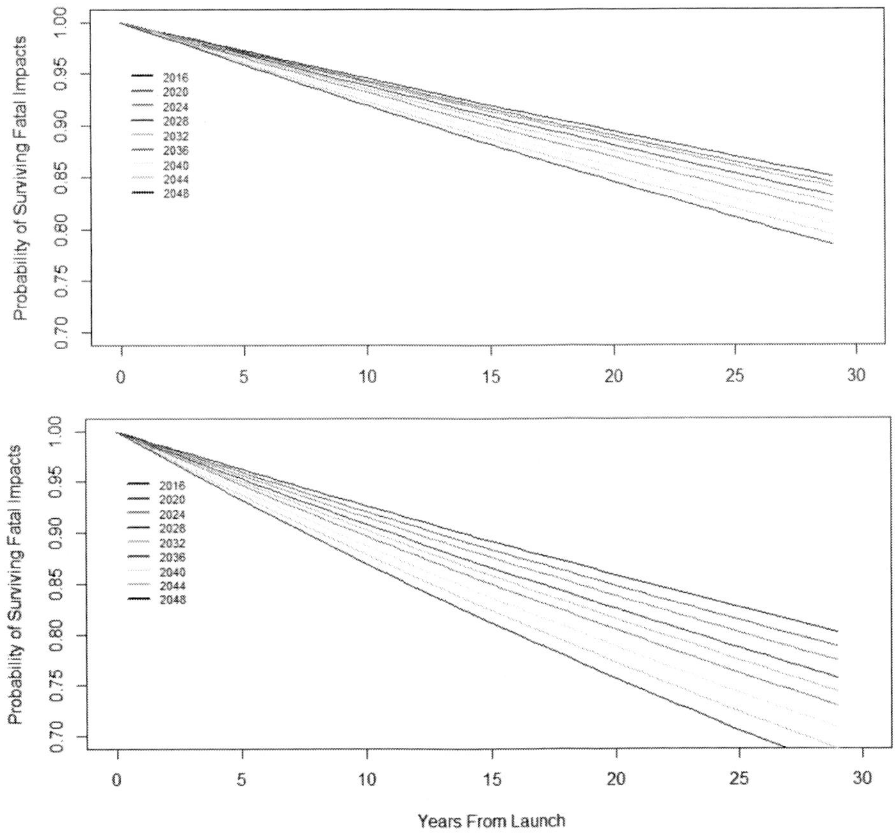

Figure 10. Survival probability for large satellite at 850 km with solar panels as function of launch date and years from launch with Baseline Flux (top) and FCM Flux (bottom)flux.

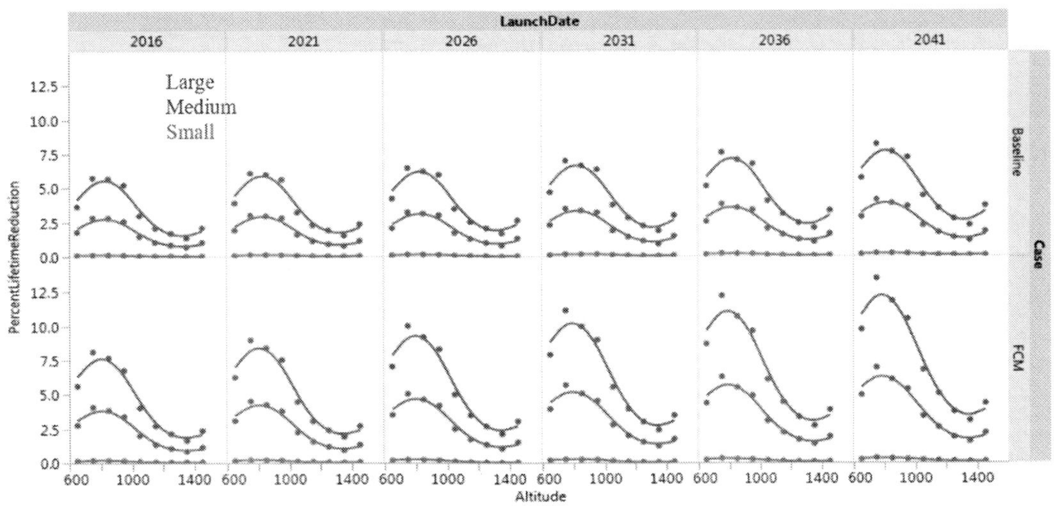

*Figure 11: Variation of lifetime of a satellite **without solar panels** in the Baseline (top) and FCM (bottom) flux environment as a function of size, operational orbit altitude, and launch date.*

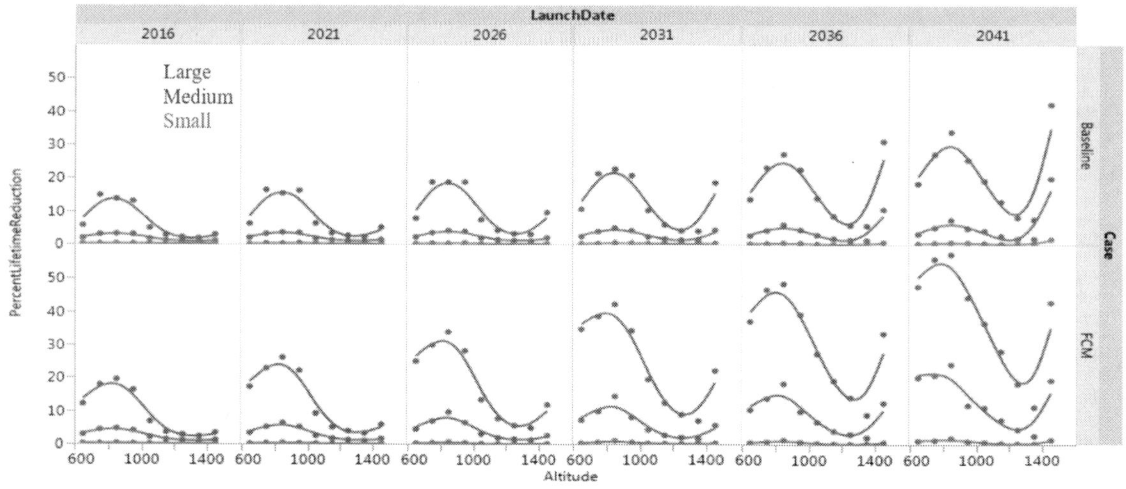

Figure 12. Variation of lifetime of a satellite ***with solar panels*** *in the Baseline (top) and FCM (bottom) flux environments as a function of size, operational orbit altitude, and launch date.*

3. CONCLUSIONS

Comparing the results for the Baseline Flux with those for the FCM Flux, it is evident that

1) Independent of the existence of the large constellations, the lifetime of satellites using current designs will gradually decrease due to the effects of the debris environment projected using the current launch rate and disposal strategies.
2) If the large constellations are added, the maximum lifetime reduction for the next 50-60 years occurs for satellites operating in orbits between 750 and 950 km, where the lifetime reduction is essentially double that which would occur if the large constellations did not exist.
3) The largest reduction in the mean lifetime of satellites used for this study occurs for the largest satellite with solar panels, where the mean lifetime drops from 20 years in a no-debris environment to ~12 years without the proposed large constellations and to ~8 years when the environment includes debris related to the large constellations.
4) The decrease in satellite lifetime at these altitudes results from debris from collisions involving satellites in and being moved to and from the large constellations accumulating in LEO and their orbits gradually decaying through lower altitudes.
5) The addition of solar panels dramatically affects the overall lifetime of operating spacecraft, primarily because of their increased exposure to debris flux and sensitivity to effects of small debris impacts.

The results of this study show that the space debris environment will have major effects on lifetime of satellites in low Earth orbits independent of the presence of the large commercial constellations that have been proposed. Adding these constellations and their related maintenance activities will amplify these effects.

Possible approaches to minimizing the growth of debris and reducing effects on satellite lifetimes include:

- Strict adherence to debris reduction guidelines and standards
- Minimizing the time from end-of-life to disposal (e.g., requiring that spacecraft be immediately disposed into the atmosphere at end of life).
- Requiring satellites and space hardware to remain operational and under control (and able to perform collision avoidance maneuvers) until their final demise in Earth's atmosphere.
- Using solar array designs that are more robust to impacts of small debris (e.g., more solar panel strings).
- Requiring that satellite designs include tracking aids to reduce the uncertainty in their orbits.
- Improving the overall quality of spacecraft tracking and conjunction warning services to enable operational satellites to avoid collisions.

Increases in the population of large and small debris in Earth orbit are inevitable, but the rate of increase will depend on how future satellites are maintained and disposed. The emergence of new constellations of satellites presents an opportunity for developing and implementing new disposal, collision avoidance, and possibly active debris removal techniques and strategies to help minimize interference with satellite operations for the long term.

4. REFERENCES

1. Ailor, W.H., Womack, J.M., Peterson, G.E., Lao, N.Y. (2010). "Effects of Space Debris on the Cost of Space Operations." IAC-10.A6.2.1-, International Astronautical Congress, Prague, Czech Republic.

2. Ailor, W.H., Womack, J.M., Peterson, G.P., Murrell, E.B., (2010). "Space Debris and the Cost of Space Operations," 4th International Association for the Advancement of Space Safety Conference, Huntsville, Alabama.

3. Peterson, G.E., Jenkin, A.B., Sorge, M.E., McVey, J.P. (2016). "Implications of Proposed Satellite Constellations on Space Traffic Management and Long Term Debris Growth in Near-Earth Environment," *IAC-16,A6,7,8,x32389,* 67th International Astronautical Conference, Guadalajara, Mexico.

4. Jenkin, A.B., Sorge, M.E., Peterson, G.E., McVey, J.P., Yoo, B.Y. (2011). "100-Year Low Earth Orbit Debris Population Model," Paper No. AAS11-410, AAS/AIAA Astrodynamics Specialists Conference, Girdwood, AK.

5. Jenkin, A.B., Sorge, M.E., Peterson, G.E., McVey, J.P. (2015). "Recent Enhancements to ADEPT and Resulting Sample Projections of the Future Debris Environment," IAC 15,A6,2,2,x28103, 66th International Astronautical Conference, Jerusalem, Israel.

ELROI: A LICENSE PLATE FOR YOUR SATELLITE TO MAKE SPACE SAFER

David M. Palmer(1), Rebecca M. Holmes(2)

(1)*Los Alamos National Lab, MS B244, PO Box 1663, Los Alamos, NM, USA, palmer@lanl.gov*
(2)*Los Alamos National Lab, MS D466, PO Box 1663, Los Alamos, NM, USA, rmholmes@lanl.gov*

ABSTRACT

Space object identification is vital for the safety and operation of spacecraft, but it is increasingly difficult and expensive due to the rapid increase in the number of satellites, especially CubeSats. An identification beacon, or 'license plate' on each object would greatly simplify the task, but radio beacons are power-hungry and can cause interference. The Extremely Low Resource Optical Identifier or ELROI [1] is a milliwatt-scale self-powered autonomous optical beacon that can be attached to any space object to transmit a persistent identification signal to ground stations. An ELROI beacon can fit in a package with the area of a postage stamp and a few millimeters thick, and requires no power, data, or control from the host object. The concept has been validated with ground tests, and the first flight test unit is scheduled for launch in 2018.

1. INTRODUCTION

Space is getting increasingly crowded. Knowing where satellites are is essential for preventing collisions, and useful for satellite owners and operators, but space objects are not included in the publicly-available catalogs unless they can be identified (or at least ascribed to a specific launch). After launch it is often difficult to determine which orbital observations correspond to which objects, especially with launches that can release more than a hundred CubeSats [2, 3], making it difficult for an operator to contact and control a specific satellite. After a satellite is identified it must thereafter be frequently and continually observed to maintain track custody, and this can be lost if there are interruptions in observations, solar flares, maneuvers, or confusingly similar satellite orbits. When a lost satellite is subsequently detected again, it can be impossible to tie it back to its original identity.

Space Object Identification (SOI) is easier if a satellite carries an identifying beacon that can be read from the ground. Ideally, this beacon would consume few resources and would last the entire orbital lifetime of the satellite. There is currently no standard or widely accepted technology to provide such a beacon, but several options are under consideration, including conventional radio beacons combined with GPS, radio tags activated by a ground signal, and passive reflectors [4]. Radio is the most developed option, but radio transmitters that can be received at orbital distances tend to be heavy and power-hungry. Furthermore, if they operate continuously they can cause radio frequency interference (RFI), so they are turned off when a satellite is decommissioned. Thus, radio beacons do not enable SOI for post-mission satellites, and they are also not suitable for passive debris objects such as rocket boosters and interstages.

The Extremely Low Resource Optical Identifier (ELROI) beacon is a new concept for a low-power optical "license plate" that can be attached to anything that goes into space [1, 5, 6] (Figs. 1 & 2). The ELROI beacon isotropically emits flashes of light with an average power of a few milliwatts. This light can be collected on the ground by a small telescope and recorded by a single-photon detector. The characteristics of the signal allow multiple background rejection techniques that permit the unambiguous recovery of the satellite identification number (ID) at very low power levels, as described below.

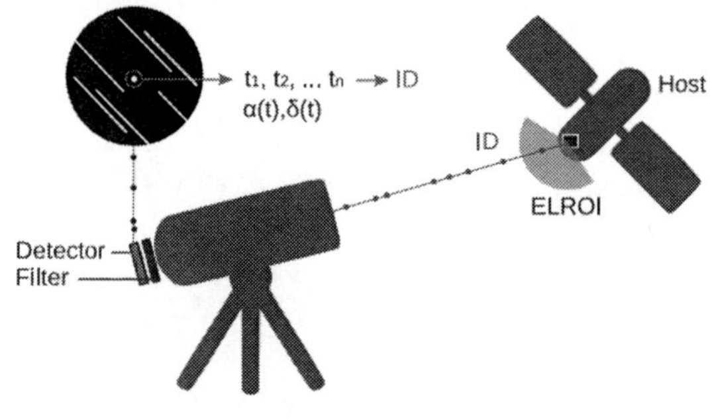

Figure 1 Overview of the ELROI system. The beacon is attached to a satellite and continuously emits its optical signal—encoding a unique ID number—over a wide solid angle. A ground telescope collects a small portion of the emitted photons, which are detected with a photon-counting sensor. A narrowband filter centered on the beacon wavelength rejects background light. The recorded data (circular inset) consists of a list of photon detection times at a tracked location

(green circle). Streaks represent background stars. The data analysis technique uses the timing characteristics of the ELROI signal to eliminate more than 99% of background photons, making it possible to read the ID in a single pass even if the signal is only a few photons per second.

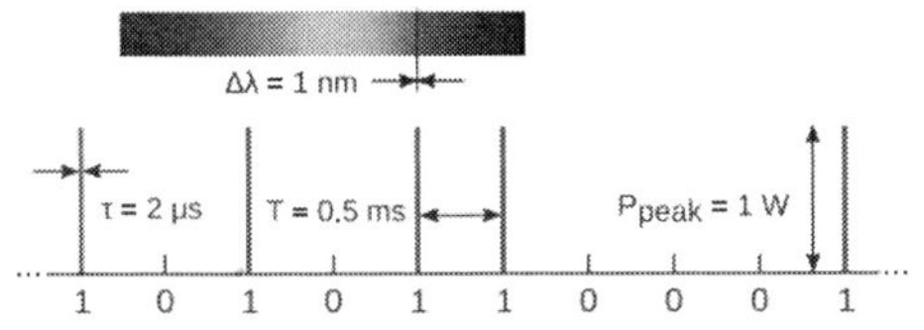

Figure 2: Representative characteristics of the ELROI signal. The spectral bandwidth of the signal is about 1 nm, and the example wavelength shown is 638 nm. The width of the optical pulses is $\tau = 2\mu s$, and the interval between pulses is $T = 0.5$ ms. If the instantaneous peak power is 1 W and half the bits contain a pulse, the average power with this duty cycle is 2 mW, and a 128-bit ID repeats every 64 ms.

This strategy allows the beacon to be low cost, small, lightweight, and low power, so it fits the $SWaP (Cost, Size, Weight, and Power) budget, even at the CubeSat level. The power requirements are low enough that the beacon can be self-powered with a small solar cell, which makes system integration simple, and allows the beacon to be flown on inert debris objects without power, communications, data, or attitude control systems. The beacon does not produce RFI, and may be operated continuously from launch to re-entry of the satellite.

The ELROI signal is intentionally open and accessible to anyone with a ground station that can be built from Commercial Off The Shelf (COTS) components. The ID of each beacon will be stored in an open registry, along with contact information for its operator and other information. This allows ELROI to be adopted as an international standard, read by ground stations around the world to assist in the worldwide problem of space traffic control (STC) and space situational awareness (SSA). The beacon can transmit additional data beyond the ID, giving satellite operators a backup channel for anomaly resolution and other diagnostic purposes. This, along with the benefits to the spacecraft operator of being able to identify their own satellite, can drive adoption of ELROI even in the absence of international norms or mandates.

The ELROI concept has been verified in ground-to-ground tests at a distance of 15 km and at expected operational received signal levels.

The first ELROI prototype has been delivered and integrated with a spacecraft (Fig. 3a) scheduled for launch in 2018.

A fully autonomous module (Fig. 3b) suitable for either attaching to a host satellite or launchable as a 1/2U CubeSat free flyer, is currently being fabricated and will be produced in quantity (6 for the initial production run) to support multiple flight opportunities as they present themselves. An optimized beacon will eventually be miniaturized to a package the size of a thick postage stamp—a few square centimeters in area and a few millimeters thick—that can be mass produced in quantity to support the global launch rate (Fig. 3c).

ELROI has been developed at Los Alamos National Laboratory (LANL), growing out of our experience both with space systems and with single-photon imaging. Initial testing will use a pre-existing ground station consisting of a COTS telescope,

a) NMTSat b) ELROI-UP c) ELROI 1.0

Figure 3: ELROI hardware development stages. ***a.*** *ELROIPC104 unit to be incorporated into a CubeSat in the PC-104 form factor.* ***b.*** *CAD Drawing of ELROI-UP unit that can operate fully autonomously when attached to host object, or as a free-flyer.* ***c.*** *Illustrative mock-up of a miniaturized ELROI unit, suitable for any LEO CubeSat or other small LEO space object.*

a Celestron 35 cm aperture telescope on a Software Bisque Paramount), and a LANL-developed single-photon imaging camera [7, 8]. However, any additional groups that wish to use their own systems to identify the ELROI beacons are invited and encouraged to do so.

In this paper, §2 describes the ELROI concept, and briefly describes the design and operations of the beacon and the ground station. §3 goes into more details about some of the design details that allow a low power signal to be transmitted, what trade-offs and optimizations are available to make this a practical system, and what else can be done with ELROI. Our progress so far in developing ELROI, with both ground tests and current and future flight hardware, is discussed in §4. In the Conclusion, §5, we discuss how ELROI can develop to become a global standard. More detailed information is in [1], which also shows how to computationally extract the ELROI ID from the ground station's data, and works through an example link budget to demonstrate the feasibility of this concept.

2. CONCEPT

ELROI uses a small optical beacon that can be attached to any object that goes into space (Fig. 1). This beacon produces flashes of light that encode an ID number, providing a "license plate" that uniquely identifies the satellite. The flashing light can be detected and read by a small ground telescope with a single-photon detector, allowing anyone to identify the satellite. The specific characteristics of the light flashes allow the ID to be read even if only a few photons are detected per second, and in the presence of background.

The optical signal is diffused to a large solid angle (almost a full hemisphere) so that spacecraft attitude control or pointing is not required. The receiver telescope tracks the satellite, using its known orbital elements. The light is spectrally filtered to select only those photons at the beacon wavelength. A single-photon detector records the arrival time of each photon, and the list of times is then analyzed to reconstruct the ID by determining which time bins contain 1's and 0's. This data analysis is described in more detail in [1].

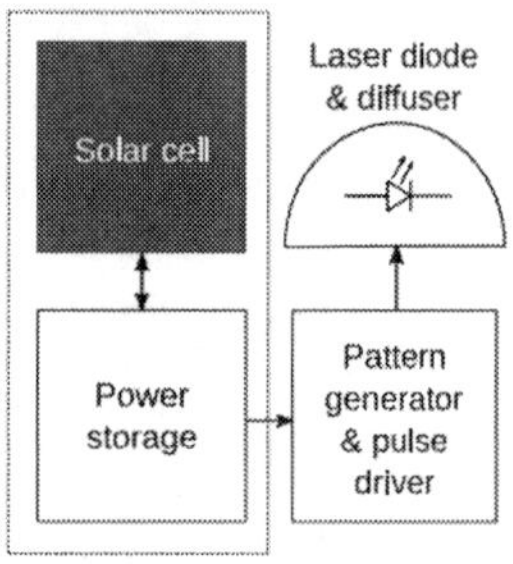

Figure 4: Components required for the beacon. A small solar cell provides power independent of the host satellite, with the battery allowing operation during orbital night. A pattern generator supplies the pulse sequence to a pulse driver, which powers the laser diode that emits the optical signal. The light is diffused into a wide solid angle so that the beacon does not need to be pointed directly at the ground station.

The combination of the beacon signal characteristics and the ground station design enable background rejection and signal extraction techniques that allow an extremely low-power signal to be recovered. Unlike a radio antenna, an optical telescope's imaging capability can reject all background sources more than a small fraction of a degree from the beacon. The spectral purity of the laser diodes allows narrow-bandwidth filters to reject almost all of the non-beacon light from the sky and the satellite itself. The use of single-photon detection and timing brings the signal into the digital domain, eliminating analog noise and allowing signal integration over the entire observation time.

The extremely low duty cycle and strict periodicity of the beacon allows a phase cut that rejects the background light that doesn't precisely match the timing of the beacon. Finally, the use of an error-correcting coding scheme for the ID makes it very unlikely that one ID will be mistaken for another, even when the values of some bits are uncertain.

2.1 System Design Overview

More detailed design considerations are discussed in §3.

The electronics required for the beacon are quite simple, as shown in the block diagram Fig. 4. The power system is mostly remarkable for its low capacity requirements compared to conventional satellite systems. The pattern generator can be a very simple microcontroller, FPGA, or ASIC. The pulse driver circuitry drives a laser diode, which provides highly monochromatic light with good efficiency. This light is diffused to cover a wide solid angle so that light reaches any ground station in its field of view without any pointing requirements placed on the host.

The ground station, shown in Fig. 5, uses a telescope to concentrate the light from the satellite and reject light from other parts of the sky. This telescope must be mounted so that it can track satellites, based on supplied orbital elements. The collected light is further filtered by wavelength, using a narrow-band spectral filter at the beacon wavelength. This rejects the majority of background light at other wavelengths (from the sky and from sunlight reflecting from the satellite).

The filtered light is then recorded with a "photon-counting detector." This is a detector that, minimally, emits a pulse that triggers a time-recording circuit every time it detects a photon. The detector may also have some position-sensitivity or imaging capability which, as we shall see in §3.2, affects the requirements on the telescope and its tracking system. The data from the photon-counting detector—a list of photon times, including both beacon photons and background photons—is then analyzed to determine the ID of the beacon, as described in [1].

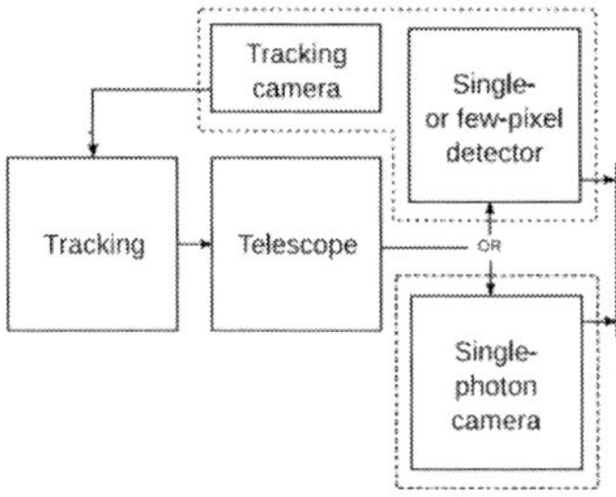

Figure 5: Components required for the ground station. A tracking system drives a telescope equipped with a photon-counting sensor. If a single- or few-pixel detector is used, tracking requirements are more stringent than for an imaging sensor, and a separate tracking camera may be required. A time-tagger records the time of each photon detection. (For a single-photon camera, the location of each detection is also recorded.) The signal is decoded by acquiring the carrier frequency, rejecting out-of-phase background photons, and assigning a value to each bit based on the number of photons detected. The resulting bit values are compared to the entries in a registry to find the best match to an existing ID.

2.2 Operations

The ELROI beacon requires a targeted observation to read an ID. It is intended as an adjunct to persistent satellite tracking operations by radar and optical methods that maintain and update a self-consistent catalog. It will typically be used for objects where the identity is unknown or questionable, such as immediately after launch when multiple satellites must be matched to multiple object detections. The existence of an ELROI beacon on an object relaxes the demands on the persistent tracking systems, because the consequences of a lost track are diminished if the identity can be recovered.

Due to sky background, the typical ground station can only operate at night. Observations will require clear skies, at least over a significant portion of the apparent path of the satellite, but atmospheric turbulence (seeing) is unlikely to have much of an effect on receiver sensitivity. The narrow field of view of the telescope will require that the satellite's orbit be known in advance. The required orbit prediction accuracy is determined by the field of view of the photon counting detector, and the presence or absence of sunlight on the satellite to allow closed-loop tracking. For larger satellites, observations in eclipse will reduce the background and allow for faster identification, at the cost of requiring accurate "blind" pointing without seeing the satellite itself in reflected sunlight.

A full implementation of ELROI for global Space Traffic Control (STC) will require a worldwide network of ground stations to provide prompt identifications for a variety of orbital geometries, regardless of the weather and other conditions at individual ground stations. Different locations, or different ground stations at a single location, can have different design choices in terms of telescope, mount and detector, to cover the diversity of situations expected.

3. DESIGN AND TRADE-OFFS

The capabilities of the beacons and the ground stations will have trade-offs that affect the total cost of the global ELROI system, both in the early days when only a few satellites carry beacons to the anticipated time when they all do.

3.1 Beacon Design

The optimal ELROI beacon signal characteristics—wavelength, peak power, pulse width and separation, etc.—will depend on trade-offs between the resources that will be devoted to each beacon ($SWaP), the resources for each ground system (primarily cost), the characteristics of the satellite, and the required reliability of identification. Some of these characteristics must be standard for all ELROI units, while others can vary from satellite to satellite.

The strictest technical requirements for an ELROI standard are merely that the beacon produce sufficient light at an agreed upon wavelength, so that a telescope with an arrow-band filter and a photon-counting detector can record the photons. Variation in the details of the timing and coding scheme can be drawn from a small set of standards, which can evolve with time as more experience is gained and more capabilities are required. (The computational power required to process a list of photon time-tags is minimal, allowing the data to be tested against each of the standards in use to find the correct one.)

Semiconductor laser diodes at the watt scale are readily available in a variety of technologies. Because this application does not require coherence, multiple independent emitters can be combined to provide the desired optical power. This allows P_{peak} to be scaled to arbitrary levels merely by adding more emitters. Larger satellites will tend to require more P_{peak} to compete against the increased sunlight reflection background, while small satellites at larger distances may use longer pulses to increase P_{ave} without requiring extra emitters.

If the beacon is a single unit mounted on a flat surface, this restricts emission to a hemisphere ($\Omega = 2\pi$ sr) which for a randomly-oriented craft leads to a 50% probability of being observable from a ground station at any given time. This can be addressed in a number of ways: the emitter may be mounted on a corner of the satellite or at the end of a solar panel to provide a wider view; multiple emitters or multiple beacons can be mounted around the spacecraft; or the restricted emission can be accepted as a factor that reduces but does not eliminate the probability of a valid ID. If, for example, the satellite is expected to be freely tumbling so that its emitter is visible to the ground station for half of a typical pass, then the aspect and attitude variation of a satellite during a given pass may be enough to provide a valid ID acceptably often.

Although Fig. 4 shows the components of an autonomous beacon, some of the functions can be integrated with other spacecraft systems. A system that taps a small amount of power from the existing spacecraft bus and is controlled by surplus capacity of a processor or FPGA could be incorporated into a spacecraft during the design stage, requiring nothing more than the laser diodes, drivers, and diffusers to be added to the spacecraft hardware. However, an autonomous beacon as a separate package can be added to a spacecraft much later in the design cycle with a much lower engineering expense.

The design of an error-correcting coding scheme that selects the ID numbers is beyond the scope of this paper. One possibility is a Bose-Chaudhuri-Hocquenghem (BCH) code, which for a 127-bit ID length would provide 2 million different IDs while being robust to up to 21 erroneous bits. Another possibility is a 32-of-128 code, which has less error correction capability but (due to the lower number of '1' bits) requires less power to transmit.

3.2 Ground Stations Design

The major design trade-offs for the ground station are among the capabilities and costs of the detector, the telescope, and its tracking system.

The sensitivity of the ELROI ID measurement relies on the use of a "photon-counting detector". That is, each photon registered by the detector must produce a discrete signal whose timing can be measured with resolution better than the pulse width τ. The detector will have a quantum efficiency, ε_{DQE}, with which it registers the photons that hit it, and a detector dark count rate, R_{dark}, which is the rate of false photon detections.

This detector can be a single pixel, or it can be an imaging detector that provides the position as well as the time of each incident photon. As discussed in [1], imaging detectors tend to be more expensive and complicated, but they simplify the task of tracking a satellite with the required accuracy.

Thus an optimal design will balance telescope and mount cost against detector cost. This balance may depend on the orbital characteristics of the satellites, the expertise and legacy equipment available to the ground station operator, and the advance of technology.

3.3 Additional Capabilities

The ID signal consists of low-duty-cycle periodic flashes confined to an in-phase time slice of width $_\tau$ in each period T. This permits additional signals to be interleaved as extra flashes out of phase with the ID bits. The additional signal must have low enough spectral power at the period of the ID signal to avoid interference with the clock recovery, but this can be achieved by a careful choice of coding scheme.

One possible use for this auxiliary data channel is to provide "black box" information for diagnosing and recovering from spacecraft anomalies. The operation of an autonomous ELROI unit will be unaffected by most failure modes of its host satellite, and so this can provide an independent low-bandwidth channel for small amounts of status information when other information is not available. The ELROI unit may have its own sensors, such as MEMS accelerometers, gyroscopes, and shock sensors. Even with no additional hardware, variations in the solar cell voltage can be analyzed to detect sudden changes in satellite attitude or spin state.

The cost in $SWaP for implementing this optional capability is relatively low, and includes an increase in complexity of the pattern generator, any interfaces or internal sensors that are required for the status information, and an increase in overall power proportional to the additional light generated to transmit the additional bits.

4. CURRENT RESULTS AND FUTURE DEVELOPMENT

4.1 Horizontal Range Test

We have verified the ELROI concept, and validated our link budget calculations, with a Horizontal Range Test (HRT). This was a ground-to-ground test at a distance of 15km using two transmitter units at a wavelength of $\lambda = 638$nm in the red with an equivalent isotropic power of $P_{ave} = 0.3$mW. A photon-counting sensor was fitted to a lens with aperture control to let us cover a wide range of conditions, including the expected operational count rates. The optical path from transmitter to detector has an atmospheric depth in excess of that from a sea-level observatory looking straight up into space.

The signal count rates measured in this test (covering a range of 0.1–40 counts/second depending on the aperture used and additional attenuation added at the transmitter) agreed with our model calculations. This gives us confidence in the accuracy of our calculations for a nominal LEO system at 1000 km using the existing ground station (predicting a signal of 3.3 counts/s).

The IDs for the two transmitters were successfully extracted from the data. When the required observations times were scaled from observed count rates to those predicted for the nominal system, we found that we could confidently read a satellite ID in 107–157 seconds (*i.e.*, well within the time available during a LEO satellites pass over a ground station), with a probability of an incorrect result of less than one in a billion.

The HRT therefore successfully demonstrates the ELROI concept and lays the groundwork for the first on-orbit test, ELROI-PC104, described in the next section. See [1] for more details.

4.2 ELROI-PC104

ELROI has been developed in a PC-104 form factor for integration into a CubeSat (Fig. 3a). It has 4 laser diode emitters projecting through two diffusers on opposite sides of the CubeSat. One diffuser has two $\lambda = 638$nm red laser diodes, at $P_{peak} = 1$W and 0.7W, from two different manufacturers. The other diffuser has an identical 1W $\lambda = 638$nm laser diode and a $\lambda = 450$nm 1.6W blue laser diode.

This unit relies on the spacecraft bus for power, and can be controlled by the spacecraft computer over an Inter-Integrated Circuit (I^2C) interface. Each laser emitter can be independently controlled with its own combination of pulse width τ, pulse interval T, and ID number (up to 128 bits). The relative phases of the diodes can be controlled so that they do not flash simultaneously, even if they have the same T.

In the absence of any commands from the spacecraft, the unit waits for 45 minutes after power is applied, then automatically powers up the three red laser diodes, each with its own ID and phase. The 45 minute delay before transmission is required for 13U CubeSats by the CubeSat standard [9]. Only the red diodes are automatically activated because the blue laser requires a higher operating voltage and more power from the spacecraft bus. In this three-diode, red-only mode, the entire ELROI unit draws only 56mW.

Many CubeSats fail after launch before providing their first telemetry [10]. Although the failure rate is declining as the community gains greater experience, the lack of feedback can prevent anomalies from being diagnosed, making recovery or learning from the experience very difficult. The autonomous start-up of ELROI-PC104 will improve this situation. If the ELROI signal is detected, it will demonstrate that the spacecraft power systems are active. As the spacecraft processor boots up and starts activating systems, it can update the transmitted ID, allowing progress to be monitored. Any failures detected by the processor can also be transmitted by modifying the ELROI ID. Other status information, such as the number of radio commands received from the ground, can be transmitted in the same way.

The unit shown in Fig 3a has been integrated into the New Mexico Institute of Technology's NMTSat, a 3U CubeSat with launch expected in 2018 [11].

4.3 ELROI-UP

A unit that can be attached to any satellite, the ELROI Universal Prototype (ELROI-UP), is currently being designed and built (Fig. 3b). This model includes its own solar cell and battery, and is capable of fully autonomous operation. It can accept commands or power from the host spacecraft, but does not require them. The unit can also be mounted in a passive mechanical structure and launched as a free-flying CubeSat in 1/3U, 1/2U or larger form factor.

ELROI-UP supports up to four laser diodes. The first test flight units will be populated with four $\lambda = 638$nm red laser diodes, at $P_{peak} = 2.5$W. Multiple codes and timings with different combinations of the emitters will be pre-programmed to allow testing at up to $P_{peak} = 10$W.

The first production run will be of 6 units at a marginal cost of \$4,000 each. The size is $9.8 \times 9.2 \times 3.1 = 280\text{cm}^3$, the mass is 300g and the power (if externally supplied instead of provided by the solar cell) is 50mW. This \$SWAP qualifies it as a Low Resource Optical Identifier. We are willing to provide these beacons to appropriate launch opportunities.

4,4 ELROI-1.0

Test results from ELROI-PC104 and ELROI-UP will allow ELROI to be standardized and adopted by the entire space community. CubeSats are the class of space object that have the most pressing need for an identification system, and are also the most constrained by \$SWaP. Thus we need to develop the lightest, smallest, cheapest ELROI beacon that can be mass-produced and attached to any CubeSat with minimal integration cost.

The size of beacon will be dominated by its largest components, which in this case will be the power system. A given system power mandates the solar cell area needed to supply it, and the battery volume required to maintain power both through the orbital night, and through extended periods when a randomly oriented inert or non-attitude-controlled satellite might keep the solar cell pointed away from the Sun.

For the LEO CubeSat beacon used as an example in the link budget in [1], P_{ave} is 2mW of optical power. Assuming a 50% conversion efficiency from electricity to light and allocating an additional 1mW for other components such as the pulse pattern generator, this gives a total system power of 5mW. We can assume a photovoltaic cell that is 30% efficient, is randomly oriented on the relevant timescales (50% of the time it is pointed away from the Sun, and the remaining time has a slant-adjusted area that averages to 50%) and is in eclipse for 50% of its orbit. These benchmark values combined with a 1.36kW/m^2 solar constant give an average yield of 5mW/cm^2 of photovoltaic cell area. LiFePO4 cells store $\sim 0.22\text{Wh/cm}^3$ of power (*e.g.*, [12]) so 0.55cm^3 of this chemistry could power the beacon for an entire day on a full charge. Thus, if the size of the beacon is dominated by the power system, this implies a size scale of $\sim 1\text{cm}^3$. Engineering margins, derating, and other considerations are likely to moderately increase this size. Fig. 3c shows a mock-up of what such a beacon could look like. Attaching one or more of these to every CubeSat would be technically and economically feasible and, given the benefits that it provides to the spacecraft operator, likely to be widely adopted.

5. CONCLUSION

ELROI is a simple, low-cost optical beacon that allows a modest ground station to provide unambiguous Space Object Identification. The beacon is cheap and small enough for use on CubeSats, and ground stations can be built around the world using commercially-available technology. It is suitable for use on everything that goes into space to simplify the important task of Space Traffic Control.

The ELROI concept has been verified in long-range ground tests that simulate the link budget and atmospheric path-length of a LEO system. A beacon has been integrated into a spacecraft for launch in the near future, and will be observed with an existing ground station. Multiple copies of an autonomous beacon that can be attached to other spacecraft with minimal integration are being built. There are no technical barriers to miniaturizing ELROI to the size of a thick postage stamp for easy application to even small satellites.

We encourage other groups to develop their own ground systems or use existing facilities to track the ELROI beacons after launch. We can also provide prototype beacons to spacecraft developers for integration into their satellites.

Based on the data from these initial flights, international standards can be developed to allow any ground station to identify an unknown satellite that has a beacon. Producing an optimal set of requirements on the beacon and ground stations will also require extensive physical modeling and simulation, vendor surveys, technology forecasts, economic and policy consideration, and value judgements to balance the competing interests that will eventually arise.

The advantages that ELROI provides to spacecraft operators, combined with the low expenditure required to incorporate the beacons in satellites, means that they can become widely used even in the absence of international norms or mandates.

Note: This conference proceedings paper incorporates material from [1], which describes ELROI in much greater depth.

6. REFERENCES

[1] D. M. Palmer and R. M. Holmes, "ELROI: A License Plate For Your Satellite," *Journal of Space and Rockets*, Submitted 2018.

[2] J. Foust, "India sets record with launch of 104 satellites on a single rocket - SpaceNews.com," 2017. [Online]. Available: http://spacenews.com/india-sets-record-with-launch-of-104-satellites-on-a-single-rocket/

[3] "Soyuz launches 73 satellites," 2017. [Online]. Available: http://spacenews.com/soyuz-launches-73-satellites/

[4] R. M. Ewart, "Enhanced Space Object Identification: Taking the Guesswork out of LEO Cubesats," *AiaaSpace2016*, no September, pp. 1–10, 2016. [Online]. Available: 10.2514/6.2016-5287

[5] D. Palmer, "OPTICAL IDENTIFICATION BEACON, provisional patent application 62/218,232," September 14, 2015.

[6] "OPTICAL IDENTIFICATION BEACON, patent application 15/232,857," August 10, 2016.

[7] W. Priedhorsky and J. J. Bloch, "Optical detection of rapidly moving objects in space," *Applied Optics*, vol. 44, no. 3, p. 423, jan 2005. [Online]. Available: 10.1364/AO.44.000423

[8] D. C. Thompson, "Imaging One Photon at a Time," Los Alamos National Laboratory, LA-UR-13-22617, Tech. Rep., 2013. [Online]. Available: http://permalink.lanl.gov/object/tr?what=info:lanl-repo/lareport/LA-UR-13-22617

[9] "Cubesat design specification rev. 13," *The CubeSat Program, California Polytechnic State University*, 2015. [Online]. Available: http://www.cubesat.org/s/cds rev13 final2.pdf

[10] M. Swartwout, "The first one hundred CubeSats : A statistical look," *Journal of Small Satellites*, vol. 2, no. 2, pp. 213–233, 2013.

[11] M. Landavazo, B. Cooper, A. Jorgensen, C. Bernson, S. Chesebrough, C. Dang, D. Guillette, T. Hall, A. Huynh, R. Jackson, J. Klepper, J. MacGillivray, D. Park, V. Ravindran, W. Stanton, C. Yelton, and A. Zagrai, "New Mexico Tech Satellite Design and Progress," *AGU Fall Meeting Abstracts*, dec 2012.

[12] "A123 Systems 18650 Cylindrical Cell." [Online]. Available: http://www.a123systems.com/Collateral/Documents/English-US/18650 2017 2page.pdf

LANDING SITE REACHABILITY ANALYSIS FOR LEO MISSIONS USING LIFTING BODY REENTRY VEHICLES

A. Gonzalez-Puerta[1], E. Mooij[2]

Delft University of Technology, Faculty of Aerospace Engineering, Section Astrodynamics and Space Missions, Kluyverweg 1, 2629 HS Delft, The Netherlands
[1]MSc Graduate Student, Email: agzpuerta@gmail.com.
[2]Assistant Professor, Email: e.mooij@tudelft.nl

ABSTRACT

This paper presents a novel reentry guidance architecture that aims to improve current mission safety by enabling zero-wait orbital aborts. To do so, an on-board guidance algorithm based on Adaptive Multivariate Pseudospectral Interpolation (AMPI) is developed, which generates mathematical approximations of the reference trajectory to be tracked by comparing the vehicle state information at the entry point against a pre-computed trajectory database that is stored on-board. This database is populated by systematically solving a number of optimal-control problems, which ultimately generate a set of reference trajectories that target multiple pre-defined emergency landing sites. Performing a reachability analysis for these sites shows that a maximum in-orbit wait time of about 10 hours can be achieved, having all sites share over 50% of their entry-window with at least one alternative site. Finally, such database supports the proposed on-board guidance, which generates reference commands in the order of a few milliseconds, thus proving the ability to run the system in current flight-hardware.

1 INTRODUCTION

To ensure the survivability of a reentry vehicle and its payload, the changes in the vehicle's altitude and velocity are carefully controlled during the descent using a guidance system. Such a system ensures that the planned nominal trajectory is followed closely and thus removes any deviations caused by external disturbances. Although reentry vehicles mostly fly a nominal trajectory, one should account for those trajectories that are required to be flown in case an unplanned event occurs. Such trajectories are referred to as abort trajectories and they must be part of the guidance design process.

Abort trajectories differ significantly from each other, where such differences are driven by the abort scenario that triggered them, the mission phase when the abort occurs and the capabilities of the vehicle flying the abort trajectory, as discussed in Ref. [1]. The guidance system discussed in this paper mainly considers orbital aborts, which are those scenarios that originate once the reentry vehicle is on a stable operational orbit. Orbital aborts begin with a de-orbit burn that puts the reentry vehicle in a descent path towards the Entry Interface Point (EIP). Once atmospheric flight begins, the vehicle is guided towards the neighborhood of an emergency landing site.

In the past, guidance systems have been designed in such a way that they accept limited vehicle state errors at a few nominal EIPs, such as those presented in Refs. [2, 3, 4]. In such systems, on-board switching from one nominal EIP to another would be limited by any preliminary mission analysis and would likely require significant in-orbit wait time to schedule a de-orbit burn that brings the vehicle to one of the limited EIPs available. For these reasons, the available guidance systems are not fully suitable for abort scenarios, where a long planning and execution time may compromise the mission and crew safety. Consequently, an adaptive abort guidance system must accept a wide range of EIPs, only limited by the geometry of the operational orbit at which the abort is called. Furthermore, such a system could be designed in such a way that global coverage is guaranteed, which would ensure that a landing site can be safely reached by immediately executing a de-orbit burn right after the abort calling.

To ensure such capabilities, the system's trajectory planner architecture must be revisited. The architecture presented in this paper is largely based on Adaptive Multivariate Pseudospectral Interpolation (AMPI), which generates mathematical approximations of the reference trajectory to be tracked by comparing the vehicle state information at the EIP against a pre-computed trajectory database that is stored on-board, as in Ref. [5]. Such architecture results in a robust guidance system that is capable of running in current flight hardware and uses proven software building blocks, such as off-board trajectory optimization and interpolation algorithms. In terms of abort trajectory-planning, AMPI requires the generation of a sufficiently large trajectory database that accounts for all possible abort scenarios. The flexibility of this architecture is only limited by the information stored in the database, allowing for multiple landing sites to be targeted and a wide range of entry conditions. Once the planned trajectories are generated on-board, these are fed to the trajectory tracker, which safely steers the vehicle during descent.

Sec. 2 presents the mathematical models that govern the reentry mechanics. Sec. 3 presents the method used to generate the optimal trajectory database that feeds the trajectory-planner together with the main features of such database. In Sec. 4, the top-level architecture that builds the abort guidance system is presented. Next, the guidance system is tested with the goal of assessing its performance. Finally, Sec. 5 summarizes all the findings of this paper and presents a number of recommendations for future research.

2 THE RE-ENTRY PROBLEM

This section presents the mathematical description of the reentry problem, including a description of the reference vehicle, the reentry dynamics, the trajectory constraints and the vehicle controls.

2.1 Reference Vehicle

The selected reference vehicle for the orbital abort mission, the Subscale Precursor Hypersonic X Vehicle (SPHYNX), was conceived by the European Space Agency (ESA) as a 1:3 scaled version of the X-38. The goal of SPHYNX was to demonstrate European reentry capabilities as well as validating the multiple technologies developed for the X-38. The development of SPHYNX concluded in 2003 once a feasibility consolidation study was completed at system level, as discussed in Ref. [6]. Tab. 1 indicates the most relevant vehicle dimensions. Note that the nose radius of the vehicle has been artificially increased to 0.5 m from the nominal value of 0.12 m. The reason behind such increase arises from the fact that the nominal nose radius is simply too small to comply with the heat-flux constraint established in Sec. 2.3. Finally, the SPHYNX vehicle is intended to perform the final leg of the descent via the use of a steerable parafoil.

Table 1: Dimensions of the SPHYNX reference vehicle[1]

m [kg]	R_N [m]	L_{ref} [m]	W_{ref} [m]	S_{ref} [m^2]	X_m [m]	Z_m [m]
550	0.5*	2.804	1.657	2.408	1.547	0.268

2.2 Dynamics

The entry dynamics that govern the vehicle motion throughout the descent are described by the following three equations, which relate the external forces to variations in the vehicle's velocity as discussed in Ref. [7]:

$$\dot{V} = \frac{F_V}{m} + \omega_E^2 R\cos\delta(\sin\gamma\cos\delta - \cos\gamma\sin\delta\cos\chi) \tag{1}$$

$$V\dot{\gamma} = \frac{F_\gamma}{m} + 2\omega_E V\cos\delta\sin\chi + \frac{V^2}{R}\cos\gamma + \omega_E^2 R\cos\delta(\cos\gamma\cos\delta - \sin\gamma\sin\delta\cos\chi) \tag{2}$$

$$V\cos\gamma\dot{\chi} = \frac{F_\chi}{m} + 2\omega_E V(\sin\delta\cos\gamma - \cos\delta\sin\gamma\cos\chi) \tag{3}$$

where V is the vehicle's groundspeed, m is the vehicle's mass, ω_E is the rotational velocity of Earth, R is the vehicle's radial position, δ is the geocentric latitude, γ is the flight-path-angle and χ is the heading-angle. The external forces along each axis of the vertical frame, F_V, F_γ, F_χ, are obtained by summing the aerodynamic forces and the gravitational forces:

$$F_V = -D - mg_R\sin\gamma - mg_\delta\cos\gamma\cos\chi \tag{4}$$

$$F_\gamma = L\cos\sigma - mg_R\cos\gamma + mg_\delta\sin\gamma\cos\chi \tag{5}$$

$$F_\chi = -L\sin\sigma + mg_\delta\sin\chi \tag{6}$$

where D is the aerodynamic drag force, L is the aerodynamic lift force, σ is the vehicle's bank-angle and g_R is the component of the gravitational acceleration pointing radially towards the center of the central body and g_δ is the component of the gravitational acceleration that lies in the local horizontal plane and points positive in a northern direction. The equations of motion are completed with the three kinematic relations shown in Eqs. (7) to (9). Such equations relate the vehicle's velocity to variations in the vehicle's position.

$$\dot{R} = V\sin\gamma \tag{7}$$

$$\dot{\tau} = \frac{V\cos\gamma\sin\chi}{R\cos\delta} \tag{8}$$

$$\dot{\delta} = \frac{V\cos\gamma\cos\chi}{R} \tag{9}$$

where τ is the geocentric longitude.

2.3 Path Constraints

Throughout this paper a number of trajectory constraints are enforced that ensure the safety of the vehicle and its payload during descent.

The heat-flux constraint defines the maximum heating rate per unit area that the vehicle can sustain as a consequence of aerothermodynamic heating. The heat-flux is computed using the hot-wall approximation, which computes a wall temperature T_w by equating the incoming convective flux with the outgoing radiative flux q_{rad} for

[1]European Space Agency, "SPHYNX Flight Mechanics", Private Communication, 2016.

an assumed adiabatic wall temperature T_{aw}.

$$q_c = \left(5.28137 \cdot 10^{-5}\right)\sqrt{\frac{\rho}{R_N}}V^{3.15} \cdot \left[1 - \frac{T_w}{T_{aw}}\right] \leq 670\ \text{kW/m}^2 \quad (10)$$

$$q_{rad} = \epsilon\sigma T_w^4 \quad (11)$$

where q_c is the convective heat-flux at the vehicle's stagnation point, R_N is vehicle's nose radius, ρ is the free-stream air density, ϵ is the emissivity of the applied material and σ is the Stefan-Bolzmann constant. Eqs. (10) and (11) can be combined together into Eq. (12). A root-finding method is used to compute the wall temperature, which is then used to obtain the convective heat-flux using Eq. (10).

$$\left(5.28137 \cdot 10^{-5}\right)\sqrt{\frac{\rho}{R_N}}V^{3.15}\left[1 - \frac{T_w}{T_{aw}}\right] - \epsilon\sigma T_w^4 = 0 \quad (12)$$

The aerodynamic load constraint defines the maximum aerodynamic acceleration experienced during descent:

$$n_g = \frac{\rho V^2 S_{ref}}{2mg_0}\sqrt{C_D^2 + C_L^2} \leq 3\ \text{g} \quad (13)$$

where n_g is the aerodynamic load, S_{ref} is the vehicle reference area, g_0 is the Earth's gravitational acceleration at sea level, C_D is the drag coefficient and C_L is the lift coefficient.

2.4 Boundary Constraints

The vehicle state at the EIP and the guidance termination point are given in Tab. 2. Since the abort mission start could occur anywhere throughout the operational orbit, the entry latitude and longitude are left free, where the entry latitude is bounded by ±52 deg as derived from the inclination of the operational orbit. As for the entry heading-angle, this is derived analytically from the entry latitude and longitude using Eqs. (14) to (16).

$$\chi_{EIP} = \pi - \cos^{-1}\left(-\cos\left(\tau_{EIP}\right)\cos\left(\frac{\pi}{2} - i\right)\right) \quad (14)$$

$$\delta_{EIP} = f\left(\chi_{EIP}\right) = H\sin^{-1}\left(\frac{\pi}{2} - \left(\frac{\sin\left(\frac{\pi}{2} - i\right)}{\sin\left(\pi - \chi_{EIP}\right)}\right)\right) \quad (15)$$

$$H = \begin{cases} 1 & \text{if } \tau_{EIP} < \pi \\ -1 & \text{elsewhere} \end{cases} \quad (16)$$

where χ_{EIP} is the heading angle of the EIP, i is the orbit inclination and δ_{EIP} and τ_{EIP} are the latitude and longitude angles of the EIP, respectively. As for the terminal conditions, these are mostly defined by the parafoil-deployment requirements provided by ESA. These require the parafoil opening to occur at an altitude between 10 and 13 km in the subsonic regime with a maximum vertical speed of 140 m/s. Note that the terminal latitude and longitude are given by the location of the targeted emergency landing site as established in Tab. 3. Finally, the vehicle must be brought within a 10 km radius from the landing site.

2.5 Controls

SPHYNX is modeled as a point-mass, meaning that rotational dynamics are not considered. Thus, the vehicle is controlled by means of the bank-angle σ and the angle-of-attack α. It is assumed that the reentry is performed in coordinated flight, consequently the angle-of-sideslip β is zero and no side-forces are exerted on the vehicle. Note that the bank-angle is limited to ±85 deg to avoid lift-down entries, whereas the angle-of-attack is limited by the regions of the $\alpha - M$ where vehicle trimming is possible. The angle-of-attack boundaries are imposed by the aerodynamics of SPHYNX, which are discussed in more details in Ref. [8].

3 LANDING SITE REACHABILITY STUDY

In the framework of entry mission planning, a reachability study can be defined as the process of determining the landing footprint of a vehicle. Performing a thorough reachability study is essential to determine the possible abort opportunities of an entry vehicle. Most importantly, the results of such a study explicitly conform the trajectory database required by the AMPI trajectory planner discussed in Sec. 4.

Table 2: State variable values at the EIP and the guidance termination point.

State Variable	Initial Value	Terminal Value
h [km]	120	10 - 13
τ [deg]	*Free*	*Landing Site*
δ [deg]	*Free*	*Landing Site*
V [m/s]	7640	M_{lim} *Dependent*
γ [deg]	-1.28	$V_{r,max}$ *Dependent*
χ [deg]	*Entry Latitude*	*Free*

Table 3: Emergency landing sites for SPHYNX.

ID	Landing Site	Latitude [deg]	Longitude [deg]
1	Azores	36.974	-25.165
2	Gran Canaria	27.933	-15.388
3	Ben Guerir AFB	32.128	-7.878
4	Perth	-31.938	115.967
5	Darwin	-12.411	130.878
6	Brisbane	-27.394	153.121
7	Sidney	-33.940	151.176
8	Woomera	-31.145	136.825

3.1 Optimal Control Problem

The reachability study proposed in this paper is completed by solving the following optimal control problem:

What are the attitude controls required to steer the re-entry vehicle from an arbitrary entry point to one of the designated emergency landing-sites, while ensuring that the trajectory constraints are satisfied?

The optimal control problem is solved using the ASTOS commercial software package[2]. Note that such problem is governed by the reentry dynamics and thus the software settings are adjusted such that the environment models, vehicle models and trajectory constraints match those described in Sec. 2.

A summary of all the trajectory constraints applied in the reachability study is given in Tab. 4. In particular, the following constraints are added:

- **Max. flight-path angle:** Limits the phugoid motion of the vehicle in the lower atmosphere. A positive value is allowed to enable solutions that use a skip entry approach to achieve more downrange.
- **Max. / min. bank angle:** Avoids entry trajectories in a lift-down configuration.

The performance index is selected in such a way that the angle of attack and bank angle controls are as smooth as possible. This was found necessary for the trajectory interpolation to work, since doing otherwise would results in discontinuous control profiles at the extremals that are too dissimilar from each other. Furthermore, discontinuous control profiles at the extremals would introduce artificial oscillations at such discontinuities. To achieve such smoothing, the performance index J_{sm} is defined as the integral of control derivatives over time, as shown in the following equation:

$$J_{sm} = \int_{t_0}^{t_f} w_{sm} \boldsymbol{q}(\boldsymbol{u}(t))\, dt \qquad (17)$$

where $\boldsymbol{q}$ is the quadrature function vector and $\boldsymbol{u}$ is the control vector. The trajectory smoothing factor w_{sm} is set to $1.0 \cdot 10^{-4}$ and was found sufficient to produce smooth trajectories for the interpolation to succeed.

[2]ASTOS AeroSpace Trajectory Optimization Software, Astos Solutions, `https://www.astos.de`

Table 4: Summary of all the optimization constraints employed in the reachability study. Here, EQ refers to equality constraints, IQ to inequality constraints, IB to an initial boundary constraint and FB to a final boundary constraint.

ID	Constraint	Value	Type
1	Entry Altitude	120 km	EQ / IB
2	Entry Latitude	Variable	EQ / IB
3	Entry Longitude	Variable	EQ / IB
4	Entry Groundspeed	7640 m/s	EQ IB
5	Entry Flight-Path Angle	-1.28 deg	EQ / IB
6	Entry Heading Angle	Variable	EQ / IB
7	Max. Distance To Target	10 km	IQ FB
8	Terminal Altitude Range	10 - 13 km	IQ / FB
9	Terminal Mach Number Range	0.0 - 1.0	IQ / FB
10	Max. Terminal Vertical Speed	-140 m/s	IQ / FB
11	Max. Heat Flux Density	670 kW/m^2	IQ / Path
12	Max. Aerodynamic Load	3.0	IQ / Path
13	Max. Flight Path Angle	1.0 deg	IQ / Path
14	Top $\alpha - M$ Boundary	Variable	IQ / Path
15	Bottom $\alpha - M$ Boundary	Variable	IQ / Path
16	Max. Bank Angle	85 deg	IQ / Path
17	Min. Bank Angle	-85 deg	IQ / Path

3.2 Trajectory Database Generation Strategy

The optimal control problem described in Sec. 3.1 generates the attitude commands that steer the vehicle from a given entry point to a chosen landing-site, in other words, a single trajectory. To complete the reachability analysis, such optimal control problem must be solved for every possible EIP and for all the landing-sites listed in Tab. 3. The reachability study is completed by performing the following approach for every landing-site:

1. **Discretize entry window:** The global entry window groups all the possible EIPs that the vehicle can encounter following the deorbitation burn, where the characteristics of the resulting grid must agree with the choice of the parameter space discussed in Sec. 4.1.
2. **Probe entry window:** A script loops through all the EIPs obtained from the discretization of the entry window and probes for a possible solution to the optimal control problem.

2.1 Compute range-to-go given the probed EIP and the targeted landing-site.

2.2 Set a constant bank-angle initial guess based on an approximation of downrange capability of the SPHYNX vehicle and the current range-to-go.

2.3 Simulate the initial guess trajectory and evaluate the proximity of the terminal point to the targeted landing-site. This is established by an area of allowable terminal points delimited by a [-100, 50] deg longitude boundary and a [-30, 30] deg latitude boundary with respect to the landing-site.

(A) *Unfeasible Distance to Landing Site:* Solution of the optimal control problem is avoided and the loop continues with the next EIP.

(B) *Feasible Distance to Landing Site:* An attempt to solve the optimal control problem is made.

(a) *Optimal Solution Found:* The guidance commands and supporting trajectory information is saved to a text file.

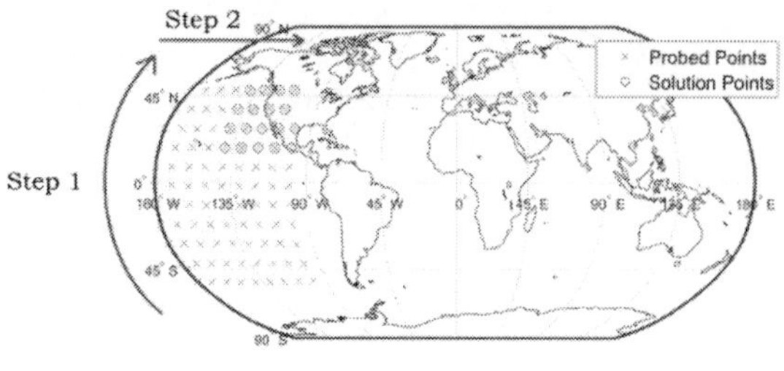

Figure 1: Strategy used to generate an entry window for an arbitrary landing-site.

(b) *No Solution Found:* If the maximum number of iterations is exceeded or divergence occurs, the solution is discarded and the loop continues with the next EIP.

As shown in Fig. 1, the discussed approach begins by probing all the possible EIP latitudes at the west-most possible EIP longitude (Step 1). This process is repeated for all the discretized longitude values (Step 2), ultimately generating worldwide entry window. In Fig. 1, probe EIPs are distributed with a spacing of 10 deg, running from -180 deg to -100 deg longitude and -52 deg to 52 deg in latitude. In this example, only 18 solution EIPs are found, which together delimit the entry window area. Note that by enlarging the area of allowable end-points, the probability of neglecting potential solution points lowers at the expense of computational time. Since all the entry-window maps are visually inspected after completion, any missing solution points can be quickly identified and executed. In summary, the feasible trajectories to a particular landing-site are determined by systematically probing all the possible EIPs, while keeping the terminal conditions the same.

The discussed approach is implemented using the ASTOS batch-loop functionality, allowing for unattended optimization tasks and thus limiting time wasted as a consequence of user interaction with the software. All the optimization tasks were run on a high-performance Xeon®E5-2687W processor clocked at 3.10 GHz, taking approximately 2 minutes to complete a single optimization task. Note that parallelizing the above strategy for multiple landing-sites achieves savings in the total computational time.

3.3 Trajectory Database Characteristics

In this section, the coverage of the trajectory database is assessed, which refers to the entry-window area from which a particular landing-site can be reached at nominal entry conditions and taking into account all of the path constraints. The entry-windows are presented in the form of "coverage maps", which display a point cloud with all the solution EIPs found in the reachability study. The entry-window is formed by the "area" which such point

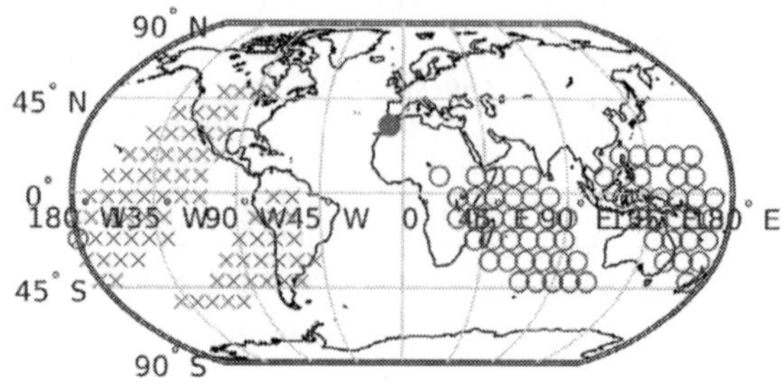

Figure 2: Entry window for the Ben Guerir site. Orange crosses: ascending cluster, blue circles: descending cluster, filled point: site location

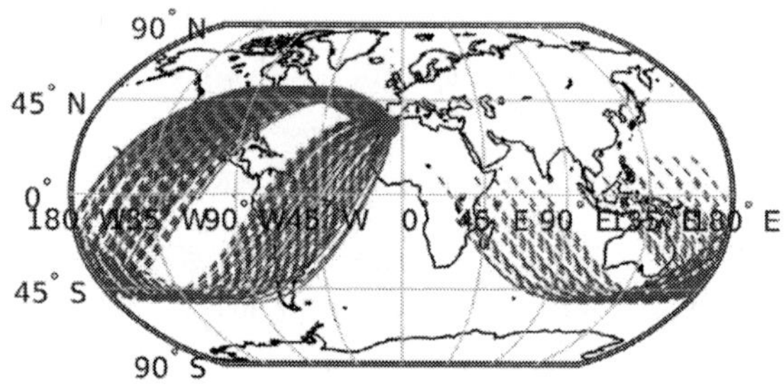

Figure 3: Groundtracks of trajectories targeting the Ben Guerir site. Orange lines: ascending cluster, blue lines: descending cluster, filled point: site location

cloud draws on the map.

For instance, the coverage map for the Ben Guerir site is shown in Fig. 2. The associated trajectory groundtracks are shown in Fig. 3, where such groundtracks originate from the EIPs shown in the point cloud. Furthermore, close examination of the entry-windows shows two EIP clusters. On the one hand, the cluster marked by orange crosses represents those EIPs that lay on an ascending leg, in other words, trajectories whose entry heading angle spans the [0, 90] deg range. On the other hand, the cluster marked with blue circles represents those EIPs laying on a descending leg. Note that for all landing-sites, one of the clusters is located significantly further west from the landing-site, meaning that the trajectories originating from such cluster travel notably further downrange. Fig. 2 also shows areas where no EIPs are found, indicating that trajectories originating from such areas would violate at least one of the path constraints or would exceed SPHYNX's downrange/crossrange capabilities.

As shown in Figs. 5 and 6, SPHYNX displays two types of entries depending on how far the EIP is from the targeted landing-site. Note that these trajectories are representative of the whole database since they exhibit entry profiles that are common to all the targeted sites.

If the EIP is close to the landing-site, the vehicle performs a regular gliding entry, characterized by a direct shallow descent towards the targeted final altitude. On the contrary, if the EIP is far from the landing-site, the vehicle performs a skipping entry characterized by a rebound in the upper atmosphere that leads to a sub-orbital flight followed by a regular gliding entry. By performing a skipping entry, the distance traveled by SPHYNX is significantly increased as shown by Fig. 6, ultimately allowing to reach the targeted landing-site. The type of entry is primarily controlled by the bank-angle, where a near full-lift-up configuration during the first 400 seconds puts the vehicle in a skipping entry. If a gliding entry is to be performed, a large bank angle magnitude is commanded shortly after the EIP, effectively reducing the vertical component of the lift vector and thus allowing the vehicle to sink into the atmosphere. The bank-angle direction controls the cross-range error whereas the angle-of-attack is primarily used to modulate the lift and drag coefficients and thus the energy dissipation.

Database Entry Coverage

Given the feasible entry trajectories found for the available landing-sites, full global time-coverage is not guaranteed. This means that a safe landing-site cannot be guaranteed irrespectively of the epoch at which the de-orbit burn is initiated. Fig. 4 shows that with the current sites, approximately 50% global time-coverage is provided, meaning that landing-sites can be targeted in 50% of the time. It is important to emphasize that to ensure full global coverage both the ascending and descending leg solutions must be available at every EIP. The area enclosed by the two black curves represents the region where no common ascending and descending EIPs are found. A rough estimate of the maximum in-orbit wait time can be made by computing time required for the groundtrack to shift across the boundaries of such area. Such approximation is given by Eq. (18):

$$t_{wait} \approx \frac{\zeta}{\omega_E} \tag{18}$$

where is the standby time in-orbit, is the required groundtrack shift and is angular velocity of Earth. In this case, the required groundtrack shift is approximate 150 deg which yields a maximum in-orbit wait-time of approximately 10 hours.

For the set of landing sites discussed in this paper, the reachability study reveals that overlapping entry-windows can be found for multiple landing sites. Any EIPs enclosed by such shared windows offer the possibility to re-target the vehicle to an alternative landing site. Such capability of the database is referred to as shared coverage, which is discussed in the remainder of this section. The shared coverage results are summarized in Tab. 5, which details the percentage of shared EIPs per site.

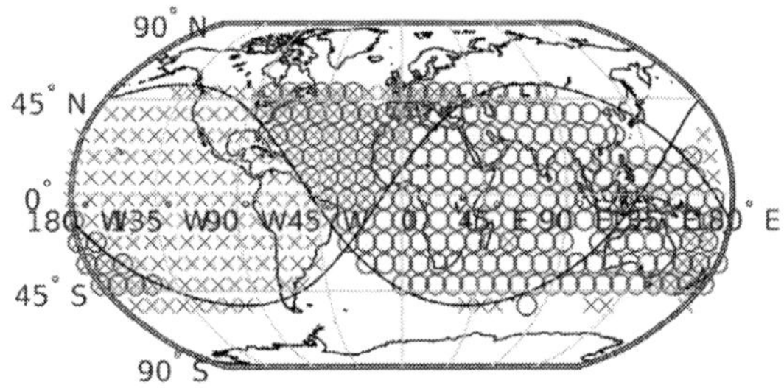

Figure 4: Global entry-window that combines the ascending and descending clusters.

Tab. 5 is split in four quadrants, where the top-left quadrant shows the percentage of shared EIPs among the northern Atlantic sites and the bottom-right quadrant shows the share among the Australian sites. The top-right and top-left quadrants give the percentage of shared EIPs across the two groups. It becomes clear that although entry-windows are expected to be mostly shared among members of the same landing group, there is a significant cross-group entry-window sharing, where landing sites of the other group can be reached by putting the vehicle in a long skip entry. It is important to emphasize that all sites share 50% of its entry-window with at least one alternative site, proving the abort capabilities of the system. Furthermore, there are entry-window areas where more than one site can be reached. This means that while clustering the sites together increases the re-routing opportunities, it significantly decreases the global coverage.

Note that the ability to re-route to a different landing site is highly dependent on the remaining energy available, which is depleted as the vehicle descends from the EIP. For example, rerouting from Azores to Perth may only be possible within the first 500 seconds of the flight. In summary, the any analysis done in this section concerns the capability of routing prior the EIP is reached. Any analysis regarding how far down the trajectory any re-routing is possible is considered beyond the scope of this paper.

Constraint Compliance

The constraint compliance characteristics of the trajectory database indicate that on average, approximately 7.4% of the trajectories stored overshoot the aerodynamic load constraint. The mean value of such overshoots ranges from 3.05 g to 3.24 g, which is moderately above the 3 g limit imposed in 2.3. The peak aerodynamic load however, can reach up to 4.26 g which is considered unfeasible. Similarly, the number of heat-flux constraint overshoots are within the same order of magnitude at an average of 7.2%. The average and peak values however, are marginally above the 670 kW/m^2 constraint. Finally, the percentage of trajectories that terminate outside the 10 km radius from the landing site is 2.6%, which is on

Table 5: Percentage of EIPs shared among landing sites.

Shared With:	Azores	Gran Canaria	Ben Guerir	Perth	Darwin	Brisbane	Sidney	Woomera
Azores [%]		42	63	14	1	23	44	23
Gran Canaria [%]	68		75	12	1	24	38	27
Ben Guerir [%]	82	61		11	2	24	45	24
Perth [%]	18	10	11		35	30	52	59
Darwin [%]	3	3	4	62		50	46	71
Brisbane [%]	34	23	28	34	32		90	72
Sidney [%]	42	22	33	38	19	58		62
Woomera [%]	28	20	22	54	36	59	79	

average lower than the percentages of aerodynamic load and heat-flux constraint overshoots. The mean overshoot distance-to-target ranges between 12.2 km to 58.1 km and the maximum overshoot value reaches up to 113.8 km, where these values exceed significantly the 10 km limit.

Overall, it becomes clear that a limited number of trajectories stored in the database exceed the limits established by the system requirements. These trajectories must be identified and re-processed, since using them to generate reference interpolated trajectories as discussed in Sec. 4 would force the guidance system to track a trajectory that violates a number of constraints in the first place. As will be discussed in Sec. 4, the guidance performance is assessed using a small subset of the database, where the trajectories in such subset are ensured to satisfy all the constraints.

4 ADAPTIVE ABORT GUIDANCE

The architecture of a guidance system can be classified under two different blocks: the *trajectory planner* and the *trajectory tracker*. On the one hand, the trajectory planner is in charge of determining the magnitude of the attitude commands and schedule these as a function of a particular independent variable. Such planning takes into account the vehicle state at the EIP, the trajectory constraints and the desired vehicle state at the terminal point. On the other hand, the trajectory tracker ensures that the reference attitude profiles produced by the planner are carefully followed during the flight, despite any disturbances or errors on the vehicle state at the EIP. Note that the guidance commands are scheduled in terms of energy instead of time, since it is a better descriptor of the entry dynamics while being a monotonic variable.

The goal of this section is to present the top-level view of the abort guidance algorithm developed for this research, while focusing on how the trajectory database design parameters impact the guidance performance. A detailed description of the mathematics behind the abort guidance algorithm can be found in Ref. [8].

4.1 AMPI On-board Trajectory Planner

The on-board trajectory planner discussed in this paper is based on the AMPI algorithm presented in Refs. [5] and [9]. As shown in Fig. 8, the AMPI algorithm receives two inputs: a set of "physics-following" trajectories and the on-board vehicle state the EIP as provided by the navigation subsystem. The set of "physics-following" trajectories refers to those trajectories that have been produced by an off-board trajectory optimization software. They are referred as "physics-following", because they have been computed by solving a set of equations of motion, and thus represent trajectories that a reentry vehicle would encounter in a potential flight scenario. The role of AMPI is to produce a "mathematical" reference trajectory via interpolation, where the "physics-following" trajectories are used as the supporting nodes of such interpolation and the on-board vehicle state at the EIP provides

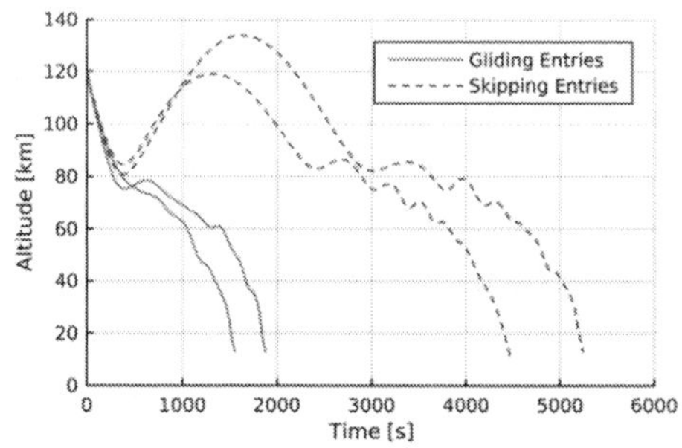

Figure 5: Altitude profiles for four example trajectories of the database that target the Azores site.

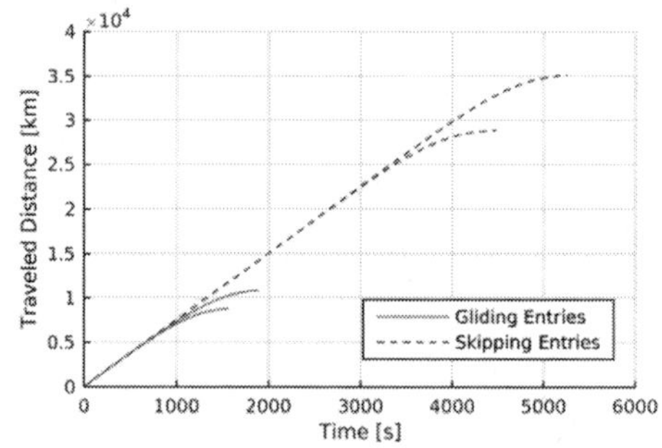

Figure 6: Traveled distance profiles for four example trajectories of the database that target the Azores site.

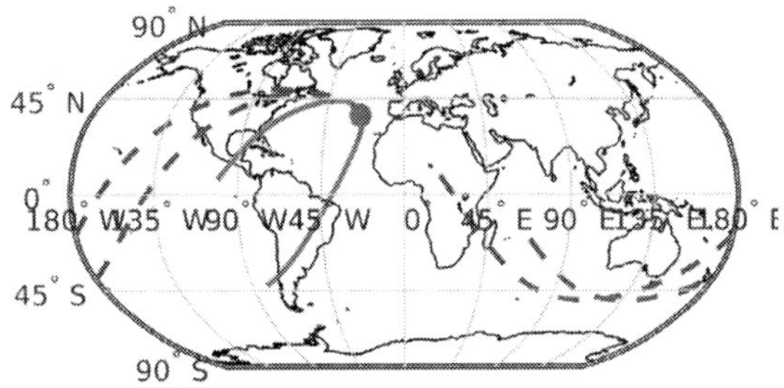

Figure 7: Example entry trajectories towards the Azores site. Orange lines: ascending cluster , blue lines: descending cluster , filled point: site location

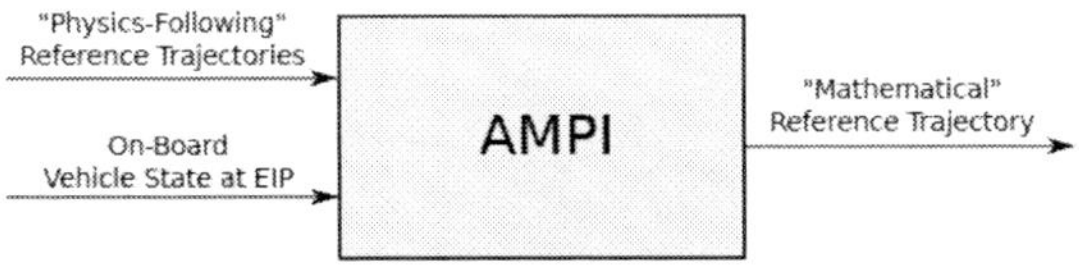

Figure 8: Sketch the concept behind on-board trajectory generation via AMPI.

the interpolation evaluation point. Note that the output of the AMPI algorithm does not represent a formal solution of the equations of motion, but does provide a good real-time approximation, as discussed in Ref. [5].

The AMPI algorithm discretizes all the parameters that are expected to be off-nominal and that can be provided during the flight by the vehicle's navigation system. In this paper, the trajectory planner feeds off a database featuring only two parameters: *entry geodetic latitude* and *entry longitude*. The entry longitude is discretized from -180 deg to 180 deg, where the entry geodetic latitude is discretized -52 deg from to 52 deg, thus encapsulating all the EIPs originating from the operational orbit. Such discretization results in grid of EIPs that resembles the coverage maps shown in Sec. 3, where each node in this grid contains the "physics-following" trajectories that guide the vehicle to the emergency landing sites.

Once the parameter space has been established and the trajectory database has been generated, the grid cell that encloses the interpolation evaluation point x_0 is identified. As mentioned above, x_0 is constructed on-board, and it contains the vehicle's longitude and geodetic latitude at the expected EIP. The trajectories that delimit the parameter subspace enclosing the interpolation evaluation point are referred to as "extremals", where e_{TL} is the top-left extremal, e_{TR} is the top-right extremal, e_{BL} is the bottom-left extremal and e_{BR} is the top-right extremal.

Finally, the AMPI algorithm performs a bilinear interpolation on the interpolation evaluation point x_0, using the four extremals as support points. The key assumption under AMPI is that the relative distance of the interpolation evaluation point to the extremals is valid not only at the entry point and for latitude and longitude, but also for all the other trajectory variables and throughout the whole time frame. Consequently it is possible to generate reference trajectories knowing only EIP information.

4.2 Adaptive Guidance Performance

As with all interpolation algorithms, reducing the relative distance from the support points to the evaluation point will generally improve the fidelity of the resulting interpolation. In AMPI, this is reflected when the database density is increased.

By bringing the extremals delimiting the interpolation cell closer to a test EIP, the cell area is reduced and the fidelity of the interpolated solutions is improved, as sketched in Fig. 9. Due to the large coverage area that the database spans, it is unfeasible to recompute the entire database for smaller grid spacings due to the limited computational time. Consequently, the tests performed in this section are restricted to a single arbitrary landing site and EIP, thus serving as a proof-of-concept. The selected EIP is placed at -125 deg longitude and 13 deg geodetic latitude, where the targeted landing site is the Ben Guerir AFB. The extremals used in this test are placed at various spacing levels, where the coarsest spacing is 10 deg and the finest spacing is 0.1 deg.

Figs. 10 and 11 show the altitude profiles for the 10 deg cell and the 1 deg cell, respectively. It becomes clear that the entry which feeds off a 10 deg cell shows extremals differ significantly in shape and trajectory length, contrary to the case where a 1-deg spacing is used. This indicates the quality of the approximation to the solution of the equations of motion will be poor, potentially leading to a trajectory that is not flyable in the first place. To quantify how the database density impacts the guidance performance, six 1000-case Monte-Carlo simulation campaigns are performed according to the dispersion characteristics given in Ref. [8].

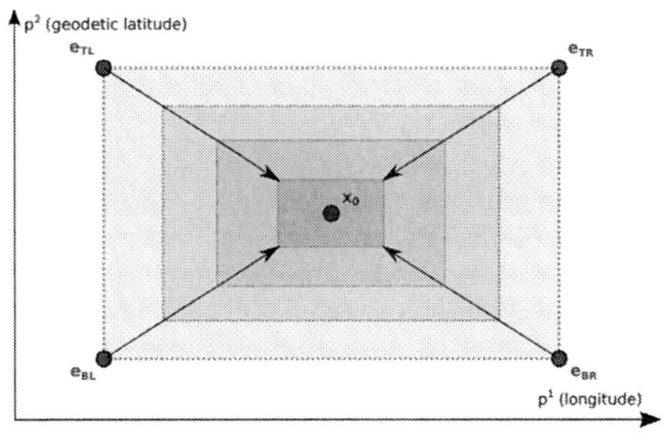

Figure 9: Sketch showing how the grid cell area reduces as a consequence of limiting the parameter spacing.

Tab. 6 shows significant dispersions in the "landing" accuracy. A trend is revealed which shows how decreasing the parameter spacing leads to a wider spread of the terminal position dispersions and an an increased mean value of such dispersions. Note that a limited number of cases falls significantly further from the landing site as compared to all the other cases of the same Monte-Carlo campaign. It is important to emphasize that these trajectories are removed from the computation of the statistical values given in Tab. 6, but will be considered in the conclusions drawn in Sec. 5. Either way, the distance-to-target dispersions exceed the 10 km radius established in Sec. 2.4, all in all suggesting that a revision of the trajectory tracker could lead to performance improvements in this aspect.

Contrary to the distance-to-target dispersions, the guidance algorithm effectively tracks the altitude and groundspeed in almost every case tested, as shown by Fig. 12. Here, three trajectories that differ significantly in altitude-airspeed profile from those in the main group are highlighted in red, which coincide with those trajectories that fall excessively far from the landing site as discussed above. Consequently, the guidance algorithm is capable of bringing the vehicle to the required altitude and Mach number in most cases. Furthermore, the parameter spacing does not make a significant impact neither on the final Mach number nor in the final vertical speed, since the mean and the standard deviation of both quantities remain practically unchanged as shown in Ref. [8].

In terms of constraint compliance, the 10 deg spacing leads to the worst performance in terms of aerodynamic-load, which is violated in a third of the cases tested with a mean overshoot of 3.42 g and a maximum overshoot of 4.11 g. Note that, although the spacings of 5 deg to 1 deg result in a large number of overshoots, the mean and maximum values of these overshoots are significantly lower that for 10 deg and within acceptable bounds. The finest spacings of 0.5 deg and 0.1 deg lead to just three overshoots and are very close to the maximum of allowable value of 3.0 g.

Contrary to the aerodynamic-load overshoots, the heat-flux constraint of 670 kW/m^2 is complied with in most of the cases. Despite this, some of the tested cases show significantly high maximum heat-flux values, in particular the 10 deg and the 0.5 deg cases. These very high constraint violations correspond to those trajectories that deviate too far from the altitude-groundspeed profile as highlighted in Fig. 12. Finally, it is important to emphasize the large number of heat-flux overshoots encountered for the finest spacing. This is related to the fact that extremals that support the interpolation closely track the heat-flux constraint and consequently any errors in the reference profile could easily lead to the violation of the constraint. Despite this, the overshoot values are still within acceptable limits and are expected to improve with a revision of the trajectory tracker.

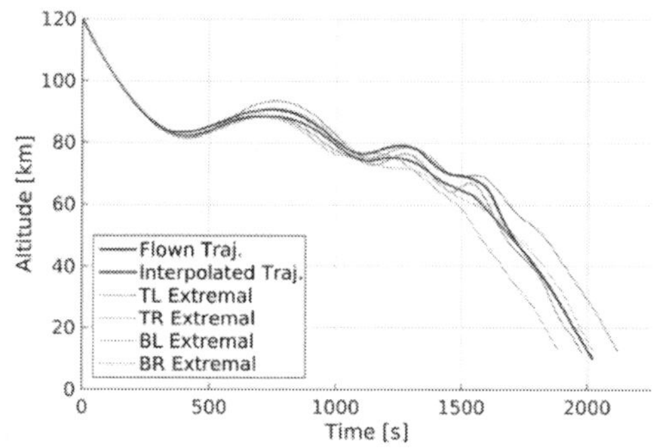

Figure 10: Altitude profiles for the test entry with extremals separated 10 deg apart from each other.

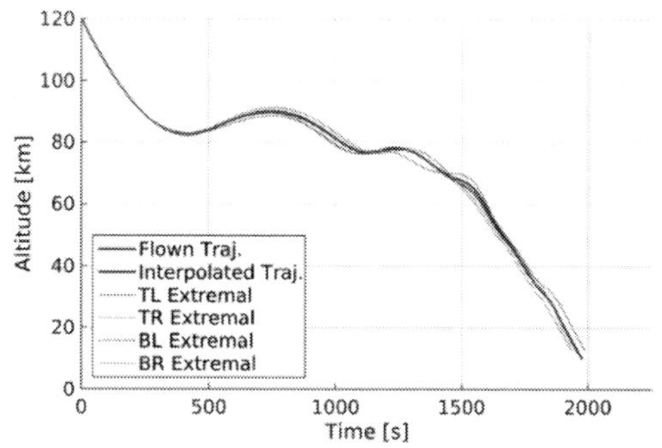

Figure 11: Altitude profiles for the test entry with extremals separated 1 deg apart from each other.

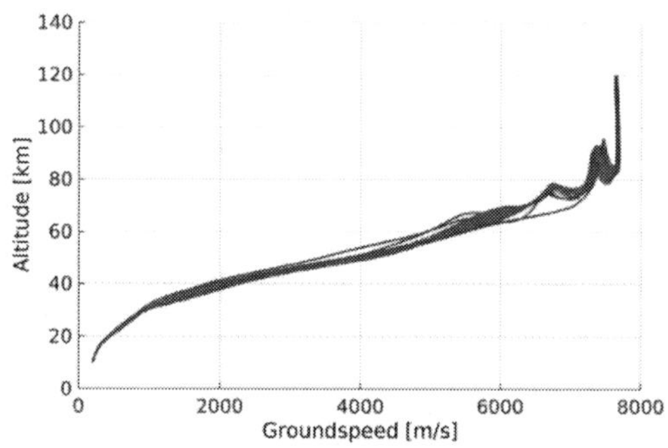

Figure 12: Altitude-Groundspeed dispersions for a parameter spacing of 2 deg. Marked in red are example trajectories that differ significantly from the main group.

In summary, the constraint violations are in line with the expected trend for the performance of the system as a function of the parameter spacing. Coarser spacings lead to reference trajectories that do not approximate the solutions of the equations of motion sufficiently well, thus forcing the guidance system to track a trajectory that is not flyable in the first place.

Table 6: Dispersion properties of the terminal position as a function of parameter spacing.

Parameter Spacing:	**10.0 deg**	**5.0 deg**	**2.0 deg**	**1.0 deg**	**0.5 deg**	**0.1 deg**
Mean [km]	209.91	82.71	35.69	34.47	19.53	13.15
Standard Deviation [km]	118.15	45.03	23.70	21.26	21.90	13.59

5 CONCLUSIONS & RECOMMENDATIONS

The goal of this paper was to present a dedicated guidance system for abort entry missions, focusing on flights that originate from a low Earth orbit. The abort guidance architecture presented in this paper features an Adaptive Multivariate Pseudospectral Interpolation (AMPI) on-board trajectory planner, which feeds off a large database of optimal trajectories computed off-board.

The reachability analysis performed in this paper shows that the choice for the emergency landing sites plays a critical role when providing entry opportunities. The current site distribution guarantees a maximum in-orbit wait time of approximately 10 hours and allows to target at least one site during 50% of the time. In addition, all sites share over 50% of their entry-window with at least one alternative site.

The developed guidance system is capable of planning a reference trajectory in the order of a few milliseconds, proving its capability to run on-board. Furthermore, the system can handle multiple abort scenarios, since the trajectory planner is capable of selecting the most appropriate reference trajectories from the database based on information provided by the navigation system. Thorough testing of the system showed that while the constraint compliance is marginally affected by the density of the trajectory database, the position dispersions at the terminal point are strongly correlated with the chosen density. Results show that increasing the spacing between the stored EIPs in the database increases the mean and spread of the terminal position dispersions.

In summary, the results shown in this paper serve as solid proof that a lightweight autonomous adaptive abort guidance system can be successfully implemented by means of combining a trajectory planner based on trajectory interpolation and a trajectory tracker. Despite this, testing showed that in a limited number of cases the guidance fails to track well the altitude and groundspeed in the early portions of the flight, causing the vehicle to land excessively far from the landing site. Implementing a trajectory tracker where the traveled range is directly tracked could improve the robustness to dispersions. Furthermore, a full system test where the choice for landing site is left free is recommended, where a similar dispersions analysis to the one presented in this paper would have to be made.

ACKNOWLEDGEMENTS

The authors would like to thank Celia Yabar Valles and Guillermo Ortega from the European Space Agency for providing support throughout the whole research project. Thanks to Marco Sagliano from DLR for his kind help deciphering the inner workings of his AMPI algorithm. A special thanks goes to Francesco Cremaschi from ASTOS Solutions for his unattended help and for providing an academic ASTOS license to work with. Last but not least, the authors thank the International Space Safety Foundation (ISSF) for showing faith in this research project with the award of the Graduate Fellowship Grant.

REFERENCES

[1] Muratore, J. P. E., Musgrave, G. E., Larsen, A. M. and Sgobba, T., "Emergency Systems", Safety Design for Space Operations, Butterworth-Heinemann, Chap. 7, 2009, pp. 225-266.

[2] Harpold, J. C. and Graves, C. A., "Shuttle Entry Guidance", Journal of the Astronautical Sciences, Number 3, Vol. 27, 1979, pp. 239-268.

[3] Saraf, A., Leavitt, J. A., Chen, D. T. and Mease, K. D., "Design and Evaluation of an Acceleration Guidance Algorithm for Entry Journal of Spacecraft and Rockets", Vol. 41, 2004, pp. 986-996.

[4] Xue, S. and Lu, P. "Constrained Predictor-Corrector Entry Guidance", Journal of Guidance, Control, and Dynamics, Vol. 33, 2010, pp. 1273-1281.

[5] Sagliano, M., Mooij, E. and Theil, S., "Onboard Trajectory Generation for Entry Vehicles via Adaptive Multivariate Pseudospectral Interpolation", Paper no. 2016-2115, AIAA Guidance, Navigation, and Control Conference, 2016.

[6] Gerard, Y. and Tumino, G., "FLPP Re-Entry In-Flight Experimentation: The Intermediate Experimental Vehicle (IXV) Technical and Programmatic Objectives", International Astronautical Congress, 2005.

[7] Mooij, E., "The Motion of a Vehicle in a Planetary Atmosphere", Delft University Press, Chap. 4, 1997, pp. 50-51.

[8] Gonzalez-Puerta, A., Mooij, E., "Reachability Analysis to Enable Zero-Wait Entry Guidance", AIAA Guidance, Navigation, and Control Conference, 2018, TO BE PUBLISHED.

[9] Sagliano, M., Mooij, E. and Theil, S., "Onboard Trajectory Generation for Entry Vehicles via Adaptive Multivariate Pseudospectral Interpolation", Journal of Guidance, Control, and Dynamics, Vol. 40, No. 2, 2016.

[10] Larson, W. J. and Wertz, J. R., "Space Mission Analysis and Design", Microcosm Press, Chap. 16, 2005, pp. 669.

AN EXERCISE IN PLANETARY DEFENSE

William Ailor

Affiliated with The Aerospace Corporation
2310 East El Segundo Blvd, El Segundo, CA 90245 william.h.ailor@aero.org

ABSTRACT

A highlight of the recent 2017 IAA Planetary Defense Conference was a tabletop exercise that challenged conference participants to develop responses to a fictitious asteroid threat. The threat began with the discovery of the fictitious 100 to 250-meter asteroid, which at the time of discovery had a 1% chance of impacting the Earth in ten years. The threat and possible responses were considered by seven focus groups representing expert advisors to leadership on the nature of the threat, methods of deflection, consequences of impact, disaster mitigation, design of missions to observe and deflect the object, and other areas. The "advisors" presented their recommendations to a panel of conference attendees who volunteered to be "World Leaders." World Leaders made decisions that affected the evolution of the threat, which culminated with a successful deflection action three years before impact. The paper provides details of the exercise, the threat, discusses issues that will arise as the threat evolves, and highlights some of the tough decisions that must be made when an object presenting a real threat is discovered.

1. INTRODUCTION

Fig. 1 shows how the number of known Near Earth Objects (NEOs), asteroids and comets whose orbits bring them into the Earth's neighborhood. The figure also notes that there are over 1800 known Potentially Hazardous Objects (PHOs), objects that have the potential to make threatening close approaches to the Earth, with 157 of these larger than 1 km. Fig. 2 shows the number of NEOs that have passed within one Lunar distance of Earth over the last 20 years. Some of these were not discovered until after their close approaches.

The surface of the moon, recent events such as the asteroid entry and explosion over Chelyabinsk, Russia, as well as known impacts to Jupiter prove that planets in our solar system have been impacted many times and impacts continue. The threat of new impacts is with us always.

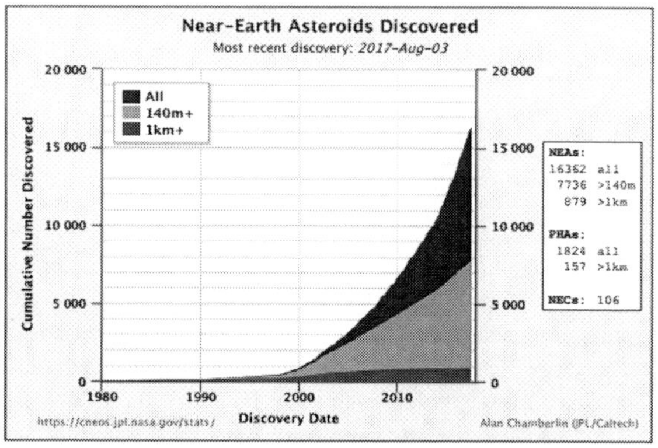

Figure 1. Known Near Earth Objects as of August 3, 2017 (NEAs—Near Earth Asteroids; NPHAs—Potentially Hazardous Asteroids; NECs—Near Earth Comets [1].

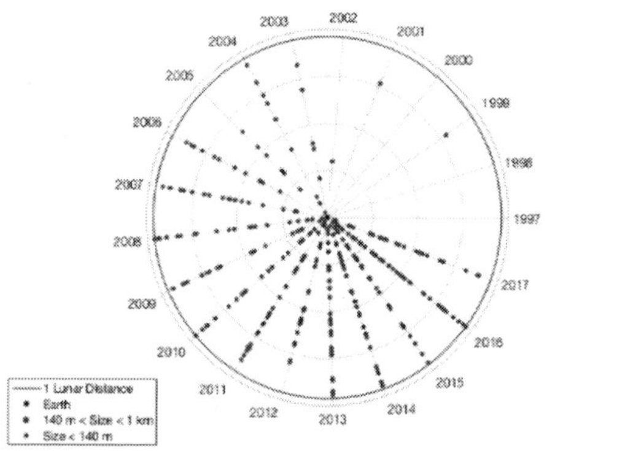

Figure 2. Asteroid close approaches to Earth (within one distance), 1997-2017.

Fortunately, for the first time in our planet's history, we have the ability keep an impact from happening: we can move a threatening asteroid away from an impact trajectory—IF we find it in time and are able to make timely decisions to take defensive actions.

In 2004, the first of a series of international conferences was held in Anaheim, California to focus on what we know about threatening asteroids and what we might do if one is discovered. Six conferences have followed in the series: 2007 in Washington, DC; 2009 in Granada, Spain; 2011 in Bucharest, Romania; 2013 in Flagstaff, Arizona; 2015 in Frascati, Italy; and 2017 in Tokyo, Japan. Fig. 3 shows attendees of the 2017 conference.

Figure 3. Attendees at the 2017 IAA Planetary Defense Conference.

All conferences have brought world experts together to discuss many aspects of planetary defense, including what is currently known about asteroids and comets; characteristics such as asteroid mass, shape and rotation rate that might affect a deflection effort; techniques that might be used to move an object off of an impact trajectory; design of deflection missions and campaigns; consequences of an asteroid impact; public education; disaster mitigation and management; and things that would affect the decision to take action.

On this last point, beginning in the 2013 conference, organizers included in the program a hypothetical threat exercise designed to elicit responses from attendees from several perspectives on what might be done given data available in various stages as the threat evolved. Details of the threats and responses for each conference scenario are included in the summary reports for each conference [2].

This paper describes the threat and the resulting discussion for the 2017 IAA Planetary Defense Conference.

2. THREAT SCENARIO

The asteroid threat response exercise for the 2017 IAA Planetary Defense Conference began with publication of a press release before the conference began. The first release highlighted the International Asteroid Warning Network (IAWN) announcement that an asteroid had been discovered that would pass very close to Earth in ten years, on July 21, 2027 (IAWN is a United Nations-endorsed organization that links existing NEO discovery teams, NEO observers, orbit computation centers, and communication experts to assist governments in understanding asteroid impacts, their frequency, consequences, and possible responses). The body was estimated to be between 100 and 250 meters in size, and IAWN estimated there was a small probability, 1%, that the object would strike Earth on that pass. If it were to impact, the impact would be somewhere along the line defined by the red dots in Fig. 4. Details of the threat and copies of the press releases are available at https://cneos.jpl.nasa.gov/pd/cs/pdc17/.

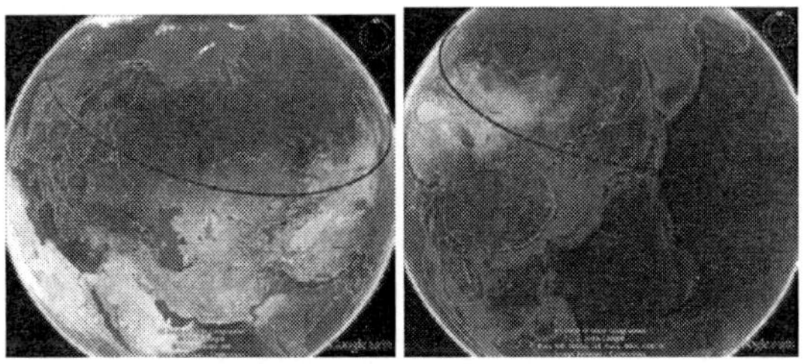

Figure 4. Impact threat corridor shortly after asteroid 2017 PDC was discovered (image courtesy Paul Chodas, NASA/Jet Propulsion Laboratory).

That release and related material was briefed to all conference attendees at the end of the first day. Each attendee then selected and joined one of eight expert groups to consider the threat information. The expert groups were representative of the types of groups that would be established to advise leaders in the event of a real threat. The initial expert groups were:

Group 1: Represented the point of view of **International Asteroid Warning Network (IAWN)** and rovided information on what is known about the asteroid and the threat.

Group 2: **Impact Effects**—Provided information on possible consequences should that object actually strike Earth.

Group 3: **Deflection and Disruption**—Techniques that might be used to deflect or disrupt the object.

Group 4: Represented the point of view of the **Space Mission Planning Advisory Group (SMPAG)**, a United Nations-endorsed organization that brings together space agencies to prepare for an international response to the NEO threat through the exchange of information, development of options for collaborative research and mission opportunities, and to conduct NEO threat mitigation planning activities.

Group 5: **Decision to Act**—Overviewed information on the threat and provided recommendations to leadership on what actions were possible and whether the information was sufficient to take action.

Group 6: **Communication**—Provided information on what and how information should be communicated to the public and what the public was saying about the threat.

Group 7: **Disaster Planning and Management**--Provided recommendations on preparations for managing an impact disaster should it occur.

Group 8: **World Leaders**—Received input from other groups and made final decisions as to what was to be done as the threat evolved.

Press releases presented at the end of each day provided new details on the evolution of the threat. Discussions by exercise group followed and each group developed specific recommendations to the World Leaders. World Leaders discussed the recommendations and authorized actions to be taken (e.g., launch of an observation mission, launch of deflection payloads). After the conference ended for the day, the exercise development team incorporated World Leader decisions into the development of the scenario and press release for the next day.

IAWN released the second press release on November 30, 2018, nine years to potential impact (the object was behind the Sun prior to this date, and observations were not possible). Based on new observations, the probability of collision with Earth rose to 96% and the threat corridor shortened to that shown in Fig. 5 Measurements by the Hubble Space Telescope and ground-based instruments updated the size range to be between 200 and 280 meters. The object was headed away from Earth and the press release noted that the object will be too faint to be visible again for 11 months.

The refined prediction of possible impact points enabled prediction of possible consequences should the object actually impact. Fig. 6 shows the population that would be affected should the object impact at a spot along the threat corridor. Note that an impact in the Pacific would affect a smaller population that a land impact, and this fact led to possible consideration of moving the object in that direction as an option.

At this point, the Deflection and Disruption working group recommended launch of two rendezvous spacecraft, with each possibly carrying a nuclear explosive, followed by launch of six kinetic interceptor spacecraft. The working group noted that only three impacts were required for deflection, the rest provided redundancy. The goal was to move the object westward into the Pacific to minimize hazards to the public. The World Leaders accepted the recommendation to launch two rendezvous spacecraft to orbit the asteroid, followed by six kinetic impactors—three were estimated to be sufficient for deflection, the extra three were for redundancy. The leaders also ordered that each of the rendezvous spacecraft include a nuclear explosive that would be used only as a last resort.

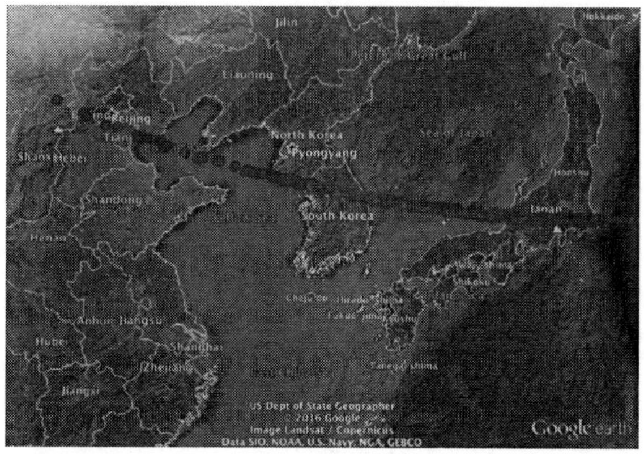

Figure 5. Possible impact locations from second press release (image courtesy Paul Chodas, NASA/Jet Propulsion Laboratory).

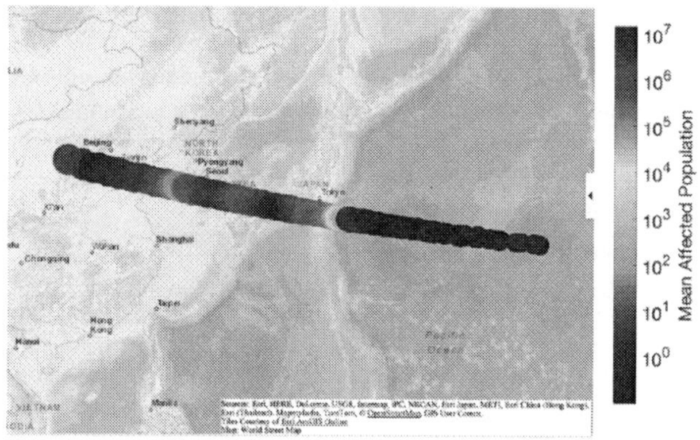

Figure 6. Mean affected population (figure courtesy L. Wheeler, D. Mathias, NASA Ames Research Center).

IAWN released the third press release on May 15, 2020, which showed that the object was on a collision course with Earth and would impact somewhere along the shorter corridor shown in Fig. 7. Leadership-mandated flyby missions had found that the object was about 270 meters in size and that the density was about 20% higher than previously estimated. In addition, the flyby spacecraft showed that the object had a ~100-meter moon, which itself would cause serious damage when it impacted. The release noted that six kinetic impactors were on the way to the asteroid, but given the updated size, the worst-case analysis suggested that five kinetic impactors were required to deflect the object away from Earth. The two rendezvous spacecraft with nuclear explosives would reach the object in three years, in May 2023.

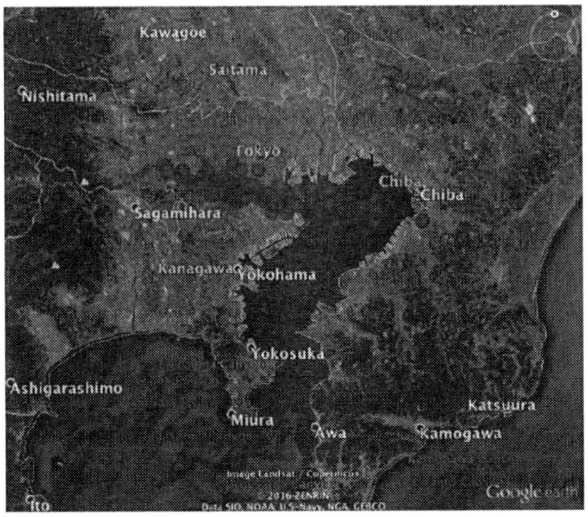

Figure 7. Threat corridor May 15, 2020 (3 years to impact).

The final press release was released on June 15, 2023. The release noted that only one rendezvous spacecraft successfully reached and was orbiting the asteroid; the second experienced a reaction-wheel failure a year after launch and was abandoned.

3. OUTCOME

Discussion of the fourth press release centered on two options: using the six incoming kinetic impactors as the primary method of deflecting the asteroid and its moon, or activating the nuclear explosive on the spacecraft currently orbiting the asteroid first, with the kinetic impactors as backup. During the discussion, it was noted that the nuclear explosive was felt to have the highest probability of success given that all five of the kinetic impactors would need to successfully hit the larger object, with the sixth impactor targeted for the secondary object. In addition, the larger object was about 20 times smaller than the 7.6 km comet Temple 1 (the target for the successful Deep Impact mission in 2005), so both it and the smaller object would present targeting challenges for the incoming vehicles. It was also noted that each impact would likely create a debris cloud that could affect targeting for subsequent impactors. The impact velocity would be 8 km/sec, below the 10 km/sec impact velocity for Deep Impact mission. Other considerations were:

- A nuclear explosive has not been used in deep space, and the long resident time for the device in deep space raised some uncertainty about its reliability.
- A successful impact of one kinetic impactor would move the asteroid's impact point into the Pacific Ocean, yielding a tsunami as a near-immediate threat. Consensus from experts was that the tsunami would be of a size that nations bordering the Pacific prepare for already, and given the warning, damage would be sustainable. There was no discussion of possible longer-term effects on the atmosphere, weather, etc.
- Given the public's concern about using nuclear explosives, there was strong support for making that option a last resort if all else failed.
- The public was concerned about the spacecraft with the nuclear explosive device following the asteroid into the atmosphere and possibly adding a "dirty bomb" incident to the consequences, so they felt that all devices left in orbit should be activated to eliminate that threat.

As a final consideration, members of the leadership group noted that constituents were very unsettled by the possibility of an impact, that the world economy was experiencing a severe downturn, and that an actual impact would be even more consequential. All parties wanted the threat eliminated and a return to regular order.

As a result, and with resigned acceptance that use of the nuclear device offered the quickest and most certain option for eliminating the threat of impact of both objects, leadership approved activating the nuclear explosive at a point that would move the primary and destroy the secondary. The exercise ended at that point.

4. LESSONS LEARNED

Despite the fictitious nature of the threat, conference attendees were very engaged in their discussion and evaluation of the threat. Key points raised during discussions were:

- The use of a nuclear explosive was quite controversial with great public concern about the launch and accidental activation of such devices. Several attendees suggested that the name be changed to Atomic Deflection Device or some similar description to help remove the image of these devices as "weapons."
- Different nations will have different opinions on what should be done and when it should be done, and more work should be done on managing public impressions and the decision-making process and factors that will affect that process. A good example was discussion of the use of nuclear explosives as a deflection technique. This has been a recurring issue for all conferences to

date and suggests that work needs to be done to raise public awareness of the realities involved with accidents involving nuclear explosives and options for moving large objects.

- The exercises have not had significant involvement by "real" decision makers--political leaders who could bring the perspective of national budgets, competing interests for funding, etc., into the discussions. In a real scenario, how would decisions be made as to the need for assigning significant resources toward planning deflection options when the warning might subsequently be found to be a false alarm? Who would be authorized to make the decision? What information should be provided to increase the acceptability of using atomic deflection devices?
- The cost and economic implications and consequences of an asteroid threat, even a threat that is determined to be a false alarm, must be considered. Areas that might be affected should an object impact will likely be public information shortly after the object is discovered, and work should be done to assess the response of the public and elected officials in the initial and evolving threat corridor and provide appropriate, authoritative information to the public, nations, and decision makers as the threat evolves.

5. SUMMARY

The threat of an impact of an asteroid or comet is real and ongoing. We continue to find threatening objects that pass very close to our planet. The community recognizes that the most likely threat is that posed by objects smaller than 1-kilometer in size that could obliterate a city or a region and might be discovered with only a few years warning.

Three planetary defense conferences have highlighted the threat and have simulated the decision-making process for responding to the threat and an ensuing disaster. The exercises make it clear that work on understanding the potential physical, psychological, political, environmental, and economic consequences of a serious threat should be prioritized before a serious threat is discovered. Education of the public and decision makers should highlight planetary defense options. Exercises of the type described in this paper can be used to increase understandings of the difficult decisions leaders and disaster responders will be required to make as a threat evolves.

6. REFERENCES

1. "Discovery Statistics," https://cneos.jpl.nasa.gov/stats/totals.html
2. Conference reports for 2013 and 2015 are available at http://pdc.iaaweb.org. The 2017 report is in preparation.

7. ACKNOWLEDGEMENTS

The asteroid threat response exercise for the 2017 IAA Planetary Defense Conference was developed by the individuals below:

William Ailor, The Aerospace Corporation, Exercise Coordinator
Brent Barbee, NASA Goddard Spaceflight Center
Mark Boslough, Sandia National Laboratories
Paul Chodas, NASA Jet Propulsion Laboratory
Barbara Jennings, Sandia National Laboratories
Nahum Melamed, The Aerospace Corporation

Dr. Paul Chodas of JPL was the "Threat Master" and developed details of the threat for the first and subsequent press releases.

PROCEEDINGS OF THE 9TH INTERNATIONAL SPACE SAFETY CONFERENCE

SESSION 24 RE-ENTRY SAFETY – III

A FIRST STEP TOWARD FRAGMENTATION PROCESS ASSESSMENT OF RE-ENTERING SPACECRAFT: MECHANICAL STRESS ANALYSIS WITH THE SPACECRAFT ORIENTED SIMULATION TOOL PAMPERO

G. Prigent [(1)], J. Carro [(2)], B. Crusson [(2)], L. Stainier [(3)]

[(1)] CNES, 18 av. Edouard Belin, 31401 Toulouse, FRANCE, Email: guillaume.prigent@cnes.fr
[(2)] GMV, 17 Rue Hermès, 31520 Ramonville-Saint-Agne, FRANCE, Email: jcarro@gmv.com
[(2)] GMV, 17 Rue Hermès, 31520 Ramonville-Saint-Agne, FRANCE, Email: Bcrusson@gmv.com
[(3)] GEM laboratory, 1, rue de la Noë, 44321 NANTES cedex, FRANCE, Email: Laurent.Stainier@ec-nantes.fr

ABSTRACT

In the frame of the French Space Operation Act, the French space agency works continuously to improve the accuracy of predictive numerical simulations of spacecraft that operates atmospheric re-entry.

Some preliminary sensitivity analysis have shown that the outcomes of the fragmentation process of spacecraft have a major influence on risk assessment resulting from survivability analysis performed for each fragment.

To get a deeper understanding of the physic at play during the fragmentation process, the French Space Agency has initiated the development of a new feature of its spacecraft oriented simulation tool called PAMPERO. Besides computing aerodynamics forces, trajectory and parietal heat flux, a new module has been added to PAMPERO to compute the stress resulting from the aero-thermal loads applied on the re-entering object. This new mechanical module relies on the validated simulation tool Code_ASTER developed by Electricité de France.

1. INTRODUCTION

To be able to get a better understanding if the underlying physics of fragmentation, CNES has developed a new feature to its in-house spacecraft oriented code PAMPERO that already deals with the aerothermodynamics phenomenon of the re-entry [1]. Indeed, it is expected by improving our understanding of the fragmentation process to provide more accurate initial conditions for survivability computations and therefor for safety assessment. This is of important matter also while checking the compliance with regulation such as FSOA.

Other codes have been developed in Europe to perform destruction analysis of a spacecraft during re-entry. Among them some can handle the fragmentation process taking into account mechanical stress.

The Ariane Group (AG) fragmentation tool suite is developed within a co-funded R&T program with CNES [2]. The suite is composed of several modules coupling the main phenomena: aerothermodynamics, aerodynamics, trajectory, tank internal thermodynamic, structure thermal and thermo-mechanical response. The fragmentation chain provides input for the code ADRYANS, used at the end of the process for the survivability simulation of the debris. The chaining between the modules is automatic but the fragmentation decision and debris definition are still controlled manually by experts.

SAM [3] is a spacecraft oriented simulation tool designed to be accommodating to uncertainty analyses. The spacecraft is composed by component assembled using an unstructured mesh. The components are linked with the weakest join where the fragmentation could occur. Fragmentation model consider that fragmentation occurs at specified pair of temperature and force. The values used have been provided by preliminary experiments and depend on the join considered.

SCARAB [4] is an integrated software composed of several modules destruction of spacecraft during re-entry. The spacecraft is constructed using panellised geometry primitives. Destruction is handled through analysis by local melting or fracture.

The CNES Spacecraft oriented tool PAMPERO designed for re-entry analysis has been developed in order to provide the ability to assess fragmentation. In a first step features of PAMPERO have been enriched by the ability to compute mechanical stress within a single monolithic object given the thermal and mechanical loads applied on its boundaries. Forces induced by deceleration, rotation and gyroscopic effects are now taken into account. This paper describes the model and assumptions taken in the first part followed with a section devoted to the development and verification process. Finally, conclusions on this preliminary work and short term perspectives are presented.

2. MODELLING

2.1. Aerothermodynamics

PAMPERO is a spacecraft oriented simulation tool developed by CNES [1] and used for the simulation of spacecraft re-entry and breakup.

The code computes the trajectory, attitude and thermal ablation of the spacecraft provided the aerodynamics forces and parietal heat fluxes obtain form correlation. PAMPERO can perform simulations in all the regimes encountered during re-entry ranging from the free molecular to the continuum regime.

Distribution of the heat flux and pressure on the external surfaces of the spacecraft takes into account the real geometry of the spacecraft through the curvature and the shadowing effects.

2.2. Mechanical stress

The developed mechanical stress module will have to take into account forces acting on a spacecraft that operates a re-entry.

As a first approach, only small displacements are considered. The modelling of the problem is described below.

The general motion of the spacecraft can be decomposed in rigid-body translation $x_0(t)$ and rotation $Q(t)$ superposed with local displacements:

$$\vec{x}(\vec{X},t) = \vec{x}_0(t) + \overline{\overline{Q}}(t) \cdot (\vec{X} + \vec{u}(\vec{X},t) - \vec{X}_0) \quad (1)$$

The deformations induced by local displacements are considered small and their impact in the computation of aerodynamic forces can be neglected. Local displacements are also considered quasi-static, allowing separating the problem of global motion and the problem of deformation in a co-rotational frame:

$$\begin{gathered} \rho\ddot{\vec{x}} = \vec{\nabla} \cdot \overline{\overline{\sigma}} + \vec{f} \quad \Longrightarrow \quad \vec{\nabla}_0 \cdot \overline{\overline{\overline{\sigma}}} + \vec{\tilde{f}} = \vec{0} \\ \vec{\tilde{f}} \approx \rho_0 \left[\overline{\overline{Q}}^T \cdot \left(\vec{g} - \ddot{\vec{x}}_0 \right) - \dot{\vec{\omega}} \wedge (\vec{X} - \vec{X}_0) - \vec{\omega} \wedge \left(\vec{\omega} \wedge (\vec{X} - \vec{X}_0) \right) \right] \end{gathered} \quad (2)$$

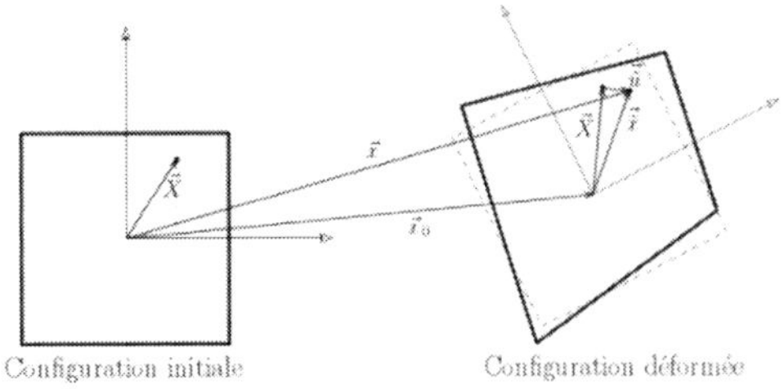

Figure 1: Configuration evolution

As can it be seen in the Eqs. 1-2, apparent body forces in the co-rotational frame can be computed from the knowledge of the global motion. Boundary conditions correspond to applied aerodynamic loads, which have been computed for the global motion problem.

For the problem of deformation in the co-rotational frame, we have assumed small perturbations and linear elastic behavior, yielding a linear problem which can be solved by FEM. In order to connect deformations in the co-rotational frame and the global rigid-body motion, we also have to block rigid-body modes: two methods have been evaluated in the prototype. Both methods shall give similar numerical solutions allowing performing an additional validation of the implementation.

- Definition of local axis
- Definition of link equations

To solve the set of equation, FEM will be used and a trade-off of the existing software has been performed.

3. Trade-off to choose the implementation of the solution

A list of possible solutions proposed by all the team members was built-up and analyzed using analytic hierarchy process [5].

Proposed solutions included:

- Development of a FEM engine from scratch.
- Coupling with existing simulation code, such as Code_ASTER, NASTRAN_MARC, Impact, FreeFEM++, Calculix.

Several selection criteria have been considered, among them we can mention:

- Possibilities of evolutions towards modelling multi-material/multi-objects and transitional problems.
- Level of validation of the solution.
- Possibility to perform sensivity analysis.
- Open source / Non Open source license.
- Portability.

After analysis, it has been concluded that Code_Aster [6] was the solution matching the best the criteria, even if it is not portable and can only be used on Linux. Nevertheless, this limitation has been circumvented using a virtual machine with the advantage that it simplifies the management of the environment between the different teams.

4. DEVELOPMENT AND VERIFICATION PROCESS

4.1. Development: Prototyping of the interface with Code_Aster

The selection of Code_Aster has allowed instancing the preliminary architecture into the following working prototype:

Output interface

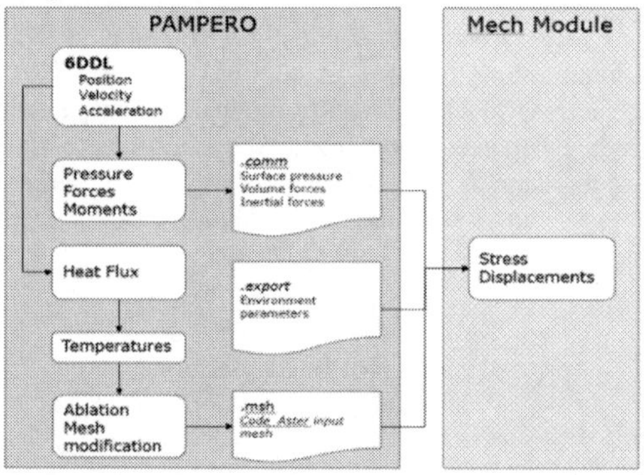

Figure 2: Output interface, PAMPERO →Code_ASTER

Input interface:

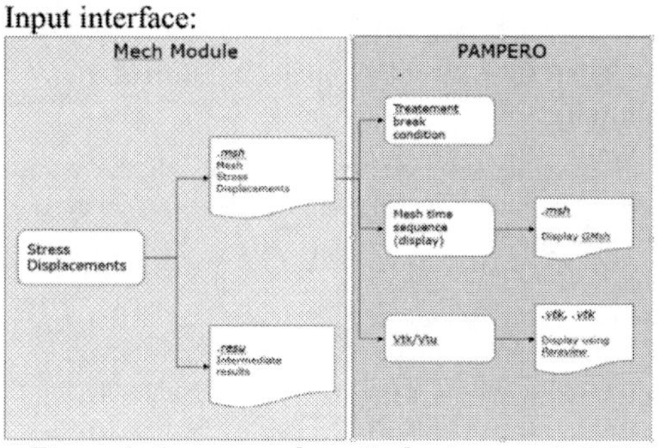

Figure 3: Input interface, Code_Aster → PAMPERO

For the moment the coupling is strong from PAMPERO to Code_Aster (Figure 2). The other direction is weaker (Figure 3), since the mechanical result is not needed to perform the aerothermodynamics computation.

To compute the mechanical constrains, Code_Aster needs inputs, and even if the data is available in PAMPERO, it has to be shaped to fit Code_Aster requirements. For instance, the pressure on the faces has to be put in the good frame, and the inertia forces have to be deducted from the solid's global trajectory.

As explained previously, Code_Aster needs an isostatic problem, so we decided to use two algorithms to block some points in the solid:

- Local axis method: we define an orthonormal base using 3 points close to the CoG (Center of Gravity). We can then express constrain on those 3 points in the new base to block rigid modes of the solid.

- Link equations method: we find the points of the surface of the solid using principal vectors from the CoG. We can then express constrain on those 3 points in the principal base to block rigid modes of the solid.

4.2. Verification strategy:

We have chosen to perform the verification with 2 types of cases:

Unit test cases (UT) that compare the obtained results against reference ones (obtained from analytical solutions or using other software), the objectives of this cases are to validate the hypotheses and to verify the numerical result of each of the contributions (deceleration forces, rotational forces, etc).

Verification test cases (VT) that allow to perform more complex and realistic cases to validate the integrated results and the robustness of the software when adding other phenomena as the modification of the solid geometry due to thermal ablation.

The following table summarizes the test cases realized

Type	Name	Goal
UT	Cube	▪ Pressure field ▪ De-acceleration forces ▪ CoG blocking
UT	Cylinder	▪ Pressure field ▪ De-acceleration forces ▪ CoG handling ▪ Validity of the Co-rotational approach
UT	Disc	▪ Rotation inertial forces
VT	Pipe	▪ Code robustness
VT	Box	▪ De-acceleration forces ▪ Co-rotational approach
VT	ATV	▪ Realistic complex test ▪ Consistency
VT	Ablation	▪ Geometry modification robustness (thermal ablation)

Table 1: Unit and verification test cases

Cube

This first case allows verification of the coupling between PAMPERO and Code_Aster, the application of the effect of the deceleration forces and its associated pressure field and finally the independence of the results from the algorithm being used to block the points to create and isostatic problem.

The following images Figure 4, Figure 5, shows that the field of efforts obtained with both point blocking algorithms are the same

Figure 4: Stress field computation using the Local axis method

Figure 5: Stress field computation using the Link equations method

The comparison of the result obtained PamperoMech/Code_Aster against the analytic solution (between Figure 6, Figure 7) show that difference in the Von Mises criteria is below 2%, in our opinion the origin of the difference comes from the roughness of the mesh.

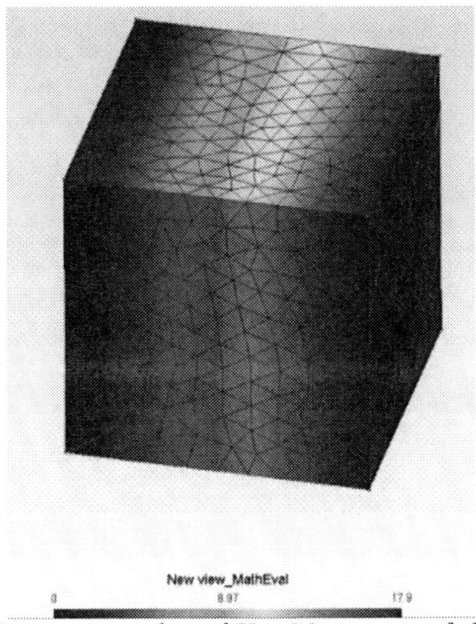

Figure 6: Analytical Von Misses norm field

Figure 7: Von Mises norm field computed with PAMPERO

Cylinder

With this case, we compare our result with the result of an explicit transient dynamics simulation (performed with Abaqus/Explicit, 265000 steps required to reach a stationary regime).

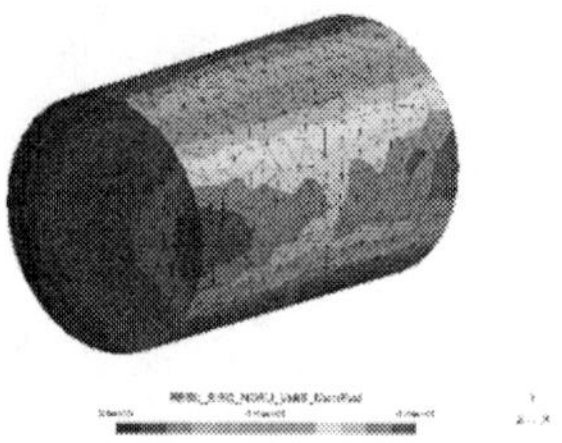

Figure 8: Von Mises norm field computed with Abaqus

As can be seen by comparison of figures 8 and 9, the results obtained in terms of stress distribution, are in quite good agreement. A precise quantitative comparison is difficult since the aerodynamic loading is not exactly the same (simplified model in Abaqus). Nonetheless this comparison provides some validation of the co-rotational approach. It also gives an indication about the savings in computational effort allowed by this approach.

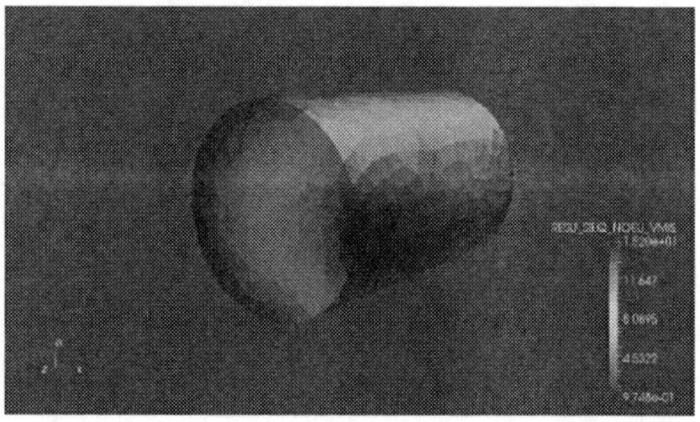

Figure 9: Von Mises norm field computed with Pampero

Disc

The objective of this case is to verify the efforts and the corresponding displacement induced by the rotational inertia forces on a rotating disk against the corresponding analytical solution:

$$u(r) = \frac{(1-\nu)}{8E}\rho\omega^2 R^3 \left[(3+\nu)\frac{r}{R} - (1+\nu)\left(\frac{r}{R}\right)^3\right] \quad (3)$$

The displacements computed are shown in Figure 10.

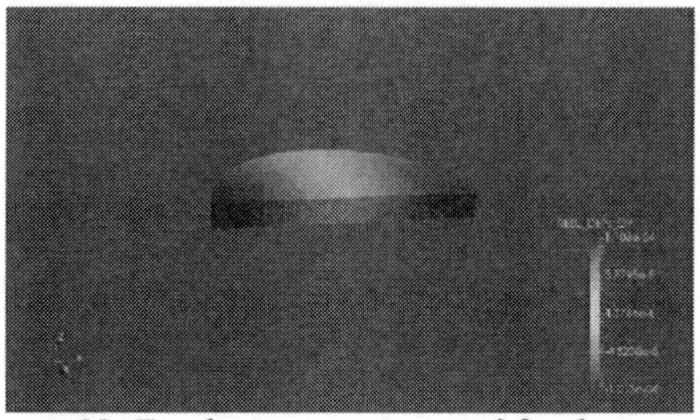

Figure 10: Displacements computed for the rotating disk

The errors with regard to the analytical solution are between 3% and 6% from edge to edge. The origin of the differences comes probably from the accuracy of the mesh.

Pipe

The objective of this test is to verify the robustness of the algorithm used to block the points of the solid to obtain an isostatic problem when the CoG of the solid is not within the solid boundaries.

The results of the computation are shown in Figure 11.

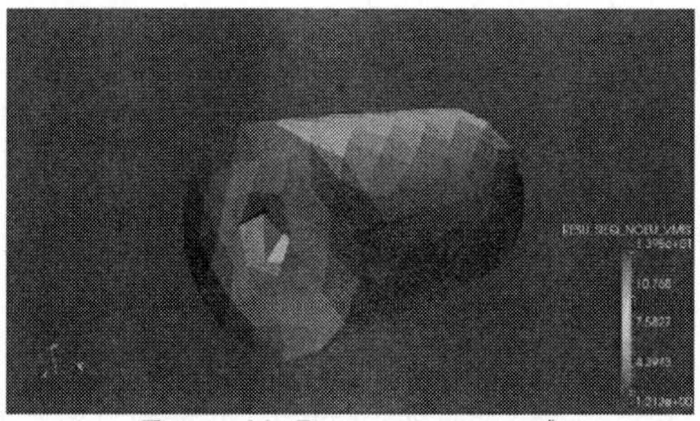

Figure 11: Pipe test case result

The conclusion of this test is that the application of the link equations method is more general as can be used to cases where the solid's CoG is not within its boundaries.

Box

The objective of this case is the qualitative verification of the effort induced by the deceleration forces in the parts of the solid that are not exposed to the flow, meaning that they are not subjected to the application of a pressure field.

As expected, it can be observed in Figure 12, the stress field is more important at the base of the inner beam.

Figure 12: Stress field in the decelerating boxe

ATV

The objective of this test case is to perform a qualitative verification of the software using a complex object as the ATV vehicle.

The vehicle is receiving the flow produced by its advancement through the atmosphere along its longitudinal axis (Angle of Attack equals to zero).

We can observe on Figure 13 and Figure 14 that the stress and temperature fields are different; the stress field is maximal in the junction between the body and the solar panels of the vehicle whereas the temperature field is maximal in the nose that is receiving the thermal flux and in the solar panel attack edge.

Nevertheless, the stress field is also important at the nose of the vehicle with regard to the rest of the body and can't be seen on the Figure 13 because of the very important stress (few greater order of magnitude) located on the panel junction.

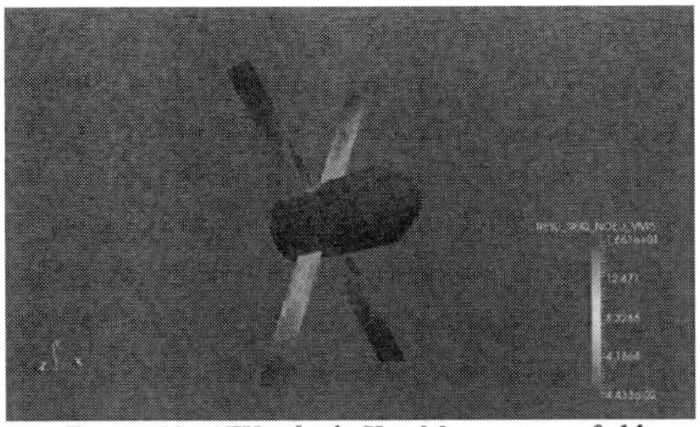

Figure 13: ATV vehicle Von Misses stress field

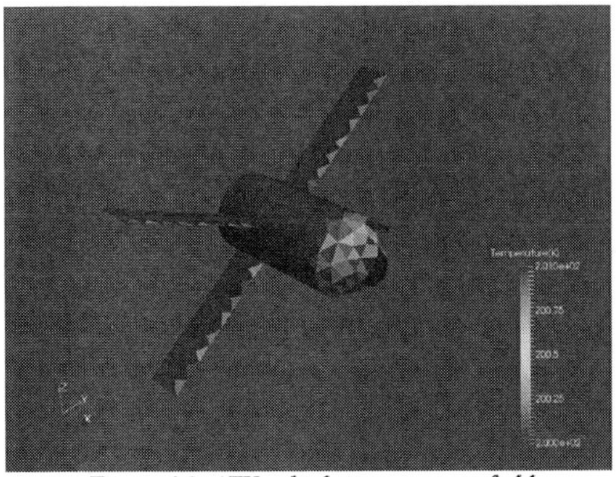

Figure 14: ATV vehicle temperature field

Ablation

The objective of this test case is to verify the qualitative operation of the PAMPERO-Code_Aster coupling when the geometry of the mesh degraded due to the ablation phenomena.

As we can observe Figure 15 that Code_Aster is able to compute a solution for irregular meshes resulting from the elimination of ablated tetrahedrons even when the number of surviving elements is minimal as we can see in the last image of the series.

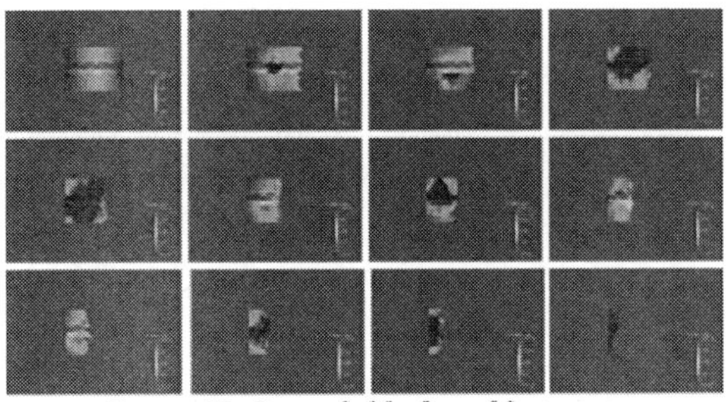

Figure 15: Stress field of an ablating pipe

5. CONCLUSION AND PERSPECTIVES

Developments presented in this paper have provided the CNES with a simulation tool still license free that could take into account mechanical stress analysis on top of trajectory and aerothermodynamics. This new features complete the code PAMPERO that could be used for preliminary breakup analysis.

A verification strategy has been followed and ensures that the coupling of PAMPERO and Code_Aster works effectively and provides the expected results. In addition, the assumptions and modelling of the problem has also been validated.

Nonetheless, an important validation work has to be performed on real configurations. This work represents the next step along with further developments. Indeed, multi-material, multi-object and thermo-mechanical effects are currently being implemented. This will provides the ability to perform validation test on real tests cases.

Finally, a suitable database that supplies models with the required physical properties is also of important matters. More specifically, characterization and modelling activities on composite and ceramic materials would be relevant.

REFERENCES

1. J. Annaloro J., Galera S., Kärräng P., Prigent G., Lips T., Omaly P. (2017). Comparison between two spacecraft-oriented tools: PAMPERO & SCARAB. The Journal of Space Safety Engineering 4 (2017) 15–21.

2. Finzi C., Bertorello C., Pinaud G., Bouilly J.-M. (2017). Simulation of the Ariane 5 EPC Reentry with the fragmentation tool suite. Proc. 7th European Conference on Space Debris, Darmstadt.

3. Beck J, Holbrough I., Merrifield J., Joiner N., Bainbrdige S. (2017). Progress in hybrid spacecraft/object oriented destructive re-entry modelling using the SAM code. Proc. 7th European Conference on Space Debris, Darmstadt.

4. Koppenwallner G., Fritsche B., Lips T., Klinkrad H. (2005). SCARAB – A multi-disciplinary code for destruction analysis of space-craft during re-entry. Proceeding of the 5th European symposium on aerothermodynamics for space vehicles, 8-11 November 2005, Cologne, Germany.

5. Saaty, Th.L., Peniwati, K.(2008). Group Decision Making: Drawing out and Reconciling Differences. Pittsburgh, Pennsylvania: RWS Publications. ISBN 978-1-888603-08-8.

6. Electricité de France, (1989-2017). Finite element Code_ASTER, Analyses de Structures Thermo-Elastiques pour des Etudes et des Recherches, Open source on www.code-aster.org

EXTRAPOLATION OF POPULATION GRIDS FOR RISK ANALYSIS

A. Bellucci[1], N. Tholey[2], Mathias Studer[3], J.F. Goester[4], N. Fuentes[5]
[1][4][5]CNES, 18 av. E. Belin 31401 Toulouse Cedex 9, FRANCE,
[2][3] ICube/SERTIT, Université de Strasbourg, 300 bd Sébastien Brant, CS 10413, 67412 Illkirch Cedex, FRANCE
[1]+33561287506, aurelie.bellucci@cnes.fr,
[2] nadine.tholey@icube.unistra.fr,
[3] mathias.studer@unistra.fr,
[4]+33561281678, jean-francois.goester@cnes.fr,
[5]+33561282714, nathalie.fuentes@cnes.fr,

ABSTRACT

To fulfill its mission of control within the framework of the FSOA, including in particular the monitoring of safety requirements for people and property, CNES has developed and has been using since height years the model and tool ELECTRA, especially designed to estimate human risk during launch and controlled or uncontrolled re-entry. The major contributor in probabilistic assessment of human risk is the population density on areas where some re-entry fragments may fall. In this framework, information about population spatial distribution at a global scale is a key parameter. In particular, CNES needs to have an estimation of the evolution of population for future decades to estimate risk during launch phase and re-entry for future space missions presently in preparation.
The objectives of the CNES POPSCAN study, started in 2011 with SERTIT, include census of available databases about worldwide population, qualitative and quantitative analyses of these data and proposals to improve population models used in risk analysis.

In this paper, we present a method to extrapolate until 2050, at different time steps, the world population grid provided by CIESIN (Columbia University), i.e. the GPW-V4-UN-adjusted geo-referenced raster dataset. Reasons having led to the choice of this particular database are presented as well as some of the main existing extrapolation methods. The method exploits existing geo-spatial population information at a pixel level provided by past population distribution grids and computes the successive future population layers: it is based on an exponential approach with an adjustment to UN projections (United Nations World Population Prospects, the 2015 Revision). This approach combines two advantages: keeping the gridded representation of population distribution from geo-data bases such as GPW and benefiting from the tabular data computed by UN and their sophisticated methods for world population projections.

1. INTRODUCTION

CNES initiated the POPSCAN study in 2011 with the help of SERTIT (Service Régional de Traitement d'Image et de Télédétection), the Image Processing and Remote sensing service of the University of Strasbourg. This valorisation service is specialized in geographic information production and analysis derived from Earth Observation data. Its Rapid Mapping Service is an operator of the Copernicus Emergency service and delivers also crisis geo-information and maps within the International Charter "Space and Major Disasters", which helps rescue operations and emergency management.
Each year, available population databases are reviewed, analysed and updated if necessary. Specific researches have also been led on air and maritime traffic. The risk on air traffic is detailed in another CNES study presented in [1]. In 2017, the POPSCAN study focuses on the extrapolation of population grids until 2050 in order to estimate the risk caused by space operations for future missions. A new extrapolation method is described and implemented for the generation of future population grids for the period 2020-2050.

2. POPULATION DATABASES

Within the perspective of generating global population grids for the next decades, an update of available population geo-databases has been performed. We focused on databases existing at a global scale, i.e. having a worldwide coverage.

- GPWv4, the fourth version of the Gridded Population of the World [2], is gridded with an output resolution of 30 arc-seconds (approximately 1 km at the equator). It provides a set of population layers for population count, population density (GPWv4 standard), UN-adjusted population counts, UN adjusted population density (GPWv4_UN_adj) available for the years 2000, 2005, 2010, 2015 and 2020. GPWv4 is based on censuses conducted around 2010 (i.e. between 2005 and 2014). Data were collected at various administrative levels from country to municipality. Then, data have been extrapolated per administrative unit for the different years of the GPWv4 collection (using annual exponential growth rates, cf part.3) and disaggregated into a grid using a uniform distribution.

- Landscan ™ dataset [3] provides each year a global population distribution data at 30 arc seconds (approximately 1 km resolution at the equator) representing an ambient population (average over 24 hours). The modelling process uses administrative census data which are disaggregated and ventilated at pixel level using spatial data and Earth Observation imagery analysis technologies to measure daytime movements. Landscan products are available since 2000, on a yearly update basis. Last version is the Landscan 2015 Global Population Database.

- The GHS-POP-GLOBE [3] is a global grid at 250 m and 1 km resolutions for 1975, 1990, 2000 and 2015. This raster dataset provides the distribution and density of population per cell. It is based on the population count per census units, used for GPWv4, which has been disaggregated using the distribution and density of agglomerations and built-up areas, derived from the interpretation of multi-temporal satellite images datasets (Landsat collection).

3. REVIEW OF AVAILABLE METHODS

To guide the design of a new extrapolation method for population grids until 2050, a review of the main existing methods was carried out. For population growth calculation, two approaches coexist. On one hand, global mathematical models calculate trends in the size of population. On the other hand, more detailed projection methods are based on demographic assumptions related to population categories, fertility, mortality and migrations assumptions (e.g. the cohort component method used by UN or INSEE (Institut National de la Statistique et des Etudes Economiques)). Since these projections methods require detailed demographic parameters which are difficult to assess for the next decades, these methods are not appropriate and easily exploitable for the production of future population grids. Thus are not going to be detailed here.

The most immediately interesting, easy to use, and widespread mathematical model to extrapolate the population value is the exponential approach, which is used also for the GPW grids calculation. This model assumes that the population is growing at an annual rate *r* depending on past population values:

$$r = \frac{\left(\ln\left(\frac{P_2}{P_1}\right)\right)}{t}$$

with r: annual growth rate,
P_1, P_2 : past population census values
t : number of years between P_1 and P_2 census

The estimated population value $P(x)$ for a given year x is then :

$$P(x) = P_2 e^{rt}$$

In the GPWv4 UN-adjusted population count grids (UN World Population Prospects, the 2015 Revision), this population value is adjusted at country level by a factor:

$$a = \frac{P_{UN} - P(x)}{P_{UN}}$$

with a: adjustment factor
P_{UN}: UN population estimated value
$P(x)$: estimated population for the target year x

The UN adjusted population value, *Padj*, is then:

$$Padj = P(x) * a$$

4. SELECTION OF INPUT DATASET

The Landsat and GPWv4 time series of grids have been analysed and tested in order to select the most appropriate input datasets for the extrapolation process. Due to the lack of multi-temporal data between 2000 and 2015, the GHS-POP-GLOBE database has not been retained. The multi-temporal population values provided by Landscan and GPWv4 have been analysed at pixel level and at country level. First, we evaluate if those grids can be extrapolated with available methods. Then we measured the shift over time of extrapolated population values using one of the extrapolation methods. The 2015 UN World Population Prospects figures were used as 'reference' values for the country level analyses, in addition to INSEE population values for France.

At a pixel level, the Landscan and GPWv4 databases were tested on a series of sample cells representative of a variety of landscapes (Oceania, Africa, Europe). The population values at pixel levels were extrapolated using the exponential model (amongst others), and then compared to the values provided by actual grids. For both databases, extrapolations were made with a 30 arc-seconds resolution (native resolution) and with a 1 arc degree resolution (aggregation of pixels) :

- The Landscan 2012 and 2013 data were used to calculate the 2014 values. Those results were compared to the 2014 values provided by Landscan.

- The GPWv4 data from 2005 and 2010 were used to compute 2105 values. Those results were compared to the 2015 values provided by GPWv4.

These pixel level tests have shown that:

- The 2014 evolution of Landscan cells is not deduced from a classic model of population growth. Indeed, Landscan data are constructed from census values collected each year by administrative units and then redistributed spatially by cells according to the occupation of ground, the zones of activities, etc.). Therefore, the Landscan grids can hardly be retained as input time series data for an extrapolation process over decades.

- The 2015 evolution of the GPWv4 cells corresponds to an exponential method applied to the previous grid cells over time. The GPWv4 data appear to be more appropriate to define and initialize a mathematical process of cell population values extrapolation. This process could be derived from the exponential method.

At a country level, further tests were performed over France using the two GPWv4 grids, i.e. the GPWv4 standard one and the GPWv4_UN-Adj. Population values for each pixel were extrapolated using the exponential method. As before, 2005 and 2010 data were used to compute 2015 year and following ones. A GIS (geographic information system) summation of all pixels extrapolated values within the country limits gives the extrapolated value for the country to be compared with tabular United Nations figures and INSEE reference data as well as GPWv4 data from 2015 and 2020 (GPW obs in Fig 1). The GPWv4_UNadj grid turns out to be the most appropriate input for such a process. The use of 2005 and 2010 UNadj grids to initialize the exponential growth rate for the 2020-2050 epoch, generates less drifts of population values from UN and INSEE reference than the use of the standard GPWv4 grids. For example, for the total population of France, the discrepancies between the calculated GPWV4_UNadj_exp and the UN value are less than those generated by the extrapolation of GPWv4 standard grids and range from 1.3% in 2020 to 10.5% in 2050, as illustrated by fig.1.

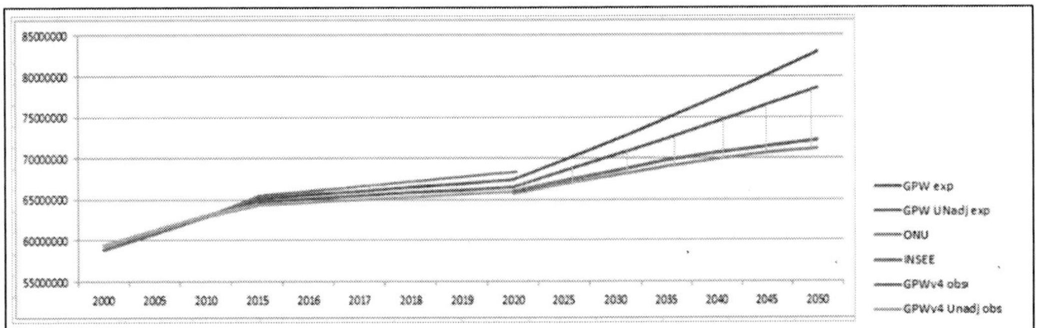

Difference between GPWv4_exp and UN (%) :	2020	2025	2030	2035	2040	2045	2050
	2,6	4,3	6,2	8,3	10,7	13,5	16,6

Difference between GPWv4_Unadj_exp and UN (%) :	2020	2025	2030	2035	2040	2045	2050
	1,3	2,3	3,5	4,7	6,3	8,3	10,5

Figure 1 : Extrapolation of the GPWv4 grids and 2020-2050 results comparison with UN and INSEE values for France

5. PROPOSED EXTRAPOLATION METHOD

Previous tests have shown that the exponential method applied to the GPW4_UNadj data is the method which generates less drift over time. Thus, to generate the future population grids for the period 2020-2050, we develop a method based on an exponential approach applied to the data GWPv4_Unadj at a pixel level.

However, in order to reduce the gap between UN reference and exponential extrapolation of GPWv4_UNadj grids, an improvement of this approach is proposed. This improvement will reduce the discrepancies by adjusting the extrapolated grids to the UN values provided every 5 years between 2020 and 2050.

This improvement could be achieved through a two-step process applied to each cell (pixel) for each 5-year interval. This method maintains a gridded representation of population values, different for each pixel, such as existing in the original GPW grids. And it offers the advantage of being consistent with the UN projected global population values issued from component or cohort methods. To compute the population grid of year n+5 using input grids from years n-5 and n, the process could be summarized as follow:

• Step 1: Computation with the classic exponential method of the population value (first values) for each cell for year n+5 (grid '*_calc n+5*'). We use for each pixel an annual growth rate (*taux_exp_init*) computed with cell values for years n-5 and n.

• Step 2: Calculation of the grid adjustment factor to UN Global Values (every 5 years) by summation of the first values of the grid cells (grid '*_calc n+5*' from step 1) and comparison to the UN values. This global adjustment factor is applied to each pixel of the grid grid '*_calc n+5*' computed in first step to produce grid '*calc_Unadj n+5*'. Then, a new growth rate per pixel (*taux_exp_adj*) is computed with an interpolation method between years n to n+5 to recalculate the successive values of population grids over the considered 5-years period.

This process is described in fig.2 and detailed at a pixel level in fig.3.

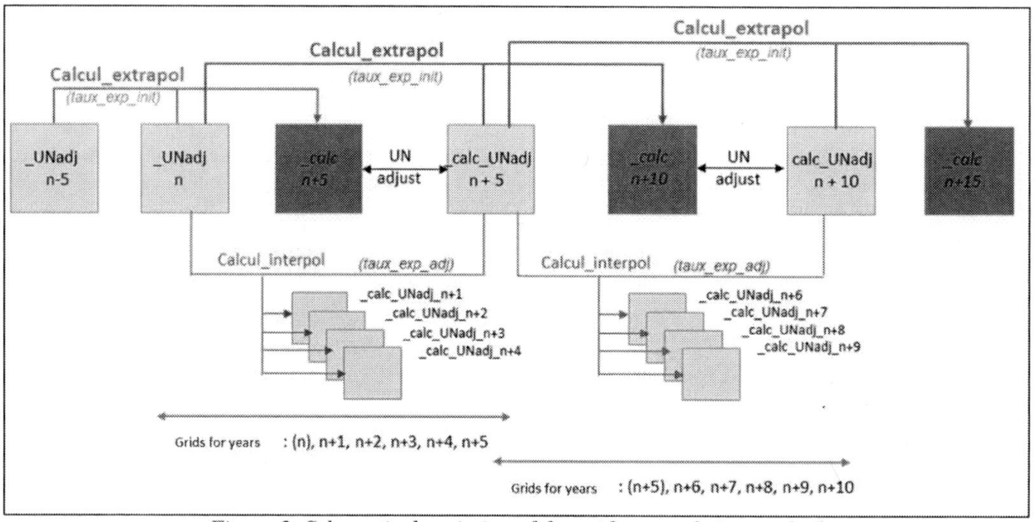

Figure 2: Schematic description of the grid extrapolation method

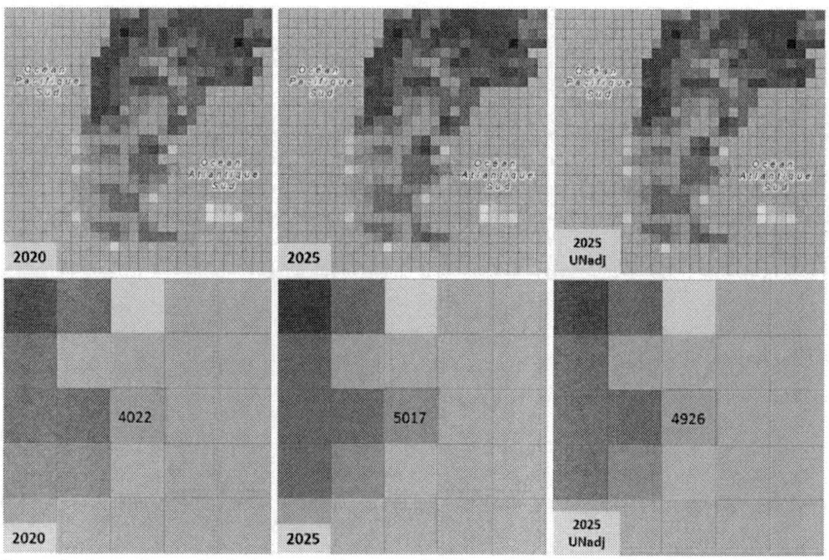

Figure 3: Pixel population value calculation steps between 2020 and 2025: initial value in 2020 (4022), intermediate value in 2025 (5017, calculated in the 1rst step) and final value in 2025 (4926, obtained in the 2nd step after UN re-adjustment)

6. RESULTS SYNTHESIS

The proposed method has been used to generate the 2020-2050 population grids at 1 deg. arc resolution (ELECTRA main format) but could of course be used for other resolutions. The initial inputs are GPWv4 grids resulting from the aggregation of the 30 arc sec. resolution layers. The GPWv4 grids are manipulated under GIS. The method is implemented using the attributes of each cell which describe the successive population values taken by the cell during the extrapolation calculation process as well as several pixel level specific computational information, for each year to calculate, such as the initial increment, the UN readjustment factor, and re-adjusted growth rate used for each pixel. Associated to the population value grids (fig. 4), population density grids have also been generated for the target years: 2020, 2025, 2030, 2035, 2040, 2045 and 2050 (fig. 4). Those population models will be available in next versions of ELECTRA to be used for risk computation relative to long-term missions.

The results obtained for the various extrapolated grids have been compared to the UN world population count of the associated years. The overall shift is situated between 0.01 % and 0.1% (fig.5).

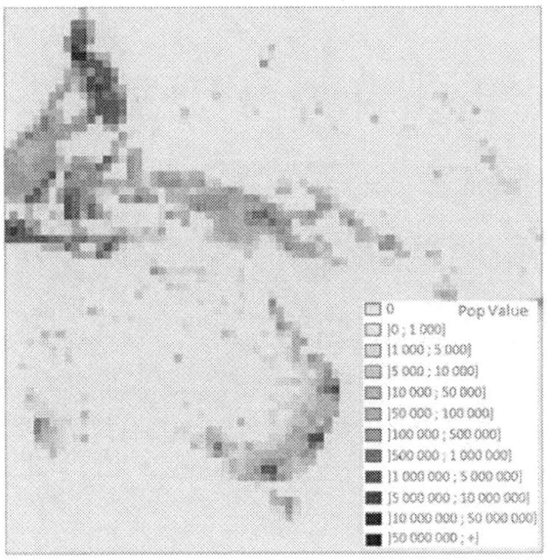

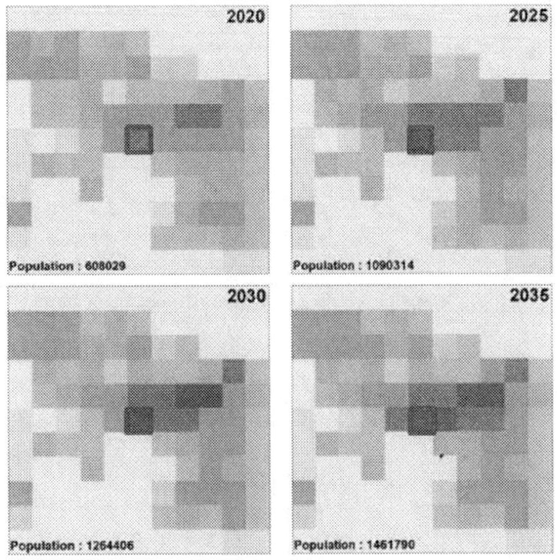

Figure 4: Examples of 2020, 2025, 2030 and 2035 extrapolated grids at 1 deg.-arc resolution

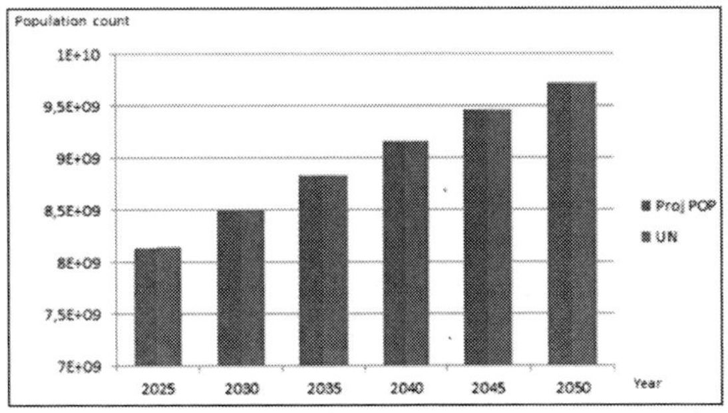

Figure 5: Comparison of the UN population count and the calculated grids for the 2025-2050 time period

7. PERSPECTIVES

This extrapolation method for the generation of the population grids for the 2020-2050 time range is a first approach. Several refinements and improvements could be considered such as:

- a synchronised update of the grids according to the next revisions of the UN World Population Prospects,
- a spatial refinement of the adjustment process at regional or national levels,
- an integration of the low and high level estimation scenarios figures of the UN World population prospects within the grids which are actually reflecting the median figures of the population trends described in the UN projection scenarios.

Moreover, the pertinence of integrating outputs of urban growths models which are currently being developed could also be studied, in order to evaluate their effects and potential improvements in the construction of future world population grids, especially those being at 1 deg cell resolution.

8. REFERENCES

1. Fuentes N, Tholey N., Studer M. (2014) Risks of atmospheric re-entries on aircrafts, CNES progress non studies. *7th IAASS (International Association for the Advancement of Space Safety)*

2. GPWv4, Gridded Population of the World, version4, CIESIN, Columbia University, http://sedac.ciesin.columbia.edu/data/collection/gpw-v4

3. Landscan, global population distribution data, http://web.ornl.gov/sci/landscan/

4. GHS POP GLOBE, Global Human Settlement Population Grid, http://ghsl.jrc.ec.europa.eu/ghs_pop.php

RISK ANALYSIS BETWEEN AIRCRAFT AND SPACE DEBRIS DURING ATMOSPHERIC RE-ENTRY

A. Bellucci(1), N. Fuentes(2) A. Guerra-Algaba(3), M. Cointe-Fourrier(4), J.F. Goester(5),
(1)(2)(5)CNES, 18 av. E. Belin 31401 Toulouse Cedex 9, FRANCE,
(3)(4)APSYS,- 37 av. Escadrille Normandie Niemen 31700 Blagnac, FRANCE
(1)+33561287506, aurelie.bellucci@cnes.fr,
(2)+33561282714, nathalie.fuentes@cnes.fr,
(3)Ana.Guerra-Algaba@apsys-airbus.com,
(4) Morgan.Cointe-Fourrier@apsys-airbus.com,
(5)+33561281678, jean-francois.goester@cnes.fr,

ABSTRACT

To fulfill its mission of control within the framework of the FSOA (French Space Operation Act), including in particular the monitoring of safety requirements for people and property, CNES has developed and has been using since 2010 the model and tool ELECTRA, especially designed to estimate human risk during launch and controlled or uncontrolled re-entry. The major contributor in probabilistic assessment of human risk is the population density on areas where some re-entry fragments may fall. However, airplanes fly over unpopulated regions such as oceans, currently identified as safe regarding the fall of space debris. Then, what is the risk for air traffic over North Atlantic in case of a launch failure? What is the risk for air-traffic over equatorial regions regarding the recurrent falls of space debris?

This paper presents a model to estimate these risks. Development of this model relies on three independents facets: characterization of space debris (through dedicated CNES tool: DEBRISK), vulnerability of aircraft, data on air traffic. Those inputs are combined to estimate risk on aircrafts using ELECTRA and its associated tools with specific parameters.

1. INTRODUCTION

For nearly 8 years, the Electra tool has been used by CNES and French Space operators to compute the risk for populations during space operations (launch failure, random or controlled re-entry) to comply with FSOA requirements ([1], [3]). In the assessment of human risk, the risk induced on aircrafts is considered as much lower than the precision required for estimations in the frame of FSOA. Moreover, the population under risk is globally considered, whatever the activity and place (outside, in a building, in a car, train or plane). The method followed does not allow to separate the risk for people on ground from the risk for people in airplanes.

Thus, as the air traffic has been constantly increasing for many years, CNES carries out studies on this topic. Firstly, during launch phases, rockets fly over oceans, which are inhabited areas but with potentially high air traffic, in North Atlantic for instance. It could be necessary to estimate the risk to aircrafts in case of launch failures. Secondly, with the expanding population of debris in space, the number of falling debris has been (and will be) increasing. Random re-entries of rocket bodies or satellites parts may now occur anywhere on Earth between [-i, +i], i being the inclination of the orbit. We know that many re-entries will concern upper stages with a low inclination and a high eccentricity, so that they will take place around the equatorial circle. Finally, although the risk to aircraft is considered as very low compared to the risk to people and properties on ground, consequences could be high. So, it is necessary to study it precisely.

A primary study was presented in 2014 ([2]) dealing with the construction of air traffic density grids and a simplified model of plane vulnerability. In 2017, CNES has worked with the APSYS company (Airbus Group) to develop a model to compute the risk for air traffic of space operations. This work is articulated in four main steps presented here after: knowledge of debris population, vulnerability of airplanes, use of air traffic databases and computation of the risk using Electra additional tools.

2. DEBRIS POPULATION

In 2015, about 20 000 objects with sizes over 10 cm orbit the Earth. It's a risk for operating satellites and for populations on Earth in case of re-entry (about four fallouts of large debris each month). Falling debris are usually rocket bodies or satellites parts. Parts of those objects ablate during re-entry but some parts are very resistant and fall on Earth such as titanium spheres or fuel tanks.

3. AIRCRAFT VULNERABILITY

Airplanes are designed to minimize the safety repercussion of various well defined trajectory hazards that can affect several systems or components. Looking for similarity between those particular risks and the impact of a debris, the vulnerability of the aircraft is evaluated.
The vulnerability of an aircraft being hit by a space debris is given by three main factors: A/C (aircraft) dimensions, A/C protection against trajectory hazards and size of the space debris.

3.1. A/C Dimensions

It has been defined an average aircraft in terms of dimensions and proportions, in order to use it as model for our study. Based on main aircraft manufacturers (Airbus and Boeing represent more than 90% of the world fleet) and considering long-distance flights models (those that are most likely to be found over the oceans and deserts), and considering the longest distance between length and span, the average characteristic length has been defined as **65,18 m.** This is the diameter of the circle surrounding the average aircraft. The aircraft occupies 27% of this surrounding circle (green circle in Fig 1), rounded up to 30%.

The vulnerable surface is defined as the surface of the aircraft likely to be hit by a space debris: horizontal view + frontal view (estimated pessimistically at 15% of its horizontal surface). Based on previous defined average aircraft, and the proportions of Airbus A350 this surface has been quantified at **1150 m²**.

Figure 1. Aircraft model

3.2. A/C protections against trajectory hazards

As part of the aircraft safety studies, the PRA (Particular Risk Analyses) are implemented in order to mitigate the effects of any Particular Risk, which are defined as those transverse risks which may affect several systems, structural components, areas and/or aircraft integration. They can be internal (hazard intrinsic to equipment design, technology, etc, such as engine burst), or external (hazards from environment, such as birdstrike). PRA aims at analysing the consequences of such an event at aircraft level and make sure that they are acceptable from a safety point of view.

Three main types of protections are implemented in order to reduce the aircraft vulnerability:

- Redundancy: duplication of equipment and energy and information channels.
- Segregation: appropriate physical separation between redundant systems and wiring.
- Dissimilarity: Different means to achieve the same function in order to avoid common modes.

For the potential impact of space debris with flying aircrafts, analogy with trajectory PRA has been analyzed, focused on the vulnerability of aircrafts. This analysis is performed based on different parameters:

- Affected area/zone depending on possible trajectories.
- Energy of the impact (capability to go through A/C surfaces and equipment).
- Protections to assure compliance with airworthiness requirements.

The different aircraft areas affected by possible trajectories have been defined based on the average aircraft (Fig.1), and thus been divided into 7 different zones:

A	Cockpit
B	Fuselage (fwd & bwd)
C	Wings
D	Engines
E	Fuselage (central)
F	Tail
G	Stabilizers

Table 1. A/C zones

Table 2 shows the list of the PRA analyzed with a short description, the affected zones, the energy of the PRA debris and the severity of worst consequences with regards to human beings:

PRA title		Affected areas	Description	Debris max energy [KJ]	Trajectory schema	Worst Consequences
UERF	Uncontained Engine Rotor Failure	C D E	Part of the engine rotor (3rd of the disk) is detached and its possible trajectories may impact the aircraft.	Infinite	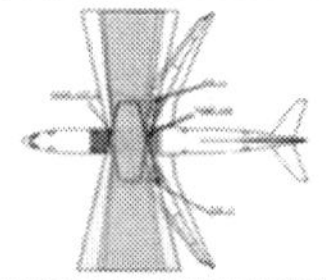	Potential deaths
UARF	Uncontained APU Rotor Failure	F G	Part of the Auxiliary Power Unit rotor is detached and its possible trajectories may impact the aircraft.	3,2	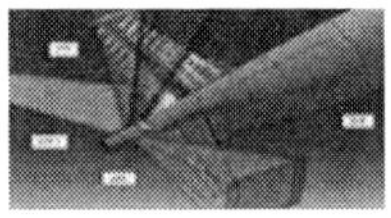	No human consequences
RBRF	RAT Blade Release Failure	C B E	A blade of the Ram Air Turbine (emergency power generator) is detached and its possible trajectories may impact the aircraft.	15,8		No human consequences
WTF	Wheel & Tire Failure	A C D E	Some of the wheels rim and tires come loose (rupture burst) and their possible trajectories may impact the aircraft.	24,7	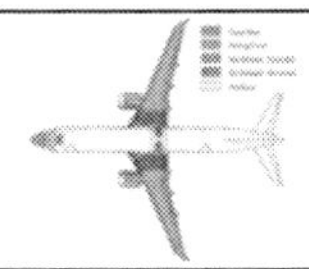	Reduction of safety margins. Potentially vulnerable
BS	Bird Strike	A C D G	Impact of a bird with aircraft.	128	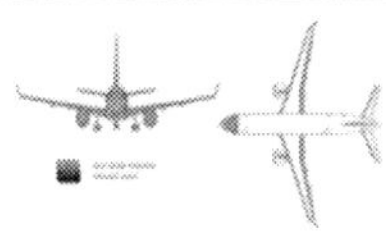	Reduction of safety margins. Potentially vulnerable
HAB	Hydraulic Accumulator Burst	E	Part of the hydraulic pressure accumulators is detached and its possible trajectories may impact the aircraft.	53		Potential deaths

Table 2. PRA Summary

3.3. Size of the space debris

It has been considered that a space debris larger than a square of 10cm x 10cm carries sufficient energy to perforate the structure of the aircraft. This threshold of 100cm2 is chosen to estimate the consequences on A/C of an eventual impact with a space debris in free fall.

The limit of 10 cm is also compliant with most of segregation principles installed in commercial aircrafts. So that a debris smaller than 10cm won't hit simultaneously two redundant or segregated systems. Beside, CNES studies using DEBRISK tool have shown that debris smaller than this size are ablated during reentry.

The trajectory of the space debris has been considered as vertical at the time of the impact, then slightly deviated due to the A/C speed and resistance as shown in Fig 2.

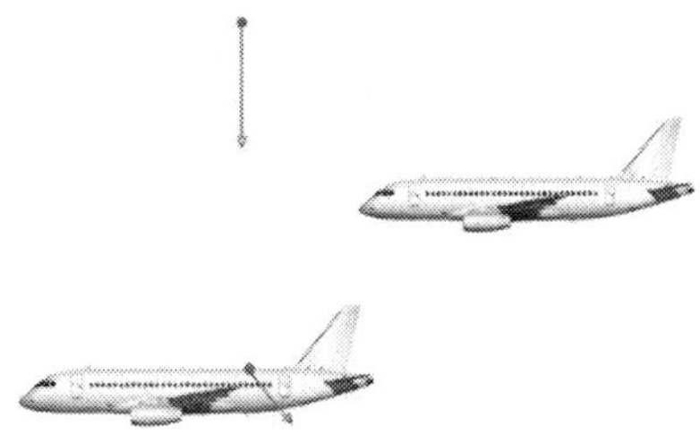

Figure 2. Space debris trajectory

3.4. Conclusion for aircraft vulnerability

Results of this analysis are summarized in Table 3. It presents, for each zone of the average aircraft, the vulnerability to the impact of a debris. It appears that 79% of the vulnerable surface might lead to the loss of at least one passenger/crew member in case of impact of a space debris bigger than 100cm^2 in section.

Zone		% A/C surface	Surface [m²]	Worst consequences
A	Cockpit	**2%**	23,02	Potential deaths
B	Fuselage (fwd & bwd)	**19%**	218,69	Potential deaths
C	Wings	**48%**	552,49	Potential deaths
D	Engines	**5%**	57,55	No human consequences
E	Fuselage (central)	**10%**	115,10	Potential deaths
F	Tail	**3%**	34,53	No human consequences
G	Stabilizers	**13%**	149,63	No human consequences

Table 3. Consequences per A/C zone

This analysis proves that the impact of a space debris on a flying aircraft would not automatically lead to a crash of the aircraft with the loss of passengers. Protections already would existing in aircrafts partly protect them from some impacts depending on the energy of the debris and the impacted area.

However, the death of at least one person is a catastrophic situation in FSOA rules. Consequently, it has been retained the conservative approach that states that "any impact of a space debris with a flying aircraft would lead to a catastrophic situation".

4. AIR TRAFIC DATABASES

Air traffic is characterized by great disparities depending on geographical areas and time of the day. The identification and the use of aeronautical databases make it possible to model the density of the air traffic and match the most possible with reality.

4.1. First analysis provided in 2014

In 2014, SERTIT worked with ICAO (International Civil Aviation Organization) database. This database contains for all flights listed the origin and arrival airport and the number of flight between each pair of airports. These vectors are transposed and represented by lines on earth globe. A grid of airspace density appears. This map has been arranged to show density of aircraft in cells of 1° and density around airports.

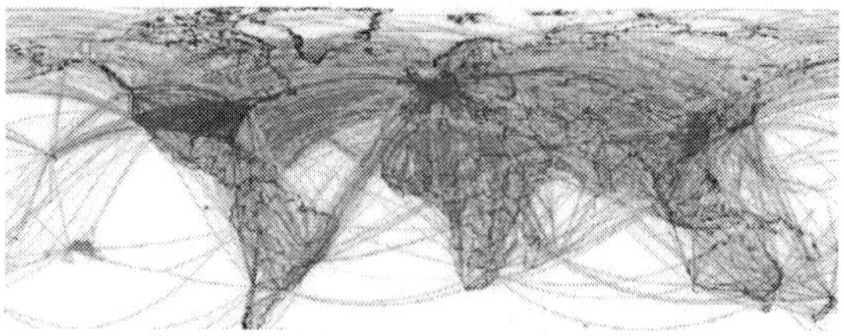

Figure 3. ICAO airspace density map

Main limits of this model are:

- Real aircraft paths are not straight lines.
- Computed density is an annual average. Though, density variations are important following the time in the day
- Only two mean speeds are taken into account: speed around airports and cruise speed.

The possible improvements to this first model are:

- Database choice and use of up to date database.
- Integration of air tracks (North Atlantic for instance).
- Integration of ETOPS (Extended-range Twin-engine Operational Performance Standards) rules.
- Real flight trajectories (take-off and landing procedures ...).
- Realistic speeds in cruise and around airports.
- Integration of traffic real time.

4.2. Databases

Several commercial flights databases are available with different sources, coverage, completeness and costs. Apsys has chosen to work with the ICAO and Flightradar24 databases.

ICAO inventories all scheduled flights and can provide a list of flights for a defined year. Air traffic doubles every fifteen years. SERTIT provided an analysis with ICAO database with 2012 data. Today, database containing 2015 flights is available and would allow updating airports list and number of flights (around 15-20% higher).Global density and completeness will also be more accurate.

Flightradar24 mostly contains data (aircraft model, airline, ground speed, vertical speed, altitude, position, scheduled trajectory) provided by aircraft themselves through ADS-B (Automatic dependent surveillance – broadcast). This signal is periodically broadcasted by aircraft transponders during the flight and received by ground radar. Departure and arrival time are taken from ICAO database and airports. Figure 4 shows results of computed data, each yellow dot represents an aircraft, moving in real time and providing parameters described previously.

Figure 4. FR24 air trafic map

4.3. Methodology and evolutions

Depending on the choice of database, the methodology differs. But the two methods produce a density map of air traffic to be used in Electra to compute the risk related to the fall of a space debris.

- Model update with ICAO 2015 database:

The method developed by SERTIT with a 2012 database can be updated with:

- the use of ICAO 2015 database,
- the adjustment of parameters such as cruise speed and speed around airports.

However, there is no possible evolution regarding:

- real trajectories,
- density around airports,
- tracks location,
- real time features.

A lot of non-accuracy remains with this approach, mainly due to the representation of aircraft path as straight lines. The result of this model is a density map that can be integrated in ELECTRA tool to compute the risk.

- Model update with Flightradar24 2016 database:

Flightradar24 gives a real evolution of this model. All needed parameters are provided in this database and allow integrating real trajectories, density around airports, tracks location and time notion.

The main interest of this database is to provide a map with the real routes followed by aircraft for area covered by ground radar. However, ADS-B signal can only be received until 300 miles from coasts. Beyond that limit, oceanic zones are not covered

Likewise, African continent is poor in ADS-B ground radar. , Information of real aircraft paths is lost for trajectories over uncovered area. In those cases, missing information is represented with a straight line as for the use of ICAO database.

Figure 5 shows the whole set of ADS-B signals received by ground radars over a 2 year period (2014 & 2015). Each point contains enough identification information to link together points corresponding to the same flight. The number of aircraft flying over each 1° cell of the grid during a defined period produced the expected density map (Fig 6).

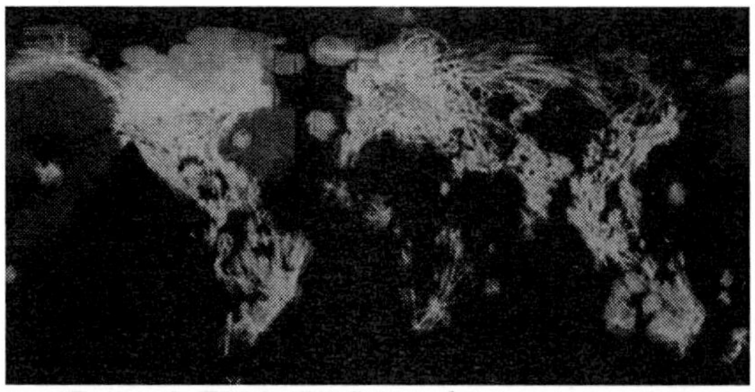

Figure 5. FR24 ADS-B signal transmission map

Figure 6. FR24 air trafic density map

Flightradar24 provides real time information showing for instance the variation of air traffic over the day. For instance, at 0h00 UTC, air traffic density is very high over North America and very low in Europe, while at 10h00 UTC the situation is reversed.

As ELECTRA doesn't manage temporal variation of density map, several density maps can be produced not to lose this important information:

- A set of 2 maps for 'worst case' or high density (for instance from 06h00 to 22h00) and for 'low density' (for instance from 22h00 to 06h00).
- A set of N density maps for required time frame (for instance, one density file every 2 hours).

4.4. Conclusion for air traffic database

The use of updated ICAO database will improved previous SERTIT model for the number of flights and aircraft speeds (cruise speed and around airports). The use of flightradar24 would include in the model all parameters enhancement.

Final choice between these two databases has yet to be done, but high cost of FR24 might put a brake on this choice.

5. RISK MODEL USING ELECTRA TOOLS

ELECTRA software has been designed by CNES to support French Space Operation Act. It computes the risk of making a victim due to atmospheric re-entries in several contexts ([1], [4]):

- RA mode for uncontrolled space objects (satellite or launcher part). The method considers only the latitudes the object flies over, meaning the risk depends on the inclination of the orbit.

- RL mode for launcher trajectory eventually dispersed. The method computes the risk due to failures occurring during this launching phase.

- RC mode for controlled trajectories following deorbit maneuvers. The method evaluates the risk associated with maneuver failures.

- RF mode: in this case, we consider uncontrolled re-entries but only some days before the final fallout. It is then possible to compute the risk more precisely than in RA mode. This mode is only available since the V4.1 version.

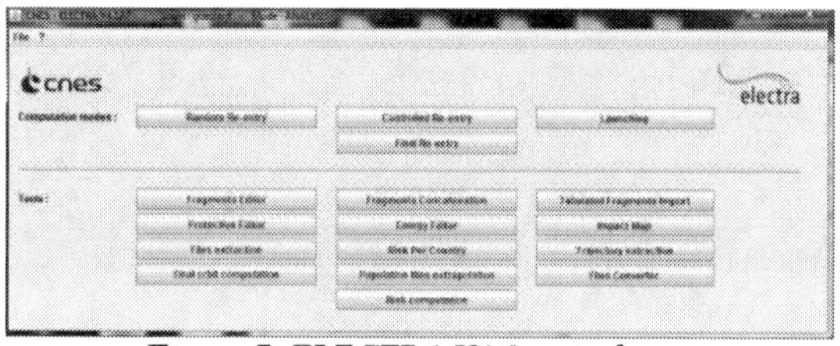

Figure 7. ELECTRA V4.1 main frame

To compute the probability of incurring at least one victim, ELECTRA uses, among others parameters, the vulnerable surface of a person (surface of the disk containing a stand-up human being), the density of population (population grids), the estimated energy of the impacting space debris, a protection coefficient (which considers roofs, buildings, etc.) and the type of re-entry space debris.

Changing some of the parameters, ELECTRA can be used to compute the risk for aircrafts during re-entry of space debris.

- The diameter of a person will be replaced by the diameter of the circle with surface equal to the vulnerable surface of the average aircraft computed above (section 3.1): 1150 m^2. This corresponds to a diameter of 38.28 m.
- The population grid will be replaced by an air traffic density map computed using one of the method presented in section 4.3.
- As we assume that any impact of a space debris with an aircraft would lead to a catastrophic situation, protection coefficients will be set at 1 (maximum probability).

This study will benefit from a new tool included in last version of ELECTRA V4.1, to be released by the end of 2017. This "a posteriori risk computation" tool allows re-computing risk much faster than the nominal modes. Indeed, in nominal modes, the Monte Carlo simulations with thousands of trajectory propagations require a lot of CPU time. But using the resulting impact file, the new tool can compute in a few seconds the risk for the same trajectories with different inputs for population density, protection coefficients or diameter of a person.

To compute the risk of atmospheric re-entries for aircraft, this new tool will be used with existing trajectories and the parameters specified above for the vulnerable surface, the air traffic density map and protection coefficients. This analysis is still in progress. Numerical results will be available by the end of 2017.

6. CONCLUSION

So far, the risk for aircrafts to be hit by a space debris in its re-entry phase has been considered negligible in terms of probability. This study should provide qualitative and quantitative arguments to validate this assumption, considering the increase of both the number of space debris and the air traffic.
The vulnerability of aircrafts is mitigated by already installed protections to external hazards (PRAs). The fall of a debris on a flying aircraft might not induce a crash but could lead to the death of at least one person, which is a catastrophic situation in FSOA rules. Hence, the analysis uses the conservative approach that any collision with a space debris would make a victim.
To compute the risk using ELECTRA tool, a density map of air traffic is required. Two databases can be used to build this map: updated ICAO as in previous study or Flightradar24 with its exhaustive information from aircraft real-time tracking. As the choice has yet to be done, numerical results about risk computation are not available. ELECTRA tools will be used for that computation using data on aircraft vulnerability and air traffic grid in place of usual data on populations.

7. REFERENCES

1. Hourtolle C., Gaudel A., Blazquez A. (2012). ELECTRA: Launch and re-entry risk analysis tool. *5th ICATT (International Conference on Astrodynamics Tools and Techniques)*

2. Fuentes N, Tholey N., Studer M. (2014) Risks of atmospheric re-entries on aircrafts, CNES progress non studies. *7th IAASS (International Association for the Advancement of Space Safety)*

3. Goester J.F., Bellucci A. (2017) JELECTRA : new features of the Cnes launch and re-entry risk analysis tool, *9th IAASS (International Association for the Advancement of Space Safety)*

BENCHMARK OF JAXA AND CNES RE-ENTRY SAFETY ANALYSIS TOOLS FOR ACCURATE HEAT-FLUX PREDICTION

Keiichiro Fujimoto(1), Hideyo Negishi(1), Saito Yasuhiro(1), Martin Spel(2), Guillaume Prigent(3)

(1) *Japan Aerospace Exploration Agency, 2-1-1, Sangem, Tsukuba, Ibaraki, Japan, fujimoto.keiichiro@jaxa.jp*
(2) *R-Tech, Parc Technologique Delta Sud, 09340 Verniolles, FRANCE, martin.spel@rtech.fr*
(3) *Centre National d'Etudes Spatiale, 18 Avenue Edouard Belin, 31400 Toulouse, France, Guillaume.Prigent@cnes.fr*

ABSTRACT

Comprehensive consideration on the design and the disposal operation based on the spacecraft-oriented re-entry risk analysis is effective to minimize the space debris-related risk. JAXA and CNES have been working on the joint research project especially for the benchmark study of the aerodynamic and heat flux models to improve those accuracies. Basic shape such as cylinder and box, and the rocket upper stage was considered. Computed results by the correlation models were compared with those of the computational fluid dynamics (CFD) analysis. Correlation model results were agreed well with the CFD results for basic shapes. Correlation model results for rocket upper stage were also agreed well in the pitch angle zero case. While, the discrepancy was found to be larger for the high incidence cases. It was shown that the large heat flux near the sharp corner can be quantitatively predicted by considering the small curvature radius effect. On the other hand, the heat flux discrepancies were larger for the concave shapes such as rocket engine nozzle, and larger when the down-stream object is inside the up-stream object's wake flow or in case of the shock wave impingement. It is essential to consider these key effects for further improvements in model accuracy.

1. INTRODUCTION

Space debris problem is a growing concern to be tackled internationally to keep our space activity sustainable. Debris mitigation is one of the essential approaches, which includes a set of strategies to reduce the number of orbital debris in the critical orbital zones for long term sustainability. The derelict spacecraft and launch vehicle orbital stages those are passing through the protected low earth region should be removed by controlled re-entry or the reduction of orbital lifetime followed by natural re-entry. The destructive re-entry of satellites and rockets and the related ground risk due to fragments reaching the ground are getting increased interest in recent years. The related risk should be minimized as much as possible by the comprehensive consideration on the design for demise and the disposal operation based on the mission analysis and the risk assessment. In the re-entry risk analysis, detailed demise and destruction process should be analysed based on the multidisciplinary physics-based models such as the attitude and trajectory, the aerodynamic characteristics, heat flux, the thermal transfer, the mechanical and thermal destruction under the wide range of environmental conditions. Shapes of the spacecraft and the launch vehicle upper stage are complicated and it changes in time, numerous fragments are generated after the break-up. Such behaviour is directly modelled to predict the risk without the simplification in the spacecraft-oriented approaches, so that the design for demise consideration is applicable. Remained challenge to establish the spacecraft-oriented approach is the validation of the physics-based models for various shapes under the wide range of conditions. Aerodynamic characteristics and the heat flux model have significant effects on the resulting risk, and further research efforts should be made to improve those accuracies.

JAXA and CNES have been working on the joint research activity wherein benchmark study of the aerodynamic characteristics and the heat flux models. At first, the basic shapes such as spheres, boxes and cylinders are considered. Then, the rocket upper stage is considered. As already discussed in the previous study [1], the empirical correlation models are validated by the comparison with the high-fidelity numerical simulation and the ground and flight experiment data. In this study, the heat flux correlation models are validated by comparing with CFD data, and the flow mechanism is investigated. Finally, further research items are identified to achieve the ultimate model accuracy to establish the spacecraft-oriented risk analysis methodology.

2. OBJECTIVES

In order to establish the spacecraft-oriented re-entry risk analysis, research efforts to improve the model accuracy of the aerodynamic characteristics and the heat flux models are essential. Model uncertainty should be quantified for various shapes under the wide range of flow conditions. Heat flux distributions over the fragment's surface is generally evaluated by the reference heat flux magnitude and the distribution function. This formulation is employed in most of the re-entry risk analysis methods. Those are formulated based on the experimental and analytical work for the basic shapes such as spheres, cylinders and bluff-bodies. Since most of the previous investigations have been carried out basically for the usual re-entry capsule and vehicles, so that further research effort should be taken to understand the high-enthalpy flow physics for the complicated shapes.

Uncertainties on the heat flux correlation models used for the spacecraft-oriented re-entry risk analysis are quantified by the comparing with CFD results in this study. At first, the basic shapes such as cylinder and box are considered to cover important shape features such as smooth curved surface, the flat faces, and the sharp corners. Secondary, the rocket upper stage is considered for the benchmark study to cover the flow interactions between multiple components and the concave shape. Investigation to understand important flow mechanism and to identify the main cause of the discrepancy between the correlation model result and the CFD result. Finally, further research topics to establish quantitative re-entry safety analysis are summarised.

3. NUMERICAL APPROACHES

In this section, the numerical methods employed at JAXA and CNES to predict the aerodynamic characteristics and the heat flux distributions are described respectively. Since probabilistic analysis is essential for the expected casualty prediction, high fidelity numerical simulations are not practical. High fidelity numerical simulation is used for the uncertainty quantification of the correlation models and the investigation on the assumption employed in the re-entry risk analysis.

3.1. LS-DARC (JAXA)

Orbital re-entry survivability analysis tool of Japan (ORSAT-J) is object-oriented re-entry safety analysis tool used in JAXA, which is derived from NASA-ORSAT ver. 4.0, which has been widely used for the risk analysis of spacecrafts and rocket upper stages. It is still used as the research code, however, investigation and development of the spacecraft-oriented re-entry risk analysis tool, LS-DARC is under the way at JAXA [1]. Current capability and the physics-based formulations such as the aerodynamic characteristics and the heat flux are almost the same with the other risk analysis tools [2, 3]. In the heat flux model, the surface local curvature effect can be considered [4].

3.2. PAMPERO (CNES)

PAMPERO is a spacecraft oriented simulation tool developed by CNES [5] and has been used for the simulation of spacecraft re-entry and breakup. The trajectory, attitude, and the thermal ablation of the spacecraft are evaluated based on the aerodynamic forces and moments and heat fluxes obtained from the correlation models. It covers all the environmental conditions encountered during re-entry ranging from the free molecular to the continuum regime. For the evaluation of the surface pressure and the heat flux distributions, the complicated real shape effects are considered such as the shadowing effect and the local curvature effect on the heat flux level. Further development of the destruction model for the thermos-mechanical loads is under the way [6].

3.3. LS-FLOW (JAXA)

At the moderate altitude, roughly bellow 90 km, the flow can be treated as continuous. For the prediction of the surface pressure and heat flux distributions, the compressible Navier-Stokes flow solver for the arbitrary polyhedral unstructured grids, named LS-FLOW [7] is employed in this study. It is the core CFD code of JAXA, which has been applied to various flow problems including the aerodynamic analysis [8], high temperature reactive flows and the cryogenic flows [9] for the liquid rocket engines. It was initially developed by the present author [7] and has been improved continuously, various numerical schemes have been implemented [8]. LS-FLOW has been validated by comparing with the experimental data for the various geometries under the wide range of flow conditions [10, 11]. Then, it has been applied to various aerodynamic design problems such as the aerodynamic analysis for the expendable and re-usable rockets and the re-entry capsules, and the capsule outer shell separation in the chaotic freestream conditions.

The system of equations is discretized and solved by the cell-centered finite volume method based on the arbitrary polyhedral unstructured grids. Various numerical schemes have been employed such as for the gradient evaluation of flow variables, the gradient limiter functions, Euler flux, and the viscous flux. Some of the schemes have been developed in JAXA in order to establish the robust and still spatially accurate aerodynamic analysis methodology for the turbulent high Reynolds number flows with including the strong shockwaves and the massive separated flows. Various turbulence models and the high-order reconstruction scheme for the unstructured grids [12] are available. Further detailed descriptions on the key features can be followed in the previous publications [11].

For the capability to handle moving multiple complicated objects, LS-FLOW has the capability to handle the arbitrary polyhedral unstructured grid, and the overset grid method with including 6DoF motion. For the computational grid, body-fitted Cartesian grid generated by LS-GRID [13] is used in this study, in which Cartesian grid is used as the volume cells and the body-fitted layer grid is used to resolve high Reynolds number boundary layers. Since volume cells are topologically Cartesian grid, and thus the AMR method is available, which is essential to achieve practical turn-around time to consider the various types of geometries under the wide range of flow conditions.

3.4. MISTRAL (CNES)

For the evaluation of the aerodynamic characteristics and the heat flux distributions at the continuum regime, three-dimensional Navier-Stokes solver, MISTRAL is employed. The gas behaviour can be described as the calorically perfect gas, the thermally perfect gas, the real gas under the assumption of equilibrium conditions, or a mixture of thermally perfect gases in chemical or thermochemical non-equilibrium. The weak ionization model is available for the chemical non-equilibrium flow. In addition, various catalicity effect models are available for the different materials.

A finite volume method on domain decomposed block structured grids is used to exploit the benefit from massively parallel computing architectures. For the inviscid fluxes, the AUSM-type scheme is used in the main flow regions where second order spatial accuracy is obtained using MUSCL interpolation. In regions of shocks the accuracy is reduced to first order and the AUSM scheme is blended with the more diffusive Van Leer fluxes. During the last 15 years the code has been validated on several national and European projects such as MSRO, IXV [5][14] and ExoMars [15] under which new models such as re-radiation, catalicity, energy deposition (SINAPS).

4. RESULT AND DISCUSSIONS

At first, basic shapes such as box and cylinder, and the rocket upper stage are considered. In the object-oriented approach, the spacecraft shape is treated as the combination of the base shapes such as sphere, box, and cylinder. Since most of the correlation models to predict pressure distributions, the stagnation heat flux, and heat flux distributions are derived from the experimental or analytical studies for these kinds of basic shapes. These correlation models are also used for the reference value evaluation for the spacecraft-oriented analysis. Therefore, the correlation models had already been validated and expected to have sufficient accuracy for the basic shapes. Therefore, the basic shapes are employed for the benchmark case for the easier identification of the prediction discrepancy with considering less numbers of uncertainty factors.

At second, the rocket upper stage is considered, which is the key driver of the ground risk. It is multiple shape problems including the concave shape, the shadowing effect, and the flow interactions. These effects are generally neglected in most of the re-entry risk analysis, so that it should be worthful case to identify the research items for the model accuracy improvement.

4.1. BASE SHAPES

The aerodynamic characteristics and the heat flux distributions are evaluated for the box and the cylinder under the hypersonic flows, free-stream Mach number M_∞ is 11.72 for the box, 14.35 for the cylinder. Considered pitch angles are 0, 30, 60 and 90 degrees. LS-FLOW of JAXA is used to obtain reference distributions to compare with the correlation model results. The dimension of the box is $0.11 \times 0.085 \times 0.085$ m, and length of cylinder is 1.97 m and its radius is 1.665 m. Computational grids are shown in *Figure 1.* Minimum grid spacing size for the box is 3.36×10^{-6}, and that for the cylinder is 1.39×10^{-5} those are corresponding to the cell Reynolds number 0.5. Total cell number is 2.64 million for both shapes.

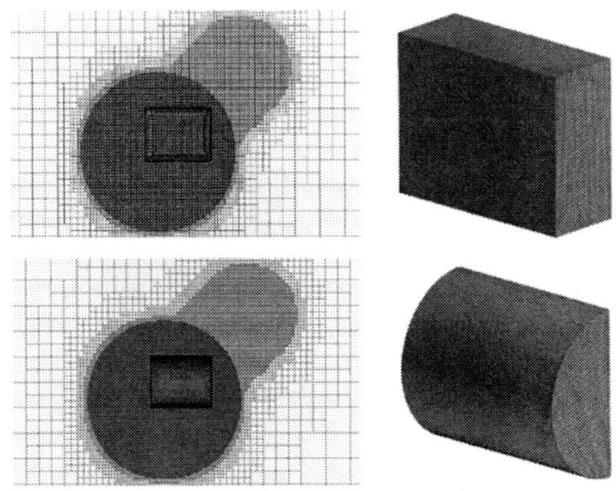

Figure 1. Computational grids for box and cylinder.

Computed heat flux distribution on the surface and the temperature distribution in the symmetric plane are shown in Figure 2. Strong detached shockwave is observed in front of the body, and the temperature level near the surface is more than 9000 K. Local maximum of the heat flux is observed near the sharp corner.

Comparisons of the heat flux distribution in the symmetry plane are shown in Figure 3. Global trend and the magnitude level of the heat flux are predicted quantitatively by using the correlation model. Especially the heat flux distribution by the correlation models for windward surface is in good agreement with CFD result.

As shown in the CFD results, the significant increase of the heat flux magnitude at the sharp corner is observed. This effect is also successfully predicted by both correlation models by considering the local curvature effect in the formulation. Peak values predicted by the correlation model are generally smaller than CFD results.

Heat flux level for the shadowed leeward surface evaluated by the correlation models are smaller than that of the CFD results. Although heat flux magnitude is assumed to be zero in the correlation model, some amount of heat flux is observed in the CFD results as shown in *Figure 3*. This fact is also observed in the previous experimental works [16].

Computed heat flux distributions for cylinder evaluated by the correlation models are compared with the CFD results in *Figure 4*. Evaluated heat flux distributions in the smooth side surface of the cylinder are agree well with the CFD results. As also already discussed in the discussion for the box, the increase of the heat flux at the sharp corner is successfully predicted by the correlation models. The value is generally underpredicted by the correlation model as comparing with CFD results.

It is shown that the prediction accuracy of the correlation models of both JAXA and CNES are generally good agreement with the CFD results, especially in the windward surface and the sharp corners. Further research efforts to improve prediction accuracy for the shadowed leeside surface should be made in the future research. In addition, the heat flux distribution over the flat surface becomes larger for the larger pitch angle cases as already discussed in the present author's study [4]. It was shown that the temperature distribution change due to the turbulent flow effect is the root cause. In most correlation models, this effect is not considered, therefore, it is also one of the research topics in the future research.

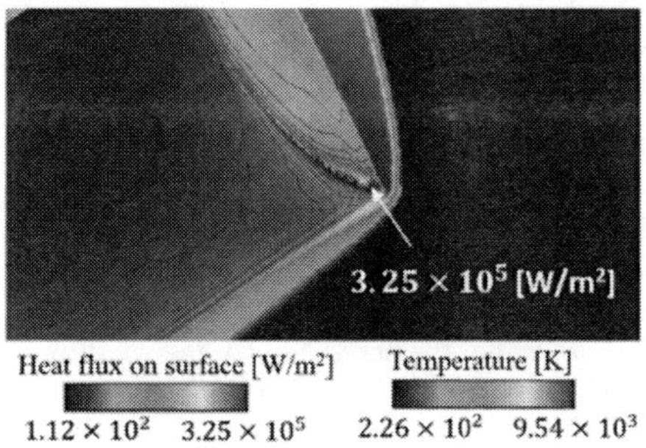

Figure 2. Heat flux and temperature distribution in the symmetric plane for $\boldsymbol{M_\infty} = \mathbf{14.35}$ *and pitch angle 30 deg.*

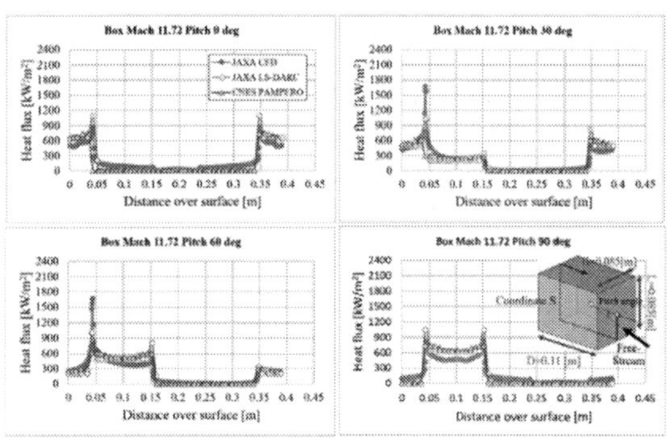

Figure 3. Comparison of heat flux distribution in the symmetry plane for the box at $\boldsymbol{M_\infty} = \mathbf{11.72}$.

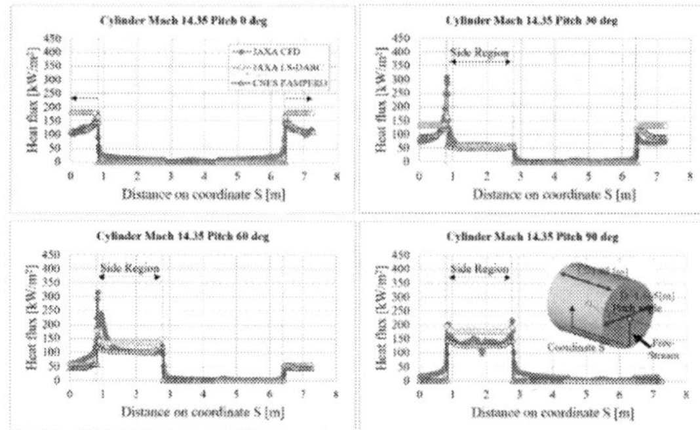

Figure 4. Comparison of heat flux distribution in the symmetry plane for the cylinder at $\boldsymbol{M_\infty} = \mathbf{14.35}$.

4.2. ROCKET UPPER STAGE

Risk mitigation on the rocket upper stage is getting increased interest in recent years. Uncertainties on the aerodynamic and heat flux models are one of the dominant factor on the resulting expected casualty of the re-entry safety analysis. In this study, simplified geometry of the rocket upper stage is considered, and the its aerodynamic characteristics and the heat flux distributions are evaluated by the CFD and the correlation models. Considered altitude is 45 km, the flow speed is 3467.8 m/s which corresponds to the Mach number $\boldsymbol{M_\infty}$ 10.64.

Half of the simplified upper stage shape is considered in this study. Computational grid for the LS-FLOW is the unstructured body-fitted Cartesian grid as shown in *Figure 5*. Total number of cells is 3.97 million. Minimum grid spacing size in the normal direction at the wall is 1.85×10^{-6} m, which corresponds to cell Reynolds number 5.0. Computational grid for the MISTRAL is three-dimensional structured grid which has 7.94 million cells as shown in *Figure 6*. Minimum grid spacing size in the normal direction at the wall is 1.0×10^{-6} m.

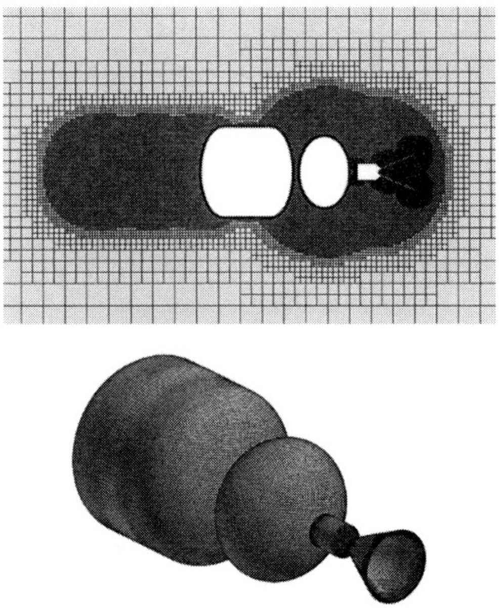

Figure 5. Computational grid of the rocket upper stage for LS-FLOW of JAXA (3.97 million cells).

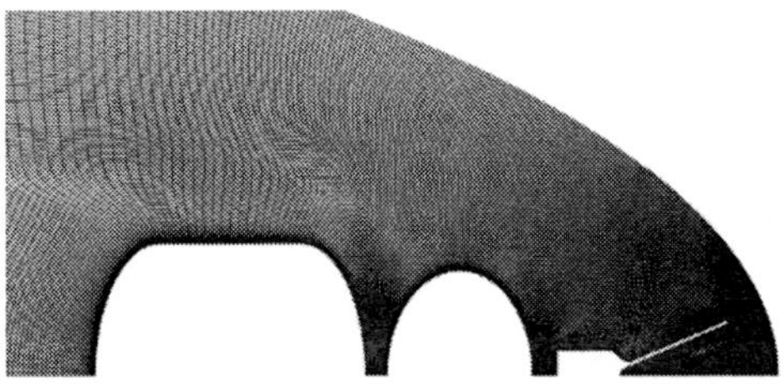

Figure 6. Computational grid of the rocket upper stage for axisymmetric computation of MISTRAL of CNES (7.94 million cells).

Time-averaged computed temperature distributions in the symmetry plane and the heat flux distribution are shown in *Figs. 7* and *8*, respectively. Angle of attack zero corresponds to the condition the free-stream direction is coincide with the vehicle's body axis as shown in *Fig. 7*. At these conditions, the high speed free-stream is directly flowing into the nozzle, and thus recirculating flow is observed in both cases. It is resulting in the unsteadiness of the up-stream detached shock-wave, and the unsteady strong shear layers initiated from the nozzle rip as shown in *Fig. 7*.

This shear layer and the high-speed flow outside the shear layer hits on the lee-side LOX and LH2 tanks respectively. As a result, local peak of the heat flux is observed at the wind-side region A for LOX tank and the region B for LH2 tank as shown in *Fig. 7*.

At the 30 degrees pitch angle case, the magnitude of the heat flux is significantly larger than that of zero pitch angle case as shown in *Fig. 8*. At high incidence conditions, the free-stream directly impinges on the windward surface of the LOX and LH2 tanks as shown in *Fig. 8*. In addition, the shear-layer and the shock-wave interaction are observed and it is resulting in the large heat flux as shown region C and D in *Figure 8*. Furthermore, the hot temperature region is observed behind the strong detached shock-wave in front of the tanks. Large local heat flux is observed at the nozzle rip region which has larger local curvature in both cases. Maximum heat flux for zero pitch angle case is observed at the nozzle rip region, that for 30 degrees pitch angle case is observed at the windward surface of LOX tank.

Time-averaged computed heat flux distributions in the symmetry plane obtained by LS-FLOW are shown in *Fig. 9* for the engine, LOX tank, and LH2 tank, respectively. All origins for horizontal plot axis are corresponding to the windward point on the body axis. As described above, the heat flux magnitude levels for pitch 30 degrees are much larger than that for the pitch zero case. Heat flux magnitude level for pitch 30 degrees is roughly 5 to 10 times larger than as comparing with that for zero pitch angles case. Temperature level behind the frontal detached shock-wave is almost the same, however, near wall temperature gradient is different due to the velocity distributions in the re-circulating flow inside the nozzle. Temperature gradient in the surface tangential direction is small at inside the nozzle in both cases. Local peak heat flux can be observed at the nozzle rip region, this is due to the smaller curvature radius. In the pitch 30 degrees case, there is also larger peak at the chamber outer shell area E as shown in *Figure 8*. Magnitude of the peak heat flux is more than 1 MW/m^2 as shown in *Fig. 9*.

In the zero pitch angle cases, there are local peaks at the wind-ward shoulder A and B shown in Figure 7 and Figure 9. Heat flux levels at the windward dome of LOX and LH2 tanks for pitch 30 degrees case are much larger than that for zero pitch case. In the zero pitch angle case, most extent of the LOX and LH2 tank is located inside the wake generated by the upstream objects. This is the main cause of the smaller heat flux level in this case. On the other hand, in pitch 30 degrees case, the heat flux level becomes very large mainly due to the higher temperature at near surface just behind the shock wave and the larger flow rate coming toward the wall. This is the main cause of the larger heat flux level for the high incidence conditions. Furthermore, large heat flux is observed at region C in the pitch 30 degrees case. This is due to that the shock wave generated by the engine impinges on the LOX tank's dome part as shown in *Fig. 8*.

Computed heat flux distributions by the correlation models for the re-entry risk analysis, LS-DARC and

PAMPERO are compared with CFD results in *Figs. 10* and *11*. Predictions by the correlation model are in good agreement with the CFD results for the zero pitch angles cases. It was shown that the heat flux distributions even for the concave nozzle and the sharp corner of nozzle rip can be quantitatively predicted.

Heat flux distributions inside the nozzle evaluated by CFD is almost constant. This is mainly due that the recirculating flow inside the nozzle which is confined and the free-stream cannot directly get inside the nozzle. Recirculating flow inside the nozzle is compressed by the dynamic pressure of the high-speed free-stream coming inside the nozzle. There is still some discrepancies between the correlation model and the CFD. The heat flux gradient inside the nozzle is almost the constant and its level would be described by small number of the parameters such as the dimensions of the nozzle and the free-stream conditions, so that there is a chance to establish the general formulation of the empirical reduced order model. It is one of the research targets in the future.

In the evaluation by the correlation models for the 30 pitch angle case, the same formulation and the same model parameters are used. Resulting evaluated heat flux level is much smaller than the CFD result. Heat flux level predicted by CFD is roughly 5 times larger than that for the correlation model. As already discussed previously, it is mainly due to the flow interaction effects between the multiple objects. The flow interaction effects such as the shock interaction and the higher flow rate are not considered in the current correlation model. It is essential to investigate the appropriate formulation to decide the reference heat flux level depending on the flow mechanism.

Finally, it is also clarified that some amount of the heat flux is observed even at the lee-ward region behind the body. In most of the re-entry risk analysis models, the heat flux magnitude at the shadowed surface behind the body assumed to be zero. Improvement in the model to consider this effect can be another research topic.

5. CONCLUSION

It was clarified that the accuracy of the correlation model used in the re-entry safety analysis is sufficient for the basic shapes such as box and cylinder. It is mainly due to that the formulations of the empirical models are derived from the analytical reduction for the simple shapes. The local heat flux peak near the sharp corners are also quantitatively predicted by the correlation model, in which the local curvature radius effect is considered.

In the benchmark studies on the rocket upper stage, the concave shape and the flow interaction between the multiple objects were considered. At the lower incidence case, the leeward objects were found to be surrounded by the wake flow of the windward objects. Therefore, it was resulting in the lower dynamic pressure for the leeside objects, so that the heat flux level of the leeside objects becomes very small. At the higher incidence case, the impingements of the shock wave and the strong shear layer on the surface of the leeside tank have been observed. These effects are not considered in the correlation models. Therefore, the predicted heat flux level by the correlation model is found to be much smaller than the CFD results.

It has shown that the predicted heat flux distributions by the correlation models in good agreement with the CFD results. On the other hand, the discrepancies are larger for the high incidence cases. In order to reduce the uncertainty to establish the accurate correlation model for the heat flux evaluation, the further research effort should be made to consider the lower dynamic pressure wake effect, the concave shape effect, the shock wave and the shear layer interaction effect, and the turbulence flow effect.

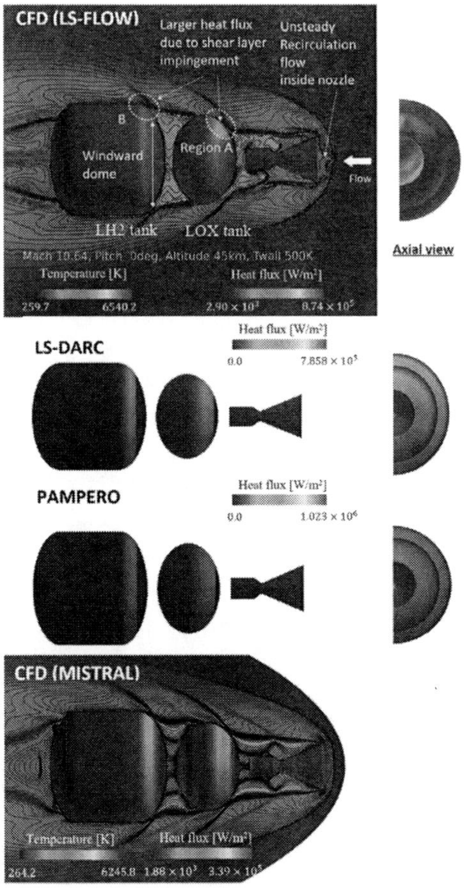

Figure 7. Computed temperature distribution in the symmetry plane and the heat flux distribution for $M_\infty = 10.64$ *and* $\alpha = 0$ *degree.*

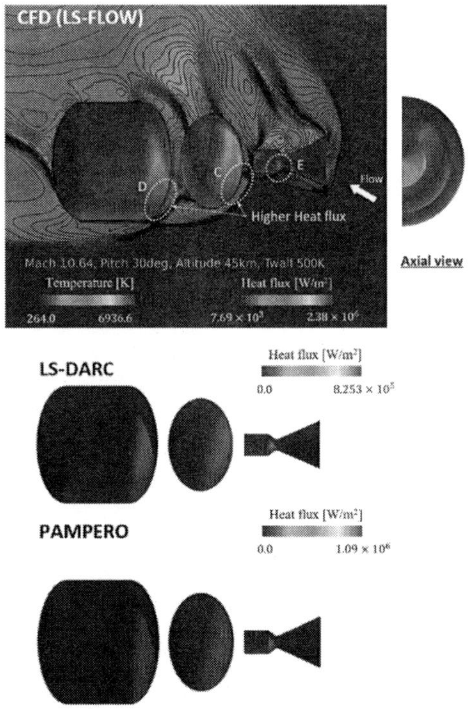

Figure 8. Computed temperature distribution in the symmetry plane and the heat flux distribution for $M_\infty = 10.64$ *and* $\alpha = 30$ *degrees.*

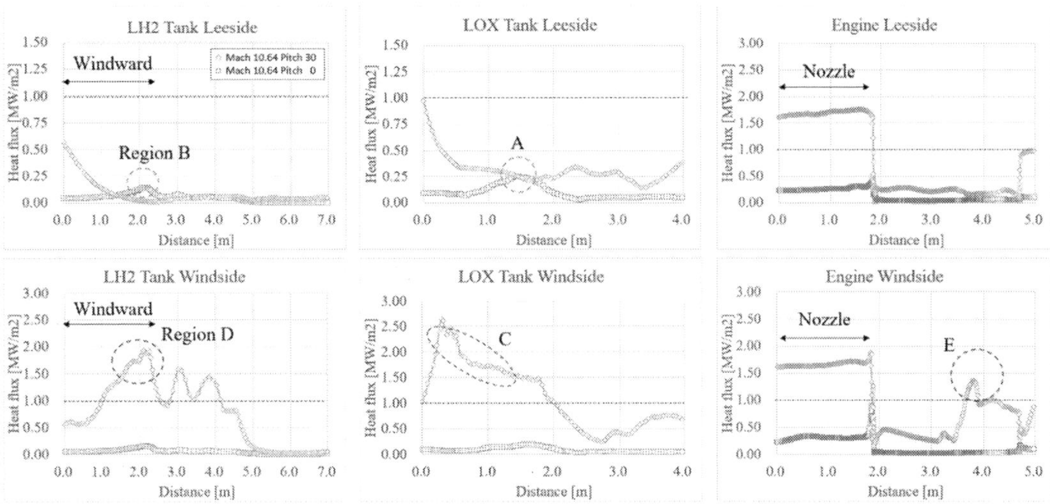

Figure 9. Computed heat flux distribution for rocket upper stage at $M_\infty = 10.64$.

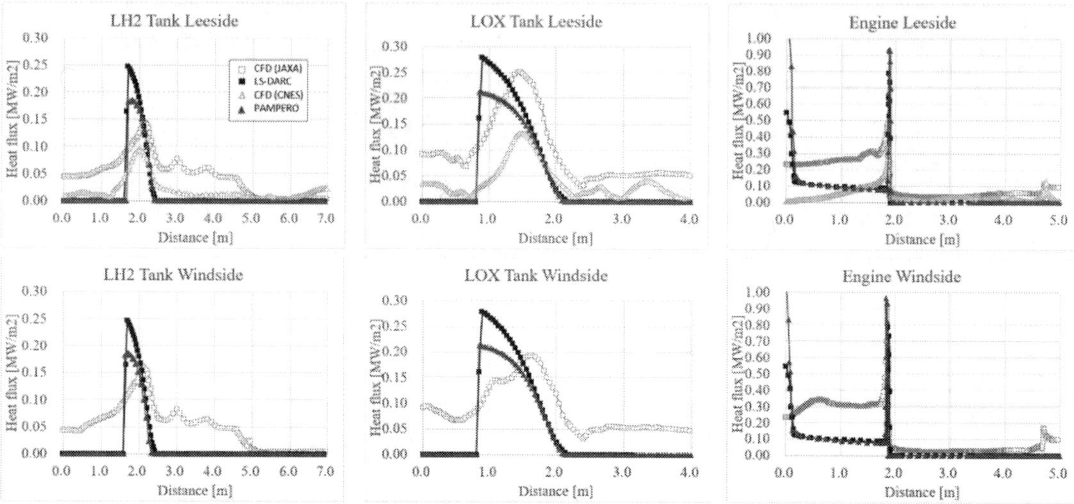

Figure 10. Computed heat flux distribution for rocket upper stage at $\boldsymbol{M_\infty = 10.64}$ *and* $\boldsymbol{\alpha = 0}$ *degree.*

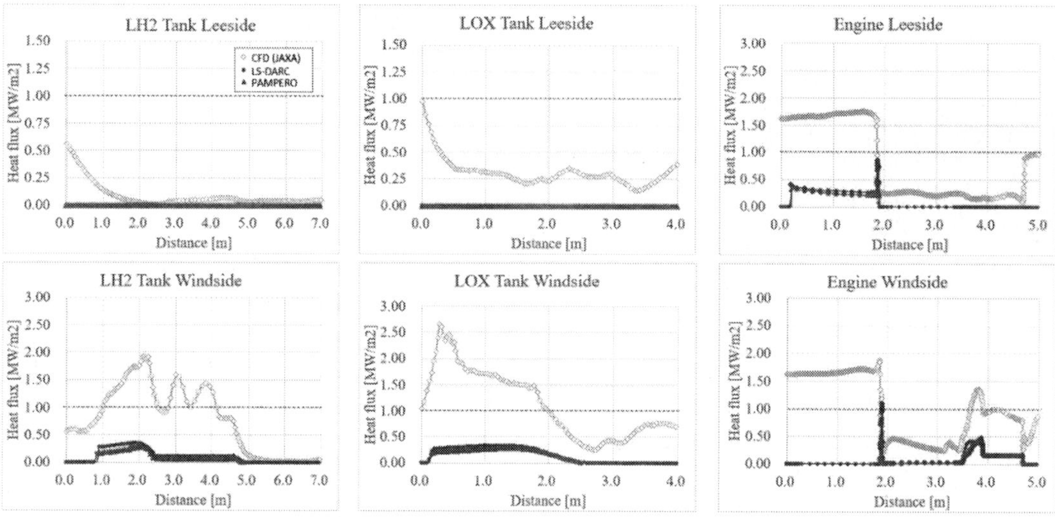

Figure 11. Computed heat flux distribution for rocket upper stage at $\boldsymbol{M_\infty = 10.64}$ *and* $\boldsymbol{\alpha = 30}$ *degrees.*

6. REFERENCES

1. Fujimoto, K., Tani, H., Negishi, H., Saito, Y., Iizuka, N., and Okita, K., "Update of aerodynamics and heat flux model for ORSAT-J", Proceedings of the 8th IAASS Conference, 2016.
2. T. Lips, B. Fritsche, M. Homeister, G. Koppenwallner, H. Klinkrad and M. Toussaint, "Re-entry Risk Assessment for Launchers – Development of the New SCARAB 3.1L", Proceedings of the 2nd IAASS Conference, 2007.
3. Annaloro, J. "Elaboration of New Spacecraft-orientated Tool: PAMPERO", 8th European Symposium on Aerothermodynamics for Space Vehicles, Lisbon, 2-6 March 2015.
4. Fujimoto, K., Tani, H., Negishi, H., Saito, Y., Iizuka, N., and Okita, K., "High-Fidelity Numerical Simulations for Destructive Re-entry of Upper Stages", 7th European Conference on Space Debris, 2017.
5. Tribot, J.P., Dutheil, S., Viguier, P., Vérant, J-L., Charbonnier, D., Vos J., Van Hauwaert, P., Spel, M., Ferrarella, D., Mareschi, V., Rufolo, G., "Intermediate Experimental Vehicle: Extrapolation Ground to Flight Wind Tunnel and CFD Approach", IAC-13,D2.6,5x17026, 64th International Astronautical Congress, Beijing, China, 2013.
6. Prigent, G., Carra, J., Crusson, B., Stainier, L., Omaly, P., "A first step toward fragmentation process assessment of re-entering spacecraft: mechanical stress analysis with the spacecraft oriented simulation tool PAMPERO", Proceedings of the 9th IAASS Conference, 2017.
7. Fujimoto, K., "Study on the Automated CFD Analysis Tools for Conceptual Design of Space Transportation Vehicles," Ph.D. Dissertation, Univ. of To-kyo, Tokyo, Japan, 2006.
8. Kitamura, K., Nonaka, S., Kuzuu, K., Aono, J., Fujimoto, K., and Shima, E., "Numerical and Experimental Investigations of Epsilon Launch Vehicle Aerodynamics at Mach 1.5," Journal of Spacecraft and Rockets, Vol.50, No.4, pp.896-916, 2013.
9. Negishi, H., Aono, J., Shimizu, T., Sunakawa, H., Sezaki, C., Nagao, N., Nan-ri, H., "Mixing Characteristics of Transcritical Hydrogen Flows around a Mixer in a Liquid Rocket Engine," JSASS-2016-2117-A, 2016 (in Japanese).
10. Kitamura, K., Fujimoto, K., Shima, E., Kuzuu, K., and Wang, Z. J., "Validation of Arbitrary Unstructured CFD Code for Aerodynamic Analyses," Trans-actions of the Japan Society for Aeronautical and Space Sciences, Vol.53, No.182, pp.311–319, 2011.
11. Fujimoto, K., Nambu, T., Negishi, H., and Watanabe, Y., "Validation of LS-FLOW for Reentry Capsule Unsteady Aerodynamic Analysis", AIAA Paper 2017-1411, 55th AIAA Aerospace Sciences Meeting, AIAA SciTech Forum, 2017.
12. Haga, T., and Kawai, S., "Toward accurate simulation of shockwave-turbulence interaction on unstructured meshes: A coupling of high-order fr and lad schemes," AIAA Paper 2013-3065, 2013.
13. Fujimoto, K., Fujii, K., and Wang, Z. J., "Improvements in the Reliability and Efficiency of Body-Fitted Cartesian Grid Method," AIAA Paper 2009-1173, 2009.
14. Vérant, J-L., Pelissier, Ch., Sourgen, F., Fontaine, J., Garcon, F., Spel, M., Van Hauwaert, P., Charbonnier, D., Vos, J., Vallee, J-J., Pibarot, J., Tribot, J-P., Mareschi, V., Ferrarella, D., and Rufolo, G., "Intermediate eXperimental Vehicle, ESA program, Aerothermodynamics – Transition and Steps and Gaps Assessment", 7th European Symposium on Aerothermodynamics for Space Vehicles, Bruges, Belgium, 2011.
15. Vérant, J.-L., Spel, M., and Bonnal, B., "EXOMARS Capsule Aerodynamics Analysis", AIAA 2010-4646, 10th AIAA/ASME Joint Thermophysics and Heat Transfer Conference, Chicago, 2010.
16. Stoney, W. E. and Swanson, A. G., "Heat Transfer Measured on a Flat-Face Cylinder in Free Flight at Mach Numbers up to 13.9," NACA RM-L-57E13, June 1957.

PROCEEDINGS OF THE 9TH INTERNATIONAL SPACE SAFETY CONFERENCE

SESSION 25 DESIGNING SAFETY – III

DERIVATION OF THE FRENCH SPACE OPERATION ACT REQUIREMENTS IN THE SPECIFICATIONS OF THE FUTURE EUROPEAN LAUNCHER ARIANE 6

N.Dias

ArianeGroup - 51/61 Route de Verneuil - BP 71040, 78131 Les Mureaux Cedex, France
Email : nathalie.dias@ariane.group

ABSTRACT

Ariane 6 is the future European Launcher launched from The Guiana Space Center. Designed to answer to launcher market expectations for the full range of satellites, any orbits and missions, the Ariane 6 launcher comes in 2 configurations, A62 and A64 corresponding to a design with 2 or 4 side Solid Rocket Motors. The first launch is planned in 2020 with a launch rate objective at full operational capability of 12 over a year.

Besides, the Ariane 6 launcher program corresponds to the 1st application of the French Space Operation Act from the beginning of an European Launch System development. The Space Debris Mitigation for European Space Agency projects is also considered as prerequisite for the Ariane 6 launcher development.

1 INTRODUCTION

This paper presents an overview of the integration of the safety requirements into the Ariane 6 Launcher System Specifications implemented by ArianeGroup as Prime Contractor of the Ariane 6 Launcher System. It explains how the French Space Operation Act and ESA Space Debris Mitigation requirements have been derived from Launcher System down to products in terms of reliability and safety objectives for both ground and flight phases. It identifies how these requirements are followed all along the development until the operational phase.

Some new technologies and systems by comparison with Ariane 5 launcher, participating to safety or implemented in safety mitigation devices are also introduced in the current paper.

2 ARIANE 6 LAUNCHER DESCRIPTION

The Ariane 6 launcher is the future European Launcher to be launched in 2020. It is composed of:

- a Central Core providing thrust with a Lower Liquid Propulsion Module (LLPM) equipped with a Vulcain engine and an Upper Liquid Propulsion Module (ULPM) equipped with a Vinci engine;
- two (A62) or four (A64) Equipped Solid Rockets (ESR) depending on the Ariane 6 Launcher configuration;
- an Upper Part including the fairing, the Launch Vehicle Adaptor, the Payload Adaptor Fitting (to fit with payload interface diameter) and any system allowing to perform dual launch or multiple launch.

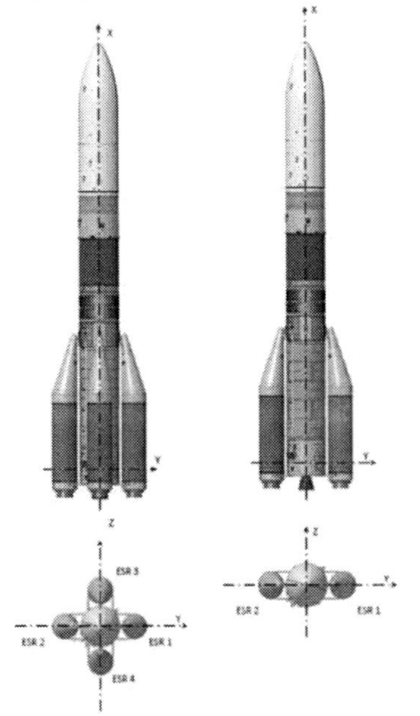

Figure 1. Ariane 6 - A64/A62 configurations

The Ariane 6 Mission Domain has been established by the European Space Agency and derived to ArianeGroup through the System Requirement Description document. It covers a large panel of missions in order to come up launcher market expectation both commercial and institutional.

3 FRENCH SPACE OPERACTION ACT AND SPACE DEBRIS MITIGATION

Besides, the Ariane 6 launcher program corresponds to the 1st application of the French Space Operation Act (FSOA) from the beginning of an European Launch System development. For the 1st time Ariane launchers will comply with the whole safety requirements without interim provisions identified in article 55 of the Technical Regulation.

Both orders of the French Space Operation Act aiming at mitigating the risks for safety of people and its property, and for public health and environment are fully taken into account:

- the decree regulating of the operation of the Guiana Space Centre Facilities, so-called REI, covering the risks during ground operations and in-flight as long as the neutralization of the launcher from ground is possible;
- the Technical Regulation related to the risks during the flight continuation up to disposal and the Space debris mitigations in orbit.

The Technical Regulation meets the Space Debris mitigation for European Space Agency projects adopting the 2nd edition of ISO 24113.

As Launcher System Prime contractor, ArianeGroup is responsible for the application of laws and regulations related to safety and space debris mitigation application at Launcher System level. It shall guarantee that all the requirements coming from the French Space Operation Act orders and ESA Space Debris Mitigations are considered within the Ariane 6 development programme as well as during its operational phase.

4 DERIVATION AND FOLLOWING OF SAFETY REQUIREMENTS

At the early step of the Ariane 6 development, the requirements coming from FSOA orders and applicable to Ariane 6 Launcher System have been introduced into its specification documentation.

They are derived from Launcher System level down to products. They have been taken into account since the Preliminary Definition of the Ariane 6 Launcher System. The safety requirements are thus taken into account as any other Ariane 6 Launcher System requirement.

The safety requirements are allocated into the whole specification documentation including Launcher System Technical Specification, Design Rules and also operational and working constraints. The analyses to be carried out in order to assess the conformity of Ariane 6 launcher with the French Space Operation Act have also presented into the Launcher System Specifications.

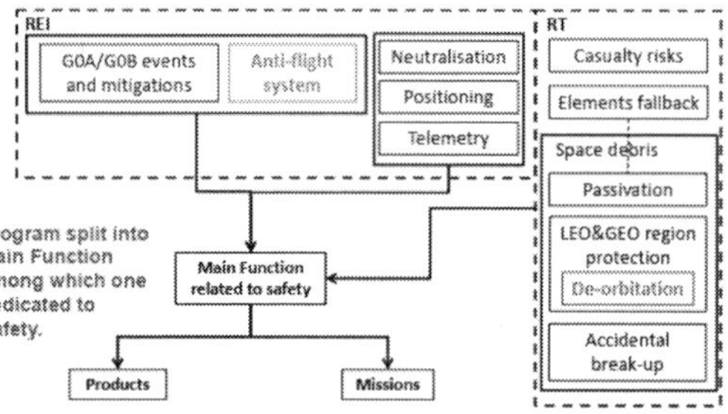

Figure 2. FSOA requirement derivation

The safety requirements on ground are related to the risks during the launch campaign with the identification of all hazardous systems whether pyrotechnical, electrical, fluidic, mechanical… or related to hazardous atmosphere (confinement and explosion risks, toxic hazards). Measures and rules are identified in order to mitigate the risks related to these hazardous systems. Qualitative and quantitative requirements related to ground-based activities during the launch campaign are thus taken into account and assessed within the whole development.

At the beginning of the flight a neutralization system in the close-range zone is required in order to interrupt the flight before the flight conditions became hazardous. This neutralization system independent from functional system is remote-controlled from the ground.

The Technical Regulation specifies requirements with regards to the risk related to fall-down of launcher elements separated from the launcher before being orbitated in order to avoid their fallback over inhabited territories and even territorial waters. The evaluation of casualty risks is expected and the risks limited during ascent phase and at re-entry. Moreover space debris mitigations are identified with the protection of LEO and

GEO regions and passivation in order to prevent space debris production risk.

Reliability, Availability, Maintainability and Safety (RAMS) activities are part of the Launcher System Group organization. They are fully involved in the allocation and assessment of safety objectives. Hazardous systems both on ground and in flight are identified and mitigations evaluated through the Launcher System Hazard Analysis all along the development. A Preliminary Hazard Analysis has been performed at the moment of the Preliminary Definition phase enabling to identify all hazardous systems of the Ariane 6 Launcher and the foreseen mitigations. The risk mitigation measures have thus been characterized and derived into the Launcher System Technical Specification since the beginning of the development. The implementation of the mitigations into Launcher System Technical Specification enables to check the progress of their application and to perform the evaluation of remaining risks. Non compliances are identified through Critical Point process. In the end, it also takes into account the hazards related to operations and human failures during operations. RAMS Robustness analysis of the current design (product and processes) have also been considered on the basis of lessons learned. The Launcher System Risk Analysis is then enhanced at each main milestone of the development phase. All the mitigations are integrated into launcher specifications and their achievement assessed within the Ariane 6 development.

This Risk Analysis at Launcher Level is also an input to risk analysis at Launch System level held by ESA.

Moreover, the Ariane 6 Launcher System has been split into several Main Functions. Among these Launcher System Main Functions, one is dedicated to Safety ensuring the end-to-end acceptance of the launcher operations on ground and the flight up to disposal with regards to safety requirements.

Considering safety requirements since the early phase of the Ariane 6 development enables implementing suitable safety mitigations since the beginning of the project and to fully integrate them into the Launcher Design. It also supports the Safety submission and Conformity process, held by ESA at Launch System level, as a continuous activity throughout the development.

5 SAFETY ON GROUND

The general qualitative requirements on ground and also in-flight are:

- Fail Safe/ Fail Safe criteria for activities with catastrophic consequences (G0A) and requires thus 3 safety barriers for all G0A hazardous systems
- Single Fail Safe criteria for activities with severe consequences (G0B) with 2 safety barriers to be considered

These barriers are split into Ground and Launcher System. RAMS analysis at Launcher System enables to identify the safety barriers available on-board and to highlight additional needs required on ground means.

5.1 Opto-pyrotechnology

Ariane 6 is a new launcher with challenging commitments on mission domain, recurring costs and launch rates.

Challenging topics have already been investigated as the use of new opto-pyrotechnical technology implemented on the launcher for functional and neutralization pyro-devices. This new technology which facilitates all integration and monitoring aspects questions the safety requirements to be considered.

Figure 3. Ariane 6 – Opto-pyrotechnology

New devices have to be developed and qualified to apply the optical technology to the Ariane 6 launcher. A Pyro Firing Unit (PFU or NFU for neutralization function) is under development. The optical signals generated by this device are sent to the pyrotechnic initiators trough an optical harness to the end pyrotechnic functions. The

Laser Initiation Device (LID) ensures the conversion of the optical signal into a detonation wave. Moreover, an Optical Safety Barrier (OSB) is located between the NFU/PFU and the pyrotechnic devices and acting as a mechanical barrier ensures ground security with regards to opto-pyro chain untimely activation. Associated to this new technology, new feared events related to optical domain have to be identified and taken into account to derive mitigations and constraints on equipment and operations (use of laser sources).

5.2 ESR anti-flight device

With comparison to Ariane 5 launcher, one important evolution is related to the ground safety with regards to the risk of solid propellant booster untimely ignition. The Equipped Solid Rockets (ESR) are fit with an anti-flight system during ground phases. A counter thrust process has been selected enabling to fulfil the operational duration constraint and appears to be the best solution for an Ariane 6 anti-flight system. This anti-flight system is implemented at Solid Rocket Motor (SRM) production and aims at remaining as late as possible before the launch. The initial anti-flight system was foreseen to remain in flight. Because of the risk of unreliability (when shifting from safe to flight mode on ground, and during flight), it was replaced by a removable system. The main matter associated to this new device is its strong operational impact of its removal before launch to fulfill operation duration constraint.

6 SAFETY IN FLIGHT AND SPACE DEBRIS MITIGATION

6.1 Neutralization system

As Ariane 5, the Ariane 6 launcher comprises a means of neutralization remote-controlled in the close-range zone from the ground.

The current baseline of the neutralization system is based on pyro-cord activation to neutralize the Solid Rocket Motor, while the Liquid Propulsive Modules are neutralized by stopping the engine. The acceptability in terms of safety (consequences in case of neutralization) is being estimated (see below launcher fragmentation assessment).

A neutralization system similar to Ariane 5 one based on the opening of all propulsive tanks by pyro-cord activations (including Lower and Upper Propulsive Module tanks) is considered as an alternative.

In the event of failure, the launcher is neutralized so that it cannot create any hazards in the Protected Area (ZP) enabling to protect Sinamary and Kourou inhabitants. This area and other protected zones are defined in the article 61 of the decree regulating of the operation of the Guiana Space Centre Facilities (REI).

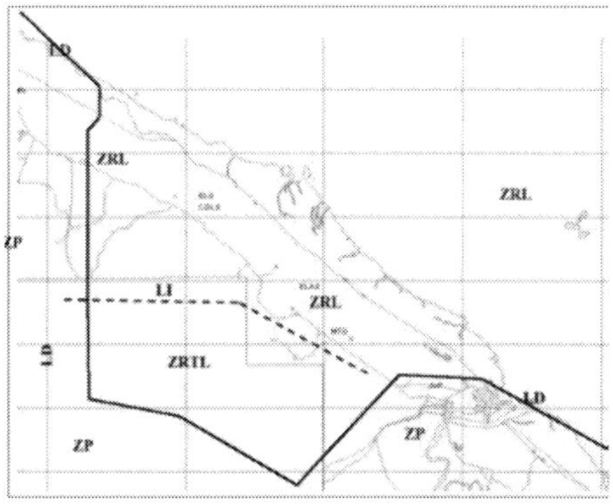

Figure 4. Protected zones in near-field of French Guyana Space Center

In case of neutralization the launcher debris fallback and the impact associated to launcher fallback on ground (as detonation) shall not override the Danger Limit (LD). The launcher effects on ground in case of neutralization are a major topic and have to be considered early within the project. Indeed in case the fallout zone is not acceptable, the beginning of launcher trajectory has to be modified (modification of launch azimuth or evolution of launcher altitude) with an impact more or less important on launcher performance. It is thus necessary to characterize the sizing failure scenario related to launcher neutralization and the associated launcher fragmentation and effects.

Launcher fragmentation assessment at an early step of the development phase appears to be a technical challenge. This assessment in case of neutralization is being established with the support of CNES (French Space Center) experts in charge of past Ariane 5 safety analysis.

6.2 Risks on ground in ascent phase and at de-orbitation

Prevention of the risks related to the fallout of elements designed to separate from the launcher before being orbited or after de-orbitation is another major topic related to launcher fragmentation. The article 21 of the Technical Regulation requires that the fallback zone, associated with a probability of 99.999%, must not impinge on the territory, including the territorial waters, of any State.

As previously in case of launcher neutralization, the outcomes of launcher element fallout could have an impact on launcher performance in case it is necessary to divert from optimal trajectory in order to avoid inhabited areas. An early evaluation enables to anticipate potential issues and to take them into account in the launcher performance evaluation.

A preliminary assessment has been performed within the mission domain to estimate the foreseen fallout zones of Equipped Solid Rockets (ESR), Lower Liquid propulsive Module (LLPM) and fairing. No major issue has been identified within the analyzed trajectories. This evaluation is being consolidated along the development considering the whole mission domain and the updated Ariane 6 data in terms of fragmentation.

Moreover an early evaluation of fallout zone for Upper Liquid Propulsive Module (ULPM) after de-orbitation has been performed in order to assess the feasibility of required de-orbitation within the whole mission. The appraisal is achieved taken into account the fallout zone opportunity and the delta-V to be provided in order to ensure ULPM de-orbitation with the required constraints on the fallout zone. A large scrutiny of the Mission Domain has been performed and the outcomes are being further analyzed. As for other launcher element fallback, this evaluation has to be comforted along the development considering updated Ariane 6 data in terms of fragmentation.

Launcher fragmentation assessment and debris survivability requires technical improvements of the associated tools, especially at an early step of the development phase. The Ariane 6 launcher fragmentation in ascent phase and at re-entry in case of de-orbitation is currently being estimated considering the aerodynamic and aerothermo-dynamic loads faced by the launcher during falldown. Thermal and thermodynamic fluxes are also considered to identify the general deformation of the launcher and assess launcher fragmentation. Launcher debris survivability is then calculated to identify the launcher parts reaching the ground and entering in fallout zone and casualty risk evaluations.

7 SPACE DEBRIS MITIGATION

As requested in the article 21 of the Technical Regulation and in the ESA Space Debris Mitigation, the LEO&GEO region protection is considered in the definition of the Ariane 6 mission domain and on the mission definition.

LEO region (so-called Protected Region A) is a spherical region extending from the surface of the Earth up to an altitude (Z) of 2,000 km

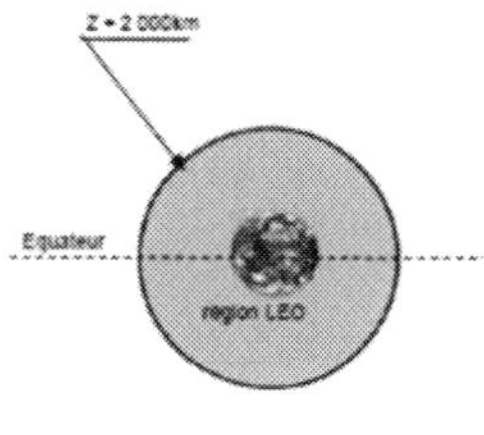

Figure 5. LEO protected region

GEO region (so-called Protected Region B) is a segment of the spherical envelope defined as follows:

- lower limit = geostationary altitude (Z GEO = 35,786) minus 200 km
- upper limit = geostationary altitude plus 200 km
- -15 degrees (South) ≤ latitude ≤ +15 degrees (North)

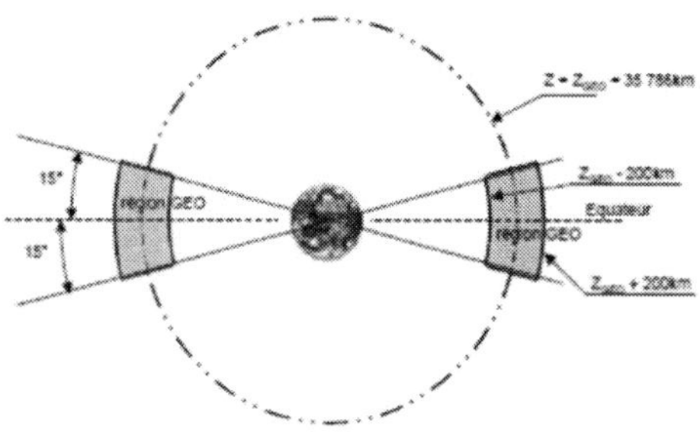

Figure 6. GEO protected region

The Ariane 6 Mission Domain covers a large panel of missions in order to come up launcher market expectation both commercial and institutional. It comprises:

- GTO and sub-GTO missions which orbit crosses the protected LEO region (low perigee) and with an apogee close to protected GEO region;
- GTO+ missions with an apogee close to protected GEO region;
- Circular/quasi-circular SSO and LEO missions which orbits are inside the protected LEO region;
- L1/L2 transfer missions with an orbit perigee crossing the protected LEO region;
- Direct GEO missions with altitude close to protected GEO region;
- MEO Galileo and equatorial missions.

As required in the Technical Regulation the Upper Liquid Propulsive Module is de-orbited when its orbit passes through the protected region A. This maneuver is a major evolution with regards to Ariane 5 missions.

The feasibility of this de-orbitation in terms of fallout zone but also in terms of launcher capability has been evaluated since the beginning of the Ariane 6 project. The Ariane 6 launcher performance assessment is thus estimated considering required de-orbitation boost.

The ability of the new Vinci engine (ULPM motor) to re-ignite enables to perform this de-orbitation boost when high delta-V is required.

Moreover, the introduction of a smart innovation with new APU (Auxiliary Power Unit) function which initial concern has been transformed into a new capability allowing better use of propellant has been retained for de-orbitation of upper module on GTO/sub-GTO missions (smaller delta-V).
The APU (Auxiliary Power Device) is composed of a gas generator equipped with heat exchangers which allow to pressurize the tanks in some dedicated flight phases, additionally to the Helium pressurization systems, and to the GH2 pressurization system of the LH2 tank (engine GH2 tap-off).

The thrusts levels produced by the gas of the combustion chamber are also used to provide acceleration to the stage in different flight phases.

The APU system participates thus in several functions:

- the pressurization of the Lox tank with Gox (heated Lox) during the propulsive phases and ballistic phase before Vinci re-ignition;
- the re-pressurization of the LH2 tank with GH2 (heated Lh2) in ballistic phase before Vinci re-ignition;
- the de-orbitation of the stage at the end of the mission, using the thrust produced by the APU System;
- the effect of the impulse scattering created by the Vinci thrust decay by ending the boost using only the APU (increasing of the P/L injection precision);
- the reverse engine chill-down;
- the passivation of the ULPM by permanently opening the APU feeding circuit.

Figure 7. Auxiliary Pressurization Unit

The Dual Launch Structure (derived from Ariane 5 SYLDA structure without propulsive system) has no propulsive system and cannot be de-orbited. As a result, the natural re-entry duration of Dual Launch Structure has been evaluated when its orbit passed through the protected LEO region. The final orbit has been selected to fulfil the 25 years maximal duration of uncontrolled re-entry requited by French Space operation Act. The computation has been performed using STELA CNES tool of orbit propagation. On GTO orbit the limit of perigee altitude to ensure the 25 years duration criteria without any constraint on launch date or launch windows has been evaluated. Above this perigee with the assessment has to be complete considering launch windows constraints. For circular orbit a preliminary assessment has been performed leading to estimate the maximal altitude enabling to fulfill the 25 year criteria requirement.

Moreover ULPM and Dual Launch System remaining in orbit have been analyzed with respect to the protected GEO region. The final orbit has been defined in order to not interfere with this region for more than one year and to not return to this region within one hundred years following the end of the operation .On elliptic orbit the apogee limit has been identified. For circular orbit the maximal altitude has been characterized.

This analysis enables to have a preliminary overview of possible constraints on the mission domain. It will be comforted with the evolution of launcher characteristics.

8 CONCLUSION

The French Space operation Act and Space Debris Mitigation requirements are implemented in the Ariane 6 launcher project since the beginning of the development. This early consideration enables to derive the safety needs within the Launcher System Specification documents and to take them into account since the launcher system preliminary definition. The impact of safety requirements on launcher definition can thus be optimized.

9 REFERENCES

1. Arrêté CNES/P n°2010-1 du 09 Décembre 2010 portant Réglementation de l'Exploitation des Installations du Centre Spatial Guyanais

2. Arrêté du 31 mars 2011 relatif à la Réglementation Technique en application du décret N° 2009-643 du 9 juin 2009 relatif aux autorisations délivrées en application de la loi N° 2008-518 du 3 juin 2008 relative aux opérations spatiales (NOR : ESRR1103737A)

3. RNC-LS-GM-0-X-3-CNES Version 1 du 06 juillet 2012
   REGLEMENTATION
   Guide des bonnes pratiques pour la démonstration du respect de la réglementation Technique (RT) Opérations de lancement

4. ESA/ADMIN/IPOL(2014)2 on 28/03/2014
   Space Debris Mitigation for Agency Projects

RISK MANAGEMENT FOR DYNAMIC RADIOISOTOPE POWER SYSTEMS

Christopher S. R. Matthes[(1)], David F. Woerner[(2)], Terry J. Hendricks[(2)]

[(1,2)]*NASA Jet Propulsion Laboratory, California Institute of Technology, 4800 Oak Grove Dr., Pasadena, CA, 91109*
[(1)]*+1-818-354-3255, cmatthes@jpl.nasa.gov*

ABSTRACT

The implementation of dynamic power conversion technology in Radioisotope Power Systems (RPS) for spaceflight has potential for improved specific power and efficiency, compared with existing Radioisotope Thermoelectric Generators (RTGs). This proposed expansion of current RPS technology necessitates a full exploration of the requirements, goals, and concerns related to risks in developing and deploying such systems. The nature of dynamic systems also presents a new set of challenges related to the presence of moving machinery not intrinsic to traditional RTG units.

A general RPS risk management methodology is outlined, which is used to identify and assess the variables and operational scenarios introducing risk throughout the design, fabrication, and system integration processes. This paper will demonstrate how mission requirements for Dynamic RPS concepts (DRPS) drive decisions throughout the development process. This work will define the standard practices for decision making within the scope of the risks associated with DRPS hardware development and deployment in sensitive spacecraft near sensitive payloads. Examples of potential risk areas are analyzed for the proposed dynamic systems, and compared to those associated with traditional RTG technologies. This analysis shows the promise for DRPS systems to elevate and extend the capabilities for power systems used in future NASA missions.

1. INTRODUCTION

For the past several decades, NASA deep space missions have employed Radioisotope Thermoelectric Generators (RTGs) to power spacecraft. These power systems operate using static thermoelectric (TE) designs, which are advantageous in terms of their stability and reliability. RTGs have drawbacks to their effectiveness, however, namely in their conversion efficiency. By contrast, dynamic power conversion in Radioisotope Power Systems (RPS) could achieve up to four times the efficiency of available RTGs[1]. This improvement would offer the ability to produce equal power levels using far less radioisotope fuel, or significantly greater amounts of power per unit fuel. Dynamic RPS (DRPS) would possess additional advantages over RTGs by eliminating sublimation mechanisms within the convertor that result in efficiency degradation.

While a static RTG system converts energy from a heat source into electricity with no moving parts, a dynamic system would involve a heat engine that converts heat into mechanical energy using pistons, dynamic seals, magnetics and linear alternators, and a working fluid, which in turn produces electricity. In recent years, NASA and the Department of Energy (DOE) have pursued dynamic system development with an Advanced Stirling Radioisotope Generator (ASRG), shown in Fig. 1, which would use the Stirling cycle to produce electricity from the radioisotope heat provided by two General Purpose Heat Source (GPHS) modules [2]. The ASRG system design includes two Stirling generators with separate heat sources, which are operated synchronously opposed, in order to significantly minimize vibrations. This effort allowed for substantial testing and research to evaluate reliability and performance, as well as the exploration of other issues related to the development of a dynamic system.

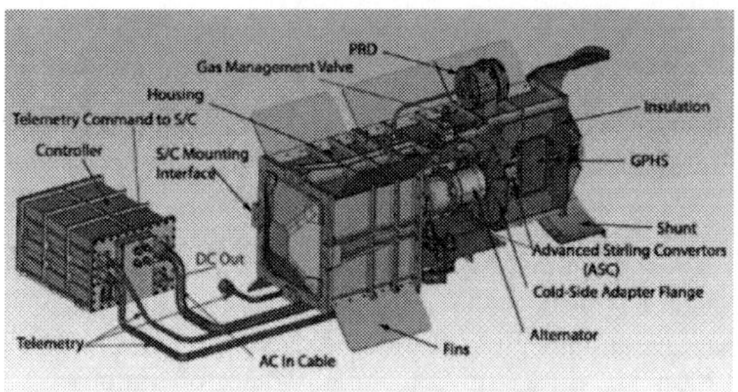

Figure 1. Diagram of the ASRG system

While DRPS technology would offer a number of valuable characteristics, there do exist challenges in qualifying this type of system for spaceflight, arising from its differences with conventional static technology. The fundamental nature of moving parts introduces an entirely new set of variables to a RPS, largely from the resulting vibrations. Electrical integration for a DRPS presents challenges due to an alternating current output from a single, or several generators, as well as the need for a generator controller. New convertor degradation and failure mechanisms also offer further areas to monitor and diagnose[3]. In addition, the desired qualities of these types of systems would come with the need to practically demonstrate their improved

efficiency, reliability, performance, and longevity, while proving them to be viable for integration on a spacecraft.
More recently, NASA has partnered with the DOE to further the development of a DRPS system concept, and is currently exploring general concepts for dynamic generators. This proposed implementation of a new type of power system necessitates a robust analysis of associated risks. Effective risk management is critical to mission success, which begins with defining the system requirements and recognizing the constraints related to the power system design. This paper will first provide insight into the design considerations that drive the system requirements, and offer context to the technology development. The Risk Management Plan (RMP) for the RPS Program will then be outlined by describing the components and standards of the Risk Management Process. Lastly, the procedure will be explained for providing conclusions and recommendations resulting from the risk analysis.

2. SYSTEM CONCEPT REQUIREMENTS

2.1 Design Considerations

The Multi-Mission Radioisotope Thermoelectric Generator (MMRTG), used for NASA's Mars Science Laboratory mission, is currently the only space-qualified RPS available for future missions, and in many ways acts as a baseline for future RPS designs. A number of constraints are applied by existing technology. Development of a new heat source would not be feasible or necessary, so a DRPS would utilize the Step 2 GPHS modules used in the MMRTG. For the same reasons, the 9904 shipping cask used for the GPHS RTG would be reused, which applies certain sizing constraints on the system.

Dynamic convertor designs would either produce axial vibrations or rotating imbalances, which must be addressed through one of a number of approaches. One option is the system can be designed to pair convertors such that they are either positioned opposite, or out of phase with one another. In this case, convertor failure may have a greater impact on the system. Another strategy is to employ active dampening or balancing to minimize movement. The generator may also be isolated from the spacecraft so vibrations do not affect operations or other equipment.

Recent work on DRPS following the termination of the ASRG as a flight hardware project in late 2013 has focused on using Stirling convertor technology in a modularized dynamic RPS concept[4]. The proposed design configuration for each individual module involved four convertors evenly positioned around a common GPHS stack. In the case of a convertor failure, less heat would be drawn from the GPHS module, and the remaining convertors would be exposed to elevated temperatures exceeding their given operational limits. Thermal modeling was performed to demonstrate that in the case of a single failed convertor, adjacent converters experienced a rapid increase in temperature before achieving a steady-state elevated temperature level. In this situation, a controller would be needed to detect convertor failure and adjust the remaining convertors to draw a greater amount of heat from the GPHS module and avoid damaging thermal levels. This solution would necessitate running the convertors below their peak power level under nominal conditions in order to accommodate adjustments in the case of convertor failure. The system may also need to be designed to manage elevated temperature levels.

Additional design constraints may be driven by a number of factors, including:

- Power output level of the generator
- Power level of individual convertors
- Quantity and configuration of GPHS modules
- Methods to utilize and reject heat
- Force balancing strategies
- System-level fault tolerance.

These quantities and considerations would be important to analyze within the context of the mission objectives throughout the design process. Similarly, associated risks would need to be extensively explored to support decision-making and to best ensure mission success.

2.2 Envisioned Operational Capabilities

A number of notional design requirements are derived from the generator's generic function of supplying power to the space vehicle. For the design life, the conceptual DRPS shall meet its operating specifications for at least 10 years after Beginning of Mission (BOM), with a goal of 14 years. Related to this requirement, its net electric power degradation shall not exceed a rate of 1.5% per year during the specified lifetime. This degradation figure includes the plutonium dioxide fuel decay of approximately 0.8%. The generator is to provide a steady-state electric DC power output of at least 300 W, with a goal of up to 500 W. The conversion efficiency should reach at least 20% while providing the specified power output. In the case of significant degradation or component failure, a fault-tolerant design should be utilized to allow for graceful degradation levels by adjusting to lower power output levels, without complete failure.

Mission-specific capabilities are also important to consider to accommodate future mission objectives. Potential future missions that could benefit from DRPS technology include those destined for relatively nearby bodies such as Mars and Earth's moon, as well as deep-space ocean and ice worlds such as Titan, Triton, Enceladus, and Europa. There are a number of environmental conditions that must be endured by the DRPS during liftoff, flight, entry, descent, and landing (EDL), and throughout planetary operations. These conditions include thermal transients and disturbance forces such as quasi-static g-loads, parachute deployments, and pyrotechnic actuations, among others. The system must be capable of withstanding launch vibrations and acceleration of particular magnitudes associated with all planned launch vehicles for RPS missions. The DRPS must be designed to operate under peak g-loading during planetary EDL of up to 20 g for a duration on the order of minutes, and up to 5 g for at least five days during spin-stabilization of the space vehicle. Since the DRPS would idealy be able to be mounted in any orientation on the space vehicle, it should have the ability to withstand these acceleration forces along any axis. Flight and operational conditions may require the ability to withstand micrometeorites and dust particles, surface pressure, atmospheric gases, as well as radiation environments.

The DRPS must accommodate requirements relating to its interface with space vehicles. Therefore, the generator would be expected to interface with standard space vehicle communications architectures used for other NASA missions. Fault tolerance for avionics must be admissible by supporting redundant data interfaces between the vehicle and the generator. In addition, the DRPS must minimize disturbance forces transmitted to the host vehicle due to its operation. These forces may include torque, vibration, or angular momentum from moving parts. A maximum housing temperature and/or maximum interface mounting temperatures must also be maintained to prevent excessive thermal loads on instrumentation or other space vehicle systems within close proximity of the DRPS.

2.3 Reliability

For a newly developed power system to be adopted for use in NASA missions, it must be shown to possess the highest degree of reliability that is feasible. Subsystems and their individual components must be rated as highly reliable and configured to last sufficiently long. This way, the power system can be best ensured to provide mission operation and completion of its goals over the course of the mission lifetime.

A dynamic system would possess key differences from static RTG systems, which impact the way reliability is considered. The most obvious distinction lies in the number of convertors, the number of moving components, and their nature of power generation within the RPS configuration. The MMRTG has a total of 768 TE elements, with each contributing a small amount to the entire system power. Performance issues or failure of an individual TE convertor has little impact on the overall power output level. In the case of a DRPS, the entire system may consist of one or several individual convertors, depending on the design. In the first case, degradation or failure of a single convertor may be catastrophic to the ability of the power system to meet mission needs. Losing one DRPS among several could at least represent a large "step" drop in power output, jeopardizing mission power needs, rather than a graceful power degradation.

To achieve a sufficient level of reliability in a mission concept, there are various redundancy and reliability strategies available. A single convertor/controller string may be used within a generator if it is considered to be highly reliable. Alternately, a generator may contain multiple convertor/controller strings if an additional level of reliability is necessary. Generator redundancy may also be employed on a mission if necessary and feasible. Figure 2 shows three different convertor/controller configurations for a four-convertor generator, with each configuration offering different levels of reliability. The converters are labeled 1-4, and the controllers are shown as the empty boxes. According to standard practice, the controllers are implemented redundantly to offer improved reliability.

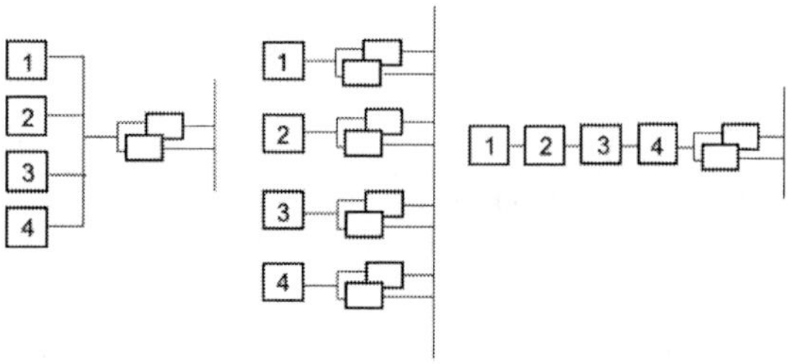

Figure 2. Convertor/Controller Configuration Diagrams

The respective system reliabilities for each of the wiring configurations in Figure 2 are modeled using the following expressions[5]:

$$R_{(a)} = [1-(1-R_{conv})^4] \times [1-(1-R_C)^2] \quad (1)$$

$$R_{(b)} = 1-\{1-R_{conv} \times [1-(1-R_C)^2]\}^4 \quad (2)$$

$$R_{(c)} = R_{conv}{}^4 \times [1-(1-R_C)^2]. \quad (3)$$

Figure 3 shows the modeled reliability behavior characteristic of the three convertor/controller wiring configurations depicted in Fig. 2. Convertor reliability, R_{conv}, is represented across the x-axis, and the system reliability is shown for two different controller reliabilities, R_C=0.5 and R_C=1. These reliability metrics represent the likelihood of each component meeting mission requirements by the end of mission. The top plot (Fig. 3a) corresponds to Fig. 2a for convertors wired in parallel, sharing a common dual controller. Fig. 3b is the same configuration, but with each convertor possessing its own dual controller (Fig. 2b). Fig. 3c is for a single string of converters in series, corresponding to Fig. 2c. The figures show that the most reliable configuration depends on the component reliabilities. For a system with high reliability converters and relatively low reliability controllers, option (b) will clearly result in the highest system reliability. Alternately, when controller reliability is high, there may not be a great difference between the options (a) and (b), and the number of controllers may be limited to a single pair to save weight and cost. The third plot shows the value of parallel wiring compared to series, which exhibits very low system-level reliability overall. The most advantageous configuration will be selected by taking into account cost and weight considerations, among other factors, needed to achieve the necessary system reliability.

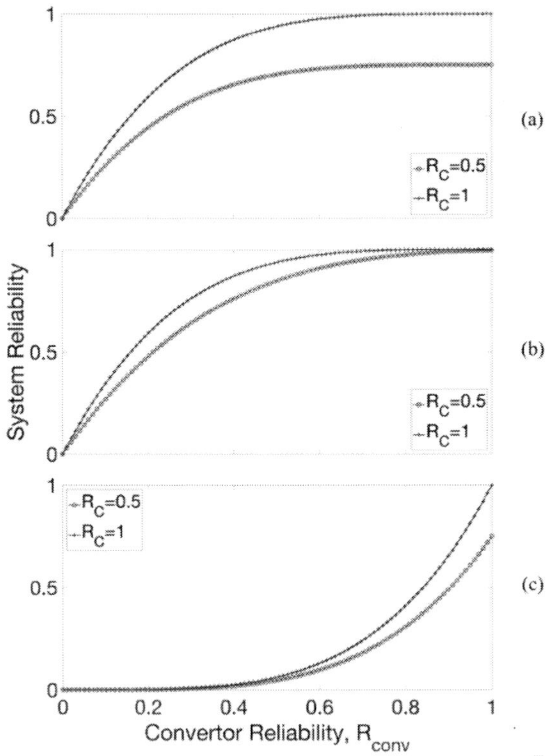

Figure 3. Reliability of Different Convertor/Controller Configurations

3. RISK MANAGEMENT PROCESS

3.1 RIDM and CRM

The Risk Management (RM) process at JPL aims to integrate system requirements, constraints, and design considerations, with an understanding of the associated risks of implementing any new technology. The goal is to minimize or mitigate risks, and when necessary, accept risks responsibly and methodically. This practice uses a complementary combination of two integrated procedures:

a) Risk-Informed Decision Making (RIDM)
b) Continuous Risk Management (CRM).

These processes work together to improve the probability of achieving the RPS Program objectives and goals within the defined schedule and budgetary constraints. The purpose of the RM process is to explicitly address performance shortfalls related to safety, cost, schedule, technical, and programmatic needs.

RIDM initiates the RM process by providing decisions arrived upon through extensive consideration of alternative outcomes. RIDM is comprised of three sequential steps:

1. Identification of Alternatives
2. Risk Analysis of Alternatives
3. Risk-Informed Alternative Selection.

Firstly, "Identification of Alternatives" involves recognizing a range of opportunities and configurations that fit within the context of the program objectives. This stage begins with understanding stakeholder expectations, which are decomposed into distinct performance objectives. Quantifiable performance measures are associated with each individual objective in order to assess all viable alternatives that fit within the imposed constraints. An example of a performance objective may be to minimize cost. The associated performance measure for a particular alternative would be the project cost. Any alternatives that fit within the imposed budgetary constraint may be considered, where the effectiveness of that alternative in achieving the objective is quantified by the performance measure. That is, alternatives with lower cost more effectively fulfil the objective for that particular criterion. In the case of DRPS, performance objectives may lead to consideration of various design configurations that minimize vibrations, or redundancy options in controllers that aim to enhance system reliability. This process results in a compiled set of feasible alternatives, which are obtained through a combination of stakeholder input and evaluation of the various candidate alternatives.

Next, "Risk Analysis of Alternative" requires the development of an integrated perspective, in which the entire scope of the program is taken into account. A technical basis is established for the various options, and deliberation provides further insight into each alternative. Specifically, Risk Analysis of a particular alternative consists of pairing performance assessment, measured according to the previously identified objectives, with probabilistic modeling, which quantifies uncertainties in a specific alternative's effectiveness at achieving program objectives. This practice begins with establishing a methodology for the analysis, followed by a quantification of the established performance measures probabilistically. The result is a technical basis for deliberation and decision making. An example may be the use of a reliability model for various convertor configurations and redundancies. Based on standard practice for reliability estimates based on a convertor's probability of failure, different wiring configurations can be considered relative to associated system considerations such as cost and weight.

Lastly, "Risk-Informed Alternative Selection" refers to the consideration and risk-informed selection of a specific alternative, such as a particular level of controller redundancy. In this process, performance commitments are defined, being informed by, but not solely based on, the aforementioned risk analysis. Consistent levels of risk tolerance are applied across the range of contending alternatives to normalize the field using either a self-consistent quantitative process or a qualitative assessment. Subsequently, the responsible parties deliberate on the existing options and provide a selection with well-documented rationale. This step is iterative, and may require additional analysis or revisions to the chosen performance commitments. The RIDM process culminates in a selected alternative to be inputted into the CRM process.

True to its name, CRM is implemented continuously throughout all program phases to mitigate risks as they become apparent. The purpose of this requirement is to identify risks as they arise so they may be collected and captured. This process is iterative, containing the steps illustrated in Figure 4 and summarized as follows, according to JPL's risk management process:

1. Identify: Identify contributors to risk (shortfalls in performance relative to the baseline performance requirements).
2. Analyze: Estimate the probability and consequence components of the risk through analysis, including uncertainty in the probabilities and consequences and, as appropriate, estimate aggregate risk
3. Plan: Decide on risk disposition and handling,

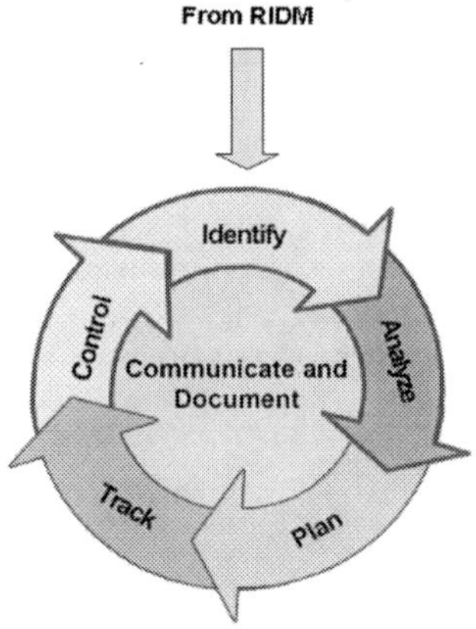

Figure 4. CRM Process Flow, Fed by the Results of the RIDM Process

develop and execute mitigation plans, and decide what will be tracked.
4. Track: Track observables relating to performance measures (e.g., technical performance data, schedule variances).
5. Control: Control risk by evaluating tracking data to verify effectiveness of mitigation plans, making adjustment to the plans as necessary, and executing control measures.

In conjunction with each of these steps, communication and documentation is practiced throughout the lifecycle of the RPS Program and Project.

An example of CRM for a DRPS may be related to assessing risks associated with modularity in the design. The RIDM process may result in a decision to pursue modular layers for the RPS, each with a particular power output, for greater flexibility in its usage for various mission scopes. A higher degree of complexity in the design, and a greater number of interfaces, may be "identified" as contributors to risk. "Analysis" may assign reliability statistics to each interface, and provide an understanding of the increased risk each size level offers to the previous one. That is, if a lower-level assembly is assigned a particular measure of risk, subsequently larger sizes will multiply the number of mating tasks, and the associated risk level. The "planning" phase may result in adjustments and impose design constraints that will minimize the mating tasks, such as prescribing a maximum level of modularity. The effectiveness of these adjustments may be monitored in the "tracking" phase, perhaps by testing the power levels across mating assemblies. Based on these results, certain measures may then be taken to "control" risks through additional adjustments. It is at this point that the risk may either be closed, accepted, or the CRM cycle restarted to mitigate the risk by "identifying" further shortfalls in performance of the assembly mating points.

3.2 Risk Analysis (NASA 5x5 Risk Matrix)

All items on the project risk list will be ranked according to a qualitative risk assessment. This task is accomplished by characterizing the likelihood and consequence of each risk on a scale of 1 to 5. This combined assessment provides the probability of a particular event occurring, and measures the impact in the case of such an event. Table 1 defines the ratings of consequence from both an implementation and mission standpoint. A cost evaluation is also included in Table 1 to approximate the resources necessary to address the risk, as a percentage of remaining project reserves. Resources, as they relate to implementation risk, may refer to performance, contingency, budget, safety, and schedule, among other categories.

Table 1. Risk Consequence Category Definitions

Rating	Consequence	Implementation Risk	Mission Risk
5	Very High	Cannot achieve flight readiness with remaining resources	Mission Failure
4	High	Consume all (100%) of remaining resources	Significant reduction in return
3	Moderate	Consume significant (26-99%) remaining resources	Moderate reduction in return
2	Low	Consume little (10-25%) of remaining resources	Small reduction in return
1	Very Low	Consume minimal (<10%) remaining resources	Minimal reduction in return

The percentage definitions related to the likelihood of a risk occurring, and the associated qualitative rating, are displayed in Table 2.

Table 2. Risk Likelihood Category Definitions

Rating	Likelihood	Definition
5	Very High	Almost Certain (> 90%)
4	High	More Likely than Not (75 < P < 90%)
3	Moderate	Significant but Not Assured (30 < P < 75%)
2	Low	Unlikely (10 < P < 30%)
1	Very Low	Very Unlikely (< 10%)

The likelihood and consequence ratings are tabulated in a traditional 5x5 risk reporting matrix shown in Figure 5. The corresponding rating position on the Risk Matrix categorizes the risk as either "low," positioned in the green section, "moderate" in the yellow, or "high" risk in the red.

3.4 DRPS Concept Risk Mitigation

As part of the communication and documentation for CRM, a Risk Focus Chart is compiled for each risk, which identifies the status of risk mitigation actions. This information includes a risk ID, title, statement, context, and ratings at each status point in the process. From a mission concept perspective, a number of risks associated with using a DRPS may be identified. For example, risk ID "DRPS-M-05" (see Table 3) possesses the title "DRPS Reliability Risk Due to Convertor Failure." The associated risk statement explains that in a DRPS composed of one or more convertor units, there exists a possibility that any of these units may become inoperable, resulting in a loss of power and excess

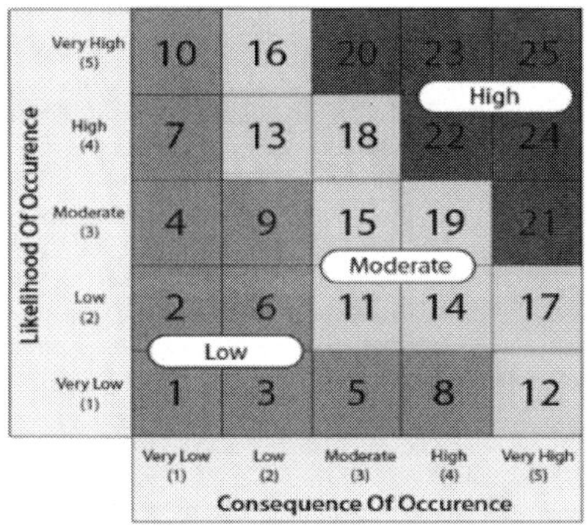

Figure 5. NASA 5x5 Risk Matrix

unused thermal energy input from the GPHS. The context of the risk is such that the power system must meet power output and thermal dissipation requirements over the entire mission. In the event of a single convertor failure in a multi-convertor configuration, the generator needs to meet these minimum requirements while operating with one less converter. A fault-tolerant design must be adopted such that no single credible fault condition renders the entire system inoperable. If the approach for addressing this risk is designated as "mitigate," then steps must be taken to lower the risk rating to an acceptable level such that it can be either accepted or closed. Table 3 shows the mitigation process for the stated risk, in which the rating is reduced from a high-level risk to low-level, to the point where it may be acceptable for flight.

Table 3. Risk Mitigation Process for DRPS-M-05

	Description	L	C	Rating
1	*Initial Rating*	5	5	25
2	Use parallel convertor wiring for improved system reliability rating	2	5	17
3	Initially operate convertors below peak power output	2	2	6
4	Perform reliability tests to improve component reliability rating	1	2	3

For a generator composed of 4 convertor units, the likelihood of convertor failure and the corresponding performance consequence can be tabulated. The initial configuration shown in step 1 of Table 3 corresponds to the series convertor/controller arrangement shown in Figure 2c. The risk likelihood can be quantified using Eqs. 1-3 presented in Section 2.3, initially assigning a conservative reliability estimate of 0.5 for all components. The series arrangement yields a reliability R=0.047, meaning nearly certain failure will occur, corresponding to a likelihood rating of 5 shown in Table 2. This reliability estimate evaluates the probability of each individual component meeting mission requirements by the end of mission. The consequence of a convertor unit outage in a series arrangement would prevent all upstream convertors from successfully contributing to the power output, meaning there exists a higher likelihood of losing each convertor with this design. Assuming that the loss of a single convertor will result in mission failure, the performance consequence in this case is characterized by a 5 rating, for an initial overall "high risk" rating of 25.

To mitigate the risk, the first step is to adopt a different convertor/controller wiring configuration. The arrangement shown in Fig. 2a is selected, resulting in a reliability estimate of R=0.703, lowering the likelihood rating to 2, meaning "low" probability of system failure, and producing an overall rating of 17, within the "moderate risk" regime. The wiring arrangement in Fig. 2b would result in a slightly higher reliability, but the added weight and cost make other risk mitigation options more advantageous. At this point, the likelihood has been lowered to an acceptable level, but the consequence of a convertor failure would result in excessive thermal levels and unacceptable reduction in power output. To address this issue, the convertors used in the DRPS may be rated for a higher power output than utilized during initial operation. In the case of a single convertor failure, the remaining units can ramp up to draw the excess heat from the GPHS and output the required amount of power. In this case, a single failure will have a low performance consequence, limited to a slightly diminished power output and an increased thermal output within reasonable margins. This change reduces that performance consequence rating to a 2, producing an overall risk rating of 6, now in the "low risk" regime. Finally, reliability tests and experimentation can be performed on the convertor units to raise the component reliability rating to a higher level. If the component reliability is raised from 0.5 to 0.7, the overall system reliability reaches R=0.903, allowing the performance consequence rating to drop to 1, for an overall rating of 3. At this point, the risk can be reviewed and closed. Procedurally, if a risk is rated at 4 or above, it may be accepted and watched, whereas if it reaches 3 or below, it may be closed.

4. PLANS & CONCLUSIONS

Dynamic RPS technology offers great potential for revolutionizing the way future NASA planetary missions would be powered. Key characteristics, potential mission constraints, and notional requirements for developing this type of system concept have been presented. A number of key design elements will need to be fully considered and analyzed throughout development to arrive upon a flight-qualified DRPS generator. These considerations include convertor technologies, converter configuration and design, component layout, mission concept needs, budget, and schedule, among many others. As DRPS development proceeds, each of these areas and the associated criteria will need to be well specified and understood. Risk management will be a critical component to decision-making throughout this undertaking.

The JPL Risk Management process was outlined as it relates to DRPS development. The risk management process should be matched to DRPS design considerations and requirements in order to determine viability. There exists great upside to advancing dynamic RPS technology. With this upside comes uncertainty related to generator and spacecraft integration, as well as viability for future use. A full understanding, evaluation, and quantification of the risk associated with implementing these new types of systems will be critical as NASA pursues future decisions about DRPS.

5. ACKNOWLEDGEMENT

The research was carried out at the Jet Propulsion Laboratory, California Institute of Technology, under a contract with the National Aeronautics and Space Administration.

6. REFERENCES

1. Cataldo, R. L., & Bennett, G. L. (2011). US Space Radioisotope Power Systems and Applications: Past, Present and Future.

In *Radioisotopes-Applications in Physical Sciences*. InTech.

2. Lange, R. G., & Carroll, W. P. (2008). Review of recent advances of radioisotope power systems. *Energy Conversion and Management*, *49*(3), 393-401.

3. Qualls, A. L., Schmitz, P., Rusick, J., Zakrajsek, J. F., Woerner, D. F., & Cairns-Gallimore, D. (2017, March). Dynamic Radioisotope Power System development for space exploration. In *Aerospace Conference, 2017 IEEE* (pp. 1-7). IEEE.

4. Schmitz, P. C., Mason, L. S., & Schifer, N. A. (2016). *Modular Stirling Radioisotope Generator*. National Aeronautics and Space Administration, Glenn Research Center.

5. Ely, N., & O'Brien, T. P. (1995). Space logistics and reliability. *Space mission analysis and design (2nd edition), Dordrecht, Netherlands, Kluwer Academic Publishers*, 693-714.

MARS SPACE SUIT SAFETY

Joao Lousada(1)

(1)*GMV Insyen, Münchener Straße 20, 82234 Weßling, Germany, Email: jlousada@gmv-insyen.com*

ABSTRACT

The Human exploration of Mars is often seen as the next step in human exploration of the Solar System. Decades of scientific explorations related to Mars recorded data on environment of the planet, including geological aspects, atmospheric composition, life aspects, and helped to create a foundation defining main hazards and security risk assessment. Based on data from decades of observation and robotic missions to Mars, many agencies and organizations around the world started planning the first human mission to Mars. Extravehicular Activities (EVAs), EVAs will play a crucial role in any mission long or short term as human participation in it is an essential step in the future. In a short term mission EVAs will allow to collect additional data about the planet and its environment, to run external experiments, to work on setting up a part of future habitats if needed etc. In the long term mission EVAs will help to run maintenance activities supporting the whole operation of the habitat or station and continue all needed scientific activities. Previous studies were dedicated to the analysis of possible hazards for the habitat, station and spacesuit. The most relevant hazards for the spacesuit design were defined including structural failures, power, thermal control and life support systems off-nominal situations (ONS), communication problems, loss in data management and problems related with humans in the crew. This paper will examine the hazards associated with thermal control and possible mitigation strategies and design options.

1. INTRODUCTION

Due to the development of technologies and increasing interest, the idea of Mars mission is becoming more and more feasible. Based on data from decades of observation and robotic missions to Mars, many agencies and organizations around the world started aiming for the first human mission to Mars.

In a initial mission, Extravehicular Activities (EVAs) will allow to collect additional data about the planet and its environment, to run external experiments, to work on setting up a part of future habitats if needed etc. In the long term mission EVAs will help to run maintenance activities supporting the whole operation of the habitat or station and continue all needed scientific activities.

EVA is defined as any activity completed by an astronaut outside a spacecraft beyond Earth's atmosphere or a planetary surface other than Earth's. EVA systems comprised of the following components: a spacesuit; a portable life support system (PLSS) that provides a breathable atmosphere and removable of Carbon Dioxide; subsystems providing stable pressurization, temperature regulation, mobility, power, and communications; waste collection; particle and radiation protection; rovers and mobility aids; and EVA specific tools that enable crewmembers to accomplish mission tasks.

Previous studies [1, 2] were dedicated to the analysis of possible hazards for the habitat, station and spacesuit. The most relevant hazards for the spacesuit design were defined including structural failures, power, thermal control and life support systems off-nominal situations (ONS), communication problems, loss in data management and problems related with humans in the crew. This paper will examine past experience in spacesuit design and its restrictions including aspects of mobility and operability, the conditions they are created to work in and its applicability in Mars conditions. Special attentions will be paid to the thermal subsystem and associated design concepts that could be used in the spacesuit design, spacesuit design for Mars.

2. DEVELOPMENT OF SPACE SUIT

Spacesuits can generally be divided in 3 groups:

1. Rescue (IVA) suits (Intra-Vehicular Activities spacesuits) – special suit used by astronauts in case of decompression of the module or in case of crucial ONSs related to environment onboard;
2. EVA suits used for activities in outer space, on the surface of the spacecraft or around it;
3. EVA suits for operation on the surface of different celestial bodies.

The first spacesuit of USSR was called "SK-1" and was used for "Vostok" series. NASA's first spacesuits were developed for the Mercury program. SK-1 and Mercury fell under the category of rescue suits and were based on simple pressurized suits. Next step was related with development of a new spacesuit that would allow human to perform EVA. The main issues needed to be solved were: assuring human safety, thermoregulation, radiation protection, low weight of the equipment and possibility to perform actions (flexibility). Solutions were realized in Berkut and Gemini spacesuits. These suits did not contain their own life support systems. The Gemini spacesuit was connected with the spacecraft by the cord/tube called umbilical and Berkut had an open life support system placed in a special "backpack" and also an additional systems connected by a cord/tube.

Next generation of spacesuits, which replaced the Berkut suit was the Yastreb. It had a new life support system, which had a regenerative respiratory system. Early Apollo suits were analogues of Yastreb suits. Further development of these spacesuits was dedicated to Moon programmes (3rd category). Main challenges of the realization became maneuverability (including the option of working with additional devices needed for exploration or repair) and integration of Life support systems. Apollo suit had a special inside suit equipped with special bio-telemetry sensors. The Russian prototype of Moon suit was called Krechet [3, 4].

More modern examples of "rescue suits" are Sokol-KV2 used in the Soyuz spacecrafts or the ACES used in space shuttle missions. EVA suits used for spacewalks are NASA's Extravehicular Mobility Unit (EMU) or the Russian Orlán-MK.

Two companies, Boeing and SpaceX, will use new spacesuits for astronauts riding on their respective Starliner and Dragon spacecrafts. Boeing unveiled its Starliner spacesuit, while SpaceX continues to work on its Crew Dragon suit [5].

At the moment spacesuits are not optimized and long-distance missions require a new generation of IVA and EVA spacesuits. Most modern EVA spacesuits using nowadays are designed to operate in microgravity and vacuum, since they are only being used in Low Earth Orbit (LEO). A Martian spacesuit will require a novel design, with specific features that mitigate the hazards of the Martian surface.

3. INITIAL CONDITIONS

To define requirements for a Martian Spacesuit, it is necessary to specify conditions of the environment and to list expected actions taking during explorations and general operation.

Climate and thermal conditions depend largely on the distance to the Sun. Mars temperature can vary from -125oC to 20oC. Giant dust devils routinely kick up the oxidized iron dust that covers Mars' surface. The dust storms of Mars are some of the largest in the solar system, capable of blanketing the entire planet and lasting for months. Mars also has "snow/ice" poles that are made of carbon dioxide rather than water [6].

Martian atmosphere is 100 times thinner than Earth. The air consists mostly of CO_2. Chemical component distribution is:

- 95.32% CO_2,
- 2.7% N_2,
- 1.6% Ar,
- 0.13% O_2,
- 0.08% CO,
- minor amounts of: H_2O, NO, Ne, HDO, Kr and Xe.

These atmosphere conditions make several liquids, such as saliva, nostrils and lacrimal liquids and even blood, boil almost immediately. In any human, this would quickly lead to death.

Based on this, the spacesuit has to protect against the low temperatures and low pressure of the Martian atmosphere. Technical requirements for a Mars EVA suit include mitigation of the thermal fluctuations resulting from the Martian atmosphere, limitation of dust contamination, prevention of organic venting, and radiation protection. The main biomedical requirements for a Mars EVA suit include a low EVA system mass, high mobility and dexterity of spacesuit enclosure with optimized joint design, provision of water, waste collection, and a life support system that is able to work continuously for 8-10 hours, thermal control system, and sun visor assembly.

The spacesuits used so far by astronauts in space missions and the Moon weigh more than 90 kg, which in the reduced lunar gravity (1/6 of the terrestrial gravity) represent about 15 kg, but the same suit on Mars (3/8 of Earth gravity) would weigh more than 33 kg which would result in great discomfort and lack of dexterity for activities in the Martian atmosphere.

Operation ability of astronaut in spacesuit is a key parameter in Extra-vehicular activities. It has got 3-jointed mechanical measuring arm, which combined force-torque sensors in the handle, and magnetic brakes installed in every joint. All data measured were reported in form of a data file and 3-dimensional graphics. The arm has three main parts: elevator, mechanical arms, and motors (DC, servo.)

In Fig. 1, we can identify the areas of operation when performing an EVA. Additionally, we can here identify some operations that are expected to be performed while on EVAs on Mars:

- External system check after landing on the site
- Support the solar panels deployment to have power for laboratory and provide power to critical systems
- Determine the locations worth of further investigation
- Perform geology and mineralogy investigations of the surrounding area (e.g. searching for evidence of life, either past or present)
- Look for water on mars or evidences of water, valleys, reservoirs
- Drill deep below the surface to check the find water resources and what beneath the land
- Keep continuous contact with the habitat to receive general instructions for further experiments [7]

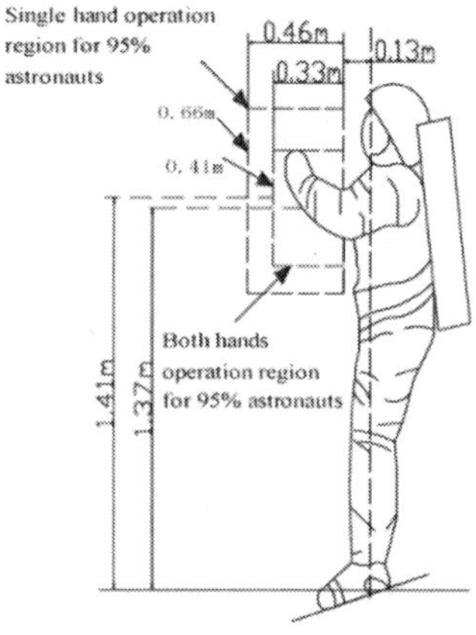

Figure. 1. Operation in the spacesuit.

4. DESIGN CONSIDERATIONS

The generally accepted [15, 16, 24] Martian space suit design includes the following:

1. Primary Life Support Subsystem – PLSS: Contains water cooling equipment's,· removes the exhaled CO2, fan to circulate the oxygen, two-way radio transmitter
2. Upper Torso: hard upper torso, covers chest part and support arm assembly, also covers the back part. It also contains tubes to drain the water and to allow oxygen flow.
3. Arms: arms available in different custom sizes, sizing rings can make the length longer or shorter
4. Gloves: it is very important that astronauts can do movements of their fingers with ease while performing EVA operations like collecting rock samples. Simultaneously it is very important that gloves save from wear and tear in harsh environment of the planet. Bearings can fix the gloves to arm.
5. Displays and Control Module: an inter-vehicular crew member can help operate the different suit systems remotely from this module
6. In-Suit Drink Bag: valve can be set near to mouth so that while performing any activity on planet one can bite the valve to open the water tube. Releasing the bite will again close the tube. The water bag will be attached in the upper torso with the Velcro.
7. Lower Torso Assembly: this assembly has hard material seal closure which can be attached to the upper body torso. This assembly covers the pants, boots and joints. Within this assembly one can take turns either way. D ring is very important which has the tethers which can keep a constant connection to the habitat.
8. Helmet: this assembly has vent pad which will control the flow of the oxygen from the primary life support system to hard upper torso and front of the helmet. The clear plastic bubble is the main part of the helmet which is covered by the visor assembly. The visor assembly will protect the spacewalker from the direct sun light. A video camera and additional lights can be attached to the suit.
9. Communications Carrier Assembly: this assembly made up of plastic cap which has microphone and earphones with which the astronauts can talk to others peers and also hear the warnings/cautions.
10. Liquid Cooling and Ventilation Garment: The underwear garment made up of stretchy spandex material, which uses almost 91.5 meters, or 300 feet, of narrow tubes throughout. Water is pumped through the tubes near the spacewalker's skin. The chilled water removes extra heat as it circulates around the crewmember's entire body. The vents in the garment draw sweat away from the astronaut's body. Sweat is recycled in the water-cooling system. Oxygen is pulled in at the wrists and ankles to help with circulation within the spacesuit.
11. Maximum Absorption Garment: one spacewalk can last long for five-six hours so specially made garments can be used to absorb the sweat and urine.
12. Wrist Mirror: this mirror will help astronauts to see the control module readings. The information in the suit is written in backward so that with the mirror it can read properly.
13. Layers: spacesuit has total 14 layers. First three layers covered by liquid cooling and ventilation garment. The next layer is bladder layer which will control the pressure of the suit. It also holds the oxygen underneath for breathing. The next layer will hold the bladder layer to make it flexible per astronaut's movement and it is made up of same materials as camping tents. The next seven layer will make the suit act as a thermos so temperature of inside will remain same without changing.
14. Cuff Checklist: holds the checklists of tasks.

5. THERMAL SUBSYSTEM

The thermal control system keeps the temperature inside the suit within acceptable limits for the astronaut. This ranges from eliminating excess heat from the other systems and physical activity, to providing heat for the astronaut when operating in shadowed or colder areas. Contrary to thermal control in a spacecraft, the systems in EVA suits have to operate with the relatively unpredictable thermal output from the astronaut, and in an almost randomly changing environment while the astronaut moves.

The Technical University of Munich (TUM) has been developing the Virtual Habitat (V-HAB) project at the Institute of Astronautics (LRT) at the Technical University of Munich (TUM) aiming to develop a modular, flexible simulation environment to dynamically model and simulate thermal and life support systems (LSS) for space missions [111].

Recently, a fully configurable and modular thermoregulation layer was implemented. It models all mass, mass flows, heat capacities and heat flows using the default V-HAB elements and solvers, which are also used to simulate e.g. physio-chemical subsystems. This allows a close interface between the human thermal layer and the environment, in this case the liquid cooling garment (LCG) and ventilation gas within the suit. [112]
Additionally, TUM has been working since 2012 on V-SUIT, a spin-off from V-HAB that focuses specifically on thermal and life support systems of space suits. [113] Simulation tools like the V-SUIT help us model and better understand how a space suit design would operate under Mars conditions, taking also into account the human physiological and thermal output.
Nonetheless, the thermal subsystem has to adapt quickly and operate within a wide range of thermal outputs. The biggest concerns regarding the thermal control subsystem relate to failures due to the changing operating conditions is the most relevant risk to the thermal control system.
The unpredictable level of physical activity of the astronaut during the EVA in the surface causes a constant variation on the amount of heat that needs to be eliminated from the system by the thermal control system. Extra physical activity, due to complicated terrain for example, would produce peaks in the amount of generated heat, which would disappear almost instantly when the complicated terrain has been cleared. This can create issues on thermal control, potentially causing overheating for the astronaut.
It is almost certain that the astronauts will have to face unpredictable changes in their levels of physical activity. Thermal control issues during EVAs can reduce the comfort of the astronaut inside the suit and limit physical capabilities, sometimes putting the astronaut at serious risk of harm. In this subsystem, it is important to find a reasonable tradeoff between lightweight and well insulating materials, and the structural stability of the suit, taking also into account the system reliability.

6. THERMAL CONTROL SUBSYSTEM DESIGN

Based on the possible hazards identified in the thermal subsystem, it is important to consider what design solutions can be taken, in order to eliminate or mitigate these issues.
It is clear that, any spacesuit design will need to employ a redundant critical thermal system: while protecting the astronaut with a higher degree of safety, having a double-fault tolerant or an auxiliary system (such as the Advanced EMU [114] mounted on the suit means increasing its complexity, mass and volume. It is important to find a trade-off between safety and weight minimization.

Additionally, it is important to define procedures and protocols to follow proper in case the suit temperature is outside the nominal range, it is important that the astronaut is promptly informed with an alarm. In case higher values of temperature are reached, it is important to distinguish its reason. If it is caused by a fault in the system, the astronaut should be able to reach promptly the habitat or the necessary tools that are needed to fix it. A value outside the nominal range can also be caused by too much workload. In this case, the astronaut should reduce the workload until the temperature lowers. In the opposite situation, the astronaut should have available insulating material to put on top of the suit in case of too low temperature inside the suit. In general, it is recommended to reduce the distance between the astronaut's workplace and the required tools.
Finally, special attention must be given to the chouce of materials in the spacesuit design, particularly for the material used in the insulating layers. There is a possibility to use novel materials that, being able to change their overall surface properties, could help in regulating the temperature, with a variation in their absorption and reflection coefficients. Electronic ink, for example, is made of tiny microcapsules that change color (white or black) when there is a change in the electric field. When it is white, the suit would reflect more light, maintaining a lower temperature in the suit. The opposite happens if the suit is black [115]. However, advances in e-ink wearables are necessary to be used for this case.

7. BIOMEDICAL ISSUES

It is also important to consider biomedical issues and how they influence the thermal dynamics of the human-suit system and its thermal regulation.
Biomedical telemetry is aimed to collect various physiological parameters. Depending on a mission, biomedical monitoring included [3, 4, 11]: electrocardiogram (ECG), blood pressure measurement, respiratory rate, galvanic skin response and resistance, rectal and core body temperatures, passive-dosimeter for radiation exposure measurement, cardio tachometer for measuring heart rate, impedance pneumogram, electroencephalogram, electromyography. This is of particular important for the thermal control.
Simulation missions [12, 13] (e.g. operation with AoudaX spacesuit) showed that additional weight burden and thermal stress of the suit can severely strain the crewmember, leading to adverse health events that can impact the overall mission. A stable core body temperature is required to ensure optimal functioning of the human body; therefore, astronaut core temperature should be a measured value. Helmet humidity and temperature measurements would provide insight into the actual interior temperature of the spacesuit. It is evident that EVA exposes crewmembers to substantial

physical stress through intense activity involving increasing workloads and stress levels.

ECG also holds a very important information providing insight into the cardiac health status of the astronaut during the mission. In combination with other physiological parameters (i.e., heart rate, respiratory rate, and arterial oxygen saturation), the ECG can assist in evaluating the workload capability and fatigue level of the astronaut. To monitor thermal stress levels of the astronaut during the mission it would be best to have measurements of core body temperature and in-suit temperature and humidity.

Previous software models to monitor physiological responses during EVAs could measure heart rate, respiratory rate, O2 volume consumption, CO2 volume generation, and mean core body temperature. Design of new monitors should allow wider range of measurements allowing a better insight into the condition of the astronaut during EVAs. As an example, a metabolic rate could now be estimated during EVAs by measuring the inlet and outlet temperatures of the liquid cooling and ventilation garment (LCVG) as well as heart rate and oxygen consumption measurements [14].

7.1. Spacesuit Data Systems and Sensors

The On-Board Data Handling (OBDH) system is a key component of the spacesuit as it provides the suit's power supply and data acquisition, processing, and communication. The various sensors with the spacesuit facilitate physiological and environmental monitoring of the astronaut. Helmet temperature is measured using an absolute air pressure sensor and helmet humidity is measured using a sensor module [17, 18]. Dew point and potential condensation within the helmet can be surmised from helmet temperature and humidity data. With regards to suit gas concentrations, CO2 is measured using a module located at the apex of the helmet's inner surface. Thus, the measurement is an air mix between inhalation and exhalation CO2 levels. In contrast, O2 is measured using an oxygen sensor located in the left side of the helmet. Biomedical telemetry is collected through an ECG device consisting of a 1-channel ECG amplifier and a 3-channel accelerometer that transmits signals to either a PC via Bluetooth or stored locally on a SD card for later analysis. Additionally, the ECG raw data is used to calculate heart rate variability.

Astronauts should have an access to all the data to analyze their own condition. That is why the system requirements should include indication of heart rate parameters, helmet CO2 level, helmet O2 pressure, core body temperature, body skin temperatures (3 minimum), inner and outer suit temperatures (4 minimum), suit outlet dew point, suit outlet gas temperature, liquid cooled garment inlet and outlet temperature, and liquid cooled garment flow rate. Requirements should include warnings and alerts for any of the following instances: low suit pressure, high heart rates (> 175 bpm), high core body temperature, high partial pressure CO2 (> 15 mmHg), or excessive sweat rate (suit outlet dew point > 27°C). Also warnings should be calculated taking into account environmental and physiological responses to thermal stress, workload, and fatigue (Fig. 2).

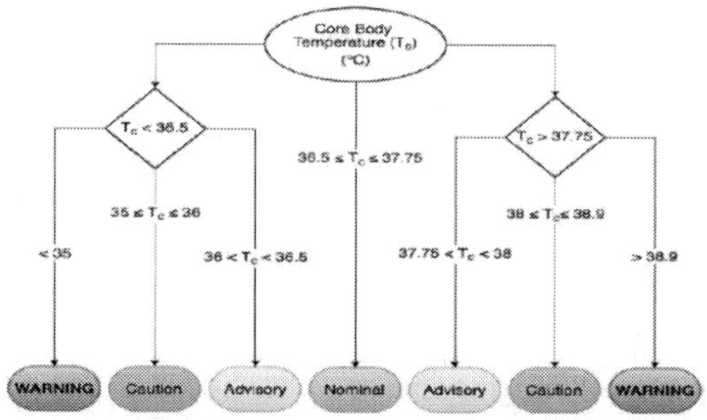

Figure. 2. Example of alerting algorithm.

7.2. Space Suit Sensors

A major spacesuit systems requirement is the ability to measure all critical data. Critical data, with respect to health monitoring, the systems requirement document as the following parameters: heart rate, helmet CO2 level, core body temperature, body skin temperatures (minimum 3), suit internal and external surface temperatures (minimum 4), suit outlet dew point, and suit outlet gas temperature.

The top priority sensor should be integrated into the spacesuit should be core body temperature because the majority of thermal stress research utilizes this measurement. It is important for human temperature regulation and can influence human performance. Sensors for measuring core body temperature have become more common in recent years. For example, an ingestible telemetric sensor has been used successfully in sport and occupational applications such as diving, military training, and long-duration exercise. Body temperature measurement should be taken with a minimum of 3 skin locations. It would facilitate monitoring of thermal stress exposure and provide insight to the body's thermoregulation practices in the spacesuit. Body skin temperature has traditionally been measured using hardwired devices such as thermistors, but new iButton® wireless sensors do not require receivers for data collection and are not subject to cross-talk interference found that the wireless iButton® sensors can effectively measure body skin temperature during laboratory and field investigations where other methods for body skin temperature measurement were problematic.

Body-worn monitors for measuring physiological health have been developing by space agencies such as NASA for use in extreme environments. Life Guard is a physiological monitoring system designed for astronauts that provides measurements for the following parameters: ECG, respiration, skin temperature, 2-axis acceleration, pulse oximetry, pulse rate, and blood pressure [19].

8. CONCLUSIONS

Following previous studies that were dedicated to specification of major hazards associated with Martian manned missions, this work focused on the spacesuit design, with particular focus to the thermal control system. A short review of a general background of spacesuit design and development was given, followed by the identification of challenges for the long-term mission suits. It was concluded that even with many currently available studies and prototypes of Martian spacesuits as well as analog spacesuits, there is no optimized approach to the perspective spacesuit design. The thermal control subsystem was introduced and its major hazards and challenges identified. Based on a literature review, the main design aspects were discussed and proposed solutions were analyzed. Correlating initial data and considering hazards with the proposed solutions, a set "best practices" for design of this system was given for future discussion. This paper paid special attention to the thermal control system, including analysis of the human biomedical issues and needed telemetry sensors. The main goal for the future work should be on further analyzing the proposed solutions and "best practices" as well as to further investigate the remaining subsystems that are a part of the overall spacesuit design.

9. REFERENCES

8.1 Sample References

1. J. Lousada, Guzel Kamaletdinova, Pierre Bertrand, Matteo Emanuelli, Seyed Ali Nasseri, Rajendrasing Rajput, Hamed Gamal & Charlotte Kiang, Technical Safety Analysis of a One-Way Human to Mars Mission, Proceedings of the 66th International Astronautical Congress, Jerusalem, Israel, 2015.
2. C.M. Entrena Utrilla, Laura Bettiol, Jessica Piness, Istvan Revesz, Funmilola Oluwafemi & João Lousada, Safety Analysis of Spacesuit Design for Martian Surface, Proceedings of the 67th International Astronautical Congress (IAC), Guadalajara, Mexico, 2016.
3. Sovkos, http://www.sovkos.ru/ (accessed June 2017)
4. Astronautix, http://www.astronautix.com/ (accessed June 2017)
5. Florida Today, http://www.floridatoday.com/story/tech/science/space/2017/04/21/florida-institute-technology-space-suit-human-spaceflight-spacex-boeing-nasa-blue-origin-virgin-galactic/100705536/ (accessed June 2017)
6. Space.com, http://www.space.com/16903-mars-atmosphere-climate-weather.html, (accessed June 2017)
7. NSSDC, https://nssdc.gsfc.nasa.gov/planetary/mars/marmarsl.html (accessed June 2017)
8. M. Gernhardt, Integrating Life Sciences, Engineering and Operations Research to Optimize Human Safety and Performance in Planetary Exploration. National Aeronautics and Space Administration. Washington, D.C., 2009
9. A.F.J. Abercromby et al., Integrated Extravehicular Activity Human Research Plan: 2016. 46th International Conference on Environmental Systems. Vienna, Austria, 2016
10. NASA, https://www.nasa.gov/audience/foreducators/spacesuits/home/clickable_suit_nf.html, (accessed June 2017)
11. M. Czupalla (2011), The Virtual Habitat - Integral Modeling and Dynamic Simulation of Life Support Systems, " Munich: Verlag Dr. Hut.
12. J. Schnaitmann & C. Olthoff (2017). Validation and Use Cases for the new Thermal Layer of the V-HAB Crew Model, 47th International Conference on Environmental Systems, Charleston, USA.
13. C. Olthoff (2017). Validation of the Virtual Spacesuit using Apollo 15 Data, 47th International Conference on Environmental Systems, Charleston, USA.
14. C. Campbell, Advanced EMU Portable Life Support System (PLSS) and Shuttle/ISS EMU Schematics, a Comparison. AIAA 2012-3411, 42nd International Conference on Environmental Systems, San Diego, California, 2012, 15-19 July.
15. E Ink Corporation, Ink Technology: Electrophoretic Ink, explained. http://www.eink.com/technology.html (accessed 07.09.2016)
16. V.P. Katuntsev, Y.Y. Osipov & S.N. Filipenkov, Biomedical problems of EVA support during manned space flight to Mars. Acta Astronautica 64, 682–687, 2008
17. AMADEE-15 Mission Report. Osterreichisches Weltraum Forum, 2015. Innsbruck, Austria
18. Aouda System Handbook. Osterreichisches Weltraum Forum, 2015. Innsbruck, Austria.
19. Committee for the Decadal Survey on Biological & Physical Sciences in Space & National Research Council, Recapturing a Future for Space Exploration: Life and Physical Sciences Research, New Era The National Academies Press, Washington D.C., 2011

20. C. Byrne & C.L. Lim, The ingestible telemetric body core temperature sensor: a review of validity and exercise applications, Br J Sports Med 41, 2006
21. S. Cheung et al., Thermal stress, human performance, and physical employment standards, Appl Physiol Nutr Metab 41, 2016
22. NASA, https://www.nasa.gov/hrp/bodyinspace (accessed June 2017)

ASTRONAUT COGNITION

Bettina L. Beard

NASA Ames Research Center, MS 262-4, Moffett Field, CA 94040, USA, Tina.Beard@nasa.gov

ABSTRACT

Spaceflight mission success and safety depends on a crewmember's ability to attend to, consolidate and interpret information. If spaceflight stressors were to compromise the crew's ability to think clearly, the mission may be in peril. Research to date does not support consistent changes in cognitive processes during spaceflight. Small sample sizes, insufficient baseline practice, individual differences, inconsistent methodologies and test sensitivity concerns plague the literature. There are results that suggest that spaceflight stressors affect the central executive function of working memory. The practical implication is that during adaptation to gravity transitions and high stress periods, crewmembers may be more susceptible to resource competition that could affect performance on a primary task. These dual-task effects should be further explored and correlated with known spaceflight stressors including noise, workload, visual phenomena, CO2, caffeine use, light therapy, back pain, headache, concussive jolts, use of hypnotics, current health, etc.

Tasks performed by astronauts require both fine motor and cognitive processes. NASA Crew Notes include entries when crewmembers miss procedure steps, schedules run long and mistakes are made. Cosmonauts and astronauts have reported instances when they could not think clearly during missions. Stressors can detrimentally affect motor and cognitive performance terrestrially including high workload [92], fatigue [108], vibration [26,96], concussive forces [3,40,90,94,95], danger [9], high CO2 levels [2,53,98], headaches [5], decompression sickness [99,110] and extreme environments [80]. Space missions include all of these stressors. For this reason, existing standardized tests have been adapted for use in microgravity environments. These tests uncover glimpses of the covert processes underlying overt behavior. Although many of the tests lack face validity, they generally show sensitivity to different insults in clinical populations. Many tests show large individual differences and significant practice effects.

The two fundamental metrics used in fine-motor and cognitive tests are reaction time and response accuracy. Reaction times provide clues about the time required to process information, if several stages are needed to process the information or whether separate stimuli are being processed serially or in parallel [74]. Processing speed is extremely vulnerable to environmental stressors and neurological insult. For example, reaction times are faster with moderate doses of caffeine [73] or when the person is anxious [109]. On the other hand, reaction times increase the longer a person is awake [113], with background noise [107] and with brain injury [10]. Tests of simple and choice reaction times do not reveal significant degradation except in a few novice crewmembers or crew suffering from space adaptation syndrome (14,46,91,35). However, when complex tasks are involved, reaction times have been shown to increase during spaceflight.

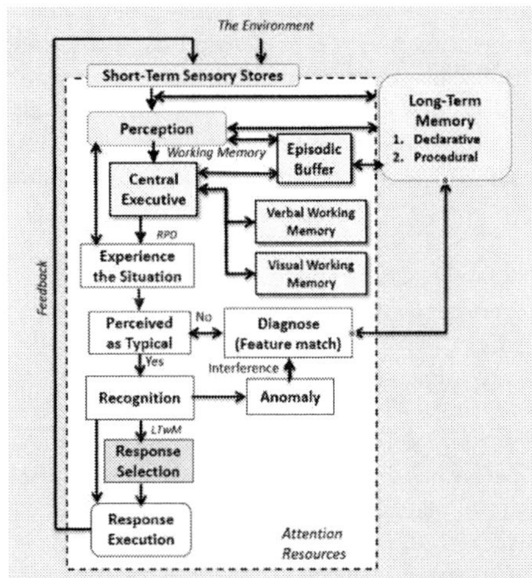

Figure 1. Schematic of how humans process information [11]. Note that attention resources are required for almost every process and that there are many feedback loops.

Current models of the human cognitive process can help inform our knowledge about astronaut cognition. While specific cognitive constructs vary in the literature, Fig. 1 compiles several leading behavioral theories of cognition. Sensory receptors (e.g., rods, cones, hair cells, baroceptors, etc.) are portals for incoming data from the environment. Sensory receptors transduce light waves and particles, sound waves, arterial pressure, etc. into neural signals. These data are stored briefly in sensory memory (i.e., the Short-Term Sensory Store) where features are extracted.

To perceive sensory signals requires paying attention. Fig. 1 depicts attention resources by a large rectangle encompassing almost the entire cognitive processing scheme. A signal's salience and context, and the individual's experience (i.e., stored memories in Long-Term Memory) guide attentional focus. Visual perception has evolved to use the pull of gravity as a frame of reference and so has been assessed during launch and on orbit. As examples, during launch, positive Gz acceleration reduces contrast sensitivity in the mid-frequency range [23]. On orbit, intra-ocular pressure is elevated [66], near visual acuity declines [85] and vertical lines are perceived as shorter [25]. After flight, there is a reduction in dynamic visual acuity [89].

A lifetime of experiences on Earth has led to internal models, i.e., templates, of what to expect in our environment. In microgravity, these learned patterns may no longer be applicable. The incoming perceptions from the different perceptual systems are integrated for robust interpretations. When exposed to gravity transitions a crewmember's visual, vestibular and somatosensory systems provide discordant information. This results in some compelling self-orientation and kinetic illusions [38,43,59,83]. For example, simply seeing another crewmember that is oriented upside-down relative to oneself can trigger a sudden change in perceived body orientation [20].

Approximately twenty years ago, [1,12,13] introduced a new experimental tool into visual psychophysics, the classification image. These studies have led to the widespread use of classification images in visual psychophysics [77]. Classification images reveal the underlying template (or stored internal model) used in visual detection and discrimination. These templates demonstrate plasticity, depending on experience [63] and disease [62]. Using the classification image technique, space agencies could determine if the stimulus features used in visual perception are altered by microgravity.

The boldly outlined boxes in Fig. 1 represent a theoretical construct referred to as working memory. Working memory actively maintains, manipulates and organizes perceived information. Information is maintained through rehearsal. Working memory maintenance is often assessed using a visual object learning task [37]. Drugs taken for space motion sickness affect the working memory maintenance process [87]. The rehearsal process is also easily disrupted by distractions (irrelevant stimuli) and interruptions (stimuli that require attention). Interruptions have a more detrimental effect on performance than distractors likely because the interrupting stimulus has disrupted the memory trace [24]. In aviation and in medicine, distractions and interruptions have contributed to aircraft maintainers [81], aircraft pilots [64], air traffic controllers [67], nurses [44,48] and anesthesiologists [42] inadvertently skipping critical procedure steps, sometimes leading to disaster [65,78].

Although the effect of distractions and interruptions during everyday space mission tasks and specialized payload or maintenance procedure execution has not been quantified, Mission Control Center capsule communicators should time transmissions so that astronauts who are stepping through a procedure or otherwise engaged are not interrupted [32,47,49]. This is because distractions and interruptions create prospective memory tasks [33]. Prospective memory is the process necessary to fulfill an intention to perform a specific action in the future. Deferred tasks, particularly safety critical tasks, require additional cognitive resources because the individual must repeatedly remind themselves about the deferred task while performing the current task. People are vulnerable to forgetting to resume interrupted tasks [31], are more prone to errors [106] and if they do remember, it can take extra effort and time to mentally reconstruct the status of the interrupted task [33]. Prospective memory performance has been affected by 45-days of head-down bed rest [22]. A common form of prospective memory error is referred to as a post-completion error that are often seen after an individual has achieved a task (e.g., a procedure) but forgets to perform the last step before moving on to the next task. Everyday examples of post-completion errors include forgetting to turn off a stove burner after cooking a meal or forgetting to retrieve the original document after photocopying it [6]. The best countermeasure for reducing prospective memory errors is to avoid creating situations that require deferring a task (i.e., interruptions) and when that is not possible there needs to be noticeable or conspicuous reminders in the environment that indicate what task needs to be performed and when it needs to be performed [72].

A cognitive test used to assess distractions to working memory is the n-back. An n-back task consists of the presentation of a stream of letters, numbers, or shapes. The software presents each item for a brief amount of time. The study participant is to report whether the item just presented matched the item presented "n" items back. Multiple executive processes are involved in the n-back task besides actively maintaining the previous "n" items including updating working memory with new items, binding each item to its serial position and inhibiting interference from items that appeared earlier in the list. The n-back is sensitive to the effects of headache [76] but it is unclear what aspects of the task are affected [79]. Reference [100] did not find a decrement in 1-back performance during either 60 or 90 days of bed rest.

The central executive (see Fig. 1) of working memory directs attention to relevant stimuli and suppresses irrelevant information. Spaceflight cognitive tests have included a Stroop test that taps into the ability to suppress irrelevant information. In the original Stroop task, subjects are asked to provide the color of a "color word". For example, the word "RED" is presented in green ink and the task is to say "green". It is amazingly difficult to inhibit the automatic behavior of word reading in favor of the less practiced task of naming the ink color. Three crewmembers showed significantly more difficulty inhibiting information during an 11-day spaceflight [86] when the test stimuli were emotionally relevant to the astronauts. Virtual reality versions of the Stroop Test are effective at detecting decrements in executive function [4].

In a Sternberg [103] task, the study participant first memorizes a short list of stimuli (letters, numbers or symbols). Following presentation of a test stimulus (the probe), the participant is required to decide if it is one of the items in the memory list. The latency of the response reveals something about the working memory retrieval process. The number of items in working memory slows the retrieval time. Quite a few spaceflight studies use a Sternberg memory task, with an active hypothesis that spaceflight stressors compromise the retrieval process. The results are variable. Some results show no change [35,70,80] while other studies show a slowing of response times [56,68]. Methodological factors in several of the studies make interpretation of the results difficult.

The central executive directs, focuses, switches and divides attention when automated processes won't suffice, giving priority to certain activities. Attention serves to integrate the distributed information from the various stores into unified objects [29,111].

The central executive controls directed attention. Directed attention refers to the capacity for mental flexibility that allows individuals to shift their focus of attention between tasks having different cognitive requirements. A directed attention task was used on two Shuttle flights. When asked to randomly switch between spatial and mathematical processing tasks, some, but not all crewmembers showed a performance decrement in the beginning and the end of the flight [35].

The ability to read, solve problems, navigate and perform two tasks taps into working memory manipulation and organizational processes. Dual-task paradigms require the participant to perform two tasks simultaneously. A dual-task test is tapping into the attentional selectivity aspect of the working memory central executive [82]. The performance on both tasks is measured relative to performance on each task alone. Concurrent tasks requiring processing in similar cognitive networks show greater declines (i.e., increases in response latency and/or decreases in accuracy) than in dissimilar networks because of the increased cognitive demand of the concurrent task [112].

Motor skills are poorer just after entering microgravity and upon returning to Earth's gravity [15,17,18,45,60,68,69,80,97,101]. Reference [19] found deficits in three astronauts throughout a six-month ISS mission using a two dimensional (versus 1 dimensional) unstable tracking task. When a motor task is combined with a Sternberg memory task, or when tapping a pre-learned rhythm, motor deficits are even greater [19,70,71,102]. Other studies have failed to find dual-task deficits during spaceflight [35]. When a motor task was combined with a reaction time task, and the astronauts were told to consider the motor task as the primary task, then motor deficits were not observed [36]. Instead, reaction time errors of commission became more variable.

Long-term memory is thought to encode information as semantically related memory traces [34] referred to as schemas. Rehearsal transfers information from working memory into long-term memory, as does linking new information to existing schemas. Novice crewmembers require well-designed instruction since they lack the schema-based knowledge. In contrast, crewmembers already possessing the schema-based knowledge can use their own internal guidance. External guidance becomes redundant and adds to their working memory load [55].

To prevent forgetting the information requires recalling the information repetitively [41]. There is some evidence that, during non-REM sleep, memory traces are reactivated to promote their transfer into storage [16]. The sleep deprivation experienced by many crewmembers [8] may be slowing the formation of long-term memory schemas.

Long-term memory schemas may explain expertise. When experts in a certain field are required to make a decision, they will use past experience to govern the verbal or motor response [58]. The recognition-primed decision-making model proposes that experts use a combination of intuition and analysis to make decisions. The recognition-primed decision-making (RPD) process (see Fig. 1) begins with experiencing the situation [57]. Using mental simulation to imagine how a situation will play out, the current situation is matched to a recognized course of action. If the current situation will work, a response is initiated and then executed. If it would almost work, they will adapt that course or consider other actions choosing the first workable option, rather than trying to find the best possible option. The process allows the decision maker to build a story from a combination of the incoming information, knowledge and experience. In recurring situations, experienced people will generally be able to come up with a decision more quickly because the situation may match a prototypical situation they have encountered before. A seasoned decision-maker will spend much more time and energy trying to diagnose a problem, and distinguishing between different explanations. This is where decision aiding could be beneficial, not in providing options for the person, but in problem diagnosis. Decision aids could also help with simulating the course of action chosen by the expert. The aid could help to identify weaknesses in the plan and provide improvement options. Novices, lacking this experience, must cycle through different possibilities, and tend to use the first course of action that they believe will work.

Our understanding of the neural basis of cognition is considerably richer today. The role of the prefrontal cortex, the frontoparietal cortex, and sensory cortices has recently been embellished by an important correlation between psychomotor and cognitive abilities [21,88,93]. There has been an increasing interest in the cerebellum's role in cognition [7,30,61,75]. Anatomically, much of the cerebellum is linked with motor systems, but a specific zone is linked with the cerebral cortex and therefore different aspects of cognition [39,52,104]. Just as sensorimotor practice guides the development of internal models within the cerebellum that are then sent to the motor cortex for implementation, sensorimotor practice also forms internal models within the cerebellum that are projected to the cerebral cortex for executive functions, e.g., mental manipulations, rehearsal or pulling information from long-term memory storage. Different zones of the cerebellum develop internal models in the same way (i.e., similar sets of neural fibers [51]). New internal models develop in adjacent locations rather than overwriting previously stored models. This modular organization of internal models may support flexible adaptation depending on the context [50]. Studies are underway to unravel these complex interrelationships.

9. REFERENCES

1. Ahumada Jr, A. J. (1996). Perceptual classification images from Vernier acuity masked by noise. *Perception*, *25*(1_suppl), 2-2.

2. Allen, J. G., MacNaughton, P., Satish, U., Santanam, S., Vallarino, J., & Spengler, J. D. (2016). Associations of cognitive function scores with carbon dioxide, ventilation, and volatile organic compound exposures in office workers: a controlled exposure study of green and conventional office environments. *Environmental health perspectives*, *124*(6), 805.

3. Arlinghaus, K. A., Shoaib, A. M., & Price, T. R. (2005). Neuropsychiatric assessment. In J. M. Silver, T. W. McAllister, & S. C. Yudofsky (Eds.), Textbook of traumatic brain injury (pp. 59-78). Arlington, VA: American Psychiatric Publishing.

4. Armstrong, C. M., Reger, G. M., Edwards, J., Rizzo, A. A., Courtney, C. G., & Parsons, T. D. (2013). Validity of the Virtual Reality Stroop Task (VRST) in active duty military. *Journal of Clinical and Experimental Neuropsychology*, *35*(2), 113-123.

5. Attridge, N., Crombez, G., Van Ryckeghem, D., Keogh, E., & Eccleston, C. (2015). The Experience of Cognitive Intrusion of Pain: scale development and validation. *Pain*, *156*(10), 1978.

6. Back, J., Cheng, W., Dann, R., Curzon, P., & Blandford, A. (2007). Does being motivated to avoid procedural errors influence their systematicity?. *People and Computers XX—Engage*, 151-157.

7. Baillieux, H., De Smet, H. J., Paquier, P. F., De Deyn, P. P., & Mariën, P. (2008). Cerebellar neurocognition: insights into the bottom of the brain. *Clinical neurology and neurosurgery*, *110*(8), 763-773.

8. Barger, L. K., Flynn-Evans, E. E., Kubey, A., Walsh, L., Ronda, J. M., Wang, W., ... & Czeisler, C. A. (2014). Prevalence of sleep deficiency and use of hypnotic drugs in astronauts before, during, and after spaceflight: an observational study. *The Lancet Neurology*, *13*(9), 904-912.

9. Barnard, K. E., Broman-Fulks, J. J., Michael, K. D., Webb, R. M., & Zawilinski, L. L. (2011). The effects of physiological arousal on cognitive and psychomotor performance among individuals with high and low anxiety sensitivity. *Anxiety, Stress, & Coping*, *24*(2), 201-216.

10. Bashore, T. R., & Ridderinkhof, K. R. (2002). Older age, traumatic brain injury, and cognitive slowing: some convergent and divergent findings. Psychol Bull., 128, 151-198.

11. Beard, B. L. (pending). Cognitive functions and human error. In *Space Safety and*

Human Performance, T. Sgobba, B. Kanki, J. Clervoy and G. Sandal (eds.), International Association for the Advancement of Space Safety, IAASS, Elsevier, Paris, FR.

12. Beard, B. L., & Ahumada Jr, A. J. (1997). Relevant image features for vernier acuity. *Perception*, *26*(1_suppl), 86-86.
13. Beard, B. L., & Ahumada Jr, A. J. (1998). A technique to extract relevant image features for visual tasks." In Human Vision and Electronic Imaging, 3(3299), 79-85.
14. Benke, T., Koserenko, O., Watson, N. V., & Gerstenbrand, F. (1993). Space and cognition: The measurement of behavioral functions during a 6-day space mission. Aviation, Space, and Environmental Medicine, 64, 376 – 379.
15. Berger, M., Mescheriakov, S., Molokanova, E., Lechner-Steinleitner, S., Seguer, N., & Kozlovskaya, I. (1997). Pointing arm movements in short-and long-term spaceflights. *Aviation, space, and environmental medicine*, *68*(9), 781-787.
16. Bergmann, T. O., Mölle, M., Diedrichs, J., Born, J., & Siebner, H. R. (2012). Sleep spindle-related reactivation of category-specific cortical regions after learning face-scene associations. *Neuroimage*, *59*(3), 2733-2742.
17. Bock, O., Abeele, S., & Eversheim, U. (2003). Sensorimotor performance and computational demand during short-term exposure to microgravity. *Aviation, space, and environmental medicine*, *74*(12), 1256-1262.
18. Bock, O., Fowler, B., & Comfort, D. (2001). Human sensorimotor coordination during spaceflight: an analysis of pointing and tracking responses during the" Neurolab" Space Shuttle mission. *Aviation, space, and environmental medicine*, *72*(10), 877-883.
19. Bock, O., Weigelt, C., & Bloomberg, J. J. (2010). Cognitive demand of human sensorimotor performance during an extended space mission: a dual-task study. *Aviation, space, and environmental medicine*, *81*(9), 819-824.
20. Brandt, T., & Dieterich, M. (1999). The vestibular cortex: its locations, functions, and disorders. *Annals of the New York Academy of Sciences*, *871*(1), 293-312.
21. Carretta, T. R. (1997). Group differences on US Air Force pilot selection tests. *International Journal of Selection and Assessment*, *5*(2), 115-127.
22. Chen, S., Zhou, R., Xiu, L., Chen, S., Chen, X., & Tan, C. (2013). Effects of 45-day-6° head-down bed rest on the time-based prospective memory. *Acta Astronautica*, *84*, 81-87.
23. Chou, P. I., Wen, T. S., Wu, Y. C., Horng, C. T., & Liu, C. C. (2003). Contrast sensitivity after+ Gz acceleration. *Aviation, space, and environmental medicine*, *74*(10), 1048-1051.
24. Clapp, W. C., & Gazzaley, A. (2012). Distinct mechanisms for the impact of distraction and interruption on working memory in aging. *Neurobiology of aging*, *33*(1), 134-148.
25. Clément, G., Skinner, A., Richard, G., & Lathan, C. (2012). Geometric illusions in astronauts during long-duration spaceflight. *Neuroreport*, *23*(15), 894-899.
26. Conway, G. E., Szalma, J. L., & Hancock, P. A. (2007). A quantitative meta-analytic examination of whole-body vibration effects on human performance. *Ergonomics*, *50*(2), 228-245.
27. Cowings, P. S., & Toscano, W. B. (1982). The relationship of motion sickness susceptibility to learned autonomic control for symptom suppression. *Aviation, space, and environmental medicine*, *53*, 570-575.
28. Cowings, P. S., & Toscano, W. B. (2000). Autogenic-Feedback Training Exercise Is Superior to Promethazine for Control of Motion Sickness Symptoms. *The Journal of Clinical Pharmacology*, *40*(10), 1154-1165.
29. Delvenne, J. F., & Bruyer, R. (2004). Does visual short-term memory store bound features?. *Visual cognition*, *11*(1), 1-27.
30. Diamond, A. (2000). Close interrelation of motor development and cognitive development and of the cerebellum and prefrontal cortex. *Child development*, *71*(1), 44-56.
31. Dismukes, R. K., & Nowinski, J. (2007). Prospective memory, concurrent task management, and pilot error. *Attention: From theory to practice*, 225-236. In A. Kramer, D. Wiegmann, & A. Kirlik (Eds.), *Attention: From theory to practice*. New York: Oxford University Press.

32. Dismukes, R. K., Young, G. E., Sumwalt III, R. L., & Null, C. H. (1998). Cockpit interruptions and distractions: Effective management requires a careful balancing act. Downloaded from: https://ntrs.nasa.gov/search.jsp?R=20020062698

33. Dodhia, R. M., & Dismukes, R. K. (2009). Interruptions create prospective memory tasks. *Applied Cognitive Psychology*, *23*(1), 73-89.

34. Dudai, Y. (2004). The neurobiology of consolidations, or, how stable is the engram?. *Annu. Rev. Psychol.*, *55*, 51-86.

35. Eddy, D. R., Schiflett, S. G., Schlegel, R. E., & Shehab, R. L. (1998). Cognitive performance aboard the life and microgravity spacelab. *Acta Astronautica*, *43*(3-6), 193-210.

36. Fowler, B., Comfort, D., & Bock, O. (2000). A review of cognitive and perceptual-motor performance in space. *Aviation, space, and environmental medicine*. Aviation, Space, and Environmental Medicine 71(9 Suppl), A66–8.

37. Glahn, D. C., Gur, R. C., Ragland, J. D., Censits, D. M., & Gur, R. E. (1997). Reliability, performance characteristics, construct validity, and an initial clinical application of a visual object learning test (VOLT). *Neuropsychology*, *11*(4), 602.

38. Glasauer, S., & Mittelstaedt, H. (1998). Perception of spatial orientation in microgravity. *Brain Research Reviews*, *28*(1), 185-193.

39. Glickstein, M., Sultan, F., & Voogd, J. (2011). Functional localization in the cerebellum. *Cortex*, *47*(1), 59-80.

40. Grabherr, L., & Mast, F. W. (2010). Effects of microgravity on cognition: The case of mental imagery. *Journal of Vestibular research*, *20*(1, 2), 53-60.

41. Greene, R. L. (1987). Effects of maintenance rehearsal on human memory. *Psychological Bulletin*, *102*(3), 403.

42. Grundgeiger, T., & Sanderson, P. (2009). Interruptions in healthcare: theoretical views. *International journal of medical informatics*, *78*(5), 293-307.

43. Harm, D. L., Reschke, M. F. & Wood, S. J. (2015). Spatial Orientation and Motion Perception in Micrograyity. In Hoffman, R. R., Hancock, P. A., Scerbo, M. W., Parasuraman, R., & Szalma, J. L. (Eds.). *The Cambridge handbook of applied perception research*. Cambridge University Press.

44. Hayes, C., Jackson, D., Davidson, P. M., & Power, T. (2015). Medication errors in hospitals: a literature review of disruptions to nursing practice during medication administration. *Journal of clinical nursing*, *24*(21-22), 3063-3076.

45. Heuer, H., Manzey, D., Lorenz, B., & Sangals, J. (2003). Impairments of manual tracking performance during spaceflight are associated with specific effects of microgravity on visuomotor transformations. *Ergonomics*, *46*(9), 920-934.

46. Hideg, J., Bognár, L., Remes, P., Kozarenko, O. P., Miasnikov, V., & Ponomareva, I. P. (1982). Psychophysiological performance examination onboard the orbital complex Salyut-Soyuz. In *International Astronautical Federation, International Astronautical Congress, 33 rd, Paris, France* (p. 1982).

47. Ho, C. Y., Nikolic, M. I., Waters, M. J., & Sarter, N. B. (2004). Not now! Supporting interruption management by indicating the modality and urgency of pending tasks. *Human Factors*, *46*(3), 399-409.

48. Hopkinson, S. G., & Jennings, B. M. (2013). Interruptions during nurses' work: A state-of-the-science review. *Research in nursing & health*, *36*(1), 38-53.

49. Huckels-Baumgart, S., Niederberger, M., Manser, T., Meier, C. R., & Meyer-Massetti, C. (2017). A combined intervention to reduce interruptions during medication preparation and double-checking: a pilot-study evaluating the impact of staff training and safety vests. *Journal of Nursing Management*.

50. Imamizu, H. (2014). Internal Models for Dexterous use of Tools: The Border Between Cognitive and Motor Skills. In Koziol, L. F., Budding, D., Andreasen, N., D'Arrigo, S., Bulgheroni, S., Imamizu, H., ... & Pezzulo, G. (2014). Consensus paper: the cerebellum's role in movement and cognition. The Cerebellum, 13(1), 151-177.

51. Ito, M. (1993). Movement and thought: identical control mechanisms by the cerebellum. *Trends in neurosciences*, *16*(11), 448-450.

52. Ito, M. (2014). What can we learn from computational models? In Koziol, L. F., Budding, D., Andreasen, N., D'Arrigo, S., Bulgheroni, S., Imamizu, H., ... & Pezzulo, G. (2014). Consensus paper: the cerebellum's role in movement and cognition. The Cerebellum, 13(1), 151-177.

53. James, J. T., & Zalesak, S. M. (2013). Surprising effects of CO2 exposure on decision making. *AIAA Paper*, (2013-3463).

54. Jozsvai, E. E., & Pigeau, R. A. (1996). The effect of autogenic training and biofeedback on motion sickness tolerance. *Aviation, space, and environmental medicine*, *67*(10), 963-968.

55. Kalyuga, S. (2007). Expertise reversal effect and its implications for learner-tailored instruction. *Educational Psychology Review*, *19*(4), 509-539.

56. Kelly, T. H., Hienz, R. D., Zarcone, T. J., Wurster, R. M., & Brady, J. V. (2005). Crewmember performance before, during, and after spaceflight. *Journal of the experimental analysis of behavior*, *84*(2), 227-241.

57. Klein, G. (2008). Naturalistic decision making. *Human factors*, *50*(3), 456-460.

58. Klein, G. A. (1993). *A recognition-primed decision (RPD) model of rapid decision making* (pp. 138-147). New York: Ablex Publishing Corporation.

59. Kornilova, L. N. (1997). Orientation illusions in spaceflight. *Journal of Vestibular Research*, *7*(6), 429-440.

60. Kubis, J. F., McLaughlin, E. J., Jackson, J. M., Rusnak, R., McBride, G. H., & Saxon, S. V. (1977). Task and work performance on Skylab missions 2, 3, and 4: Time and motion study: Experiment M151.

61. Küper, M., Dimitrova, A., Thürling, M., Maderwald, S., Roths, J., Elles, H. G., ... & Timmann, D. (2011). Evidence for a motor and a non-motor domain in the human dentate nucleus—an fMRI study. *Neuroimage*, *54*(4), 2612-2622.

62. Levi, D. M., & Klein, S. A. (2003). Noise provides some new signals about the spatial vision of amblyopes. *Journal of Neuroscience*, *23*(7), 2522-2526.

63. Li, R. W., Levi, D. M., & Klein, S. A. (2004). Perceptual learning improves efficiency by re-tuning the decision'template'for position discrimination. *Nature neuroscience*, *7*(2), 178.

64. Loukopoulos, L. D., Dismukes, R. K., & Barshi, I. (2001, March). Cockpit interruptions and distractions: A line observation study. In *Proceedings of the 11th international symposium on aviation psychology* (pp. 1-6). Columbus: Ohio State University Press.

65. Loukopoulos, L. D., Dismukes, R. K., & Barshi, I. (2003, April). Concurrent task demands in the cockpit: Challenges and vulnerabilities in routine flight operations. In *Proceedings of the 12th international symposium on aviation psychology* (pp. 737-742). Dayton, OH: Wright State University Press.

66. Mader, T. H. (1991). Intraocular pressure in microgravity. *The Journal of Clinical Pharmacology*, *31*(10), 947-950.

67. Malakis, S., & Kontogiannis, T. (2008, October). Cognitive strategies in emergency and abnormal situations training: implications for resilience in air traffic control. In *Proceedings of the 3rd Symposium on Resilience Engineering*.

68. Manzey, D., Lorenz, B., & Poljakov, V. (1998). Mental performance in extreme environments: results from a performance monitoring study during a 438-day spaceflight. *Ergonomics*, *41*(4), 537-559.

69. Manzey, D., Lorenz, B., Heuer, H., & Sangals, J. (2000). Impairments of manual tracking performance during spaceflight: more converging evidence from a 20-day space mission. *Ergonomics*, *43*(5), 589-609.

70. Manzey, D., Lorenz, B., Schiewe, A., Finell, G., & Thiele, G. (1993). Behavioral aspects of human adaptation to space analyses of cognitive and psychomotor performance in space during an 8-day space mission. *The clinical investigator*, *71*(9), 725-731.

71. Manzey, D., Lorenz, B., Schiewe, A., Finell, G., & Thiele, G. (1995). Dual-task performance in space: results from a single-case study during a short-term space mission. *Human factors*, *37*(4), 667-681.

72. McDaniel, M. A., Guynn, M. J., Einstein, G. O., & Breneiser, J. (2004). Cue-focused and reflexive-associative processes in prospective memory retrieval. *Journal of Experimental Psychology: Learning, Memory, and Cognition*, *30*(3), 605.

73. McLellan, T. M., Kamimori, G. H., Bell, D. G., Smith, I. F., Johnson, D., & Belenky, G. (2005). Caffeine maintains vigilance and marksmanship in simulated urban operations with sleep deprivation. *Aviation, space, and environmental medicine*, *76*(1), 39-45.

74. Miller, J. O., & Low, K. (2001). Motor processes in simple, go/no-go, and choice reaction time tasks: a psychophysiological analysis. *Journal of experimental psychology: Human perception and performance*, *27*(2), 266.

75. Molinari, M., Restuccia, D., & Leggio, M. G. (2009). State estimation, response prediction, and cerebellar sensory processing for behavioral control. *The Cerebellum*, *8*(3), 399-402.

76. Moore, D. J., Keogh, E., & Eccleston, C. (2013). Headache impairs attentional performance. *Pain*, *154*(9), 1840-1845.

77. Murray, R. F. (2011). Classification images: A review. *Journal of Vision*, *11*(5), 2-2.

78. National Transportation Safety Board. (1988). Northwest Airlines, Inc., McDonnell Douglas DC-9-82, N312RC, Detroit Metropolitan Wayne County Airport, Romulus, Michigan, August 16, 1987. (Report No. AAR-88-05). Washington, DC.

79. Nee, D. E., Brown, J. W., Askren, M. K., Berman, M. G., Demiralp, E., Krawitz, A., & Jonides, J. (2012). A meta-analysis of executive components of working memory. *Cerebral cortex*, *23*(2), 264-282.

80. Newman, D. J., & Lathan, C. E. (1999). Memory processes and motor control in extreme environments. *IEEE Transactions on Systems, Man, and Cybernetics, Part C (Applications and Reviews)*, *29*(3), 387-394.

81. Ng, M. W., & Li, S. Y. (2016, September). An analysis of aircraft maintenance incidents using psychological and cognitive engineering knowledge. In *Proceedings of the Human Factors and Ergonomics Society Annual Meeting* (Vol. 60, No. 1, pp. 1676-1680). Sage CA: Los Angeles, CA: SAGE Publications.

82. Oberauer, K., Süß, H. M., Schulze, R., Wilhelm, O., & Wittmann, W. W. (2000). Working memory capacity—facets of a cognitive ability construct. *Personality and individual differences*, *29*(6), 1017-1045.

83. Oman, C. (2007). Spatial orientation and navigation in microgravity. In Spatial processing in navigation, imagery and perception, Mast, F. & Jancke, L. (eds.), 209-247.

84. Oman, C. M., Lichtenberg, B. K., Money, K. E., & McCoy, R. K. (1986). MIT/Canadian vestibular experiments on the Spacelab-1 mission: 4. Space motion sickness: symptoms, stimuli, and predictability. *Experimental Brain Research*, *64*(2), 316-334.

85. Oneal, M. R., Task, H. L., & Genco, L. V. (1992). Effect of microgravity on several visual functions during STS shuttle missions. Downloaded from: https://ntrs.nasa.gov/search.jsp?R=19920013088

86. Pattyn, N., Migeotte, P. F., Morais, J., Soetens, E., Cluydts, R., & Kolinsky, R. (2009). Crew performance monitoring: Putting some feeling into it. *Acta astronautica*, *65*(3), 325-329.

87. Paule, M. G., Chelonis, J. J., Blake, D. J., & Dornhoffer, J. L. (2004). Effects of drug countermeasures for space motion sickness on working memory in humans. *Neurotoxicology and teratology*, *26*(6), 825-837.

88. Penner-Wilger, M., Fast, L., LaFevre, J. A., Smith-Chant, B. L., Skwarchuck, S. L., Kamawar, D., & Bisanz, J. (2007, January). The foundations of numeracy: Subitizing, finger gnosia, and fine motor ability. In *Proceedings of the Cognitive Science Society* (Vol. 29, No. 29).

89. Peters, B. T., Miller, C. A., Brady, R. A., Richards, J. T., Mulavara, A. P., & Bloomberg, J. J. (2011). Dynamic visual acuity during walking after long-duration spaceflight. *Aviation, space, and environmental medicine*, *82*(4), 463-466.

90. Piponnier, J. C., Forget, R., Gagnon, I., McKerral, M., Giguère, J. F., & Faubert, J. (2016). First-and second-order stimuli reaction time measures are highly sensitive

to mild traumatic brain injuries. *Journal of neurotrauma*, *33*(2), 242-253.

91. Ratino, D. A., Repperger, D. W., Goodyear, C., Potor, G., & Rodriguez, L. E. (1988). Quantification of reaction time and time perception during space shuttle operations. *Aviation, space, and environmental medicine, 59(3), 220-224.*

92. Recarte, M. A., & Nunes, L. M. (2003). Mental workload while driving: effects on visual search, discrimination, and decision making. Journal of experimental psychology: Applied, 9(2), 119.

93. Ree, M. J., & Carretta, T. R. (1992). *The correlation of cognitive and psychomotor tests* (No. AL-TP-1992-0037). ARMSTRONG LAB BROOKS AFB TX.

94. Ross, H. E. (1991). Motor skills under varied gravitoinertial force in parabolic flight. *Acta Astronautica*, *23*, 85-95.

95. Saatman, K. E., Duhaime, A. C., Bullock, R., Maas, A. I., Valadka, A., & Manley, G. T. (2008). Classification of traumatic brain injury for targeted therapies. *Journal of neurotrauma*, *25*(7), 719-738.

96. Sanes, J. N., & Evarts, E. V. (1983). Effects of perturbations on accuracy of arm movements. *Journal of Neuroscience*, *3*(5), 977-986.

97. Sangals, J., Heuer, H., Manzey, D., & Lorenz, B. (1999). Changed visuomotor transformations during and after prolonged microgravity. *Experimental brain research*, *129*(3), 378-390.

98. Satish, U., Mendell, M. J., Shekhar, K., Hotchi, T., Sullivan, D., Streufert, S., & Fisk, W. J. (2012). Is CO2 an indoor pollutant? Direct effects of low-to-moderate CO2 concentrations on human decision-making performance. *Environmental health perspectives*, *120*(12), 1671.

99. Sausen, K. P., Wallick, M. T., Slobodnik, B., Chimiak, J. M., Bower, E. A., Stiney, M. E., & Clark, J. B. (2001). The reduced oxygen breathing paradigm for hypoxia training: physiological, cognitive, and subjective effects. *Aviation, space, and environmental medicine*, *72*(6), 539-545.

100. Seaton, K. A., Slack, K. J., Sipes, W. A., & Bowie, K. E. (2009). Cognitive functioning in long-duration head-down bed rest. *Aviation, space, and environmental medicine*, *80*(5), A62-A65.

101. Semjen, A., Leone, G., & Lipshits, M. (1998). Temporal control and motor control: Two functional modules which may be influenced differently under microgravity. *Human movement science*, *17*(1), 77-93.

102. Shehab, R. L., Schlegel, R. E., & Hwang, B. C. (2000, January). Cognitive performance of astronauts during two space shuttle missions. In *Proceedings of the Human Factors and Ergonomics Society... Annual Meeting* (Vol. 3, p. 110). Human Factors and Ergonomics Society.

103. Sternberg, S. (1969). The discovery of processing stages: Extensions of Donders' method. *Acta psychologica*, *30*, 276-315.

104. Strick, P. L., Dum, R. P., & Fiez, J. A. (2009). Cerebellum and nonmotor function. *Annual review of neuroscience*, *32*, 413-434.

105. Stroud, K. J., Harm, D. L., & Klaus, D. M. (2005). Preflight virtual reality training as a countermeasure for space motion sickness and disorientation. *Aviation, space, and environmental medicine*, *76*(4), 352-356.

106. Trafton, J. G., Altmann, E. M., Brock, D. P., & Mintz, F. E. (2003). Preparing to resume an interrupted task: Effects of prospective goal encoding and retrospective rehearsal. *International Journal of Human-Computer Studies*, *58*(5), 583-603.

107. Trimmel, M., & Poelzl, G. (2006). Impact of background noise on reaction time and brain DC potential changes of VDT-based spatial attention. *Ergonomics*, *49*(2), 202-208.

108. Van der Linden, D., Frese, M., & Meijman, T. F. (2003). Mental fatigue and the control of cognitive processes: effects on perseveration and planning. Acta Psychologica, 113(1), 45-65.

109. Vasterling, J. J., & Bremner, J. D. (2006). The impact of the 1991 Gulf War on the mind and brain: findings from neuropsychological and neuroimaging research. *Philosophical Transactions of the Royal Society B: Biological Sciences*, *361*(1468), 593-604.

110. Webb, J. T., & Pilmanis, A. A. (2011). Fifty Years of Decompression Sickness Research at Brooks AFB, TX: 1960–2010. *Aviation, space, and environmental medicine*, *82*(5), A1-A25.

111. Wheeler, M. E., & Treisman, A. M. (2002). Binding in short-term visual memory. *Journal of Experimental Psychology: General*, *131*(1), 48.

112. Wickens, C. D. (2002). Multiple resources and performance prediction. Theoretical issues in ergonomics science, 3(2), 159-177.

113. Wyatt, J. K., Cajochen, C., Cecco, A. R. D., Czeisler, C. A., & Dijk, D. J. (2004). Low-dose repeated caffeine administration for circadian-phase-dependent performance degradation during extended wakefulness. *Sleep*, *27*(3), 374-381.

PROCEEDINGS OF THE 9TH INTERNATIONAL SPACE SAFETY CONFERENCE

SESSION 26 NEO HAZARDS

SIMULATED RESPONSE TO FICTITIOUS ASTEROID THREAT

Nahum Melamed[(1)]

[(1)]The Aerospace Corporation
2310 E. El Segundo Blvd., El Segundo, CA 90245, USA
Email:Nahum.Melamed@aero.org

ABSTRACT

An interactive, online Near-Earth Object (NEO) deflection app, or NDA, has been developed jointly by The Aerospace Corporation and NASA's Jet Propulsion Laboratory (JPL). NEOs are asteroids or comets whose orbits are close enough to Earth's to pose a collision risk. At the 2017 IAA Planetary Defense Conference, a hypothetical threat scenario was exercised for a fictitious 100 to 250-meter asteroid termed 2017 PDC. The asteroid was designed to impact Earth about ten years after its discovery. Conference participants used the NDA to assess possible responses to 2017 PDC. Had this had been a real threat, decisions would have needed to be made whether to launch an expensive deflection campaign immediately, launch an observer spacecraft, or wait for additional measurements to either confirm or rule out the impact. New phases of the threat were revealed each day of the conference, and participants applied the NDA to construct an approximate assessment of deflection feasibility using today's launch vehicle technology. This paper describes the decision-making process exercised at the conference, the role the NDA played in informing participants of key considerations for campaign design, and achievable Earth miss distances. The paper highlights planned upgrades to the NDA that will account for uncertainties in a threatening object's orbit, physical properties, launch failures, and other considerations that will be key to designing a realistic deflection campaign.

1. INTRODUCTION

New large, more than a kilometer in size, Near-Earth Objects (NEOs) are discovered at a slowing rate with estimates that over 90% have now been discovered. Fortunately, none have been discovered that pose a threat of collision with Earth this century1. However, the discovery of medium and small size asteroids whose orbits bring them close enough to Earth's to pose a collision risk is accelerating, with many only discovered after a very close approach. Therefore, the risk of collision has shifted to the smaller objects, which are more numerous and likely to inflict local or regional damage as opposed to damage on the planetary scale. Discovery efforts of the smaller objects are intensifying, and the discussion on how to deal with an actual threat is growing in importance.

The NEO hazard spans a wide range of orbits, sizes, compositions and discovery times, requiring different types of responses. When the time before a potential impact is adequate, preparation of a deflection campaign can be started, resources allocated, and systems built. When the warning time is short (less than a few years), more powerful measures are required and systems must be pre-built. Such short warning situations must be addressed proactively and decisions must be made ahead of a potential emergency.

Since 2004 International planetary defense conferences (PDCs) have gathered world experts to discuss the nature of the NEO threat, present current knowledge and analyses, and propose solutions2. The last three conferences included an asteroid threat response exercise in order to help better characterize the problem and assess solutions. Similar processes were applied as part of three recent NASA/FEMA tabletop exercises3. These exercise scenarios involved fictitious objects spanning a range of orbits, sizes and compositions that are representative of the real NEO population. A special fictitious object was designed by NASA/JPL for each exercise by varying the discovery time and calculating an orbit to ensure its impact with Earth at the end of the period. Size and composition were assumed parameters.

The most recent tabletop exercise was played out at the 2017 IAA Planetary Defense Conference, where an asteroid estimated to be between 100 and 250 meters in size with an initial 1% chance of impacting the Earth was discovered ten years before potential impact4. The initial parameters of the scenario were released before the conference, and additional press releases were issued throughout the conference showing that the object was on a collision course with Earth. The object's orbit and appropriate parameters were loaded into the NEO Deflection App (NDA)5 and used before and during the conference to estimate the types and numbers of launch vehicles required to deliver deflection payloads to the asteroid.

The key parameter defining the 2017 PDC impact scenario was the 10-year span from discovery to potential impact, which allowed for several opportunities to respond. As shown in Fig. 1, with a heliocentric orbital period of about 3 years, the object presents three distinct opportunities to launch kinetic impactors to deflect the asteroid. Flight time from launch to deflection is shown on the vertical axis, and the time of deflection before impact on the horizontal axis. The sum of any combination of these times constitutes possible times before impact when kinetic impactors could be launched.

Estimates produced by the NDA for the 2017 PDC object assisted the decision process exercised at the conference. Conference attendees considered the threat information6 and were encouraged to use the NDA and issue recommendations to a group of world leaders who made decisions and shaped the progression of the scenario throughout the conference7. The use of the NDA in this and in prior exercises revealed its utility, but also its limitations, in estimating deflection requirements reliably. The purpose of this paper is to highlight the usefulness of the NDA for informing mitigation decisions, and point to planned improvements that will facilitate more realistic NDA results.

2. PLANNED UPDATES TO THE NEO DEFLECTION APP

At present, the NDA is an approximation that calculates the minimum number of launch vehicles required to deliver kinetic energy deflectors to a fictitious asteroid designed to impact Earth. In tabletop exercises, leaders rely on such information to know what resources should be allocated and when. Although the NDA exclusively deals with fictitious NEOs and is used mainly for training, the NDA could be used to provide first estimates of feasible solutions for deflection of real objects when appropriate information becomes available.

Several enhancements are planned that will add realism to the calculations. First and foremost is the orbital uncertainty which is inherent to newly discovered objects before adequate observations or in-situ information are available. In its current form, the NDA does not account for orbital uncertainty so it can only produce estimates for moving an object on a "nominal" orbit. Additional launchers would have to be integrated into the campaign in order to achieve a desired level of deflection assurance if uncertainties in the orbit were included. Credible campaign size relies on systematically accounting for orbital uncertainty in the NDA, and this is a highly desired enhancement to the tool. The five Atlas V launch vehicles calculated by the NDA for the 2017 PDC scenario was increased to six to account for orbit and other uncertainties, but this increase may have been well below what a more rigorous analysis would have provided.

Another source of uncertainty currently not modeled in the NDA is the object's mass and composition. These parameters dictate the deflection effort, so their uncertainties should be accounted for in the NDA to reflect our lack of knowledge about the object. Also, when a kinetic impactor collides with the asteroid, the amount of mass ejected out of the crater, which enhances the performance of the deflection, depends on the physical properties of the object. The physical composition is also unknown and is estimated from theoretical and experimental data.

Launch vehicle and space vehicle reliability metrics should also be taken into account due to their effect on overall campaign reliability. Historical data and up-to-date analyses can be used to estimate the likelihood that a fleet of deflection systems will perform as desired and adjust the size of the campaign to meet success confidence objectives. Basic reliability metrics could be integrated into the NDA and combined with other uncertainties to compute higher fidelity estimates. The PDC17 scenario did not take vehicle reliability considerations into account, and it was assumed that the increase from five to 6 launch vehicles would be adequate to cover possible launch and other failures.

Standoff nuclear detonation remains a viable deflection technique when short warnings and object sizes do not afford other options. The 2017 PDC ten-year scenario enables many deflection strategies, but the nuclear option might be the only one available if proactive decisions to build and launch other types of deflection systems are not made in time, or if the time to do that is too short, or the overall number of launches including uncertainties is too high. Therefore, inclusion of standoff detonation is a desired and useful NDA feature.

The NDA estimates the payload mass and calculates the transfer orbit from Earth to the NEO, but it does not verify that deflection systems can be built with existing technologies. To increase its realism, the NDA is being integrated with concurrent engineering tools to perform studies incorporating current and forecasted technologies to determine the feasibility and limitations of kinetic impact and nuclear standoff deflection payloads.

Generally, more than a single mission is required to deflect an oncoming object, and as noted above, launching more than the minimum number of systems will boost campaign reliability and likelihood of success. The NDA in its current form supports multiple launches, but with all launches occurring on the same day. It is more realistic to stagger the launches and carry them out as a time series to reflect the reality of launch operations. This upgrade will make NDA predictions more credible when more than very few launches are required. At the 2017 PDC, the NDA and other simulation tools were used, and a one week gap was randomly inserted between launches. Therefore, adding a staggered launch schedule capability to the NDA will make launch window calculations more realistic.

The transfer trajectory from Earth launch to NEO interception is approximated by the solution of a Lambert problem in the NDA. The Lambert solution is limited in that it represents a two-body, Keplerian arc between Earth and the object. Other simulation tools employing low thrust approximations at the 2017 PDC produced a richer set of transfer trajectories, which were not supported by the NDA in its current form. It is therefore a desired capability for the NDA to model guiding the deflection payload continuously as it is in route to the NEO.

In summary, planned upgrades to the NDA include:

1. Accounting for detection and observation limitations and modeling NEO orbital uncertainty
2. Sizing the deflection campaign according to our best launch and space vehicle reliability data
3. Modeling NEO deflection uncertainties (object mass, impact ejecta)
4. Modeling standoff nuclear detonation
5. Allowing breaks between launches in a deflection campaign to reflect the reality of launch operations
6. Including a low thrust approximation to expand the range of feasible solutions

3. APPLICATION OF THE NEO DEFLECTION APP TO THE 2017 PDC NEO IMPACT SCENARIO

The addition of the NDA features discussed above might have modified the outcome of the 2017 PDC exercise in two primary ways. Accounting for uncertainties would have required more powerful deflections over nominal NDA predictions, manifested by either additional launchers in a kinetic deflection campaign, or an increased yield of a standoff explosion. Consequently, the NDA did reduce the impact likelihood of the PDC17 asteroid, but not to an acceptable level of confidence. The second affect would have been the benefits realized by an expanded range of deflection solutions due to the addition of low thrust approximations. Other tools employing such techniques at the conference computed late-scenario deflection opportunities not identifiable by current NDA capabilities.

The first opportunity to launch kinetic deflectors in the 2017 PDC scenario is shortly after discovery and is predicated on the preexistence of deflection systems. From discovery on 6 March 2017 to potential impact on 21 July 2027 there are 3789 days, or 10.38 years. Opportunities to launch kinetic deflectors shortly after discovery are depicted in region A in Fig. 1. Assuming the initial maximum estimated object diameter of 250 meters and density of porous rock (1.5 g/ccm), the NDA calculates that a flight time of 66 days and deflection time of 3713 days before impact, totaling 3779 days before impact, allowing 10 days of preparation for a single ready-to-launch Atlas V 551. This would yield a small miss distance of 0.007 Earth radii depicted by the green dot moved just outside the red capture circle in Fig. 2. The red dot marks the impact point with no deflection.

At present, no such deflection systems exist and no actionable decision has been made to build such capabilities, so the option to launch at short notice was ruled out in the discussions to reflect today's reality. However, even if launch on demand had been possible, it would result in the very small miss distance of 0.007 Earth radii, marked by the green dot in Fig. 2. That is inadequate because the NDA in its current form does not account for the large initial orbital uncertainty that is inherent to the scenario. When the initial orbital uncertainty is projected forward in time, it creates a region of impact likelihood rather than a dot, requiring larger deflection to move this region off the Earth.

The planned NDA upgrades will include orbital uncertainty calculations in order to estimate the deflection effort necessary to miss Earth with an acceptable level of confidence. An updated NDA with orbital uncertainty data would display an uncertainty ellipse around the nominal point. The deflection effort required to move the uncertainty ellipse outside the impact zone is larger than moving just the nominal impact point, and that may require an increase in the campaign size with additional launches. A notional uncertainty ellipse that must be moved outside the capture zone is depicted in Fig. 2. An upgraded NDA with orbital uncertainty information will inform decisions with improved realism.

In addition to orbital uncertainty considerations, system reliability data would also have to be factored in to add further realism to NDA campaign size estimates. Ref. 8 provides a possible approach for sizing a deflection campaign based on historical data. The methodology is then applied to the PDC 2017 object assuming worst case 250 meters diameter and 3.0 g/cm3 density (slightly heavier than the ~2 g/cm3 density assumed here). Fig. 3 shows the increase in the campaign size required to attain 90% campaign reliability. The NDA in its current form is lacking actual launch systems performance data. Hence, its predictions produced smaller campaign sizes than are required for a desired overall reliability.

As stated above, launch of a deflection system shortly after the PDC17 discovery was ruled out. Other mission options and key dates considered at the conference are depicted in Fig. 4. Considering that the asteroid will not be visible to optical telescopes for about two years from late 2017 to late 2019, an attractive approach is to build and launch a reconnaissance spacecraft to perform a flyby and gather important information in-situ. The three opportunities for doing that are depicted at the top of Fig. 4 with their respective construction times, launch times, and arrival times. An obvious benefit of this approach is information on the orbit, size and other factors critical to increasing the effectiveness of future deflection missions that is impossible to gather from the ground. A reconnaissance spacecraft would refine the orbit and could either rule out the impact, or confirm the need for deflection. An upgraded NDA with the ability to compute reconnaissance missions will increase its value.

In addition to high velocity kinetic impact, standoff nuclear detonation was also a considered a viable deflection technique for 2017 PDC. This can be done by installing a deployable nuclear explosive device (NEDs) on the spacecraft, and detonating it at the right time and distance from the asteroid. The NDA used in the 2017 PDC scenario does not model standoff detonation.

Use of a NED in a flyby mission enables a short transfer time, but the large relative approach velocity presents technical challenges with fusing timing and terminal guidance. Instead, a hybrid rendezvous mission was considered in which the spacecraft carrying a nuclear device would orbit the asteroid, observe a series of kinetic impactor deflections, and could carry out a controlled deflection if needed. The travel time associated with rendezvous compared with a flyby will be estimated by the NDA as it is expanded to approximate this range of possibilities. In the 2017 PDC scenario, the two rendezvous options 4 and 5 in Fig. 4 were proposed.

Three proposed kinetic impact mission options are enumerated 6, 7 and 8 in Fig. 4. The main difference between the missions is the time for construction and testing of the deflection systems. It is possible, in some cases, that information from an earlier reconnaissance flyby mission is available before the launch time, and if it would rule out an impact, the kinetic impactor missions would be aborted and repurposed. The NDA supported nominal (no uncertainties) kinetic impact missions at the 2017 PDC. Finally, a post deflection flyby was proposed as option 9 in Fig. 4 to monitor the outcome of a deflection operation. The assumed time for systems development is a minimum of 2 years.

Following daily discussions by the attendees, and deliberations by the decision makers, mission options 3 and 5 were selected for implementation. One fast flyby characterization spacecraft was built and launched in October 2019, successfully encountering the asteroid in May 2020 and reported that 2017 PDC was a binary asteroid. The primary was then estimated at about 270 meters in size, and the secondary at about 100 meters. Masses were then known to within ~20%. With this new information, the NDA calculated that kinetic impactor payloads launched by two Atlas V launch vehicles could successfully deflect the primary in December 2020 if launched in November 2019 (region B in Fig. 1). This option was deemed infeasible to carry out because there was not enough time to build and test two launchers.

Further along the scenario, two rendezvous characterization and observer spacecraft were built and launched in June 2020, arriving in May 2023, according to Option 5 in Fig 4. Both were installed with deployable nuclear explosive devices and would be in place to either carry out a controlled nuclear deflection or observe a series of kinetic impactor deflections of the primary in Feb. 2024. Approximation of rendezvous missions is a planned NDA capability that was not available at the 2017 PDC.

Due to the highly controversial use of nuclear explosive devices for asteroid deflection, world leaders decided to select Option 8 in Fig. 4 (in addition to Option 5) and build and launch a campaign of kinetic impactors to deflect the primary in February 2024. The NDA calculated that a minimum of five Atlas V launch vehicles were required to deflect the 270-meter primary (worst case) (region C in Fig. 1). To improve its reliability, the campaign was randomly increased to six kinetic impactors that were built and launched in March 2020 aimed for deflection in February 2024. An upgraded NDA with reliability analysis would estimate the 2017 campaign more accurately.

4. SUMMARY

The NDA has successfully supported planetary defense exercises to date at three Planetary Defense Conferences and three NASA/FEMA workshop discussions, each having unique impact threat characteristics. It has proved helpful with the design and implementation of simulated NEO impact threat scenarios at conferences, workshops and classes. The experience and user feedback gained points to its usefulness, as well as to its limitations. It is clear that upgrades in several areas will make NDA solutions more credible and realistic. These include: accounting for detection and observation limitations and modeling NEO orbital uncertainty; sizing the deflection campaign according to our best launch and space vehicle reliability data; modeling NEO deflection uncertainties; modeling standoff nuclear detonation; allowing breaks between launches in a deflection campaign to reflect the reality of launch operations; and including a low thrust approximation to expand the range of feasible solutions. The proposed enhancements will allow the upgraded NDA to create a more realistic and full set of NEO impact mitigation possibilities and more effectively inform training and decisions.

5. REFERENCES

1. "Discovery Statistics," https://cneos.jpl.nasa.gov/stats/
2. "5th IAA Planetary Defense Conference - Gathering for Impact!," http://pdc.iaaweb.org/
3. "Hypothetical Impact Scenarios," https://cneos.jpl.nasa.gov/pd/cs/
4. "Planetary Defense Conference Exercise - 2017," https://cneos.jpl.nasa.gov/pd/cs/pdc17/
5. "NASA/JPL NEO Deflection App," https://cneos.jpl.nasa.gov/nda/nda.html
6. "How to Participate in the Asteroid Threat Exercise," http://planetary.jp/PDC/exercise.html
7. Ailor H. W., "An Exercise in Planetary Defense," 9th International Association for the Advancement of Space Safety Conference, Toulouse, France, 18-20 October 2017
8. Iwata C., et al, "NEO Mitigation Mission Assurance," IAA-PDC-17-05-14, Planetary Defence Conference, Tokyo, Japan, 15-19 May 2017
9. Barbee B., "Preliminary 2017 PDC Mission Design Considerations," https://cneos.jpl.nasa.gov/pd/cs/pdc17/2017PDC_Mission_Considerations_Day1.pdf

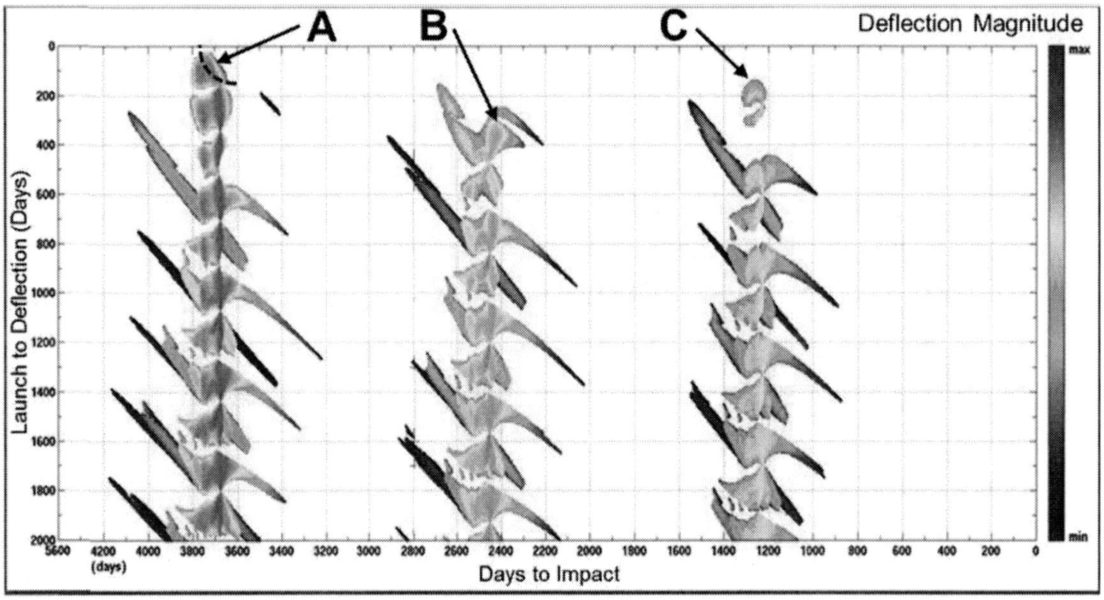

Figure 1. PDC17 Launch and deflection windows[5]

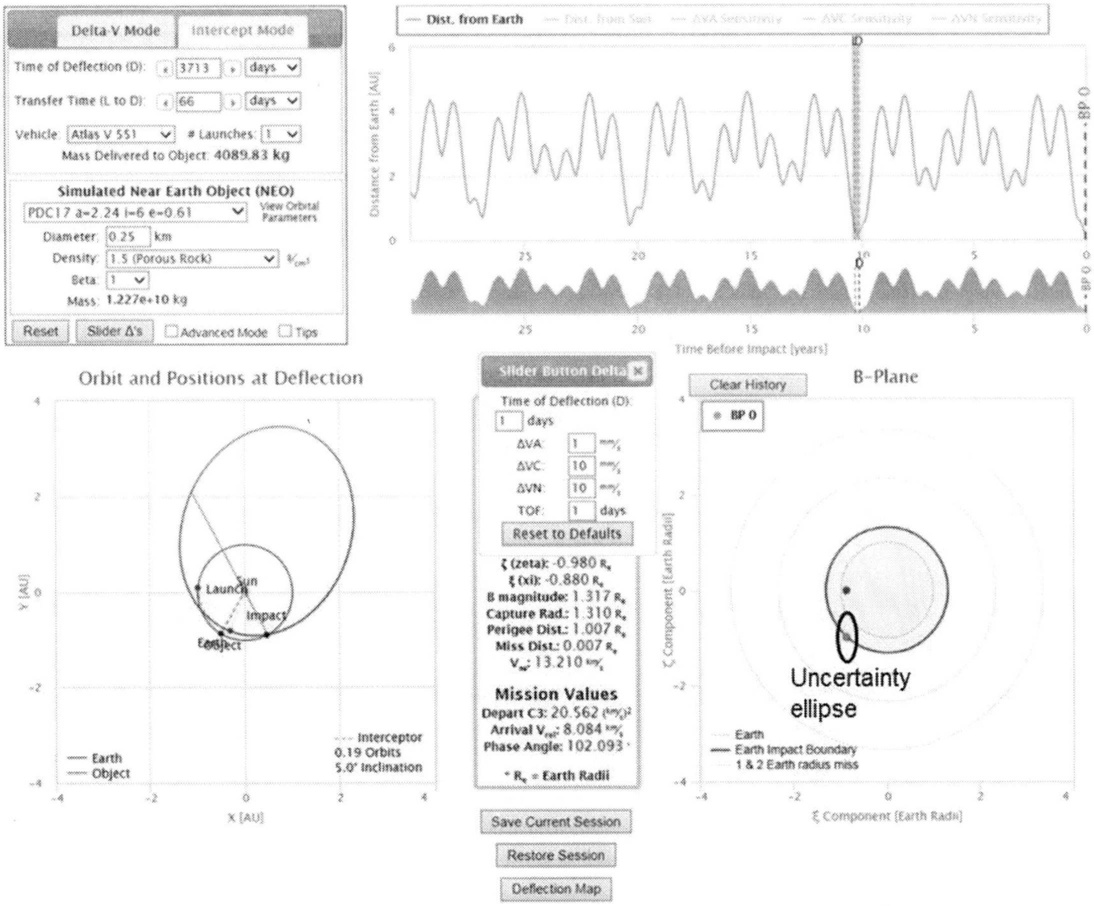

Figure 2. Kinetic impact deflection solution shortly after discovery[5]

		Number of Delta IV Heavy	
	Launch Window	Minimum Req.	90% Campaign Reliability
L1	Soon after discovery	2	5
L2	~7 years before impact	2	5
L3	~3.5 years before impact	3	7

Figure 3. Number of Delta IV Heavy launches required for deflection of PDC 2017[8]

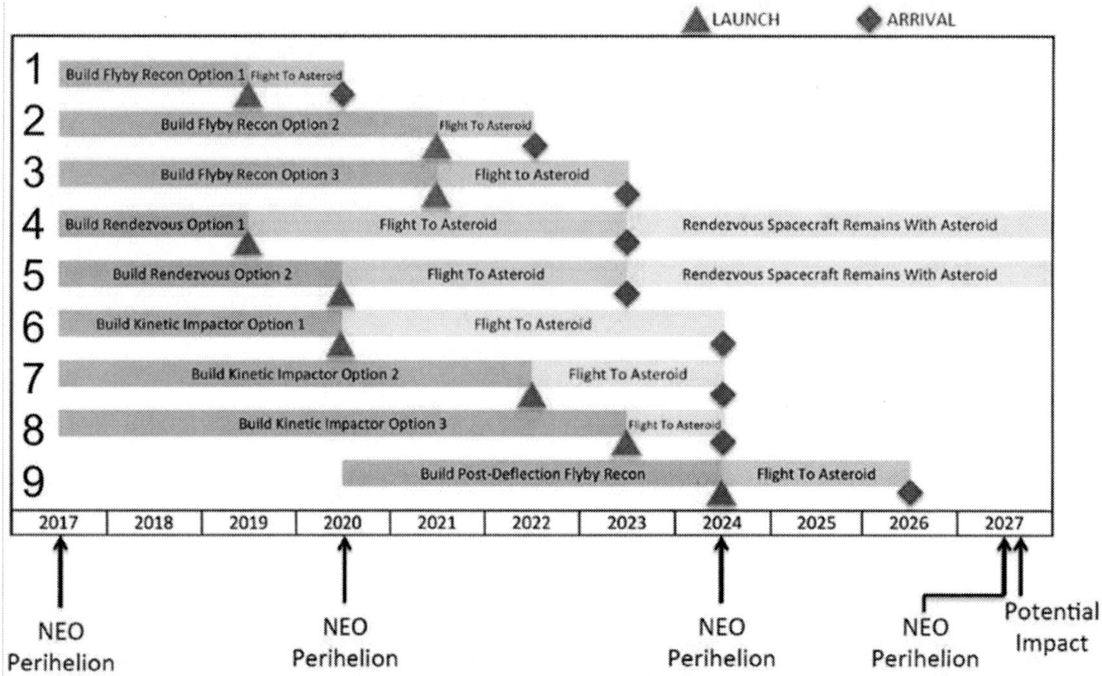

Figure 4. Mission options and key dates[9]

PROCEEDINGS OF THE 9TH INTERNATIONAL SPACE SAFETY CONFERENCE

SESSION 27 HUMAN PERFORMANCE FOR SAFETY & ORGANIZATIONAL CULTURE

COMPARISON BETWEEN TWO TRI-AXIAL ACCELEROMETERS TO DETECT GAIT CHARACTERISTICS AFTER 60 DAYS OF BED REST

Grassi M.(1), Mittag U. (2), Lau P. (2), Mulder E. (2), Daumer M.(1), Gruber M. (3), Rittweger J.(2)

(1) SLCMSR e.V. – The Human Motion Institute, Hohenlindener Str. 1, 81677 Munich (Germany), Email:info@slcmsr.org
(2) Deutsches zentrum für Luft und Raumfahrt (DLR), Linder Höhe, 51147 Cologne (Germany), Email: joern.rittweger@dlr.de
(3) Sport Science Department, Unviersität Konstanz, 78464 Constance (Germany), Email: m.gruber@uni-konstanz.de

ABSTRACT

It has been demonstrated in previous studies that bed rest is able to induce physiological modifications comparable with those seen after a long-term space mission. The aim of this paper is to explore the effects of 60 days of bed rest (Cologne, 2016) on habitual gait characteristics. To that purpose, reliable instrumentation is needed to correctly answer the various research questions. Thus, a comparison between two tri-axial accelerometers to detect gait characteristics (e.g. gait speed, number of steps, etc.) has been carried out. Subjects wore two portable tri-axial accelerometers on their waist for 14 consecutive days immediately after the bed rest. We hypothesized that for walking characteristics, the two tri-axial accelerometers will show a high level of agreement, while the same level of agreement will not be reached for running due to the nature of the activity. Results show a high level of disagreement between the two devices in walking speed calculation with a mean difference of -0.22 m/s and 95% limits of agreement ranged from -0.3 to -0.1 m/s ($p < 0.001$) and running speed with a mean difference of -0.09 m/s and 95% limits of agreement ranged from -0.4 to 0.3 m/s ($p = 0.004$). Moreover, results show that one device (*actibelt*) in both activities reported greater values compared with the second device.
This study highlights the importance of proper device selection. Moreover, proper calibration and verification is needed beforehand in order to reduce and minimize device-dependent errors.

1. INTRODUCTION

Nowadays, the negative effects of bed rest on the human body systems are well known [1]. Thus, 8 weeks of bed rest decrease calf muscle size by approximately 20% [2] with related loss in the plantar flexor muscle strength by circa 20%. As a consequence of this deconditioning, a decline of 30% has been observed in peak jumping power [3]. It is evident how such decreases in the musculoskeletal system affects a person's ability to locomote. In considering this in a wider range, changes in gait speed are an indicator of decreased functionality and even mortality at old age [4]. Thus, it seems important to understand how musculoskeletal de-conditioning, e.g. by bed rest, will impact on gait characteristics, and how the sub-sequential recovery behaves. Such scenarios might also represent an extraordinary "playground" for rehabilitative medicine research and to further study the recovery process after bed rest.

Measuring and correctly quantifying variables is not always as easy and straightforward as it might seem; a good measurement tool plays a key role in obtaining valid and reliable outcomes throughout the entire study. As suggested by Kimberlin [5], when choosing a measurement device, several questions must be taken into consideration such as the existence of a device that answers your research questions or the need to build a new one from scratch; how well the chosen device matches the definition of your outcomes; or how well the instrument will be well accepted by test subjects. All considered, tri-axial accelerometers are used to assess and quantify gait characteristics. Thanks to their easy handling, they are considered as one of the preferred objective methods to assess physical activity [6].

The aim of this paper is to quantify gait characteristics after bed rest with the aid of two tri-axial accelerometers and compare the outcomes from the two different devices to see from a methodological perspective the differences in using similar measurement tools.

The first device, named *GCDC X16-4* (Gulf Coast Data Concepts, LLC. Mississippi, United States) was chosen because previously used in other studies, and the second device named *actibelt* (Trium Analysis Online GmbH, Munich, Germany) was introduced while the study was ongoing to use an opportunity to collect data that allow to quantify the uncertainty in the measurement accuracy [7].

2. MATERIALS AND METHODS

2.1. Study design

A 60-day bed rest study was commissioned by the European Space Agency (ESA) and performed at the German aerospace center (DLR) in Cologne, Germany. The study was performed from February to April 2016 and lasted for 60 consecutive days plus an in-house recovery period of two weeks before and after the confinement to perform experiments on the subjects. This extended period of 3 months was the second campaign of a bigger 60-day bed rest study (details of the study have been published elsewhere [8]).

2.2. Participants

After the screening process (see [8] for details) 12 healthy male participants were recruited for the second campaign of the bed rest study. One subject interrupted the study for medical reasons not related to the study. Subjects were randomly assigned to one of two groups; the control group (n = 4; age 31±7 years; height 179±2.1 cm; weight 73±8.7 kg) and the exercise group (n = 7; age 29.5±6 years; height 184±4.7 cm; weight 81.5±4.1 kg). The control group did not perform any physical activity during the bed rest period, while the exercise group performed daily training sessions on a custom device named Sledge Jump System (SJS) that allows mimicking reactive jumps at different gravity loads [9].

During the recovery period (two weeks) after the bed rest, subjects were asked to wear two tri-axial accelerometers placed on the waist to allow continuous recording. Subjects were instructed to wear them as much as possible throughout the whole day.

2.3. Gait course

During the in-house recovery period, a series of 3 gait courses were performed 1, 7 and 14 days after the end of the bed rest period (R+1, R+7 and R+14, respectively) by each subject. The gait course was performed in a 74.36 m corridor at the DLR facilities. During gait course, subjects were asked to walk at a self-paced speed along the corridor, take a break and run back at a self-paced speed at a pace they would choose if they had to run 10 km. Each subject performed the gait courses in single sessions to avoid being influenced by the other subjects and resulting in an alteration of the gait speed. The walking bouts were repeated four times within one gait course, resulting in a distance covered of 297.44 m. Running bouts were repeated two times within a gait course resulting in a covered distance of 171.54 m.

Throughout the entire gait course session, participants were always under the supervision of a supervisor.

2.4. Accelerometers

As mentioned before, two different types of tri-axial accelerometers were used. *actibelt* was worn with the aid of a belt on the waist, in the frontal region below the umbilicus. *GCDC X16-4* was worn on the back, in the lumbar region.

Both devices recorded with a sample rate of 100 Hz and were rechargeable; *actibelt* is equipped with a 1000mA rechargeable lithium-ion battery allowing a recording period up to 2 months, while *GCDC X16-4* is equipped with a 500mA rechargeable lithium polymer battery. Data from both devices can be downloaded to a computer via an USB interface.

2.5. Data processing

From the walk and run sessions during the gait course, the following parameters were extracted: (i) gait speed, (ii) number of steps needed to complete the activity, (iii) time to complete the activity, (iv) step frequency, and (v) stride length.

Data were first downloaded to a computer and then processed:

actibelt. Data were processed with purpose-built packages (SLCMSR e.V. The Human Motion Institute, Munich, Germany) using R Studio (R Foundation for Statistical Computing). Algorithms from these packages allow extraction of walking and running segments. Moreover, speed is estimated at each step with a support vector regression machine trained by gold standard measurements (detailed methodology described elsewhere [10]).

GCDC X16-4. Data collected with this device were processed using MATLAB R2015a (The MathWorks, Inc., Natick, Massachusetts, United States). To extract the variables from this device, an easier approach, compared with the approach used for *actibelt* data processing was used: the number of steps needed to complete each activity was calculated finding the peaks in the signal in the interested period window. A peak was considered as such

"Table 1". Mean, standard deviation, and median for steps detected and speed estimated during walking and running using *actibelt* and *GCDC X16-4*.

	Steps detected		Speed estimated (m/s)	
	Mean ± SD	*Median*	*Mean ± SD*	*Median*
Walking				
actibelt	374.84 ± 33.43	374	1.67 ± 0.2	1.65
GCDC X16-4	362.88 ± 39.95	350.5	1.45 ± 0.19	1.42
Running				
actibelt	142.35 ± 27.26	142	3.32 ± 0.69	3.24
GCDC X16-4	131.38 ± 28.92	137	3.26 ± 0.6	3.23

if the threshold set using the standard deviation of the magnitude (calculated with Eq. 1) was crossed.

$$mag_i = \sqrt{x_i^2 + y_i^2 + z_i^2} \quad (1)$$

Where x is the vertical axis, y the antero-posterior axis and z is the medio-lateral axis.

Knowing the length of the corridor, the number of steps and the time needed to cover that distance, all the other variables (gait speed, step frequency and stride length) were derived.

2.6. Statistical analysis

Statistical analysis focused on two variables: number of steps detected and gait speed estimated during walking and running.

To graphically visualize the level of agreement between these paired variables from the two devices, Bland-Altman plots [11] were constructed between the mean value of the outcome on the *x-axis* and the differences between the paired observations on the *y-axis*. The upper and the lower limits of agreement were plotted with 95% confidence interval (mean ± 1.96 *x* standard deviation). Wilcoxon signed-rank test [12] was used to compare the medians of the variables from the two devices. All the analysis was carried out using R Studio (R version 3.4.1).

3. RESULTS

Walking. Bland-Altman plots for comparison of number of steps and gait speed estimated is shown in Fig. 1.a and Fig. 1.b, respectively. The mean of the difference for steps detected was -12 steps/297.44 m and 95% limits of agreement ranged from -72 to 48 steps/297.44 m (Fig. 1.a). The Wilcoxon signed-rank test showed that the median (mdn) of number of steps detected with *GCDC X16-4* (mdn = 350.5) statistically differs from those detected with *actibelt* (mdn = 374), $V = 103.5$, $p = 0.014$. The mean of the difference for walking speed was -0.22 m/s and 95% limits of agreement ranged from -0.3 to -0.1 m/s (Fig. 1.b). Wilcoxon signed rank-test showed that the median of walking speed estimated with *GCDC X16-4* (mdn = 1.42) statistically differs from the speed estimated with *actibelt* (mdn = 1.65), $V = 0$, $p < 0.001$.

Running. Bland-Altman plots for comparison of number of steps and gait speed estimated is shown in Fig. 1.c and Fig. 1.d, respectively. The mean of the difference for steps detected was -10 steps/171.54 m. and 95% limit of agreement ranged from -38 to 18 steps/171.54 m (Fig. 1.c). The Wilcoxon signed-rank test showed that the median (mdn = 137) of steps detected with *GCDC X16-4* statistically differs from those detected with *actibelt* (mdn = 142), $V = 24$, $p < 0.001$. The mean of the difference for running speed was -0.09 m/s and 95% limit of agreement ranged from -0.4 to 0.3 m/s (Fig. 1.d). Wilcoxon signed rank-test showed that walking speed estimated with *GCDC X16-4* (mdn = 3.23) statistically differs from the speed estimated with *actibelt* (mdn = 3.24), $V = 91$, $p = 0.004$.

Tab. 1 shows the descriptive statistics for the parameters mentioned above.

4. DISCUSSION

Many methods have been used to determine and assess physical activity levels; from subjective methods, even if not consistent [13], to objective ones, which are more consistent although the outcomes sometimes might be misinterpreted or the equipment might not be practical for measurement in the real-world, but only under laboratory settings. For these reasons, in the last years accelerometers became very popular for measuring physical activity levels [6] and pattern recognitions [14]. Thus, within a

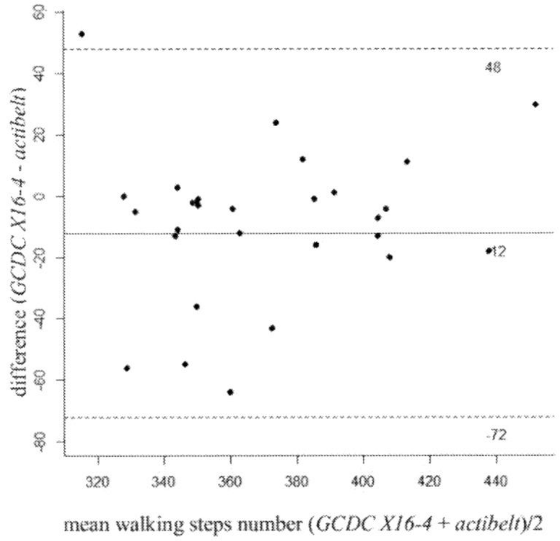

1.a. Bland-Altman plot of differences in number of steps detected during walking.

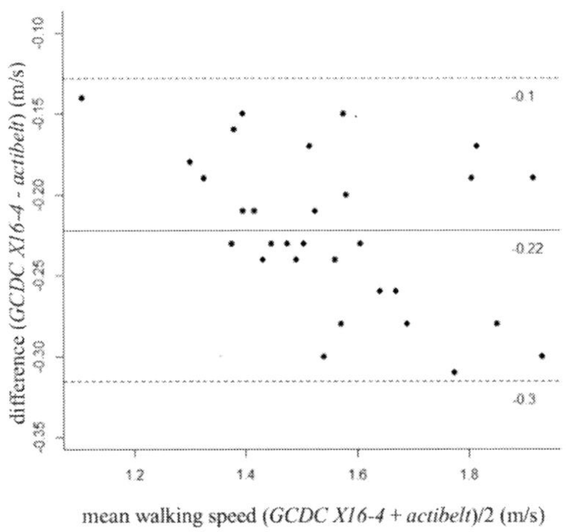

1.b. Bland-Altman plot of differences in gait speed estimated during walking.

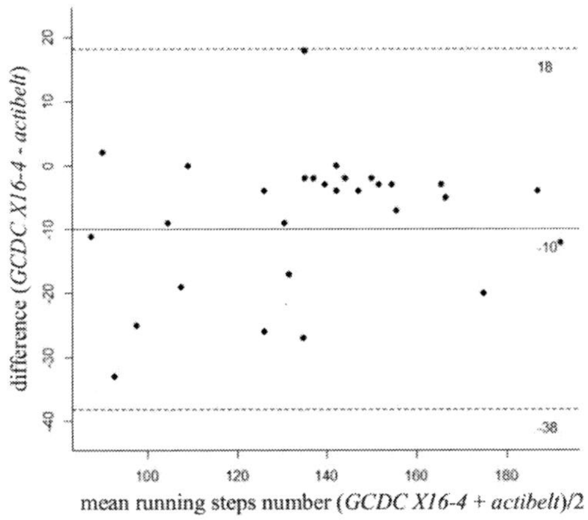

1.c. Bland-Altman plot of differences in number of steps detected during running..

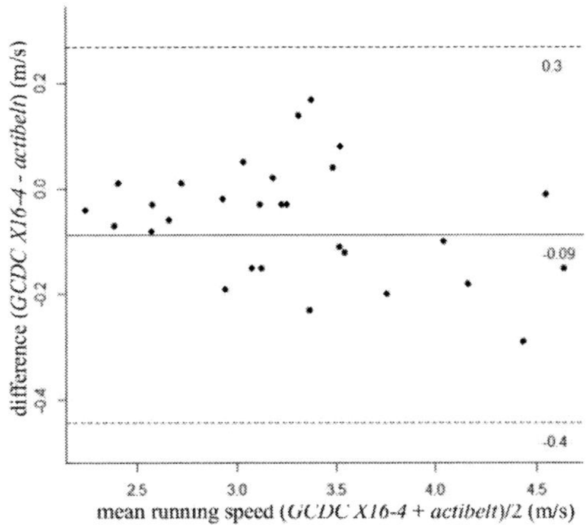

1.d. Bland-Altman plot of differences in gait speed estimated during running.

Figure 1. Bland-Altman plots. Dashed lines: 95% limit of agreement. Continued lines: mean difference.

bigger research study, authors wanted to compare outcomes from two similar tri-axial accelerometers.

Tendency of over/underestimate walking parameters (e.g. daily number of steps) is well known in the field of mobile accelerometry; [15] reported an underestimation of 7.1% and an overestimation of 6.9% in two accelerometers respectively in a 100-step accuracy test. Similar results have been reported by other authors [16, 17] when comparing different accelerometers and pedometers to detect number of steps and physical activity. Two previous studies [10, 18] tested the accuracy of *actibelt* showing reliable outcomes. While on the other hand, there is no such documentation about the accuracy of steps/speed detection algorithms used by *GCDC X16-4*; the main reason being that algorithms used to extract the parameters have been developed and tested internally.

To further explore the differences between these two devices, the nominal sample frequency (*sf*: 100 Hz) was tested over 210 consecutive minutes resulting in a mean and median *sf* of 100 and 100.02 Hz respectively for *actibelt* and a mean and median *sf* of 102.54 and 102.57 respectively for *GCDC X16-4*.

From our findings over a 30-minute gait course, a discrepancy of 3.3% and 8.4% has been found in the number of steps detected between both devices during walking and running. A higher level of agreement or a lesser percent error (calculated with Eq. 2) during walking activity between the two devices was expected, even though the difference between both conditions resulted to be statistically significant.

$$\frac{\frac{1}{n}\sum_{i=1}^{n} GCDC_i - \frac{1}{n}\sum_{i=1}^{n} actibelt_i}{\frac{1}{n}\sum_{i=1}^{n} GCDC_i} * 100 \qquad (2)$$

Where $\frac{1}{n}\sum_{i=1}^{n} GCDC_i$ and $\frac{1}{n}\sum_{i=1}^{n} actibelt_i$ represent the average of the values between all the subjects obtained with *GCDC X16-4* and *actibelt* respectively.

Although it is very challenging to determine gait speed in real-world scenarios, [10] showed various models used by *actibelt* to predict walking speed in real-world situations, and the best predictor seems to be a support vector regression model with a concordance correlation coefficient of 0.93 and 95% confidence intervals at 0.92 and 0.94. Moreover, [18] showed that *actibelt* tends to overestimate speeds compared to gold-standard measurements. In the above-mentioned study, authors showed that in a sample of 51 subjects with clinically definite multiple sclerosis, *actibelt* overestimates walking speed by 0.12 m/s. When dividing subjects into three levels of disability (mild disability, moderate disability and severe disability), *actibelt* tends to perform better with faster gait speeds (mild disability group gait speed: 1.45 m/s; *actibelt* estimated gait speed: 1.47 m/s; difference: 0.02 m/s).

For calculating gait speed from the *GCDC X16-4* accelerometer, a more uncomplicated approach was used. Speed was calculated as: $v = s/t$, where v is velocity, s is distance and t is time.

With this method it is possible to derive velocity, or speed, as a function of time defining the length of the corridor as a constant. Even though in theory this is a correct method, in practice there are several issues. First of all, the assumption that the length covered is always constant and respected by every subjects might not be always correct. Second, the supervisor of the gait course might lead to a boundary effect due to manual supervision and altering the actual time spent to cover the given distance. Lastly, having an average as outcome does not depict the intra-gait variability of the subject, so part the information is going to be lost.

As a result of different algorithms used in the two accelerometers to calculate walking speed, the mean difference was of 0.22 m/s. This deviation in the outcome is not negligible since changes of 0.1 m/s in walking speed are considered to be clinically meaningful [19].

As during the study gold-standard measurements of gait parameters were not obtained, it is now not possible to know which method predicted values closer to the actual values after the 60-day bed rest.

5. CONCLUSION

To conclude, in this paper it was shown that it is not possible to rely only on a good piece of hardware or only on good software, but the right combination is needed in order to obtain accurate and reliable measurements. The findings of this study are similar to what others showed with regard to level of agreement between different devices to monitor and quantify physical activity levels or gait characteristics.

However, a more detailed comparison of the raw data and of the algorithms from both devices is needed in order to have a better comparison of the measurement systems [7].

Future studies in the field of metrology that want to compare different devices should also include gold-standard data to rely on.

6. ACKNOWLEDGMENTS

Authors want to express their gratitude to all participants involved in the bed rest study and the staff who continuously took care of them.

We want to say thanks to all those persons who helped and contributed with their time and energies to the data collection first at DLR, and the production of this manuscript then.

Lastly, an owed big thank you to ESA and also to the DLR's institute of Aerospace Medicine for having made this possible.

This bed rest was funded by the European Space Agency (400011387-15-NL).

7. REFERENCES

[1] A. Pavy-Le Traon, M. Heer, M. V. Narici, J. Rittweger, and J. Vernikos, "From space to Earth: advances in human physiology from 20 years of bed rest studies (1986-2006)," *European journal of applied physiology*, vol. 101, no. 2007. Springer-Verlag, pp. 143–194, 13-Apr-2007.

[2] J. Rittweger and G. A. G Beller E.Mulder..., "Prevention of bone loss during 56 days of strict bed rest by side-alternating resistive vibration exercise," 2010.

[3] B. Buehring, D. L. Belavý, I. Michaelis, U. Gast, D. Felsenberg, and J. Rittweger, "Changes in lower extremity muscle function after 56 days of bed rest," *Journal of applied physiology*, vol. 111, no. 2011. American Physiological Society, pp. 87–94, 28-Apr-2011.

[4] S. E. Hardy, S. Perera, Y. F. Roumani, J. M. Chandler, and S. A. Studenski, "Improvement in usual gait speed predicts better survival in older adults," *Journal of the American Geriatrics Society*, vol. 11, no. 2007. pp. 1727–1734, 2007.

[5] C. L. Kimberly and A. G. Winterstein, "Validity and reliability of measurement instruments used in research," *American Journal of Health-System Pharmacy*, vol. 65. American Society of Health-System Pharmacists, pp. 2276–2284, 01-Dec-2008.

[6] C.-C. Yang and Y.-L. Hsu, "A review of accelerometry-based wearable motion detectors for physical activity monitoring," *Sensors*, vol. 10. pp. 7772–7788, 20-Aug-2010.

[7] M. Sené and J. J. I Gilmore, "Metrology is key to reproducing results," *Nature*, vol. 547, pp. 397–399, Jul. 2017.

[8] A. Kramer, J. Kümmel, E. Mulder, A. Gollhofer, P. Frings-Meuthen, and M. Gruber, "High-Intensity Jump Training Is Tolerated during 60 Days of Bed Rest and Is Very Effective in Preserving Leg Power and Lean Body Mass: An Overview of the Cologne RSL Study," *PLoS ONE*, vol. 12. pp. 1–18, 12-Jan-2017.

[9] A. Kramer, R. Ritzmann, A. Gollhofer, D. Gehring, and M. Gruber, "A new sledge jump system that allows almost natural reactive jumps," *Journal of biomechanics*, vol. 43. Elsevier, pp. 2672–2677, 22-Jun-2010.

[10] M. Schimpl, C. Lederer, and M. Daumer, "Development and Validation of a New Method to Measure Walking Speed in Free-Living Environments Using the actibelt® Platform," *PLoS ONE*, vol. 6, no. 8. p. e23080, 05-Aug-2011.

[11] M. J. Bland and D. G. Altman, "Measuring agreement in method comparison studies," *Statistical methods in medical research*, vol. 8. Arnold, pp. 135–160, 1999.

[12] F. Wilcoxon, "Individual comparisons by ranking methods," *Biom. Bull.*, vol. 1, pp. 80–83, 1945.

[13] G. A. Meijer, K. R. Westerterp, F. M. Verhoeven, H. B. Koper, and F. ten Hoor, "Methods to assess physical activity with special reference to motion sensors and accelerometers.," *IEEE Trans. bio-medical Eng.*, vol. 38, no. 3, pp. 221–229, Mar. 1991.

[14] A. G. Bonomi, B. Yin, K. R. Westerterp, and A. H. Goris, "Detection of type, duration, and intensity of physical activity using an accelerometer," *Medicine and science in sports and exercise*, vol. 41. pp. 1770–1777, 01-Sep-2009.

[15] K. L. Storti, K. K. Pettee, J. S. Brach, J. B. Talkowski, C. R. Richardson, and A. M. Kriska, "Gait Speed and Step-Count Monitor Accuracy in Community-Dwelling Older Adults," *Medicine and science in sports and exercise*, vol. 40. American College of Sports Medicine, pp. 59–64, 01-Aug-2007.

[16] G. C. Masurier, S. M. Lee, and Tudor-Locke Catrine, "Motion Sensor Accuracy under Controlled and Free-Living Conditions," *Medicine and science in sports and exercise*, vol. 36. American College of Sports Medicine, pp. 905–910, 01-Dec-2004.

[17] G. C. Le Masurier and C. Tudor-Locke, "Comparison of Pedometer and Accelerometer Accuracy under Controlled Conditions," *Medicine and science in sports and exercise*, vol. 35. pp. 867–871.

[18] R. W. Motl *et al.*, "Accuracy of the actibelt accelerometer for measuring walking speed in a controlled environment among persons with multiple sclerosis," *Gait & Posture*, vol. 35, no. 2012. Elsevier, pp. 192–196, 04-Sep-2011.

[19] M. Schimpl *et al.*, "Association between walking speed and age in healthy, free-living individuals using mobile accelerometry--a cross-sectional study.," *PloS one*, vol. 6, no. 8, p. e23299, Aug. 2011.

GLOBALIZATION OF NEW SPACE INDUSTRY INTO DEVELOPING COUNTRIES AND ITS VERY SIGNIFICANT IMPACTS ON SAFETY

Norul Ridzuan Zakaria(1), Nasri Nasrun(2), Azizi Aziz(3), Mohd Jamil Mohd Nor(4), Ashwar Aziz(4), Amluddin Yusof(4)

(1)SOLVES, Corso Regio Parco, 24, Torino,Italy, ikam290200@hotmail.com
(2)Space City, Shah Alam, Malaysia, nasrinasrun@gmail.com
(3)Spaceport Malaysia, Shah Alam, pepngeri@@yahoo.co.uk
(4)Space Ventures, Shah Alam, Malysia, jamil@spaceventure.com.my, ashwar@spaceventures.com.my, amluddin@spaceventures.com.my

ABSTRACT

Developing countries like Malaysia which are keen on new industries define the industry as the space industry that is practicable to be developed without much investment on ground infrastructures and organizational structures as required by the conventional space industry that have been established in developed countries for decades if not half a century. This view and understanding is guiding the organizational and cultural behavior and planning involved in developing such industry in developing countries, which is affecting the safety issues pertaining to it.

The issues of safety become more significant as the new space industry itself is in the definition phase with various interpretations from various industries with variety of interests. As for developing countries like Malaysia that have its own tactical and strategic development plan and policy, will always try to suit new space industry in their existing development agenda, and always view the industry with their very own perspective to serve the national agenda.

Spaceport Malaysia is a very specific case where an effort championed by local private agencies with guides from international space organizations including IAASS is evolving into a national agenda with official government agencies involvement contributing and shaping it to fulfill the local public and governmental objectives. As demonstrated by this case, the definition phase which has been completed was very much important with guides from established international partners, and now the evolution phase which is very vital to ensure the effective implementation of significant issues such as safety has started.

Spaceport Malaysia is an example of a new space industry carefully planned and executed to fulfill both the professional objectives in the industry, which among them is the professional safety planning at the level of organizational and culture, and also the national interest, which wishes to see the project to be successful at the national level.

Spaceport Malaysia has fully evolved from a private venture supported by professional organizations into a national strategic space project with participation from various national government agencies. At this stage, the safety issues are very important to be established and such status is actually an opportunity not to be missed as all parties involved have realized its significance.

For almost a decade the authors and co-authors have been planning Spaceport Malaysia and presented papers related to it at conferences including the IAASS conferences with safety issues discussed. This paper summarizes all those discussions and lay down a plan for execution of safety agendas within the project. This paper discusses common issues on safety that will be raised for implementation of new space industry in developing countries with globalization understanding and views both from the perspective of professionals and the countries involved with their national and international interests. As such, this paper proposes few pro-active steps to be taken both by the professional and the host country on safety issues to make effective the globalization of new space industry in developing countries.

1. GLOBALIZATION OF NEW SPACE INDUSTRY INTO DEVELOPING COUNTRIES

Malaysia most probably is the first developing country in the world that has been reported to have a plan to establish a commercial spaceport. On May 1999, it was reported by a major local newspaper that the Department of Aerospace Engineering of the University of Malaysia in Tronoh, Perak has organized a public lecture, where a spaceport was proposed for the first time in the country. Later on August of the same year, a very popular caricature on another local newspaper has depicted a conversation about the Kuala Lumpur International Airport (KLIA) becoming a spaceport by 2020 (1).

Figure 1. The caricature (in Malay) dated 24 August 1999 published by a major local newspaper depicting the conversation about KLIA becoming a spaceport by 2020.

In those years, that particular caricature depicted issues that were very major if not the headline news of the particular day. Therefore, KLIA to become a spaceport by 2020 was a major national news reported on that month of 1999.

So, commercial spaceport is not a new interest in Malaysia. It has been on the news in the country for the past 20 years, and since those early years, it is championed by private and non-governmental organizations (NGOs).

Surprisingly, these pioneers brought the issue of Malaysian spaceport to the international space community particularly the space safety community at an early stage, when they presented a paper addressing the spaceport at the 2nd IAASS Conference in Chicago, the USA in 2007 (1).

Since then, these pioneers have got a little attention from the few related government agencies, compared to previous which was almost nil. Today, the situation is totally different, because there are already quite a number of parties including private, NGOs and government agencies in Malaysia that are initiating the establishment of a commercial spaceport in Malaysia. The most established initiative among these is Spaceport Malaysia, which today is well known to the international space community and supported by the Malaysian Space Agency (ANGKASA).

At the time of writing, ANGKASA is waiting for an approval from the Malaysian Ministry of Science, Technology and Innovation (MOSTI) to sign a memorandum of understanding (MOU) with the Italian Space Agency (ASI) to collaborate in various activities in the development of space science and technology including suborbital flight and commercial spaceport establishment and operation. This MOU between the Malaysian and Italian governments is projected to serve as a quantum leap for the spaceport and other new space industry initiatives

To developing countries like Malaysia, new space industry is defined as space industry activities that are affordable to their economy and practical to the "space commodities" of the countries. As for Malaysia, a tropical developing country, its space commodities are the steady high intensity of sunlight and the equatorial geographical factors.

Being an equatorial country that lie between 1 and 8 degrees north, Malaysia receives a steady stream of sunlight the whole year. In such countries, it is summer all year.

As commonly as anywhere else, space tourism was an early understanding of new space industry in Malaysia. A popular space tourism idea that is seem to be very suitable for countries like Malaysia is a holiday package where the tourists and the companions including family members and friends are trained, prepared and accommodated in tropical seaside resorts, where they enjoy themselves partying, traveling, shopping and blending with the local cultures, before the space tourists fly to space and those who accompany them watch them flying to space. This is the very "sunny" commercial idea of space tourism that influences the design of space tourism industry in these countries, including in sites selection and architectural works for spaceport.

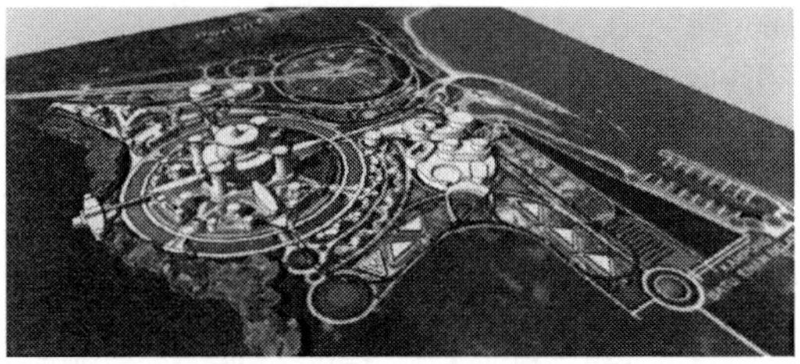

Figure 2. An illustration of Space City by Space Ventures. It is to be a seaside city in Malaysia where the Spaceport Malaysia is to be located (2).

The extension of this idea is that space tourists who fly from these tropical seaside resorts will be able view exotic sceneries on Earth, be them coral reefs, micro-archipelagoes or even smoking volcanoes in the ocean. For the space tourists, not only being in space is an achievement, but being able to view exotic sceneries on Earth from space will also be very significant (3).

This extension of the sunny space tourism idea has influenced the Malaysian works in suborbital vehicles design leading to few ideas of VTOL (vertical take off and landing) suborbital vehicles such as SOLVES (Sub-Orbital Low cost & low risk Vehicle to the Edge of Space) (4) and Aerocosmos Spacejet. The later was a result from a study completed in October 2015, by ALTEC SpA, Italy for a private company in Malaysia.

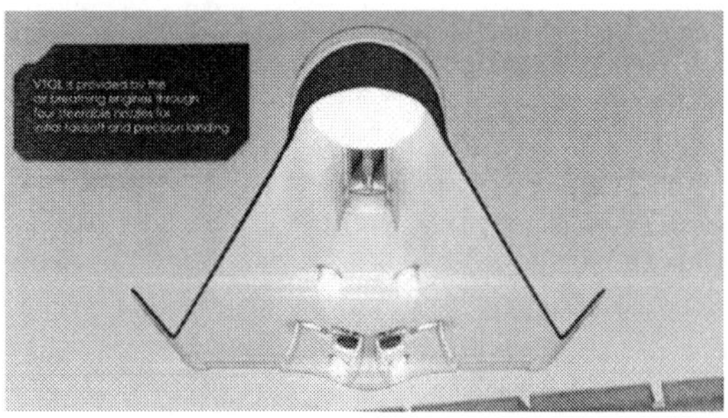

Figure 3. An llustration of Aerocosmos Spacejet indicating its VTOL capability.

The study however did not describe in detail the VTOL propulsion system, except it did say that the VTOL is to be provided by air-breathing engines, but focused more on its rocket propulsion that is to be dedicated to ascend the suborbital aircraft to an altitude of 100km from sea level. However, now there is already an idea of having an electric propulsion system to provide the VTOL capability. Aeroscosmos Spacejet most probably is the world's 1st concept of suborbital tourism aircraft with VTOL capability.

The Malaysian pioneers even went further by presenting a proposal of VTOL spaceplane operating from a seaborne spacepad, which actually is super yacht or ocean liner with spacepad, very similar to helipad but designed for such spaceplane. They reasoned that a seaborne spacepad will transport space tourists to exotic destinations in the middle of the ocean such as a smoking volcano where the spaceplane will take them to 100km from sea level for them to view such exotic event from space (5).

During the sea voyage, which will consume days aboard the ship, the space tourists can be trained, prepared and qualified for the flight. The VTOL spaceplane concept was proposed instead that of VTOL vehicle as in vertically ascend and descend rocket, because having a proven horizontal flight capability besides vertical thrust as demonstrated by Harrier operating aboard an aircraft carrier is a vital safety requirement (5).

In the perspective of scientific research, a sunny country like Malaysia is a suitable site for solar science, as long as the weather permits. This is where space weather, which is the study of solar activities which influence the global weather, becomes a preferred scientific activity under the new space industry for tropical developing countries.

Observing the sun from a terrestrial observatory in a tropical country may suffer from the disadvantage of heavy rains and/or cloudy sky. To overcome this, the idea of near space solar observatory becomes viable, where near space aircraft will be used to carry solar observation payload.

Near space aircraft operates in the regions where it is too high for conventional aircraft, but too low for orbital satellites, commercially in the region between 20km and 100km from sea level. Since it operates as an aircraft instead of satellite, its operational cost will be lesser than that of satellite.

To perform such mission effectively, a near space aircraft will have to fly not only at very high altitude above all the clouds, but also to remain airborne for a very significant time, up to weeks and months. To do so, the aircraft is to be electric powered with solar rechargeable capability, so that it will not need to be landed or descended to a lower altitude for refueling. Therefore, solar electric near space aircraft suits very much the sunny spaceport concept of tropical countries.

Such solar-powered near space aircraft can be unmanned or manned. The unmanned version has the advantage of more available space and mass for carrying payloads instead of a pilot, but the manned one, if it can carry a passenger, it can offer a cost-effective "edge of space tourism". Typically these aircraft has long wing span to fly effectively at high altitudes and accommodate maximum number of solar cells. The author has designed a 2-passenger near space aircraft for both; carrying payloads and edge of space tourism called, INSECT (Integrated Near Space Electric Carrier Transporter).

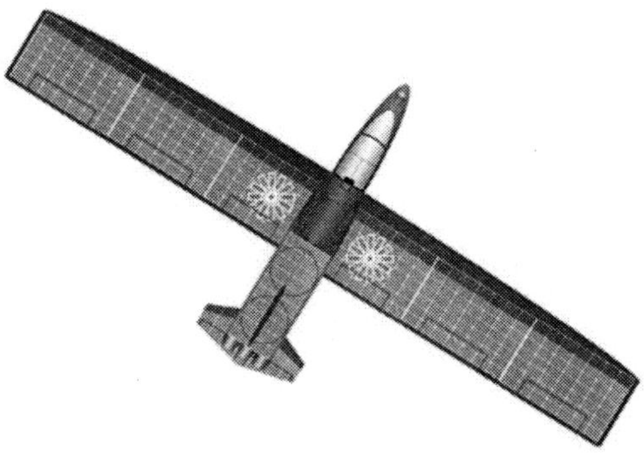

Figure 4. An illustration of INSECT.

Besides near space flight, another activity which is considered as practical and affordable new space industry in developing countries is commercial micro gravity flight. This activity seems to be lesser related to the sunny spaceport concept since it does not technically related to solar electric propulsion as in near space flight. However, taking into the consideration that a country with summer all year may provide more economic operation of such flight, microgravity flight may also be related albeit not that strongly to the "sunlight commodity".

There are 2 types of platform being considered, one is the existing large platform being operated in Europe and Russia and another one is new small platform innovated in Malaysia. The second one is being studied as an "immediately available cost effective" opportunity. Microgravity flights, including both are considered as the most practical embryonic activity for the operation of commercial spaceport (6).

The large platform is in the form of large passenger jet being flown in Europe and Russia carrying up to 30 to 40 "zero gravity" passengers. There have been plans to transfer such platform to be operational in Malaysia for a significant period of time. As in suborbital space tourism, this is to be offered as a "holiday under the Sun" package, perhaps justifying its relationship with the sunny spaceport concept.

The smaller platform planned ranges from an aerobatic plane to a combat aircraft to a small executive jet. Even though these aircraft are small platforms, they are to be equipped with an on board microgravity-boxes that are to function similar to that of black-boxes. This microgravity payload has been developed, tested and certified and is waiting for the availability of such platform.

As for the plan to employ a combat jet as the microgravity platform, the microgravity-box is to be installed inside the cockpit on the rear seat of the aircraft. Few models of 2-seater combat aircraft are already under study including MB-339 jets that are within the inventory of Royal Malaysian Air Force. Such installation does not require any significant modification of the combat aircraft.

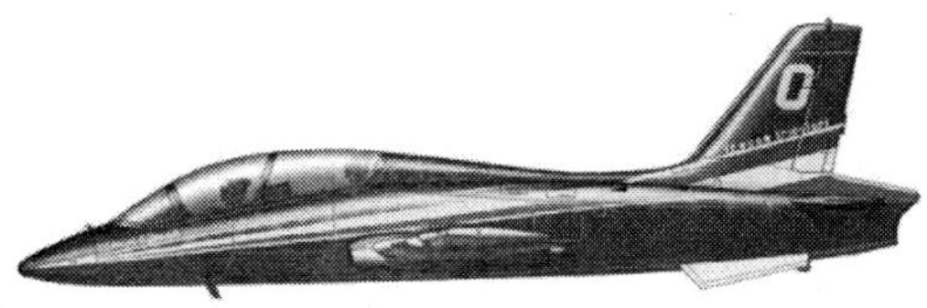

Figure 5. An illustration of MB-339.

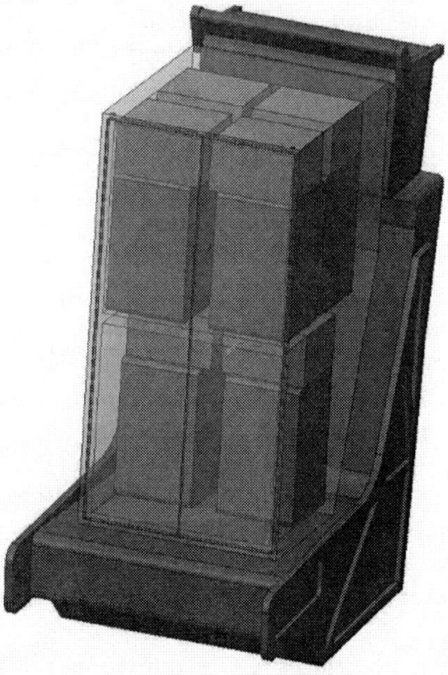

Figure 6. An illustration of the microgravity-boxes installed on an ejection seat of a 2-seater combat aircraft (7).

Related to the application of small executive jets as a microgravity flight platform, is air launch services by the same aircraft. Therefore there is also a consideration where a small executive jet is to be used both for microgravity flight and limited mass air launch, but large enough to insert nanosatellites to orbit.

Those are considered the activities of new space industry that can be implemented within a relatively short time and can be operated from any existing airport without having to wait for the completion of a commercial spaceport. The suborbital flight however is considered a medium term new space industry activity that is to be developed in parallel with the spaceport.

Recently Space Ventures has explored an alternative to realize the new space industry sooner than originally planned in Malaysia by promoting a strategic collaboration between ANGKASA and ASI involving a specific job to be executed by ALTEC SpA (Aerospace Logistic Technology Engineering Company), a subsidiary of ASI.

This collaboration will enable Malaysia to participate even if not directly, but through the subsidiaries of ASI, such as ALTEC, in existing ASI's space activities which are part of the bigger space programs under ESA (European Space Agency). Such space program that is under the consideration of this collaboration is Space Rider and its outcomes including spinoffs such as future manned and unmanned spaceplane programs.

Space Rider evolved from IXV (Intermediate Experimental Vehicle) which was successfully launched and landed in 2015 by ASI. An element that became the interest of the author and co-authors is the fact that IXV was considered as the most successful point-to-point suborbital spaceplane that had flown over the equator.

During the IXV mission, ALTEC had served the role of being the operator of the ground stations and mission control center or the contractor for the ground segment of the mission. The author and co-authors see that having such ground segment capabilities in Malaysia will enable the country to play a significant role in similar suborbital missions including Space Rider. This will enable Malaysia to participate in a proven suborbital spaceplane program without having a suborbital spaceplane development program, a kind of "short-cut" in not only suborbital, but also in point-to-point flight. This plan however is more related to the equatorial geographical factors, the other commodity besides high intensity sunlight.

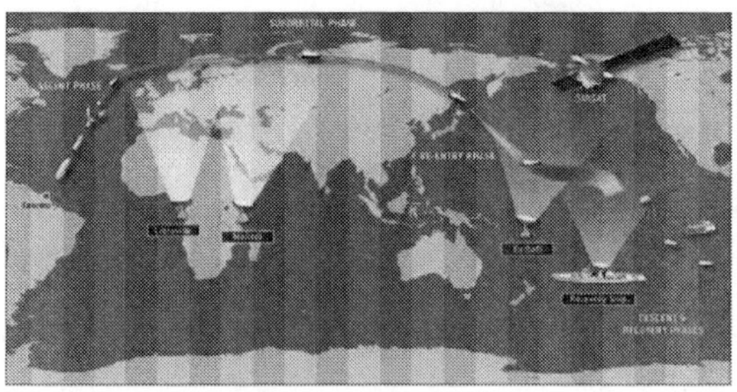

Figure 7. An illustration of IXV point-to-point suborbital flight along the equator from its launch in Guiana and its "splash-down" in the equatorial Pacific.

There is however a long term plan by the author and co-authors promoted since 2013, that is very much related to the equatorial geographical factors, generally describing the intercontinental point-to-point suborbital flight, between Spaceport Malaysia and Spaceport Puerto Rico called, TEST (Twin Equatorial Spaceport) (8).

The distance between Malaysia and Puerto Rico is strategically and approximately 20,000km by a straight line almost parallel with the equator, which is approximately half the circumference of Earth. In other words, Malaysia and Puerto Rico is located on the equator almost exactly on the opposite side of the globe.

Being at the equator, both countries are also in the middle of their region. Malaysia is located at a point at a center surrounded by Far East (China and Japan) on its northeast, Near East (Middle East and North Africa) on its northwest and Oceania (Australia and New Zealand) on its southeast. On the other half, Puerto Rico is between North America and South America. If there is a spaceport in Malaysia and Puerto Rico respectively, connecting flights will transfer the passengers to and from these regions to and from the spaceports in Malaysia and Puerto Rico.

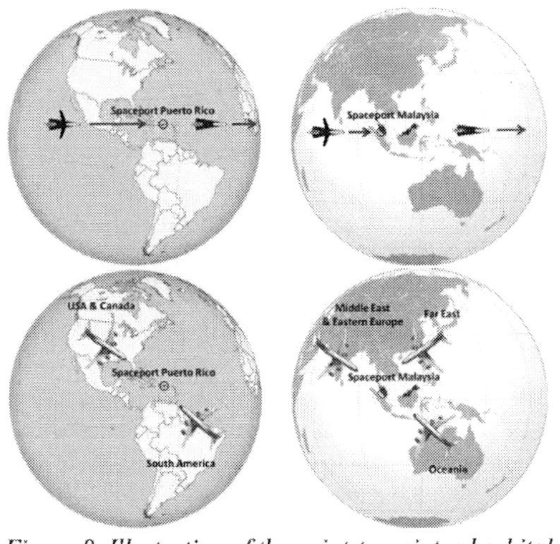

Figure 8. Illustration of the point-to-point suborbital service between Malaysia and Puerto Rico and the connecting flights to both spaceports.

With the current rocket technology, the distance of 20,000km between Malaysia and Puerto Rico can be travelled in 2 hours. Therefore, even if there is a connecting flight that takes 8 hours to travel, the total flight between Malaysia and Puerto Rico will still be less than 12 hours if the transit at the spaceport is less than 2 hours.

This geographical location for both countries is very strategic for a point to point suborbital flight between America and Asia serving not only the 2 continents but also neighboring regions. The author believe that this route along the equator with the distance of 20,000km is more effective than the more commonly proposed rapid point-to-point aerospace transportation such as between London and Sydney across the equator at the distance of 17,000km. Therefore, in 2013, the author proposed ICES (Inter-Continental Equatorial Spaceplane) to provide this service (9).

Figure 9. Illustrations of ICES.

ICES is a concept of 2-stage vehicle consisting of a universal carrier and a variable geometry wing spaceplane with electric propulsion for horizontal landing as the "suborbiter". The universal carriers are the common large passenger jets being used in inter-continental long-haul flights. This application of universal carriers as the 1st stage of ICES will be more economic than having a specially designed and developed 1st stage, because they can also be used as conventional passenger jets and they have readily facilities and trained crews to serve them at airports. This concept was presented to the government of Puerto Rico in 2013 and the suborbiter was the largest suborbital vehicles ever proposed by SOLVES.

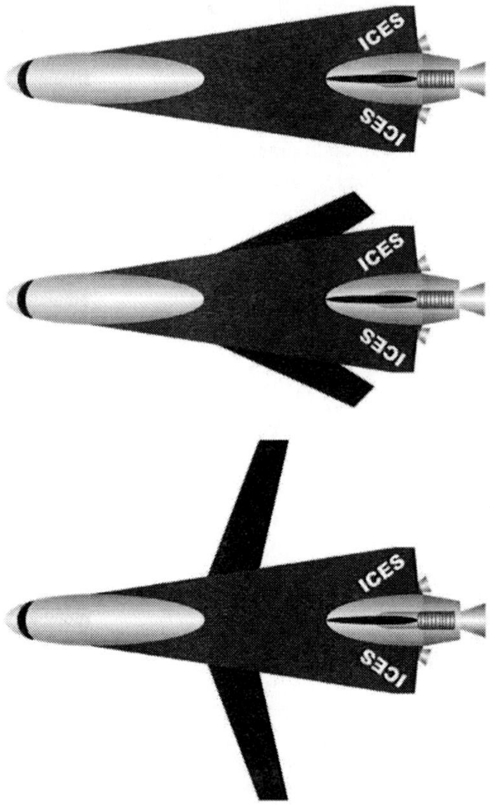

Figure 10. Illustrations of ICES variable-geometry suborbiter.

ICES was also proposed to be fueled by biofuel developed from agricultural industry in existence or that can be developed both in Malaysia and Puerto Rico. A candidate agro product for this fuel was palm oil, which Malaysia is a major global producer and exporter.

At one time, ICES had become an interest of Sierra Nevada that had seen it as a program for the development of their Dream Chaser spaceplane. ICES is also similar to IXV because both are "point-to-point equatorial suborbital spaceplane".

To summarize, the immediate term plan of the author and co-authors now is to start the new space industry in Malaysia by having a ground segment capability to participate in a proven and strategic point-to-point suborbital spaceplane mission. The medium term plan is to develop a spaceport and suborbital vehicles, and the long term plan is to develop point-to-point suborbital spaceplane. This long term plan may also be the continuation of its immediate term plan, which although it is concerning only the mission control center and ground tracking facility, it is also actually a ground segment of a point-to-point suborbital flight industry.

2. THE BIOLOGICAL DNA TERMINOLOGY –THE VERY SIGNIFICANT IMPACTS ON SAFETY AND ACTIONS TAKEN

The Malaysian experience is a case of a developing country trying to establish new space industry. The effort was influenced by equatorial geographical factor, national agenda and professional understanding of new space industry. In this case, the professional understanding is actually the cultural and organizational interpretation of the industry.

Now the government of Malaysia has a policy that is meant to transform the country into a fully developed nation by 2050. Each ministry is assigned to identify industries that are relevant to them and the methodology and technology involved to develop those industries. Perhaps the immediate term now being promoted by the author and co-authors, which is the establishment of ground segment capability for point-to-point suborbital flight is the first part of the right methodology and technology required for TN50, as the policy to achieve the fully developed nation is known.

New space industry is a result of the globalization of space industry. It is not only the spreading of the industry to various parts of the world outside the regions where the industry was originally established, but also the evolution of the industry with local factors at the new regions where the industry is trying to reach. So, at an equatorial developing country like Malaysia, the industry evolves into new industry with local factors, which are, as earlier mentioned, the equatorial geographical factor, national agenda and the cultural and organizational understanding of new space industry. The cultural and organizational interpretation is where and how it affects safety.

Developing countries with weak cultural and organizational understanding of safety in new industrial development will place other elements as priority, particularly that are related to short term and low cost but high risk. They wish to witness physical development the shortest term at the lowest cost possible. So they may develop certain activities on new space industry without enough attention given to safety issues.

The Malaysian case however is quite unique, because the authors and co-authors who lead the effort in establishing new space industry in the country are quite aware of the important of safety. Unfortunately, they are not the regulators or representing official organizations with authorities in such industry. This is where their decision to pursue their short term objective counts, because this objective is already part of an establish space mission with established safety contents. This is a case of "bringing safety with new space industry"- a safe approach, as far as safety is concerned.

With the complexity of a space system, safety can be viewed as an "atom" of the "DNA" of the space system, while the space system is the "organism". So, if the "safety atom" is not fit enough to a standard, the result will be an inferior DNA, - the space system will develop, but without a safety culture. It will grow, as DNA is providing the information for the growth, but without the capability to develop safety parallel with its development, resulting in an "inferior organism".

Such phenomenon can easily occur in the globalization of new space industry into developing countries with weak cultural and organizational value of safety. And if we stick to this "biological DNA" terminology, finally the system will be outcompeted, come to an end and become extinct due to the inferiority of its "DNA without the fit safety atom". This is expected, due to "natural selection/survival of the fittest" as determined by market and policy.

The existence of such "inferior organism" space system in developing countries is very dangerous, because in those countries such project is "protected" by the government to ensure its success. The government most probably will be willing to spend extra budget to make sure that the image of such project is not tarnished as its image of being a successful project is so much very important to the government. This is where the impact to safety will be most affected due to the "inferior organism" phenomenon.

Safety is a culture, and culture is built up of both "DNA" and experiences. For developing countries that do not yet have established space agencies and space industries and at the stage of initiating the new space industry due to its globalization, most probably they lack both, the DNA and experiences. But, this is also an opportunity to develop a good culture of safety if the right DNA and experiences can be provided by "those that already have the safety culture".

An effective method to provide that culture is by relocating related activities to the developing countries, rather than waiting for such activities to be initiated by those countries. So, taking into the consideration the equatorial geographical factor, certain space activities with safety culture can be located at developing countries along the equator. Even a major new space industry similar to ICES can be located at those countries.

ICES is an example of a possible mega space project involving collaboration between several developing countries sharing the common interest and that is directly useful to the economics of those countries. The countries involved will not only be Malaysia and Puerto Rico (a developing region of the United States of America), but also countries along the equator that will provide not only tracking services, but also emergency landing plus search and rescue services, in case the ICES suborbiter need to make an emergency landing before reaching its destination. Therefore ICES can easily be seen as a project that will involve tremendous effort in flight safety over a large geographical area in the equator among the developing countries involved.

In the case of ICES, 4 major locations along the flight path were proposed to provide the services of ground tracking, emergency landing and search and rescue services. They are Praia in Cape Verde, Sao Tome in Sao Tome and Principe, Jizan in Saudi Arabia and Male in Maldives.

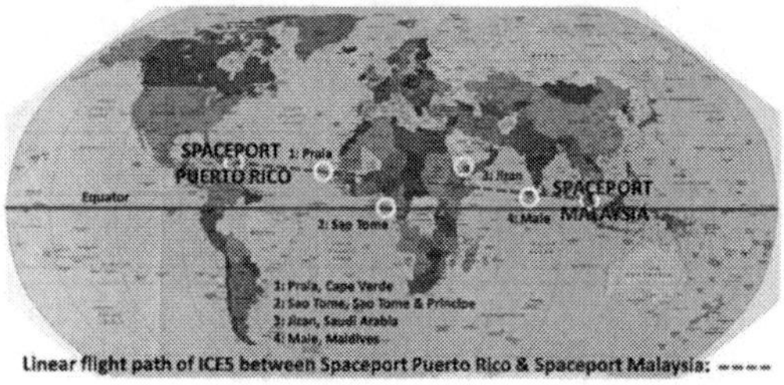

Figure 11. Illustrations of tracking stations and emergency landing points between Spaceport Malaysia and Spaceport Equator along the equator.

ICES is also an example of new space industry that will directly bring the benefits to the developing countries that provide the services to it because they will get paid for the services they provide from the customers who will use ICES as passengers or people who transport their cargoes. It is not like sending a probe to Mars that nobody will pay for the trip, but the taxpayers and the taxpayers themselves have to reach out to receive the indirect benefit from the space program.

From the installation of ground segment capability to the planning of operating an equatorial point-to-point suborbital flight, these are the activities of the new space industry that is spreading to the developing countries due to the globalization of the industry. The spreading must come with the culture of safety, inserted within its DNA and strengthen through experiences. Such globalization has very significant impact on safety. Realizing this, both parties, the professional bodies, such as the IAASS and the developing countries that plan to develop new space industry must move toward each other and agree to meet at a point. The space agencies of those countries, including ANGKASA, the Malaysian space agency should starts seriously looking at the safety issues as there are already activities in the new space industry being planned in Malaysia.

Among the efforts done by the author and co-authors with very good collaboration from the IAASS is bringing senior ministers and officers from Malaysia to participate in the IAASS Conferences and meet with the IAASS professionals. This has been done since the 2nd IAASS Conference, when 2 very senior officers from the Malaysian Perak state government participated in the conference. At the 6th IAASS Conference, the Malaysian High Commissioner to Canada has delivered a keynote speech and the Malaysian Chief Minister of Perak is delivering a keynote speech at this 9th IAASS Conference. They are fully aware of the new space industry initiatives in Malaysia and the Chief Minister of Perak now is a driving force of a specific new space industry in Perak focusing on near space flight. This has made them and their organization aware of the IAASS and its missions and vision in space safety.

3. PLANS

A year after ICES was proposed, in 2014, the author has improved its concept particularly for better safety (10). The newer ICES was low wing (unlike the high wing earlier version) so that its large wing box to accommodate the variable geometry wing mechanism below its fuselage will also serve as sponson in case of emergency landing in the sea. This is a major safety issue since most of the flight will be above the sea. It has emergency exits on top of its fuselage so that after it has landed and halted safety on the surface of the sea, the passengers and crews can exit through those exits on the roof. It has jettison-able fuselage cone, where the cone accommodates the pilot cockpit and a mini cabin for passengers in "semi-standing-seating" so that the cone can be separated from the fuselage in flight in case of emergency and safely landed.

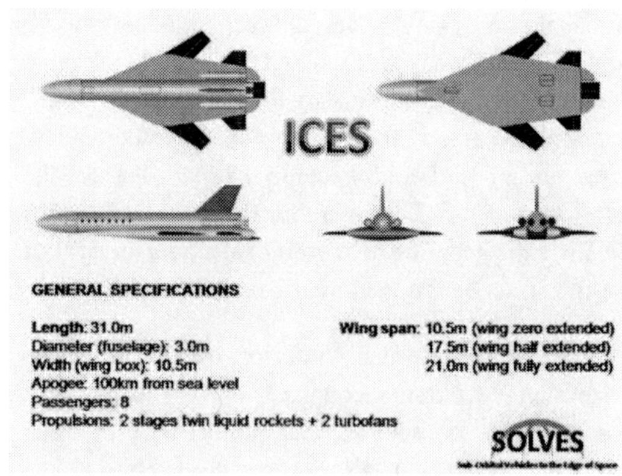

Figure 12. The New (2014) ICES.

The author plan to re-launch this newer ICES program in 2018 with further improvement in propulsion and safety technology as the author can see that ICES is a program that is suitable to promote new space industry in the developing countries along the equator.

The author also plan to bring many more Malaysian ministers and very senior government officers to the next and following IAASS Conference so that the Malaysian government will always be aware of the significance of space safety in its globalization and particularly the development of new space industry in Malaysia.

As for 2017, the author is organizing an official relationship between the IAASS and Perak state government of Malaysia with the participation of an official delegation of the state government led by the state chief minister in the 9th IAASS Conference.

Finally, the author and co-authors plan to continue promoting the establishment of Spaceport Malaysia at Space City with more effective strategy and safety and market concern.

Figure 12. another illustration of Space City where Spaceport Malaysia is to be located (2).

4. REFERENCES

1. Zakaria, N.R., Zahari, R., Abd. Majid, A.A., Othman. (2007). The Symbiotic Relationship between Atronaut Program and Space Tourism Development – A 3rd World Perspective. 2nd IAASS Conference.
2. Space Ventures Sdn Bhd. (2016). Space City Proposal, Malaysia.
3. Zakaria, N.R., Mettauer, A., Abu, J., Ismael, A.T, Othman, J., Shaari, C.Z., Nasrun, N. (2010). Human Factors Engineering in Designing the Passengers' Cockpit of the Malaysian Commercial Suborbital Spaceplane. 4th IAASS Conference.
4. Zakaria, N.R., Nasrun, N., Zulkifli, M.R.,Yamin, M.I., Othman, J., Zakaria, N.R. (2013). The Design and Operation of Suborbital Low Cost and Low Risk Vehicle to the Edge of Space (SOLVES), 6th IAASS Conference.
5. Zakaria, N.R., Nasrun, N., Abu, J., Jusoh, A., Azim, L., Said, M.A., Ishak, M.S., Zakaria, N.R. (2011). The Advantages, Potentials and Safety of VTOL Suborbital Space Tourism Operations. 5th IAASS Conference.
6. Abu Samah, S.K., Zakaria, N.R. Nasrun, N., Abu, J., Shukor, S.M. (2013). Zero Gravity Flights as the Most Effective Embryonic Operation for Planned Commercial Spaceport. 6th IAASS Conference
7. ALTEC SpA. (2016). Microgravity Experiment Module on Ejection Seat.
8. Space Ventures Sdn Bhd. (2013). Twin Equatorial Spaceport (TEST).
9. SOLVES. (2013). Inter-Continental Equatorial Spaceplane (ICES).
10. SOLVES. (2014). The New ICES (Inter-Continental Equatorial Spaceplane).

HEMODYNAMIC PUSH-PULL EFFECT DURING SIMULATED SUBORBITAL SPACEFLIGHT

Arjan J.H. Meskers [1,*], Eric Groen [1], Mark M.J. Houben [1], Ries M. Simons, Erik Frijters [2]

(1) TNO Human Factors, Kampweg 5, 3769 DE, Soesterberg, The Netherlands,
**Corresponding author.*
(2) Center for Man in Aviation, Royal Netherlands Air Force, Kampweg 3,3769 DE, Soesterberg, The Netherlands
Email addresses: arjan.meskers@tno.nl (A.J.H. Meskers), eric.groen@tno.nl (E.L. Groen), mark.houben@tno.nl (M.M.J. Houben), simons-aeromed@ziggo.nl (R. Simons), E.Frijters@mindef.nl (E. Frijters).

ABSTRACT

From military and aerobatic flight it is known that a pilot's tolerance for positive Gz loading is lowered when preceded by a negative Gz load; this is termed the 'push-pull' effect. The result is an increased risk of G-induced Loss of Consciousness (i.e. G-LOC). The same holds for astronauts in orbit where the absence of gravity results in a continuous headward fluid shift (i.e. analog to negative Gz). During suborbital spaceflight the changes in Gz occur over a timespan of several minutes instead of seconds (i.e. military and aerobatic flight), or hours to months (i.e. orbital spaceflight). For this study a centrifuge-based simulator was used to investigate whether a push-pull effect is present in response to a simulated suborbital spaceflight profile. Each of the 22 male participants underwent two different flight profiles: one conventional centrifuge profile, which started with an upright cabin so that the participant was only exposed to a 'pull' phase (max. +3.16 Gz); and one profile which started with a five-minute period of head-down tilt so as to induce a headward fluid shift ('push') followed by a 'pull' phase (max. +3.16 Gz). The outcome of the subjective measures indicates the presence of a 'push-pull' effect during simulated suborbital spaceflight.

1. Introduction

Suborbital spaceflights will bring their participants to an altitude above 100 km (i.e. 62 mi, 328.000 ft) above sea level, which is defined 'space' according to the definition of the *Fédération Aéronautique Internationale*. Although flight trajectories may differ between different concepts, a suborbital spaceflight typically involves G-loads up to 4-6 G during the ascent and reentry phase, and offer about five minutes of microgravity upon apogee. Total flight duration varies from less than one hour up to a few hours. The large variations in physical loading during these flights may trigger specific physiological responses with potential medical risks. On the one hand, microgravity or '0 G', causes a headward fluid shift which could result in decreased cardiac output, cardiac rhythm disturbances, increased intra-ocular pressure, and possibly intra-cranial hypertension [1,2]. On the other hand, positive G-loads – oriented along the longitudinal body or Z-axis (i.e. Gz) – induce a footward fluid shift which leads to decreased cerebral oxygenation, which may result in loss of color vision ('gray out'), peripheral light loss, and ultimately G-induced Loss Of Consciousness (G-LOC) [3,4,17].

Of particular interest is the rapid transition phase between microgravity and positive Gz loads, which takes place when moving from apogee to reentry; moving from 'push' to 'pull', causing a 'push-pull' effect. This is especially the case in a suborbital spacecraft where the participants are seated upright rather than restrained in supine position.

The headward fluid shift that occurs in microgravity may – to a lesser extent – have the same effects as negative Gz experienced by pilots [4,5,11]. Several seconds of exposure to a negative Gz load causes a state of enhanced parasympathetic tone, resulting in bradycardia, decreased cardiac contractility, and decreased total peripheral resistance [10]. A subsequent transition to positive Gz can then cause a profound drop in cerebral blood pressure and may take 8-10 s to compensate. The reduced tolerance to positive Gz loads due to the hemodynamic push-pull effect has been estimated to be +1.27 Gz, but some individuals have shown a reduction at as large as +3.9 Gz [2,13]. This push-pull effect has been identified as a possible cause of 30% of G-LOC events [2,16]. Up till now, it remains unclear whether a push-pull effect occurs upon transition from the microgravity to the reentry phase of suborbital spaceflight. The effect has been described in parabolic flight [4] and concerns have been expressed that it could occur in suborbital flight as well [3,12].

Current training programs for future suborbital spaceflight participants include exposure to increased G-loads in human centrifuges [8]. The objective of the present study was to investigate the existence of the push-pull effect during a simulated suborbital spaceflight. The study was performed in a centrifuge in which the cabin's orientation can be controlled independent of the centrifugation. This allowed for the application of head down tilt (HDT) to simulate the headward fluid shift occurring in microgravity [9]. Physiological and subjective responses during simulated 'push-pull' flights were compared to conventional centrifugation not preceded by a HDT phase.

2. Hemodynamic 'push'

Reports from Soyuz cosmonauts in the 70's [22] stated that a 6° head down tilt (HDT) in supine posture produced a similar sensation of blood flow toward the head compared to what occurs in a micro-gravity environment. This information formed the basis for the HDT angle commonly used in bedrest studies to mimic a microgravity environment for the human cardiovascular system.

HDT causes a headward fluid shift, leading to an increase of blood pressure in the upper body. This is sensed by the baroreceptors at the aortic arch (i.e. near the heart), and at the carotid sinuses (i.e. at the neck), triggering the baroreflex. This mechanism activates the autonomous nervous system which leads to increased parasympathetic nervous system activity and results in decreased heart rate and increased vasodilation (i.e. widening of the blood vessels); in effect reducing the increase in blood pressure.

For the current experiment we also used a 6° HDT. It should be noted that our participants had a seated posture, so that during HDT the legs were relatively above the head, which may have increased the cardiovascular load compared to HDT in supine position.

3. Desdemona motion simulator

The Desdemona simulator, Fig. 1, features a fully gimbaled cabin to allow for unlimited rotations in three degrees of freedom. The cabin can be moved vertically over 2 m and horizontally over 8 m. With the cabin positioned 4 m away from the planetary rotation axis a Gx-load of 3 G can be achieved during centrifugation [7], totaling about +3.16 Gz (i.e. including Earth's gravity).

4. Participant posture

The cabin consisted of an F16 mockup. The seat had a five-point belt and was adjustable in height such that the upper legs were fully supported and the participant's heels could well touch the cabin's floor. The angle between the calves and upper legs amounted 85° and between the back and the vertical it was 30° (i.e. between the calves and the back). The participant was always facing 'outward', i.e. having his back directed toward the simulators' centrifugal center.

Figure 1. The Desdemona motion simulator during operation.

5. Flight profiles

For the experiment each participant underwent two different flight profiles: one 'pull' profile and one 'push-pull' profile.

During the onset phase (i.e. angular acceleration of the simulator) the Gz vector was kept oriented 'head to feet' by means of active cabin tilt.

5.1. The 'pull' flight profile

This flight profile – comparable to conventional centrifugation – started with an upright seated participant who remained idle in this orientation for five minutes. This period was the 'baseline' phase. After this phase the simulator started centrifugation with a maximum onset rate of 0.25Gz/s [2], Fig. 2 *(top)*.

During the G-onset phase the cabin rotated backward, from 0° toward 72°, leading to a maximum G load of +3.16 Gz. Upon reaching maximum G-load the simulator maintained constant centrifugation for an average duration of 22.9 s. After this period centrifugation was decelerated at a maximum rate of 0.125 Gz/s (i.e. half of the onset rate, to minimize tumbling sensations and thereby nausea).

5.2. The 'push-pull' flight profile

The 'push-pull' profile was similar to the 'pull' profile, with the only difference that the participant was tilted backward by 66° (i.e. the backrest already had a 30° tilt when sitting 'upright'). This resulted in a 6° HDT during the five minute baseline phase. After this phase the simulator followed a similar motion profile as during the 'pull' flight, Fig. 2 *(bottom)*.

It is important to note that the change in cabin tilt angle required to remain aligned with the Gz-vector during a 'push-pull' flight was 6° (i.e. 72° - 66°) further backward during onset, whereas for the 'pull' profile this rotation amounted 72° (i.e. 72° - 0°). Hence, a 'pull' flight involved more Coriolis-related false cues.

6. Data logging

During the experiment the G-loads and orientation of the cabin were continuously logged. In addition, various physiological recordings were made to monitor the participant's state and to investigate the physiological responses to the push-pull effect. The physiological recordings included electrocardiography (ECG), electromyography (EMG) of the upper right leg, and Near Infrared Spectroscopy (NIRS) and photoplethysmography (PPG) at the head. All sensors logged their data time synchronously for dataset overlap. The results of the physiological measures will be reported elsewhere, in this paper the focus is on the psychophysical responses (i.e. subjective measures).

Besides local monitoring the participants were also monitored remotely with cameras, and they stood in direct contact with the test-leader via a headset.

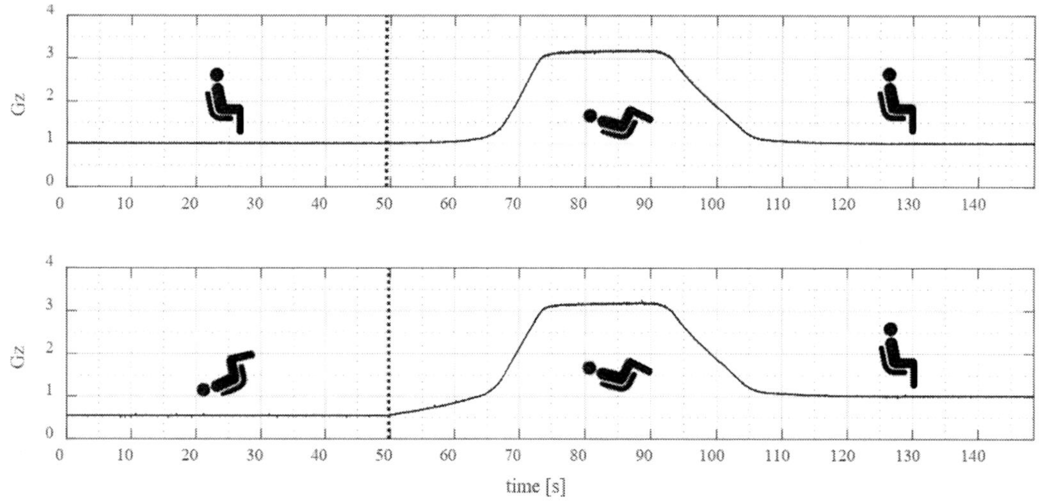

Figure 2. (top) 'pull' flight profile where the participant starts seated upright, (bottom) 'push-pull' flight profile, where the participant is tilted head down by 6° before Gz-loading. The black dotted line marks the start of centrifugation.

7. Subjective measures

After each flight the participants were asked to report the subjective effects that they experienced. This included grey-out (i.e. loss-of-color), peripheral light loss (PLL, i.e. tunnel vision), black-out, and nystagmus. The participant's state involving (almost) loss of consciousness (A-LOC), and G-induced loss of consciousness (G-LOC) were determined by the flight surgeon who continuously monitored the participants.

Grey-out was derived from observed changes in the colors of a color wheel and/or post-its mounted in front of the participant, Fig. 3.

Peripheral light loss was rated by using six symmetrically installed post-its, numbered from '1-3' outward. The angular separation ranged from 25° between the number '1' post-its (orange), 55° between the number '2' post-its (magenta), and 80° between the number '3' post-its (yellow).

Nystagmus was retrieved using the reported direction(s) that the color wheel seemed to 'jump' to (i.e. cabin tilt during G-onset and G-downset resulted in a combination of horizontal and vertical nystagmus, leading to an apparent clockwise or counter-clockwise rotation of the color wheel).

8. The course of the experiment

Participants were selected based upon a received questionnaire containing an individual medical background, which was reviewed by a flight surgeon.

The experiments took place over several days. Each test-day was divided into a morning and an afternoon session. Each session involved two participants (i.e. A and B) who were simultaneously instructed and equipped with sensors at the session-start. During the session the participants took alternately part in the experiment.

The 'push-pull' and 'pull' flight profiles were alternated to exclude external influences such as the time of day or any other sequential impact of the tests. When participant A was in the simulator, participant B rested (and vice versa); in between 'flights' each participant had a resting period of at least 30 minutes to recover.

9. Participant instructions

In preparation for the experiment the participants were instructed not to perform intense exercise, not to drink alcohol, not to use narcotics, and to rest as much as possible. On the day of the experiment the participants were instructed not to drink coffee or tea to reduce the risk of nausea. The participants were in advance briefed about the forces and physiological effects/symptoms to expect, for familiarization and to explain what to recognize and report. They were furthermore advised to keep their head fixed to the headrest during centrifugation to minimize nausea due to Coriolis stimulation. After the briefing the sensors were attached and checked for operation.

With the participant sitting in the simulator the markings inside the cabin were addressed and explained how to use them to recognize and to report experienced visual symptoms. This was only done before the first flight of each participant.

The participants were allowed to briefly strain the upper legs and buttocks when they had the feeling that the effects of having a reduced amount of blood in the head – resulting from +Gz – were 'too much' (i.e. when feeling 'fading away' or going towards unconsciousness). Continuous EMG sensing of the right upper leg allowed for a post-hoc check if a participant had used this technique.

The final instruction concerned the use of a sensor held in the hands. This sensor was to be 'tapped' upon

Figure 3. During the experiment cabin temperature was kept at minimum to minimize nausea. Instead of using projectors colored post-its were hung up together with a 'color wheel' at the visual center of the participant.

any event of which the participant found worthwhile to mention. Upon each 'tap' the event had to be remembered by the participant until the simulator came at standstill, then the content of the taps was asked for by the test-leader who logged them. This way events were logged without having to talk with a participant during G-loading (i.e. as talking would interfere with the NIRS recording).

10. Results

Table 1 summarizes the results of the subjective measures of each profile. Fifteen participants reported to suffer a gray-out in both conditions. This indicates that this is a quite sensitive measure, although it did not discriminate between the two profiles.

PPL was more frequently reported in the 'push-pull' condition (15) than in the 'pull' condition (9). Furthermore, the average viewing angle during PLL, Fig. 4, amounted 54.7° in the 'pull' condition, and 32.5° in the 'push-pull' condition. The more frequent presence and the levels of PLL indicate that the 'push-pull' condition was physiologically more demanding.

Brief blackouts were reported by three participants: two in the 'push-pull' and one in the 'pull' condition. The same numbers apply for the report of 'stars' in the field of view. The severity of the latter effect was reported *"mild"* in the 'pull' condition, whereas *"quite impactful"* and *"very present"* in the 'push-pull' condition. Related to this symptom was the feeling of losing consciousness (i.e. not A-LOC), reported twice during 'push-pull'.

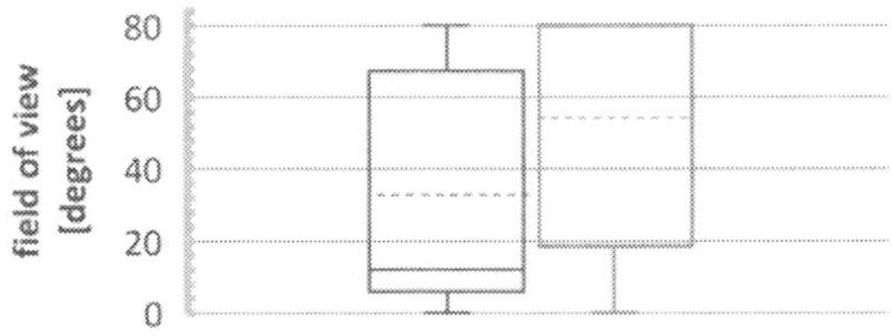

Figure 4. Viewing angle based on the post-its upon report of PLL, blue = push-pull, average field of view 32.5°, orange = pull, average field of view 54.7°. Note: 0° viewing angle is the result from a visual blackout

In the 'push-pull' condition for two participants PLL was nearly instantly followed by an almost-loss of consciousness (A-LOC). None of the participants experienced G-LOC in either condition.

In the push-pull condition eight participants found it necessary to strain, compared to four in the pull condition, indicating that the latter was experienced as less strenuous than the former.

Table 1. Results of subjective measurements.

Type of Flight	push-pull	pull
no. of runs	22	22
# grey-out	15	15
# PLL	15	9
# stars in field of view	2	1
# blackout	2	1
# A-LOC	2	1
# G-LOC	0	0
# straining	8	4

11. Conclusion

This preliminary analysis showed a higher number of reported symptoms (PLL, stars in field of view, blackout, and straining) in the 'push-pull' condition compared to the 'pull' condition. The magnitude of the reported symptoms were also larger for the 'push-pull' condition. This is indicative of the occurrence of an hemodynamic push-pull effect due to a prior HDT. In a future paper the physiological measures (ECG, NIRS, and PGG) will be reported to further substantiate this conclusion.

Acknowledgement

The authors thank Mark Wentink, Bernd de Graaf, and Martijn Rambonnet for their input and time creating the 'flight' profiles and operating the Desdemona simulator.

References

1. Antuñano MJ, Hobe S, Gerzer R. *Position paper on medical safety and liability issues for short-duration commercial orbital spaceflights.* Paris: International Academy of Astronautics; 2009.

2. AsMA *Commercial Spaceflight Working Group. Position paper: suborbital commercial spaceflight crewmember medical issues.* Aviat Space Environ Med 2011; 82:475–484.
3. Banks RD, Brinkley JW, Allnutt R, Harding RM. *Human response to acceleration.* In: Davis JR, Johnson R, Stepanek J, Fogarty JA. *Fundamentals of aerospace medicine*, 4th ed. Amsterdam: Lippincott Williams & Wilkins; 2008.
4. Banks RD, Grissett JD, Saunders PL, Mateczun AJ. *The effect of varying time at -Gz on subsequent +Gz physiologic tolerance (push-pull effect).* Aviat Space Environ Med 1995; 66:723–7.
5. Banks RD, Grissett JD, Turnipseed GT, Saunders PL, Rupert AH. *The "push-pull effect".* Aviat Space Environ Med 1994; 65: 699–704.
6. Benni PB, Li JK, Chen B, Cammarota J, Amory DW. *NIRS monitoring of pilots subjected to +Gz acceleration and G-induced loss of consciousness (G-LOC).* Adv Exp Med Biol. 2003; 530:371–9.
7. Bles W, Groen EL. *The DESDEMONA motion facility: applications for space research.* Microgravity Science and Technology, 2009: 21(4): 281-286.
8. Blue RS, Riccitello JM, Tizard J, Hamilton RJ, Vanderploeg JM. *Commercial spaceflight participant G-force tolerance during centrifuge-simulated suborbital flight.* Aviat Space Environ Med 2012; 83:929–34.
9. Buckey JC. *Effect of spaceflight on the cardiovascular system.* Space physiology. New York: Oxford University Press; 2006; 149-54.
10. Goodman LS, Banks RD, Grissett JD, Saunders PL. *Heart rate and blood pressure responses to +Gz following varied-duration -Gz.* Aviat Space Environ Med 2000; 71:137 – 41.
11. Green NDC. *Effects of long duration acceleration.* In: Rainford DJ, Gradwell DP, editors. Ernsting's Aviation Medicine. 4th ed. New York: Edward Arnold Ltd; 2006:137–58.
12. Haerkens MH, Simons R, Kuipers A. *Doctor, may I travel in space?* Aeromedical considerations regarding commercial suborbital spaceflights.[Article in Dutch] Ned Tijdschr Geneeskd. 2011; 155(48):A3904.
13. Howard P. *The physiology of positive acceleration.* In Gillies, JA. *A text book of Aviation Physiology.* Oxford: Pergamon press; 1965: 551–687.
14. Kobayashi A, Kikukawa A, Onozawa A. *Effect of muscle tensing on cerebral oxygen status during sustained high +Gz.* Aviat Space Environ Med. 2002; 73(6): 597–600.
15. Ledderhos C, Rall G, Gens A. *Objective determination of the efficiency of anti-G suits using different pulse wave components.* Presentation at the 3rd European Conference in Aerospace Medicine, 9-11 November 2012, London, UK.
16. Liu Y, Zhang LF, Zhang KL, Lu HB. *Role of carotid baroreflex and sympathetic responses in the push-pull effect: a simulation study.* Aviat Space Environ Med 2012; 83: 841–9.
17. Prior A. *Negative to positive Gz Acceleration transition.* In: Current concepts on G protection Research and Development AGARD Report LS-202, France: NATO 2-1-2-8(1995).
18. Schneider S, Abeln V, Askew CD, et al. *Changes in cerebral oxygenation during parabolic flight.* Eur. J. Appl. Physiol. 2013 Jun;113(6):1617-23.
19. Shender BS, Forster EM, Hrebien L, Ryoo HC, Cammarota JP Jr. *Acceleration-induced near-loss of consciousness: the "A-LOC" syndrome.* Aviat Space Environ Med. 2003;74(10): 1021–8.
20. Tran CC, Florence G, Tinet E, et al. *Cerebral hemodynamics and brain oxygen changes related to gravity induced loss of consciousness in rhesus monkeys.* Neuroscience Letters 2003; 338: 67–71.
21. Wertheim AH, Ooms J, Regt GP de, Wientjes CJE. *Incidence and severeness of seasickness: Validation of a rating scale (Report IZF 1992 A-41).* Soesterberg, The Netherlands: Institute for Perception TNO; 1992.
22. Smith JD, Cromwell RL, Kundrot CE, Charles JB, *Six-degree head-down tilt bed rest: forty years of development as a physiological analog for weightlessness.* American Society for Gravitational and Space Biology Conference 2011, San Jose, CA, USA.

PROCEEDINGS OF THE 9TH INTERNATIONAL SPACE SAFETY CONFERENCE

SESSION 28A LECTURE

INSIGHTS GAINED FROM THE MASSIVE COLLISION MONITORING ACTIVITY

Darren McKnight(1), Sophie Behrend(2), and Patrick Casey(3)
(1)IAI, 15020 Conference Center Drive, Chantilly, VA 20120 (USA), dmcknight@integrity-apps.com
(2) IAI, sbehrend@integrity-apps.com
(3) IAI, pcasey@integrity-apps.com

ABSTRACT

Over the last two years a monitoring and characterization experiment has been conducted to assess the conjunction dynamics of massive derelict objects in low Earth orbit (LEO). Four separate clusters of massive derelicts were examined whose constituents represent about 500 objects and 800,000kg of mass; this equates to about a third of all derelict mass in LEO. The encounter statistics for these four clusters are reviewed and the expansion of this experiment, in both scope and functionality into a regular space safety operation, is detailed. Based on the last year of conjunction history, a "worst offenders" analysis was conducted identifying the top ten massive derelicts in LEO that are recommended for removal to minimize the debris-generating potential of the existing population. The Massive Collision Monitoring Activity (MCMA) recently was enhanced so it now daily scrutinizes about 800 objects constituting over ~1.6Mkg of abandoned mass; this is over 2/3 of the derelict mass in LEO.

1. BACKGROUND

The aerospace community has largely focused on how to prevent the potential for the cascading of the debris environment that was identified decades ago (i.e., the Kessler Syndrome). [1] The traditional response to combatting the Kessler Syndrome is the steady, long-term removal of abandoned rocket bodies and payloads (i.e., five large objects a year for 100yrs). [2] The typical ADR analysis assumes that we are unable to determine which events might occur first. As a result, the overall effect is that one must remove 35-50 objects on average for each collision prevented. While this scenario is predicted to prevent nearly 30 collisions, around 15 collisions are anticipated to still occur. These events would erode space flight safety significantly even while preventing environmental instability.

The alternative approach of assuring near-term space flight safety as the primary objective has encouraged us to examine the possible short-term, highly consequential events (i.e., create very large amounts of debris). This focus led us to examine debris "hot spots" in Earth orbit where the most massive objects might interact with each other at hypervelocity speeds (i.e., well above 6km/s).

It was shown previously that two SL-16 rocket bodies colliding around 850km would result in the reduction of the operational lifetime of all satellites in the 650-1050km altitude range by an average of ~10%. [3] An examination of the satellites residing within this altitude range amount to at least $10-15B of operational space systems; 10% of this asset pool is indeed significant. [4] It should be noted that the cost of space assets may not be the most complete means to quantify loss since ongoing revenue and/or replacement cost might be more relevant. However, these values are much more difficult to estimate.

A focus on individual close approaches within specific clusters of massive derelicts has produced empirical data documenting the number and proximity of conjunctions between these massive objects. In the spirit of establishing a culture of safety, it is imperative to monitor, characterize, and share information about these individual events in order to motivate responsible action. This experiment is called the Massive Collision Monitoring Activity (MCMA).

A recent review of aerospace and business catastrophes clearly show the importance of such communications: *"Multiple near misses preceded (and foreshadowed) every disaster and business crisis we studied and most of the misses were ignored or misread.* Our work also shows that cognitive biases conspire to blind managers to near misses. Relevant cognitive biases are normalization of deviance (i.e., over time accept anomalies) and outcome bias (i.e., focus on results more than, often unseen, complex processes)." [5] More pointedly, the experts in the psychology of anomaly reporting and risk acceptance state that "viable approaches to preventing [such] catastrophes is to observe near-misses and use them to identify and eliminate problems before they produce large failures." [6]

2. MCMA BASICS

The clusters identified for the initial MCMA effort were selected considering (1) proximity of apogees and perigees (i.e., clumped in altitude); (2) common inclination values (causes greater interaction rates); (3) total mass involved in a potential collision; and (4) altitude of the center of the cluster (i.e., high enough so that debris produced is long-lived).

These criteria resulted in the selection of four clusters in low Earth orbit (LEO) that include nearly 500 objects that have an aggregate mass of nearly 800,000kg. The table below provides the salient characteristics of the four clusters.

Table 1. The four clusters that MCMA have been monitoring are the "hot spots" in LEO as the collisions will be the most consequential and the probability of collision operationally relevant.

Cluster Name	Cluster Member	Mass (kg)	Number	Apogee (km)	Perigee (km)	Inclination (deg)
C775	SL8 RB	1,434	44	793	733	74
	SL8 PL	850	44	802	742	74
C850	SL16 RB	8,300	18	860	814	71
	SL16 PL	3,250	18	868	823	71
C975	SL8 RB	1,434	144	1020	935	83
	SL8 PL	800	142	1024	934	83
	Other PL	1500	15	997	905	64
C1500	SL8 RB	1,434	17	1660	1330	74
	SL14 RB	1,407	24	1530	1363	83
	SL14 PL	2477	24	1507	1381	83
TOTAL		**~756k**	**490**			

Overall, the four clusters have distinctly different characteristics in mass span (total mass of derelicts divided by altitude in which their orbits reside, kg/km), number span (total number of objects divided by the altitude in which their orbits reside, #/km), and annual collision rate (CR/yr) as shown below, respectively:

	Mass Span	Number Span	CR/yr
C775:	1393kg/km,	1.39#/kg,	1/413
C850:	**3850kg/km,**	0.67#/km,	1/578
C975:	2610kg/km,	**2.55#/km,**	**1/92**
C1500:	539kg/km,	0.32/km,	1/1345

The maximum value in each category is bolded above emphasizing that a collision in C850 will be the most catastrophic, but a collision is most likely to occur in C975 (i.e., highest CR/yr).

2.1. Encounter Statistics

The table below depicts the monthly encounter rates while Appendix A contains a monthly accounting of interactions for each cluster and statistics on the total set of encounter data summarized in the table below.

As would be expected from examining the data above, C975 has the greatest number of encounters as it has the greatest concentration of objects (2.55 objects/km) which yields the greatest predicted collision rate between the members of the cluster (1/92 per year).

Table 2. The encounter rates for 1km and 5km miss thresholds do not directly provide an absolute measure of probability of collision but they do provide a means to compare measured data against models used to estimate orbital debris interactions.

	C775	SL-16 R/Bs	C850	C975	C1500
#<5km/mon	~200	~20	~70	~1800	~36
#<1km/mon	~10	~1	~2	~80	~2

There have been nearly 27,000 conjunctions less than 5km recorded in the last two years and 1,275 less than 1km for all of these clusters; the vast majority of these occured in C975 over the last year.

As of 1 September 2017, there have been 287 events (19 in C775, 9 in C850, 257 in C975, and 2 in C1500) that if either object were an operational satellite, it would have triggered the issuance of a Conjunction Data Message (CDM) by the Joint Space Operations Center (JSpOC) (i.e., total miss <1km and radial miss <200m).

3. WORST OFFENDERS ANALYSIS

A major objective of the MCMA effort is to determine the efficacy of selecting the "worst offenders" (i.e., identify primary objects for ADR) before they are on a collision course with another massive derelict. First, each cluster is evaluated generally to identify the worst "hot spot" in LEO. C975 has a slightly higher immediate risk (~43 to ~28 for C850) but it also will have a greater persistence as debris produced in C975 will have ~3.5x longer orbital lifetime than debris from a C850 event. While C1500 has the lowest immediate risk, debris liberated in C1500 will persist over ~10x longer versus C975 (and ~40x versus C850).

Table 3. Cluster ratings show that C975 and C850 are the worst clusters.

	Probability (CR/yr)	Consequence (Trackable Debris Created)	**"Risk" (Probability x Consequence)**	Persistence (lifetime, yrs, for 0.01m²/kg)
C775	1/413	3,500	**~9**	300
C850	1/578	16,000	**~28**	500
C975	1/92	4,000	**~43**	1,800
C1500	1/1345	6,000	**~5**	20,000

For individual objects, the worst offenders scoring is done by (1) frequency of interactions (34%), (2) potential consequence of collision (33%), and (3) potential probability (33%, combination of low eccentricity and overlap with largest number of fellow cluster members). The results using the weightings above produced the top ten worst offenders listed in the table below. [7]

Table 4. The worst of the worst offenders are all rocket bodies from C975 (SL-8) and C850 (SL-16).

Satellite Number	Cluster
15056	C975
32053	C975
31793	C850
22285	C850
22220	C850
23088	C850
23405	C850
25407	C850
28353	C850
9044	C975

However, in examining the worst offenders results, it was determined that possibly the individual categories should be weighted differently. The research team proceeded through three new weightings as detailed in the table below with Scenario A being the one used previously (i.e., the table above).

Scenario B added a NASA simulation to the Potential Probability portion of the test since it more accurately portrays the long-term interaction dynamics than the original simple two-step evaluation (i.e., altitude span of orbit and number of cluster members with which it overlaps). [7] Scenario C weighted the NASA simulation even more since our two-step test actually provided an alternative to the NASA simulation results.

Table 5. The MCMA Team experimented with four different worst offenders tests.

Factors & Thresholds	A	B	C	D
ACTUAL FREQUENCY	34%	34%	34%	25%
#<500m	0.18	0.18	0.18	0.16
#<1km	0.14	0.14	0.14	0.08
#<5km	0.02	0.02	0.02	0.01
POTENTIAL CONSEQUENCE	33%	33%	33%	50%
Mass x Involvement				
P/L = 1.5; R/B = 2.5				
POTENTIAL PROBABILITY	33%	33%	33%	25%
Apogee – perigee	0.165	0.11	0.0825	0.0625
# overlapping sats	0.165	0.11	0.0825	0.0625
NASA simulation	---	0.11	0.165	0.125

Scenario D changed the weightings of the families so that the consequence was weighted at 50% and the probability (cOmbined actual frequency and potential probability) was weighted 50%.

Each of these tests had their own drawbacks and benefits so instead of only using the last scenario, Scenario D, it was decided that the worst offenders would be decided upon by using the results of all of these tests. All of the objects within the four clusters were scored against all four tests. Each object was given a score based upon its position in each list and these scores were summed. For example, if an object was rated the third worst offender in each test then it would receive a total score of 12 (i.e., four scores of 3 = 12). The lowest score is the worst offender.

This process uses the concept of the diversity prediction theorem whereby a diverse suite of models (i.e., model of models) will always be more accurate than its average member. [8, 9] More analytically, the suite's squared error is equal to the average of the models' individual errors squared error minus the diversity of the models' legacies:

$$(\text{model of models error})^2 =$$

$$\sum(\text{model error})^2/\#\text{ of models} - \sum(\text{diversity of models})$$

Therefore, as we get better models (i.e., less average error) and more diversity in them, the smaller the "model of models" error will be. **The table below depicts the list of the top 10 worst offenders using each scenario with the last column showing the composite final list of worst offenders.**

*Table 6. The four individual scenario results along with the final composite worst offenders list (last column) are shown. **Members of C975 are bolded** while all other objects come from C850. Payloads have "p" after the number while all others are rocket bodies.*

A	B	C	D	Final
15056	**15056**	**15056**	31793	31793
32053	**32053**	**32053**	22285	**15056**
31793	**10537**	**10537**	23088	**32053**
22285	**9044**	**9044**	28353	22285
22220	**6708**	**6708**	23405	23088
23088	31793	**12092**	22220	22220
23405	**12092**	**8421**	25407	28353
25407	**8421**	**11804**	22566	**6708**
28353	**11804**	**15597p**	19650	23405
9044	22285	**11735p**	16182	25407

It is expected that Scenario D, where the consequence factor had the greatest weighting, produced only C850 rocket bodies in the top ten worst offenders list.

The very subtle variations between the scenarios provided some drastic differences in the ratings. It is also noteworthy that only a couple of payloads from C975 showed up at all and no payloads from C850. The final column has a color coding to highlight that the top four worst offenders were significantly riskier than the rest; within this top four, there are two rocket bodies from each of our two primary clusters. The next five entries, highlighted in yellow, were also fairly tightly clumped in scores.

3.1. Unique Encounter Signatures

Upon close examination, pairs of objects within these clusters appear to interact in repeatable ways. The encounters normally happen in a sequence of conjunctions that we call a "walk-in." The rate at which these close approaches get closer each orbit is called the closure rate. These rates can vary from 10's of kilometers per orbit to as little as 10's of meters per orbit. However, it has been found that when two objects encounter each other they have a set closure rate that they follow whenever they encounter each other. Thirty sets of walk-in sequences were examined and 28 of them had closure rates that were the same as previous encounters. This observation might be useful in determining the likelihood of a "very close" approach (i.e., a collision) in the future.

4. MCMA: FROM EXPERIMENT TO OPERATIONS

The motivation for MCMA was to monitor and characterize a small, but consequential, subset of the cataloged population with the hypothesis that they pose the majority of the risk for large debris-generating events. Indeed, it is without debate that encounters within the four clusters monitored within MCMA produce the vast majority of the most consequential collision events in Earth orbit. However, while these four clusters account for ~500 massive derelicts amounting to ~800,000kg of mass, there are still hundreds of massive derelicts that are not in "clusters" and, thus, their debris-generating potential are not accounted for. It would be prudent to consider more objects in order to try to be more complete in the assessment of the massive derelict population. As a result, ~300 more objects amounting to an additional ~700,000kg of mass have been added to the first four clusters.

The threshold for inclusion is that any object (1) whose orbit crosses the 700-2000km altitude range (i.e., LEO) and (2) has a mass of over 2,000kg will be monitored.

The figure below depicts the state of the massive objects being considered by MCMA in the initial "compelling" version and the new "complete" framework. The original four clusters are the dark rectangles in the figure. Each of the four clusters had derelicts within their altitude range (but previously excluded since they did not have the inclination value(s) for each cluster); these were added to each cluster.

A new cluster, C1200, with only eight objects was also created to cover massive objects that fall between C975 and C1500. Cleo includes 22 objects that stay within 700-2000km but spans more than one of the clusters. Finally, Chigh contains many massive objects that reside in highly elliptical orbits nominally being rocket

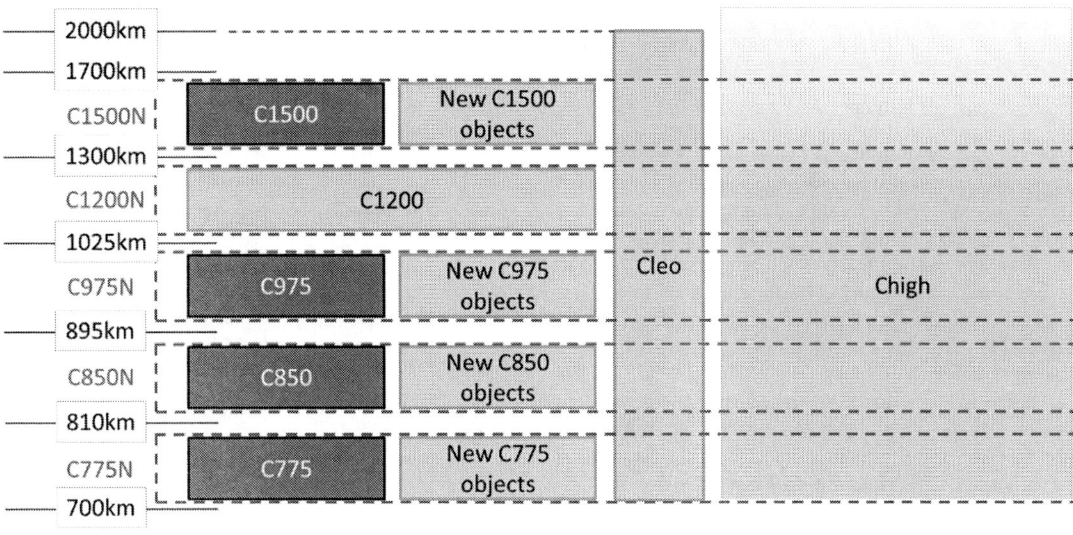

Figure 3. The new set of objects to be monitored and characterized by MCMA provides a more complete coverage of massive derelict encounter possibilities: ~800 objects amounting to ~1,600,000kg.

bodies abandoned in geosynchronous orbit (GEO) transfer orbits (GTO) or medium Earth orbit (MEO) transfer orbits (MTO). Chigh contains nearly 200 objects amounting to over 570,000kg (about 2/3 of the total mass contained within the original four "pure" clusters). Chigh is a unique population subset since if one of these objects were to fragment, the debris cloud would transit between LEO and GEO/MEO creating a potential cross-contamination hazard.

The new clusters analyzed for a complete characterization of debris-generating collisions are now C775N, C850N, C975N, C1200N, and C1500N where each one considers interactions amongst themselves and with Cleo and Chigh. It is our hypothesis that there will be few "new" encounters in the new clusters as compared to the four original clusters. However, it is imperative that this process be completed, and over the next year, the characterization of these new clusters will help to answer the question as to whether they warrant daily scrutiny. In 60 days of monitoring (1 May 2017 to 31 July 2017) these new objects, that contributed a ~60% increase in number of objects monitored, resulted in only ~20% increase in the number of conjunctions. More pointedly, the ~40% number increase due to Chigh only yielded ~1% more conjunctions. This result is what was anticipated but this larger screening population reduces the chance that some highly consequential, low probability conjunction will be missed.

The five new clusters in total now contain nearly 2/3 of the derelict mass that crosses LEO in its ~800 objects. Expanding the objects to be monitored by MCMA not only expanded the mass (~1.6M kg out of ~2.5M kg in LEO) and number of objects under examination but also now contains derelicts from all major spacefaring countries. Over the next year, the research team will provide updates on the entire population being monitored.

4.1. Severity and Probability of Collision

In addition to considering a more complete population for monitoring, the research team is also implementing enhanced characterization features to include severity of event (i.e., estimate number of fragments) and deterministic probability of collision for events with miss distances smaller than a few hundred meters.

Severity refers to the quantity of debris generated from a collision event (i.e., consequences) determined through the use of fragmentation models. To estimate the consequences of a potential collision, two models were used: the NASA Orbital Debris Program Office's (ODPO) fragmentation model and the Mass Involvement empirical fragmentation model, specifically developed for MCMA. Both are described in more detail in Appendix B.

The Mass Involvement model predicts only the number of catalogued and lethal nontrackable (LNT) debris generated. Catalogued debris in LEO refers to debris with diameters greater than 10cm and LNT debris is between 5mm and 10cm. The ODPO model provides a continuous cumulative mass distribution, where characteristic lengths of 10cm and 5mm are used to find the number of catalogued and LNT debris generated, respectively. The severity is given as a range for both cataloged and LNT determined by the values from the two models.

The probability of collision (PC) is determined through the use of three separate models and, like the severity modelling, the PC result is given as a range of the maximum and minimum PC provided by these models. These models are by (1) Knowles, (2) Patera, and (3) Alfano. The PC models are described in more detail in Appendix B. These algorithms primarily take into account positional uncertainty, the miss distance, and the object sizes. The table below provides a brief overview of how these parameters are used in these three PC models.

It is important to repeat again that the number of encounters in the 1-5km range does not reflect any measure of actual probability of the objects colliding. These encounter rates are instructive to examine relative to accepted statistical models for these interactions. However, once the miss distance gets on the order of a ~300m or less then there is a measurable probability of a collision which is what MCMA now calculates. Finally, a risk value is calculated by multiplying the severity with the probability (of collision).

Table 7. The three PC equations selected have diverse philosophies and use the key terms differently.

Method	Positional Uncertainty, PU	Miss Distance, D	Object Sizes
Knowles	R, I, and C components	Only calculate PC when D < PU	Hard body radius is average of maximum dimension and minimum dimension of two conjuncting objects
Patera	Positional uncertainty in "collision" plane (i.e., major and minor)	Function of encounter angle	
Alfano	N/A	Use miss distance directly	

The table below provides the current output for potential conjunctions amongst the five "new" MCMA clusters. The objects are identified by their common name and satellite number. The conjunction date and time are provided in zulu time. The total miss distance in kilometers (i.e., range) is provided plus the miss distance in the radial, in-track, and cross-track components (i.e., R, I, C). Next, the total relative velocity and its components are provided followed by the longitude, latitude, and altitude of the conjunction point.

The "Notes" section is provided in case there are some data flags triggered such as "breakup altitude is above the apogee of one of the objects," etc. The Two-Line Element (TLE) age in days is provided for both objects since TLEs older than 1-2 days may provide a weak mathematical basis upon which to base the PC calculations. The mass of each object and the cluster within which each object resides is also provided.

The last row of information provides the probability of collision, consequence, and risk values that have been added this year to provide more characterization details. As stated earlier, the probability and severity values are generated using outputs from a variety of models described in Appendix B. The risk value is simply the product of the maximum PC and maximum catalogued debris produced and the same for the two minimum values. The risk values have little meaning in an absolute sense but provide a valuable way to compare between potential conjunctions monitored by MCMA.

5. SUMMARY

The MCMA is an ongoing space risk experiment which has yielded several insights:

- Clusters of massive derelicts in LEO pose a uniquely subtle but significant risk to space flight safety that is obfuscated by averaging these objects into the overall LEO population.

- Over the last two years, nearly 300 near misses have occurred between these derelicts that would have warranted the issuance of a Conjunction Data Message (CDM) if either had been an operational payload.

- The "ten worst offenders" in LEO have been identified, providing a draft target list for ADR systems to pursue; they are all in C975 or C850.

- The expanded list of objects being monitored by MCMA creates a much more complete assessment of potential massive debris-generating events.

Table 8. The output from the enhanced characterization of the clusters provides new information vital to short-term response and long-term modeling efforts.

Object 1	Object 2	Conjunction Date/Time	Range, km	Range R	Range I	Range C
C1452 13991	C1486 14240	8/19/17 09:43:33 PM	0.242	0.236	0.013	-0.048

Vrel, km/s	Vrel R	Vrel I	Vrel C	Longitude, deg	Latitude, deg	Altitude, km
13.763	0.000	-12.675	-5.364	51.707	45.353	779.318

Notes	TLE age Object 1	TLE age Object 2	Object 1 Mass	Object 2 Mass	Object 1 Cluster	Object 2 Cluster
None	0.052	0.015	750	750	C775	C775

Minimum PC	Maximum PC	Minimum Cataloged Debris Produced	Maximum Cataloged Debris Produced	Minimum LNT Debris Produced	Maximum LNT Debris Produced	Minimum Risk	Maximum Risk
4.63E-04	1.00E-02	3750	8830	37500	1473200	1.74	88.48

6. REFERENCES

1. Kessler, D. "Collision Frequency of Artificial Satellites: The Creation of a Debris Belt", Journal of Geophysical Research, Vol. 83, No. A6, 1978.
2. Liou, J.-C., Johnson, N.L. A sensitivity study of the effectiveness of active debris removal in LEO. Acta Astronautica 64, 236–243, 2009a.
3. McKnight, D., Di Pentino, F., and Knowles, S., "Massive Collisions In LEO - A Catalyst To Initiate ADR," 65th International Astronautical Congress, Toronto, CA, September 2014.
4. Approximate cost of existing satellite systems was determined by examining databases at http://www.satview.org and http://uscusa.org.
5. Robin Dillon Merrill: Tinsley, C., Dillon, R., and Madsen, P., "How to avoid catastrophe," HBR, April 2011.
6. Dillon, R.; Tinsley, C.; Madsen, P.; and Rogers, E.; "Organizational Correctives for Improving Recognition of Near-Miss Events" Journal of Management, Vol.42, No.3, March 2016 671-697.
7. McKnight, D., Walbert, K., Casey, P., Behrend, S., and Speaks, S., "Preliminary Analysis of Two Years of the Massive Collision Monitoring Activity," 68th Astronautical Congress, Adelaide, Australia, September, 2017.
8. Ioannides, Y. and Page, S., "The Difference: How the Power of Diversity Creates Better Groups, Firms, Schools, and Societies," Journal of Economic Literature, 2007.
9. Page, Scott (2011). *The Hidden Factor: Why Thinking Differently is Your Greatest Asset.* Chantilly, VA: The Great Courses.
10. Patera, R., "Calculating Collision Probability for Arbitrary Space-Vehicle Shapes via Numerical Quadrature," Journal of Guidance, Control, and Dynamics, Vol.28, No.6, Nov-Dec 2005.

APPENDIX A: Encounter data for all clusters

	C775			SL16 R/B Only			C850			C975			C1500		
	<5km	<1km	Closest	<5km	<1km	Closest	<5km	<1km	Closest	<5km	<1km	Closest	<5km	<1km	Closest
Jun 2015				18	0	1146									
Jul 2015				20	0	1971									
Aug 2015				44	3	425									
Sep 2015				12	0	1066									
Oct 2015				9	0	2260									
Nov 2015				25	0	1140									
Dec 2015				15	2	477									
Jan 2016				21	3	447									
Feb 2016				12	0	1178									
Mar 2016				32	2	778									
Apr 2016				19	0	1609									
May 2016				4	0	1113									
Jun 2016				25	0	1080				1786	77	27			
Jul 2016				17	2	946				1762	81	73			
Aug 2016				9	0	1115	49	0	1115	1796	73	64			
Sep 2016				114	2	634	144	3	634	1710	55	65			
Oct 2016				18	0	1936	54	1	750	1803	56	102			
Nov 2016				17	0	1013	38	1	867	1911	89	122			
Dec 2016	211	8	565	18	1	815	50	1	815	2135	94	116	33	3	694
Jan 2017	194	11	176	14	0	1492	48	0	1402	1798	77	190	39	1	459
Feb 2017	155	11	436	32	2	501	82	1	501	1735	76	160	31	3	208
Mar 2017	211	7	200	12	0	1653	46	2	759	1874	76	207	35	1	718
Apr 2017	200	13	135	12	0	2137	54	3	812	1836	71	56	36	2	512
May 2017	212	14	472	14	2	169	46	3	260	1745	75	101	39	0	1358
Jun 2017	201	9	329	26	0	1202	75	3	652	1786	78	122	30	1	202
Jul 2017	245	11	260	13	0	1737	48	3	778	1764	74	47	15	1	782
Aug2017	180	11	241	12	0	1052	46	1	898	1963	80	105	24	2	668
TOTAL	1809	95		435	19		688	22		23690	1132		282	14	
Min	155	7	135	4	0	169	46	0	260	1710	55	27	15	0	202
Mean	201	11	313	18	1	1146	63	2	751	1822	75	103	31	2	622
MEDIAN	201	11		16	0		49	1		1786	76		33	1	

Total:
26,742: #<5km
1,275: #<1km

Appendix B: Fragmentation and PC models

Fragmentation Models

The NASA ODPO fragmentation model predicts the number of objects N as a function of characteristic length L_C.

$$N(L_C) = 0.1 * (P)^{0.75} * L_C^{-1.71}$$

Where $P = (M + m) * V_{rel}$ when $\frac{m*V_{rel}^2}{2M} > 40$

$P = m * V_{rel}$ when $\frac{m*V_{rel}^2}{2M} \leq 40$

M = mass of larger object, target
m = mass of smaller object, impactor

To determine the number of catalogued objects, an L_c of 10cm is used and L_c is 5mm for LNT. The number of catalogued objects is subtracted from the result of an L_c of 5mm so that the LNT count does not include catalogued objects.

The "Mass Involvement" model shown below uses the mass of the objects and a structural coefficient (SC) based on the type of interacting objects, their relative velocities, and the type of debris of interest (trackable or LNT).

$$\text{Number of Fragments} = (SC_1 * M_1) + (SC_2 * M_2)$$

Where M_1 and M_2 are the object masses and SC is a structural coefficient. A table of SC values is found below:

Table 1: List of SC numbers for various circumstances.

Relative velocity of Impact in km/s	Object Type	SC for Trackable Fragments	SC for LNT
V>12	PL	2.50	25
	RB	2.00	20
8<V<12	PL	2.25	22.5
	RB	1.75	17.5
4<V<8	PL	2.00	20
	RB	1.50	15
1<V<4	PL	1.00	10
	RB	0.75	7.5
V<1	PL	0.75	7.5
	RB	0.50	5.0

For example, to determine the number of LNTs that will be generated by a collision between a 3000kg rocket body and a 900kg payload at a relative velocity of 10km/s, the result is:

#LNT = (17.5*3000) + (22.5*900) = 72,750

PC Models

All of the following models take object size into account through the use of the hardbody radius (HR). Since the object's orientation is unknown, the object's HR is found by adding the longest and shortest dimension of the pair of objects and dividing by two, e.g., a conjunction between two rocket bodies of length 12m and diameter 3m has a HR of 7.5m.

1: *Knowles' method* takes into account only the positional uncertainties (PU) in the R, I, and C dimensions (i.e., PU_R, PU_I, and PU_C), and the HR. If the miss distance is smaller than the smallest positional uncertainty, Knowles' method, shown below, can provide a rough but accurate estimation of the PC.

$$PC = (HR/PU_R)*(HR/PU_I)*(HR/PU_C)$$

2: *Alfano's method* uses the ratio between HR and the miss distance, D. Depending on the value of this ratio, r, a different equation is used, as seen below.

$$r = \frac{HR}{D}$$

$$PC_{Max} = 0.48394 * r \qquad 0 < r < 0.8$$
$$PC_{Max} = 0.21329e^{1.01511*r} - 0.0925 \qquad 0.8 < r < 1$$
$$PC_{Max} = 0.5 \qquad r = 1$$

3: *Patera's method* looks at the conjuncting objects in the two dimensional collision plane using the closest miss distance, the angle between the line of sight of the objects, and the major axis of the uncertainty ellipse, as seen below [10]:

$$P = -\frac{1}{2\pi} * \int_{\theta_{min}}^{\theta_{max}} e^{-\alpha * r^2} d\theta$$

Where $r(\theta) = R * cos(\theta) - \sqrt{\rho^2 - R^2 * sin(\theta)^2}$

α = angle between the line of sight and major axis of the uncertainty ellipse based on positional uncertainty

θ = angle of encounter

ρ = normalized function of HR and positional uncertainty

R = miss distance

PROCEEDINGS OF THE 9TH INTERNATIONAL SPACE SAFETY CONFERENCE

PLENARY CLOSING SESSION

HUMAN-CENTERED DESIGN OF UPCOMING HUMAN MISSIONS TO MARS

Guy André Boy, Ph.D.[1]

ABSTRACT

Human missions to Mars are not only a matter of technology-centered engineering, it is foremost a matter of human-centered design (HCD) where technology, organizations and people roles should be co-designed in a participatory and agile way. Major human centered components are trajectory and propulsion (defining HCD in terms of time), life support (physiological, psychological and social factors), exploration (jobs during flights and on the planet) and anticipation of a Martian civilization (governance, natural and artificial resource management). I will review a few experiments that were conducted so far, such as Mars-500, MDRS and what we are doing in HCD, such as key cognitive and social concepts of operations that rely on situation awareness and decision-making, complexity analysis, dealing with the unexpected, automation and autonomy, and collaborative work. I will develop the perspective that human-systems integration (HSI) is a matter of integrating HCD and systems engineering. We then need to orchestrate the design of the various structures and functions involved in human Mars operations. This orchestration can be made using modeling and human-in-the-loop simulations looking for emerging cognitive and physical functions towards the definition of underlying cyber-physical systems. HSI can be managed in terms of design flexibility, resource commitments, and human-systems knowledge.

1. INTRODUCTION

For the last forty years or so, I worked on cockpit analysis, design and evaluation. Today, we talk about "interactive cockpit", that is people in charge interact with computers that interact with other things (e.g., physical entities such as engines and wings, as well as cognitive entities such as ground control and crews). Even if I carried out research and development on aircraft and spacecraft cockpits most of the time, I use the term "workspace" or "life-space" in a generic way in this paper to denote a location where people control and manage a complex system that includes people and systems. The concept of workspace or life-space considerably changed during the last four decades. Workspace or life-space design requires an orchestration of technology (i.e., hardware and software), organizations and people.

This paper addresses the difficult question of function allocation among people and systems in the scope of future human missions to Mars. Before providing solutions, it is crucial to focus on stating the overall problem correctly. Problem stating is a matter of creativity as well as risk taking. Creativity will be defined as integration. As a metaphor, a painter who wants to create a new color, a kind of orange for example, mixes red and yellow. He or she incrementally integrates red and yellow until a satisfactory orange comes up. Dealing with technology, the same schema applies. A new technology is basically an integration of several components that already exist, or have to be designed (i.e., being themselves integration of components). Integration is key and a recursive process. Now, components can be technological or human. This is why human-systems integration (HSI) emerges as a fundamental goal in design. Such design will be denoted human-centered design (HCD).

For that matter, HCD of upcoming human missions to Mars requires creativity and design thinking, involving various kinds of expertise and experience owned by aerospace engineers, computer scientists, astronauts and ground operators, for example. On May 30-31, 2017, I participated in a Mars Mission Social Sciences Workshop at NASA Kennedy Space Center, where we addressed and discussed a set of key research questions such as technology, time, environment, number of colonists, simulation, highest priority problems and showstoppers, Earth-crew connectivity, budget and government decisions, task allocation, privacy, technology lifecycle and HSI. We also addressed and discussed methods and approaches for the design of support systems such as analogs (e.g., Mars 500), not only artificial reality, and actual scenarios, going to the Moon before Mars, look at the government, industries

[1] Dr. Boy is a former University Professor and Dean of the School of Human-Centered Design, Innovation and Art at Florida Institute of Technology, U.S.A. He was also Chief Scientist for Human-Centered Design at NASA Kennedy Space Center. He is the Chair of the Human-Systems Integration Working Group at the International Council on Systems Engineering (INCOSE) and Fellow of the Air and Space Academy.

and transportation, and life support systems (e.g., how do we produce the food) in such an extreme environment. One of the recommendations was to test prototypes at different levels that lead to Human Missions to Mars, and use analogs and simulations both virtual and physical.

Challenges and barriers are expected to be the followings. The first question is: Why is going to Mars important? People on earth need to be convinced why we need to go to Mars and what are the benefits of the technology that gets us there. Going to Mars is foremost a political decision at the level of planet Earth. An important challenge is communication between crews and Earth mission control. It seems that a causal eventdriven approach based on classical predictions does not work. Conversely, an anticipatory goal-driven approach that will attempt to demonstrate the relevance of potential futures is more appropriate. We then need to focus on the creation of possible futures on Mars, as well as travels to and from Mars.

2. HUMAN MISSIONS TO MARS STATING THE PROBLEM

Going to and living on Mars requires us to state the overall problem correctly. One way to frame the problem is considering the duration of a mission. If it is one year, then the following requirements become important (non-exhaustive list): food production; clothing (i.e., what should we wear?); healthcare and life support systems; education knowledge base. If it is five years, then the following requirements can be considered: updating of unanticipated needs in technology; analog simulations. If it is ten years, then the following requirements can be considered: continue long duration human space flight missions at International Space Station and beyond, like around the moon; lifecycle of technologies.

Five time contexts will have to be studied: getting on board; pre-departure preparation; transit to Mars; establishment of a colony; and maintenance of the colony.

Stating the problem of human missions to Mars: what does it mean to live in a spacecraft for more than six months, and in extreme environments where almost everything will need to be artificially constructed both a priori and during the mission lifecycle of technologies?

3. WHAT IS HUMAN CENTERED DESIGN?

The human-systems integration goal can be supported by human-centered design. HCD supports the orchestration of technology, organizations and people (Figure 1).

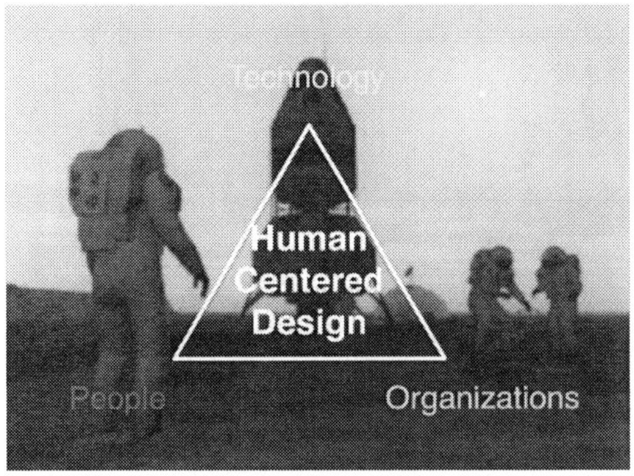

Figure 1. Using the TOP model (technology, organizations and people) for framing the overall problem of human missions to Mars.

We will consider four major human-centered components:

- trajectory and propulsion systems;
- human and life-support systems; • Mars and exploration systems;
- Mars and Martian civilization.

Three types of Human Earth-Mars trajectories can be investigated. Framing the mission for Trajectory 1, which is "short stay with Venus flyby" (Figure 2) requires the following mission time periods: Outbound for 224 days; Stay for 30 days; Return for 291 days. Total time would then be 545 days.

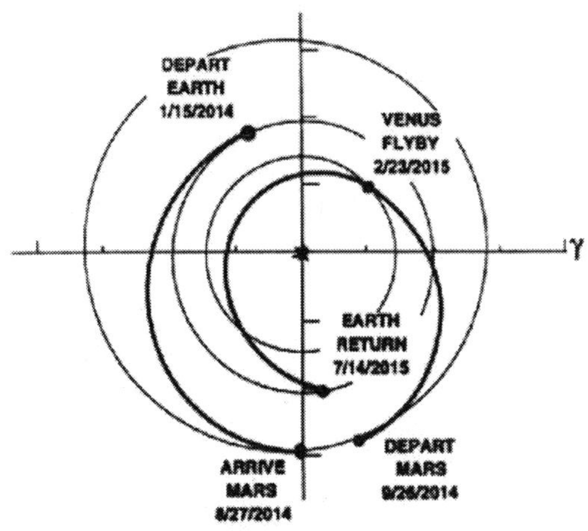

Figure 2. Trajectory 1: short stay with Venus flyby.

Framing the mission for Trajectory 2, which is "minimum energy" (Figure 3) requires the following mission time periods: Outbound for 224 days; Stay for 458 days; Return for 237 days. Total time would then be 919 days.

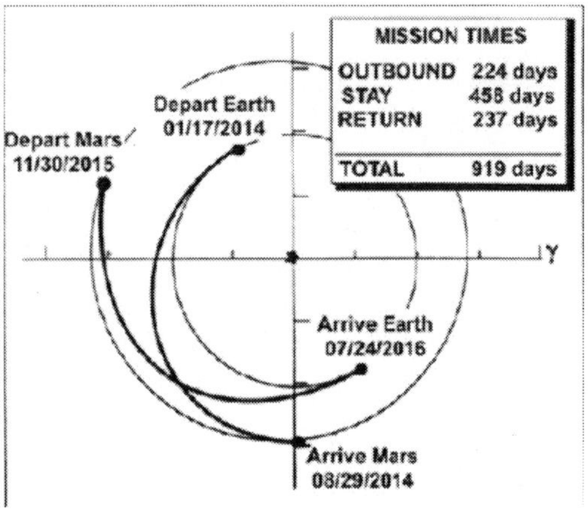

Figure 3. Trajectory 1: minimum energy.

Framing the mission for Trajectory 3, which is "long stay, fast transit" (Figure 4) requires the following mission time periods: Outbound for 150 days; Stay for 619 days; Return for 110 days. Total time would then be 879 days.

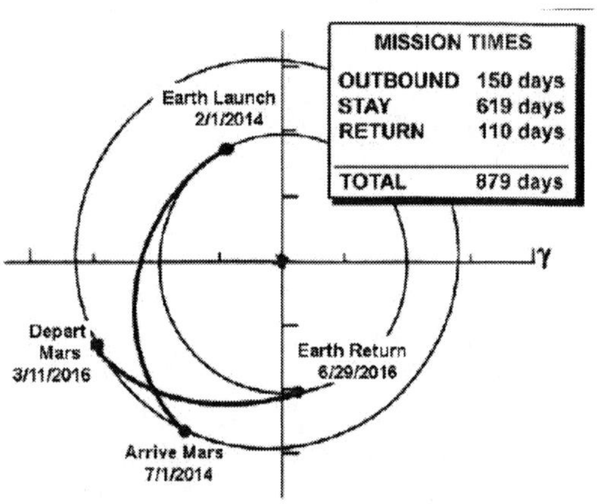

Figure 4. Trajectory 3: long stay, fast transit.

HCD should be used to incrementally define and better understand the fundamental issues of what a human mission to Mars is about. It is based on several agile development cycles that include participatory design, prototyping, tests (lots of tests) and redesign. It is now time to define related concepts and processes. **4.**

4. PARTICIPATORY DESIGN AND AGILE DEVELOPMENT

Participatory design started to develop in Scandinavian countries during the nineteen sixties and seventies, as cooperative design, often called today co-design. It involves in the design process all stakeholders, such as end-users, designers, engineers, certifiers, maintainers, workers, suppliers and so on. It enables design teams to incrementally consider all kinds of requirements related to culture, practice, experience and expertise to improve final product usefulness and usability. It is a great democratic approach for creativity and innovation. In life-critical system design, we typically focus on three main principles: safety, efficiency and comfort.

Agile development was first developed in software engineering. It is based on the Agile Manifesto that defines its values and principles[2]. In this paper, agile development will refer to human-systems integration, which involves both software and hardware. In complex systems design, we need to consider integration at three levels: technology integration, human-systems integration, and organizational integration. The TOP model supports this design approach (Boy, 2013). Instead of designing and developing engineered systems in isolation from each other and then integrated when all of them are finished as it is currently done, it is better to have a holistic approach from the beginning using both existing components, which could be extended or even changed during the agile design process, or/and still-abstract components, which will be further developed with respect to formative evaluations of activities emerging from operations design simulations.

5. ORCHESTRATING COGNITIVE AND PHYSICAL FUNCTIONS

One of the first attempt to anticipate and simulate a human mission to Mars was the Mars500 experiment carried out at the Institute of Biomedical Problems of the Russian Academy of Sciences in Moscow, Russia, from 2007 to 2011. Three different crews of volunteers lived and worked in a spacecraft prototype. A first 15-day experiment was carried out in November 2007. Its purpose was purely technical (i.e., testing equipment, facilities and operations procedures. The second 105-day experiment was carried out from March to July 2009 involving six volunteers who lived in isolation in the spacecraft prototype. The third 520-day experiment was carried out from June 2010 to November 2011. The crew included three Russians, a Frenchman, an Italian/Columbian and a Chinese citizen. The simulation facility included a spacecraft simulator, an

[2] http://agilemanifesto.org

ascent/descent craft and a Martian surface. Participant backgrounds were in engineering, medicine, biology, and human spaceflight. Physiological, psychological and social data were recorded and analyzed. The purpose was to improve understanding on long-term isolation in a limited room space.
Mars500 mission (Figure 6) required the following mission time periods: Outbound for 250 days; Stay for 30 days; Return for 240 days. Total time would then be 520 days. Three major issues were elicited: time effects; cultural influences; and individual differences (Tafforin, 2013). During outbound, it was observed common activity and normal feelings. During return, good spirit and fatigue or stress was observed. Stay was expected to be a traumatic period, but it was not.

A study was carried out by Wang et al. (2014) on mental issues due to confinement of long-term spaceflight mission, using Mars 500 data. They analyzed psycho-physiological and psycho-social data. They used the International Affective Pictures System (IAPS), the Profile of Mood States (POMS) questionnaire and the levels of four types of plasma hormones (i.e., cortisol, 5-hydroxy tryptamine, dopamine, and norepinephrine). They concluded that the confined crew tended to assign positive ratings to negative pictures with time, which might be driven by a defensive system, which was probably activated by the aggravating psychological stress with time. There was a stage-changing pattern of psychological adaptation of the Mars 500 crew.

Ushakov et al. (2014) carried out psychological studies in the Mars500 experiment to find out the integrated impact of interplanetary spaceflight factors on the behavior and activities of an isolated small group. Main findings are about the necessary changes in the selection, training, and medical and psychological support of the future crew of a human interplanetary mission. They identified positive (salutogenic, i.e., related to successful coping with space stress) and negative (in particular, the separation phenomenon) effects of human participation. This raised a number of fundamentally new issues of space medicine such as the problem of performing complex operator activities on another planet after prolonged exposure to weightlessness and technological aspects of an adequate terrestrial simulation of operations on other space objects.

Mars500 is not the only attempt to simulate and anticipate a Mars mission. The Mars Desert Research Station (MDRS) experiments in Utah desert are interesting attempts also. They are carried out by the Mars Society (http://mdrs.marssociety.org/) and based on Robert Zubrin's stories.

More generally, these experiments that consist in prototyping possible futures for human missions to Mars require the development of tangible scenarios and effective human and system function analysis. Elicited functions can be both cognitive and physical, and to be orchestrated (Boy, 2013). I tried to illustrate such emergent functions in this short section of the paper. The main goal of such a human-centered approach to design and develop successful human missions to Mars is to gain maturity at three levels: technology maturity, maturity of practice (at the individual level), and organizational maturity (the TOP model again). This orchestration is necessarily incremental and should be carried step by step. This is why it seems that the next step could be to start training on the Moon before going to Mars. Risk taking is mainly a matter of incremental preparation and accumulation of experience that could take time (Boy & Brachet, 2010).

6. THE TANGIBILITY ISSUES

During the 20th century, engineering was based on the assembly of tangible pieces of hardware. We needed to wait the end of the century to see electronics and software invading hardware, to the point that our lives shifted from (structural) hardware issues at design and development time to (functional) software issues at operations time. One of the major issues was automation and its appropriation by people. From the beginning of the 21st century, we observe the opposite trend, i.e., almost all industrial projects start as a piece of software. Typically, we start by using PowerPoint, or equivalent software to present ideas and concepts. We continue by developing models, animations and simulations. Even better, we can create and further refine human-in-the-loop simulations (HITLS) that enable us to very early discover behaviors, properties and functions emerging from interactions between technology, organizations and people. Consequently, functional issues can be discovered very early during design and development. Problems may arise at the (structural) hardware level since structures can be typically deduced from functions (i.e., 3D printing is now possible). Of course, this is not always linear (i.e., from software to hardware) because structures and functions are intimately related, and therefore co-designed. Even if automation remains a typical issue caused by software, it can be better understood at design time. Conversely, the tangibility issue emerges from immersion into real physical and organizational structures. This is why we need to make progress by implementing experimental simulations incrementally. In systems engineering, this is called an agile approach.

Automation commonly denotes automation of machines, i.e., automating machine functions (Figure 5).

However, it is important to figure out that operations procedures tend to automate people, who are required to follow them. In contrast, when neither procedures nor automation provide appropriate solutions, people need to solve problems by themselves using their knowledge and experience. They need to be autonomous. More generally, organizations that include humans and systems need to be autonomous to deal with unexpected events.

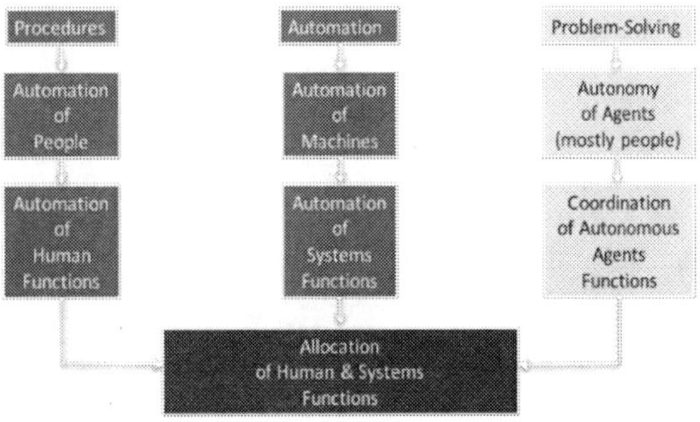

Figure 5. Human-systems function allocation for procedure following, automation use and problem solving.

Autonomy requires constant construction deliberative functions and discovery of emerging functions (Figure 6). Modeling and HITLS are important support to such function construction and discovery. Tangibility of the resulting multi-agent network depends on the definition of such functions. Tangibility can be tested using metrics related to multi-agent stability, maturity, flexibility and sustainability (Boy, 2016).

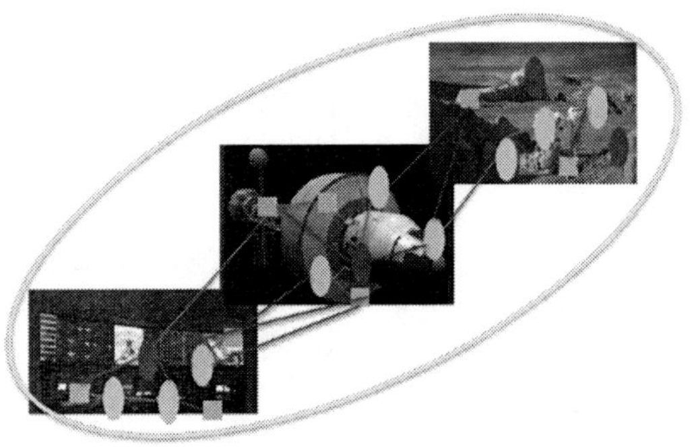

Figure 6. Human and system functions among Earth control center, spaceship and Mars operations.

7. CONCLUSION

This paper provided a few concepts and perspectives on design and development of human missions to Mars. Human-centered design puts forward context awareness as a key problem, which can be incrementally solved by defining scenarios, using human-in the-loop simulations, and eliciting emergent cognitive functions. Scenario-based design requires strong conceptual models. This is why expertise and experience are required, as well as creativity and design thinking. In other words, HCD of a human mission to Mars requires participation of astronauts and experienced space personnel, as well as human centered designers and human-systems integration architects. Of course, designing a

human mission to Mars today is a brand-new endeavor that both astronauts and experienced space personnel never experienced; this is the reason why we need to work on analogs. Mars500 is a good example. There are many others. These experiments are therefore mandatory to accumulate experiences that enable people involved to take reasonable risks.

Such experiences contribute to the reinforcement of our capacity to deal with the unexpected. This should be understood in terms of situation awareness and decision making. Being aware of a situation requires better understanding of distinctions between possible, probable and improbable risks. Any decision cannot be correctly made if the current situation is not well perceived, understood and projected. This requires appropriate technology and organization, as well as excellent training. More specifically, collaborative work needs to be better understood among the various agents (i.e., Earth control center, spaceship and Mars operations). For all these reasons, cognitive function analysis is mandatory (Boy, 1998, 2013, 2016).

Acknowledgements

I would like to thank the people who helped me produce this paper. More specifically, discussions I had with Anthony Genova, Ondrej Doule and Buzz Aldrin contributed to improve the quality of the paper. I am also thankful to the organizers and participants of the Mars Mission Social Sciences Workshop at NASA Kennedy Space Center in May 2017, for deeper discussions and insights.

8. REFERENCES

Boy, G.A. (1998). *Cognitive Function Analysis*. Praeger/Ablex, USA; ISBN 9781567503777.

Boy, G.A. (2013). *Orchestrating Human-Centered Design*. Springer, U.K.

Boy, G.A. (2016). *Tangible Interactive Systems: Grasping the Real World with Computers*. Springer, U.K. ISBN 978-3-319-30270-6.

Boy, G.A. & Brachet, G. (2010). *Risk Taking*. Dossier of the Air and Space Academy, Toulouse, France, ISBN 2-913331-47-5.

Tafforin, C. (2013). Adaptive behavior of an isolated and confined crew during a Mars mission simulation. *Human Ethology Bulletin*. Volume 28, Number 1. ISSN 2224-4476.

Ushakov, I.B., Morukov, B.V., Bubeev, Y.A., Gushin, V.I., Vasil'eva, G.Y., Vinokhodova, A.G. & Shved, D.M. (2014). Main findings of psychophysiological studies in the Mars 500 experiment. Original Russian Text © published in Vestnik Rossiiskoi Akademii. 3316, *Herald of the Russian Academy of Sciences*, 2014, Vol. 84, No. 2, pp. 106–114. © Pleiades Publishing, Ltd., 2014. ISSN 1019.Nauk, 2014, Vol. 84, No. 3, pp. 212–221. *(PDF Download Available)*. Available from: https://www.researchgate.net/publication/271632265_Main_findings_of_psychophysiological_studies_in_the_Mars_500_experiment [accessed Sep 25, 2017].

Wang, Y. Jing, X., Lv, K., Wu, B., Bai, Y., Luo, Y., Chen, S. & Li, Y. (2014). During the Long Way to Mars: Effects of 520 Days of Confinement (Mars500) on the Assessment of Affective Stimuli and Stage Alteration in Mood and Plasma Hormone Levels. *PLoAS One*; 9(4): e87087. Published online 2014 Apr 2. doi: 10.1371/journal.pone.0087087.

PROCEEDINGS OF THE 9TH INTERNATIONAL SPACE SAFETY CONFERENCE

POSTER SESSION

SAFETY ANALYSIS AND RISK CONTROL FOR SATELLITE NAVIGATION SYSTEM BASED ON BAYESIAN NETWORK

Peipei Gong, Liming Ren, Yingchun Ran, Zhuopeng Yang

China Astronautics Standards Institute, Astronautics Standards Building, No.89 Xiaotun Road, Fengtai District, Beijing, China, 100071, gongpeipei2005@163.com

ABSTRACT

Aiming to solve safety (integrity and continuity) concerns of satellite navigation system raised by satellite-station interactive operation, coupling faults of hardware and software, severe space environment effects, multiple factors such as hardware malfunctions, software failures, environment effects, etc, are analyzed and considered synthetically. Integrity risk model and continuity risk model of satellite navigation system based on Bayesian Network are constructed. The weak components of satellite navigation system are identified through fault diagnosis, and risk control measures of satellite navigation system addressing those weak components are given. Technical as well as managerial considerations are all consolidated toward safety enhancement.

The results in this paper show that the quality of space broadcasting signals and the receiver robustness are both significantly improved. The risk of satellite navigation system integrity and continuity are also reduced tremendously. This method can be further extended to be used in safety analysis and risk control, such as integrity monitoring of autonomous navigation, continuity guarantee of integrated navigation system, etc.

1. INTRODUCTION

With the satellite navigation system in the international civil aviation, maritime and other life safety services in-depth application. Guaranteed navigation services safety for safety critical applications, such as the International Civil Aviation Organization (ICAO) and the International Maritime Organization (IMO) and so on, has become increasingly important. Navigation service safety is mainly concerned with the user positioning error over the alarm limit without timely alarm, or other events that cause the user service interruption. Navigation service safety is driven by integrity & continuity requirements [1], and relevant factors that lead to integrity and continuity risks include whether the user positioning error exceeds the limit, whether the system sends a timely alarm or not.

User positioning error mainly depends on two factors, the Dilution of Precision (DOP) and the User Equivalent Range Error (UERE). DOP are mainly related to constellation design. There are many factors that affect UERE, roughly from the space segment, operation control segment, environment segment and user segment 4 aspects.

- The factors related to the space segment include Code/Carrier Divergence, Navigation message data errors, Clock jump, ramp, and/or acceleration errors, Signal Deformations/ Distortions ("Evil Waveforms"), Low Signal Power, etc. [2]
- The factors related to the operation control segment include Delayed/Missed/Bad Upload, Ephemeris calculations errors (GA or MS induced errors), Operational errors (Health Settings and Data Content), etc. [3]
- The factors related to the environment segment include Ionospheric Delay, Tropospheric Delay, Multipath Effects, etc.
- The factors related to user segments include hardware failures, software failures, human operations errors, etc.

The main factors affecting the alarm capability are related to the satellite, control station and receiver alarm design. The three are:

- Satellite autonomous integrity Monitoring (SAIM).
- Ground Integrity Channel (GIC).
- Receiver autonomous integrity Monitoring (RAIM).

Aiming to solve safety (integrity and continuity) concerns of satellite navigation system raised by satellite-station interactive operation, coupling faults of hardware and software, severe space environment effects, multiple factors such as hardware malfunctions, software failures, environment effects, etc, are analyzed and considered synthetically. In this paper, integrity risk model and continuity risk model of satellite navigation system based on Bayesian Network are constructed. The weak components of satellite navigation system are identified through fault diagnosis, and risk control measures of satellite navigation system addressing those weak components are given.

2. MODELING AND ANALYSIS ALGORITHM FOR NAVIGATION SERVICE SAFETY RISK BASED ON BAYESIAN NETWORK

According to the analysis of the influence factors of navigation system safety, and the rules and experiences of fault and risk propagation, the models of integrity and continuity risk analysis based on Bayesian network are established.

2.1. Navigation Service Integrity Risk Analysis Model

Navigation Service integrity failure mechanism includes: all satellites are in non-fault mode and the satellite is marked as "healthy" but fails, as well as user segment failure. The "integrity risk ('IR' in Fig.1)" analysis model based on Bayesian network and some intermediate nodes CPT are shown in Fig.1.

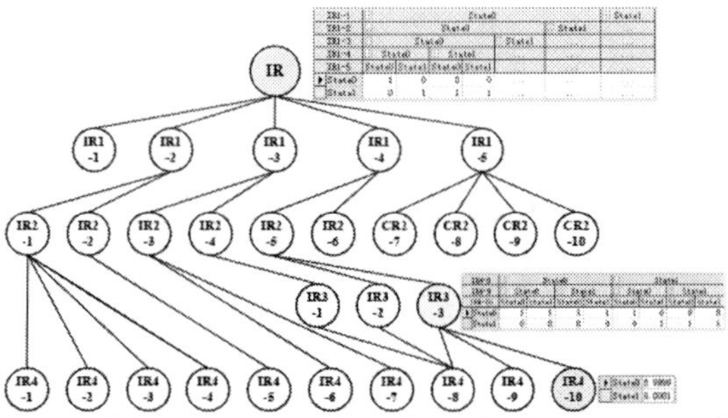

Figure 1. The Bayesian Network model for integrity risk analysis

The probability of integrity failure when all satellites are in "Fault-Free ('IR1-1' in Fig.1)"mode is very low[4], and the probability is set to 0 in the model, corresponding to the first branch in Fig.1.

The satellite is marked as "healthy" but fails includes the following:

(1) Single Satellite Failure Missed Protection ('IR1-2' in Fig.1)
The signal in space (SIS) is abnormal and is not detected, ie "a single SIS is failing ('IR2-1' in Fig.1)" and "Single failure missed detecting ('IR2-2' in Fig.1)". SIS failure modes include "Code/Carrier Divergence ('IR4-1' in Fig.1)", "Navigation message data errors ('IR4-2' in Fig.1)", "Clock jump, ramp, and/or acceleration errors ('IR4-3' in Fig.1)" and "Signal Deformations/ Distortions ('IR4-4' in Fig.1)". "Single failure missed detecting" means that "the satellite autonomous integrity monitoring dose not detect the failure ('IR4-5' in Fig.1)". The above corresponds to the second branch in Fig.1 and the downward branches.

(2) Single Satellite Failure Missed Protection ('IR1-2' in Fig.1)
"Ephemeris calculations are abnormal ('IR2-3' in Fig.1)", and are not detected by the ground monitor station (GMS), "the satellite signal information is wrong but set to 'healthy' ('IR2-4' in Fig.1)". Factors that cause Ephemeris Anomaly include "the OSPF (Orbit and synchronization Processing Facility) output contains misleading Information ('IR4-6' in Fig.1)" or "insufficient satellites ('IR4-7' in Fig.1)". Factors that cause Health setting Anomaly are mainly caused by "Ground IPF (Integrity Processing Facility) failure ('IR4-8' in Fig.1)" that causes "SIS integrity flag setting incorrectly ('IR3-1' in Fig.1)". The above corresponds to the third branch in Fig.1 and the downward branches.

(3) Hazard Misleading Information (HMI) in the SIS occurs according to a failure in the integrity information dissemination ('IR1-4' in Fig.1)
It is caused by two factors that are "GSM processing failure ('IR2-5' in Fig.1)" and "RAIM missing detecting ('IR2-6' in Fig.1)".Factors that cause "GSM processing failure" are mainly caused by "Ground IPF (Integrity Processing Facility) failure ('IR4-8' in Fig.1)" that causes "SISA processing failure ('IR3-2' in Fig.1)" and "SISMA processing failure ('IR3-3' in Fig.1)". The above corresponds to the fourth branch in Fig.1 and the downward branches.

The influence factors in the user segment are generally complicated, including "receiver hardware failure('IR2-7' in Fig.1)", "software failure ('IR2-8' in Fig.1)", "human operation error ('IR2-9' in Fig.1)" and "multipath effects ('IR2-10' in Fig.1)" , etc. The above corresponds to the fifth branch in Fig.1 and the downward branches.

2.2. Navigation Service Continuity Risk Analysis Model

Navigation Service continuity failure mechanism includes: Intact loss of integrity message, critical satellite signal interruption, as well as user segment failure. The "continuity risk ('CR' in Fig.2)" analysis model based on Bayesian network and some intermediate nodes CPT are shown in Fig.2.

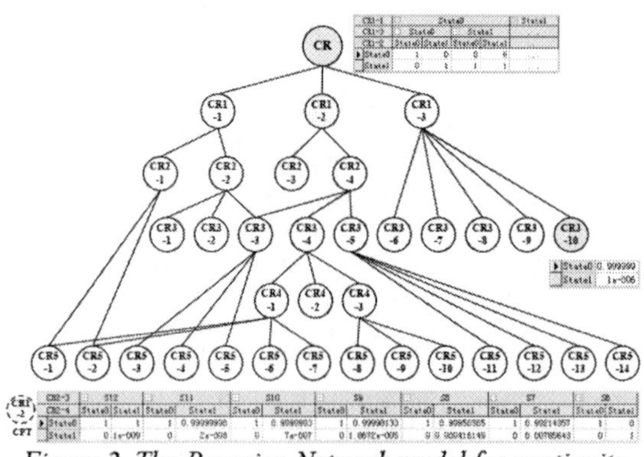

Figure 2. The Bayesian Network model for continuity risk analysis

(1) Discontinuity due to loss of integrity message ('CR1-1' in Fig.2)[4]

The satellite signal is not interrupted but the information is not guaranteed to be correct, including "Control section fails to generation and propagation integrity information ('CR2-1' in Fig.2)" and "SIS fails to propagation integrity information ('CR2-2' in Fig.2)". Control section failures include "IPF Outage in IM determination ('CR5-1' in Fig.2)" and "GMS Outage in IM dissemination ('CR5-2' in Fig.2)". SIS failures include "Satellite fails to receive IF ('CR3-1' in Fig.2)", "Satellite fails to transmit IF ('CR3-2' in Fig.2)" and "SIS containing IF message is not received ('CR3-3' in Fig.2)". Factors that affect 'CR3-3' include "All bus and payload Failures ('CR5-3' in Fig.2)", "Navigation message data errors ('CR5-4' in Fig.2)" and "Low Signal Power ('CR5-5' in Fig.2)". The above corresponds to the first branch in Fig.2 and the downward branches.

(2) Discontinuity due to non-local SIS Failure or Exclusion ('CR1-2' in Fig.2)
This depends on whether the number of satellites is sufficient at the initial stage, as well as whether there is a satellite interruption during the navigation phase. It includes "The probability of Visible Satellites ('CR2-3' in Fig.2)" and "SIS interruption ('CR2-4' in Fig.2)". Here, "the probability of visible satellites" is derived from a single global navigation system constellation simulation (the number of visible satellites is in 6-12, and the simulated values are shown Fig.1).

Factors that cause SIS interruption include "SIS containing IF message is not received ('CR3-3' in Fig.2)", "Correctly flagged Down signal - True Alert ('CR3-4' in Fig.2)" and "Good signal is not used - False Alert ('CR3-5' in Fig.2)".

"The True Alert" includes "Satellite Alerts correctly ('CR4-2' in Fig.2)", and "Alert due to ground anomaly ('CR4-1' in Fig.2)" or "Alert due to SIS accuracy jump ('CR4-3' in Fig.2)". 'CR4-1' include "IPF Outage in IM determination ('CR5-1' in Fig.2)", "GMS Outage in IM dissemination ('CR5-2' in Fig.2)", "Delayed/Missed/Bad Upload ('CR5-6' in Fig.2)", and "Ephemeris calculations are abnormal ('CR5-7' in Fig.2)". 'CR4-3' includes "Signal Deformations/ Distortions ('CR5-8' in Fig.2)", "Clock jump, ramp, and/or acceleration errors ('CR5-9' in Fig.2)", and "Code/Carrier Divergence ('CR5-10' in Fig.2)".

"CR3-5-False Alert" includes "Incorrect transmission of valid IF by satellite ('CR5-11' in Fig.2)", "Unintentional Health Bit set to Unhealthy ('CR5-12' in Fig.2)", "IPF False Alert triggering ('CR5-13' in Fig.2)", and "Integrity data dissemination corruption ('CR5-14' in Fig.2)".

The above corresponds to the Second branch in Fig.2 and the downward branches.

(3) Discontinuity due to User Segment ('CR1-3' in Fig.2)
As similar with IR, The influence factors in the user segment including "receiver hardware failure ('CR3-6' in Fig.2)", "software failure ('CR3-7' in Fig.2)", "human operation error ('CR3-8' in Fig.2)", "multipath effects ('CR3-9' in Fig.2)" and "False Alert from RAIM ('CR3-10' in Fig.2)", etc. The above corresponds to the third branch in Fig.2 and the downward branches.

2.3. Probabilistic Reasoning Algorithm for Bayesian networks

2.3.1. Joint Probability Calculation Algorithm

To calculate the security risk, we need to obtain the joint probability distribution of all nodes According to chain rules for Bayesian network [5], we can get integrity and continuity risk by calculating the joint probability of all nodes. For example, a node X_{N+1} of a Bayesian network has N parent nodes X_k (k = 1, 2, ... N), the conditional probability of node X_{N+1} is $P(X_{N+1}|\pi(X_{N+1}))$, where $\pi(X_{N+1})$ stands for parents of node X_{N+1}, then the node X_{N+1} joint probability is

$$P(X)=\prod_{k=1}^{N}P(X_k|\pi(X_k)) \quad (1)$$

2.3.2. Diagnostic Reasoning Algorithm

Bayesian network diagnosis reasoning is generally the process of solving Maximum a Posteriori assignment (MAP) problem. MAP is the problem of finding the most probable instantiation of a set of variables given the partial evidence on the other variables in a Bayesian network.

In this paper, we propose the Annealed MAP algorithm, which uses Markov Chain Monte Carlo (MCMC) methods and the simulated annealing technique to solve MAP problem in Bayesian network[6].

Formally, the MAP problem in a Bayesian network is defined as followed. Let X be the set of MAP nodes whose most probable configuration we are interested in. Let E be the set of evidence nodes. The remainders of the nodes are denoted by Y. Let Z=X ∪ Y ∪E, then MAP is the instantiation that satisfies

$$\hat{x}=\max_{x} p(X|E) \quad (2)$$

Using the law of total probability and the chain rule for Bayesian networks, we have

$$\hat{x} = \max_x \sum_Y p(X,Y|E)) = \max_x \sum_Y \prod_{Z_i \in Z} p(Z_i|\pi(Z_i)) \quad (3)$$

We introduce MCMC methods to sample from the target distribution $p(x)$, and applies the simulated annealing technique to simulation a non-homogeneous Markov chain whose invariant function is $p^{1/T_i}(x|E)$, which is a probability density that concentrates itself on modes of $p(x)$ as $\lim_{i\to\infty} T_i = 0$.

Let T be the transition matrix of constructed Markov chain. The chain will converge to the invariant function $p(x)$ when T satisfies irreducibility and aperiodicity, and make sure that

$$p(x^*)T(x|x^*) = p(x)T(x^*|x) \quad (4)$$

The Markov chain of the Metropolis-Hasting(MH) algorithm moves to a new state x^*, with acceptance probability, where $q(x^*|x)$ is proposal distribution.

$$A(x,x^*) = \min\left\{1, \frac{p(x^*)q(x|x^*)}{p(x)q(x^*|x)}\right\} \quad (5)$$

Suppose we have n-dimensional Vector $X = x_1, x_2, \ldots x_n$, and the expressions for the full conditionals $p(x_j|x_{-j})$ for each j, where $X_{-j} = x_1, \ldots x_{j-1}, x_{j+1}, \ldots x_n$ .Gibbs sampling uses the fowling proposal distribution for j=1, …,n

$$q(x|x^*) = \begin{cases} p(x_j^*|x_{-j}), & if \ \ x^*_{-j} = x_{-j} \\ 0, & otherwise \end{cases} \quad (6)$$

We can use simulated annealing technique to simulate non-homogeneous Markov chain whose invariant function is no longer $p(x)$, but

$$p_i(x) \propto p^{1/T_i}(x) \quad (7)$$

Where $\lim_{i\to\infty} T_i = 0$, $p^{\infty}(x)$ is a probability density that concentrates itself on modes of $p(x)$.
Then, the corresponding acceptance probability is

$$\begin{aligned} A(x,x^*) &= \min\left\{1, \frac{p^{1/T_i}(x^*|E)q(x|x^*)}{p^{1/T_i}(x|E)q(x^*|x)}\right\} \\ &= \min\left\{1, \frac{p^{1/T_i}(x_j^*,x_{-j}^*|E)p(x_j|x_{-j}^*,E)}{p^{1/T_i}(x_j,x_{-j}|E)p(x_j^*|x_{-j},E)}\right\} \\ &= \min\left\{1, \frac{p^{1/T_i-1}p(x_j^*|x_{-j},E)}{p^{1/T_i-1}p(x_j|x_{-j},E)}\right\} \end{aligned} \quad (8)$$

When $T_i = 1$, the above algorithm is just Gibbs sampling within variant function $p(X|E)$. By cooling down T_i to 0, we eventually get a density function which concentrates itself on the global maxima of $p(X|E)$.

3. EXAMPLES OF RISK ANALYSIS AND WEAK LINK IDENTIFICATION FOR NAVIGATION SERVICE SAFETY

3.1. Navigation Service Continuity Risk Analysis Model

Navigation service safety risk analysis is to calculate the integrity and continuity risk of navigation services under a given prior probabilities of Bayesian network root node.

3.1.1. The Assumptions of Prior Probability in Bayesian Network Model

According to the actual experience of GPS, GLONASS and BDS, the basic assumptions about the prior probability values of the Bayesian network root nodes in the integrity and continuity risk analysis models are given, as shown in Tab.1 and Tab.2. The value of State1 in the table indicates that the probability of an undesired event occurs.

Table1. The assumption of the prior probability of the root node in the integrity risk model

Node Name	State1	Node Name	State1
IR1-1	0.0E+00	IR4-3	1.0E-04
IR2-6	1.0E-03	IR4-4	1.0E-04
IR2-7	1.0E-05	IR4-5	1.0E-03
IR2-8	1.0E-06	IR4-6	1.0E-07
IR2-9	1.0E-07	IR4-7	1.0E-09
IR2-10	1.0E-04	IR4-8	1.0E-06
IR4-1	1.0E-04	IR4-9	1.0E-04
IR4-2	1.0E-04	IR4-10	1.0E-04

Table2 The assumption of the prior probability of the root node in the continuity *risk model*

Node Name	State1	Node Name	State1	Node Name	State1
CR3-1	1.0E-06	CR5-1	1.0E-07	CR5-9	1.0E-04
CR3-2	1.0E-06	CR5-2	1.0E-07	CR5-10	1.0E-04
CR3-3	1.0E-08	CR5-3	1.0E-04	CR5-11	1.0E-06
CR3-4	1.0E-05	CR5-4	1.0E-06	CR5-12	1.0E-06
CR3-5	1.0E-08	CR5-5	1.0E-04	CR5-13	1.0E-06
CR3-6	1.0E-05	CR5-6	1.0E-04	CR5-14	1.0E-06
CR3-7	1.0E-06	CR5-7	1.0E-06		
CR4-2	1.0E-05	CR5-8	1.0E-04		

3.1.2. Navigation service integrity and continuity risk analysis results

Based on the Bayesian network model and the conditional probability in Chapter 3, we can get the joint probability of navigation service integrity and continuity risk, as well as the intermediate nodes, by inputting the prior probabilities of the root nodes given above. Navigation service integrity and continuity risk are in the order of 10^{-5}. Among the several influencing factors, the safety risk posed by the user segment exception is the highest.

3.2. Navigation Service Safety Weak Links Identification and Risk Control

3.2.1. Weak Links Identification

Based on the input values above, the evidence of the top event (node "IR" and node "CR") in the integrity risk and continuity risk analysis model is set to "state1=1", which means risk even happened. By reasoning calculations, the posteriori probabilities of the most important factors that affect the risk of navigation service safety are shown in Tab.3 and Tab.4. As can be seen from the analysis results, when the navigation service integrity or continuity risk event occurs, most likely due to the receiver hardware failure, whose posteriori probabilities is more than 50%. Without considering the impact of user segments, SIS failure and SAIM capability are the weak links in the navigation system itself that affect navigation service safety.

Table3 The integrity risk of weak links analysis results with evidence

NO.	Evidence		Posterior Probability			
	Node Name	Risk	Node Name	State1	Node Name	State1
1	IR	1	IR1-5	0.953	IR2-7	0.934
			IR1-2	0.037	IR2-1	0.038
			IR2-2	0.039		
2	IR	1	IR1-2	0.798	IR2-1	0.798
	IR1-5	0	IR2-2	0.798		

Table4 The continuity *risk of weak links analysis results with evidence*

NO.	Evidence		Posterior Probability			
	Node Name	Risk	Node Name	State1	Node Name	State 1
1	CR	1	CR1-3	0.665	CR3-7	0.603
			CR1-2	0.322	CR2-4	0.324
			CR3-3	0.126	CR3-4	0.196
			CR4-3	0.187		
2	CR	1	CR1-2	0.960	CR2-4	0.968
	CR1-3	0	CR3-3	0.377	CR3-4	0.583
			CR4-3	0.559		

3.2.2. Navigation service safety risk control

According to the analysis of the weakness of the navigation service security, the user segment is the main factor affecting the service integrity and continuity. The impact of the user segment is mainly reflected in the receiver hardware failure and environmental impact. Increased receiver robustness design and multipath interference suppression techniques can significantly reduce the integrity and continuity risk, see line NO.1 in Tab. 5 and Tab. 6. For the navigation system itself, the main factors affecting service integrity and continuity are single satellite failure / interruption, mainly reflected in the quality of the broadcast signal (Clock jump, Signal Deformations, Low Signal Power, etc.) and SAIM. Strengthen the reliability of satellite design and space environment protection design, improve the quality of space signal broadcast, can effectively reduce the integrity and continuity risk, and see line NO.2 in Tab.5 and Tab.6.

Table5 The results of the integrity risk analysis were compared before and after the improvement of the influencing factors

NO.	Node Name	Risk Before	Risk After	Node Name	State1 Before	State1 After
1	IR*	1.1E-5	1.7E-6	IR2-7	1.0E-5	1.0E-6
2	IR**	1.1E-5	1.3E-6	IR2-7	1.0E-5	1.0E-6
				IR4-1	1.0E-4	1.0E-5
				IR4-2	1.0E-4	1.0E-5
				IR4-3	1.0E-4	1.0E-5
				IR4-4	1.0E-4	1.0E-5

Table6 The results of the continuity risk analysis were compared before and after the improvement of the influencing factors

NO.	Node Name	Risk Before	Risk After	Node Name	State1 Before	State1 After
1	CR*	1.7E-5	7.6E-6	CR3-7	1.0E-5	1.0E-6
2	CR**	1.7E-5	2.9E-6	CR3-7	1.0E-5	1.0E-6
				CR5-3	1.0E-4	1.0E-5
				CR5-5	1.0E-4	1.0E-5
				CR5-7	1.0E-4	1.0E-5
				CR5-8	1.0E-4	1.0E-5
				CR5-9	1.0E-4	1.0E-5

At present, satellite navigation systems have been applied in depth to areas related to life safety services. ICAO, IMO and the railway applications have made clear requirements for the safety risks of Satellite navigation system services [4][7][8]. According to the above analysis results, the satellite navigation system can basically meet the safety requirements of sailing and railway users' navigation services, without the help of other auxiliary and enhanced systems. But any single existing satellite navigation system cannot meet the aviation users precision approach to the navigation requirements. It is recommended that the satellite navigation system be used as an auxiliary navigation means or to increase the safety performance of the auxiliary or enhanced system in the case of high performance requirements for life safety services.

4. CONCLUSION

This paper comprehensively analyzes the influence factors of space, ground segment, user segment and environment segment on UERE, DOP and alarm capability of satellite navigation system. The Bayesian network is used to construct the satellite navigation system integrity and continuity risk model respectively. The risk analysis and the weak link identification are carried out, and the risk control measures are given to provide the technical and management basis for improving the safety of the satellite navigation system. Technical as well as managerial considerations are all consolidated toward safety enhancement.

The results in this paper show that the quality of space broadcasting signals and the receiver robustness are both significantly improved. The risk of satellite navigation system integrity and continuity are also reduced tremendously. This method can be further extended to be used in safety analysis and risk control, such as integrity monitoring of autonomous navigation, continuity guarantee of integrated navigation system, etc.

5. REFERENCES

1. Andrew Brown (2010). Space Born Support for Safety Critical Applications. Navagation solutions powered by Europe.
2. Karen Van Dyke, Karl Kovach, John Lavrakas (2003). GPS Integrity Failure Modes and Effects Analysis. ION NTM 2003, 22-24 January 2003, Anaheim, CA
3. U.S. Department of Defense (2008). Global Positioning System Standard Positioning Service Performance Standard[S]. Washington. DC, 4th Edition.
4. Veit Oehler, Jörg Hahn (2004). The Galileo Integrity Concept. ION GNSS 17th International Technical Meeting of the Satellite Division, 21-24 Sept. 2004, Long Beach, CA
5. LIU Shufen, YANG Shuangshuang, WANG Hui (2014). Fault Diagnosis of Equipment Based on Fault Tree Combined with Bayesian Network. Journal of Jilin University (Science Edition), 2014, 52(9): 982-983.
6. Changhe Yuan, Tsai-Ching Lu, Marek J. Druzdzel (2004). Annealed MAP. In Proceedings of the 20th Annual Conference on Uncertainty in Artificial Intelligence (UAI-2004), AUAI Press, Arlington, Virginia, pages 628-635.
7. Alan Grant, Paul Williams, George Shaw, Michelle De Voy & Nick Ward (2011). Understanding GNSS availability and how it impacts maritime safety, International Technical Meeting of The Institute of Navigation, San Diego, CA, January 24-26
8. Ing. G. Schaenzer (2005). Safety Critical Applications of Satellite Navigation. CERGAL 2005, 14.04.2005, Braunschweig.